HUMAN *Anatomy*

LABORATORY MANUAL

Second Edition

CHRISTINE M. ECKEL
*West Virginia School of
Osteopathic Medicine*

Mc
Graw
Hill

Connect
Learn
Succeed™

HUMAN ANATOMY LABORATORY MANUAL, SECOND EDITION

Published by McGraw-Hill, a business unit of The McGraw-Hill Companies, Inc., 1221 Avenue of the Americas, New York, NY 10020.

Some ancillaries, including electronic and print components, may not be available to customers outside the United States.

This book is printed on acid-free paper.

1 2 3 4 5 6 7 8 9 0 DOW/DOW 1 0 9 8 7 6 5 4 3 2 1

ISBN 978–0–07–352566–2
MHID 0–07–352566–9

Vice President, Editor-in-Chief: *Marty Lange*
Vice President, EDP: *Kimberly Meriwether David*
Senior Director of Development: *Kristine Tibbetts*
Executive Editor: *Colin H. Wheatley*
Senior Developmental Editor: *Kristine A. Queck*
Marketing Manager: *Denise M. Massar*
Senior Project Manager: *April R. Southwood*
Senior Buyer: *Sandy Ludovissy*
Senior Media Project Manager: *Tammy Juran*
Senior Designer: *David W. Hash*
Cover Illustration: *Electronic Publishing Services Inc., NY*
Cover Image: *Marili Forastieri/Digital Vision/Getty Images, RF*
Lead Photo Research Coordinator: *Carrie K. Burger*
Photo Research: *Danny Meldung/Photo Affairs, Inc*
Compositor: *Electronic Publishing Services Inc., NYC*
Typeface: *9.5/12 ITC Slimbach Std*
Printer: *R. R. Donnelley*

Some of the laboratory experiments included in this text may be hazardous if materials are handled improperly or if procedures are conducted incorrectly. Safety precautions are necessary when you are working with chemicals, glass test tubes, hot water baths, sharp instruments, and the like, or for any procedures that generally require caution. Your school may have set regulations regarding safety procedures that your instructor will explain to you. Should you have any problems with materials or procedures, please ask your instructor for help.

www.mhhe.com

CHRISTINE ECKEL, Ph.D., is Associate Professor of Anatomy at West Virginia School of Osteopathic Medicine (WVSOM) in Lewisburg, WV. She is the Course Director for Medical Gross Anatomy and Director of the Human Gift Registry at WVSOM, where she teaches courses in medical gross anatomy and medical microanatomy (histology). Dr. Eckel received her B.A. and M.A. degrees in Integrative Biology and Human Biodynamics from the University of California at Berkeley, where she engaged in research in muscle physiology and locomotion biomechanics and taught undergraduate courses in human anatomy, general biology, and biomechanics. She taught courses in human anatomy, human physiology, and general biology at Salt Lake Community College (SLCC) for eleven years (1998-2009), where she advanced to Associate Professor and earned tenure. While continuing to teach full time at SLCC, Dr. Eckel returned to graduate school to earn her Ph.D. in Neurobiology & Anatomy at the University of Utah School of Medicine. There, she taught gross anatomy to first-year medical and dental students, pathology residents, and orthopedic surgery residents. Her Ph.D. work includes educational outcomes research and development of a cadaver autopsy project, an innovative approach to teaching that integrates courses in gross anatomy, histology, and pathology for first-year medical students. She continues to engage in educational outcomes research and is adapting the cadaver autopsy project to curricular innovations at WVSOM. She is an acclaimed cadaver prosectionist and biomedical photographer, with her animal and human dissections and photographs appearing in numerous publications, including this laboratory manual. She has also authored ancillary materials for both human anatomy and human physiology textbooks, and has developed electronic media for teaching gross anatomy to both undergraduate and medical students.

Christine has received several teaching honors, including an Outstanding Graduate Student Instructor award from U.C. Berkeley and a Teaching Excellence award from Salt Lake Community College. She was awarded the Frank L. Christensen Endowed Fellowship from the University of Utah and was named the Betty Cook Karrh Endowed P.E.O. Scholar for 2004–2005. She is an active member of the Human Anatomy and Physiology Society (HAPS), and served as Western Regional Director from 2004-2008. She is also an active member of the American Association of Anatomists (AAA), where she was a member of the Advisory Committee of Young Anatomists from 2007-2009. She currently serves on the Educational Affairs Committee for the AAA.

ontents

Preface

Human anatomy is a complex but fascinating subject, and is perhaps one of the most personal subjects a student will encounter during his or her education. Yet it is also a subject that can create a great deal of anxiety for students because of the sheer volume of material, and a misconception among students that it's "all about memorization." Too often, confusion in the anatomy laboratory only enhances this misconception and enhances student frustration with the subject. My goal in writing this laboratory manual was to create a manual that guides students through their laboratory experience in a very focused way, and to provide them with tools that make the material more relevant to the student's daily experiences with their own bodies and the world around them.

The study of human anatomy really comes to life in the anatomy laboratory. Here is where students get hands-on experience with human cadavers and bones, classroom models, preserved and fresh animal organs, and histology slides of human tissues. Yet, most students are at a loss as to how to proceed in the anatomy laboratory. They are given numerous lists of structures to identify and histology slides to view, but comparatively little direction as to *how* to recognize structures or *how* to relate what they encounter in the laboratory to the material presented in lecture. In addition, most laboratory manuals on the market contain little more than material repeated from anatomy textbooks, which provides no real benefit to a student. All of these things lead to student frustration and take away from the joy that comes from discovering the beauty of the human body through touch, observation, and dissection. This laboratory manual is designed to be a user-friendly *manual* to the human anatomy laboratory that addresses just such issues.

New in the Second Edition

The most exciting change in this edition is the inclusion of an all-new chapter. Chapter 1: The Human Anatomy Laboratory introduces safety procedures, use of cadavers in the laboratory, common laboratory equipment, and dissection techniques. A multitude of new photos that familiarize students with laboratory equipment are presented in the context of step-by-step exercises demonstrating proper usage. All subsequent chapters have been renumbered to accommodate the placement of this new chapter at the very beginning of the book.

Another major change is the substantial overhaul of bone and histology photos. Of particular note, the author photographed a series of new skull images using a first-grade skull. The high-quality specimen and improved contrast provide much clearer detail than seen in the previous edition. In addition, the author photographed several new tissue slides to obtain photos that coincide with specific steps in the histology activities so students see exactly what is described in the exercises.

Finally, much effort has been spent fine-tuning content accuracy and clarity. Every chapter in the book was carefully scrutinized, resulting in numerous tweaks to table entries, adjustments in terminology, strategic addition and/or reconfiguring of labels and leader lines in the figures, and general polishing to improve readability. Other book-wide changes include the addition of new questions to nearly all of the Pre- and Post-Laboratory Worksheets to make them more challenging and more complete, and numbering the chapter learning objectives for easier reference and tracking.

In addition to the general changes above, the following list provides a chapter-by-chapter overview highlighting some (but by no means all) of the updates in the second edition.

1. Introduction—All-new chapter introducing laboratory equipment and procedures.
2. Orientation to the Human Body—Added body planes to figure 2.1 and directional terms to figure 2.2.
3. The Microscope—More tips regarding microscope usage added to Focus Box: Caring for the Microscope and to Table 2.2.
4. Cellular Anatomy—Added illustrations of each cell organelle to table 4.1. Replaced micrograph of interphase in Table 4.2.
5. Histology— New micrographs added to the following figures: 5.3e (dense irregular connective tissue), 5.17 (elastic connective tissue), 5.19 (hyaline cartilage), 5.27 (smooth muscle). Added multiple new "Study Tip!" boxes, including clarification of basement membrane vs. basal surface, and suggestions for relating tissue appearance to everyday items. Elaborated on description of bone tissue.
6. Integument— Added new micrograph for figure 6.2 (pigmented skin) and a new Study Tip! regarding the tissue types found in leather. Expanded the description of keratinocytes and made various terminology updates for consistency.
7. Skeletal System Overview: Bone Anatomy—Enlarged photos in Exercise 7.5 to make figures easier to label.
8. Axial Skeleton—New photos using better specimens and coloration provided for the following figures: 8.1 (anterior skull), 8.5 (lateral skull), 8.6 (posterior skull), 8.8 (inferior skull), 8.11 (cranial floor), 8.12 (hyoid bone). Improved contrast of existing photos for figures 8.7 (superior skull), 8.13 (fetal skull), 8.16-8.19 (vertebrae), 8.24 (rib).
9. The Appendicular Skeleton— Improved contrast of photos in figure 9.3 (humerus) and added close-up view of ulnar notch to figure 9.4. New Study Tip! with mnemonic for remembering the carpal bones.
10. Articulations—New Focus box and accompanying photo on hip replacement.
11. Muscle Tissue and Introduction to the Muscular System-- New Study Tip! clarifying the difference between flexion and contraction.
12. Axial Muscles-- Modified figure 12.11 to indicate location of the linea alba. Provided innervations in table 12.1. New Study Tip! regarding conventions of muscle naming.
13. Appendicular Muscles—Revised muscle actions to refer to limbs rather than joints--for example, "flexes elbow" was changed to "flexes forearm."
14. Nervous Tissues—Minor terminology and table updates.

15. The Brain—Modified layouts of several figures and enlarged figure 15.6 (superior brain) for ease of labeling.
16. Cranial Nerves—Reformatted text and enlarged several figures for easier labeling. Shaded the branches of the trigeminal nerve in figure 16.7a to match the sensory distribution map of 16.7b.
17. The Spinal Cord and Spinal Nerves—New Study Tip! for distinguishing anterior and posterior horns.
18. The Nervous System: General and Special Senses— New micrographs for figures 18.3d (vallate papilla), 18.3e (foliate papilla), 18.8c (cochlea cross-section). Added close-up photo of cochlea model to figure 18.16. Expanded structures covered in tables 18.7 and 18.8. Clarified descriptions of anterior cavity vs. anterior chamber.
19. The Endocrine System—Added illustrated reference icons to figures 19.7 (adrenal glands) and 19.8 (pancreas), and a new photo showing the hypothalamus and pituitary to figure 19.9 (endocrine organs).
20. The Cardiovascular System: Blood—Minor tweaks for clarity and correctness.
21. The Cardiovascular System: The Heart—Increased size of figure 21.3 (heart in thoracic cavity) for easier labeling. Improved contrast of figure 21.4 (pericardial sac).
22. Vessels and Circulation—New micrographs for figures 22.1b (blood vessel wall), 22.2 (elastic artery). Increased size of figure 22.18 for ease of labeling. Modified shading in figure 22.21 (lower limb circulation) to differentiate superficial and deep vessels. Shaded capitate and cuboid bones for clarity in figures 22.20 and 22.23 respectively
23. The Lymphatic System—Two new micrographs for figure 23.4 (Peyer patches). Added illustrated reference icons to figures 23.6 and 23.9.
24. The Respiratory System—Included a new exercise (Exercise 24.3) and table on the histology of the bronchi and bronchioles. Added illustrated reference icons to figures 24.3-24.5. New photomicrograph of bronchus in Post-Laboratory Worksheet.
25. The Digestive System—Modified several anatomic descriptions for clarity. Added illustrated reference icon to figure 25.8, and a new micrograph of the ileum in Post-Laboratory Worksheet.
26. The Urinary System— Reorganized layout of several figures for ease of labeling. New micrographs (ureter, kidney cortex, urinary bladder wall) for Post-Laboratory Worksheet questions.
27. Reproductive System—New micrographs for figure 27.7b (epididymis), 27.8b (ductus deferens), and 27.11 (penis cross-section). Increased size of micrographs in table 27.6 (menstrual cycle phases) and figure 27.13 (model of female pelvic cavity) to improve visibility. Added new table 27.11, The Female Breast. New micrographs (ductus deferens and ovary) for Post-Laboratory Worksheet questions.

Distinguishing Features

Overall Approach

First and foremost, this laboratory manual was designed *not* to repeat textbook material. However, students still need critical information to proceed in the laboratory. Thus, as much as possible, reference information necessary for completing laboratory activities is presented in summary tables that act as a concise resource for students.

Laboratory exercises are presented in steps that guide students precisely through each activity. Interesting and pertinent points about the structures students are observing or dissecting are provided within the text of each exercise. Detailed anatomical descriptions of structures such as individual bones of the skull are left to the main textbook. Rather, the discussions in this laboratory manual give students alternative ways to understand, organize, and make sense of the material. The text is written in a friendly, conversational tone so as to not be intimidating to students, while at the same time not being overly chatty or brief on details.

Photographs and Illustrations

The photographs in this laboratory manual are intended to truly capture the laboratory experience. The author, an accomplished prosectionist and biomedical photographer, personally prepared and shot the vast majority of the photographs of dissections, bones, human cadavers, and classroom models for this laboratory manual, as well as several of the histology images. While writing the dissection exercises, she performed the dissections herself and photographed each dissection at key stages that would be of most benefit for students as they perform the same steps. This gives each photo a unique perspective that could not be accomplished any other way.

Illustrations and photographs appearing in this manual have been tailored to the specific needs of the associated laboratory exercises, and are generally unique to avoid unnecessary repetition of lecture textbook images.

Organization

Because observation of histology slides and observation of human cadavers and classroom models are usually performed in separate physical spaces or at specific times within each laboratory classroom, chapters in this laboratory manual are similarly separated into two sections: Histology and Gross Anatomy. Each exercise within these chapter sections has been designed with the student's actual experience in the anatomy laboratory in mind. Thus, each exercise covers only a single histology slide, classroom model, or region of the human body (for example: muscles of the abdominal wall, histology of cardiac muscle, model of the human ear). In addition, organization of each chapter into a series of discreet exercises makes the laboratory manual easily customizable to any anatomy classroom, allowing an instructor to assign certain exercises, while telling students to ignore other exercises.

Pedagogy

This laboratory manual utilizes several pedagogical devices to assist students in learning human anatomy in the laboratory setting.

- **Outline and Objectives** Each chapter begins with an outline that lists the exercises within the chapter. Below each exercise is a list of objectives that conform to the activities the students are asked to complete within each exercise.
- **Pre- and Post-Laboratory Worksheets** Pre-laboratory worksheets at the beginning of each chapter are intended to

give the student a "warm up" before entering the laboratory classroom. Some questions pertain to previous activities that are relevant to the upcoming activities (for example: review questions about nervous tissues in the pre-laboratory worksheet for the chapter on the brain and cranial nerves), while others are basic questions that students should be able to answer if they have read the chapter from their lecture text before coming into the classroom. The goal of completing these worksheets is simple: have students arrive at the laboratory prepared to deal with the material they will be covering so they do not waste valuable in-class time reviewing necessary background information.

Post-laboratory worksheets at the end of each chapter help students review the material they just covered, and challenge them to apply the knowledge gained in the laboratory (for example: questions asking students to determine loss of function if a particular nerve or part of the brain is damaged). The post-laboratory worksheets contain more in-depth, critical thinking types of questions than the pre-laboratory worksheets. Post-laboratory worksheets are perforated so they can be torn out of the manual and handed in to the instructor if so desired.

- **In-Chapter Learning Activities**   The exercises in this laboratory manual are about *doing*, not just observing. Exercises offer a mixture of activities including labeling exercises, sketching activities, coloring exercises, table completion, data recording, palpation of surface anatomy structures, and the like.

- **Labeling Activities**   In the gross anatomy exercises of this manual, images of things such as cranial bones, muscles of the body, and so on are *not* presented as labeled photos because the students *already have* labeled photos in their main textbook. Instead, each image is presented as a labeling activity with a checklist of structures. The checklists serve two purposes: (1) they guide students to what items they need to be able to identify on classroom models, fresh specimens, or cadavers (if the laboratory uses human cadavers), and (2) they double as a list of terms students can use to complete the labeling activities. Answers to the labeling activities are provided in the appendix. Thus, if a student does not know what a leader line is pointing to, or cannot remember the correct term, he or she can consult the appendix to locate the correct answer. This is a bit more challenging to students than having a pre-labeled image in the lab manual. However, that is precisely the goal: challenge the students!

- **Study Tip!**   Handy "Study Tip!" boxes coach students through the more problematic areas of study. They offer tips such as mnemonic devices, points of clarification, and things to be aware of and/or careful of when they are making certain observations.

- **What Do You Think? Questions**   Placed at key points within exercises, these critical thinking questions challenge students to think beyond the "what" of the structures they are observing and start to think about the "why." Answers are provided in the appendix.

- **Tables**   Each chapter contains numerous tables, which concisely summarize necessary details. As stated previously, the goal of this manual was expressly not to repeat textual material. However, students still need the information as reference while in the laboratory classroom. Thus, critical information and key structures are covered in table format. Most tables contain a column that provides word origins for each structure listed within the table. These word origins are intended to give students continual exposure to the origins of the language of anatomy, which is critical for learning.

- **Focus Boxes**   Several chapters contain Focus boxes, which feature information related to the subject matter but not necessarily critical for understanding or performing the laboratory activities. Focus box topics include: the use of human cadavers in the anatomy classroom (chapter 1), how to prepare a human blood smear (chapter 19), and breast self-examination (chapter 26).

- *Anatomy & Physiology* | REVEALED® **Correlations** AP|R Where pertinent, optional activities indicated by the logo above direct students to where they can find related content on Anatomy & Physiology Revealed. See below for more information on this cutting-edge anatomy software.

Teaching Supplements

McGraw-Hill Higher Education and Blackboard have teamed up.

Blackboard, the Web-based course-management system, has partnered with McGraw-Hill to better allow students and faculty to use online materials and activities to complement face-to-face teaching. Blackboard features exciting social learning and teaching tools that foster more logical, visually impactful and active learning opportunities for students. You'll transform your closed-door classrooms into communities where students remain connected to their educational experience 24 hours a day.

This partnership allows you and your students access to McGraw-Hill's Connect™ and Create™ right from within your Blackboard course—all with one single sign-on.

Not only do you get single sign-on with Connect™ and Create™, you also get deep integration of McGraw-Hill content and content engines right in Blackboard. Whether you're choosing a book for your course or building Connect™ assignments, all the tools you need are right where you want them—inside of Blackboard.

Gradebooks are now seamless. When a student completes an integrated Connect™ assignment, the grade for that assignment automatically (and instantly) feeds your Blackboard grade center.

McGraw-Hill and Blackboard can now offer you easy access to industry leading technology and content, whether your campus hosts it, or we do. Be sure to ask your local McGraw-Hill representative for details.

Instructors can obtain access to the following resources by calling McGraw-Hill Customer Service at 1-800-338-3987 or contacting their local sales representative.

Assignable Questions

Pre-and Post-Laboratory Worksheet questions are available for use in online assignments via McGraw-Hill's Connect.

Textbook Images

Image files for use in presentations and teaching materials are provided for instructor use.

Instructor's Manual

A helpful manual containing materials lists, presentation ideas, and answer keys for the Pre- and Post-Laboratory Worksheets is available to instructors using this laboratory manual.

ACKNOWLEDGMENTS

I would like to thank the entire team at McGraw-Hill for their hard work on this laboratory manual. In particular, I am extremely grateful for the invaluable insights of Senior Developmental Editor Kris Queck. Kris has a unique ability to find solutions to layout "problems" and similar issues. She made working on the second edition of this book even better than the first! Managing Editor and Project Manager April Southwood and Senior Photo Research Coordinator Carrie Burger worked relentlessly to bring this project to fruition, and they also have my profound thanks. The entire team at McGraw-Hill is a class act and I am grateful to work with such a talented group of people.

Numerous reviewers (listed below) put a lot of time and thought into providing constructive feedback on the first edition of this book. I greatly appreciate their efforts, which have made the second edition of this book even more user-friendly than the first.

A special thank-you goes out to Matthew P. Cauchi and Sarah C. Shaw who worked on the Instructor's Manual for this book. Matt and Sarah are fourth-year medical students and Graduate Teaching Assistants for the Medical Gross Anatomy course at West Virginia School of Osteopathic Medicine. They are two of WVSOM's finest and are on their way to becoming outstanding physicians.

As always, I give my most sincere thanks to all the individuals who selflessly donated their bodies after death for medical education and research. Without their generous donations, none of us would have the opportunity to truly learn anatomy. They have given us the most precious gift.

To the end users of this book, thank you in advance for any feedback, suggestions for improvement, or corrections that will help improve future editions. I am dedicated to producing the highest quality laboratory manual that will help students learn and develop a love of this most beautiful and fascinating subject. Thank you!

Christine M. Eckel
Division of Biomedical Sciences
West Virginia School of Osteopathic Medicine
400 N. Lee Street
Lewisburg, WV 24901
ceckel@osteo.wvsom.edu

Reviewers

Debra J. Barnes
Contra Costa College

Ethan A. Carver
University of Tennessee at Chattanooga

Anne E. Hays
Slippery Rock University

Laura M. Juárez de Ku
Austin Community College

John W. McDaniel
Skyline College

Gavin C. O'Connor
Ozarks Technical Community College

Rachel D. Smetanka
Southern Utah University

Mark D. Tillman
University of Florida

Marcia J. Holstad Walton
Des Moines Area Community College

Michael Yard
Indiana University-Purdue University at Indianapolis

Michele Zimmerman
Indiana University Southeast

The Human Anatomy Laboratory

OUTLINE and OBJECTIVES

Module 1: SKELETAL SYSTEM

INTRODUCTION

Welcome to the human anatomy laboratory! If you are like most students, you are experiencing a lot of excitement along with a lot of anxiety about the course you have just enrolled in. The human body is a fascinating subject, and the study of human anatomy is an experience that you will surely never forget.

If this laboratory manual is required for your anatomy course, then your course is likely to be an integrated, systems-based course that combines both human gross anatomy and histology. **Gross anatomy** is the study of structures that can be seen with the naked eye. This means any structure that you do not need a microscope to see. **Histology** is the study of tissues. Unlike gross anatomy, the study of histology requires the use of a microscope. It is our hope that once you have completed this course, you will develop an understanding of and appreciation for how tissue structure relates to gross structure, and vice versa. However, in the laboratory classroom itself, you may find yourself studying the two levels of structure somewhat separately. That is, your classroom studies in gross anatomy will likely involve observing classroom models, dissecting animal specimens, or making observations of human bones and/or human cadavers, whereas your classroom studies in histology will likely involve observing histology slides with a microscope or using some sort of virtual microscopy system. To aid in your observations, the exercises in this manual are divided into two types of activities: gross anatomy exercises and histology exercises. Where applicable, each chapter will begin with a section on histology and end with a section on gross anatomy.

What should you expect to find when you enter your anatomy laboratory classroom? The human anatomy laboratory has relatively few complicated pieces of equipment as compared to most biology laboratories. The main items you will find, the "tools of the trade," mainly consist of dissection instruments, microscopes, and perhaps cadaver tables. The purpose of this chapter is to introduce you to common equipment and dissection instruments you will encounter in the laboratory, common chemicals used in the laboratory, protective equipment, proper disposal of waste materials, and common dissection techniques. Chapter 3 of the laboratory manual covers use of the microscope.

Focus | Use of Human Cadavers in the Anatomy Laboratory

Where did that body lying on a table in your human anatomy laboratory come from? Typically, the body was donated by a person who made special arrangements before the time of death to donate his or her body to a body donor program so it could be used for education or research. Individuals who donate their bodies for these purposes made a conscious decision to do so. Such individuals have given us an incredible gift—the opportunity to learn human anatomy from an actual human body. It is important to remember that what that person has given you is, indeed, a gift. The cadaver deserves your utmost respect at all times. Making jokes about any part of the cadaver or intentionally damaging or "poking" at parts of the cadaver is unacceptable behavior.

The idea that you will be learning anatomy by observing structures on what was, at one time, a living, breathing human being might make you feel very uncomfortable at first. It is quite normal to have an emotional response to the cadaver upon first inspection. Would you want it any other way? You will need time and experience to become comfortable around the cadaver. Even if you think you will be just fine around the cadaver, it is important to be aware of your first response and of the responses of your fellow students. If at any time you feel faint or light-headed, sit down immediately. Fainting, though rare, is a possibility, and people can get hurt from falling down unexpectedly. Be aware of fellow students who appear to lose color in their faces or start to look sick—they might need your assistance.

Typically the part of the body that evokes the most emotional response is the face, because it is most indicative of the person that the cadaver once was. Because of this, the face of the cadaver should remain covered most of the time. This does not mean you are not allowed to view it. However, when you wish to do so, you should make sure that other students in the room know that you will be uncovering the face. If you have a particularly strong emotional response to the cadaver, take a break and come back to it later when you are feeling better.

Those of us with a great deal of experience around cadavers had a similar emotional response our first time as well. What happens with time is that you learn to disconnect your emotions from the experience. Certainly at one time the body that is the cadaver in your laboratory was the home of a living human being. However, now it is just a body. Eventually you will become comfortable using the cadaver and will find that it is an invaluable learning tool that is far more useful than any model or picture could ever be. You will literally see for yourself that there is nothing quite like the real thing to help you truly understand the structure of the human body. Make the most of this unique opportunity—and give thanks to those who selflessly donated their bodies so you could partake in the ultimate learning experience in anatomy.

Students who are curious about the uses of cadavers in science and research are encouraged to check out the following book from the library: Mary Roach, *Stiff: The Curious Lives of Human Cadavers* (New York: W.W. Norton, 2003).

Chapter 1: The Human Anatomy Laboratory

Name:_____

Date:_____ Section:_____

PRE-LABORATORY WORKSHEET

1. Define *histology*: _____

   _____

   _____

2. Define *gross anatomy:*_____

   _____

   _____

3. What is a "sharps" container? _____

   _____

   _____

4. Give one example of an item that must be disposed of in a sharps container. _____

   _____

   _____

5. Briefly describe the process of "blunt dissection."

   _____

   _____

   _____

   _____

   _____

   _____

IN THE LABORATORY

The laboratory exercises in this chapter will introduce you to common equipment and dissection instruments found in the human anatomy laboratory classroom. It will also introduce you to the proper use of such instruments, the proper disposal of laboratory waste, and common dissection techniques. It is likely you will not perform any actual dissections of human or animal tissues until later on in your course (such as the cow bone dissection in chapter 7). Thus, you will likely find yourself coming back to this chapter when you get to activities that require you to utilize the equipment and techniques described in this chapter. However, at this time it is important that you become familiar with the equipment you will be using.

Laboratory Equipment

The typical human anatomy classroom consists of laboratory tables or benches that provide ample room for use of microscopes and for dissection. If your classroom uses human cadavers, it will also have a space dedicated to the tables where the cadavers are stored. When you enter the classroom, look around and familiarize yourself with your environment. Pay particular attention to the location of the sinks, eyewash stations, and safety equipment such as first-aid kits and fire extinguishers. Your instructor will give you a detailed introduction specific to your laboratory classroom, safety procedures, and accepted protocol. The main purpose of this chapter is to introduce you to common safety devices and dissection equipment, and to help you identify some of the items your laboratory instructor may be introducing to you. *Do not* use the information in this chapter as your sole source of information on laboratory safety, as it is not intended to be a safety manual for the laboratory.

EXERCISE 1.1   Identification of Common Dissection Instruments

Several dissection instruments are commonly found in the human anatomy laboratory classroom. **Table 1.1** describes each of these instruments and their uses.

1. Obtain a dissection kit from your laboratory instructor, or view your own dissection kit if you were required to purchase your own materials.

2. Using table 1.1 as a guide, identify the instruments listed in **figure 1.1**. Then label figure 1.1.

Table 1.1	Common Dissection Instruments		
Tool	**Description and Use**	**Photo**	**Word Origin**
Blunt probe	An instrument with a blunt (not sharp) end on it. It is used to pry and poke at tissues when you do not want to damage them. Some probes come with a sharper point on the opposite end that can be used for "picking" at tissues.		*proba*, examination
Dissecting needles	Long, thick needles that have a handle made of wood, plastic, or metal. These needles are used to pick at tissues and to pry small pieces of tissue apart.		*dissectus*, to cut up
Dissecting pins	"T" shaped pins that are used to pin tissues to a dissecting tray, thus allowing you to see a particular area more easily.		*dissectus*, to cut up

(continued on next page)

Table 1.1	Common Dissection Instruments *(continued)*		
Tool	**Description and Use**	**Photo**	**Word Origin**
Dissecting tray	Metal or plastic tray used to hold a specimen. The tray is filled with wax or plastic. The wax and/or plastic is soft enough to allow you to pin tissues to it.		*dissectus,* to cut up
Forceps	Resemble tweezers, and are used for holding objects. Some are large and have tongs on the ends that assist with grabbing tough tissues. Some are small and fine (needle-nose) for picking up small objects. Forceps may also be straight-tipped or curve-tipped.		*formus,* form + *ceps,* taker
Hemostat	In surgery these are used to compress blood vessels and stop bleeding (hence the name). For dissection they are useful as "grabbing" tools. The handle locks in place, which allows you to pull on tissues without fatiguing your hand and forearm muscles.		*haimo-,* blood + *statikos,* causing to stop
Scalpel	A sharp cutting tool. Generally the blade and the blade handle will be separate, unless you are using a disposable scalpel. See specific directions in the text regarding proper use of a scalpel, as they can be dangerous!		
Scalpel blade	Both the cutting part and the disposable part of a scalpel. The number of the blade indicates the size of the blade, and must be matched with an appropriately numbered blade handle. When the blades become dull, they may be removed and replaced with a new blade. Used blades must be disposed of in a sharps container.		*scalpere,* to scratch
Scalpel blade handle	The non-disposable part of a scalpel that is used to hold the blade. The number on the handle indicates the size of the handle and is used to match it with a particular blade size. A scalpel blade handle can be a very useful tool for blunt dissection when used *without* a blade attached.		*scalpere,* to scratch
Scissors	Some scissors come with pointed blades and some have one curved (blunt) and one pointed blade. Scissors with the curved/blunt edge are used when you need to be careful not to damage some structures. To use them, direct the curved blade toward the structures you do not want to damage. Pointed-blade scissors are particularly helpful for using "open scissors" technique (see text).		*scindere,* to cut

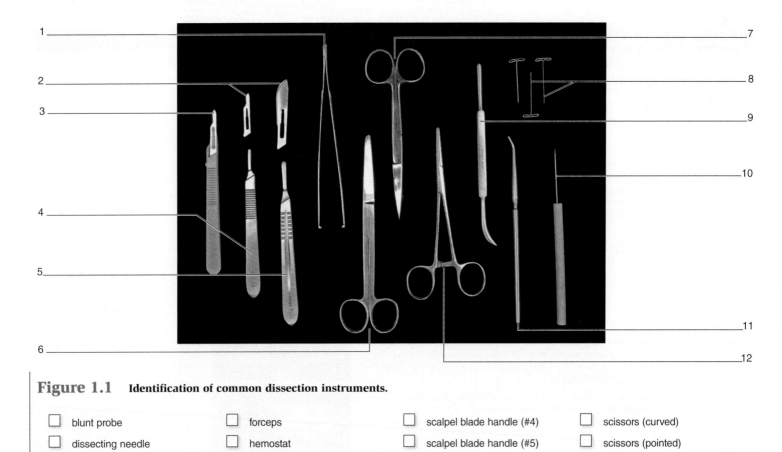

Figure 1.1   **Identification of common dissection instruments.**

☐ blunt probe   ☐ forceps   ☐ scalpel blade handle (#4)   ☐ scissors (curved)

☐ dissecting needle   ☐ hemostat   ☐ scalpel blade handle (#5)   ☐ scissors (pointed)

☐ dissecting pins   ☐ scalpel (disposable) ☐ scalpel blades

Protective Equipment

The human anatomy laboratory poses few risks, although it is important to be aware of what these are. The main risks are damage to skin or eyes from exposure to laboratory chemicals (covered in the next section) or cuts from dissection tools. As a general precaution, whenever you are working with fresh or preserved specimens (animal or human) wear protective gloves to keep any potentially infectious or caustic agents from contacting your skin. If there is a risk of squirting fluid, then wear protective eyewear (safety glasses or safety goggles). When wearing gloves, be sure to wear the correct size for your hands. If the gloves are too small, they may tear easily. If they are too big, you may not be able to handle instruments and tissues easily. When your gloves become dirty, remove them and put on a new pair. When removing gloves, start at the wrist and pull toward your fingers, turning the glove inside-out as it is removed. This will prevent any potentially damaging fluids from contacting your skin during removal of the gloves.

When using dissecting tools there is a risk of cutting yourself or others. First and foremost—*never* wear open-toed shoes to the laboratory. Dissecting tools are often dropped and will cut your feet if you are not wearing protective footwear. When using sharp tools such as scalpels, always be aware of where you are pointing the blades of those instruments. They should always be pointed away from you *and* away from others in your laboratory. If you are dissecting, be aware of where others are standing or sitting, and consider the risk posed to yourself and others if your hand were to slip. *Never* put your hands in the dissecting field when someone

else is dissecting. If another person asks for assistance holding tissues while they are dissecting, use forceps or some other device to hold the tissue so your hands are not within reach of the scalpel blade. Always be aware of the location of your scalpel, particularly when you are not using it. Individuals can be accidentally cut when they reach into a dissecting tray or table and unexpectedly find a scalpel sitting there. Whenever you are finished using dissecting pins, remove them from the specimen so they will not poke unsuspecting individuals.

Hazardous Chemicals

Relatively few chemicals are used in the human anatomy laboratory. Most of these chemicals are used to preserve, or "embalm," animal specimens or human cadavers. Generally you will not find any of these chemicals in their full-strength form in the laboratory. Instead, you will find tissues that have been previously injected with solutions containing these chemicals. Thus, safety measures in the laboratory are designed to protect you from the forms of these chemicals that you are most likely to encounter. The most common chemicals used for embalming purposes are formalin, ethanol, phenol, and glycerol.

Table 1.2 summarizes the uses and hazards of these chemicals. The majority of these chemicals are used to fix tissues and prevent the growth of harmful microorganisms, such as bacteria, viruses, and fungi. "Fixation" refers to the ability of the chemical to solidify proteins, thus preventing their breakdown. **Preservatives** both fix tissues and inhibit the growth of harmful microorganisms. Because most preservatives also dehydrate tissues, "humectants" are added to embalming solutions. **Humectants,**

Table 1.2	Common Chemicals Encountered in the Human Anatomy Laboratory				
Chemical	**Description**	**Use**	**Hazard**	**Preventing Exposure**	**Disposal**
Ethanol	Inhibits growth of bacteria and fungi.	Preservative	Flammable, so requires storage in a fire-safe cabinet. Generally safe in small quantities.	Gloves and eye protection. Rinse tissues immediately if exposed, particularly eyes. Seek medical attention if irritation persists.	Small amounts may be flushed down the sink along with plenty of water to dilute the solution.
Formalin	Fixes tissues by causing proteins to cross-link (solidify). Destroys autolytic enzymes, which initiate tissue decomposition. Inhibits growth of bacteria, yeast and mold.	Preservative	Flammable, so requires storage in a fire-safe cabinet. Toxic at full-strength. Penetrates skin. Corrosive. Burns skin. Damages lungs if inhaled. May be carcinogenic.	Gloves and eye protection. Rinse tissues immediately if exposed, particularly eyes. Seek medical attention if irritation persists.	Do not pour into sinks.
Glycerine (glycerol)	Helps control moisture balance in tissues. When used with formalin, it counteracts the dehydrating effects of formalin.	Humectant	Flammable, so requires storage in a fire-safe cabinet. Generally safe. Can pose a slipping hazard if spilled on the floor.	Gloves and eye protection. Rinse tissues immediately if exposed, particularly eyes. Seek medical attention if irritation persists.	Small amounts may be flushed down the sink along with plenty of water to dilute the solution.
Phenol	Assists formalin in fixing tissues through protein solidification. Inhibits growth of bacteria, yeast, and mold.	Preservative	Flammable, so requires storage in a fire-safe cabinet. Extremely toxic at full strength. Rapidly penetrates the skin. Corrosive. Burns skin. Damages lungs if inhaled. NOTE: when used as embalming preservative concentration (and thus toxicity) is extremely low.	Gloves and eye protection. Rinse tissues immediately if exposed. Use an eye wash station if solution gets in the eyes. Seek medical attention if irritation persists.	Do not pour into sinks.

such as glycerol, attract water. When humectants act alongside preservatives, they help keep tissues moist. Other chemicals that may be added to embalming solutions are pigments, which make the tissues look more natural, or chemicals that mask the odors of the preservative chemicals. **Formalin** and **phenol** are the most toxic and odorous preservative chemicals. Luckily, your exposure to them in the anatomy laboratory will be very low. Although it may smell as if the concentrations of these chemicals are high, the odor is often misleading because these chemicals can be detected by odor in extremely small quantities. Although the concentrations of formalin and phenol that you are exposed to may be very low, if these chemicals have been used to preserve any specimens you are working with, then you need to wear protective clothing to prevent the chemicals from contacting your skin or eyes. Use gloves whenever you handle the specimens, and use protective eyewear whenever there is a risk of chemicals getting into your eyes. If your skin is exposed, rinse it immediately. If your eyes are exposed, use the eyewash station in your laboratory to rinse your eyes thoroughly. If you have experienced contact exposure to these chemicals and your skin or eyes continue to be irritated after rinsing, consult a medical doctor.

Proper Disposal of Laboratory Waste

There are several types of waste that must be disposed of in the human anatomy laboratory. Much of this waste is "normal" waste, such as tissues, paper towels, rubber gloves, and the like. Such waste should be disposed of in the regular garbage/waste container found in the classroom. However, it is likely that at times you will find yourself needing to dispose of potentially **hazardous waste,** which must be disposed of in a special container. The general rule for determining if something is potentially hazardous or not is this: if you think someone else may be injured in any way from handling this waste, it is hazardous. Follow this rule, and be sure to ask your instructor how to properly dispose of something any time you have

a question as to whether it is hazardous or not. It is always better to err on the side of caution.

What is hazardous waste?

- Any sort of fresh tissue and/or blood is potentially hazardous
- Laboratory chemicals
- Broken glass, scalpel blades, or any other sharp item that may cut an individual who handles the waste

Sharps Containers

Sharps containers (figure 1.2) are plastic containers (often red or orange) that are used to dispose of anything "sharp," such as needles, scalpel blades, broken glass, pins, or anything else that has the potential to cut or puncture a person who handles it. Such items should NEVER go in the garbage, because they may injure anyone who handles the garbage thereafter. When in doubt, put it in the sharps container.

Biohazard Bags

Special **biohazard bags** may be available in your laboratory. These are used for biological materials such as blood or other fresh animal tissue that requires special disposal. When it comes to human blood, an item containing a small amount of blood (such as a band-aid) can be disposed of in a normal wastebasket. However, if you have a towel soaked with blood, then that must be disposed of in a biohazard bag. A biohazard bag is usually red or clear and has the symbol shown in **figure 1.3** on it. When you are dealing with tissues that must be disposed of in a biohazard bag, your instructor will generally inform you of this. Again, when in doubt, always ask before disposing of something potentially hazardous. Important note: human cadaveric tissues do *not* go into biohazard bags. They must be kept with the cadaver. Any piece of human tissue removed from a cadaver must eventually be returned to the cadaver to be cremated with the entire body.

Figure 1.2   Sharps Containers. Samples of two different models of sharps containers. Such containers allow you to place sharp objects into the container, but they cannot be removed once placed inside. Note the biohazard warning symbol on the containers.

Biohazard symbol

Figure 1.3   Biohazard Waste Symbol

EXERCISE 1.2 Proper Disposal of Laboratory Waste

1. Circle the letter (a, b, or c) of the correct waste receptacle (shown in **figure 1.4**) for each item listed below.

1.	broken scissors	A	B	C
2.	cotton swab	A	B	C
3.	dissecting pins	A	B	C
4.	glass slide	A	B	C
5.	paper towel	A	B	C
6.	rubber glove	A	B	C
7.	scalpel blades	A	B	C

(a)

(b)

(c)

Figure 1.4   Common Waste Receptacles in the Laboratory
(a) Sharps container (b) Waste basket (c) Hazardous waste bag

Dissection Techniques

The word *dissect* literally means to cut something up. Most of us have been led to think that the first thing a surgeon or anatomist does when planning to dissect is to pick up a scalpel and cut. However, skilled dissection does not always involve actually cutting tissues. In fact, the dissector's best friend is a technique called "blunt dissection." Blunt dissection specifically involves separation of tissues *without* using sharp instruments (hence the term *blunt*). When dissecting tissues, always try using blunt dissection before picking up sharp instruments such as scissors and scalpels. Sharp instruments are very handy—as they are good at cutting things. However, often students will end up cutting many things they do not wish to cut, purely by accident. Thus, being sparse and prudent in your use of sharp tools is one of the most important things you can do to perform a good dissection.

For this exercise, the demonstration of techniques will be shown using a fresh chicken purchased from a grocery store.

However, your instructor may choose another specimen for you to practice on. For now, our goal is to separate the skin from the underlying tissues such as bones and muscle (the "meat") of the specimen.

Sharp Dissection Techniques

"Sharp" dissection techniques are the techniques most familiar to most of us. These techniques involve the use of sharp instruments such as scissors and scalpels. They are "cutting" techniques. They are advantageous in that you can use them to separate tough tissues from each other, or to remove pieces of tissue from a dissection specimen. The danger in using sharp techniques is that novice and experienced dissectors alike will often end up cutting things they do not wish to cut, such as blood vessels and nerves. Thus, sharp dissection techniques should be used with care.

EXERCISE 1.3 Placing a Scalpel Blade on a Scalpel Blade Handle

Scalpels come in many forms. Some are of the disposable type, which typically means that the handle and blade come as one unit and the handle is made out of plastic **(figure 1.5)**. Often the blades and handles are separate items. This allows you to replace the blade whenever it becomes dull from use. In this exercise we will learn how to properly place a scalpel blade on a scalpel blade handle, and how to properly remove it once finished.

1. Obtain a **scalpel blade** and **scalpel blade handle** from your instructor. Scalpel blades and handles come in various sizes, and it is important to match the size of the blade to the size of the blade handle. Observe the scalpel handle and look for a number stamped on it, which will be a 3 or a 4 **(figure 1.6a)**. Next, observe the blade packet and note the number on it (figure 1.6b). A number 3 handle is used to fit number 10, 10A, 11, 12, 12D, and 15 blades. A number 4 handle is used to fit number 18, 20, 21, 22, 23, 24, 24D, or 25 blades. Larger handles and blades are generally used for making bigger, deeper cuts whereas the smaller handles and blades are generally used for finer dissection. The most commonly used combination in anatomy laboratories is the number 4 handle matched with a number 22 blade.

2. Once you have paired the scalpel handle and blade size, carefully open the scalpel blade packet halfway **(figure 1.7a)**. Note the bevel on the blade. This bevel matches the bevel on the blade handle, so that there is only one way to properly place the scalpel blade on the handle. The blade handle has a bayonet fitting that is matched to the opening on the scalpel blade (figure 1.7b), which will lock the blade in place on the handle. The safest way to place the blade on the handle is to first grasp the end of the blade using **hemostats** (figure 1.7c; table 1.1). Then, while matching the bevel on the blade to the bevel on the handle, slide the blade onto the handle until it clicks, indicating it is locked in place (figure 1.7d).

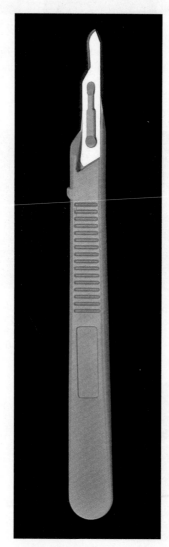

Figure 1.5 **Disposable scalpel.** The scalpel blade and scalpel blade handle are both disposable. The entire unit must be disposed of in a sharps container.

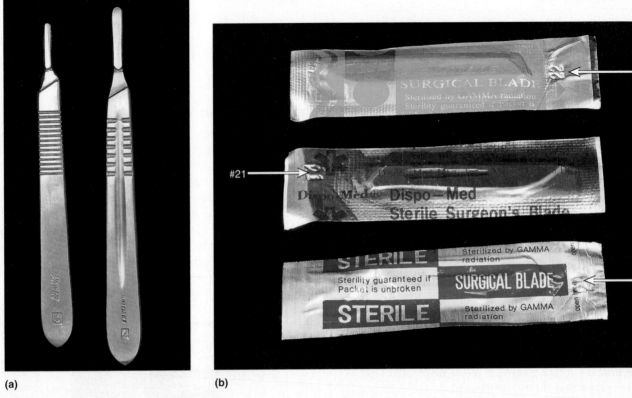

(a)   (b)

Figure 1.6   Scalpel Blade Handles and Blades.   (a) The number on the scalpel blade handle indicates what size blades will fit on the handle. (b) The number on the blade wrapper indicates the size of the blade. See text for description of what size blades fit on what size blade handles.

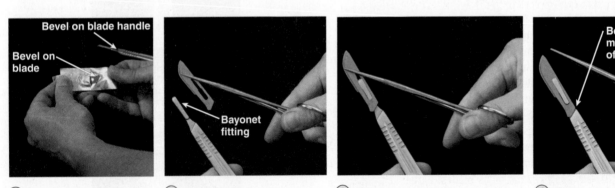

① Open the foil packet and note the bevel on the blade.

② Grasp the blade firmly using hemostat and line the blade up so that it matches the bevel on the blade handle.

③ Slide the blade onto the bayonet of the blade handle.

④ The blade should "click" as it locks in place on the blade handle.

(a)

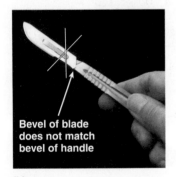

(b)

Figure 1.7   Scalpel Blade Placement.   (a) Correct procedure. (b) Incorrect placement of a blade on a blade handle. Notice that the bevel on the blade does not match up with the bevel on the blade handle. If placed in this fashion, the blade will not be secure on the handle and may slip off the handle and injure someone.

If it does not go on easily, check to make sure you haven't placed it on incorrectly (example: figure 1.7e). Now it is ready to use!

3. The safest way to remove a blade from a handle is to use a device that is both a **blade remover** and a sharps container all in one (an example is shown in **figure 1.8**). If a blade remover is not available, you will remove the blade using hemostats.

Obtain a pair of hemostats. Pointing the blade *away* from you (but not toward someone else), clamp the part of the blade nearest the handle with the hemostats **(figure 1.9a)**. Once you have a firm grip on the blade, slide it over the bayonet on the handle and away from you until the blade comes off of the handle (figure 1.9b). Using the hemostats, transport the blade to a sharps container and dispose of it in the sharps container (figure 1.9c).

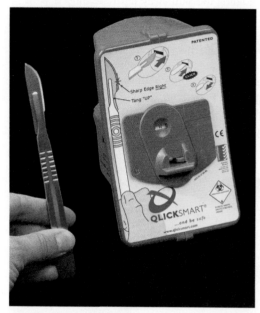

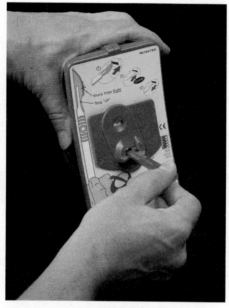

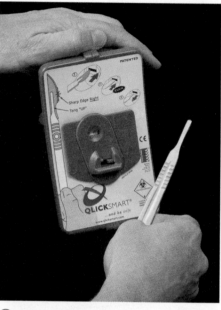

① Orient the blade and blade handle with sharp edge of the blade pointed to the right, as shown on the front of the device.

② Push the blade into the slot on the device until you hear and feel a distinct "click."

③ While holding the removal device firmly with your free hand, pull the blade handle out of the device.

Figure 1.8   **Removal of a Scalpel Blade from Handle Using All-in-One Blade Remover/Sharps Container.**

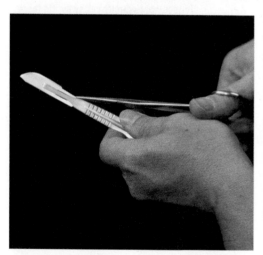

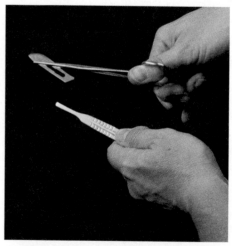

① With the blade pointed away from you and the bayonet surface of the handle also directed away from you, grasp the base of the blade with hemostats and lock the hemostats firmly to the blade.

② Slide the blade off of the bayonet on the blade handle. Again, push it away from you (and away from others in your vicinity as well).

③ Once the blade has been removed from the handle, continue to grasp it firmly with the hemostats.

Figure 1.9   **Removal of a Scalpel Blade from Handle Using Hemostats.**

EXERCISE 1.4 Dissecting with a Scalpel

1. Obtain a dissection specimen from your instructor and place it on a dissecting tray.

2. Obtain a scalpel with a blade (see exercise 1.3) and some **tissue forceps** (table 1.1). If the tissue is difficult to hold on to, you may want to use **hemostats** instead of forceps. Hemostats allow you to "lock" on to the tissue so the tissue is not dropped when you release your grip on the handle.

3. Using the forceps or hemostats, pull the skin away from the muscle on your dissection specimen (**figure 1.10a**). Carefully cut into the skin, using the tip of the scalpel blade (figure 1.10b). Note how easily a new blade cuts into the tissue. When cutting with a scalpel, you want to take care not to cut too deep, or too aggressively, or you will damage tissues. Once you have cut a small slit in the skin, observe the stringy tissue that lies between the skin and

the muscle. This tissue is a loose connective tissue called fascia (figure 1.10c), which you will learn more about in chapter 5. Because our goal is to separate the skin from the muscle, this means we want to loosen the "grip" of the fascia that holds the skin and muscle together. One way to do this is to cut into the fascia using the scalpel.

4. Next, *without* holding the skin away from the muscle with forceps, just cut into the skin using a considerable amount of pressure. Note how easily you cut through the skin directly into the muscle. This is not desirable. You may ask yourself, "If I can't manage to use forceps to easily separate skin from muscle before cutting with the scalpel, how can I avoid damaging the underlying tissues?" One technique is to push a blunt probe or scalpel handle (*without* blade attached!) into the space between the skin and muscle, thus protecting the underlying tissues. Then cut with the

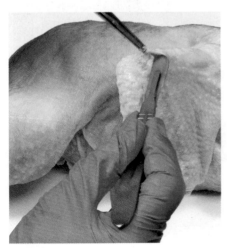

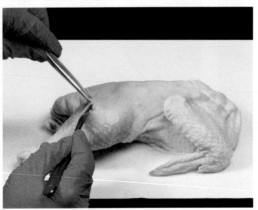

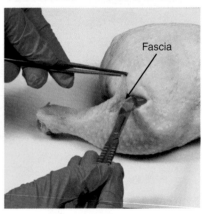

Fascia

① Pull the skin away from the underlying tissues using tissue forceps.

② Begin cutting the skin with the scalpel, taking care not to cut delicate tissues deep to the skin.

③ To assist with removal of the skin, use the scalpel to gently cut away the fascia that loosely holds the skin to the muscle. Maintain as much tension on the skin as possible and always keep the sharp end of the blade pointed toward the skin, not the underlying tissues.

Figure 1.10 **Dissecting with a Scalpel.**

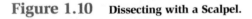

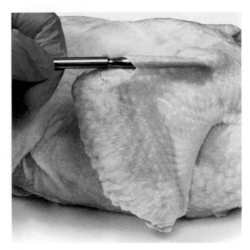

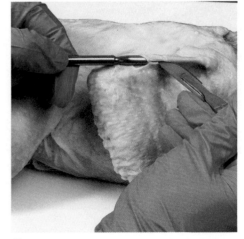

① Pull the skin away from the underlying tissues using tissue forceps and cut a small slit in the skin with the scalpel, taking care not to cut delicate tissues deep to the skin.

② Push the probe under the skin along the line where you wish to make your cut.

③ Cut the skin superficial to the probe with the scalpel. Notice how the blunt probe limits the depth at which the scalpel can cut, thus protecting underlying tissues.

Figure 1.11 **Protecting Underlying Tissues with a Probe when Dissecting with a Scalpel.**

scalpel superficial to the probe (**figure 1.11**). This way the probe limits the depth at which the scalpel blade can cut, thus protecting the underlying tissues.

5. Once you have pulled back enough of the skin to grasp it easily with forceps or hemostats, you want to put as much tension on the skin as possible, thus stretching out the fibers in the fascia (**figure 1.12**). Once the fascia is stretched, you can use the scalpel to cut the fascia and remove the skin from the specimen. When you cut with the scalpel, always point the sharp end of the blade toward the skin, not toward the underlying tissues, so as to protect those underlying tissues.

6. Practice using the forceps, hemostats, blunt probe, and scalpel to remove the skin from part of the specimen. Note areas where this is more difficult than others. As you are practicing, consider carefully whether or not the scalpel is the best instrument for the job, or if using it is causing you to damage too many tissues.

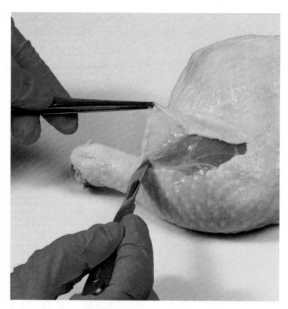

Figure 1.12   **Removing the Rest of the Skin with the Scalpel.** As you continue to cut, use the forceps to pull the skin away from the underlying tissues and be sure to keep as much tension as possible on the fascia. Cut the fascia with the scalpel, always keeping the sharp part of the scalpel blade pointed toward the skin. This way, if the blade slips accidentally, it will cut the skin, not the underlying tissues that you wish to preserve.

EXERCISE 1.5   Dissecting with Scissors

1. Using the dissection specimen you began with in exercise 1.4, you will now practice using scissors to cut tissues.

2. Obtain a pair of **pointed scissors** and **forceps** (table 1.1). Using the forceps, grasp part of the skin covering part of the specimen that you have not already dissected and pull it away from the muscle. Next, cut a small slit into the skin until you can see the fascia beneath it. Continue to lengthen your cut until it is about 2 inches long (**figure 1.13**).

3. In theory, you can begin to separate the skin from the muscle by cutting with the scissors in much the same

way as you did with the scalpel. However, there are a lot of tissues within that fascia that we may want to preserve, such as nerves and blood vessels. If we just cut away using "sharp" techniques, we may cut those vessels accidentally. For this reason, we prefer to use "blunt" dissection technique here to prevent ourselves from destroying potentially important structures. To do this we will use what is called an "open scissors" technique, so named because the dissecting action of the scissors is performed by starting with the scissors closed and then actively opening them. This is exactly the opposite of how most scissors are used.

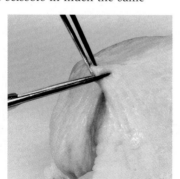

① Use tissue forceps to pull the skin away from underlying tissues.

② Make a small cut in the skin with the scissors.

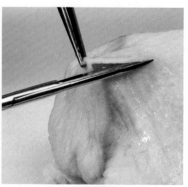

③ Now you will have created a small hole in the skin with which to begin cutting directionally along the skin.

④ Continue your cut along the skin, making sure to continue to use the tissue forceps to pull the skin away from underlying tissues before making each cut so as not to damage underlying structures.

Figure 1.13   **Dissecting with Scissors.**

4. With the scissors closed, push the tip of the scissors into the space between the skin and the muscle so that it pierces the fascia (**figure 1.14a**). Once you have the tip of the scissors within the fascia, open the scissors (figure 1.14b). Notice how this action causes the fibers within the fascia to separate from each other and loosens the hold between the skin and the fascia.

5. Continue to loosen the fascia using the open scissors technique. As you do so, notice small structures such as blood vessels and nerves that may run in the space

between the skin and the underlying tissues. Notice how the fibers in the fascia easily separate from each other without damaging those structures when you use the open scissors technique. At times, the hold of the fascia is too tight, and "open scissors" technique will no longer work effectively. At those times, switch to "normal" scissors technique and cut away the tough tissue.

6. Practice using both open and normal scissors technique to continue to remove skin from underlying tissues.

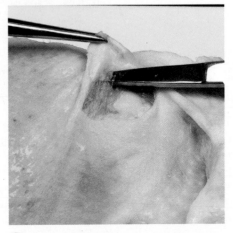

① After you have created a cut in the skin and can pull the skin away from underlying tissues using forceps, push the pointed end of a pair of scissors into the fascia deep to the skin, taking care not to pierce other underlying structures.

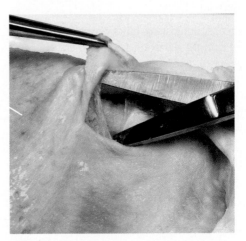

② Open the scissors, thus separating the fibers of the fascia.

Figure 1.14   Open Scissors Technique.

Blunt Dissection Techniques

As we began to see in exercise 1.3, there are times when sharp dissection technique is undesirable, because tissues might be damaged if we use sharp instruments. At these times it is best to switch to blunt dissection techniques. Blunt dissection is designed to separate tissues without damaging delicate structures.

EXERCISE 1.6 Blunt Dissection Techniques

1. Using the dissection specimen you began with in exercise 1.4, you will now practice using blunt dissection techniques.

2. Obtain a pair of **pointed scissors** and **forceps** (table 1.1). Using the forceps, grasp part of the skin of the specimen where you have not previously dissected and pull it away from the muscle. Next, make a small cut into the skin until you can see the fascia beneath it (**figure 1.15**). When you cut this way, you are using "sharp" dissection technique.

3. We have already learned that one way to separate skin from underlying tissues is to cut into the fascia. However,

there are tissues within that fascia that we may want to preserve, such as nerves and blood vessels. If we just cut away using "sharp" techniques, we may cut those structures accidentally. For this reason, we prefer to use "blunt" dissection technique whenever possible so as to prevent ourselves from destroying potentially important structures. "Open scissors" is one blunt dissection technique that you learned in exercise 1.4. Use this technique to loosen the hold between the skin and the fascia on your specimen.

4. Once you have created a space between the skin and muscle that is large enough for you to put a finger in, set

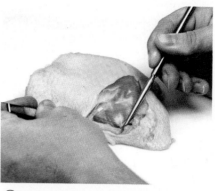

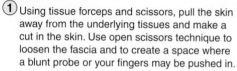 Using tissue forceps and scissors, pull the skin away from the underlying tissues and make a cut in the skin. Use open scissors technique to loosen the fascia and to create a space where a blunt probe or your fingers may be pushed in.

② Using your fingers, pull the skin away from the underlying tissues. When necessary, use a sharp instrument to cut any fascia that is very tough and won't separate using blunt techniques.

③ A blunt probe can be moved around under the skin to gently separate the connective tissue without damaging underlying structures.

Figure 1.15 **Blunt Dissection.** Blunt dissection techniques involve separating tissues with your fingers or blunt instruments such as a probe.

the scissors down. You will now proceed to separate the skin from the muscle using only your fingers (figure 1.15b). Because you are not using any sharp instruments to perform this, it is referred to as "blunt" dissection.

5. Obtain a blunt probe (table 1.1). A blunt probe can be used in place of your fingers to separate structures when your fingers are too large (figure 1.15b). Because it does not cut the tissue, this is also a "blunt" dissection technique.

6. Other items that can be used for blunt dissection are scalpel blade handles (*without* the blades on them!), or the rounded ends of forceps and the like. Spend some time practicing the use of these tools by trying out different tools to separate skin from muscle in different regions of the dissection specimen. The general rule of thumb is that you want to start with sharp dissection techniques to cut slits in the skin, but then transition to blunt techniques whenever possible so as to prevent accidental damage to underlying tissues.

Chapter 1: The Human Anatomy Laboratory

Name:_____

Date:_____ Section:_____

POST-LABORATORY WORKSHEET

1. Briefly describe what fascia is. _____

   _____

2. You have just finished dissecting a fresh cow bone as part of the day's laboratory activities. What do you think is the most appropriate way to dispose of this waste?

   _____

   _____

   _____

3. Where do you dispose of used scalpel blades?

   _____

   _____

   _____

4. You are dissecting the wing of a chicken, and the skin is held tight to the bones beneath it. You would like to remove the skin, but in doing so you would like to preserve the bone and muscle beneath the skin. What tools might you use, and how will you use them, to complete this task? (There is more than one right answer to this question; use your best judgment.)

   _____

   _____

   _____

   _____

   _____

   _____

5. List two common chemicals used to preserve human or animal specimens.

   _____

6. Formalin and phenol are potentially hazardous chemicals. What steps should be taken to prevent exposure to these chemicals?

   _____

   _____

   _____

   _____

7. What should you do if your skin or eyes are exposed to formalin or phenol?

   _____

   _____

   _____

Orientation to the Human Body

INTRODUCTION

The human body is both beautiful and complex. You are about to embark on an incredible journey in which you will develop an understanding of that beauty and complexity. To be successful in this venture, you must do a great deal of work. The good news is that if you put in the time, you *will* succeed. Anatomy is not a conceptually difficult subject, but historically it is considered quite difficult mainly because of the enormous time commitment it requires.

To put things in perspective, consider this: a beginning student in human anatomy is typically asked to learn more new words in a one-semester anatomy class than a beginning student learns in the first semester of a foreign language class. In fact, you *will* be learning a new language. This *language of anatomy* has its origins principally in Latin and Greek. One of your primary tasks will be to establish a firm understanding of the meanings of common word origins. I encourage you to look up the origins of all words that are new or unfamiliar to you, using the list of common word origins located on the inside front and back covers of this manual. This will require a small amount of work each time you look up a word, but over time you will develop an impressive

OUTLINE and OBJECTIVES

Module 1: BODY ORIENTATION

vocabulary of anatomical/medical terms. In addition, you will establish a rich knowledge base that will allow you to begin to interpret the meanings of new words as you encounter them later in the course.

Let's practice by analyzing the origin of the word *anatomy*. This word can be broken down into two parts, *ana-* and *-tomé*. The word part *ana-* means "apart." The word part *-tomé* means "to cut." Thus, the word **anatomy** literally means "to cut apart." Though you might not literally be cutting up human bodies as medical students do and early anatomists did, you will at the very least be *conceptually* "cutting up" the body to understand its component parts.

Chapter 2: Orientation to the Human Body

Name: _____

Date: _____ Section: _____

PRE-LABORATORY WORKSHEET

1. In the space below, make a simple line drawing of an individual in the anatomic position. Next to your drawing, describe the anatomic position using your own words.

\

_____

_____

_____

_____

_____

_____

_____

_____

_____

_____

2. A plane that separates the body into superior and inferior portions is a _____ plane.

3. A plane that separates the body into right and left portions is a _____ plane.

4. A plane that separates the body into anterior and posterior portions is a _____ plane.

5. Match each directional term in Column A with its definition in Column B.

Column A	*Column B*
_____ 1. anterior (ventral)	a. in front of; toward the front surface
_____ 2. caudal	b. in back of; toward the back surface
_____ 3. cranial	c. above
_____ 4. deep	d. below
_____ 5. distal	e. at the tail end of the body
_____ 6. inferior	f. at the head end of the body
_____ 7. lateral	g. toward the midline of the body
_____ 8. medial	h. away from the midline of the body
_____ 9. posterior (dorsal)	i. on the inside; beneath the surface of the body
_____ 10. proximal	j. on the outside; toward the surface of the body
_____ 11. superficial	k. closer to the attachment point of a limb to the trunk
_____ 12. superior	l. farther from the attachment point of a limb to the trunk

6. Match each regional name in Column A with its description in Column B.

   Column A

   _____ 1. axillary

   _____ 2. brachial

   _____ 3. buccal

   _____ 4. carpal

   _____ 5. cephalic

   _____ 6. cervical

   _____ 7. digital

   _____ 8. femoral

   _____ 9. gluteal

   _____ 10. inguinal

   _____ 11. lumbar

   _____ 12. mental

   _____ 13. orbital

   _____ 14. plantar

   _____ 15. umbilical

   _____ 16. vertebral

   Column B

   a. arm

   b. armpit

   c. buttock

   d. cheek

   e. chin

   f. eye

   g. fingers or toes

   h. groin

   i. head

   j. lower back; loin

   k. navel

   l. neck

   m. sole of the foot

   n. spinal column

   o. thigh

   p. wrist

7. The body cavity that encases the brain is the _____ cavity, and the body cavity that encases the spinal cord

   is the _____. These two cavities combined make up the _____ _____.

8. The body cavity that surrounds the heart is the _____ cavity.

9. The body cavity that contains most of the digestive, reproductive, and urinary system organs is the _____ cavity.

10. The abdominopelvic cavity can be divided into a total of _____ quadrants or _____ regions. The central point of reference for

    dividing the abdominopelvic cavity into quadrants or regions is the _____.

Gross Anatomy

Anatomic Terminology and the Anatomic Position

Because the human body can be placed in numerous positions and each position has the potential to change the definition of terms such as *front* and *back*, in anatomy and medicine all such terms refer to the body when it is placed in the **anatomic position**. The anatomic position (**figure 2.1**) is the position of the body when one is standing up, facing forward (anterior), with the palms of the hands facing forward (anterior). In this position, no two bones of the body cross each other. The anatomic position is similar to the natural position of an individual when standing up, except for the position of the wrist and hand. Because the position of the wrist and hand is somewhat unusual in the anatomic position, always be extra careful when using directional terms that relate to the upper limbs.

Anatomic Planes and Sections

The study of anatomy that involves viewing sections (or slices) of an organ or the body is called **sectional anatomy**. An understanding of sectional anatomy is increasingly important in clinical settings, where medical imaging techniques such as CT (computed tomography) and MRI (magnetic resonance imaging) scans are used extensively. The ability to analyze such scans requires special skill that is developed over time. At this point in your education, it is important that you simply understand the terms related to the three major planes that divide the body into sections: the coronal plane, the transverse plane, and the sagittal plane (**table 2.1**).

> ## Study Tip!
>
> A **plane** (*planus*, flat) is an imaginary two-dimensional flat surface. One way of visualizing a plane is to imagine a transparency film passing through the body; a **section** (*sectio*, a cutting) is a slice made along one of these two-dimensional planes.

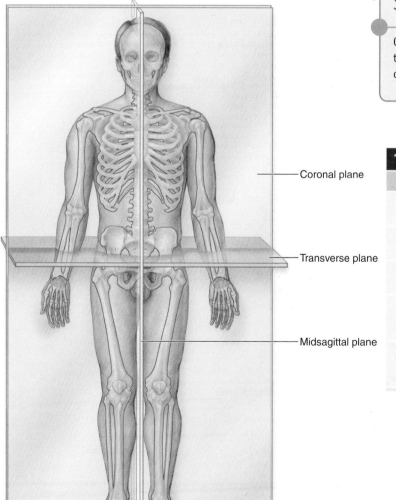

Coronal plane

Transverse plane

Midsagittal plane

Table 2.1	Anatomic Planes
Body Plane	**Description**
Coronal (frontal)	Separates anterior portions from posterior portions.
Transverse (horizontal)	Separates superior portions from inferior portions.
Sagittal (parasagittal)	Separates right portions from left portions.
Midsagittal (median)	Separates right and left portions equally; runs down the midline of the body.
Oblique	Runs at an angle to any of the three main planes of the body (coronal, transverse, or sagittal).

Figure 2.1   **Anatomic Position and Body Planes.** In the anatomic position, no two bones are crossed.

EXERCISE 2.1 Determining Anatomic Planes and Sections

1. **Figure 2.2** shows photographs of a brain that has been sectioned along different planes. Determine which plane the brain was sectioned along for each photo, then enter the information in the appropriate spaces in figure 2.2 below. Answer choices are provided in the figure legend, and each answer may only be used once. Refer to the chapter on the brain in your main textbook if you need help getting oriented.

2. *Optional Activity:* **AP|R**—1. **Body Orientation**—Review all dissections in this module to familiarize yourself with general anatomic terminology.

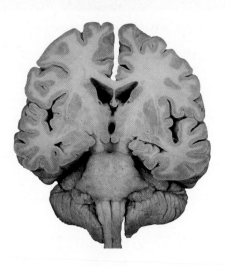
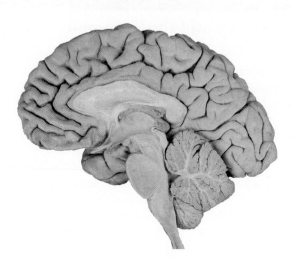

1 _____
Anterior view

2 _____
Medial view

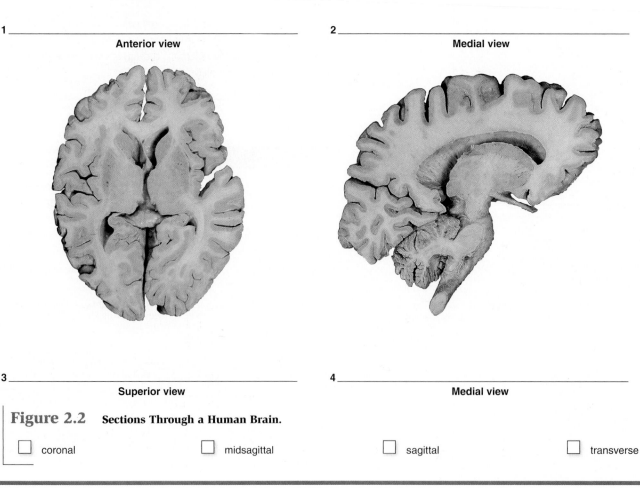

3 _____
Superior view

4 _____
Medial view

Figure 2.2   **Sections Through a Human Brain.**

☐ coronal   ☐ midsagittal   ☐ sagittal   ☐ transverse

Directional Terms

When we use everyday language we often use terms like "front" and "back," and "on top of" or "on the bottom of" to give directions. While this is perfectly appropriate in everyday language, it can create a great deal of confusion when referring to directions in the human body. For instance, when we say "on top of," we need to know what part of the body "on top of" refers to, and different people might use the term in different ways or in reference to different structures. Furthermore, they may be thinking "on top of" relative to different body positions. This becomes problematic in a medical setting because it leads to confusion, which has the potential to create severe consequences for a patient. Thus, we use an agreed-upon set of directional terms in anatomy and medicine to be as specific as possible when describing directions, but also to ensure we are all speaking the same language. **Table 2.2** lists these directional terms and gives definitions of each of them. In exercise 2.2 you will practice using these directional terms.

Table 2.2	Directional Terms
Directional Term	**Definition**
Anterior (ventral)	Toward the front of the body (the belly side)
Posterior (dorsal)	Toward the back of the body (the back side)
Superior	Above; closer to the head
Inferior	Below; closer to the feet
Cranial (cephalic)	At the head end of the body
Caudal	At the tail end of the body
Medial	Toward the midline of the body
Lateral	Away from the midline of the body
Superficial	Toward the surface of the body; on the outside
Deep	Beneath the surface of the body; on the inside
Proximal	Near; closer to the attachment point of a limb to the trunk
Distal	Far; farther from the attachment point of a limb to the trunk

EXERCISE 2.2   Using Directional Terms

Figure 2.3 shows a posterior view of a human. Three locations marked on the body are denoted with the numbers 1–3. In the spaces below, describe the locations of markings 1–3 as specifically as possible using correct anatomic terminology. When you are finished, compare your answers to those of other students in your class to see how similar your answers are.

Location 1

_____

_____

_____

Location 2

_____

_____

_____

Location 3

_____

_____

_____

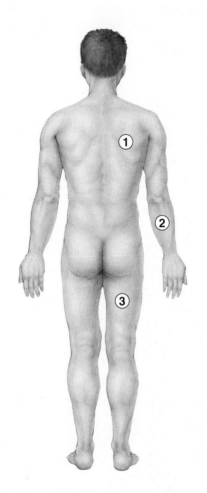

Figure 2.3   **Posterior View of an Individual with Three Reference Locations (1–3) Marked.**

☐ antebrachial   ☐ femoral   ☐ thoracic

Study Tip!

The directional terms *superior* and *inferior* are used when describing one structure with respect to another structure in the trunk of the body. The directional terms *proximal* and *distal* are used when describing the position of one structure with respect to another structure on the limbs. Thus, it is more appropriate to say the elbow is located *proximal* to the wrist, rather than to say it is superior to the wrist. On the other hand, it is quite appropriate to say the thorax is located *superior* to the abdomen.

Regional Terms

Just as with directional terms, we have common, everyday terms that we use to describe regions of the body, like *arm* or *back*. The correct anatomic terms to describe regions of the body are basically synonyms for these terms, and are closer to the Latin or Greek derivative of the term. **Table 2.3** correlates some commonly used regional terms in anatomy with words we use to describe the same region using everyday language. Your main textbook has a much more inclusive table of regional terms, as well as a figure describing these regions, which you should use as references when performing exercise 2.3.

Table 2.3	Selected Regional Terms
Regional Term	**Description**
Axillary	Armpit
Brachial	Arm
Buccal	Cheek
Carpal	Wrist
Cephalic	Head
Cervical	Neck
Digital	Fingers or toes
Femoral	Thigh
Gluteal	Buttock
Inguinal	Groin
Lumbar	Lower back
Mental	Chin
Orbital	Eye
Plantar	Sole of the foot
Umbilical	Navel
Vertebral	Spinal column

EXERCISE 2.3 Using Regional Terms

1. Identify the body regions listed in **figure 2.4** on your own body.

2. Label figure 2.4 with the appropriate regional terms.

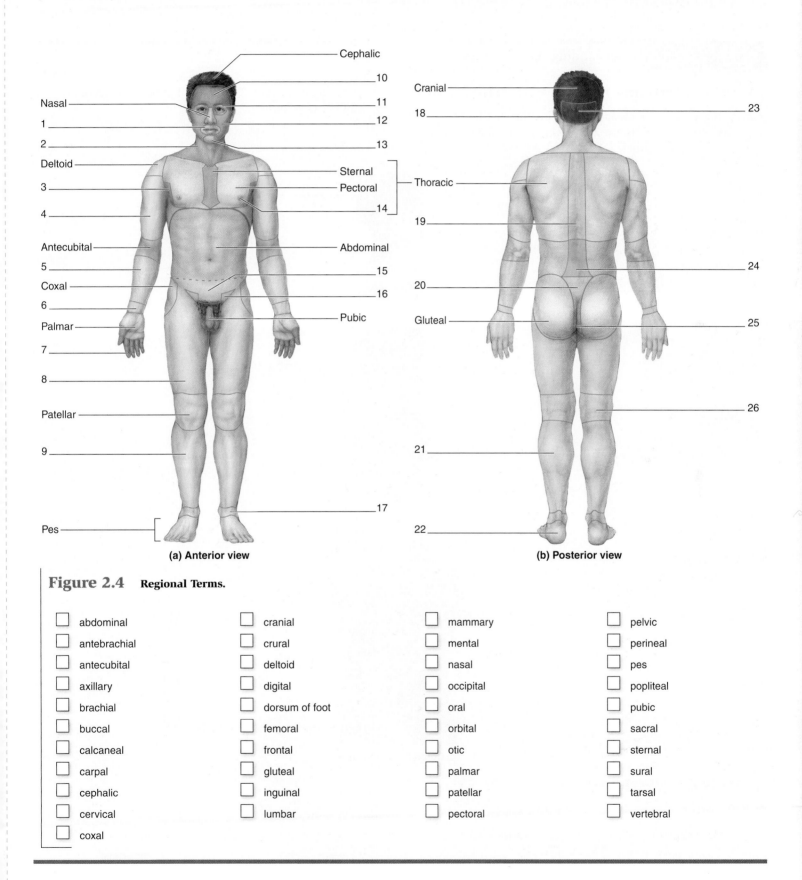

Figure 2.4 **Regional Terms.**

- ☐ abdominal
- ☐ antebrachial
- ☐ antecubital
- ☐ axillary
- ☐ brachial
- ☐ buccal
- ☐ calcaneal
- ☐ carpal
- ☐ cephalic
- ☐ cervical
- ☐ coxal

- ☐ cranial
- ☐ crural
- ☐ deltoid
- ☐ digital
- ☐ dorsum of foot
- ☐ femoral
- ☐ frontal
- ☐ gluteal
- ☐ inguinal
- ☐ lumbar

- ☐ mammary
- ☐ mental
- ☐ nasal
- ☐ occipital
- ☐ oral
- ☐ orbital
- ☐ otic
- ☐ palmar
- ☐ patellar
- ☐ pectoral

- ☐ pelvic
- ☐ perineal
- ☐ pes
- ☐ popliteal
- ☐ pubic
- ☐ sacral
- ☐ sternal
- ☐ sural
- ☐ tarsal
- ☐ vertebral

Body Cavities and Membranes

Many organs within the body are compartmentalized and separated from each other by a *body cavity*. Compartmentalizing the organs this way allows the separate organs to perform their functions without interfering with the functioning of other organs. For example, the pumping action of the heart does not interfere with the expansion and contraction of the lungs because each organ is enclosed in its own cavity. In addition, the encasement of organs within separate cavities helps to prevent the spread of infection from one region of the body to another.

EXERCISE 2.4 Understanding Body Cavities

1. Observe a human torso model or a human cadaver.

2. Using your textbook as a guide, identify the body cavities listed in **figure 2.5** on the torso model or human cadaver. Then label figure 2.5 with the appropriate terms.

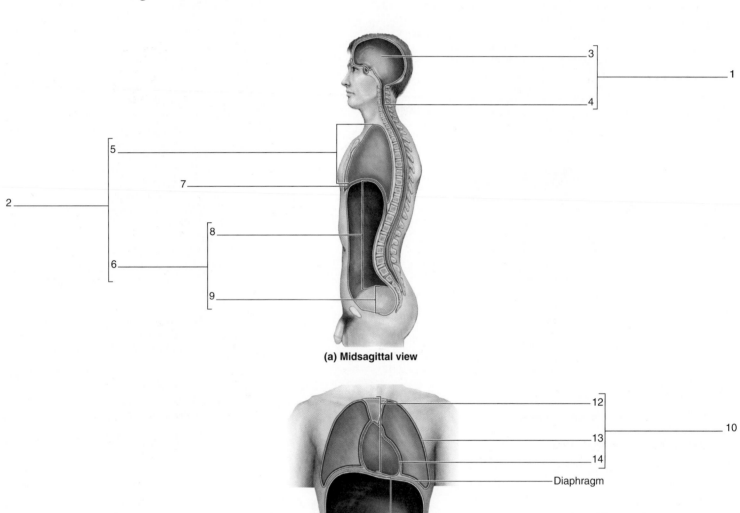

(a) Midsagittal view

(b) Coronal (frontal) view

Figure 2.5   Body Cavities.

- ☐ abdominal cavity
- ☐ abdominopelvic cavity
- ☐ cranial cavity
- ☐ diaphragm
- ☐ mediastinum
- ☐ pelvic cavity
- ☐ pericardial cavity
- ☐ pleural cavity
- ☐ posterior aspect
- ☐ thoracic cavity
- ☐ ventral cavity
- ☐ vertebral canal

Abdominopelvic Regions and Quadrants

To describe the locations of organs in the abdominopelvic cavity, we need a set of terms that allows us to specify where within the cavity an organ or tissue is located. There are two approaches used to do this. The first approach is to divide the abdominopelvic cavity into **quadrants** (*quad*, four). This is done by passing one imaginary line vertically through the *umbilicus* (belly button) and another horizontally through the umbilicus. The four resulting quadrants are: right upper, left upper, right lower, and left lower (**figure 2.6a**). Because this approach is simple, it is the approach used most often in a clinical setting. The second approach is to divide the abdominopelvic cavity into **regions**. This is done by drawing one vertical line to the left of the umbilicus and another to the right of the umbilicus (at the *midclavicular line*; a vertical line that passes through the midpoint of the clavicle), then drawing one horizontal line superior to the umbilicus and another inferior to the umbilicus. The result is a grid similar to a tic-tac-toe layout with nine regions formed (figure 2.6b). This approach allows us to be more specific about describing the locations of organs and tissues within the abdominopelvic cavity. The resulting regions are: umbilical, epigastric (*epi-*, above, + *gastēr*, belly), hypogastric (*hypo-*, below, + *gastēr*, belly), right and left hypochondriac (*hypo-*, below, + *chondro-*, cartilage; as in the cartilages that attach ribs to sternum), right and left lumbar (*lumbus*, a loin), and right and left iliac (*ilium*, flank). Exercise 2.5 is designed to introduce you to the locations of major organs within the abdominopelvic cavity and to help you become comfortable with the terms used to describe the quadrants and regions.

EXERCISE 2.5 Locating Major Body Organs Using Abdominopelvic Region and Quadrant Terminology

1. Using figure 2.6 as a guide, identify the following structures on a human torso model or on a human cadaver:

   ☐ left kidney   ☐ spleen

   ☐ liver   ☐ stomach

   ☐ pancreas   ☐ urinary bladder

   ☐ small intestine

2. Based on your observations of the cadaver or human torso model, complete the chart on p. 30 by indicating the quadrant(s) or region(s) in which each organ is found.

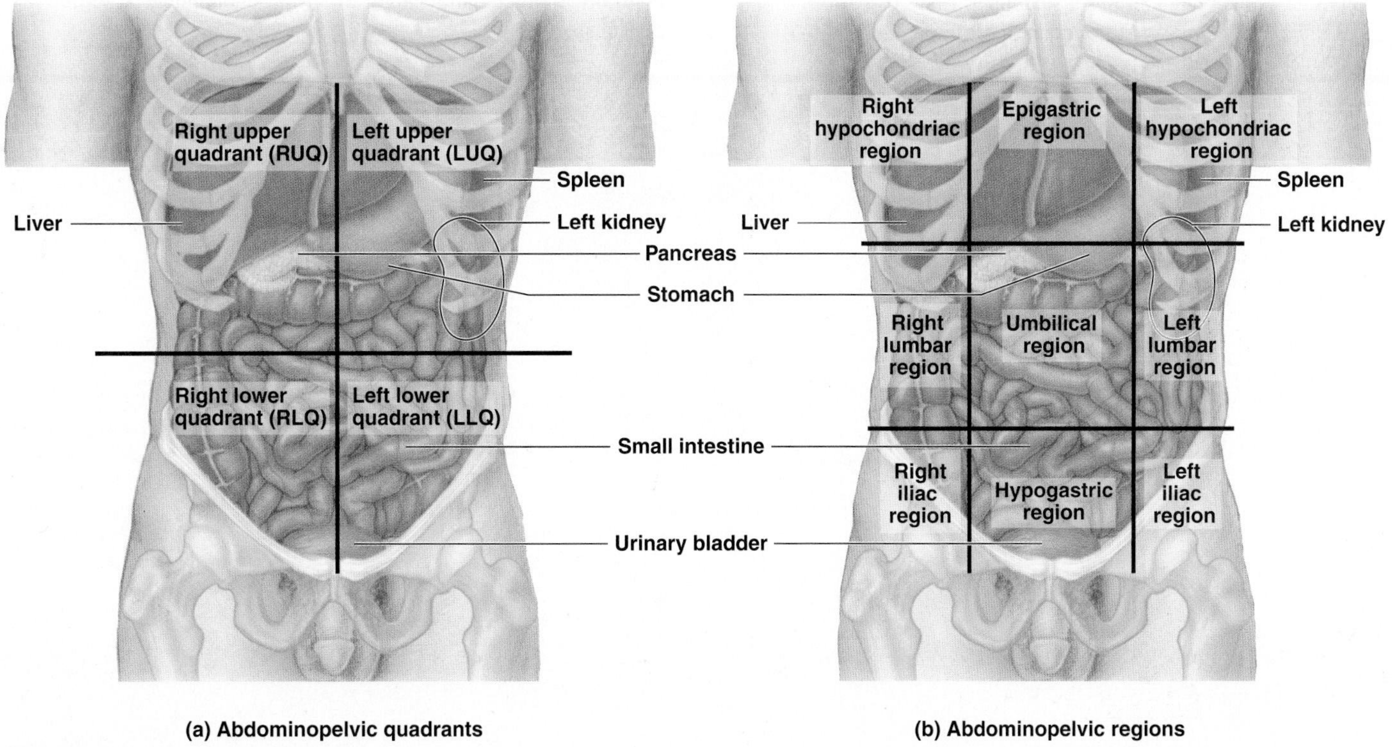

(a) Abdominopelvic quadrants

(b) Abdominopelvic regions

Figure 2.6   **The Abdominopelvic Cavity.** The abdominopelvic cavity can be subdivided into (a) four abdominopelvic quadrants, or (b) nine abdominopelvic regions.

Organ	Quadrant(s)	Region(s)
Left kidney		
Liver		
Pancreas		
Small intestine		
Spleen		
Stomach		
Urinary bladder		

WHAT DO YOU THINK?

1. A man who had been in a car accident arrived at the emergency room. He was awake and alert and was able to tell the physician that he was experiencing severe pain near his lower ribs on the left side of his body and in his abdomen. Upon palpation, the patient's abdomen was rigid, indicating the possibility of internal bleeding. The physician suspected that broken ribs may have injured an organ within the abdominal cavity. Which organ would most likely be injured in this case? What abdominopelvic **quadrant** is this organ located in? What abdominopelvic **region** is it located in?

Chapter 2: Orientation to the Human Body

POST-LABORATORY WORKSHEET

1. The anatomic position requires that the forearm be rotated to face anteriorly. Why is the forearm rotated this way in the anatomic position?

   _____

   _____

2. Fill in the figure below with the names of the appropriate abdominopelvic regions:

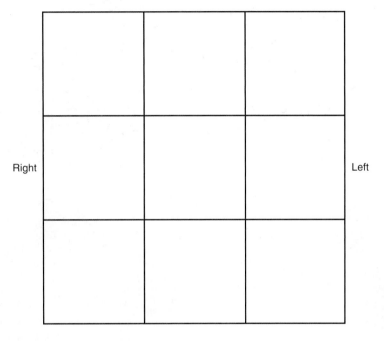

Right   Left

3. Complete the following table by listing the major body organs located within each of the listed cavities.

Body Cavity	Major Organs Within the Cavity
Abdominal Cavity	
Cranial Cavity	
Pelvic Cavity	
Pericardial Cavity	
Pleural Cavity	
Thoracic Cavity	
Vertebral Canal	

4. Label these photos of a human torso model using the terms below.

☐ brain

☐ diaphragm

☐ esophagus

☐ eye

☐ heart

☐ large intestine

☐ left kidney

☐ left lung

☐ liver

☐ pancreas

☐ small intestine

☐ spleen

☐ stomach

☐ trachea

☐ ureter

☐ urinary bladder

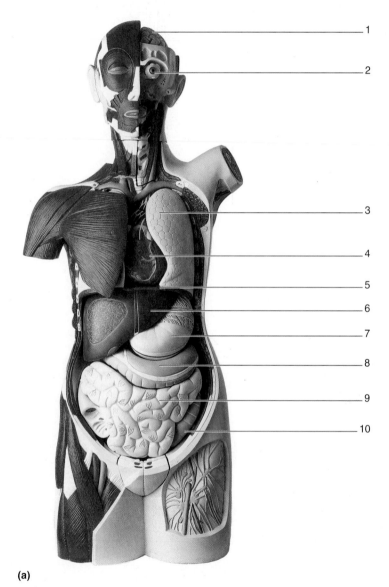

(a)

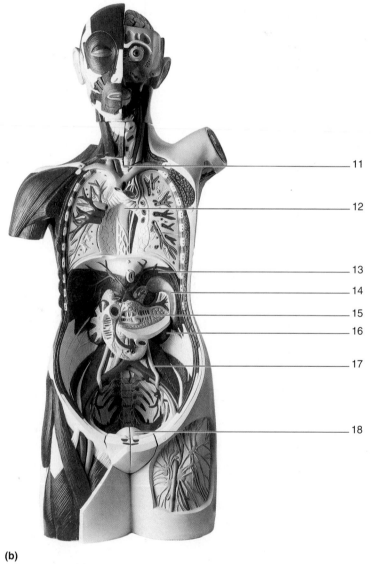

(b)

5. Using directional term(s), describe the location of the pancreas with respect to the liver.

   _____

   _____

6. A physician would like to obtain a radiographic view of both the liver and the spleen of her patient, so she orders a CT scan. What plane of section does the physician ask the radiology department to scan so she may view both of these organs in the same image?

   _____

7. The same physician would like to obtain a radiographic view of both the right lung and the liver of her patient, so she orders a CT scan. What plane of section does the physician ask the radiology department to scan so she may view both of these organs in the same image?

   _____

8. A female is suffering from menstrual cramps, which originate in the uterus. In which abdominopelvic region is the pain localized? _____

   _____

9. A teenage boy is diagnosed with appendicitis (inflammation of the appendix) and must undergo an operation. To remove the appendix, the surgeon

   will operate on the boy's _____ abdominopelvic quadrant.

10. For each of the following, insert the most appropriate directional term.
    a. The elbow is located _____ to the wrist.
    b. The umbilicus is located _____ to the sternum.
    c. The chin is located _____ to the mouth.
    d. The nose is located _____ and _____ to the ears.
    e. A scratch wound, which does not penetrate the skin, is said to be a _____ wound. In contrast, a stab wound, which penetrates the skin, is referred to as a _____ wound.

The Microscope

3

INTRODUCTION

The study of anatomy involves observation of both gross (large) and microscopic (small) structures. To get the most out of your experience observing microscopic structures, it is imperative that you know how to use the compound microscope properly. Although you have probably used a microscope before, pay very careful attention to the instructions in this chapter. Even though many students have been taught the proper use and care of the microscope, most continue to use poor technique. This, in turn, causes great frustration and is a major impediment to the learning process. Treat this laboratory exercise as an opportunity to refine and improve upon your technique.

OUTLINE and OBJECTIVES

Focus | Caring for the Compound Microscope

Microscopes are very expensive instruments. You should always use great care when you transport, use, and store them.

- *Workspace:* Keep your workspace clear of all materials except for the microscope, your laboratory manual, lens paper, and microscope slides. Remove or restrict any loose articles of clothing or jewelry so they do not interfere with your work. Long necklaces that dangle in front of your neck can damage the microscope. If you have long hair, tie it back to keep it out of the way.

- *Transport: Always* use both hands when carrying the microscope. Place one hand on the base of the microscope and the other on the arm and use care to always move deliberately when you are carrying the microscope. (**figure 3.1**).

- *Lens care:* Lenses are some of the most expensive and delicate parts of the microscope and should be treated with great care. Special lens paper is the only thing that should ever be used to clean dirty lenses. This paper is designed so it will not scratch the lenses. *Never* use facial tissues, paper towels, articles of clothing, or anything else to clean the lenses because they may scratch the lenses. A special cleaning solution can be used with the lens paper, although lens paper generally does a fine job of cleaning when used alone. *Never* use saliva, water, or other fluids to moisten the lens paper.

- *Storage:* When you are finished using the microscope, move the stage to its lowest position. Make sure that no slides are left on the stage, and then rotate the nosepiece so the lowest-power objective lens is over the stage. Wrap the power cord around the base of the microscope and replace the dust cover.

- *Handling slides:* Always hold a slide by the edges so you do not leave smudges and fingerprints, which will interfere with your ability to view the slide clearly. If the slide is dirty, clean it using lens paper before placing it on the microscope stage. If a slide becomes cracked or broken, notify your instructor so that it can be disposed of properly. Broken slides should be placed in a special broken glass container, never in the garbage can.

Other things to be aware of:

- When removing the microscope cover, be careful not to pull the ocular lens off with the cover

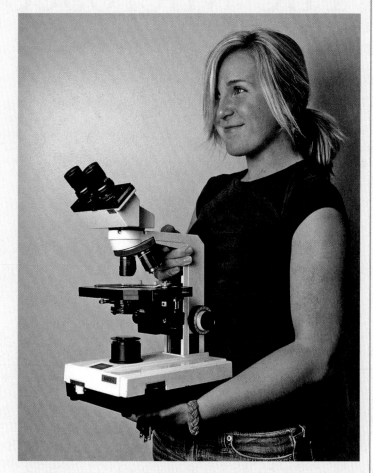

Figure 3.1   **Proper Technique for Carrying a Microscope.**

- When using the course adjustment knob, be careful not to allow the objective lens to smash into the slide, especially on high power.

Chapter 3: The Microscope

Name: _____

Date: _____ Section: _____

PRE-LABORATORY WORKSHEET

1. Match the microscope part in column A with its definition in column B.

Column A

_____ 1. condenser

_____ 2. mechanical stage

_____ 3. voltage regulator

_____ 4. iris diaphragm

_____ 5. arm

_____ 6. field of view

_____ 7. ocular lens

_____ 8. fine adjustment knob

_____ 9. diopter adjustment ring

_____ 10. working distance

_____ 11. nosepiece

_____ 12. eyepiece

_____ 13. objective lens

_____ 14. base

_____ 15. depth of field

_____ 16. mechanical stage controls

_____ 17. power switch

_____ 18. coarse adjustment knob

_____ 19. head

_____ 20. illuminator

_____ 21. total magnification

Column B

a. the light source located in the base of the microscope

b. a device on the eyepiece that is used to focus the eyepiece

c. everything that is visible when looking through the eyepiece

d. a measure of how much of the total thickness of the specimen is in focus

e. the part of the microscope that connects the head to the base

f. a part of the condenser that controls the amount of light that passes through

g. knobs that are used to move the slide on the stage

h. the platform that a slide is placed upon

i. a device that connects the objective lenses to the head of the microscope

j. lenses that are attached to the nosepiece

k. lenses that are located within the eyepieces

l. a switch that is used to turn the power on or off

m. calculated by multiplying the ocular lens power by the objective lens power

n. the distance between the mechanical stage and the tip of the objective lens

o. a knob that is used to regulate the brightness of the light

p. provides attachment and support for the objective lenses

q. a knob that moves the mechanical stage up and down in small increments

r. a tube located on the head of the microscope that a person looks into

s. a knob that moves the mechanical stage up and down in large increments

t. a device that is used to focus the light coming from the illuminator

u. the bottom support of the microscope

2. In the space below, draw or describe in words the proper procedure for holding and transporting a microscope.

3. Explain how to calculate the total magnification of the microscope. _____

_____

4. As total magnification power increases, depth of field _____.

5. As total magnification power increases, diameter of the field of view _____.

IN THE LABORATORY

In this laboratory exercise you will learn the proper use and care of the compound microscope, which is used to view slides of cells and tissues. You will also learn how depth of field and diameter of the field of view change as total magnification increases. After performing these exercises you will be prepared for the observations of histology (tissues) that you will be making in later chapters.

The Compound Microscope

A compound microscope is used to view structures that are not visible to the naked eye, and most compound microscopes can magnify images anywhere from 40 to 1000 times (40× to 1000×) normal size.

EXERCISE 3.1 Parts of a Compound Microscope

1. Obtain a compound microscope.

2. Using **figure 3.2** and **table 3.1** as guides, identify all the parts listed in table 3.1 on your compound microscope.

3. Record the magnifications of the lenses on your microscope in the spaces below.

   Magnification of ocular lenses: _____

   Magnification of objective lenses:

   Scanning _____   Low _____   High _____

4. To determine the total magnification of an image, simply multiply the magnification of the ocular lens with that of the objective lens. For example, if the magnification of the ocular lens is 10× and the magnification of the objective lens is 4×, then the total magnification is 40×. In the spaces below, calculate the total magnification of your microscope for each of these objective lenses: scanning, low-power, and high-power.

   Total magnification (ocular × objective):

   Scanning _____   Low _____   High _____

5. Use the information from steps 3 and 4 to answer question 3 of the Post-Laboratory Worksheet (p. 45).

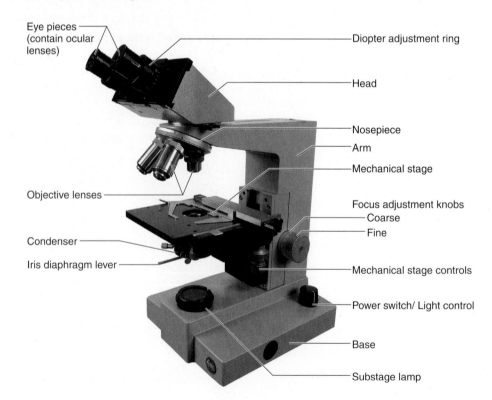

Eye pieces (contain ocular lenses)
Diopter adjustment ring
Head
Nosepiece
Arm
Mechanical stage
Objective lenses
Focus adjustment knobs
Coarse
Fine
Condenser
Iris diaphragm lever
Mechanical stage controls
Power switch/ Light control
Base
Substage lamp

Figure 3.2   **Parts of a Compound Microscope.**

Table 3.1	Parts of the Microscope
Term	**Description and Usage**
Arm	Connects the head to the base. When transporting the microscope, always grasp the arm with one hand.
Base	The bottom part of the microscope. It supports the entire microscope and encases much of the electrical wiring. When transporting the microscope, one hand should always be placed under the base for support.
Condenser	A device used to focus the light coming from the illuminator/substage lamp. It generally works best in the position closest to the mechanical stage.
Coarse adjustment knob	Moves the mechanical stage up and down in relatively large increments. It is used to position a specimen into your field of view when you first begin to scan a slide at low power. It should never be used when the microscope is on the higher powers.
Depth of field	How much of the total thickness of the specimen is in focus.
Diopter adjustment ring	Changes the focus of the eyepiece. Used to make adjustments when you have better vision in one of your eyes and you are not using corrective lenses. To use this device begin by closing the eye that looks through the ocular lens containing the diopter adjustment ring. Looking through the *other* ocular lens, bring the sample into focus. Then, open the eye that was closed and bring the sample into focus for that eye using the diopter adjustment ring.
Eyepieces	The parts of the microscope that you look into. Always use both eyepieces. The eyepieces are movable so you can adjust them to the width of your own eyes.
Field of view	Everything visible when looking through the eyepieces.
Fine adjustment knob	Moves the mechanical stage up and down in small increments. Once a specimen is brought into view using the course adjustment knob, the fine adjustment is used to bring the specimen into focus.
Head	The part of the microscope that provides attachment and support for the objective and ocular lenses. It also serves as the support for the eyepieces.
Substage lamp (illuminator)	The light source located in the base of the microscope. When it is turned on, light passes through the specimen on the stage, through the lenses of the microscope, and ultimately hits your eye, allowing you to see the specimen.
Iris diaphragm	A part of the condenser that can be opened or closed to control the amount of light passing through the condenser. More light (iris open) decreases contrast between structures. Less light (iris closed) increases contrast.
Light control (voltage regulator)	This is typically a rotating or sliding knob that alters the voltage going to the substage lamp to regulate the brightness of the light.
Mechanical stage	Platform that holds the specimen. It contains clips to hold a slide in place and knobs for positioning the slide on the stage.
Mechanical stage controls	Knobs used to position the slide on the stage.
Nosepiece	Device that connects the objective lenses to the head of the microscope, and rotates to change objective lenses.
Objective lenses	A typical microscope has three objective lenses, though some have four. Each objective lens has a label that tells how much it magnifies the image. For instance, a 4× objective lens magnifies the specimen four times normal size. The lowest power objective lens (4×) is also called the "scanning" objective.
Ocular lenses	Lenses located within the eyepieces, which generally magnify the specimen ten times normal size.
Power switch	Usually located somewhere on the base of the microscope; used to turn the power on or off.
Working distance	The distance between the mechanical stage (and the slide on it) and the tip of the objective lens.

Focus and Working Distance

The purpose of exercise 3.2 is to familiarize you with the microscope and help you to obtain an appreciation for the relationship between what you see when you look directly at a slide and what you see when you look at a slide through the microscope lenses. If you experience difficulties and cannot see anything through the microscope or cannot focus the lenses, refer to **table 3.2,** "Troubleshooting."

Table 3.2	Troubleshooting
Problem	**Solution**
"No light is coming from my illuminator."	Make sure the microscope is plugged into a working power outlet.
	Check the power switch and make sure it is turned on.
	Check the light control/voltage regulator and make sure it isn't turned all the way down.
	If the first 3 steps don't solve the problem, see your instructor. The bulb may have burned out.
"I can't find anything on my slide."	Go back to scanning power. Lower the stage as far as it will go. Look at the slide on the stage and position it so that the specimen is illuminated. Look through the eyepiece and, using the coarse adjustment knob, slowly bring the stage up until the specimen comes into view.
"I can't use both eyes to view the slide." or "I have to close one eye to view the slide."	Move your head back from the eyepieces slightly. If you are too close, it is more difficult to see a single image.
	Move the eyepieces (closer together or farther apart) until you see a single image through the microscope.
	If the first 2 steps don't work, get a classmate to help you measure the distance between your pupils using a ruler. Then use the ruler to move the eyepieces apart that same distance.
"I see a dark crescent in the view."	Make sure the objective lens is clicked into place.
	Adjust the condenser.
"I can't get the specimen in clear focus."	Make sure the slide is not upside-down on the mechanical stage.

EXERCISE 3.2 Viewing a Slide of the Letter *e*

EXERCISE 3.2A Focusing the Microscope

1. Obtain a compound microscope and a slide of the letter *e*.

2. *Always begin your observation with the lowest possible total magnification (scanning objective in place).* Make sure the mechanical stage is in its lowest position, closest to the base of the microscope, then place the slide on the stage and turn on the illuminator. Adjust the light control/voltage regulator so it is somewhere near the middle or low end of its range.

3. Once you are sure that light is coming out of the illuminator, position the microscope stage so the letter *e* on the slide is over the opening in the stage where light comes through. Do this by looking directly at the microscope stage (not through the eyepiece(s)). This ensures that your specimen will be in the field of view, or at least very close to it, when you look through the lens.

4. Check to see if the scanning objective is in place over the stage, and look into the eyepiece(s). Using the coarse adjustment knob, slowly move the stage up until the specimen comes into view, and then use the fine adjustment knob to bring the specimen into focus. Next, play with the adjustments of the iris diaphragm and light control/voltage regulator to see how each affects the clarity and contrast of your image.

5. In the circle below, draw what you see within the field of view, and record the total magnification.

Total magnification = _____ ×

6. How does the image you see through the microscope differ from what you see when you look directly at the slide?

_____

_____

_____

_____

_____

7. Without changing the position of the slide or the stage, rotate the nosepiece so the low-power objective is in place. Because you had the slide in focus with the scanning objective, you should only need to use the fine adjustment knob to bring the specimen into focus with the low-power objective. Each time you increase the magnification, you should only need to use the fine adjustment knob to focus the specimen, as long as you had the specimen in clear focus at the lower magnification.

8. Repeat step 7 using the high-power objective.

EXERCISE 3.2B Working Distance

The **working distance** is the distance between the mechanical stage (and the slide upon it) and the tip of the objective lens (**figure 3.3**). The shorter the working distance, the more likely it is that the objective lens will run into the slide or the stage. Because of this, only the fine adjustment knob should be used when viewing specimens using high-power objective lenses.

1. Begin with the microscope stage at its lowest position. Place the slide of the letter *e* on the stage and bring it into focus using the scanning objective.

2. Using a millimeter ruler, measure the **working distance**: the distance between the top of the slide on the stage and the bottom of the objective lens. Record the distance here: _____ mm.

3. Change to the low-power objective lens and repeat the process in step 2. Record the distance here: _____ mm.

4. Change to the high-power objective and repeat the process in step 2. Record the distance here: _____ mm.

5. Use your results in answering question 3 of the Post-Laboratory Worksheet (p. 45).

WHAT DO YOU THINK?

1. How does working distance change as total magnification increases? What are the practical consequences of this change in working distance?

Figure 3.3   **How to Measure Working Distance.**

Diameter of the Field of View

The **field of view** is everything that is visible when you look through the eyepiece. You have probably noticed already that as magnification increases, the field of view decreases. In this activity you will determine the diameter of the field of view at the various magnifications of your microscope. The diameter of the field of view is simply the distance (usually given in millimeters [mm], and sometimes in micrometers [μm]) across the widest part of the field.

EXERCISE 3.3 Measuring the Diameter of the Field of View

1. Obtain a compound microscope and a stage micrometer or clear ruler with mm increments.

2. Begin with the scanning objective in place. Place the stage micrometer slide on the stage and position it so that you can see the markings within the field of view when you look through the eyepiece. Then line up the first marking with the left-hand side of the field of view at the widest part. If you are using a ruler, line up one of the mm lines with the left-hand side of the field of view at the widest part.

3. Count the number of markings you see across the widest part of the field of view and record the number: _____ mm. This is the diameter of the field of view at scanning power.

4. Repeat steps 1 and 2 with the low-power objective in place, and record the diameter of the field of view at low power: _____ mm.

5. It is often difficult to use the method from steps 1 and 2 to determine the diameter of the field of view for the high-power objective. However, you can calculate the diameter based on the measurements you took for the scanning or low-power objectives using the following formula (LP = low-power objective, HP = high-power objective):

diameter at LP • total magnification at LP = diameter at HP • total magnification at HP

Example:

- Diameter of the field of view using low-power objective: _4.8_ mm

- Total magnification using low-power objective: _40_ ×

- Diameter of field of view using high-power objective: _unknown_
(this is what you will calculate)

- Total magnification using high-power objective: _400_ ×

- *Calculation:*   4.8 mm • 40× = unknown • 400×
192 = unknown • 400×
192/400 = 0.48 mm (round off the number if necessary)

The diameter of the field of view at a magnification of 400× is 0.48 mm.

6. Use the information recorded in this exercise to answer question 3 of the Post-Laboratory Worksheet (p. 45).

Study Tip!

A *millimeter* (mm) is 10^{-3} meters. A *micrometer* (μm) is 10^{-6} meters. At higher magnifications, you should record the size of structures in micrometers instead of millimeters. To convert 0.2 mm to micrometers, simply move the decimal point three positions to the right. Thus, a distance of 0.2 mm is a distance of 200 μm.

Practice: If the diameter of the field of view at a magnification of 400× is 0.48 mm, what is the diameter in micrometers? _____ μm

EXERCISE 3.4 Estimating the Size of a Specimen

Now that you know the diameter of the field of view for each of the objective lenses of your microscope, you can determine the approximate size of a specimen by comparing the specimen's size to the diameter of the field of view.

1. **Figure 3.4** shows a specimen (outlined in red) at various magnifications. In the space below each image, record the approximate diameter of the specimen. Assume that the diameter of the field of view at 200× total magnification is 0.6 mm. Calculate the diameter of each field of view before estimating the length of the object.

2. Is it possible to measure or estimate the length of the object in figure 3.4c? _____ If not, why? _____

   _____

   If yes, describe how to do it._____

   _____

   _____

WHAT DO YOU THINK?

2. If you can see 4 cells within the field of view at its maximum diameter at a total magnification of 200×, how many cells will you be able to see at a total magnification of 500×?

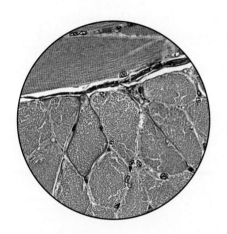

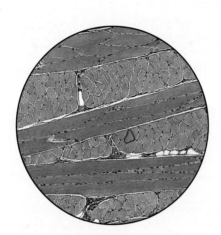

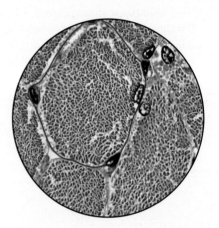

(a) Total magnification = __200x__

Diameter of field   = 0.6 mm

Diameter of object = _____

(b) Total magnification = __40x__

Diameter of field   = _____

Diameter of object = _____

(c) Total magnification = __500x__

Diameter of field   = _____

Diameter of object = _____

Figure 3.4   **Estimating Specimen Size.** The circle indicates the field of view. The specimen to be measured is outlined in red.

Depth of Field

The **depth of field** is how much of the total thickness of a specimen is in focus at each magnification. As total magnification increases, the depth of field decreases and a smaller portion of the specimen will be in focus. **Figure 3.5** demonstrates what happens to the depth of field as total magnification changes.

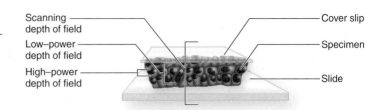

Figure 3.5   **Depth of Field.** Depth of field narrows as total magnification increases.

---

1. Obtain a compound microscope and a slide of three crossed, colored threads.

2. Place the slide on the microscope stage and observe the slide with the scanning objective in place. Once you have the three crossed threads in the field of view, use the fine adjustment knob to focus up and down and determine which of the threads is on top, which is in the middle, and which is on the bottom. Then record your answers:

    Color of top thread: _____

    Color of middle thread: _____

    Color of bottom thread: _____

3. Now observe the slide with the low-power objective in place. Do you have to move the fine adjustment knob more or less than you did with the scanning objective in place to see all three threads? _____

4. Finally, observe the slide with the high-power objective in place. At this magnification, each individual thread will take up almost the entire diameter of the field of view. Is the entire thickness of an individual thread in focus at this magnification? _____. What does this tell you about the depth of field at this magnification? _____

    _____.

Finishing Up

When you are finished, follow proper cleanup procedures.

1. Remove the slide from the microscope stage and put it back where it belongs.
2. Turn the power switch to the "off" position, unplug the power cord, and wrap the cord neatly around the base of the microscope.
3. Rotate the nosepiece so the scanning power objective clicks into place.
4. Lower the microscope stage to its lowest position.
5. Put the dust cover back on the microscope.
6. Return the microscope to its proper storage location.

Chapter 3: The Microscope

POST-LABORATORY WORKSHEET

1. Label the parts of the microscope.

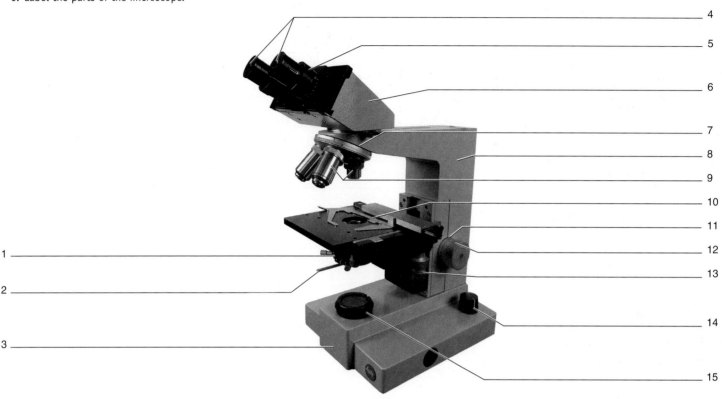

1 _____

2 _____

3 _____

2. What is the total magnification of a microscope set up with an ocular lens magnification of 10× and an objective lens magnification of 43×?_____

3. Complete the following table with the numbers you calculated in exercises 3.1, 3.2, and 3.3.

Power	Ocular Magnification	Objective Magnification	Total Magnification	Diameter of the Field of View	Working Distance
Scanning					
Low					
High					

4. Refer back to your calculations on the sample "specimens" in figure 3.4. Enter information from your calculations in the spaces below. Pay attention to the units you are asked to use because they are not all the same.

A. Total magnification = __200×__

    Diameter of field = __0.6 mm__

    Length of object = _____ mm

B. Total magnification = __40×__

    Diameter of field = _____ mm

    Length of object = _____ mm

C. Total magnification = __500×__

    Diameter of field = _____ μm

    Length of object = _____ μm

5. Record your answers to the questions about colored threads (exercise 3.5, # 2, p. 44) in the spaces below:

Color of top thread: _____

Color of middle thread: _____

Color of bottom thread: _____

6. Explain why proper microscope technique requires that you always begin viewing a slide with the scanning objective first before moving to higher

power objectives. Use the concept of *depth of field* in your explanation. _____

_____

_____

_____

_____

Cellular Anatomy 4

OUTLINE and OBJECTIVES

Module 2: CELLS & CHEMISTRY

INTRODUCTION

The cell is the basic unit of life. Organisms can be unicellular or multicellular, but they must be composed of cells to be considered living entities. Human beings are, of course, multicellular organisms. Our bodies are composed of **tissues**: groups of similar cells and associated extracellular materials that function together as a unit. The study of tissues is called **histology** [*histos*, web (tissue), + *logos*, study]. To understand the study of histology, you must first become comfortable identifying cells and cellular organelles under the microscope. Most cells are easily seen with a light microscope, but most cellular organelles are too small to be seen without the use of a more powerful electron microscope. The purpose of this laboratory session is to introduce you to the microscopic appearance of cells, give you a brief review of cellular anatomy and the stages of mitosis, and help you become more confident in your use of the compound microscope.

Most of the slides that you will view in the anatomy laboratory have been stained with hematoxylin and eosin and are labeled "H and E." This stain makes the **cytoplasm** of the cell appear pink in color and makes visible the outline of the cell where the **plasma membrane** (**cell membrane**) is located. The **nucleus** of the cell appears dark purple in color, and the **nucleolus** often appears as a dark spot within the nucleus. Though these structures have just been described using references to color, you should not generally learn to recognize cell structures based on color alone. Instead, you should learn to recognize cells and cellular organelles based on *shape*. If you do this, you will have an easier time when observing slides stained with stains other than H and E, which can produce different colors.

Chapter 4: Cellular Anatomy

Name:_____

Date:_____ Section: _____

PRE-LABORATORY WORKSHEET

1. Match the cell structures in column A with their functions in Column B.

Column A

_____ 1. centrioles

_____ 2. chromatin

_____ 3. cytoplasm

_____ 4. cytoskeleton

_____ 5. endoplasmic reticulum (ER)

_____ 6. Golgi apparatus

_____ 7. lysosomes

_____ 8. mitochondria

_____ 9. nucleolus

_____ 10. nucleus

_____ 11. peroxisomes

_____ 12. plasma membrane

_____ 13. ribosomes

Column B

a. provides a selectively permeable barrier between the intracellular and extracellular environments of the cell

b. acts to suspend cellular organelles; contains enzymes that mediate many cytosolic reactions such as glycolysis and fermentation

c. contains the cell's genetic material (DNA)

d. synthesizes rRNA and assembles ribosomes in the nucleus

e. refers to the "colored stuff" found in the nucleus; consists of uncoiled chromosomes and associated proteins

f. synthesizes new proteins destined for the plasma membrane, for lysosomes, or for secretion from the cell

g. sites of protein synthesis: may be bound to the ER ("bound") or found within the cytoplasm ("free")

h. a stack of flattened membranes that are the site where proteins from the ER are modified, packaged, and sorted for delivery to other organelles or to the plasma membrane of the cell

i. membrane-enclosed sacs that contain digestive enzymes and function in the breakdown of intracellular debris

j. membrane-enclosed sacs that contain catalase and other oxidative enzymes

k. often referred to as the "powerhouse" of the cell: these organelles are the site of cellular respiration

l. paired organelles composed of microtubules that are used to organize the spindle microtubules that attach to chromosomes during mitosis

m. composed of protein filaments called microtubules, intermediate filaments, and microfilaments; provides the main structural support for the cell

2. Define *mitosis*: _____

_____

3. List the four stages of mitosis (in order):

   a. _____

   b. _____

   c. _____

   d. _____

4. The stage of the cell cycle when cells are *not* undergoing mitosis is _____. During this phase, individual chromosomes are/are not (circle one) visible within the nucleus of the cell.

IN THE LABORATORY

Most animal cells are transparent, so when tissue samples are prepared for use in the anatomy laboratory the slides are stained so cellular details will be visible when viewed under a microscope. Different parts of a cell attract the biological stains to different degrees, which makes some parts of the cell appear dark in color, and others appear lighter in color, or even transparent. The nucleus of the cell has a high attraction for most biological stains, so it is often the most recognizable part of a cell. In this laboratory session you will identify cellular organelles that are visible using a light microscope and you will observe the stages of mitosis in a whitefish embryo.

Histology

Structure and Function of a Typical Animal Cell

Table 4.1 lists the parts of a typical animal cell and gives descriptions of their functions and microscopic features. The parts of a typical animal cell that are most readily visible under a light microscope are the nucleus, nucleolus, and the border of the cell where the plasma membrane is located. Thus, as you observe animal cells under the microscope, your focus will be on finding those structures in particular. Once you have learned to recognize what parts of an animal cell are most typically visible under the light microscope you will be ready to observe the different cell types that are presented in future laboratory exercises.

EXERCISE 4.1 Observing Cellular Anatomy

EXERCISE 4.1A: Human Cheek Cells

1. Obtain a compound microscope and a prepared slide of human cheek cells, and place the slide on the microscope stage.

2. Bring the specimen into focus on low power. Then switch to high power and bring the specimen into focus once again using the fine-focus lens.

3. Cheek cells are examples of **squamous** (*squama-*, scale) cells, which are flattened cells. Each cell has a prominent round nucleus that is generally located near the center of the cell (**figure 4.1**).

4. Using figure 4.1 and table 4.1 as guides, identify the following structures on the slide:

   ☐ nucleolus   ☐ cytoplasm

   ☐ nucleus   ☐ plasma membrane

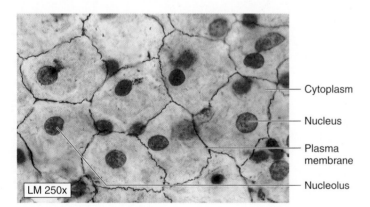

LM 250x

Figure 4.1   Human Cheek Cells.

5. In the space below, make a brief sketch of the cheek cells as seen through the microscope. Label all the structures listed in step 4 on your drawing.

_____ ×

6. *Optional Activity:* **AP|R** 2. **Cells & Chemistry**—Examine the "Generalized cell" dissection and test yourself on cell structures in the Quiz area.

Table 4.1	Parts of a Typical Animal Cell			
Organelle/Structure	**Function**	**Microscopic Features**	**Word Origins**	**Appearances**
Centrioles	Paired organelles that are used to organize the spindle microtubules that attach to chromosomes during mitosis. The area next to the nucleus that contains the centrioles is called the *centrosome*.	Visible only when a cell is actively undergoing nuclear division (mitosis).	*kentron*, center	
Chromatin	Refers to the "colored stuff" found in the nucleus; consists of uncoiled chromosomes and associated proteins.	Most of the coloration seen in the nucleus, with exception of the nucleolus, consists of chromatin.	*chroma*, color	
Cytoplasm	Acts to suspend cellular organelles; contains enzymes that mediate cytosolic reactions such as glycolysis and fermentation.	Clear and homogeneous in appearance; may contain granular substances such as glycogen in certain cells (e.g., hepatocytes).	*kytos*, a hollow (cell), + *plasma*, something formed	
Cytoskeleton	Provides the main structural support for the cell and is composed of microtubules, intermediate filaments, and microfilaments.	Not generally visible under the light microscope.	*kytos*, a hollow (cell), + *skeletos*, dried	
Endoplasmic Reticulum (ER)	Site of lipid synthesis and detoxification of drugs and alcohol (smooth ER). Additionally, rough ER synthesizes proteins destined for the cell membrane, for lysosomes, or for secretion.	Not generally visible under the light microscope. In neurons, the rough ER stains very dark and is called chromatophilic substance (Nissl bodies).	*endon*, within, + *plasma*, something formed, + *rete*, a net	
Golgi Apparatus	A stack of flattened membranes that receive proteins from the rough ER and then modify, package, and sort them for delivery to other organelles or to the plasma membrane of the cell.	Not generally visible under a light microscope.	*Golgi*, Camillo, Italian histologist and Nobel laureate, 1843–1926	
Lysosomes	Membrane-enclosed sacs that contain digestive enzymes; function in the breakdown of intracellular debris.	Not generally visible under a light microscope.	*lysis*, a loosening, + *soma*, body	
Mitochondria	Often referred to as the "powerhouse" of the cell. These organelles are the site of cellular respiration: the metabolic pathway that utilizes oxygen in the breakdown of food molecules to produce ATP.	Not generally visible under a light microscope.	*mitos*, thread, + *chondros*, granule	
Nucleolus	Synthesizes rRNA and assembles ribosomes in the nucleus.	Recognized as a small, dark, circular structure within the nucleus.	*nucleus*, a little nut	
Nucleus	Contains the cell's genetic material (DNA).	The most noticeable feature of a cell; stains very dark.	*nucleus*, a little nut	
Peroxisomes	Membrane-enclosed sacs that contain catalase and other oxidative enzymes. The enzymes break down lipids and toxic substances by first converting them into hydrogen peroxide and then breaking down the hydrogen peroxide into water and oxygen.	Not generally visible under a light microscope.	*peroxi*, relating to hydrogen peroxide, + *soma*, body	
Plasma Membrane	Provides a selectively permeable barrier between the intracellular and extracellular environments of the cell.	Visible only using an electron microscope. However, the outer border of the cell, where the cell membrane is located, is often visible under the light microscope.	*plasma*, something formed, + *membrane*, a membrane	
Ribosomes	Sites of protein synthesis: may be bound to the ER ("fixed") or found within the cytoplasm ("free").	Not generally visible under a light microscope.	*ribose*, the sugar in RNA, + *soma*, body	

EXERCISE 4.1B: Neurons

1. Obtain a compound microscope and a prepared slide of neurons (nerve cells).

2. Place the slide on the microscope stage and bring the specimen into focus using the scanning objective. Next, switch to high power and bring the specimen into focus once again using the fine-adjustment knob.

3. Using **figure 4.2** as a guide, locate the very large, branching cells called **neurons**. Neurons contain large amounts of **rough endoplasmic reticulum** (ER). The ER tends to stain very dark and appears as a dark, granular substance within the cell.

4. Using figure 4.2 and table 4.1 as guides, identify the following structures on the slide:

   ☐ rough endoplasmic reticulum   ☐ nucleolus
   ☐ nucleus   ☐ plasma membrane

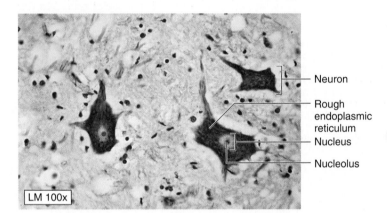

LM 100x

Figure 4.2 **Neurons.**

5. In the space below, make a brief sketch of the neurons as seen through the microscope. Label all the structures listed in step 4 on your drawing.

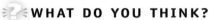
——————— ×

WHAT DO YOU THINK?

① What is the function of rough endoplasmic reticulum? Why do you think neurons might contain so much rough endoplasmic reticulum? Hint: Neurons are cells that need to be able to transport numerous ions into and out of the cell.

Table 4.2	Appearance of Cells Undergoing Mitosis During Phases of the Cell Cycle	
Stage	**Interphase**	**Prophase**
Histological View	Nucleus with chromatin	Nucleus with dispersed chromosomes
	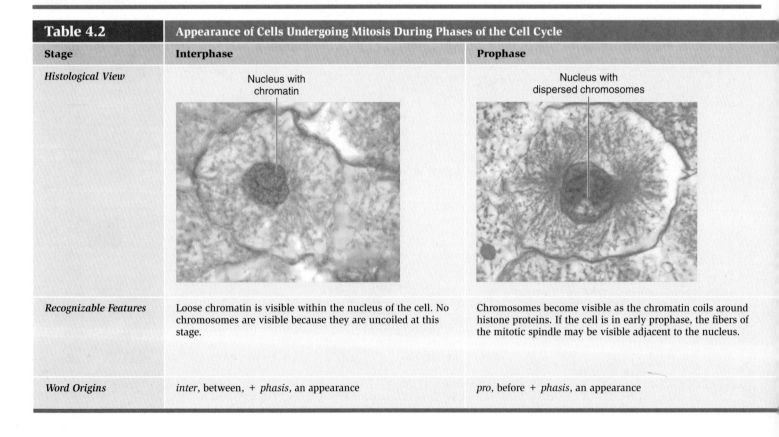	
Recognizable Features	Loose chromatin is visible within the nucleus of the cell. No chromosomes are visible because they are uncoiled at this stage.	Chromosomes become visible as the chromatin coils around histone proteins. If the cell is in early prophase, the fibers of the mitotic spindle may be visible adjacent to the nucleus.
Word Origins	*inter*, between, + *phasis*, an appearance	*pro*, before + *phasis*, an appearance

Mitosis

The **cell cycle** describes the events that occur during the life of a typical cell. The cell cycle is divided into two main phases: interphase and mitosis. **Interphase** is the time in which the cell is performing its main functions and is in a state of general maintenance. **Mitosis** (*mitos*, thread) is the process by which cells reproduce. It results in the formation of two identical daughter cells from one original parent cell. Technically, mitosis refers only to nuclear division. However, most cells undergo **cytokinesis** (*cyto-*, cell, + *kinesis*, movement) immediately following mitosis.

Thus, casual usage of the term "mitosis" usually implies that the cytoplasm and cellular organelles have also divided. Some cells undergo mitosis without undergoing cytokinesis. The resulting cell thus becomes a *multinuclear* (*multus*, much, + *nuclear*, nucleus) cell. **Table 4.2** describes the microscopic appearance of cells in each of the four stages of mitosis and interphase. In this laboratory exercise your goal is to locate cells in each of the stages of mitosis by observing whitefish embryos (blastulas). Whitefish embryos are very small and are rapidly developing, which makes them ideal specimens for observations of cells undergoing mitosis.

EXERCISE 4.2 Observing Mitosis in a Whitefish Embryo

1. Obtain a compound microscope and a prepared slide of a whitefish embryo (blastula) or other slide of cells undergoing mitosis.

2. Using table 4.2 as a guide, scan the slide and locate cells in the following four stages of mitosis and interphase. Once you locate a cell in a particular phase, switch to a higher-power objective to see the cell more clearly.

   ☐ interphase   ☐ anaphase
   ☐ prophase   ☐ telophase
   ☐ metaphase

3. Sketch cells in each of these phases (as viewed under the microscope) in the spaces provided for you in the Post-Laboratory Worksheet (question #4) on p. 57.

WHAT DO YOU THINK?

② Chemotherapy treatments are given to cancer patients in an attempt to halt or slow the growth of a tumor, which is composed of rapidly-dividing cells. Certain chemotherapy drugs exert their actions by interfering with mitosis. For example, some drugs act to prevent microtubules from lengthening or shortening. Microtubules are protein filaments that attach to chromosomes and centrioles, forming the mitotic spindle, which moves the chromosomes during mitosis. Based on this role of microtubules during mitosis, with which stage(s) of mitosis would these drugs most likely interfere?

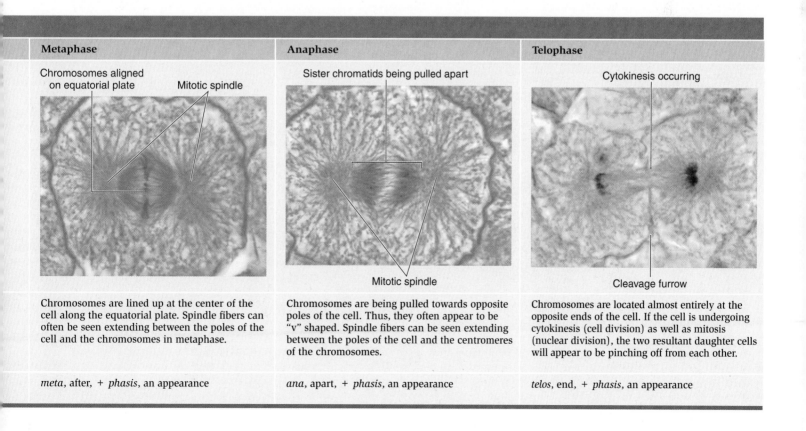

Metaphase	Anaphase	Telophase
Chromosomes aligned on equatorial plate — Mitotic spindle	Sister chromatids being pulled apart — Mitotic spindle	Cytokinesis occurring — Cleavage furrow
Chromosomes are lined up at the center of the cell along the equatorial plate. Spindle fibers can often be seen extending between the poles of the cell and the chromosomes in metaphase.	Chromosomes are being pulled towards opposite poles of the cell. Thus, they often appear to be "v" shaped. Spindle fibers can be seen extending between the poles of the cell and the centromeres of the chromosomes.	Chromosomes are located almost entirely at the opposite ends of the cell. If the cell is undergoing cytokinesis (cell division) as well as mitosis (nuclear division), the two resultant daughter cells will appear to be pinching off from each other.
meta, after, + *phasis*, an appearance	*ana*, apart, + *phasis*, an appearance	*telos*, end, + *phasis*, an appearance

Gross Anatomy

Models of a Typical Animal Cell and Stages of Mitosis

In exercise 4.3 you will observe classroom models demonstrating a typical animal cell and the stages of mitosis. **Figure 4.3** is a photograph of a classroom model of a typical animal cell, which has the organelles labeled for your reference.

EXERCISE 4.3 Observing Classroom Models of Cellular Anatomy and Mitosis

1. Obtain classroom models demonstrating a typical animal cell and the stages of mitosis.

2. Using figure 4.3, table 4.1, and your textbook as guides, identify the following structures on a classroom model of a typical animal cell.

   ☐ centrioles   ☐ nucleolus
   ☐ chromatin   ☐ nucleus
   ☐ cytoplasm   ☐ peroxisome
   ☐ Golgi apparatus   ☐ plasma membrane
   ☐ lysosome   ☐ ribosomes
   ☐ mitochondria   ☐ rough ER
   ☐ nuclear envelope ☐ smooth ER

3. Using figure 4.3, table 4.2, and your textbook as guides, identify all of the stages of mitosis on a classroom model of cells undergoing mitosis.

   ☐ interphase   ☐ metaphase   ☐ telophase
   ☐ prophase   ☐ anaphase

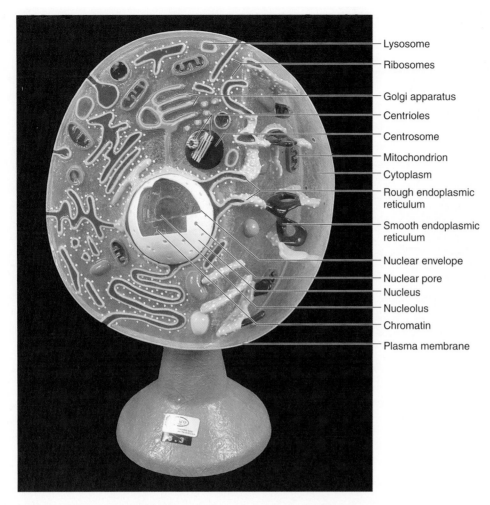

- Lysosome
- Ribosomes
- Golgi apparatus
- Centrioles
- Centrosome
- Mitochondrion
- Cytoplasm
- Rough endoplasmic reticulum
- Smooth endoplasmic reticulum
- Nuclear envelope
- Nuclear pore
- Nucleus
- Nucleolus
- Chromatin
- Plasma membrane

Figure 4.3   **Classroom Model of a Typical Animal Cell.**

Chapter 4: Cellular Anatomy

Name: _____

Date: _____ Section: _____

POST-LABORATORY WORKSHEET

1. Label the structures in this diagram of a typical cell:

1 _____   10 _____
2 _____   11 _____
3 _____   12 _____
4 _____   13 _____
5 _____
6 _____
7 _____
8 _____
9 _____

2. One major function of liver cells (hepatocytes) is to detoxify alcohol. Based on this function, what organelle(s) do you predict hepatocytes would contain large numbers of? Why?

_____

_____

3. Certain cells within the pancreas function in the synthesis and secretion of the hormone insulin, which is a protein. Based on this function, what organelle(s) do you predict these cells would contain large numbers of? Why?

_____

_____

4. In the spaces below, draw a picture of a whitefish embryo in each of the four stages of mitosis and interphase. To the right of your drawing write a brief description of the major cellular events that occur in each stage.

STAGE: Interphase

EVENTS:_____

_____

_____

_____

STAGE: Prophase

EVENTS:_____

_____

_____

_____

STAGE: Metaphase

EVENTS:_____

_____

_____

_____

STAGE: Anaphase

EVENTS:_____

_____

_____

_____

STAGE: Telophase

EVENTS:_____

_____

_____

_____

Histology

5

OUTLINE and OBJECTIVES

Module 3: TISSUES

INTRODUCTION

In this laboratory session you will begin your practical study of **histology**, or tissue biology (*histo*, tissue, + *logos*, the study of). Tissues consist of multiple cells that function together as a unit. An understanding of histology is important in the health sciences because many times the first manifestation of disease is seen at the tissue level of organization. For example, when a patient presents to his or her doctor with a tumor (*tumere*, to swell), one of the primary methods for determining the type of tumor and whether it is cancerous is to do a biopsy (take a tissue sample) and look at the tissue under a microscope. Thus, understanding normal histology is important for understanding histopathology (*histo*, tissue + *pathos*, disease).

Most beginning students find the study of histology challenging, but with perseverance and patience they eventually get the hang of it. This chapter introduces you to the key features of the four basic tissue types—epithelial, connective, muscle, and nervous. The text and figures in this chapter provide you with examples of where each type of tissue is located in the body, but at this time your primary focus will be to learn how to recognize the different tissue types.

Histology Slides

To get yourself oriented when looking at a histology slide through the microscope, one of the most important things to keep in mind is the issue of the *size of the tissue sample* you are viewing. **Figure 5.1** shows a life-size prepared slide that is similar to the slides you will find in the laboratory. Let us consider briefly how this slide was prepared. The process of making a histology slide involves five general steps:

1. Obtain a tissue sample.
2. Prepare the tissue sample for slicing.
3. Cut thin slices of the tissue using a special knife called a *microtome*.
4. Transfer the tissue slices to a microscope slide.
5. Stain the slide.

After a tissue sample has been obtained, it must be made rigid so that it will be easy to slice. This is done either by freezing the tissue or by embedding the tissue in a block of paraffin wax. The next step in preparing the histology slide involves slicing the frozen sample (or wax block) into very thin slices (on the order of micrometers—1 μm is 10^{-6} meters) using a special knife called a **microtome** (*micro*, small, + *tome* or *temmein*, to cut). The slices are cut so thin that often only a single layer of cells is contained in the slice. Once the slices are made, they are transferred to a microscope slide, which is then covered with a cover slip. Finally, the samples are stained to make intracellular and extracellular structures visible. The most common method of staining, hematoxylin and eosin (H and E), makes most structures appear pink or purple and makes the nucleus of the cell, in particular, easily visible.

Observe the life-size histology slide shown in figure 5.1. Measure the dimensions of the slide. What are they? _____ + _____ cm. These measurements will give you an idea of the size limits of what can fit on a microscope slide.

Always ask yourself the "size" question when you first view a slide. Then, when you see a slide labeled "small intestine" (for example), you will expect to see only a *portion* of the wall of the small intestine, not a cross section of the entire organ—because that would not fit on the slide. This practice will become more natural for you as you gain experience viewing slides.

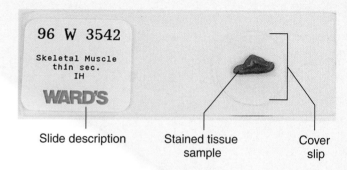

Figure 5.1   **Life-Size Histology Slide.**

WHAT DO YOU THINK?

1. For each of the examples given below, do you think the structure is likely to fit *in its entirety* on a microscope slide? (Circle yes or no for each.)

   a. A cross section of a pinky finger   Yes   No

   b. A cross section of the large intestine   Yes   No

   c. A cross section of a small blood vessel   Yes   No

   d. A cross section of the humerus (your arm bone) Yes   No

   e. A cross section of a ureter (~0.5 cm in diameter) Yes   No

Chapter 5: Histology

PRE-LABORATORY WORKSHEET

Name: _____

Date: _____ Section: _____

1. List the four basic tissue types.

    a. _____

    b. _____

    c. _____

    d. _____

2. Bone is classified as a(n) _____ tissue.

3. Which two of the four basic tissue types are excitable?

    a. _____

    b. _____

4. The basic tissue type that contains an extensive extracellular matrix (ECM) is _____.

5. List the three types of muscle tissue.

    a. _____

    b. _____

    c. _____

6. The only basic tissue type that exhibits polarity (has both apical and basal surfaces) is _____.

7. The basic tissue type that contains a basement membrane is _____ tissue.

8. The two specialized tissue types that are characterized as supporting connective tissues are _____

    and _____.

9. Fluid connective tissues include _____ and _____.

10. The only type of connective tissue that is avascular is _____.

In this laboratory session you should expect to spend nearly all of your time looking at tissues under the microscope. At first you will think you are simply looking at a lot of slides with "pink and purple stuff" on them, and you will have difficulty identifying particular tissue types and structures. This initial experience looking at tissues will probably be somewhat frustrating, but with time, patience, and *practice* you will be successful.

The exercises in this chapter address each of the four basic tissue types. You may work through the exercises in any order, or in the order your instructor chooses. However, be sure to complete all the tasks involving a particular tissue type before moving on to another type of tissue.

Histology

Epithelial Tissue

Epithelial tissues make up body coverings and linings. As such, they will have a free edge (see "Study Tip!" on p. 63). They are characteristically highly **cellular** (mostly composed of cells, with little extracellular material) and **avascular** (no blood supply). In addition they exhibit **polarity**; they have a distinct *basal* (bottom) and *apical* (top) surface. On their basal surface, they have a specialized extracellular structure called a **basement membrane**, which anchors the epithelium to the underlying tissues. The shape of the cells on the *apical* surface of the epithelium (**table 5.1**), the number of layers of cells, and presence of any surface modifications (**table 5.2**) are the characteristics used to classify epithelial tissues.

Figure 5.2 is a flowchart for classification of epithelial tissues that you can use when you are given an unknown slide of epithelial tissue. It will help you decide how to classify the epithelial tissue.

Table 5.1	Epithelial Cell Shapes			
Cell Shape	**Squamous**	**Cuboidal**	**Columnar**	**Transitional**
Micrograph	Squamous cells / Lumen / LM 130x	Cuboidal cells / Lumen / LM 165x	Columnar cells / Lumen / LM 400x	Transitional cell / Lumen / LM 180x
Description	Cells are flattened and have irregular borders.	Cells are as tall as they are wide.	Cells are taller than they are wide.	Cells change shape depending on the stress on the epithelial tissue. The cells change between a cuboidal shape and a more flattened, squamous shape.
Generalized Functions	If the epithelium is only one cell layer thick, it provides a very thin barrier for *diffusion*. If the epithelium is several layers thick, the cells specialize in *protection* (as in epidermal cells of the skin).	The shape of the cell allows more room for cellular organelles inside (mitochondria, endoplasmic reticulum, etc.). These cells generally function in *secretion* and/or *absorption*.	The large size of the cell allows even more room for cellular organelles (mitochondria, endoplasmic reticulum, etc.). These cells generally function in *secretion* and/or *absorption*.	The fact that these cells change shape means that they are good at *resisting stretch* without being torn apart from each other. They are only found lining structures of the urinary tract (ureters, urinary bladder, etc.).
Identifying Characteristics	In cross section, the nucleus is the most visible structure. It will be very flattened. In a view from the top of the epithelium, the cell borders will be irregular in shape.	Generally cuboidal cells are identified by their very round, plump nucleus, and by equal amounts of cytoplasm in the spaces between the nucleus and the plasma membrane on all sides.	The nuclei of columnar cells can be either oval or round in shape, and they generally line up in a row. If the nuclei are round, you will see more cytoplasm between the nucleus and the plasma membrane on the apical side of the nucleus than on the other three sides.	Transitional cells are located on the apical surface of the epithelium. However, they appear much more rounded or dome-shaped than typical cuboidal cells, and they are sometimes binucleate.

Table 5.2	Surface Modifications and Specialized Cells of Epithelial Tissues			
Surface Modification	**Cilia**	**Goblet cells**	**Keratinization**	**Microvilli**
Micrograph	Lumen, Cilia — LM 200x	Goblet cell, Columnar epithelial cell, Mucin within goblet cell, Lumen, Goblet cell nucleus, Location of basement membrane — LM 100x	Lumen, Keratinization — LM 40x	Brush border of microvilli, Lumen — LM 165x
Description and Function	Small hairlike structures that stick out of the apical surface of epithelial cells. Cilia actively move to *propel substances along the apical surface of an epithelial sheet*. Cilia move only in one direction.	Contain many small mucin granules and function in the *production of mucus*. The mucus is used either to *assist in transport* of substances along an epithelial sheet, to *provide a protective barrier* along the apical surface of the epithelium, or to provide *lubrication*.	Keratin is an intermediate filament produced by skin epithelial cells. When present in stratified squamous epithelium, bundles of keratin fill up entire cells and bind to desmosomes, which firmly anchor the dead squamous epithelial cells together so the layers of cells appear to be a single, homogeneous unit. Keratin imparts *strength* and *protection* to dead skin epithelial cells.	Microvilli are extremely small extensions of the plasma membrane of the apical surface of cells. Microvilli *increase the surface area* of the cell to enhance the process of *absorption*.
Identifying Characteristics	When cilia are present, and the slide is viewed at sufficient magnification, you can see what appear to be individual "hairs" on the apical surface of the epithelial cells.	Goblet cells are named for their shape. They are rounded near the apical surface and they narrow toward their basal surface. The shape is similar to the shape of a goblet (wine glass). The mucin inside the cells does not typically take up biological stains, so the cells often appear white or "empty." If the slide is stained specifically for mucin, then the goblet cells will appear dark.	Keratinization is recognized as a homogeneous, acellular-looking portion of a stratified squamous epithelium.	Individual microvilli can be seen only when the specimen is viewed under an electron microscope. Thus, you will *not* be able to see individual microvilli. Instead, the apical surface of the epithelium will appear to be "fuzzy." For this reason, epithelia containing microvilli are often said to have a "brush border."

Study Tip!

When you are viewing a slide for the purpose of identifying an epithelial tissue, remember that epithelial tissues always form linings and coverings of organs. The slides you will be observing contain more tissues than just epithelial tissues. To locate the epithelial tissue, first look for any white space, or "empty" space, on the slide. This space will typically be the outside of an organ or the lumen (inside) of the organ. The tissue that lies directly adjacent to the empty space will usually be an epithelial tissue.

Study Tip!

A **simple** epithelium is only one cell layer thick. A **stratified** epithelium is two or more cell layers thick.

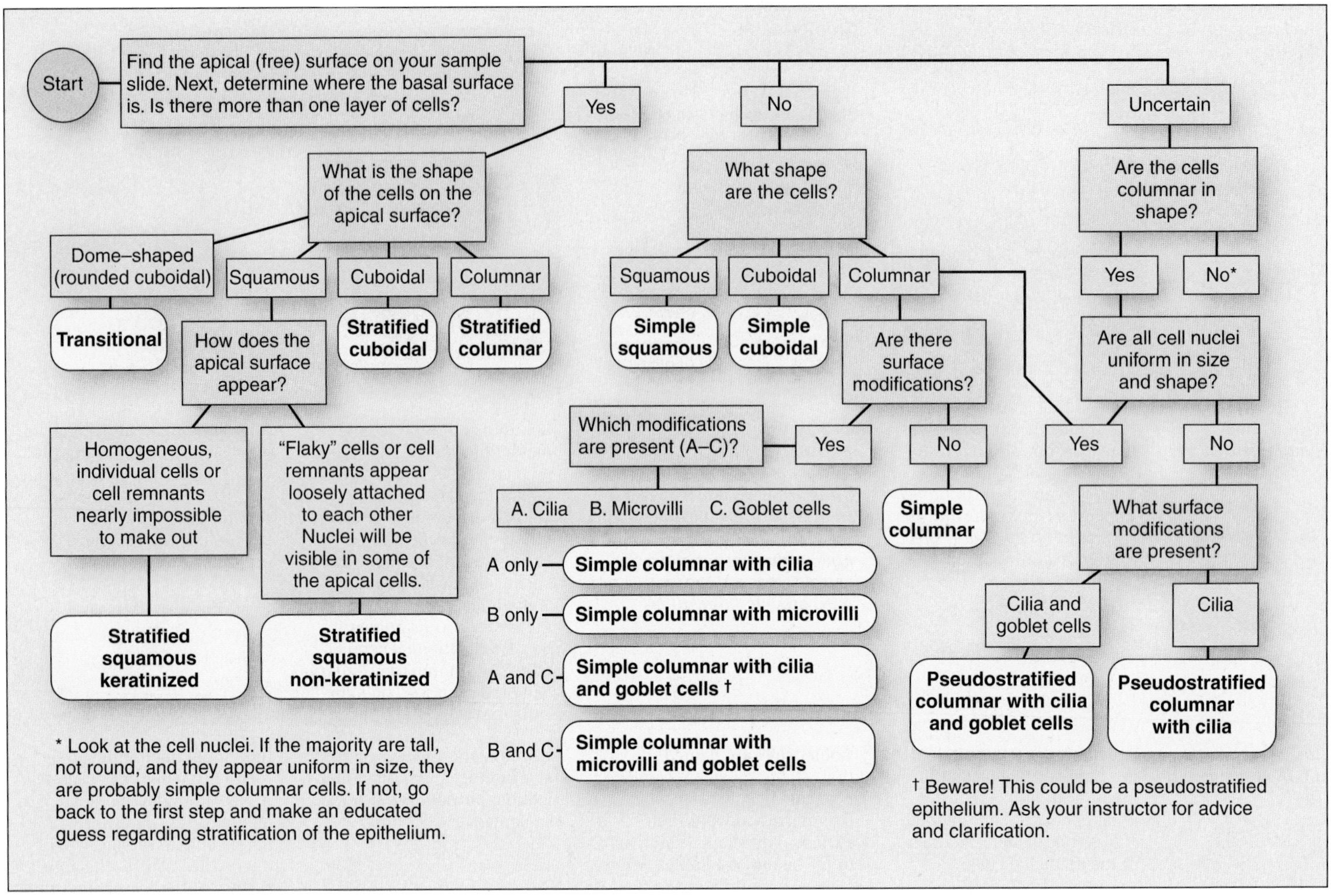
Start
Find the apical (free) surface on your sample slide. Next, determine where the basal surface is. Is there more than one layer of cells?
Yes
No
Uncertain
What is the shape of the cells on the apical surface?
Dome–shaped (rounded cuboidal)
Squamous
Cuboidal
Columnar
Transitional
How does the apical surface appear?
Stratified cuboidal
Stratified columnar
Homogeneous, individual cells or cell remnants nearly impossible to make out
"Flaky" cells or cell remnants appear loosely attached to each other Nuclei will be visible in some of the apical cells.
Stratified squamous keratinized
Stratified squamous non-keratinized
What shape are the cells?
Squamous
Cuboidal
Columnar
Simple squamous
Simple cuboidal
Are there surface modifications?
Which modifications are present (A–C)?
Yes
No
A. Cilia B. Microvilli C. Goblet cells
Simple columnar
A only
Simple columnar with cilia
B only
Simple columnar with microvilli
A and C
Simple columnar with cilia and goblet cells †
B and C
Simple columnar with microvilli and goblet cells
Are the cells columnar in shape?
Yes
No*
Are all cell nuclei uniform in size and shape?
Yes
No
What surface modifications are present?
Cilia and goblet cells
Cilia
Pseudostratified columnar with cilia and goblet cells
Pseudostratified columnar with cilia
* Look at the cell nuclei. If the majority are tall, not round, and they appear uniform in size, they are probably simple columnar cells. If not, go back to the first step and make an educated guess regarding stratification of the epithelium.
† Beware! This could be a pseudostratified epithelium. Ask your instructor for advice and clarification.

Figure 5.2 **Flowchart for Classifying Epithelial Tissues.**

EXERCISE 5.1 Identification and Classification of Epithelial Tissue

EXERCISE 5.1A: Simple Squamous Epithelium

1. Obtain a slide of a small vein in cross section (**figure 5.3**).
2. Place the slide on the microscope stage and bring the tissue sample into focus on low power.
3. Look for any "empty" space on the slide. The empty space on the slide will either be the inside of the vessel (the **lumen** of the vessel) or the outside edge of the tissue sample.
4. Locate the lumen of the vessel and move the microscope stage so the lumen is at the center of the field of view. The lumen of all blood vessels and lymphatic vessels, and the four chambers of the heart is lined with a simple squamous epithelium called *endothelium*. This type of epithelium is always somewhat difficult to see because it is extremely thin.

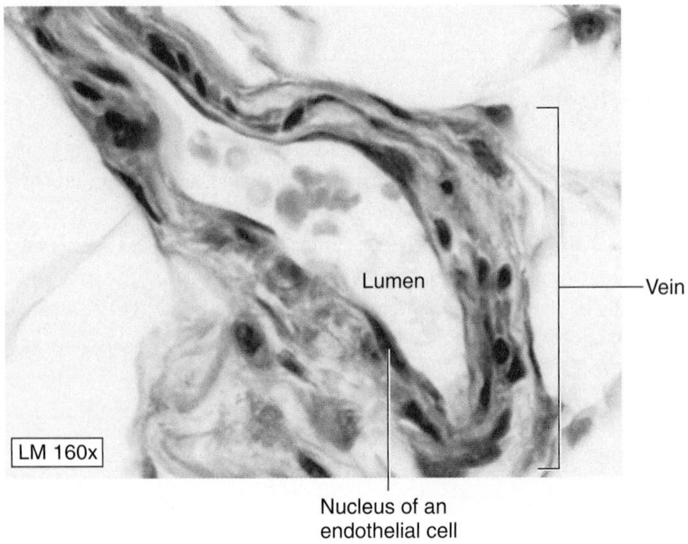
Lumen
Vein
LM 160x
Nucleus of an endothelial cell

Figure 5.3 **Simple Squamous Epithelium.** Cross section through a small vein lined with endothelial cells, which are simple squamous epithelial cells.

5. Once you have identified the lumen of the vessel and know where the epithelial tissue is located (lining the lumen), change to high power. Look for the flattened nuclei of the squamous epithelial cells that line the lumen. It is unlikely that you will be able to see much, if any, of the cellular cytoplasm because the cells are extremely thin. Because simple squamous epithelium is extremely thin, it functions in diffusion—a process that occurs over very short distances (distances of approximately 10 µm). Diffusion is the movement of particles from an area of high concentration to an area of low concentration, and is one mechanism by which substances are transported in the body.

6. Using figure 5.3 and tables 5.1 and 5.2 as guides, identify the following structures on the slide:

   ☐ lumen of vein

   ☐ nucleus of squamous epithelial cell

7. In the following space, make a brief sketch of simple squamous epithelium as seen through the microscope. Be sure to identify all the structures listed in step 6 on your drawing.

_____ ×

EXERCISE 5.1B: Simple Cuboidal Epithelium

1. Obtain a slide of the kidney (**figure 5.4**).

2. Place the slide on the microscope stage and bring the tissue sample into focus on low power.

3. Locate the lumen of a tubule in cross section (figure 5.4), and then identify the cells that lie next to the lumen. These cells should have plump, round nuclei and approximately equal amounts of cytoplasm surrounding each nucleus. These are cuboidal epithelial cells, which line the kidney tubules and function in secretion and absorption of substances across the epithelium (table 5.1).

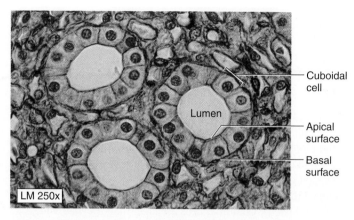

LM 250x

Figure 5.4 **Simple Cuboidal Epithelium.** Cross section of three kidney tubules demonstrating simple cuboidal epithelium.

4. Using figure 5.4 and table 5.1 as guides, identify the following structures on the slide:

   ☐ apical surface   ☐ lumen of tubule

   ☐ basal surface   ☐ cuboidal cell

5. In the following space, make a brief sketch of simple cuboidal epithelium as seen through the microscope. Be sure to identify all the structures listed in step 4 on your drawing.

_____ ×

Study Tip!

Simple squamous epithelium lining the cardiovascular system is called **endothelium**. Simple squamous epithelium lining body cavities is called **mesothelium**. Thus, these terms (endothelium and mesothelium) indicate not only the type of epithelium (simple squamous), but also the location of the epithelium.

EXERCISE 5.1C: Simple Columnar Epithelium (nonciliated)

1. Obtain a slide of the small intestine (**figure 5.5**).

2. Place the slide on the microscope stage and bring the tissue sample into focus on low power.

3. This slide will show only a part of the wall of the intestine so you will need to find some empty space on the slide first and then look for epithelium next to that empty space. Once you have the epithelium in the center of the field of view, switch to high power. Look for epithelial cells with oval, elongated nuclei that have most of their cytoplasm on the apical side of the nucleus. Columnar cells are taller than they are wide, and the nuclei generally appear to be lined up in a row. The nuclei can be either elongated or round in shape.

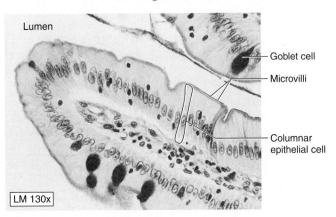

Lumen

Goblet cell

Microvilli

Columnar epithelial cell

LM 130x

Figure 5.5   Simple Columnar Epithelium with Microvilli. Simple columnar epithelium with microvilli lining the lumen of the small intestine.

4. This epithelium also demonstrates **goblet cells**, which secrete mucus, and **microvilli**, which increase the surface area of the epithelial cells for absorption (table 5.2)

5. Using figure 5.5 and tables 5.1 and 5.2 as guides, identify the following structures on the slide:

☐ columnar epithelial cell   ☐ lumen of intestine
☐ goblet cell   ☐ microvilli

6. In the following space, make a brief sketch of simple columnar epithelium as seen through the microscope. Be sure to identify all the structures listed in step 5 on your drawing.

_____ ×

7. Are any surface modifications present in the slide of the small intestine? If so, describe them: _____

_____

_____

EXERCISE 5.1D: Simple Columnar Epithelium (ciliated)

1. Obtain a slide of a uterine tube (**figure 5.6**).

2. Place the slide on the microscope stage and bring the tissue sample into focus on low power.

3. Look for a tubular structure cut in cross section (figure 5.6), and identify its lumen. Look for columnar epithelial cells next to the lumen. Once you have the epithelium in the field of view switch to high power. When cilia are present, you should be able to make out the individual "hair"-like cilia on the apical surface of the epithelial cells.

4. Using figure 5.6 and tables 5.1 and 5.2 as guides, identify the following structures on the slide:

☐ cilia   ☐ lumen of uterine tube
☐ simple columnar
   epithelial cell

5. In the following space, make a brief sketch of simple columnar epithelium with cilia as seen through the microscope. Be sure to identify all the structures listed in step 4 in your drawing.

_____ ×

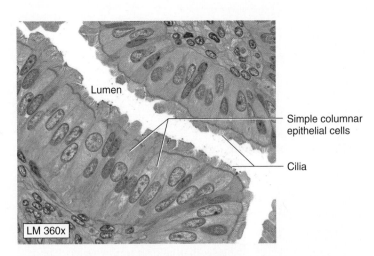

Lumen

Simple columnar epithelial cells

Cilia

LM 360x

Figure 5.6   Simple Columnar Epithelium with Cilia. Simple columnar ciliated epithelium lining the lumen of the uterine tube.

WHAT DO YOU THINK?

❷ What function do you think the cilia that line the uterine tube perform?

EXERCISE 5.1E: Stratified Squamous Epithelium (nonkeratinized)

1. Obtain a slide of the trachea and esophagus (**figure 5.7a,b**).

2. Place the slide on the microscope stage and bring the tissue sample into focus on low power.

3. This slide will contain two different organs in cross section, the trachea and the esophagus. The epithelium you are looking for lines the esophagus and consists of multiple layers of cells.

4. Locate the lumen of the esophagus (figure 5.7a), then identify stratified squamous nonkeratinized epithelium lining the lumen of the esophagus. The cells on the apical/ luminal surface of this epithelium will appear flattened. This nonkeratinized stratified squamous epithelium is sometimes referred to as stratified squamous "moist" epithelium. It lines surfaces within the body that experience friction and abrasion, but where water loss is not a problem (for example, lining the oral cavity, esophagus, and vagina).

5. Using figure 5.7b and tables 5.1 and 5.2 as guides, identify the following structures on the slide.

   ☐ lumen of esophagus
   ☐ stratified squamous nonkeratinized epithelium
   ☐ squamous epithelial cells

6. In the following space, make a brief sketch of stratified squamous nonkeratinized epithelium as seen through the microscope. Be sure to identify the structures listed in step 5 on your drawing.

_____ ×

7. Keep the slide on the microscope stage and proceed to exercise 5.1F (next page), where you will focus on the epithelium lining the trachea.

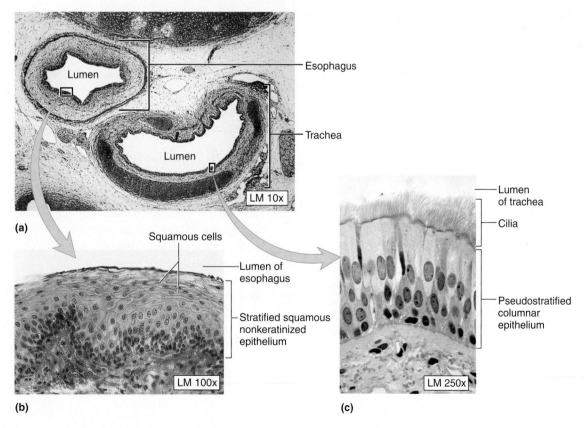

Figure 5.7 **Trachea and Esophagus.** (a) Cross section through the trachea and esophagus. (b) Stratified squamous nonkeratinized epithelium lining the esophagus. (c) Pseudostratified columnar epithelium lining the trachea.

EXERCISE 5.1F: Pseudostratified Columnar Epithelium

1. Obtain a slide of the trachea and esophagus (figure 5.7a).

2. Place the slide on the microscope stage and bring the tissue sample into focus on low power.

3. This slide will contain two organs in cross section, the trachea and the esophagus. The epithelium you are looking for lines the trachea and contains ciliated cells.

4. Locate the lumen of the trachea (figure 5.7a), then identify pseudostratified columnar epithelium (figure 5.7c) lining the trachea. Once you have the epithelium in the field of view switch to high power. This epithelium is characterized by the presence of columnar cells that are not all the same height. All cells contact the basement membrane, but not all reach the apical surface. Because not all cells reach the apical surface of the epithelium, the nuclei will *not* be nicely lined up in rows as with simple columnar epithelium. Instead, the nuclei will appear to be layered. Hence, the name of this epithelium: **pseudostratified** (*pseudo-*, false). Most, but not all, pseudostratified epithelia also contain cilia and goblet cells.

5. Using figure 5.7c and tables 5.1 and 5.2 as guides, identify the following structures on the slide:

   ☐ cilia
   ☐ lumen of trachea

   ☐ pseudostratified columnar epithelial cell

6. In the following space, make a brief sketch of pseudostratified columnar epithelium as seen through the microscope. Be sure to label the structures listed in step 5 in your drawing.

_____ ×

EXERCISE 5.1G: Stratified Cuboidal or Stratified Columnar Epithelium

1. Obtain a slide of a merocrine sweat gland (**figure 5.8**).

2. Place the slide on the microscope stage and bring the tissue sample into focus on low power.

3. Locate the lumen of a duct of the sweat gland (figure 5.8). Identify stratified cuboidal epithelium next to the lumen. Once you have the epithelium in the field of view switch to high power. Stratified cuboidal epithelium is found lining the ducts of merocrine sweat glands, which are located in the dermis of the skin. Stratified cuboidal epithelium is generally only two cell layers thick. How many layers do you see on the slide? Your laboratory may have other slides available that demonstrate stratified *columnar* epithelium lining the ducts of other exocrine glands. As with stratified cuboidal epithelium, stratified columnar epithelium is also rarely more than two cell layers thick.

4. Using figure 5.8 and tables 5.1 and 5.2 as guides, identify the following structures on the slide:

   ☐ basement membrane
   ☐ cuboidal epithelial cell

   ☐ lumen of the duct

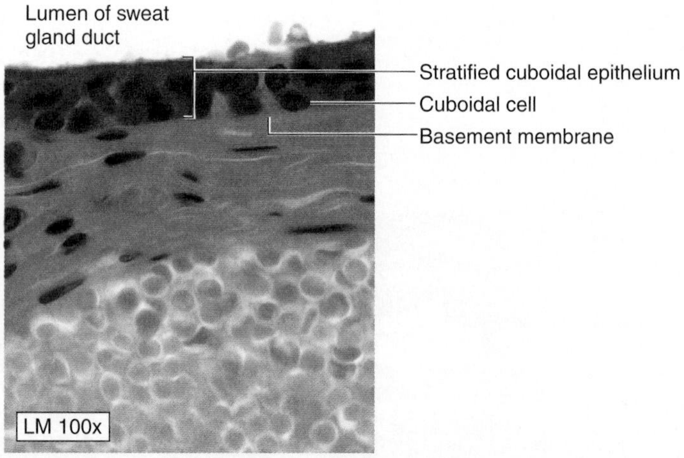

Figure 5.8 **Stratified Cuboidal Epithelium.** Epithelium lining the duct of a merocrine sweat gland.

5. In the following space, make a brief sketch of stratified cuboidal (or stratified columnar) epithelium as seen through the microscope. Be sure to label the structures listed in step 4 on your drawing.

_____ ×

EXERCISE 5.1H Transitional Epithelium

1. Obtain a slide of the urinary bladder (**figure 5.9**).

2. Place the slide on the microscope stage and bring the tissue sample into focus on low power.

3. This slide will only show a part of the wall of the bladder so you will need to locate the empty space first and then look for epithelium next to that empty space. Locate the transitional epithelium. Once you have the epithelium in the field of view switch to high power. Transitional epithelium is stratified, and it can look very similar to stratified squamous epithelium. However, the cells on the apical/luminal surface will appear more cuboidal in shape, they will be much more rounded or dome-shaped than typical cuboidal cells, and they may contain more than one nucleus per cell (these cells are sometimes referred to as "dome" or "umbrella" cells).

4. Observe the cells on the basal surface of the transitional epithelium. These cells are usually columnar in shape, in contrast to the cells on the basal surface of a stratified squamous epithelium, which tend to be more cuboidal in shape. Transitional epithelium is found lining the urinary bladder and other urine-draining structures. Its structure allows the epithelium to stretch easily to accommodate the passage or storage of urine without causing the epithelial cells to tear apart.

5. Using figure 5.9 and tables 5.1 and 5.2 as guides, identify the following structures on the slide:

   ☐ dome-shaped epithelial cells

   ☐ lumen of urinary bladder

   ☐ transitional epithelium

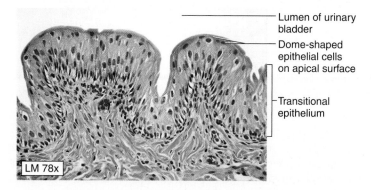

LM 78x

Figure 5.9   Transitional Epithelium. Epithelium lining the urinary bladder.

6. In the following space, make a brief sketch of transitional epithelium as seen through the microscope. Be sure to label the structures listed in step 5 in your drawing.

_____ ×

Lumen of urinary bladder

Dome-shaped epithelial cells on apical surface

Transitional epithelium

7. *Optional Activity:* **AP|R** **3. Tissues**—Watch the "Epithelial Tissue Overview" animation.

WHAT DO YOU THINK?

3 In table 5.1, cuboidal and columnar epithelial cells are described as having a shape that accommodates more cellular organelles than squamous epithelial cells. Because cuboidal and columnar epithelial cells function to transport substances across an epithelial lining for the processes of absorption and secretion, what organelles might they contain that would not be found in a simple squamous epithelial cell (which functions mainly as a thin barrier for diffusion)?

Study Tip!

Students often confuse the terms *basement membrane* and *basal surface*. To clarify, the **basement membrane** is a connective tissue structure that lies at the **basal surface** of the epithelium. The two terms are not synonyms. That is, the basement membrane is a *structure*; the basal surface is a *location*. Also note that the basement membrane is an extracellular structure that the epithelium sits on. The function of the basement membrane is to anchor the epithelium to the underlying connective tissue, provide physical support for the epithelium, and to act as a barrier to regulate the passage of large molecules between the epithelium and the underlying tissues.

Connective Tissue

Connective tissues are derived from an embryonic tissue called **mesenchyme** (see **figure 5.10**). There are three broad categories of mature connective tissues: connective tissue proper, supporting connective tissues, and fluid connective tissues. **Connective tissue proper** is the "glue" that holds things together (such as tendons and ligaments) and the "stuffing" that fills in spaces (such as the fat that fills in spaces between muscles). **Supporting connective tissues** (cartilage and bone) are specialized connective tissues that provide support and protection for the body. **Fluid connective tissues** (blood and lymph) are specialized connective tissues that function to transport substances throughout the body. The details of bone and blood are covered in chapters 7 and 20 of this laboratory manual.

EXERCISE 5.2 | Identification of Embryonic Connective Tissue

1. Obtain a slide of mesenchyme (figure 5.10).

2. Place the slide on the microscope stage and bring the tissue sample into focus on low power, then switch to high power.

3. Notice that there are no visible fibers within the extracellular matrix (mature fibers do not yet exist within mesenchyme). The **mesenchymal cells** are recognized by their large oval nuclei. Mesenchyme has the ability to differentiate into any of the mature connective tissue cell types. That is, mesenchymal cells can differentiate into any of the adult connective tissue cell types (e.g., fibroblasts, chondroblasts, osteoblasts).

4. Using figure 5.10 as a guide, identify the following structures on the slide:

   ☐ ground substance   ☐ mesenchymal cell

5. In the following space, draw a brief sketch of mesenchyme as seen through the microscope. Be sure to label the structures listed in step 4 in your drawing.

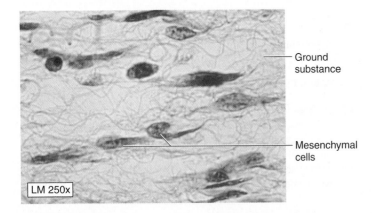

Ground substance

Mesenchymal cells

LM 250x

Figure 5.10   **Mesenchyme.** Mesenchyme is embryonic connective tissue.

——————— ×

Connective Tissue Proper

Connective tissue proper is a kind of "grab bag" category that contains all of the unspecialized connective tissues (that is, any connective tissue other than supporting or fluid connective tissues). These tissues are used either to hold things together (as with tendons and ligaments) or to fill up space (as with adipose tissue). Loose connective tissues have a loose association of cells and fibers, whereas dense connective tissues have cells and fibers that are densely packed together, which makes them much tougher than loose connective tissues. **Table 5.3** summarizes the characteristics of the different types of connective tissue proper. **Figure 5.11** is a flowchart for classification of connective tissue proper that you can use when you are given an unknown slide of connective tissue. It will help you to classify the connective tissue.

Table 5.3	Connective Tissue Proper	
Classification	**Description and Function**	**Location(s)**
Loose Connective Tissue		
Areolar (*Figure 5.12*)	Consists of a loose arrangement of collagen and elastic fibers with numerous fibroblasts. It loosely anchors structures to each other or fills in spaces between organs.	Located in the superficial fascia (hypodermis), which anchors skin to underlying muscle. It is also found surrounding many organs.
Adipose (*Figure 5.13*)	Characterized by adipocytes, which appear to be large, "empty" cells because the process of preparing the tissue removes all lipid within the cells. Collagen fibers located between the adipocytes hold the tissue together. Adipose tissue functions in insulation, protection, and energy storage.	Subcutaneous (under the skin). It also surrounds organs such as the kidneys, where it provides protection and fills in potential spaces such as within the axilla and popliteal fossa.
Reticular (*Figure 5.14*)	Composed of reticular fibers, which form an inner supporting framework for highly cellular organs such as the liver.	The inner stroma of organs such as the spleen, liver, and lymph nodes (*stroma, bed*).
Dense Connective Tissue		
Dense regular (*Figure 5.15*)	Composed of regular bands of collagen fibers all oriented in the same direction. The flattened nuclei of fibroblasts can be seen between bundles of collagen fibers. Dense regular connective tissue is good at resisting tensile forces in one direction only.	Tendons (connect muscle to bone) and ligaments (connect bone to bone).
Dense irregular (*Figure 5.16*)	Composed of bundles of collagen fibers arranged in many directions. Fibroblast nuclei can be seen between bundles of collagen fibers. Some nuclei appear round (if cut in cross section) and some appear flattened (if cut in longitudinal section). Many of the collagen fibers appear wavy because they are not all cut along the same plane. This tissue is tough and is good at resisting tensile forces applied in multiple directions.	Organ capsules, dermis of the skin, periosteum (outer covering of bone), perichondrium (outer covering of cartilage).
Elastic (*Figure 5.17*)	Consists of both collagen and elastic fibers all oriented in the same direction. Collagen fibers are thick and typically stain light pink or purple. Elastic fibers appear thin and black (if stained) or wavy. Fibroblast nuclei can be seen between the densely packed fibers. Elastic connective tissue is extensible and allows structures to stretch and recoil back to their original shape.	Walls of large arteries such as the aorta and some ligaments.

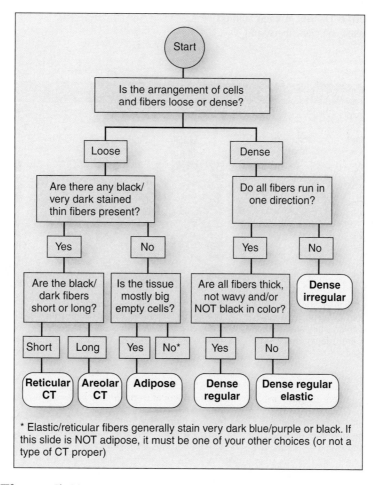

Figure 5.11   **Flowchart for Classifying Connective Tissue Proper.**

EXERCISE 5.3 Identification and Classification of Connective Tissue Proper

EXERCISE 5.3A: Areolar Connective Tissue

1. Obtain a slide of areolar connective tissue (**figure 5.12**).

2. Place the slide on the microscope stage and bring the tissue sample into focus on low power. Then change to high power.

3. You will see several small cells scattered throughout the slide. Most of these cells are **fibroblasts**, which secrete the elastic and collagen fibers (**table 5.4**). **Collagen fibers** will generally be light pink in color and somewhat thick. **Elastic fibers** will generally be long and thin, and will stain very dark or black.

4. Using figure 5.12 and tables 5.3 and 5.4 as guides, identify the following structures on the slide:

☐ collagen fibers   ☐ fibroblasts

☐ elastic fibers

5. In the following space, make a brief sketch of areolar connective tissue as seen through the microscope. Be sure to label the structures listed in step 4 on your drawing.

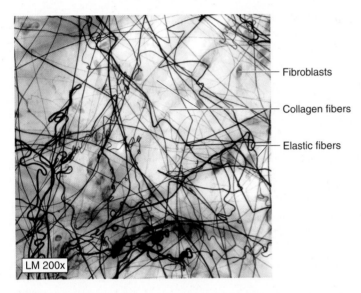

Fibroblasts

Collagen fibers

Elastic fibers

LM 200x

Figure 5.12   Areolar Connective Tissue. Areolar connective tissue contains predominantly fibroblasts, collagen fibers, and elastic fibers.

_____ ×

Table 5.4	Connective Tissue Fibers		
Fiber Type	**Collagen**	**Elastic**	**Reticular**
Micrograph			
Identifying Characteristics	Thicker than elastic or reticular fibers and are usually somewhat *pale* in color. They stain either *pink or blue*, depending on the stain used.	Appear either as fine, thin *black* fibers (when a silver stain is used) or as thin, *wavy* fibers (when silver stain is not used).	Reticular fibers are composed of a fine, thin type of collagen. The fibers can be seen only when a silver stain is applied, and they appear as an *irregular network* of thin *black* fibers (*reticular*, network). Reticular fibers are much shorter than elastic fibers.
Functions	Collagen fibers are good at *resisting tensile (stretching) forces*. They are only good at resisting stretch in one direction, along the long axis of the fiber.	Elastic fibers have the *ability to stretch and recoil*. They will often stretch to greater than 150% of their resting length without damage. Too much stretch will break the fiber.	Reticular fibers *form a delicate inner supporting framework* for highly cellular organs such as the liver, spleen, and lymph nodes.

EXERCISE 5.3B: Adipose Connective Tissue

1. Obtain a slide of adipose connective tissue (**figure 5.13**).

2. Place the slide on the microscope stage and bring the tissue sample into focus on low power. Then change to high power.

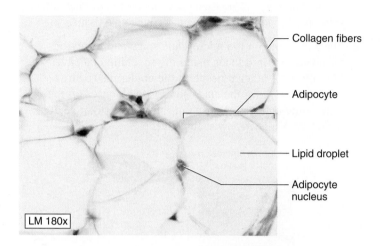

Figure 5.13   **Adipose Connective Tissue.** Adipose connective tissue is characterized by large adipocytes held together with a loose arrangement of collagen fibers.

3. Using figure 5.13 and tables 5.3 and 5.4 as guides, identify the following structures on the slide:

    ☐ adipocyte   ☐ collagen fibers

    ☐ adipocyte nucleus

4. In the following space, draw a brief sketch of adipose connective tissue as seen through the microscope. Be sure to label all the structures listed above in your drawing.

———— ×

EXERCISE 5.3C: Reticular Connective Tissue

1. Obtain a slide of reticular connective tissue (**figure 5.14**).

2. Place the slide on the microscope stage and bring the tissue sample into focus on low power. Then change to high power.

3. Using figure 5.14 and tables 5.3 and 5.4 as guides, identify the following on the slide:

    ☐ lymphocytes   ☐ reticular fibers

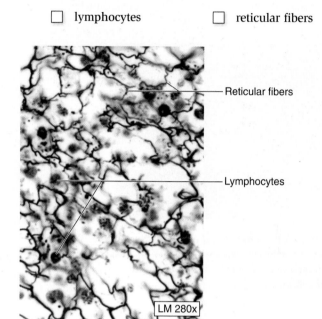

Figure 5.14   **Reticular Connective Tissue.** Reticular connective tissue consists of reticular fibers (a fine, thin form of collagen) that are visible only when a special stain is used, which makes them appear black.

4. In the following space, make a brief sketch of reticular connective tissue as seen through the microscope. Be sure to label all the structures listed in step 3 in your drawing.

———— ×

Study Tip!

Though a variety of stains are used for visualization of cells and tissues, certain tissues are visible only when special stains are used. In connective tissues, elastic and reticular fibers can be seen only when a special stain is used that stains the fibers black. Thus, if you see black fibers in a slide, you know you are looking at either elastic or reticular fibers. If the black fibers you see are very long and thin or wavy, they are elastic fibers. If they are short and form a network (*rete*, network), they are reticular fibers.

EXERCISE 5.3D: Dense Regular Connective Tissue

1. Obtain a slide of a tendon or ligament (**figure 5.15**).

2. Place the slide on the microscope stage and bring the tissue sample into focus on low power. Then change to high power.

3. Using figure 5.15 and tables 5.3 and 5.4 as guides, identify the following structures on the slide:

   ☐ collagen fibers   ☐ fibroblast nuclei

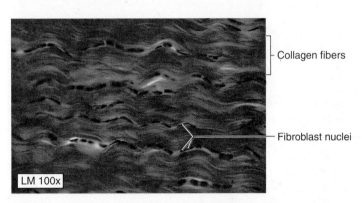

Collagen fibers

Fibroblast nuclei

LM 100x

Figure 5.15   Dense Regular Connective Tissue. Dense regular connective tissue consists of bundles of collagen fibers all oriented in the same direction, with fibroblasts located between the bundles of collagen fibers.

4. In the following space, draw a brief sketch of dense regular connective tissue as seen through the microscope. Be sure to label the structures listed in step 3 in your drawing.

_____ ×

EXERCISE 5.3E: Dense Irregular Connective Tissue

1. Obtain a slide of skin (**figure 5.16**) Skin consists of two layers, an outer epidermis, composed of stratified squamous epithelial tissue, and an inner dermis, composed of connective tissue and several skin appendages (more on this in Chapter 6). The tissue that makes the dermis of the skin very tough is dense irregular connective tissue.

2. Place the slide on the microscope stage and bring the tissue sample into focus using the scanning objective. Identify the epithelial tissue in the epidermis (figure 5.16a), then move the stage so the lens focuses on the underlying connective tissue in the dermis (figure 5.16b). Switch to a higher-power objective.

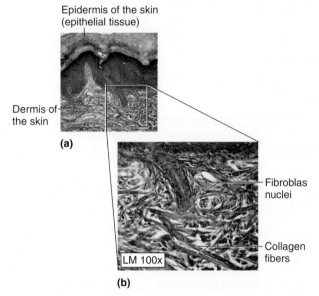

Epidermis of the skin (epithelial tissue)

Dermis of the skin

(a)

Fibroblast nuclei

Collagen fibers

LM 100x

(b)

Figure 5.16   Dense Irregular Connective Tissue. Dense irregular connective tissue is located in the dermis of the skin (a). Dense irregular connective tissue consists of bundles of collagen fibers oriented in many different directions, with fibroblasts located between the bundles of collagen fibers (b).

3. Using figure 5.16 and tables 5.3 and 5.4 as guides, identify the following structures on the slide:

   ☐ collagen fibers   ☐ fibroblast nucleus

4. In the following space, draw a brief sketch of dense irregular connective tissue as seen through the microscope. Be sure to label the structures listed in step 3 in your drawing.

_____ ×

EXERCISE 5.3F: Elastic Connective Tissue

1. Obtain a slide of the aorta or an elastic artery (**figure 5.17**).

2. Place the slide on the microscope stage and bring the tissue sample into focus on low power. Then switch to high power.

3. Using figure 5.17 and tables 5.3 and 5.4 as guides, identify the following structures on the slide:

   ☐ collagen fibers   ☐ fibroblasts

   ☐ elastic fibers

4. In the following space, draw a brief sketch of elastic connective tissue as seen through the microscope. Be sure to label the structures listed in step 3 in your drawing.

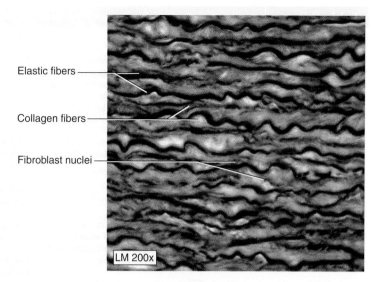

Figure 5.17   **Elastic Connective Tissue.** The wall of the aorta consists of dense regular elastic connective tissue, which contains both collagen and elastic fibers (which stain black). All the fibers are oriented in the same direction. In this slide, the collagen fibers are dark pink and the elastic fibers are black and wavy. Some fibroblast nuclei are visible in between the bundles of fibers.

Supporting Connective Tissue

Cartilage

Cartilage is a specialized connective tissue whose function is to provide strong, yet flexible, support. Cartilage is unique as a connective tissue in that it is avascular. A dense irregular connective tissue covering, called the **perichondrium**, surrounds all types of cartilage except fibrocartilage. The innermost part of the perichondrium contains immature cartilage cells called **chondroblasts**. The function of the chondroblasts is to secrete the fibers and ground substance that compose the extracellular matrix of cartilage. As a chondroblast secretes extracellular matrix, it eventually becomes completely surrounded by the matrix; at this point it is considered a mature cell, a **chondrocyte**. The space in the matrix where a chondrocyte sits is a **lacuna** (*lacus*, a lake). The function of a chondrocyte is to maintain the matrix that has already been formed. All types of cartilage contain chondrocytes in lacunae and have a ground substance that consists largely of glycosaminoglycans (GAGs). The three types of cartilage differ mainly in the type and arrangement of fibers within the matrix. **Table 5.5** summarizes the characteristics of the different types of cartilage.

Bone

Bone (table 5.5) is a specialized connective tissue whose function is to provide strong support. It protects vital organs and provides strong attachment points for skeletal muscles. Similar to cartilage, bone is surrounded by a dense irregular connective tissue covering, called the **periosteum**. The innermost part of the periosteum contains precursor bone cells called **osteoprogenitor cells**, which develop into immature bone cells called **osteoblasts**. Osteoblasts secrete the extracellular matrix (fibers and ground substance) of bone. When they become completely enveloped by the bony matrix, the osteoblasts become mature **osteocytes**.

There are two types of bone tissue: **compact bone** (dense) and **spongy bone** (cancellous). The structural and functional unit of compact bone is an **osteon**, which consists of concentric layers of bony matrix (lamellae). Along the lamellae are lacunae that contain osteocytes. The details of the two types of bone tissue will be covered in chapter 7.

Figure 5.18 is a flowchart for classification of supporting connective tissues that you can use when you are given an unknown slide of connective tissue. It will help you decide how to classify the connective tissue.

EXERCISE 5.4B: Fibrocartilage

1. Obtain a slide of an intervertebral disc (**figure 5.20**).

2. Place the slide on the microscope stage and bring the tissue sample into focus on low power. Then change to high power.

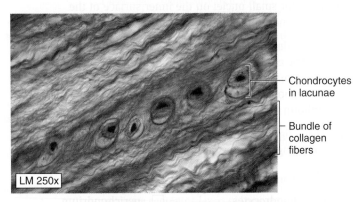

Chondrocytes in lacunae

Bundle of collagen fibers

LM 250x

Figure 5.20   Fibrocartilage. Fibrocartilage contains visible bundles of collagen fibers within its matrix. The chondrocytes (in their lacunae) often appear to line up in rows.

3. Using figure 5.20 and table 5.5 as guides, identify the following structures on the slide:

☐ chondrocytes   ☐ bundle of collagen fibers

4. In the following space, make a brief sketch of fibrocartilage as seen through the microscope. Be sure to label all the structures listed in step 3 in your drawing.

_____ ×

EXERCISE 5.4C: Elastic Cartilage

1. Obtain a slide of elastic cartilage (**figure 5.21**).

2. Place the slide on the microscope stage and bring the tissue sample into focus on low power. Then change to high power.

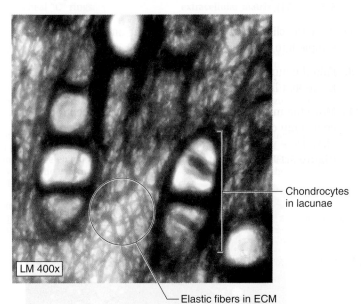

Chondrocytes in lacunae

LM 400x

Elastic fibers in ECM

Figure 5.21   Elastic Cartilage. Elastic cartilage contains chondrocytes in lacunae visible elastic fibers (which stain black) in its extracellular matrix.

3. Using figure 5.21 and table 5.5 as guides, identify the following structures on the slide:

☐ chondroblasts   ☐ lacunae
☐ chondrocytes   ☐ perichondrium
☐ elastic fibers

4. In the following space, make a brief sketch of elastic cartilage as seen through the microscope. Be sure to label all the structures listed in step 3 in your drawing.

_____ ×

WHAT DO YOU THINK?

4️⃣ Cartilage is unique as a connective tissue in that it is *avascular*. What limitations do you think avascularity imposes on cartilage as a tissue?

EXERCISE 5.3F: Elastic Connective Tissue

1. Obtain a slide of the aorta or an elastic artery (**figure 5.17**).

2. Place the slide on the microscope stage and bring the tissue sample into focus on low power. Then switch to high power.

3. Using figure 5.17 and tables 5.3 and 5.4 as guides, identify the following structures on the slide:

☐ collagen fibers   ☐ fibroblasts
☐ elastic fibers

4. In the following space, draw a brief sketch of elastic connective tissue as seen through the microscope. Be sure to label the structures listed in step 3 in your drawing.

_____ ×

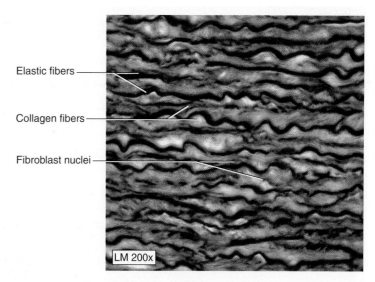

Figure 5.17 **Elastic Connective Tissue.** The wall of the aorta consists of dense regular elastic connective tissue, which contains both collagen and elastic fibers (which stain black). All the fibers are oriented in the same direction. In this slide, the collagen fibers are dark pink and the elastic fibers are black and wavy. Some fibroblast nuclei are visible in between the bundles of fibers.

Supporting Connective Tissue

Cartilage

Cartilage is a specialized connective tissue whose function is to provide strong, yet flexible, support. Cartilage is unique as a connective tissue in that it is avascular. A dense irregular connective tissue covering, called the **perichondrium**, surrounds all types of cartilage except fibrocartilage. The innermost part of the perichondrium contains immature cartilage cells called **chondroblasts**. The function of the chondroblasts is to secrete the fibers and ground substance that compose the extracellular matrix of cartilage. As a chondroblast secretes extracellular matrix, it eventually becomes completely surrounded by the matrix; at this point it is considered a mature cell, a **chondrocyte**. The space in the matrix where a chondrocyte sits is a **lacuna** (_lacus_, a lake). The function of a chondrocyte is to maintain the matrix that has already been formed. All types of cartilage contain chondrocytes in lacunae and have a ground substance that consists largely of glycosaminoglycans (GAGs). The three types of cartilage differ mainly in the type and arrangement of fibers within the matrix. **Table 5.5** summarizes the characteristics of the different types of cartilage.

Bone

Bone (table 5.5) is a specialized connective tissue whose function is to provide strong support. It protects vital organs and provides strong attachment points for skeletal muscles. Similar to cartilage, bone is surrounded by a dense irregular connective tissue covering, called the **periosteum**. The innermost part of the periosteum contains precursor bone cells called **osteoprogenitor cells**, which develop into immature bone cells called **osteoblasts**. Osteoblasts secrete the extracellular matrix (fibers and ground substance) of bone. When they become completely enveloped by the bony matrix, the osteoblasts become mature **osteocytes**.

There are two types of bone tissue: **compact bone** (dense) and **spongy bone** (cancellous). The structural and functional unit of compact bone is an **osteon**, which consists of concentric layers of bony matrix (lamellae). Along the lamellae are lacunae that contain osteocytes. The details of the two types of bone tissue will be covered in chapter 7.

Figure 5.18 is a flowchart for classification of supporting connective tissues that you can use when you are given an unknown slide of connective tissue. It will help you decide how to classify the connective tissue.

Table 5.5	Supporting Connective Tissue: Cartilage and Bone		
Tissue Type	**Description**	**Functions and Locations**	**Identifying Characteristics**
Hyaline cartilage	Chondrocytes are located within lacunae. Contains a perichondrium of dense irregular connective tissue. Extracellular matrix consists of diffuse collagen fibers spread throughout a semi-rigid ground substance, which is composed mainly of glycoproteins and water.	Hyaline cartilage provides *strong, semiflexible support* for structures such as the nasal septum, costal cartilages, articular cartilages, larynx, and tracheal "C" rings.	Hyaline cartilage is recognized by the chondrocytes in lacunae and the fact that no fibers are visible in the extracellular matrix.
Fibrocartilage	Very similar to hyaline cartilage except there is no perichondrium and the collagen fibers form thick, visible bundles.	The organization and density of the collagen fibers make this cartilage particularly good at *resisting compressive forces*. It is found in areas where compressive forces are high, such as intervertebral discs, the pubic symphysis, and the menisci of the knee joint.	Bands of fibers are easily visible. In addition, chondrocytes will appear to be lined up in rows because the thick bands of collagen fibers force them into this configuration.
Elastic cartilage	Very similar to hyaline cartilage in all aspects. However, the addition of elastic fibers to the matrix makes the cartilage much more flexible than hyaline cartilage.	The addition of elastic fibers allows the cartilage to provide *flexible support*. Locations include the epiglottis, the lining of the auditory tube, the external ear, and some cartilages of the larynx.	Elastic fibers are stained black in most preparations. The elastic fibers are generally higher in concentration near the lacunae.
Compact bone	Composed of osteons (Haversian systems), which are concentric layers of bony matrix (lamellae) surrounding a central canal. Lamellae have lacunae located along them, with osteocytes inside the lacunae. Canaliculi connect adjacent lacunae, and perforating canals run perpendicular to the central canals.	Provides extremely strong, rigid support. It is thickest in the diaphysis of long bones, but is also found as a thin layer surrounding all bone.	Multiple osteons packed tightly together. No marrow spaces.

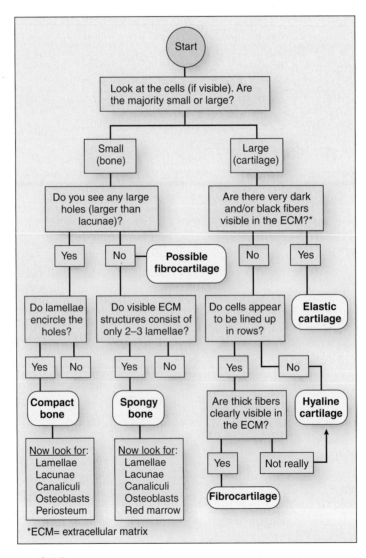

Figure 5.18   **Flowchart for Classifying Supporting Connective Tissues.**

EXERCISE 5.4 Identification and Classification of Supporting Connective Tissue

EXERCISE 5.4A: Hyaline Cartilage

1. Obtain a slide of the trachea and esophagus (see exercise 5.1g, p. 64).

2. Place the slide on the microscope stage and bring the tissue sample into focus on low power.

3. Find the lumen of the trachea and identify the epithelial tissue that lines the trachea (see exercise 5.1E, p. 67).

4. Move the microscope stage so that the tissue deep to the epithelium of the trachea is in the center of the field of view. Here you will find a plate of hyaline cartilage (**figure 5.19**).

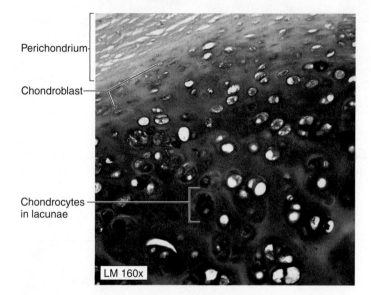

Perichondrium

Chondroblast

Chondrocytes in lacunae

LM 160x

Figure 5.19 **Hyaline Cartilage.** Hyaline cartilage contains prominent lacunae with chondrocytes. No fibers are visible in the extracellular matrix. At the top of this photograph you can see the perichondrium (light pink) with small, flattened chondroblasts on its inner surface.

5. Observe the slide on the lowest magnification possible. Identify the **perichondrium**, which surrounds the cartilage plate.

6. Look for small nuclei on the inner surface of the perichondrium. These are the nuclei of chondroblasts.

7. Next, identify the chondrocytes located within lacunae. In hyaline cartilage you will not be able to identify any fibers in the matrix because the fibers are spread very diffusely throughout the matrix. Instead, the matrix will appear uniform and smooth.

8. Using figure 5.19 and table 5.5 as guides, identify the following structures on the slide:

☐ chondroblasts   ☐ lacunae
☐ chondrocytes   ☐ perichondrium

9. In the following space, make a brief sketch of hyaline cartilage as seen through the microscope. Be sure to label all the structures listed in step 8 in your drawing.

EXERCISE 5.4B: Fibrocartilage

1. Obtain a slide of an intervertebral disc (**figure 5.20**).

2. Place the slide on the microscope stage and bring the tissue sample into focus on low power. Then change to high power.

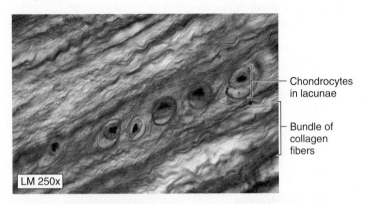

Figure 5.20   **Fibrocartilage.** Fibrocartilage contains visible bundles of collagen fibers within its matrix. The chondrocytes (in their lacunae) often appear to line up in rows.

3. Using figure 5.20 and table 5.5 as guides, identify the following structures on the slide:

   ☐ chondrocytes   ☐ bundle of collagen fibers

4. In the following space, make a brief sketch of fibrocartilage as seen through the microscope. Be sure to label all the structures listed in step 3 in your drawing.

_____ ×

EXERCISE 5.4C: Elastic Cartilage

1. Obtain a slide of elastic cartilage (**figure 5.21**).

2. Place the slide on the microscope stage and bring the tissue sample into focus on low power. Then change to high power.

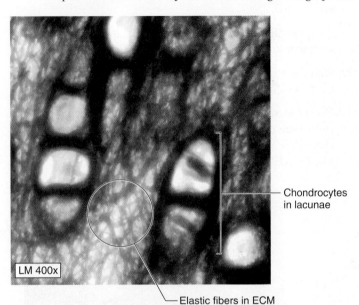

Figure 5.21   **Elastic Cartilage.** Elastic cartilage contains chondrocytes in lacunae visible elastic fibers (which stain black) in its extracellular matrix.

3. Using figure 5.21 and table 5.5 as guides, identify the following structures on the slide:

   ☐ chondroblasts   ☐ lacunae
   ☐ chondrocytes   ☐ perichondrium
   ☐ elastic fibers

4. In the following space, make a brief sketch of elastic cartilage as seen through the microscope. Be sure to label all the structures listed in step 3 in your drawing.

_____ ×

WHAT DO YOU THINK?

4️⃣ Cartilage is unique as a connective tissue in that it is *avascular*. What limitations do you think avascularity imposes on cartilage as a tissue?

EXERCISE 5.4D: Bone

1. Obtain a slide of compact bone (**figure 5.22**).

2. Place the slide on the microscope stage and bring the tissue sample into focus on low power. Then change to high power.

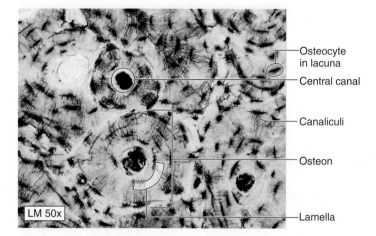

LM 50x

Figure 5.22 **Compact Bone.** Compact bone is composed of multiple osteons. Each osteon is characterized by a central canal surrounded by concentric layers of bony matrix called lamellae.

3. Using figure 5.22 and table 5.5 as guides, identify the following structures on the slide:

☐ canaliculus   ☐ lamella

☐ central canal   ☐ osteocyte

☐ lacuna   ☐ osteon

5. In the following space, make a brief sketch of compact bone as seen through the microscope. Be sure to label all the structures listed in step 3 in your drawing.

_____ ×

Fluid Connective Tissue

Fluid connective tissues (blood and lymph) are specialized in that the extracellular matrix of these tissues consists of a liquid ground substance and soluble fibers that become insoluble only in response to tissue injury. These tissues will be covered in detail in chapters 20 and 23. In this chapter we will consider only the basic characteristics of blood as a connective tissue. The cell types in blood are **erythrocytes** (red blood cells), **leukocytes** (white blood cells), and **platelets** (thrombocytes). Platelets are not actually cells. Instead they are cytoplasmic fragments of cells called **megakaryocytes**, which are found in the bone marrow. The extracellular matrix of blood consists of blood **plasma**. The ground substance is liquid, composed mainly of water and a number of dissolved substances. In addition, blood contains soluble proteins, some of which are called clotting proteins, such as **fibrinogen**. Fibrinogen becomes insoluble **fibrin** and forms fibers in response to tissue injury.

EXERCISE 5.5   Identification and Classification of Fluid Connective Tissue

1. Obtain a slide of a blood smear (**figure 5.23**).

2. Place the slide on the microscope stage and bring the tissue sample into focus on low power. Then bring the sample into focus using the oil immersion lens.

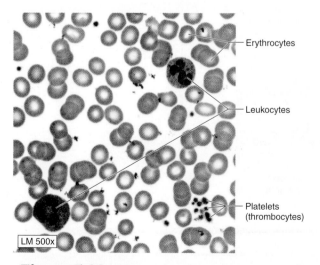

LM 500x

Figure 5.23 **Blood.** Blood is a fluid connective tissue containing erythrocytes (red blood cells), leukocytes (white blood cells), thrombocytes (platelets), and an extracellular matrix called plasma.

3. Using figure 5.23 as a guide, identify the following structures on the slide:

☐ erythrocytes   ☐ platelets (thrombocytes)

☐ leukocytes

4. In the following space, make a brief sketch of blood as seen through the microscope. Be sure to label all the structures listed in step 3 in your drawing.

_____ ×

5. *Optional Activity:* **AP|R** 3. **Tissues**—Watch the "Connective Tissue Overview" animation.

Muscle Tissue

Muscle tissue is both excitable and contractile. **Excitable** tissues are able to generate and propagate electrical signals called action potentials. As a **contractile** tissue, muscle has the ability to actively shorten and produce force. There are three types of muscle tissue: skeletal muscle, cardiac muscle, and smooth (visceral) muscle. The three types of muscle tissue are distinguished by their neural control (voluntary or involuntary),

the presence or absence of visible striations, the shape of the cells, and the number and location of nuclei. **Skeletal muscle** is found in the voluntary muscles that move the skeleton and the facial skin. **Cardiac muscle** is found in the heart, and **smooth muscle** is found in the walls of soft viscera such as the blood vessels, stomach, urinary bladder, intestines, and uterus. **Table 5.6** compares the three types of muscle tissue, and the flowchart in **figure 5.24** explains steps you can use to identify muscle tissues.

Table 5.6	Muscle Tissue		
Type of Muscle	**Description**	**Generalized Functions**	**Identifying Characteristics**
Skeletal	Elongate, cylindrical cells with multiple nuclei. Nuclei are peripherally located. Tissue appears striated (light and dark bands along the length of the cell).	Provides voluntary movement of the skin and the skeleton.	Length of cells (extremely long), striations, multiple peripheral nuclei.
Cardiac	Short, branched cells with 1 to 2 nuclei. Nuclei are centrally located. Dark bands (intercalated discs) are seen where two cells come together. Tissue appears striated (light and dark bands along the length of each cell).	Performs the contractile work of the heart. Responsible for creating the pumping action of the heart.	Branched, uninucleate cells, striations, intercalated discs.
Smooth	Elongate, spindle-shaped cells (fatter in the center, narrowing at the ends) with single, "cigar-shaped" or "spiral" nuclei. Nuclei are centrally located. No striations are apparent.	Creates movement within viscera such as intestines, bladder, uterus, and stomach. Moves blood through blood vessels, etc.	Spindle shape of the cells, no striations, cigar-shaped nuclei that are centrally located.

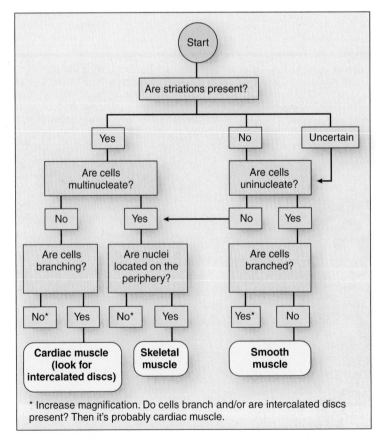

Figure 5.24 **Flowchart for Classifying Muscle Tissues.**

EXERCISE 5.6 Identification and Classification of Muscle Tissue

EXERCISE 5.6A: Skeletal Muscle Tissue

1. Obtain a slide of skeletal muscle (**figure 5.25**).

2. Place the slide on the microscope stage and bring the tissue sample into focus on low power. Then change to high power.

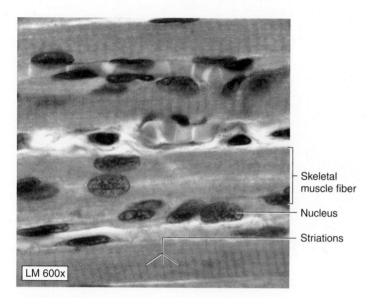

LM 600x

Figure 5.25   **Skeletal Muscle.** Skeletal muscle fibers are elongated and striated, and contain multiple peripherally located nuclei.

3. Using figure 5.25 and table 5.6 as guides, identify the following structures on the slide:

☐ skeletal muscle fiber   ☐ striations

☐ nucleus

4. In the following space, make a brief sketch of skeletal muscle as seen through the microscope. Be sure to label all the structures listed in step 3 in your drawing.

_____ ×

EXERCISE 5.6B: Cardiac Muscle Tissue

1. Obtain a slide of cardiac muscle (**figure 5.26**).

2. Place the slide on the microscope stage and bring the tissue sample into focus on low power. Then change to high power.

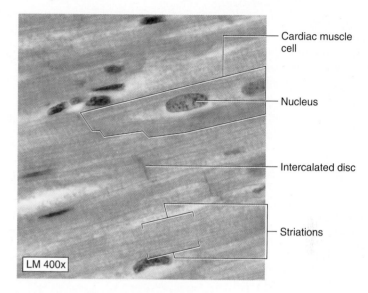

LM 400x

Figure 5.26   **Cardiac Muscle.** Cardiac muscle cells are short and branched, striated, and generally contain only one centrally located nucleus.

3. Using figure 5.26 and table 5.6 as guides, identify the following structures on the slide:

☐ intercalated disc   ☐ nucleus

☐ cardiac muscle cell   ☐ striations

4. In the following space, make a brief sketch of cardiac muscle as seen through the microscope. Be sure to label all the structures listed in step 3 in your drawing.

_____ ×

EXERCISE 5.6C: Smooth Muscle Tissue

1. Obtain a slide of smooth muscle (**figure 5.27**).

2. Place the slide on the microscope stage and bring the tissue sample into focus on low power. Then change to high power.

3. Using figure 5.27 and table 5.6 as guides, identify the following structures on the slide:

   ☐ smooth muscle cell   ☐ nucleus

4. In the following space, make a brief sketch of smooth muscle as seen through the microscope. Be sure to label all the structures listed in step 3 in your drawing.

_____ ×

5. _Optional Activity:_ **AP|R** 3. **Tissues**—Watch the "Muscle Tissue Overview" animation.

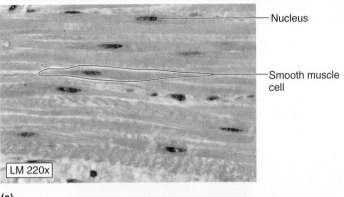

(a)

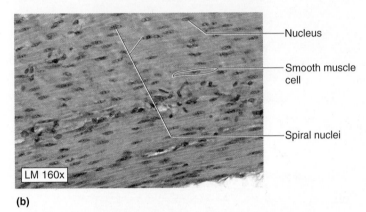

(b)

Figure 5.27   **Smooth Muscle.** Smooth muscle fibers are short and spindle-shaped, not striated, and contain only one centrally located nucleus (a). In figure 5.27_b_ you can see some nuclei that have taken on a "spiral" shape. This happens when the muscle fibers contract. As the fibers shorten, the nuclei start to coil up, or "spiral."

Nervous Tissue

Nervous tissue is characterized by its _excitability:_ the ability to generate and propagate electrical signals called action potentials. Nervous tissue is composed of two basic cell types (**table 5.7**). **Neurons** are excitable cells that send and receive electrical signals.

They have a limited ability to divide and multiply in the adult brain. **Glial cells** are supporting cells that support and protect neurons. Glial cells maintain the ability to divide and multiply in the adult brain. Glial cells constitute over 60% of the cells found in neural tissue.

Table 5.7	Nervous Tissue	
Cell Type	**Description**	**Generalized Functions**
Neurons	Though varied in shape, most neurons appear to have numerous branches coming off the cell body (soma). Neurons contain large amounts of rough endoplasmic reticulum (ER). The ribosomes and rough ER stain very dark and are collectively called chromatophilic substance.	These cells are responsible for _generating and transmitting information via electrical impulses_ within the nervous system. Thus, they are "excitable" cells.
Glial cells	Even more varied in shape than neurons, glial cells are generally much smaller than neurons with fewer (if any) branching processes.	These cells are the _general supporting cells_ of the nervous system. Their jobs are to protect, nourish, and support the excitable cells, the neurons.

EXERCISE 5.7 Identification and Classification of Nervous Tissue

1. Obtain a slide of nervous tissue (**figure 5.28**).

2. Place the slide on the microscope stage and bring the tissue sample into focus on low power. Then change to high power.

3. Using figure 5.28 and table 5.7 as guides, identify the following structures on the slide:

   ☐ axon of neuron   ☐ chromatophilic substance

   ☐ cell body of neuron   ☐ nucleus of glial cell

   ☐ dendrite of neuron   ☐ nucleus of neuron

4. In the following space, make a brief sketch of nervous tissue as seen through the microscope. Be sure to label all the structures listed in step 3 in your drawing.

_____ ×

5. _Optional Activity:_ **AP|R** **3. Tissues**—Watch the "Nervous Tissue Overview" animation.

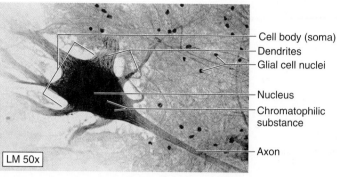

Cell body (soma)
Dendrites
Glial cell nuclei

Nucleus
Chromatophilic substance

Axon

LM 50x

(a)

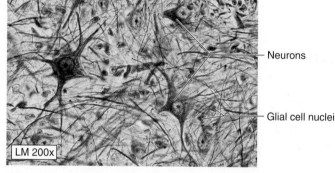

Neurons

Glial cell nuclei

LM 200x

(b)

Figure 5.28 **Nervous Tissue.** (a) Neurons are very large cells with a prominent nucleus and nucleolus, and dark-staining endoplasmic reticulum (chromatophilic substance). (b) Glial cells are much smaller than neurons, and are also more abundant.

WHAT DO YOU THINK?

5 Most tumors arise in cells that are constantly undergoing cell division, or mitosis (for example, skin cells). A patient has recently been diagnosed with a brain tumor. What type of cell do you think the tumor most likely arose from—a neuron or a glial cell?

Chapter 5: Histology

POST-LABORATORY WORKSHEET

Name: _____

Date: _____ Section: _____

1. Compare and contrast epithelial and connective tissues with respect to the following:

Characteristic	Epithelial Tissues	Connective Tissues
Cell Number and Arrangement		
Polarity		
Extracellular Matrix		
Vascularity		

2. Fully classify the epithelial tissue shown in this micrograph: _____

_____

_____

_____

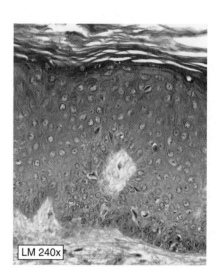

LM 240x

3. For each category of connective tissue listed in the following table, write in the major cell types, the fiber types, and the characteristics of the ground substance of the tissue. Refer to your textbook if you need help.

	Connective Tissue Proper	Cartilage	Bone	Fluid Connective Tissue
Cells				
Fibers				
Ground Substance				

4. In the following table, compare and contrast the characteristics of the various types of muscle tissue.

Characteristic	Skeletal Muscle	Cardiac Muscle	Smooth Muscle
Location and Number of Nuclei			
Cell Shape			
Presence or Absence of Striations			
Nervous Control			

5. Explain the characteristics that make blood a connective tissue. (Hints: What embryonic connective tissue is blood derived from? What is the composition of the extracellular matrix? What is the cellular component? What are the fibers?)

_____

_____

_____

_____

_____

_____

_____

6. Fully classify the epithelial tissue shown here:

_____

_____

_____

LM 200x

7. Endothelium is a _____ _____ epithelium that lines the walls of blood vessels and the heart.

8. The three types of cartilage are differentiated from each other by the type and arrangement of _____.

9. What does the presence of cilia indicate about the function of an epithelial tissue? _____

_____

10. The epithelial type that protects against abrasion is _____.

11. The only kind of muscle tissue that does NOT have striations is _____.

12. Write the name of the tissue type below each photo.

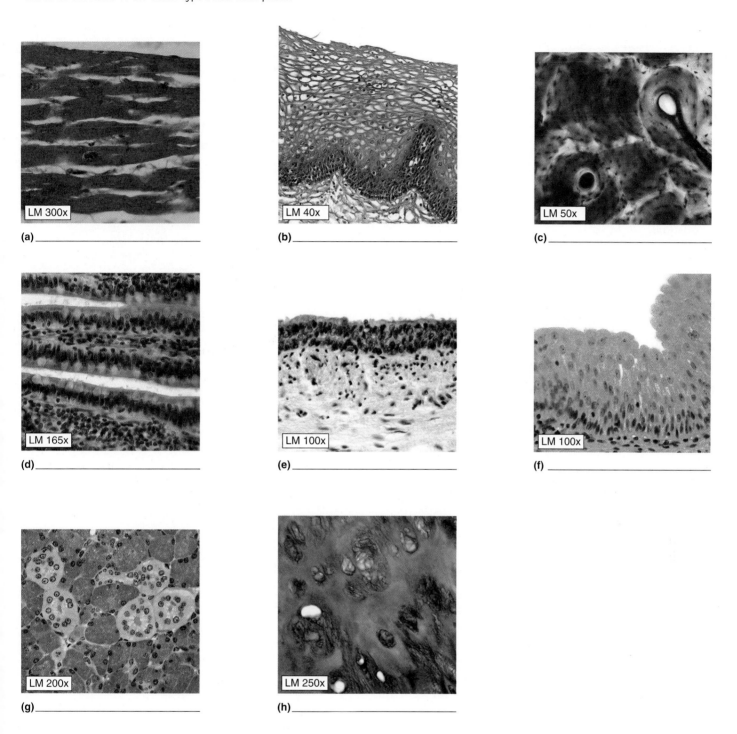

(a) _____

(b) _____

(c) _____

(d) _____

(e) _____

(f) _____

(g) _____

(h) _____

Integument

6

INTRODUCTION

In this laboratory session you will encounter the body's largest organ system, the **integument** (*integumentum*, a covering). We often mistakenly refer to the integument as the "skin." However, the two terms do not refer to the same thing. The **skin** is the outer covering of the body and is composed of an **epidermis** (*epi*, upon, + *derma*, skin) and **dermis** (*derma*, skin). The integument is an organ system that includes the skin *plus* all of the accessory structures of skin, such as sensory organs, glands, hair, and nails. The integument is a truly remarkable system, and we typically take its functions for granted—unless, of course, we happen to injure large portions of it. Damage to large areas of the integument (as might happen to a burn victim) can easily be life-threatening.

OUTLINE and OBJECTIVES

Module 4: INTEGUMENTARY SYSTEM

1. List four functions of the integument.

   a. _____

   b. _____

   c. _____

   d. _____

2. What type of epithelium composes the epidermis of the skin? _____

3. What type of connective tissue constitutes most of the dermis of the skin? _____

4. List the five layers of the epidermis of thick skin, starting with the layer closest to the basal surface.

   a. _____

   b. _____

   c. _____

   d. _____

   e. _____

5. List the two layers of the dermis.

   a. _____

   b. _____

6. List three accessory structures located in the dermis.

   a. _____

   b. _____

   c. _____

7. The main protein that composes a hair is:

   a. elastin   b. fibrin   c. keratin   d. collagen

8. The subcutaneous layer of the skin (hypodermis) is mainly composed of _____ tissue.

9. True or false: thick skin is located only on the palms of the hands and soles of the feet.

10. True or false: sweat glands are composed of modified epithelial tissue.

IN THE LABORATORY

In this laboratory session, the materials at your disposal will include histological slides of the skin and its accessory structures, and models of integument. While it is important that you learn to identify the structures of the integument under the microscope and on the models, do not forget to use your own body as an example of the real thing. For instance, while you view slides of thin skin and thick skin, be sure to also find locations on your own body that contain thin skin and thick skin. See and feel the differences between the two for yourself, and try to correlate what you see and feel on your own body with what you see under the microscope. Lastly, be sure to compare the appearance and texture of your own skin and hair with that of another student in your lab who has skin or hair of a different color or texture than your own.

Histology

The Epidermis

The epidermis is the outermost layer of the skin and consists of epithelial tissue. Thus there are no blood vessels in the epidermal layers. The main cell type composing the epithelial tissue in the skin is called a **keratinocyte**, so named because of its role in synthesizing the protein **keratin**. Keratin is an insoluble protein that imparts strength to the skin and makes it almost completely waterproof. The epidermis of *thin skin* (see figure 6.3), which is located on most of the body's surface, has four distinct layers, whereas the epidermis of *thick skin* (see figure 6.1), which is located on the palms of the hands and soles of the feet, has five distinct layers. **Table 6.1** summarizes the layers of the epidermis and their characteristics.

> ## Study Tip!
>
> The terms *thick skin* and *thin skin* refer only to the thickness of the epidermal layer. Technically, the thickest skin on the body is located on the back. However, that thickness takes into account not only the thickness of the epidermis, but that of the dermis as well. The thick skin on the palms of the hands and the soles of the feet is actually rather thin when considering the total thickness of the epidermis and the dermis combined. To understand this a little better, compare the total thickness of the skin on your back (by feel) to the total thickness of the skin on the middle arch of the sole of your foot or palm of your hand.

Table 6.1	Layers of the Epidermis	
Epidermal Layer	**Description**	**Word Origins**
Stratum Corneum	Consists almost entirely of keratin and remnants of dead keratinocytes, forming multiple layers. On the outer (apical) surface of the stratum corneum you may be able to view layers of dead cells in the process of **desquamation** (*squamosus*, scaly, fr. *squama*, scale).	*stratum*, layer, + *cornu*, horn, hoof
Stratum Lucidum	Present only in thick (glabrous) skin, which is located on the palm of the hand and sole of the foot. The layer is clear and homogeneous, and sometimes stains darker than the remainder of the stratum corneum, depending on the dye used.	*stratum*, layer, + *lucidus*, clear
Stratum Granulosum	Consists of three to five layers of cells that appear granular and darker in color than those of the underlying stratum spinosum. The graininess of the cells in this layer makes it quite distinct. Keratinocytes begin to die within the stratum granulosum as organelles and nuclei degenerate and the cells accumulate bundles of keratin.	*stratum*, layer, + *granulum*, a small grain
Stratum Spinosum	Named for the appearance of the keratinocytes when viewed at high magnification. In this layer, keratinocytes contain many cellular extensions, which make them appear somewhat "prickly" or "spiny." The "spines" are artifacts of staining and are areas where the extensions of adjacent cells interdigitate with each other and are anchored to each other by numerous desmosomes.	*stratum*, layer, + *spina*, a thorn
Stratum Basale	This layer is only one cell thick. At higher magnification, the cells appear large, with round nuclei, and you may be able to see cells undergoing cell division (*mitosis*). In addition, melanin granules are most concentrated in this layer of the skin. You will be able to view melanin granules well only in a sample of pigmented skin.	*stratum*, layer, + *basalis*, situated near the base

EXERCISE 6.1 | Layers of the Epidermis

1. Obtain a histology slide of **thick skin** (from the palm of the hand or sole of the foot); (**figure 6.1**). Thick skin is also *glabrous* skin, meaning it lacks hair (*glaber*, smooth). It does, however contain all five epidermal layers, so it is an ideal slide for demonstrating characteristics of the epidermis.

2. Place the slide on the microscope stage and scan the slide at low power. Look for any empty or clear space present in the slide, and try to find the apical surface of the skin epithelium lying next to the clear space.

3. Once you have located the skin epithelium, switch to a higher power. Scan the slide until the basal surface of the epithelium is in the center of the field of view. Here you will see the junction between the epidermis and the dermis. Notice that the basal layer is thrown into folds. These folds are called **epidermal ridges**. Epidermal ridges are located over extensions of the papillary layer of the dermis called **dermal papillae** (*papilla*, nipple). The increased area of adhesion between the epidermis and the dermis in the areas of epidermal ridges and dermal papillae helps prevent the epidermis and dermis from coming apart in areas where the skin experiences large frictional forces. In addition, epidermal ridges constitute the fingerprints and toe ridges found on the surface of the skin. Observe the fingerprints on your own hand.

4. Switch to high power and scan the epidermal/dermal junction for dermal papillae. Locate a **tactile (Meissner) corpuscle** within a dermal papilla (figure 6.4). Tactile corpuscles are sensory receptors for fine touch (table 6.3).

5. Using figure 6.1 and table 6.1 as guides, identify the following structures on the slide of thick skin:

   ☐ stratum basale   ☐ stratum lucidum

   ☐ stratum spinosum   ☐ stratum corneum

   ☐ stratum granulosum

6. In the space below, make a brief sketch of the layers of the epidermis of thick skin as seen under the microscope. Be sure to label all of the structures listed in step 5 in your drawing.

_____ ×

WHAT DO YOU THINK?

1 What layer of the epidermis represents the transition from living to dead epithelial cells? Why do keratinocytes begin to die within this layer?

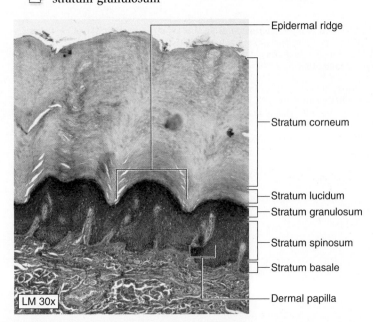

Epidermal ridge
Stratum corneum
Stratum lucidum
Stratum granulosum
Stratum spinosum
Stratum basale
Dermal papilla

LM 30x

Figure 6.1   Thick Skin. Epidermal layers in thick skin (found on the palms of the hands and soles of the feet).

The base color of a person's skin comes from the pigment **melanin** (*melas*, black). Melanin is produced by cells called **melanocytes**, which are found along the base of the stratum basale of the epidermis. Melanocytes have long dendritic (*dendrites*, relating to a tree) processes that extend between keratinocytes and transfer the melanin granules they produce to the adjacent keratinocytes. The melanin then accumulates on the apical side of the nuclei of the keratinocytes, protecting them from harmful ultraviolet light that can damage their DNA. A single melanocyte produces melanin for a number of adjacent keratinocytes (the melanocyte plus the cells served by it are referred to as an *epidermal-melanin unit*). Some areas of the body have a higher

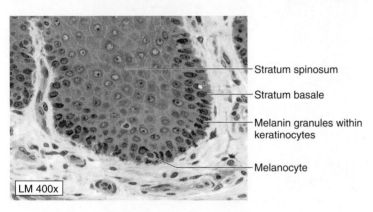

Stratum spinosum

Stratum basale

Melanin granules within keratinocytes

Melanocyte

LM 400x

Figure 6.2   Pigmented Skin. Pigmented skin containing melanin granules (brown) within keratinocytes, particularly in the stratum basale of the epidermis. Melanocytes themselves are distinguished by a halo of clear cytoplasm that surrounds the nucleus.

density of melanocytes than others. For example, the skin of the areola of the breast has more melanocytes than the skin on the rest of the breast. However, the *total number of melanocytes in light- and dark-skinned individuals is the same.* Differences in skin color have to do with the *amount of melanin* produced by melanocytes. The melanocytes of dark-skinned individuals simply produce greater amounts of melanin than those of light-skinned individuals.

1. Obtain a slide of pigmented skin and place it on the microscope stage. Observe at low power to locate the epidermis, and then change to a higher power lens.

2. Focus in on the stratum spinosum of the pigmented skin. You should be able to notice distinct black/brown melanin granules within the keratinocytes (**figure 6.2**). Can you identify any melanocytes in the slide? Melanocytes are located in the stratum basale and have small, round nuclei and pale-staining cytoplasm. If you can't find melanocytes, do not worry about it. Melanocytes are not easily identified on the standard slides found in the laboratory.

WHAT DO YOU THINK?

2 Why do you think it is important that melanin is present in its highest concentration in the keratinocytes at or near the basal layer of cells? (Hint: Did you observe cells that were undergoing cell division in this layer?)

The Dermis

The dermis of the skin is more complex than the epidermis because it contains numerous skin appendages such as hair follicles, glands, blood vessels, and nerves. The dermis consists of two layers: an outer **papillary layer**, and an inner **reticular layer** (**figure 6.3**). The papillary layer is the part of the dermis that contains "nipple-like" extensions that project into the epidermis (the dermal papillae). The papillary layer is generally quite thin and loosely defined. The reticular layer is named for

its "networked" appearance (*rete*, a net), not because it contains reticular fibers. In fact, the major fiber type found in this layer is the collagen that composes the dense irregular connective tissue. Embedded into this dense irregular connective tissue are the numerous skin appendages. The slide descriptions in this section of the laboratory manual will guide you through both general observation of the structure and function of the dermis and the specific structures and functions of the various skin appendages found within the dermis.

Study Tip!

The next time you see a piece of leather, look at it carefully and think about your own skin. The very tough tissue that composes most of the leather is dense irregular connective tissue (collagen fibers) from the dermis of the animal's skin.

WHAT DO YOU THINK?

3 Contrast the epidermal/dermal junction in thick skin with that of thin skin. Specifically note the structure of the epidermal/dermal junction in each slide. Is there a difference in the number of dermal papillae? What function do you think such differences serve?

EXERCISE 6.3 Layers of the Dermis

1. Obtain a slide of thick skin or thin skin and place it on the microscope stage.

2. Scan the slide at low power to identify the skin tissue, and then bring on the dermal layer of the skin into the center of the field of view.

3. Switch to high power and distinguish the papillary dermis from the reticular dermis (**figure 6.4**).

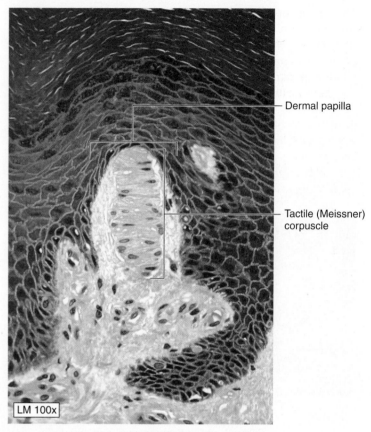

Figure 6.4 on the right shows labels: Dermal papilla, Tactile (Meissner) corpuscle, LM 100x

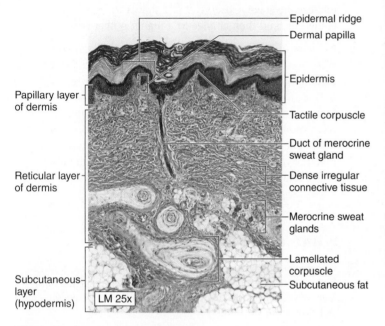

Figure 6.3 on the left shows labels: Epidermal ridge, Dermal papilla, Epidermis, Papillary layer of dermis, Tactile corpuscle, Duct of merocrine sweat gland, Reticular layer of dermis, Dense irregular connective tissue, Merocrine sweat glands, Lamellated corpuscle, Subcutaneous layer (hypodermis), Subcutaneous fat, LM 25x

Figure 6.3 **Dermis.** Thick skin demonstrating several skin appendages, such as tactile corpuscles, lamellated corpuscles, and merocrine sweat glands in the dermis.

Figure 6.4 **Tactile Corpuscle.** High-magnification view of a dermal papilla containing a tactile corpuscle.

4. Using figures 6.3 and 6.4 as a guide, identify the following structures of the dermis on the slide:

☐ dense irregular connective tissue
☐ dermal papillae
☐ epidermal ridges
☐ epidermis
☐ papillary dermis
☐ reticular dermis
☐ tactile (Meissner) corpuscle

5. In the space below, make a brief sketch of the dermis as seen through the microscope. Be sure to label all the structures listed in step 4 in your drawing.

_____ ×

WHAT DO YOU THINK?

What is the advantage of having tactile (Meissner) corpuscles located near the surface of the skin? Do you think there would be more of these sensory receptors per unit area in the skin on the palm of your hand or in the skin on your back (or would they be the same)?

EXERCISE 6.4 Merocrine (eccrine) Sweat Glands and Sensory Receptors

1. Obtain a slide of thick skin or thin skin and place it on the microscope stage. Begin by observing the epidermal and dermal layers of the skin, just as you did with the slides used to demonstrate epidermal layers and structures. As you view the current slide, focus your attention on the numerous skin appendages found within the dermis.

2. Locate the coiled, tubular glands found deep within the reticular dermis (these glands open to the skin surface, and you might see a duct traveling to the surface of the skin). These are **merocrine (eccrine) sweat glands** (*meros*, to share, + *krino*, to separate) (figure 6.3), which are located in skin covering nearly the entire body (as opposed to apocrine sweat glands, which are located only in the axillary region, pubic region, and areola). **Table 6.2** lists the types of glands located in the dermis and describes their locations and functions.

3. Focus your attention on sensory receptors found in the dermis. **Table 6.3** summarizes the characteristics of sensory receptors in the dermis. Many of the receptors are difficult to identify histologically, so you should also refer to the gross anatomy section of this manual and identify them on classroom models of integument. Observe the many dermal papillae found in the slide. Center one or more papillae in the field of view and increase the magnification.

4. Using figure 6.4 as a guide, try to locate the tiny sensory receptors called **tactile (Meissner) corpuscles** (*tactus*, to touch, + *corpus*, body) within the papillae. These sensory receptors, appropriately located near the surface of the skin, are responsible for sensing fine touch.

5. After identifying the tactile corpuscles, change back to a lower magnification and scan the lower reticular dermis and subcutaneous regions of the skin. Look for a large onion-shaped organ. This is a sensory receptor called a **lamellated (Pacinian) corpuscle** (*lamina*, plate, + *corpus*, body) (figure 6.3). Such sensory organs, located deep within the dermis, are responsible for sensing deep pressure applied to the skin. The other main sensory receptors found within the dermis, the **tactile (Merkel) cells** and **free nerve endings**, are not easily identifiable under the light microscope so you must rely on classroom models for their identification.

6. Finally, see if you can identify cross sections through the numerous small blood vessels located in the dermis.

7. In the space below, draw a sketch of the skin appendages you were able to locate.

_____ ×

Table 6.2	Glands in the Dermis				
Gland	**Location**	**Description**	**Mode of Secretion**	**Function**	**Word Origins**
Apocrine Sweat Glands	Axilla, areola of the breast, and pubic/anal regions.	Coiled tubular glands located next to hair follicles. Ducts open into the hair follicle.	Merocrine—exocytosis of vesicles containing product into the duct of the gland.	Produce a thick, slightly oily sweat that may have pheromone-like properties.	*apo-*, away from, + *krino*, to separate or secrete
Merocrine (eccrine) Sweat Glands	Most of the surface of the body.	Coiled tubular glands whose main secretory portions are found deep within the reticular layer of the dermis. Ducts open to the surface of the skin.	Merocrine—exocytosis of vesicles containing product into the duct of the gland.	Produce the thin, watery sweat that cools the body.	*meros*, share, + *krino*, to separate or secrete
Sebaceous Glands	Wherever hair follicles are found. Particularly abundant on the scalp.	Glands located next to hair follicles. Ducts commonly open into the hair follicle.	Holocrine—disintegrated whole cells filled with product are discharged into the duct of the gland.	Produce sebum, an oily substance that lubricates the skin surface, keeps it from drying out, and inhibits the growth of bacteria.	*sebaceous*, relating to sebum; oily; *holos*, whole, + *krino*, to separate or secrete

Table 6.3	Sensory Receptors in the Dermis		
Sensory Receptor	**Location**	**Structure/Appearance**	**Function**
Free Nerve Ending	At epidermal/dermal junction.	There is no special structure at the end of the nerve (hence "free" nerve ending).	General pain sensation.
Lamellated (Pacinian) Corpuscle	Deep in the dermis and hypodermis.	Multiple layers of cells wrapped around each other like an onion.	Sensation of deep pressure.
Tactile (Merkel) Cell	At epidermal/dermal junction.	Round cell located in the stratum basale of the epidermis. It associates with a sensory nerve ending in the dermis (a tactile disc).	Sensation of fine, delicate touch.
Tactile (Meissner) Corpuscle	In dermal papillae.	Oval-shaped structure with cells that appear almost layered on top of each other.	Sensation of fine, delicate touch.

EXERCISE 6.5   The Scalp—Hair Follicles and Sebaceous Glands

1. Obtain a slide of the **scalp** and place it on the microscope stage. Begin by observing the epidermal and dermal layers of the scalp, just as you did with previous slides. The scalp epithelium is thin, but the dermis is quite thick and contains numerous **hair follicles** (**figure 6.5** and **table 6.4**).

2. Scan the slide until you locate a hair follicle that is sliced longitudinally, so you can see the entire hair, from the base of the hair follicle to where it exits the skin. Notice that the color of the cells lining the hair follicle is similar to the color of the cells within the epidermis. This is because hair follicles are derivatives of the epidermis and they develop initially as down-growths of the stratum basale. There are three distinct regions to a hair: (1) the **shaft**, which is the portion of the hair that exits the skin surface; (2) the **root**, which is the portion of the hair within the skin itself; and (3) the **bulb**, which is the swelled base of the hair.

3. Observe the bulb of the hair at higher magnification (**figure 6.6**). Notice the **papilla**, a cone-shaped structure in the middle of the base of the follicle. The papilla is part of the dermis, and is separated from the hair follicle by the basement membrane of the hair follicle epithelium (this basement membrane continues external to the hair follicle as the "glassy membrane"). The papilla contains sensory nerve endings and numerous blood vessels, which are important in supplying nutrients to the developing hair.

4. Return to a lower magnification and look for the oil-secreting **sebaceous glands** (figure 6.5) that connect to the hair follicles in the region of the hair roots. Sebaceous glands secrete an oily substance, **sebum**, into the hair follicle.

5. Carefully observe several hair follicles to locate the small **arrector pili** muscles (*arrector*, that which raises, + *pili*, hair) that attach to the base of the hair follicle (not visible in figure 6.5; see figure 6.10). When these smooth muscles contract, they pull at the base of the hair follicle. This causes the hair to stand up straight, rather than lie flat against the surface of the skin. Muscle contraction pulls down on the skin, while the area where the hair shaft comes out of the skin remains elevated. Thus, in humans it gives the appearance of "goose bumps" or "goose pimples."

6. Using figures 6.5 and 6.6, and tables 6.2 and 6.3 as guides, identify the following structures related to the hair follicle:

   ☐ arrector pili muscle   ☐ root of hair follicle
   ☐ bulb of hair follicle   ☐ sebaceous gland
   ☐ hair follicle   ☐ shaft of hair follicle
   ☐ papilla of hair follicle

Table 6.4	Parts of a Hair Follicle
Structure	**Description and Function**
Connective Tissue (Dermal) Root Sheath	The connective tissue of the dermis (mainly dense collagen fibers) that surrounds the entire hair follicle.
Cortex	Constitutes the bulk of the hair. Composed predominantly of keratin.
Cuticle Layer	The outer portion of the hair itself, composed of several layers of hard plates of keratin that surround the cortex of the hair.
External Root Sheath	The outer layers of the hair follicle, which are continuous with the stratum basale and stratum spinosum of the epidermis.
Glassy Membrane	A specialized basement membrane located external to the external root sheath and internal to the connective tissue that surrounds the hair follicle (the connective tissue root sheath).
Internal Root Sheath	A sheath derived from epithelial tissue that lies between the external root sheath and the hair itself.

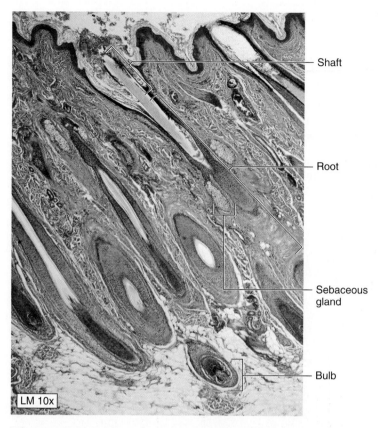

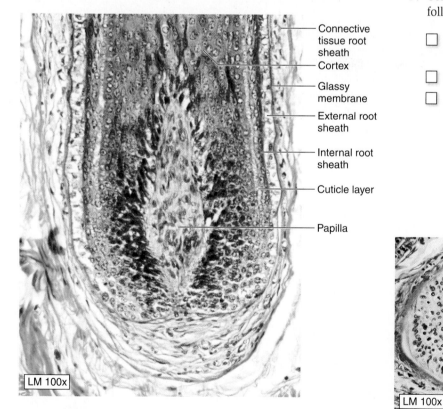

Figure 6.5 **Scalp.** Skin of the scalp, demonstrating several hair follicles.

Figure 6.6 **Close-up View of a Hair Bulb.**

7. In the space below, draw a sketch of a hair follicle and associated skin appendages as viewed through the microscope. Be sure to label all the structures listed in step 6 in your drawing.

8. Next, scan the slide until you find a hair follicle in cross section (**figure 6.7**).

9. Using figure 6.7 and table 6.4 as guides, identify the following structures:

☐ connective tissue (dermal) root sheath
☐ cortex
☐ cuticle layer
☐ external root sheath
☐ glassy membrane
☐ internal root sheath

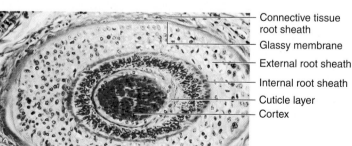

Figure 6.7 **Cross Section of a Hair Follicle.**

EXERCISE 6.6 | Axillary Skin—Apocrine Sweat Glands

1. Obtain a slide of **axillary skin** and place it on the microscope stage. Locate a hair follicle on the slide. Notice that the hair follicles in this skin are oriented at a fairly steep angle with respect to the apical surface of the epithelium. This, in part, is what makes the hairs in the axillary region curly.

2. Look for glands that open into the hair follicle. These are **apocrine sweat glands** (*apo-*, away from, + *krino*, to separate or secrete) (**figure 6.8**). Apocrine sweat glands are predominantly located in the axillary and pubic/anal regions of the body, though they are also located in the areola of the nipple and in men's facial hair. These glands produce their secretions the same way that merocrine glands do, via exocytosis. However, apocrine glands produce a secretion that is thicker and oilier than that of the merocrine glands. Thus, although the term *apocrine* historically referred to a different mode of secretion,* these glands are still referred to as apocrine glands.

3. In the space below, draw a sketch of a hair follicle from axillary skin and its associated apocrine sweat gland.

_____ ×

Opening of hair follicle
(hair is missing in this view)

Cross–section through
hair follicle

Sebaceous gland

Apocrine sweat gland

LM 10x

Figure 6.8   Apocrine Sweat Glands. Axillary skin demonstrating apocrine sweat glands, which open to the hair follicle.

*Traditionally, *apocrine* referred to a process by which the apical part of the cell discharged into the duct of the gland along with the secretion. However, we now know that this is not how apocrine glands produce their secretions.

Structure of a Nail

1. Obtain a slide of a **nail**. A nail is a very specialized skin appendage that arises from epithelial tissue. **Table 6.5** lists the parts of a nail, and **figure 6.9** shows a longitudinal section of a nail.

2. Using figure 6.9 and table 6.5 as guides, identify the following structures on the slide:

☐ body   ☐ hyponychium

☐ eponychium   ☐ lunula

☐ hidden border

3. In the space below, draw a sketch of a nail as seen through the microscope. Be sure to label all of the structures listed in step 2 in your drawing.

Table 6.5	Parts of a Nail	
Structure	**Description**	**Word Origins**
Body	The main portion of the nail.	
Eponychium	The fold of skin at the root of the nail that folds over the body of the nail.	*epi*, upon, + *onyx*, nail
Hidden Border	The portion of the nail that lies beneath the eponychium.	
Hyponychium	The skin underneath the free border of the nail.	*hypo*, under, + *onyx*, nail
Lunula	A white, curved area at the base of the nail.	*luna*, moon

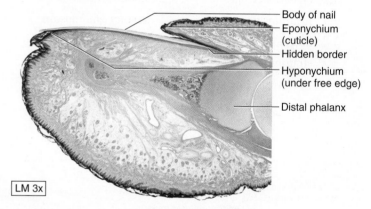

LM 3x

Figure 6.9   **Longitudinal Section of a Nail.**

_____ ×

Gross Anatomy

Integument Model

Viewing laboratory models or charts of integument will allow you to identify structures that are not easily visible under the light microscope. **Figure 6.10** demonstrates structures that are typically visible in classroom models of the integument.

EXERCISE 6.8 Observing Classroom Models of Integument

1. Locate models or posters of the integument in your laboratory.

2. Using figure 6.10 as a guide, identify the structures listed below. Pay particular attention to tactile corpuscles, lamellated corpuscles, melanocytes, and free nerve endings, because these structures are not easily seen under the microscope.

3. *Optional Activity:* AP|R 4: **Integumentary System—** Use the Quiz feature to review integumentary system structures.

Epidermis

- ☐ epidermal ridges
- ☐ keratinocytes
- ☐ melanocytes
- ☐ stratum basale
- ☐ stratum corneum
- ☐ stratum granulosum
- ☐ stratum lucidum
- ☐ stratum spinosum

Dermis

- ☐ blood vessels
- ☐ dense irregular connective tissue
- ☐ free nerve ending
- ☐ hair follicle
- ☐ tactile (Meissner) corpuscle
- ☐ merocrine sweat gland
- ☐ lamellated (Pacinian) corpuscle
- ☐ papillary layer
- ☐ reticular layer
- ☐ sebaceous gland

Subcutaneous layer (hypodermis)

- ☐ adipose connective tissue
- ☐ lamellated (Pacinian) corpuscle

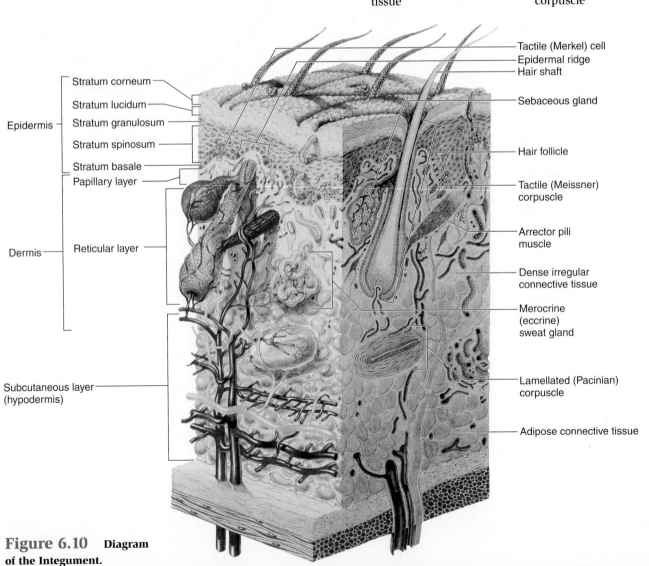

Figure 6.10   Diagram of the Integument.

Chapter 6: Integument

POST-LABORATORY WORKSHEET

1. Label the following diagram of the integument.

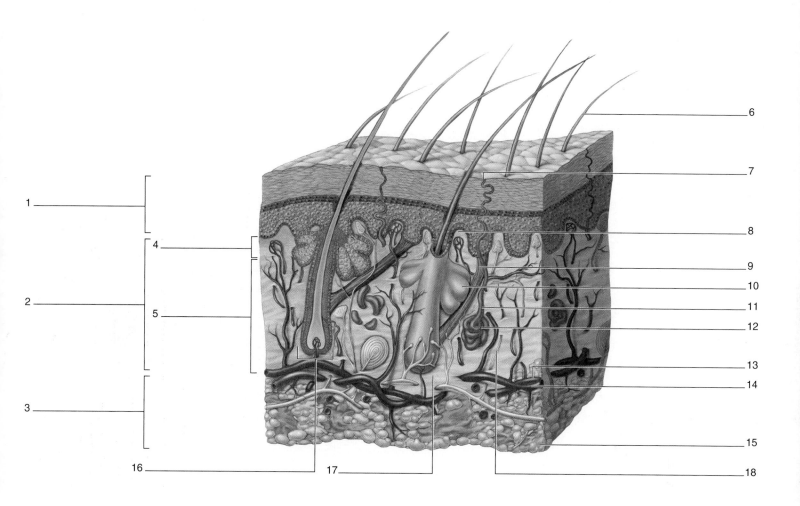

2. Label the following diagrams of a hair follicle.

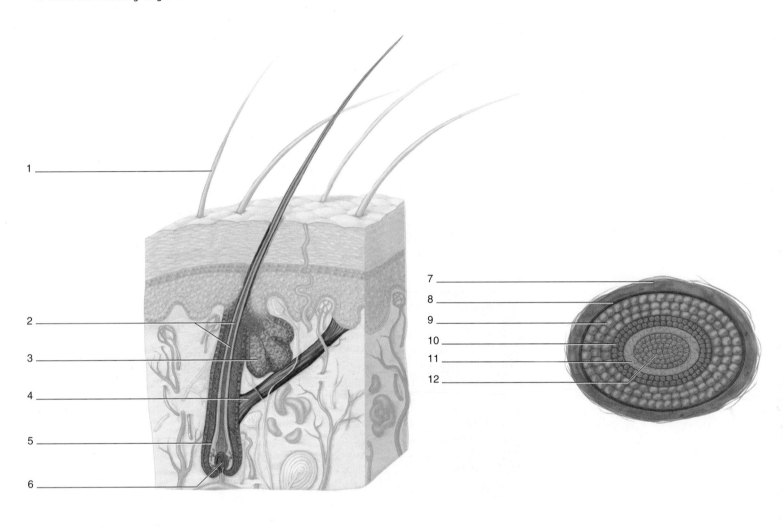

1 _____

2 _____

3 _____

4 _____

5 _____

6 _____

7 _____

8 _____

9 _____

10 _____

11 _____

12 _____

3. Define the following terms:

a. *epidermal ridge:* _____
_____

b. *keratinocyte:* _____
_____

c. *melanocyte:* _____
_____

d. *dermal papilla:* _____
_____

e. *sebaceous gland:* _____
_____

f. *merocrine sweat gland:* _____
_____

g. *apocrine sweat gland:* _____
_____

4. Label the following diagrams of a nail.

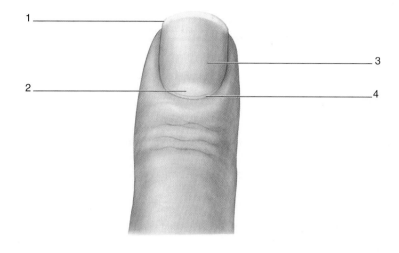

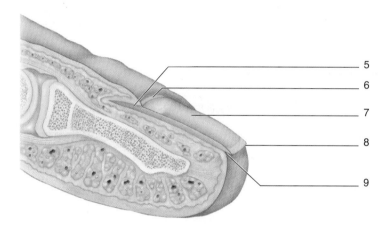

5. While sitting on the couch reading your anatomy book, you realize that you left a window open because you feel a light breeze flowing across your arm. What type of sensory receptor in your skin is responsible for detecting this sensation? _____

6. When you sit down on a bench to wait for a bus to arrive, you notice that there is a rock on the bench that you didn't see before you sat down on it. What type of sensory receptor in your skin is responsible for detecting this sensation?

_____

7. Susan accidentally tripped on the sidewalk and scraped her knee. The scrape was superficial and did not bleed. What is the deepest layer of her skin that could have been damaged without causing her to bleed?

_____

8. Compare and contrast the structure and function of merocrine and apocrine sweat glands.

_____

_____

_____

_____

_____

9. When part of the skin is exposed to a great deal of friction, such as when part of your foot rubs against your shoe, a **blister** often forms.

a. A blister is a collection of fluid that accumulates between two layers of the skin as they separate from each other. Which two layers of the skin are

these? _____ and _____ .

b. When you pull the outer layers of skin off of a blister, exposure of the underlying tissue causes a great deal of pain. What type of sensory receptor is

responsible for sensation of this pain? _____ .

c. Why is the pain worse after removal of the outer layer of the blister? _____

_____ .

10. A hypodermic needle is used to give certain types of injections.

a. Based on the name of the needle, what space is the needle usually directed into? _____ .

b. What layers of the skin must the hypodermic needle pass through, in order to get to this space? (In your answer, include all sublayers of the dermis or

epidermis that may apply. Assume the needle is passing through thin skin.) _____

_____

_____

_____

_____

_____ .

c. What structures in this space do you think are the likely targets of these needles?

_____ .

Skeletal System Overview: Bone Anatomy

7

OUTLINE and OBJECTIVES

aprevealed.com

Module 5: SKELETAL SYSTEM

INTRODUCTION

Bone is a highly specialized connective tissue that is both rigid and flexible. Most of us tend to view bone as static, nonliving tissue because the type of bones we commonly see are those left behind from the carcasses of animals or the preserved skeletons we see in the anatomy laboratory. This view of bone is quite far from reality. Living bone is dynamic, versatile tissue that is in a constant state of turnover and is one of the most metabolically active kinds of tissue within the body. It has a rich blood supply and an amazing ability to alter its structure in response to the changing stresses placed upon it. Indeed, prolonged absence of stress will cause bone to lose density and strength, whereas increased stress causes bone to increase in density and strength.

A typical long bone such as the humerus or femur (**figure 7.1**) is composed of a long shaft, called the **diaphysis** (*dia*, through, + *physis*, growth); rounded ends, called **epiphyses** (*epi*, upon, + *physis*, growth); and articulation points between the two, called **metaphyses** (*meta*, between, + *physis*, growth). Within the shaft is a large cavity called the **medullary cavity**, which in the adult is filled with **yellow bone marrow** (adipose tissue). The walls of the diaphysis are composed of a thick layer of **compact bone** tissue. The epiphyses of the bone are surrounded by a thin layer of compact bone and have **articular cartilages** on the ends. **Spongy bone** tissue is found within the epiphyses of the bone. In the fetus, the marrow spaces between the trabeculae of spongy bone are composed of **red bone marrow**. However, in the adult they are composed mainly of yellow bone marrow because of the conversion of red marrow to yellow marrow that occurs as the skeleton matures. In the adult, red bone marrow is primarily limited to the proximal epiphyses of the humerus and femur, the sternum, and the iliac crest.

The human skeleton consists of two divisions: (1) the **axial skeleton**, which includes bones of the cranium, vertebrae, ribs, and sternum, and (2) the **appendicular skeleton**, which includes bones of the pectoral girdle, upper limb, pelvic girdle, and lower limb. The **pectoral girdle** consists of the scapula and clavicle. These bones provide support and attachment points for muscles that connect the limbs to the axial skeleton. The **pelvic girdle** consists of the bones composing the **os coxae** (*os,* bone, + *coxa,* hip): the ilium, ischium, and pubis. These bones protect contents of the pelvic cavity and provide support and attachment points for muscles that connect the lower limb to the axial skeleton.

The detailed structures and functions of the bones making up the human skeleton are covered in chapters 8 and 9. The focus of this chapter is to introduce you to the general micro- and macrostructure of bone tissue, and to provide you with a brief overview of the structure and function of the human skeleton.

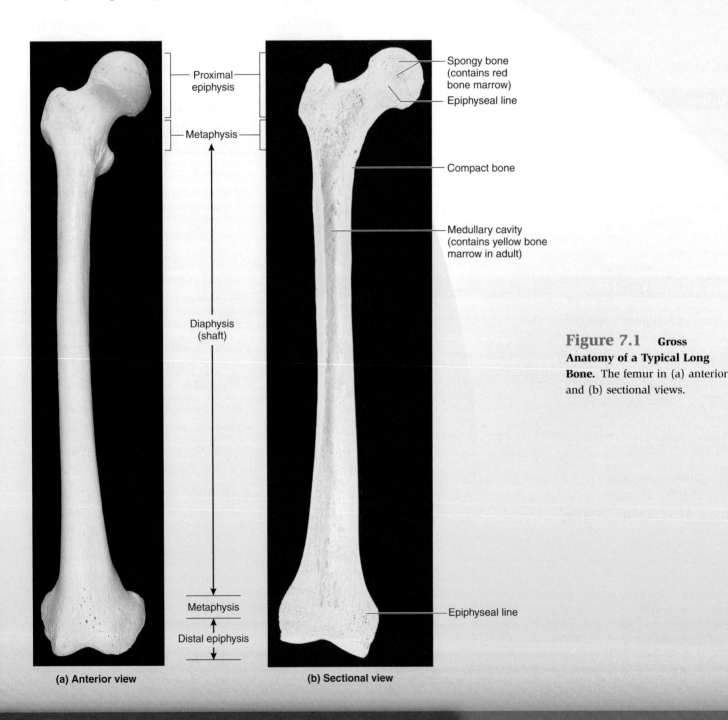

Figure 7.1   Gross Anatomy of a Typical Long Bone. The femur in (a) anterior and (b) sectional views.

(a) Anterior view   (b) Sectional view

Chapter 7: Skeletal System Overview: Bone Anatomy

Name: _____

Date: _____ Section: _____

PRE-LABORATORY WORKSHEET

1. List four functions of the skeletal system.

   a. _____

   b. _____

   c. _____

   d. _____

2. Name the two types of bone tissue.

   a. _____

   b. _____

3. Describe the following structures:

   a. *osteoprogenitor cell:* _____

   b. *osteoblast:* _____

   c. *osteocyte:* _____

   d. *osteoclast:* _____

   e. *lamella:* _____

   f. *lacuna:* _____

   g. *canaliculus:* _____

4. Fill in the blanks:

   a. The end of a long bone is the _____.

   b. The shaft of a long bone is the _____.

   c. The area where the end of a long bone meets the shaft of a long bone is the _____.

5. In an adult skeleton, _____ bone marrow fills the medullary cavity of a long bone, and _____ bone marrow is located within the proximal epiphyses of long bones such as the humerus and femur.

6. The outer covering of bone is the _____ and it is composed of _____ connective tissue.

7. List the two types of bone formation.

   a. _____

   b. _____

8. List the two divisions of the human skeleton, and describe which bones are included in each division.

   a. _____

   b. _____

IN THE LABORATORY

In this laboratory session, you will identify gross and histological features of the bony tissues that compose the human skeleton through microscopic observations of compact and spongy bone tissues and through dissection of a fresh cow bone. You will also identify the major bones that compose the human skeleton.

Histology

In these exercises you will look at the histology of compact bone tissue, spongy bone tissue, and endochondral bone development. **Table 7.1** summarizes the characteristics of the types of bone cells you will observe.

Bone Tissue

You will typically observe two types of bone histology preparations in the laboratory. In the first type of preparation, **decalcified bone** (see figure 7.2), the rigid, mineralized (calcified) matrix of the bone

Table 7.1	Types of Bone Cells		
Cell Name	**Description and Function**	**Drawing**	**Word Origins**
Osteoprogenitor Cell	A stem cell derived from mesenchyme that differentiates into an osteoblast. Located in the inner layers of the endosteum and periosteum.	Nuclei	(*osteon*, bone + *pro*, before + *genesis*, origin)
Osteoblast	A small, immature bone cell derived from an osteoprogenitor cell that functions to lay down new bone for bone growth, remodeling, and repair. These cells often appear lined up in rows next to a trabecula of spongy bone.	Nucleus	(*osteon*, bone + *blastos*, germ)
Osteocyte	A mature bone cell derived from an osteoblast that functions to maintain the matrix surrounding it. These cells are located within lacunae.	Nucleus	(*osteon*, bone + *kytos*, a hollow; a cell)
Osteoclast	A very large, multinucleate, phagocytic cell derived from bone marrow cells that also produce monocytes. Osteoclasts break down bone and are often found on the opposite side of a trabecula of spongy bone as the layer of osteoblasts.	Nuclei	(*osteon*, bone + *klastos*, broken)

is dissolved away so the tissue is soft enough to be sectioned in the traditional manner. This type of preparation preserves living cells such as osteocytes and osteoblasts, but the fine structure of the bony matrix cannot be seen because of the removal of the rigid portion of the matrix. In the second type of preparation, **ground bone** (see figure 7.3), a hard bone sample is ground down into a section that is thin enough to be viewed under a microscope. This type of preparation destroys living cells. Thus, no osteocytes, osteoblasts, or other cells can be seen. However, details of the bony matrix—such as central (Haversian) canals, perforating (Volkmann) canals, and canaliculi—are preserved.

EXERCISE 7.1 Compact Bone

1. Obtain a slide of decalcified compact bone (**figure 7.2**) and place it on the microscope stage. Bring the tissue sample into focus on low power and locate an **osteon**.

2. With the osteon at the center of the field of view, increase the magnification. The **central (Haversian) canals** will be the largest holes visible in the sample and will appear to have a lot of "junk" inside them. The "junk" consists of blood vessels and nerves. In contrast, **lacunae** will appear to be much smaller holes and there will only be one cell (an osteocyte) inside each lacuna. **Lamellae** are sometimes difficult to make out, but once you identify where the lacunae are, you will have an idea of where the lamellae are as well (lacunae are found at the border between two adjacent lamellae). As you scan toward the outer border of the bone sample, you should be able to identify **circumferential lamellae**, which surround the entire diaphysis of the bone.

3. Scan the outer edge of the bone tissue on the slide and identify the **periosteum** of the bone. The periosteum contains an outer layer of dense irregular connective tissue and an inner layer containing **osteoprogenitor cells** (between the dense connective tissue and the compact bone tissue). Can you identify this layer of osteoprogenitor cells on your slide?

4. Obtain a slide of ground compact bone (**figure 7.3**) and place it on the microscope stage. Remember, you will not be able to see any living cells in this tissue sample. The largest holes seen in the sample will be the central canals of the osteons. The lacunae appear very dark in color, and at very high magnification they may appear to be empty because of the lack of living osteocytes in the tissue.

5. Observe this slide at the highest magnification possible (without using the oil immersion lens) and notice what appear to be tiny little "cracks" or fractures that run perpendicular to the central canals. These are the tiny **canaliculi**, which in living bone contain cytoplasmic extensions of osteocytes. Within the canaliculi osteocytes connect to each other via gap junctions, which allow them to exchange substances with each other.

6. Finally, scan the slide and see if you can find any large canals that run perpendicular to the central canals. These canals are **perforating (Volkmann) canals**, which convey blood vessels from the outer periosteum into the central canals.

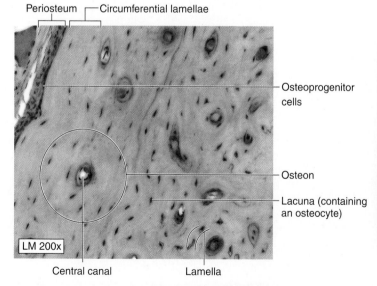

Periosteum ┌─Circumferential lamellae

- Osteoprogenitor cells
- Osteon
- Lacuna (containing an osteocyte)

LM 200x

Central canal   Lamella

Figure 7.2   **Decalcified Compact Bone.**

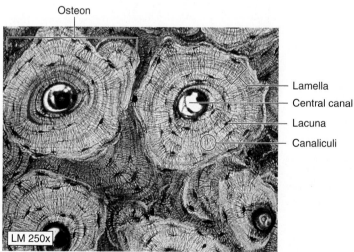

Osteon

- Lamella
- Central canal
- Lacuna
- Canaliculi

LM 250x

Figure 7.3   **Ground Compact Bone.** No perforating canals are visible in this section.

7. In the spaces below, make brief sketches of the decalcified compact bone and ground compact bone as viewed under the microscope. Using table 7.1 and figures 7.2 and 7.3 as guides, label the following on your sketches:

☐ central canal   ☐ osteocyte

☐ circumferential lamellae   ☐ osteon

☐ lacuna   ☐ osteoprogenitor cell

☐ lamella(e)   ☐ perforating canal

☐ osteoblast   ☐ periosteum

Decalcified compact bone   *Ground compact bone*

_____ ✕   _____ ✕

EXERCISE 7.2 Spongy Bone

1. Obtain a slide of decalcified **spongy (cancellous) bone** (**figure 7.4**) and place it on the microscope stage. The sample of spongy bone will appear most similar to the slide of decalcified compact bone. The main difference between the two is that the spongy bone does *not* contain osteons, so you will not see central canals or perforating canals.

2. Observe the slide at high magnification and look for areas where several small cells are lined up next to each other on the edge of a trabecula. These are **osteoblasts** (figure 7.4*a*) that actively secrete new bony matrix. Can you identify any very large, multinucleate cells on the slide? If so, these are bone-resorbing **osteoclasts** (figure 7.4*b*). Both osteoclasts and osteoblasts are actively involved in the process of bone remodeling.

3. In the space to the right make a brief sketch of the spongy bone as viewed under the microscope. Using table 7.1 and figure 7.4 as guides, label the following on your sketch:

   ☐ bone marrow space   ☐ osteoclast

   ☐ lacuna   ☐ osteocyte

   ☐ osteoblast   ☐ trabecula

_____ ×

WHAT DO YOU THINK?

1. When a bone fractures and subsequently undergoes the process of repair, what bone cells will be involved in the repair process and how will each bone cell participate in the process?

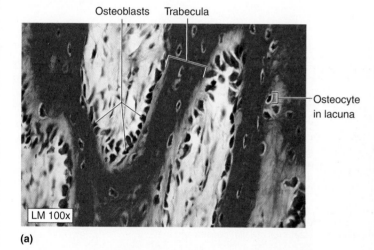

(a)

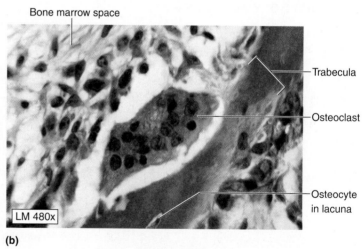

(b)

Figure 7.4   Spongy Bone. Decalcified sections of spongy bone showing (a) osteoblasts and (b) osteoclasts.

EXERCISE 7.3 Endochondral Bone Development

1. Obtain a slide of **developing long bone** (**figure 7.5**) and place it on the microscope stage. This slide will typically contain the developing femur of a young mammal. The femur develops using **endochondral ossification**, a process by which a hyaline cartilage model of the bone is gradually replaced by bone tissue.

2. Scan the slide at the lowest magnification and identify the parts of the bone—specifically, the epiphysis, diaphysis, and metaphysis. In the metaphysis of a developing long bone is an **epiphyseal (growth) plate**. Name the type of tissue you expect to find at this location: _____

   _____ .

3. Position the metaphysis at the center of the field of view and increase the magnification. Using figure 7.5 as a guide, identify the five functional layers within the epiphyseal plate.

4. In the space below make a brief sketch of the epiphyseal growth plate as viewed under the microscope. Using figure 7.5 and your textbook as guides, label the following on your sketch:

   ☐ zone of resting cartilage   ☐ zone of calcified cartilage

   ☐ zone of proliferating cartilage   ☐ zone of ossification

   ☐ zone of hypertrophic cartilage

   _____ ×

WHAT DO YOU THINK?

❷ What process must stop in order for the epiphyseal plate to close? (That is, which layer of the plate must stop its development first?)

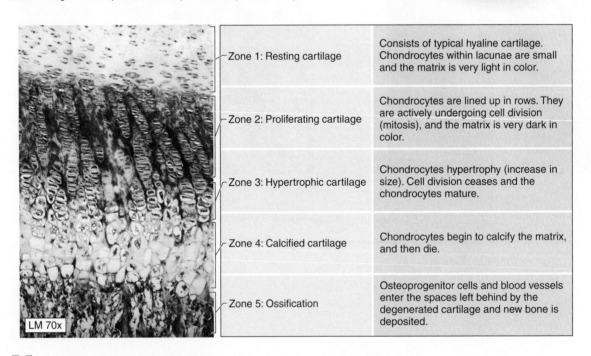

	Consists of typical hyaline cartilage. Chondrocytes within lacunae are small and the matrix is very light in color.
Zone 1: Resting cartilage	
Zone 2: Proliferating cartilage	Chondrocytes are lined up in rows. They are actively undergoing cell division (mitosis), and the matrix is very dark in color.
Zone 3: Hypertrophic cartilage	Chondrocytes hypertrophy (increase in size). Cell division ceases and the chondrocytes mature.
Zone 4: Calcified cartilage	Chondrocytes begin to calcify the matrix, and then die.
Zone 5: Ossification	Osteoprogenitor cells and blood vessels enter the spaces left behind by the degenerated cartilage and new bone is deposited.

LM 70x

Figure 7.5   **Epiphyseal Plate.** The five functional layers of the epiphyseal plate.

Gross Anatomy

Structure of a Typical Long Bone

Observing a fresh bone specimen allows you to see many of the tissues that normally associate with bones, such as periosteum, articular cartilages, muscles, tendons, ligaments, marrow, and blood vessels. In the following exercises you will first observe a fresh specimen of a long bone from a cow, and then you will study an articulated human skeleton to familiarize yourself with the major bones of the human body.

EXERCISE 7.4 Cow Bone Dissection

1. Obtain a dissecting pan, blunt probe, forceps, and a fresh cow bone cut in longitudinal section. Place the bone in the dissecting pan and begin your observations. Find the large medullary cavity that is filled with yellow bone marrow (**figure 7.6**). Notice the thick layer of compact bone that surrounds the medullary cavity.

2. If you have a dissecting microscope or magnifying glass at your disposal, use it to focus in on the compact bone tissue. Notice how "bloody" it appears. Can you see tiny dots

of blood within the compact bone? These result from the rupture of tiny blood vessels in the tissue when the bone was sectioned. This observation should reaffirm for you that living bone is a highly vascular, metabolically active tissue—quite different from the appearance of preserved bones.

WHAT DO YOU THINK?

3 Why do you suppose the medullary cavity of a long bone is filled with adipose connective tissue in the adult?

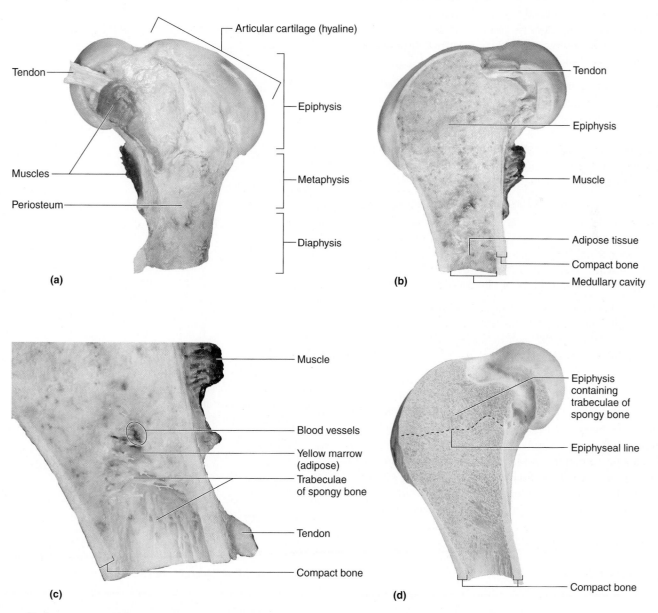

(a) — Articular cartilage (hyaline), Tendon, Epiphysis, Muscles, Metaphysis, Periosteum, Diaphysis

(b) — Tendon, Epiphysis, Muscle, Adipose tissue, Compact bone, Medullary cavity

(c) — Muscle, Blood vessels, Yellow marrow (adipose), Trabeculae of spongy bone, Tendon, Compact bone

(d) — Epiphysis containing trabeculae of spongy bone, Epiphyseal line, Compact bone

Figure 7.6 **Fresh Cow Bone.** (a) Exterior view, (b) interior view, (c) close-up of the medullary cavity. (d) The same bone, with all of the fat, muscle, tendons, cartilage, and blood vessels removed.

3. Using the blunt probe, carefully begin to clean out the adipose tissue from the medullary cavity (figure 7.6*b*). While you do this, gently probe for medium- to large-size blood vessels imbedded within the adipose tissue. If you find a blood vessel that travels from the outer surface of the diaphysis into the medullary cavity, this is most likely the **nutrient artery**, an artery that grows into the diaphysis of the bone during the initial stages of ossification of the bone.

4. As you continue to clean out the inside of the diaphysis of the bone and progress toward the inside of the epiphysis, you will begin to feel small, hard "strings" of bony tissue. These are the **trabeculae** (*trabs,* beam) of spongy bone, which are located within the epiphysis and lining the inside of the diaphysis. These trabeculae will make it difficult to clean out the adipose tissue in the epiphysis, but spend some time poking around in there anyway—this will give you a better understanding of the arrangement of the trabeculae within. You may observe a portion of the epiphysis that appears much more "bloody" than the rest of the inside of the bone. This area consists of **red bone marrow**—a **hemopoietic** (*hemo-*, blood, + *poiesis,* a making) tissue that produces blood cells. Because you are observing adult cow bones, you will observe very little red marrow; most of the spaces will be filled with yellow marrow.

5. Turn your attention to the structures on the outside of the bone. Using forceps, pick away at the dense connective tissue on the outside of the diaphysis (figure 7.6). This is the periosteum, or outer covering of the bone. The periosteum acts as an attachment point for tendons and ligaments and as an anchoring point for blood vessels and nerves that enter the bone.

6. Observe the outside of the epiphysis and look for the cut portion of a tendon or ligament where it attaches to the periosteum (figure 7.6*a,c*). If you carefully observe either one, you will see that it consists of a regular arrangement of shiny, white fibers. These are the tough collagen fibers that give the tendon or ligament its great tensile strength.

7. Observe the shiny, white cartilage on the ends of the bone. This is the articular cartilage, which is composed of hyaline cartilage. Notice that the periosteum ends where the articular cartilage begins.

8. If you are fortunate enough to have a cow tibia as your bone sample, try to identify the C-shaped pads of fibrocartilage located on top of the articular cartilages. These are the **menisci** (s. *meniscus*) of the knee joint (**figure 7.7**). If your cow bone does not have a meniscus, find another group in your laboratory whose cow bone has a meniscus so you can observe its gross structure. Within the meniscus you will see more of the shiny, white collagen fibers that you saw in the tendons and ligaments. Recall from chapter 5 that fibrocartilage contains thick bundles of collagen fibers in its extracellular matrix. Note that the meniscus is somewhat "tied in to" the connective tissues surrounding the joint.

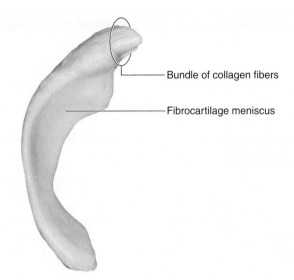

Bundle of collagen fibers

Fibrocartilage meniscus

Figure 7.7   Meniscus. One of the C-shaped meniscal cartilage pads removed from the end of the tibia of a fresh cow bone.

9. In the space below make a brief sketch of the fresh cow bone you dissected. Using figure 7.6 as a guide, label the following on your sketch:

☐ articular cartilage   ☐ periosteum

☐ compact bone   ☐ spongy bone

☐ medullary cavity   ☐ tendon/ligament

10. When you are finished with your dissection, dispose of the cow bones according to your laboratory instructor's directions, and clean your dissection instruments and work space.

WHAT DO YOU THINK?

4 Many people who have torn a ligament in the knee also have a torn meniscus. Why do you think it is so common for individuals who tear knee ligaments to also tear a meniscus?

Survey of the Human Skeleton

Learning the names of all the bones in the human body and their features can seem like a Herculean task. Often it is best to break a big task into many small tasks that can be tackled individually in short segments of time. As you begin the task of learning all of the bones and bone features of the human skeleton, an excellent "first task" is to simply learn the names of the bones that compose the skeleton. In this exercise you will do just that. The exercises in chapters 8 and 9 will then guide you in the task of learning the names of specific features unique to each bone of the skeleton.

EXERCISE 7.5 The Human Skeleton

1. Observe an articulated human skeleton.

2. Using your textbook as a guide, identify the bones listed in **figure 7.8** on the articulated skeleton. Then label them in figure 7.8.

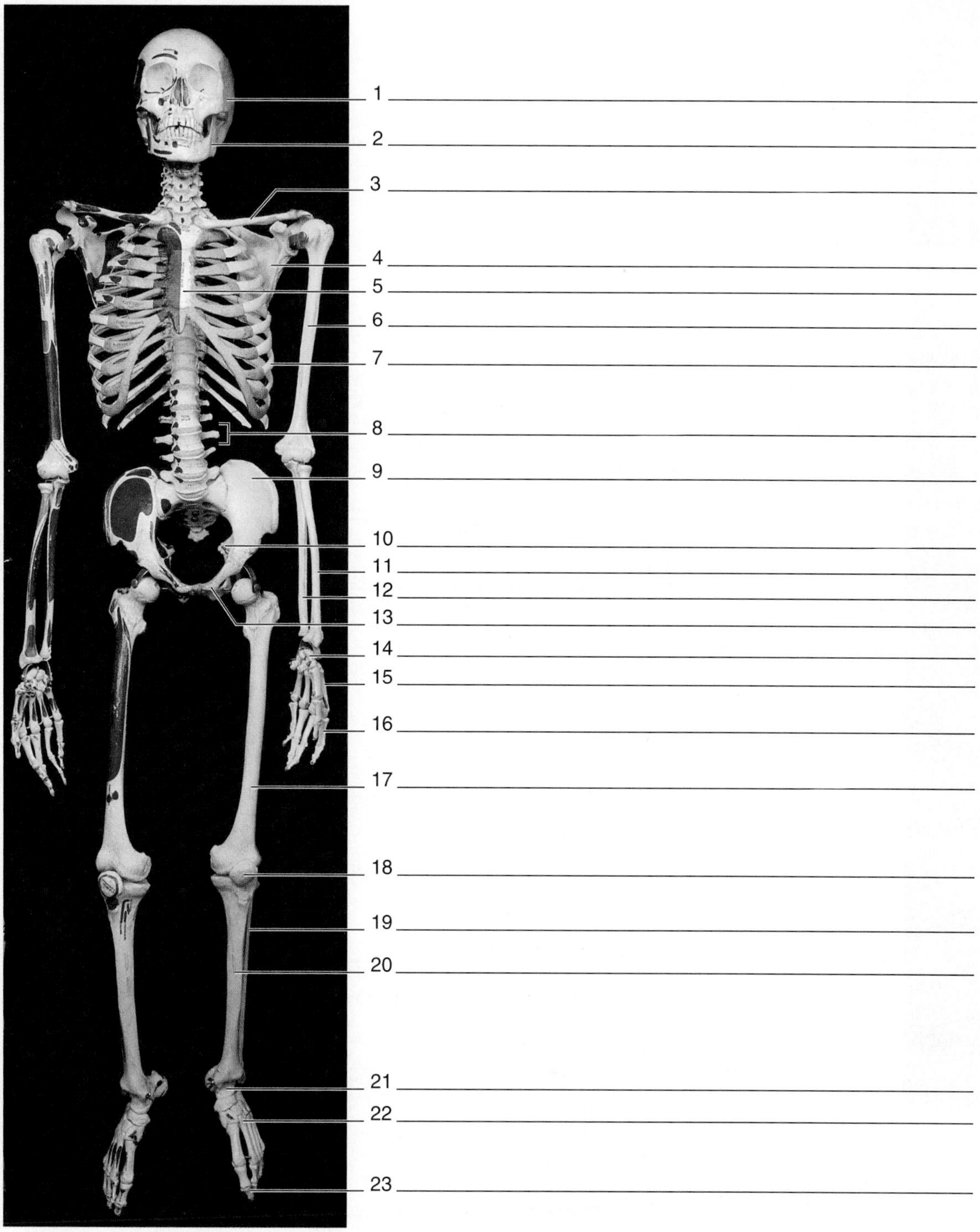

1 _____
2 _____
3 _____
4 _____
5 _____
6 _____
7 _____
8 _____
9 _____
10 _____
11 _____
12 _____
13 _____
14 _____
15 _____
16 _____
17 _____
18 _____
19 _____
20 _____
21 _____
22 _____
23 _____

☐ carpals
☐ cranium
☐ femur
☐ fibula
☐ humerus
☐ ilium
☐ ischium
☐ mandible
☐ metacarpals
☐ metatarsals
☐ phalanges
☐ pubis
☐ radius
☐ ribs
☐ scapula
☐ tarsals
☐ tibia
☐ ulna
☐ vertebrae

(a) Anterior view

Figure 7.8 **The Human Skeleton.** (a) Anterior and (b) posterior views.

Figure 7.8 **The Human Skeleton continued.** (a) Anterior and (b) posterior views.

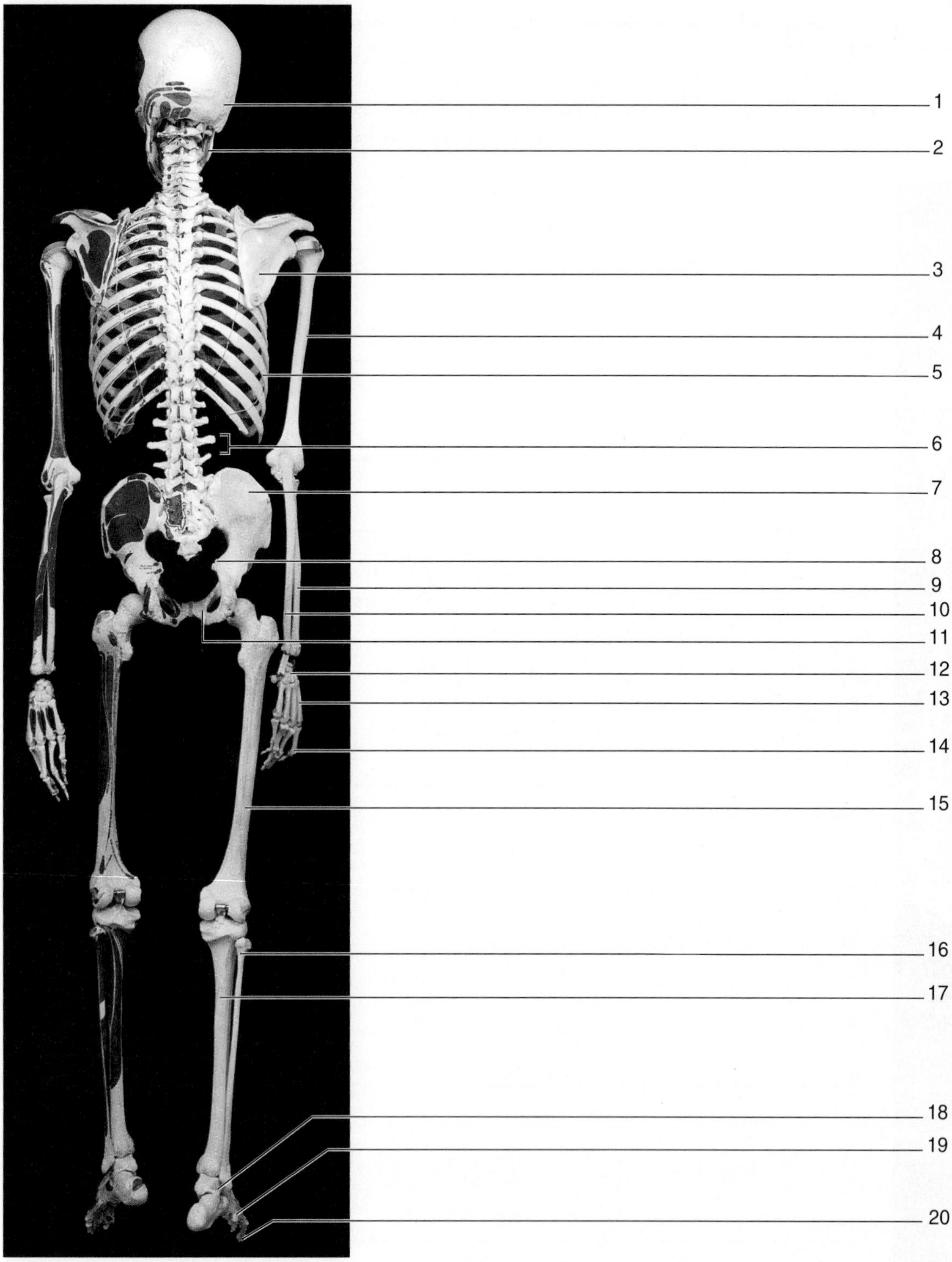

1
2
3
4
5
6
7
8
9
10
11
12
13
14
15
16
17
18
19
20

☐ carpals
☐ cranium
☐ femur
☐ fibula
☐ humerus
☐ ilium
☐ ischium
☐ mandible
☐ metacarpals
☐ metatarsals
☐ phalanges
☐ pubis
☐ radius
☐ ribs
☐ scapula
☐ tarsals
☐ tibia
☐ ulna
☐ vertebrae

(b) Posterior view

Chapter 7: Skeletal System Overview: Bone Anatomy

Name: _____

Date: _____ Section: _____

POST-LABORATORY WORKSHEET

1. Label the parts of a typical long bone on the following diagram.

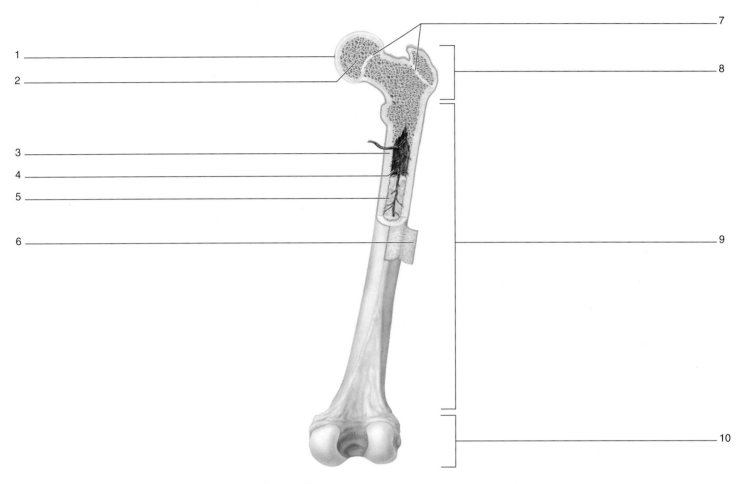

2. Label the components of compact bone on the following diagram.

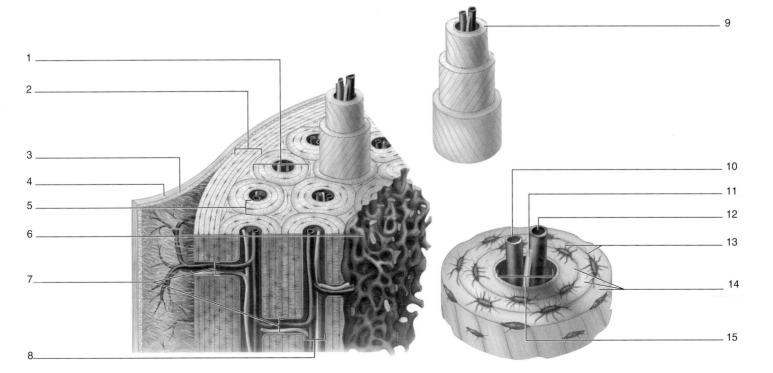

3. Label the components of spongy bone on the following diagram.

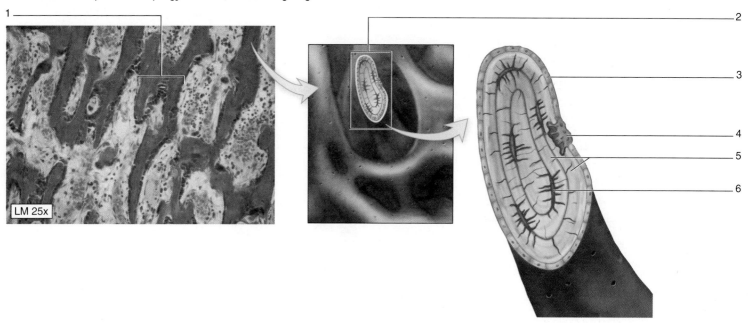

4. In the following table, list the four types of bone cells and describe the function of each.

Bone Cell	Function

5. Label all of the major bones of the human skeleton on the following figure.

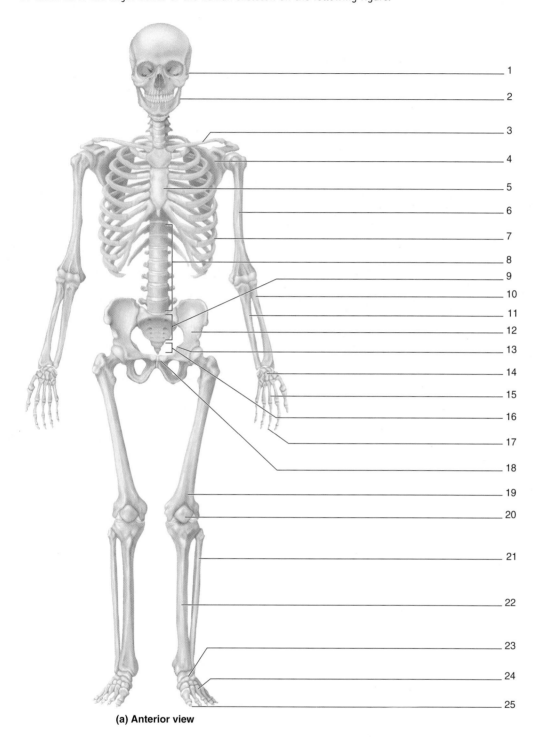

1
2
3
4
5
6
7
8
9
10
11
12
13
14
15
16
17
18
19
20
21
22
23
24
25

(a) Anterior view

6. Label all of the major bones of the human skeleton on the following figure.

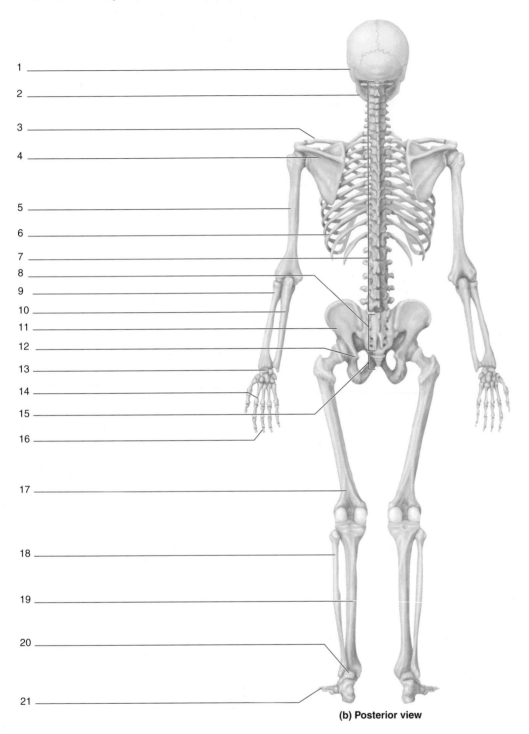

1 _____

2 _____

3 _____

4 _____

5 _____

6 _____

7 _____

8 _____

9 _____

10 _____

11 _____

12 _____

13 _____

14 _____

15 _____

16 _____

17 _____

18 _____

19 _____

20 _____

21 _____

(b) Posterior view

The Skeletal System: Axial Skeleton

8

OUTLINE and OBJECTIVES

INTRODUCTION

The term **axial** is a derivative of the term *axis*. The axial skeleton consists of bones that form the main axis of the body—the skull, vertebrae, ribs, and sternum, which collectively form the body's core structural foundation. These bones are also critical for protection of the body's most vital organs, such as the brain, heart, and lungs.

In this chapter you will begin what is often a feared task for anatomy students: the process of learning a large number of structures in a short period of time. Most students find this process of trying to learn all of the processes, projections, and foramina (holes) of the bones to be an enormous effort in memorization. It need not be if you approach it with the proper attitude. That is, as you view each structure, study it closely and contemplate why it exists. Is the process an attachment point for a muscle? Is there a nerve, artery, or vein that runs through the foramen in a living individual? The reason you are engaging in this process is to learn the answers to such questions. Although you will often view individual, disarticulated bones in the anatomy laboratory, always try to find the same bones on an articulated (complete) skeleton so you have a better idea of how each individual bone fits in with the rest of the axial skeleton.

Chapter 8: The Skeletal System: Axial Skeleton

Name: _____

Date: _____ Section: _____

PRE-LABORATORY WORKSHEET

1. List the bones that compose the axial skeleton.

   _____

   _____

   _____

   _____

2. In a typical human, how many vertebrae (individual or fused) does each section of the vertebral column contain?

   Cervical _____   Sacral _____

   Thoracic _____   Coccygeal _____

   Lumbar _____

3. All ribs articulate with _____ vertebrae.
   a. cervical
   b. thoracic
   c. lumbar
   d. sacral

4. List the three parts of the sternum, from superior to inferior.

   a. _____

   b. _____

   c. _____

5. What bone feature is unique to the transverse processeses of cervical vertebrae? _____

6. Which skull bone is the only one that is mobile (movable)? _____

7. The sella turcica is a feature of the _____ bone.

8. The foramen magnum (through which the spinal cord exits the cranial cavity) is a feature of the _____ bone.

9. What is a fontanelle? _____

10. The suture that forms between the frontal and parietal bones is called the _____ suture.

IN THE LABORATORY

In this laboratory session you will observe the bones that compose the axial skeleton on an articulated human skeleton, on disarticulated bones of the human skeleton, or on bone models. You will identify the major landmarks and identifying features of each bone, and will associate the observed structures with their associated functions.

Gross Anatomy

The Skull

The bones that make up the skull are separated into two functional categories: the **cranial bones** (frontal, parietal, temporal, occipital, sphenoid, and ethmoid) and the **facial bones** (maxilla, mandible, zygomatic, nasal, lacrimal, palatine, inferior nasal conchae, and vomer). The roof of the cranium—the **calvaria**, or skullcap—is the dome-shaped part of the skull that protects the brain. In an adult, all of the skull bones, with the exception of the mandible, are fused to each other via synarthrotic joints called sutures.

Table 8.1 describes each bone of the skull individually and lists the best view(s) for observing its features. Because your laboratory experience is practical, you will view the skull from six points of reference: anterior view, lateral view, posterior view, superior view, inferior view, and superior view of the cranial floor. Your first goal will be to identify the individual bones visible in each view. Your second goal will be to identify all processes, foramina, and major features (often formed from multiple bones) that are visible in each view of the skull. As you work on the second goal, always try to relate the bony processes, fossae, and foramina to the individual bone(s) from which they are formed.

Table 8.1	**The Axial Skeleton: Skull Bones and Important Bony Landmarks**		
Major Bone	**Bone Features**	**Description and Related Structures of Importance**	**Best View**
Ethmoid	Cribriform plate	Forms the roof of the nasal cavity and part of the floor of the cranial cavity.	Superior view of cranial floor
	Cribriform foramina	The olfactory nerve (CN I) passes through en route to the brain.	Superior view of cranial floor
	Crista galli	A projection that serves as the attachment point for the falx cerebri.	Superior view of cranial floor
	Superior nasal concha	Forms superior lateral wall of nasal cavity and causes turbulent air flow.	Anterior
	Middle nasal concha	Forms middle lateral wall of nasal cavity and causes turbulent air flow.	Anterior
	Perpendicular plate	Forms superior part of nasal septum.	Anterior
Frontal	Frontal sinus	A cavity within the frontal bone lined with respiratory epithelium.	Superior view of cranial floor
	Supraorbital foramen (notch)	A hole (that is sometimes just a notch) on the superior ridge of the orbit.	Anterior
Inferior Nasal Conchae	NA	Curved structure that forms inferior part of lateral wall of the nasal cavity and causes turbulent air flow.	Anterior
Lacrimal	Lacrimal groove	Forms the medial, inferior aspect of orbit of eye. Groove connects orbital and nasal cavities.	Lateral
Mandible	Alveolar processes	Cavities in the mandible that form the tooth "sockets."	Lateral
	Angle	The portion of the mandible connecting the body to the ramus, forming a right angle.	Lateral
	Body	The anterolateral portion of the mandible.	Lateral
	Coronoid process	Insertion point for the temporalis muscle.	Lateral
	Head	Forms a joint with the mandibular fossa of the temporal bone (temporomandibular joint).	Lateral
	Mandibular foramen	Passageway for the mandibular branch of the trigeminal nerve (CN V_3).	Lateral
	Mental foramen	Passageway for the mental artery and nerve (CN V_3).	Anterior
	Mental protuberance	An anterior projection of the mandible that forms the anterior projection of the chin.	Lateral
	Ramus	The part of the bone that forms an angle with the body of the mandible.	Lateral
Maxilla	Infraorbital foramen	A passageway for the infraorbital artery and nerve.	Lateral
	Incisive foramen (fossa)	Contains arteries and nerves passing from the nasal cavity into the oral cavity.	Inferior
	Palatine process	Forms anterior floor and part of lateral wall of the nasal cavity.	Inferior
Nasal	NA	Forms most of the bridge of the nose.	Frontal

(continued on next page)

Table 8.1	The Axial Skeleton: Skull Bones and Important Bony Landmarks *(continued)*		
Major Bone	**Bone Features**	**Description and Related Structures of Importance**	**Best View**
Occipital	External occipital protuberance	A large projection that can be palpated on the back of the head; serves as a muscle attachment point.	Posterior
	Foramen magnum	The spinal cord passes through en route to the vertebral (spinal) canal.	Superior view of the cranial floor
	Hypoglossal canal	The hypoglossal nerve (CN XII) travels through.	Superior view of the cranial floor
	Jugular foramen	The jugular vein and a number of nerves (CN IX, X, and XI) travel through.	Superior view of the cranial floor
Palatine	NA	Forms the posterior floor of nasal cavity, part of the orbit, and part of the hard palate.	Inferior
Parietal	NA	Forms the lateral, superior wall of the cranial cavity.	Lateral
Sphenoid	Foramen lacerum	Largely covered by cartilage in a living human; no one structure passes completely through it.	Inferior
	Foramen ovale	The mandibular branch of the trigeminal nerve (CN V_3) travels through.	Superior view of the cranial floor
	Foramen rotundum	The maxillary branch of the trigeminal nerve (CN V_2) travels through.	Superior view of the cranial floor
	Foramen spinosum	Middle meningeal artery and vein and a branch of the trigeminal nerve (CNV) travel through.	Superior view of the cranial floor
	Greater wing	Forms parts of the posterior orbit of the eye and the middle cranial fossa.	Superior view of the cranial floor
	Inferior orbital fissure	The maxillary branch of the trigeminal nerve (CN V_2) and the infraorbital artery and vein travel through.	Superior view of the cranial floor
	Lesser wing	Forms part of the anterior cranial fossa.	Superior view of the cranial floor
	Optic foramen	The optic nerve (CN II) travels through.	Superior view of the cranial floor
	Sella turcica	A "Turkish saddle"-shaped depression that houses the pituitary gland.	Superior view of the cranial floor
	Superior orbital fissure	The oculomotor (CN III), trochlear (CN IV), trigeminal (CN V_1), and abducens (CN VI) nerves travel through.	Superior view of the cranial floor
Temporal	Carotid canal	The internal carotid artery and associated nerves travel through.	Inferior
	External acoustic (auditory) meatus	The opening into the external auditory canal.	Lateral
	Foramen lacerum	Largely covered by cartilage in a living human; no one structure passes completely through it.	Superior view of the cranial floor
	Internal acoustic (auditory) meatus	The facial (CN VII) and vestibulocochlear nerves (CN VIII) travel through.	Superior view of the cranial floor
	Mandibular fossa	The point of articulation with the head of the mandible, forming the temporomandibular joint.	Lateral
	Mastoid process	Serves as an attachment point for muscles of the neck.	Lateral
	Petrous part	Houses mechanisms for hearing and equilibrium; separates middle and posterior cranial cavities.	Superior
	Squamous part	Forms the inferior, posterior part of the temporal fossa.	Lateral
	Styloid process	Serves as an attachment point for muscles controlling the tongue.	Lateral
	Zygomatic process	A projection that articulates with the temporal process of the zygomatic bone.	Lateral
Vomer	NA	Forms inferior and posterior part of nasal septum.	Inferior
Zygomatic	Frontal process	Articulates with the frontal bone.	Lateral
	Maxillary process	Articulates with the zygomatic process of maxillary bone.	Lateral
	Temporal process	Articulates with the zyogmatic process of the temporal bone.	Lateral

EXERCISE 8.1A: Anterior View of the Skull

1. Obtain a skull and observe it from an anterior view (**figure 8.1**). An anterior view of the skull reveals much of the detail of the facial bones. Facial bones play a role in mastication (chewing) and in the protection and support of special sensory organs such as the eye. **Table 8.2** describes bony structures of the face and lists the bones that compose each structure.

2. Using Table 8.2 and your textbook as guides, identify the structures listed in figure 8.1 on a skull or model in the laboratory. Then, label figure 8.1.

3. *Optional Activity:* AP|R **Skeletal System**—Watch the "Skull" animation to see how the bones of the skull fit together.

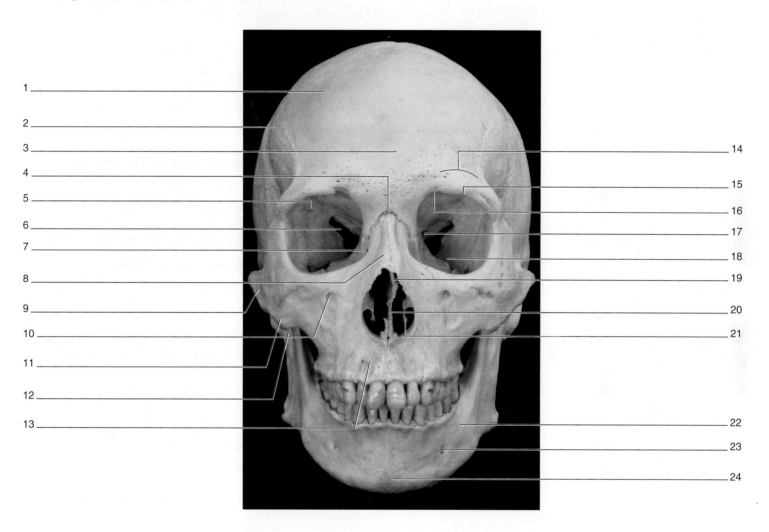

Figure 8.1 **Anterior View of the Skull.**

☐ frontal bone

☐ frontonasal suture

☐ glabella

☐ inferior nasal concha

☐ inferior orbital fissure

☐ infraorbital foramen

☐ lacrimal bone

☐ lacrimal groove

☐ mandible

☐ mastoid process

☐ maxilla

☐ mental foramen

☐ mental protuberance

☐ nasal bone

☐ optic canal

☐ parietal bone

☐ perpendicular plate of ethmoid

☐ sphenoid bone

☐ superciliary arch

☐ superior orbital fissure

☐ supraorbital foramen (notch)

☐ supraorbital margin

☐ temporal process of zygomatic bone

☐ vomer

☐ zygomatic bone

Table 8.2	The Axial Skeleton: Anterior View of the Skull		
Facial Structure	**Major Bone**	**Bone Feature**	**Description and Related Structures of Importance**
Forehead	Frontal	Squamous part	Remnant of a fetal joint between the two parts of the frontal bone.
		Coronal suture	Suture between frontal and parietal bones.
		Metopic suture	Suture between the two parts of the frontal bone; only named in the adult if the suture persists.
		Squamous suture	Suture between frontal and temporal bones.
		Glabella	Prominent bony ridge located immediately superior to the nose.
		Superciliary arch	The "brow" ridges, which are located superior to the supraorbital margin.
Orbit	Frontal	Supraorbital margin	Bony support and protection of the superior border of the orbit.
		Supraorbital foramen	Supraorbital artery and nerve travel through. Sometimes the supraorbital foramen is just a notch.
	Sphenoid	Optic canal	Optic nerve (CN II) travels through.
		Superior orbital fissure	Oculomotor (CN III), trochlear (CN IV), trigeminal (CN V_1) and abducens (CN VI) nerves travel through.
		Inferior orbital fissure	Maxillary branch of trigeminal nerve (CN V_2) and infraorbital artery and vein travel through.
	Ethmoid		Forms medial wall and part of the posterior wall of the orbit.
	Lacrimal	Lacrimal fossa	Drains tears from the surface of the eye into the nasal cavity.
	Maxilla		Forms medial and inferior walls of the orbit.
		Infraorbital foramen	Infraorbital nerve (a branch of CN V) and artery travel through.
	Zygomatic		Forms lateral border and wall of the orbit.
Nose	Nasal		Forms most of the bridge and the anterior portion of the bony skeleton of the nose.
	Maxilla	Frontal processes	Form lateral aspect of the bony skeleton of the nose.
	Frontal	Nasal spine	Forms superior aspect of the bony skeleton of the nose.
Nasal Septum	Ethmoid	Perpendicular plate	Forms superior portion of the nasal septum.
	Vomer		Forms the posterior-inferior portion of the nasal septum.
Nasal Cavity	Ethmoid	Cribriform foramina	Holes that olfactory nerves (CN I) travel through to get to the CNS.
		Cribriform plate	Forms roof of nasal cavity.
		Superior and middle conchae	Curved bony structures that form superior part of lateral wall and cause turbulent airflow.
	Inferior nasal concha		Curved bone that forms inferior part of lateral wall and causes turbulent airflow.
	Maxilla	Palatine process	Forms anterior floor and part of lateral wall of the nasal cavity.
	Palatine		Forms posterior floor of nasal cavity.
Oral Cavity (Buccal)	Palatine		Forms posterior roof of oral cavity.
	Maxilla	Palatine process	Forms anterior roof of oral cavity.
		Incisive foramen	Contains arteries and nerves passing from the nasal cavity into the oral cavity.
		Alveolar processes	Form joints with the teeth.
		Maxillary teeth	
	Mandible	Alveolar processes	Form joints with the teeth.
		Body	Forms anterior portion of lower border of oral cavity.
		Ramus	Forms lateral portion of lower border of oral cavity.
		Mandibular teeth	
Chin	Mandible	Body	The anterolateral portion of the mandible.
		Angle	The portion of the mandible connecting the body to the ramus, forming a right angle.
		Mental foramen	Mental artery and nerve (CN V_3) travel through.
		Ramus	Part of the bone that forms an angle with the body of the mandible.
		Alveolar processes	Form joints with the teeth.
		Mental protuberance	Anterior projection of mandible, forming the anterior projection of the chin.

EXERCISE 8.1B: The Orbit

1. Observe the **orbit** (**figure 8.2**) on a skull or model.

2. The orbit is the bony casing that supports and protects the eyeball. Parts of the frontal, zygomatic, maxillary, ethmoid, and lacrimal bones form the anterior border of the orbit. The ethmoid and lacrimal bones form most of the medial wall of the orbit, and the sphenoid bone forms most of the posterior wall. Identify the walls and borders of the orbit on your specimen.

3. Using Table 8.2 and your textbook as guides, identify the features of the orbit listed in figure 8.2 on a skull. Then, label figure 8.2.

EXERCISE 8.1C: The Nasal Cavity

1. Observe the **nasal cavity** (**figure 8.3**) on a skull or model.

2. The nasal cavity is a large, complex cavity that is separated into two halves by a **nasal septum** (*saeptum*, a partition). The ethmoid bone forms parts of the roof, septum, and lateral walls. The **cribriform** (*cribrum*, a sieve, + *forma*, a form) **plate** of the ethmoid bone forms most of the roof. The **palatine processes of the maxillary bones** and the **palatine bones** form the floor of the nasal cavity (and the roof of the oral cavity). The **nasal bones** form most of the bridge of the nose. Finally, the bony portion of the nasal septum is formed from the **perpendicular plate of the ethmoid bone** and the **vomer**.

3. Using Table 8.2 and your textbook as guides, identify the features of the nasal cavity listed in figure 8.3 on a skull. Then, label figure 8.3.

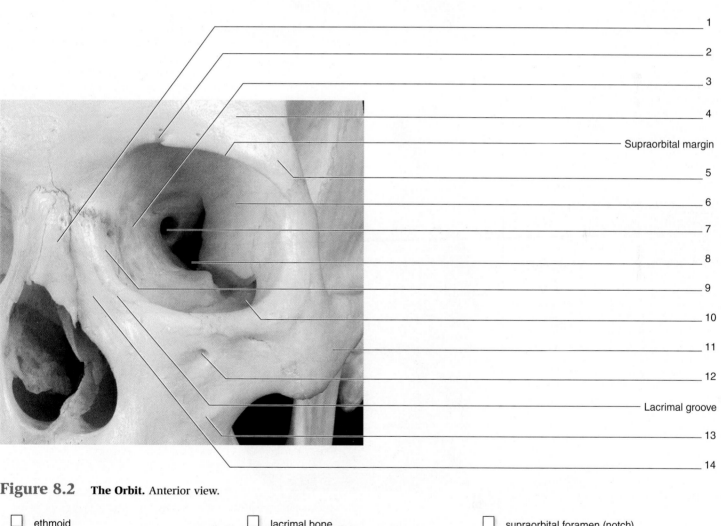

Figure labels (right side, top to bottom): 1, 2, 3, 4, Supraorbital margin, 5, 6, 7, 8, 9, 10, 11, 12, Lacrimal groove, 13, 14

Figure 8.2   The Orbit. Anterior view.

- ☐ ethmoid
- ☐ frontal bone
- ☐ frontal process of maxilla
- ☐ greater wing of sphenoid
- ☐ inferior orbital fissure
- ☐ infraorbital foramen

- ☐ lacrimal bone
- ☐ lacrimal groove
- ☐ maxilla
- ☐ nasal bone
- ☐ optic foramen
- ☐ superior orbital fissure

- ☐ supraorbital foramen (notch)
- ☐ supraorbital margin
- ☐ zygomatic bone
- ☐ zygomatic process of frontal bone

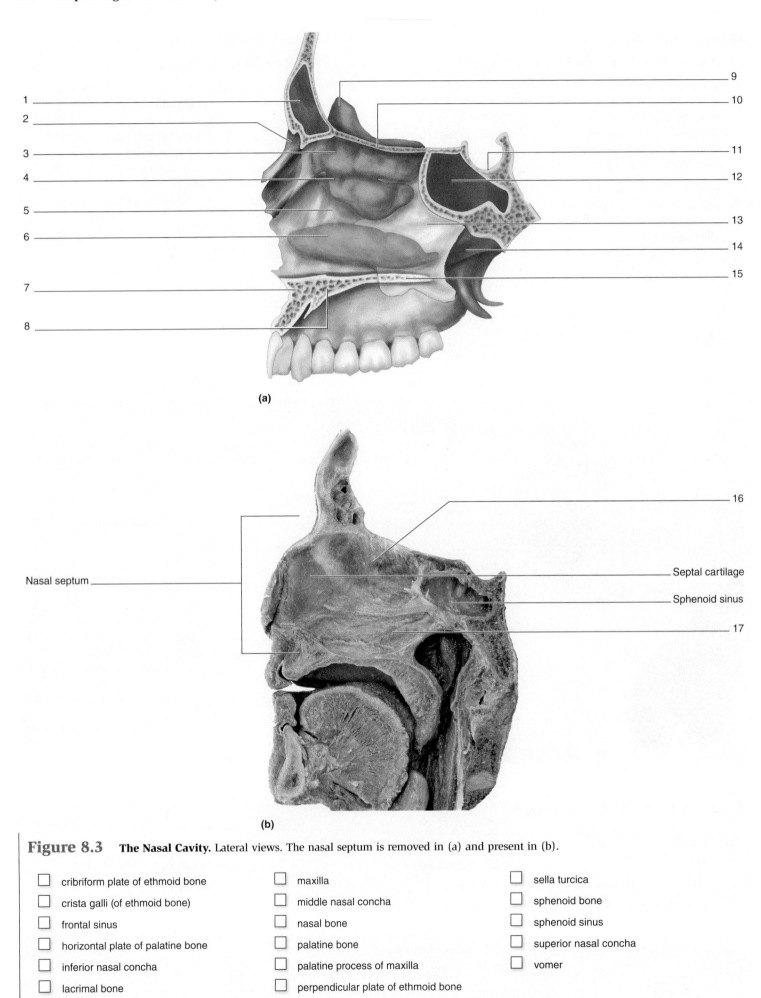

(a)

(b)

Figure 8.3   **The Nasal Cavity.** Lateral views. The nasal septum is removed in (a) and present in (b).

- [ ] cribriform plate of ethmoid bone
- [ ] crista galli (of ethmoid bone)
- [ ] frontal sinus
- [ ] horizontal plate of palatine bone
- [ ] inferior nasal concha
- [ ] lacrimal bone

- [ ] maxilla
- [ ] middle nasal concha
- [ ] nasal bone
- [ ] palatine bone
- [ ] palatine process of maxilla
- [ ] perpendicular plate of ethmoid bone

- [ ] sella turcica
- [ ] sphenoid bone
- [ ] sphenoid sinus
- [ ] superior nasal concha
- [ ] vomer

EXERCISE 8.1D: The Mandible

1. Obtain an isolated mandible or observe the mandible on an articulated skeleton or complete skull.

2. The **mandible** (**figure 8.4**) is unique among skull bones, because it is the only bone that is independently movable. It shares an articulation with the temporal bone, forming the **temporomandibular joint**. Here, the **head of the mandible** articulates with the mandibular fossa of the temporal bone. The **mandibular condyle** is the rounded projection on the head of the mandible that actually forms the joint with the temporal bone. Place your fingers just anterior to your ears and then open and close your mouth to feel the movement of the joint formed between the mandible and temporal bone (the temporomandibular joint).

3. The mandible also has a prominent **coronoid process** (*corona*, crown, + *eidos*, form, resemblance), which serves as the insertion point for the temporalis muscle.

4. The mandible contains two paired prominent foramina. The first, found on the inner (medial) surface of the ramus of the mandible, is the **mandibular foramen**. A branch of the mandibular branch of the trigeminal nerve (CN V_3) passes through this foramen into the interior of the bone. It then sends branches to the **alveolar processes** of the mandible to innervate the roots of the teeth in a living individual. When you are having dental work done on the teeth of your lower jaw, the dentist will direct a needle containing anaesthetic at the mucosa surrounding this foramen in order to bathe the nerve branches that travel into the mandible. Finally, the **mental foramen** (*mental*, chin) is located just superior to the lower border of the mandible at about the midpoint of the body of the mandible. The mental artery and nerve travel through the mental foramen in a living individual.

5. Using Table 8.2 and your textbook as guides, identify the structures listed in figure 8.4 on a mandible. Then, label figure 8.4.

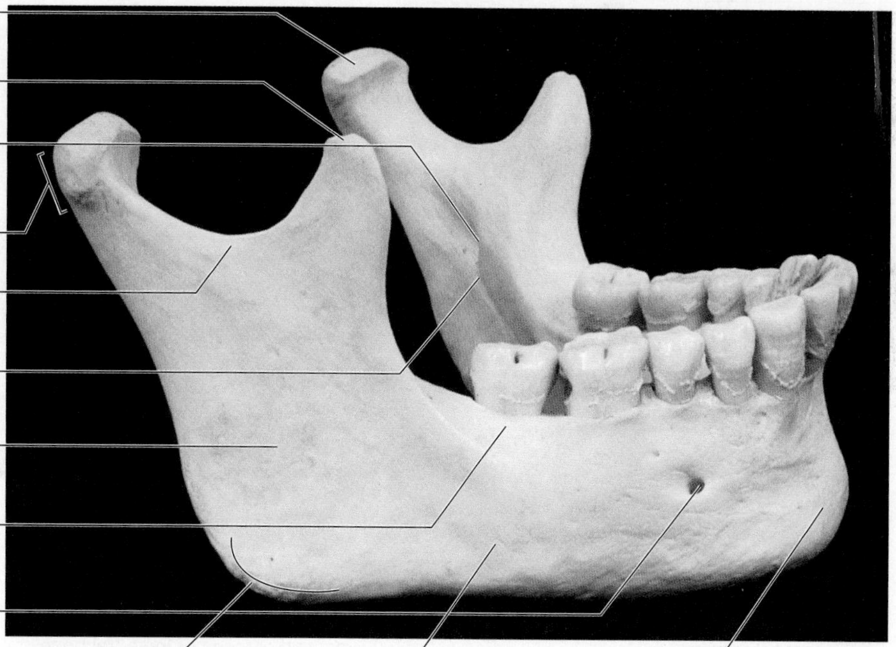

Figure 8.4 **The Mandible.** Lateral view.

- ☐ alveolar process
- ☐ angle
- ☐ body
- ☐ condylar process
- ☐ coronoid process
- ☐ head of mandible
- ☐ mandibular foramen
- ☐ mandibular notch
- ☐ mental foramen
- ☐ mental protuberance
- ☐ mylohyoid line
- ☐ ramus

EXERCISE 8.2 Additional Views of the Skull

EXERCISE 8.2A: Lateral View of the Skull

1. Obtain a skull and observe it from a lateral view (**figure 8.5**).

2. The most notable feature in a lateral view of the skull is the **zygomatic arch**, which is the bony structure that forms the superior part of a person's cheek. It is formed from the zygomatic process of the temporal bone, and the temporal process of the zygomatic bone.

3. Using Table 8.1 and your textbook as guides, identify the structures listed in figure 8.5 on a skull. Then, label figure 8.5.

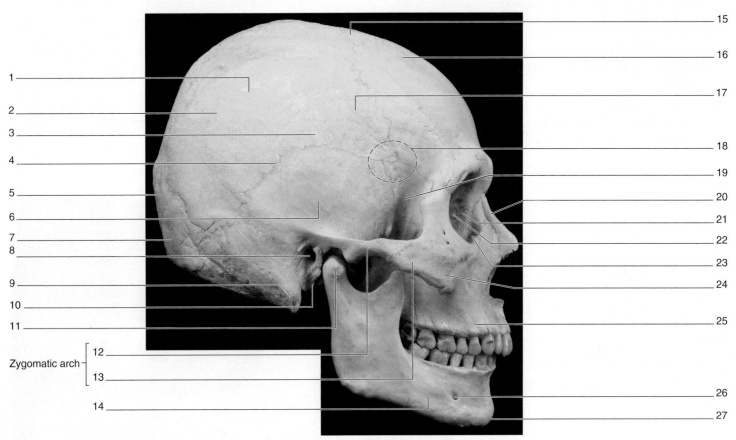

Figure 8.5   **Lateral View of the Skull.**

☐ body of mandible	☐ lambdoid suture	☐ squamous part of temporal bone
☐ coronal suture	☐ mastoid process of temporal bone	☐ squamous suture
☐ ethmoid bone	☐ maxilla	☐ styloid process of temporal bone
☐ external acoustic meatus	☐ mental foramen	☐ superior temporal line
☐ frontal bone	☐ mental protuberance	☐ temporal process of zygomatic bone
☐ greater wing of sphenoid bone	☐ nasal bone	☐ zygomatic arch
☐ head of mandible	☐ occipital bone	☐ zygomatic bone
☐ inferior temporal line	☐ parietal bone	☐ zygomatic process of temporal bone
☐ lacrimal bone	☐ parietal eminence	
☐ lacrimal groove	☐ pterion	

EXERCISE 8.2B: Posterior View of the Skull

1. Obtain a skull and observe it from a posterior view (**figure 8.6**).

2. The most notable feature in a posterior view of the skull is the **lambdoid suture** (*lambda*, the Greek letter λ, + *eidos*, resemblance), which is named for its resemblance to the Greek letter lambda. Can you see the resemblance? Draw

    the letter lambda in the space provided here: _____.

3. Using Table 8.1 and your textbook as guides, identify the structures listed in figure 8.6 on a skull. Then, label figure 8.6.

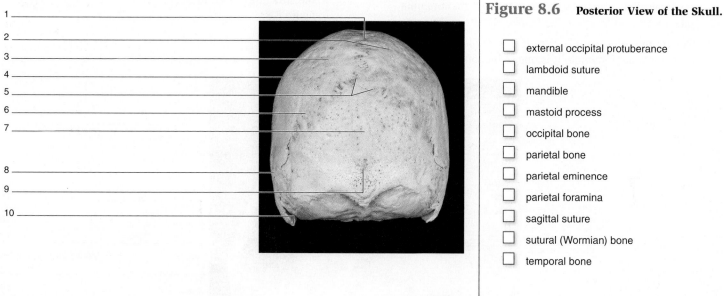

Figure 8.6   **Posterior View of the Skull.**

- ☐ external occipital protuberance
- ☐ lambdoid suture
- ☐ mandible
- ☐ mastoid process
- ☐ occipital bone
- ☐ parietal bone
- ☐ parietal eminence
- ☐ parietal foramina
- ☐ sagittal suture
- ☐ sutural (Wormian) bone
- ☐ temporal bone

EXERCISE 8.2C: Superior View of the Skull

1. Obtain a skull (with the skullcap intact) and observe it from a superior view (**figure 8.7**).

2. The most notable feature in a superior view of the skull is the **sagittal suture**.

3. Using Table 8.1 and your textbook as guides, identify the structures listed in figure 8.7 on a skull. Then, label figure 8.7.

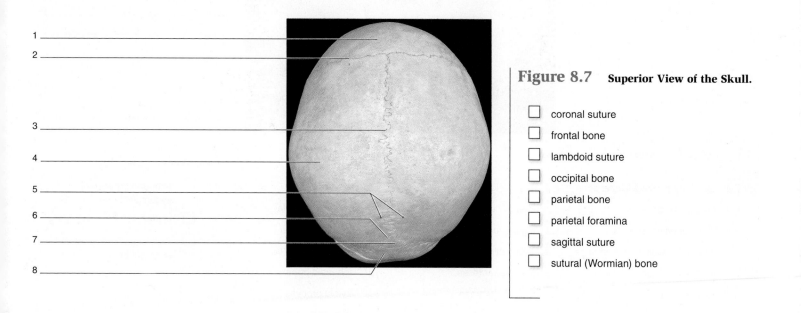

Figure 8.7   **Superior View of the Skull.**

- ☐ coronal suture
- ☐ frontal bone
- ☐ lambdoid suture
- ☐ occipital bone
- ☐ parietal bone
- ☐ parietal foramina
- ☐ sagittal suture
- ☐ sutural (Wormian) bone

EXERCISE 8.2D: Inferior View of the Skull

1. Turn the skull over and observe it from an inferior view (**figure 8.8**). Many of the structures visible in this view are foramina that nerves and blood vessels pass through to get into and out of the cranial cavity.

2. Obtain a broom straw or other nonmarking pointing device (NO pens or pencils!) from your instructor. As you identify each foramen in the inferior view, pass the broom straw through the foramen and see where it comes out in the cranial floor. This activity will help you visualize the pathways that structures traveling through the foramina take to get into or out of the cranial cavity.

3. The **foramen lacerum** (*lacero*, to tear to pieces) is unique among cranial foramina. It is one of the longest canals in

the skull (about a centimeter in length). However, no single structure passes completely through it from one opening to the other. Instead, several structures pass through small portions of the canal. Such structures include a number of nerves as well as the internal carotid artery. As the internal carotid artery travels superiorly from the thorax into the cranial cavity, it passes first through the **carotid canal** and then enters the superior portion of the foramen lacerum as it proceeds toward the brain.

4. Using Table 8.1 and your textbook as guides, identify the structures listed in figure 8.8 on a skull. Then label figure 8.8.

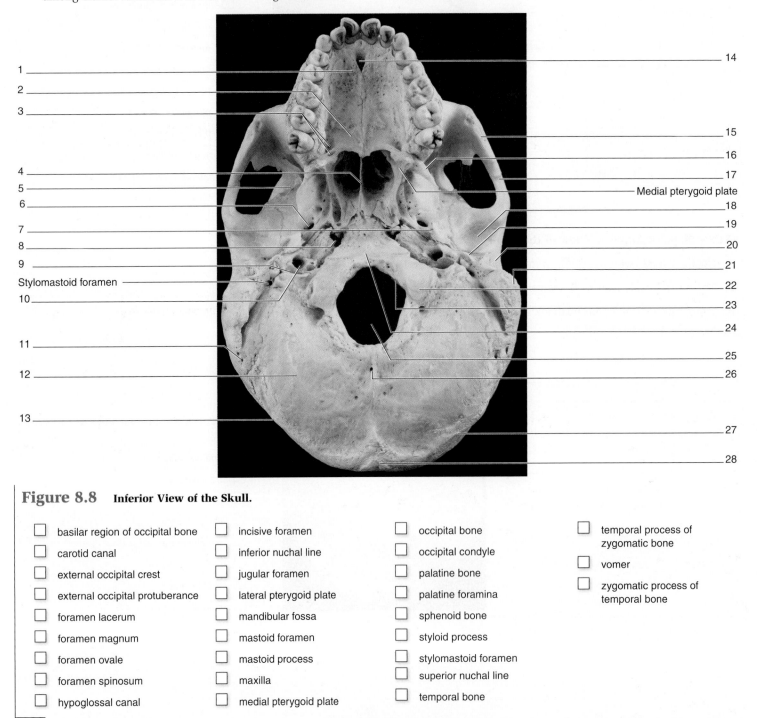

Figure 8.8   **Inferior View of the Skull.**

- ☐ basilar region of occipital bone
- ☐ carotid canal
- ☐ external occipital crest
- ☐ external occipital protuberance
- ☐ foramen lacerum
- ☐ foramen magnum
- ☐ foramen ovale
- ☐ foramen spinosum
- ☐ hypoglossal canal
- ☐ incisive foramen
- ☐ inferior nuchal line
- ☐ jugular foramen
- ☐ lateral pterygoid plate
- ☐ mandibular fossa
- ☐ mastoid foramen
- ☐ mastoid process
- ☐ maxilla
- ☐ medial pterygoid plate
- ☐ occipital bone
- ☐ occipital condyle
- ☐ palatine bone
- ☐ palatine foramina
- ☐ sphenoid bone
- ☐ styloid process
- ☐ stylomastoid foramen
- ☐ superior nuchal line
- ☐ temporal bone
- ☐ temporal process of zygomatic bone
- ☐ vomer
- ☐ zygomatic process of temporal bone

1. Obtain a skull and remove the top of the cranium so the cranial floor is visible (**figure 8.9**).

2. Notice how the floor of the cranium is separated into three fossae: anterior, middle, and posterior. The **lesser wing of the sphenoid bone** forms the border between anterior and middle cranial fossae. The **petrous part of the temporal bone** (*petrosus*, a rock) forms the "rocky" border between the middle and posterior cranial fossae.

3. Using colored pencils, color and label the anterior, middle, and posterior cranial fossae in figure 8.9. As you color in the fossae, pay attention to the structures that form the natural divisions between these fossae.

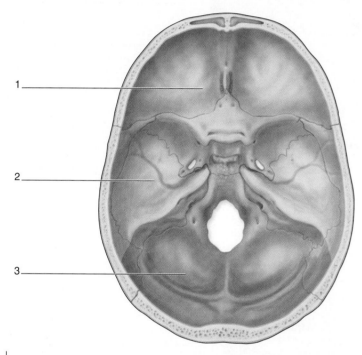

Figure 8.9   **Cranial Fossae.**

- ☐ anterior cranial fossa
- ☐ middle cranial fossa
- ☐ posterior cranial fossa

4. Observe the **sella turcica** (*sella*, saddle, + *turcica*, Turkish) in the central portion of the sphenoid bone. This structure gets its name from its resemblance to a Turkish saddle, which contains large, prominent horns (**figure 8.10**). The anterior part of the sella turcica contains a slight projection called the **tuberculum sellae** (*tuber*, a knob). Posterior to that is the **hypophyseal fossa** (*hypophysis*, an undergrowth), which houses the pituitary gland in a living human. The pituitary gland is a small pea-shaped endocrine gland that connects to the brain via a small stalk called the infundibulum (*infundibula*, a funnel). The larger projection in the posterior part of the sella is the **dorsum sellae**, which connects laterally to the two **posterior clinoid processes** (*klino*, to slope).

5. The temporal bone has two major portions; a lateral **squamous part**, which forms part of the lateral wall of the cranium, and a thick **petrous part,** which forms the border between the middle and posterior cranial fossae (although it is considered part of the middle cranial fossa).

6. Locate the petrous part of the temporal bone. Notice how it forms a sort of "rocky" ridge within the cranial floor. This portion of the temporal bone is very large and bulky because it contains the mechanisms for hearing (the cochlea) and equilibrium/balance (the semicircular canals). These structures cannot be seen from the surface of the bone. However, if you were to break open the petrous part of the temporal bone, you would find them inside.

7. Find the internal and external acoustic (auditory) meatuses. The **internal acoustic** (auditory) **meatus** is the opening into the **internal auditory canal**, which is a passageway for the nerves that carry sensory information from the cochlea and semicircular canals to the brain. The **external acoustic** (auditory) **meatus** is the opening into the external auditory canal, which is a passageway through which sound waves travel to reach the tympanic membrane (ear drum).

8. Using Table 8.1 and your textbook as guides, identify the structures listed in **figure 8.11** on a superior view of the cranial floor. Then label figure 8.11.

Study Tip!

It is useful to focus your observations on one fossa at a time as you identify bony structures within the cranial floor. This is a natural way to divide the features of the cranial floor into manageable pieces of material. If you take this approach, when you take a laboratory practical exam and are asked to identify one of the many foramina in the cranial floor, you will be able to narrow down your choices if you first identify the cranial fossa where the foramen is located.

Figure 8.10   **Photograph of a Turkish Saddle.**

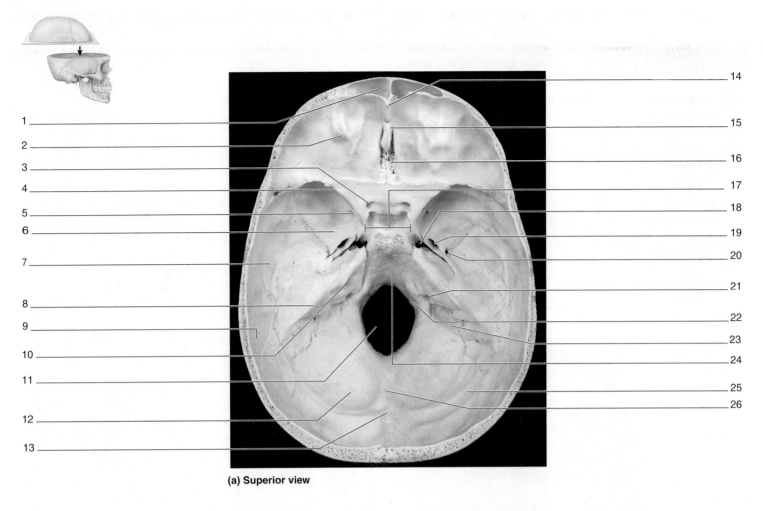

1 _____
2 _____
3 _____
4 _____
5 _____
6 _____
7 _____
8 _____
9 _____
10 _____
11 _____
12 _____
13 _____

14 _____
15 _____
16 _____
17 _____
18 _____
19 _____
20 _____
21 _____
22 _____
23 _____
24 _____
25 _____
26 _____

(a) Superior view

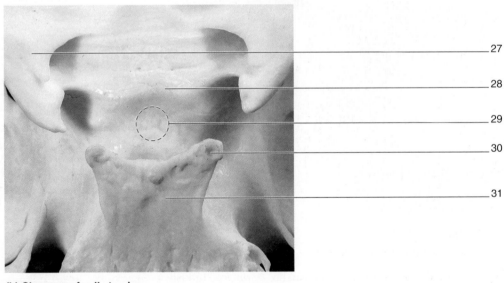

27 _____
28 _____
29 _____
30 _____
31 _____

(b) Close–up of sella turcica

Figure 8.11   **Superior View of the Cranial Floor.** (a) View of all three fossae. (b) Close-up of sella turcica.

Anterior Cranial Fossa

☐ cribriform plate of ethmoid bone

☐ crista galli

☐ frontal crest

☐ frontal bone

☐ lesser wing of sphenoid bone

Middle Cranial Fossa

☐ anterior clinoid process

☐ foramen lacerum

☐ foramen ovale

☐ foramen rotundum

☐ frontal sinus

☐ foramen spinosum

☐ greater wing of sphenoid bone

☐ hypophyseal fossa

☐ optic canal

☐ petrous part of temporal bone

☐ posterior clinoid process

☐ sella turcica

☐ temporal bone

Posterior Cranial Fossa

☐ basilar part of occipital bone

☐ dorsum sellae

☐ foramen magnum

☐ groove for sigmoid sinus

☐ groove for transverse sinus

☐ hypoglossal canal

☐ internal auditory meatus

☐ internal occipital crest

☐ internal occipital protuberance

☐ jugular foramen

☐ occipital bone

☐ parietal bone

☐ tuberculum sellae

EXERCISE 8.4 Bones Associated with the Skull

The hyoid bone and the auditory ossicles are bones of the axial skeleton that are associated with the skull, but they are not part of the skull proper. The auditory ossicles are part of the hearing apparatus and will be covered in chapter 18. This exercise will concentrate on the features of the hyoid bone.

1. Observe the hyoid bone (**figure 8.12**) on an articulated skeleton.

2. The **hyoid bone** (*hyoeidēs*, shaped like the Greek letter upsilon, υ) is the only bone in the body with no direct articulation with another bone. Muscles that move the tongue and pharynx attach to the hyoid.

3. Palpate your own hyoid by placing your thumb and index finger just medial to the angle of your mandible on either side and then moving them from side to side. Can you feel the rigid structure that you are moving?

4. Observe the articulated skeleton in your laboratory and look at the placement of the hyoid bone with respect to the mandible.

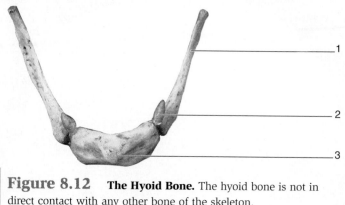

Figure 8.12   **The Hyoid Bone.** The hyoid bone is not in direct contact with any other bone of the skeleton.

☐ body   ☐ greater cornu   ☐ lesser cornu

5. Using your textbook as a guide, identify the structures listed in figure 8.12 on a hyoid bone. Then label figure 8.12.

The Fetal Skull

Like all bones in the body, the skull bones of the fetus are still developing. Recall that they develop via intramembranous ossification, which involves replacing a connective tissue membrane with bone tissue. Thus, when a fetus is born, the sutures between skull bones have not yet formed, which allows the head to distort as it moves through the birth canal. The spaces between the plates of bone in the developing skull still consist of connective tissue membranes, which are largest in places where more than two bones come together. These membranes are **fontanelles** (*fontaine*, a small fountain), and can be felt as "soft spots" on a baby's head. The fontanelles will not fill in completely until between the ages of 2 and 3.

EXERCISE 8.5 The Fetal Skull

1. Observe a fetal skull or a model of a fetal skull (**figure 8.13**).

2. Based on your observations of the adult skull, locate the major cranial bones on the fetal skull (for example: frontal, parietal, occipital, and so on).

3. Using your textbook as a guide, identify the structures listed in figure 8.13 on a fetal skull or model of a fetal skull. Then label them in figure 8.13.

WHAT DO YOU THINK?

1 Why do you think the fontanelles persist until well after the birth of the infant?

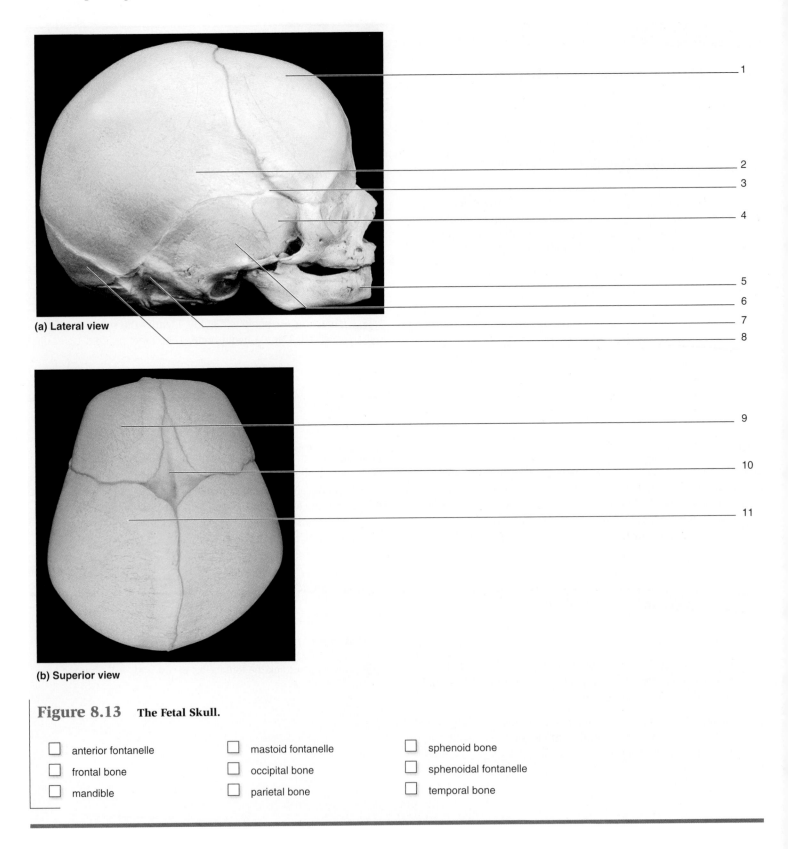

(a) Lateral view

(b) Superior view

Figure 8.13   **The Fetal Skull.**

☐ anterior fontanelle	☐ mastoid fontanelle	☐ sphenoid bone
☐ frontal bone	☐ occipital bone	☐ sphenoidal fontanelle
☐ mandible	☐ parietal bone	☐ temporal bone

The Vertebral Column

The vertebral column lies at the core of the human skeleton. It quite literally is the "backbone" that anchors nearly every major component of the skeletal support system. The vertebral column is divided into five major regions: **cervical, thoracic, lumbar, sacral,** and **coccygeal**. The vertebrae themselves change size and shape rather drastically from the cervical region to the sacral region. These changes reflect the different demands placed on the vertebrae in each region. **Cervical vertebrae** are small and light because they are not supporting a lot of weight (relatively speaking), and they are specialized to allow a lot of movement of the neck, particularly

rotation. **Thoracic vertebrae** are specialized to provide articulation points for the ribs. **Lumbar vertebrae** are very large, bulky vertebrae that are specialized for supporting the weight of the entire vertebral column and body structures above them. They do not allow much movement, but instead are designed to keep the vertebral column stable. The **sacrum**, which consists of fused vertebrae, is specialized to provide a stable anchoring point for the bones of the pelvic girdle. Finally, the **coccyx** consists of 3–5 small vertebrae, which have fused together during development. It serves as an attachment point for several ligaments and for muscles of the pelvic floor. **Table 8.3** summarizes the characteristics of each type of vertebra.

Table 8.3		**The Axial Skeleton—Vertebral Column**	
Vertebrae	**Number of Vertebrae**	**Bone Feature**	**Description and Related Structures of Importance**
Typical Vertebra	32	Lamina	Connects transverse process to spinous process.
		Pedicle	Connects body to transverse process.
		Transverse processes	Processes that are directed laterally (one on each side).
		Spinous process	A process that is directed posteriorly.
		Inferior articular process	Contains a facet that forms a joint with the superior articular process of the vertebra one level below.
		Superior articular process	Contains a process that forms a joint with the inferior articular process of the vertebra one level above.
		Vertebral foramen	Location of the spinal cord.
		Body	The largest part of the vertebra. Intervertebral discs are found between bodies of adjacent vertebrae.
		Intervertebral foramina	Formed when two vertebra come together; passageway for exit of spinal nerves.
Cervical (C)	7	Body	Small body, oval/kidney bean shape.
		Spinous process	Horizontal, bifid (forked) appearance on some (but not all).
		Vertebral foramen	Large (especially with respect to size of the body), slight oval shape.
		Transverse processes	Each contains a transverse foramina.
		Transverse foramen	Contain the vertebral artery and other structures.
		Costal facets	Located on the body and transverse processes; these form joints with the ribs.
Atlas (C1)		Body	Has no body; the body has become the dens (odontoid process) of the axis.
		Arch	Contains the articulation surface for the dens of the axis and the posterior tubercle (no spinous process).
Axis (C2)		Body	Has odontoid process (dens), which is the fused body of C1.
Vertebra Prominens (C7)		Spinous processes	Very large and blunt, not bifid, not covered by ligamentum nuchae. Therefore, is the first spinous process easily felt under the skin.
Thoracic (T)	12	Body	Heart-shaped, contains demifacets for articulation of the head of a rib.
		Spinous process	Points inferiorly.
		Vertebral foramen	Relatively small, circular in shape.
		Transverse processes	Contain facets for articulation with the tubercle of a rib.
		Costa facets	Located on the body and transverse processes; these form joints with the ribs.
Lumbar (L)	5	Body	Very large, heavy.
		Spinous process	Short and blunt, square shaped, horizontal.
		Vertebral foramen	Small (especially with respect to size of body), round.
		Transverse processes	Short and tapered at the ends.
Sacrum (S)	5 (fused)	Anterior sacral foramina	Passageway for exit of ventral (anterior) rami of sacral spinal nerves.
		Posterior sacral foramina	Passageway for exit of dorsal (posterior) rami of sacral spinal nerves.
		Median sacral crest	Represents fused spinous processes of sacral vertebrae (S1–S4).
		Auricular processes	Earlike (*auris*, ear) processes that articulate with the iliac bones.
		Superior articular processes	Contains a facet to form a joint with the inferior articular processes of L5.
		Sacral hiatus	The opening at the inferior end of the sacral canal. Formed by unfused laminae of S5.
		Sacral promontory	The anteriosuperior border of the body of S1.
Coccyx (Co)	3 to 5 (fused)	Cornu (horns)	Small projections that point superiorly (part of Co1).

EXERCISE 8.6   Vertebral Column Regions and Curvatures

1. Observe the vertebral column of an articulated skeleton (**figure 8.14**).

2. Using colored pencils, color and label the regions of the vertebral column in figure 8.14. Use a different color for each region. As you shade in each region, count the number of vertebrae that make up the region and write that number in the appropriate space in figure 8.14.

3. As the vertebral column develops, it forms several curvatures because of the stresses placed on it. The first curvatures to develop during the fetal period are **primary curvatures**. These form in the thoracic and sacral regions due to growth of the viscera. The second curvatures, which develop after birth, are **secondary curvatures**. These form in the cervical and lumbar regions. The cervical curvature forms when an infant begins to lift his head and the lumbar curvature forms when an infant begins to stand on his feet.

4. Locate all of the curvatures of the vertebral column on an articulated skeleton, and label the curvatures on figure 8.14.

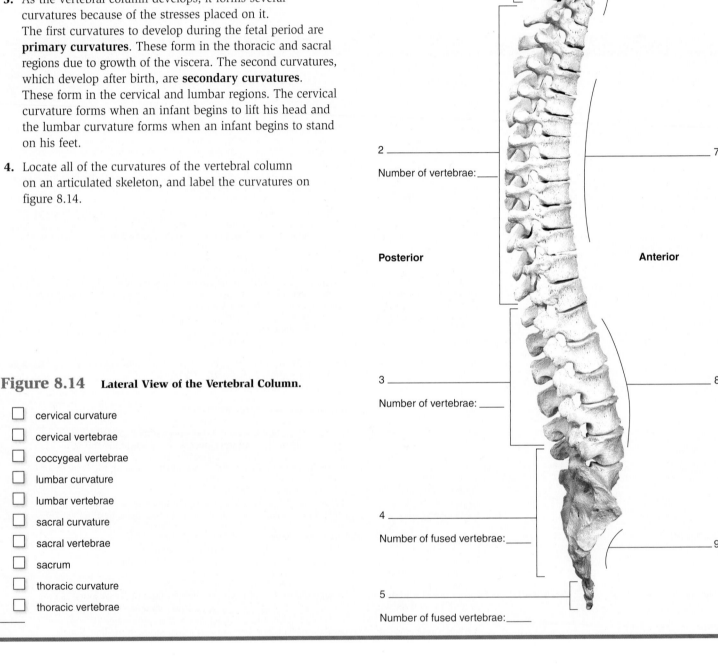

Regions

1 _____
Number of vertebrae: _____

2 _____
Number of vertebrae: _____

Posterior

3 _____
Number of vertebrae: _____

4 _____
Number of fused vertebrae: _____

5 _____
Number of fused vertebrae: _____

Curvatures

6

7

Anterior

8

9

Figure 8.14   **Lateral View of the Vertebral Column.**

☐ cervical curvature
☐ cervical vertebrae
☐ coccygeal vertebrae
☐ lumbar curvature
☐ lumbar vertebrae
☐ sacral curvature
☐ sacral vertebrae
☐ sacrum
☐ thoracic curvature
☐ thoracic vertebrae

EXERCISE 8.7 Structure of a Typical Vertebra

1. Obtain a **thoracic vertebra** (**figure 8.15**) from your instructor or bone box—this will be your example of a "typical" vertebra. It is helpful to begin your study of the vertebral column by taking a "typical" vertebra and identifying its component parts. Then as you begin observing normal variability among vertebrae you will have an easier time finding the appropriate structures.

2. Looking at the vertebra from the superior view, notice the large **body**. The body is generally the heaviest part of the vertebra, and connections between adjacent vertebral bodies (with intervertebral discs in between) provide the main support of the vertebral column. Just posterior to the body is the large foramen called the **vertebral (spinal) foramen**. The spinal cord runs through this foramen. The foramen itself is formed by a structure that is collectively referred to as the **vertebral arch**. The vertebral arch is composed of two sets of processes and the structures that connect them. We will return to a description of this arch after you have identified the major processes of the vertebra.

3. Observe the vertebral processes that project posteriorly and laterally from the vertebral arch. The largest vertebral processes are the **spinous processes**, which are directed posteriorly, and the **transverse processes**, which are directed laterally.

4. Return your attention to the vertebral arch. Notice the bony connections between the vertebral body and the transverse processes. These structures are called **pedicles** (L. *pediculus*, dim. of *pes*, foot). The word *pedicle* comes from a word meaning "foot." If you use your imagination a bit, can you imagine how the vertebral arch stands upon the body on its "feet?" Now notice the bony connections between the transverse processes and the spinous process. These structures are called **laminae** (*lamina*, layer).

5. Next, turn the vertebra so that you are observing it from an anterior view. Notice that there are two prominent structures that project superiorly from the vertebral arch and two that project inferiorly. The projections are respectively called the **superior articular processes** and **inferior articular processes**. Note that on each process there is a smooth, flat surface. These surfaces are called **facets** (*facette*, face). The term *facet* literally means "a little face." This is the same term used to describe the surfaces on a diamond. Each vertebra contains upon its superior and inferior processes a pair of **superior articular facets** and **inferior articular facets**. These facets are the surfaces that form the joints between vertebrae, as described in the next step (6).

6. Pick up another vertebra that articulates (forms a joint) with the vertebra in your hand. As you put the two together, observe how the superior facets and inferior facets articulate with each other to form a joint. These joints are much more mobile than the intervertebral joints (the joints between the vertebral bodies), and they are the sites where most of the movement is allowed by the vertebral column.

7. Now that you have the two vertebrae articulated with each other, look at them from a lateral view. Notice the foramen that forms between the pedicles of adjacent vertebrae. This is the **intervertebral foramen**. This foramen is the location where spinal nerves (nerves that come off of the spinal cord) exit the vertebral canal to travel to their destinations throughout the body.

8. Using Table 8.3 and your textbook as guides, identify the structures listed in figure 8.15 on a typical vertebra. Then label them in figure 8.15.

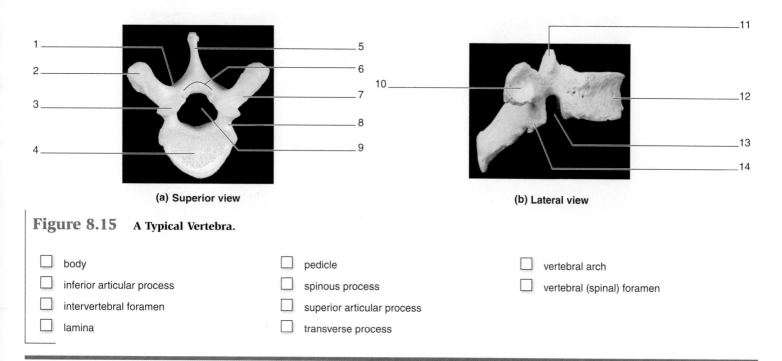

(a) Superior view   (b) Lateral view

Figure 8.15   **A Typical Vertebra.**

☐ body

☐ inferior articular process

☐ intervertebral foramen

☐ lamina

☐ pedicle

☐ spinous process

☐ superior articular process

☐ transverse process

☐ vertebral arch

☐ vertebral (spinal) foramen

EXERCISE 8.8 Characteristics of Individual Vertebrae

Study Tip!

As you observe individual vertebrae, always try to keep function in mind. Ask yourself questions, such as: Why does this vertebra have such a large/small body? Why does this vertebra have such a large/small vertebral canal? How does this vertebra "fit" with other aspects of the skeletal system? Asking yourself these questions will help you identify each type of vertebra correctly. Finally, when you are observing isolated vertebrae, always be sure to identify the same vertebrae on an articulated skeleton so you can develop an appreciation for how the vertebral column is put together.

EXERCISE 8.8A: Typical Cervical Vertebrae

1. Obtain a cervical vertebra (*cervix*, neck) (**figure 8.16**). As you identify the features of each of the individual cervical vertebrae, think about how modifications of the cervical vertebrae allow for a great deal of movement in the neck region of the vertebral column.

2. Observe the vertebra from a superior view. Typical cervical vertebrae have a small, oval body and a large triangular vertebral foramen. The spinous process of some of them is forked, or **bifid** (*bifidus*, cleft in two parts).

3. Observe cervical vertebrae on an articulated skeleton. Notice how the fork on one vertebra fits over the top of the spinous process of the vertebra below it.

4. Observe the transverse processes on a cervical vertebra. Notice that it has a hole, or **transverse foramen**, in it. This foramen protects an artery, the **vertebral artery**, as it travels from the thorax to the cranial cavity to supply the brain with blood.

5. Using Table 8.3 and your textbook as guides, identify the structures listed in figure 8.16 on a cervical vertebra. Then label them in figure 8.16.

WHAT DO YOU THINK?

② How might the bifid spinous processes of cervical vertebrae affect anterior-posterior movement in the cervical region of the vertebral column?

The Atlas (C1) and Axis (C2)

The Greek Titan, Atlas, held up the heavens on his shoulders. The first cervical vertebra is named the **atlas** because it holds up the head in much the same way. The second cervical vertebra is called the **axis** because it forms an axis of rotation for the first cervical vertebra to rotate about. Both of these vertebrae are specialized to allow for extensive flexion, extension, and rotational movements of the neck. As you observe the special modifications of the atlas and axis, try to visualize how these modifications allow extensive movement of the head and neck.

EXERCISE 8.8B: The Atlas (C1)

1. Obtain an atlas (C1) and an axis (C2).

2. Notice that the atlas is missing a body (**figure 8.17**). During development, the tissue that would normally become the body of the atlas fuses with the body of the axis, forming the *dens* (odontoid process) of the axis. This modification allows the atlas to rotate around the axis.

3. Instead of laminae and pedicles, the atlas has an **anterior arch** and **posterior arch**. Notice the **articular facet for the dens** on the inner surface of the anterior arch. Also note that instead of a spinous process there is a smaller **posterior tubercle** on the posterior arch.

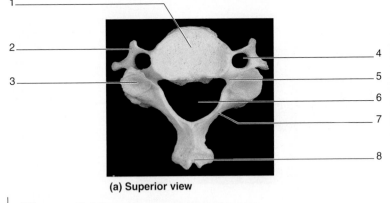

(a) Superior view

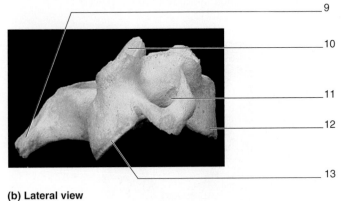

(b) Lateral view

Figure 8.16   Cervical Vertebra.

☐ body

☐ inferior articular process (and facet)

☐ lamina

☐ pedicle

☐ spinous process

☐ superior articular process (and facet)

☐ transverse foramen

☐ transverse process

☐ vertebral (spinal) foramen

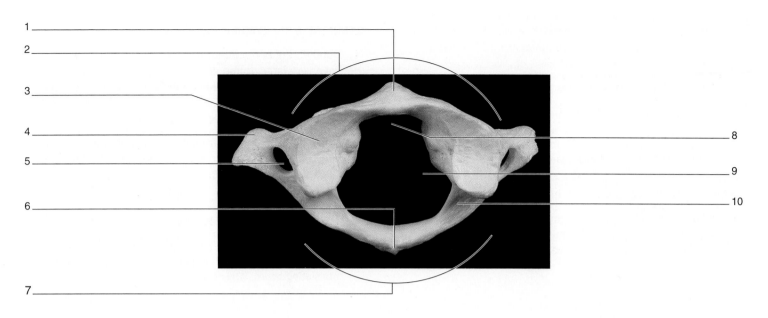

1 _____

2 _____

3 _____

4 _____   8 _____

5 _____   9 _____

6 _____   10 _____

7 _____

Figure 8.17   **The Atlas (C1).** Superior view.

☐ anterior arch   ☐ posterior arch   ☐ transverse foramen

☐ anterior tubercle   ☐ posterior tubercle   ☐ transverse process

☐ articular facet for dens   ☐ superior articular facet   ☐ vertebral foramen

☐ groove for vertebral artery

4. Observe the **superior articular facets** of the atlas. These facets are oriented horizontally in the atlas, rather than vertically as with the other vertebrae. These facets articulate with the **occipital condyles**. Observe the occipital bone and atlas on an articulated skeleton to see how these structures fit together to form the **atlanto-occipital joint**. This joint allows flexion and extension movements of the neck—as when we nod our heads to indicate "yes."

5. Using Table 8.3 and your textbook as guides, identify the structures listed in figure 8.17 on an atlas (C1 vertebra). Then label them in figure 8.17.

EXERCISE 8.8C: The Axis (C2)

1. Obtain an atlas (C1) and an axis (C2).

2. The **axis** (**figure 8.18**) is more similar to a typical cervical vertebra than the atlas. However, it has an extra appendage that no other vertebra has. This appendage/process is the **dens**, or **odontoid process** (*odont-*, tooth). Where did this process come from (developmentally)? _____

_____

3. Place the atlas upon the axis and observe their articulation with each other to form the **atlantoaxial joint.** This joint allows lateral rotation of the neck—as when we turn our heads from side to side to indicate "no". Holding the atlas (C2) in place, rotate the axis (C1) around the dens of the atlas to observe this movement.

4. Similar to the atlas (C1), the superior and inferior articular processes of the axis (C2) lie in a horizontal plane. In

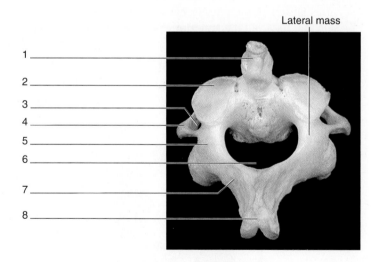

Lateral mass

1 _____

2 _____

3 _____

4 _____

5 _____

6 _____

7 _____

8 _____

Figure 8.18   **The Axis (C2).** Superior view.

☐ dens (odontoid process)   ☐ superior articular process

☐ lamina   ☐ transverse foramen

☐ pedicle   ☐ transverse process

☐ spinous process   ☐ vertebral (spinal) foramen

addition, the axis (C2) has a large bony surface where the laminae and pedicles come together called the **lateral mass.** The transverse processes connect to the lateral mass.

5. Using Table 8.3 and your textbook as guides, identify the structures listed in figure 8.18 on an axis (C2 vertebra). Then label them in figure 8.18.

WHAT DO YOU THINK?

3 How do the superior and inferior articular processes of the atlas differ from the same processes on a "typical" vertebra, and how does this difference contribute to the special movement allowed at the atlanto-occipital and atlantoaxial joints?

EXERCISE 8.8D: Thoracic Vertebrae

1. Obtain a thoracic vertebra (**figure 8.19**). **Thoracic vertebrae** are the only vertebrae that articulate with the ribs. Thus, these vertebrae have special articular surfaces (facets) in locations where the ribs and vertebrae meet and form joints.

2. Observe the thoracic vertebra from a superior view. Thoracic vertebrae typically have a heart-shaped body (medium in size), a round vertebral foramen, a spinous process that projects inferiorly, and superior and inferior articular processes with surfaces that lie in the frontal plane.

3. Look at the relationship between the ribs and vertebrae on an articulated skeleton. Notice that the **tubercle** of a rib articulates with the transverse process of a thoracic vertebra. Notice also that the head of the rib articulates at the junction between two vertebral bodies. Thus, it articulates with the **inferior costal facet** of the vertebra superior to it and the **superior costal facet** of the vertebra inferior to it.

4. Using Table 8.3 and your textbook as guides, identify the structures listed in figure 8.19 on a thoracic vertebra. Then label them in figure 8.19.

EXERCISE 8.8E: Lumbar Vertebrae

1. Obtain a lumbar vertebra (**figure 8.20**). **Lumbar vertebrae** have very large, round or oval bodies, small vertebral foramina, a short and blunt spinous process that projects posteriorly. The superior and inferior articular processes have facets that face medial and lateral, respectively.

2. Using Table 8.3 and your textbook as guides, identify the structures listed in figure 8.20 on a lumbar vertebra. Then label them in figure 8.20.

WHAT DO YOU THINK?

4 Notice that when you put two lumbar vertebrae together, little to no lateral rotation is allowed because of the shape of the articulating bones. Why do you think they are built this way?

EXERCISE 8.8F: The Sacrum and Coccyx

1. Obtain a sacrum and coccyx (**figure 8.21**), or observe them on an articulated skeleton.

2. The **sacrum** (*sacr-*, sacred) forms by the fusion of five primitive vertebrae that subsequently form a single bony structure. As you identify the features of the sacrum, one of your goals is to recognize the parts of a typical vertebra within the sacrum.

3. The **coccyx** is usually composed of three to five small bones. The vertebrae have only two prominent structures, the *cornu* (*cornu*, horn) and the *transverse processes*.

4. Using Table 8.3 and your textbook as guides, identify the structures listed in figure 8.21 on a sacrum and coccyx. Then label them in figure 8.21.

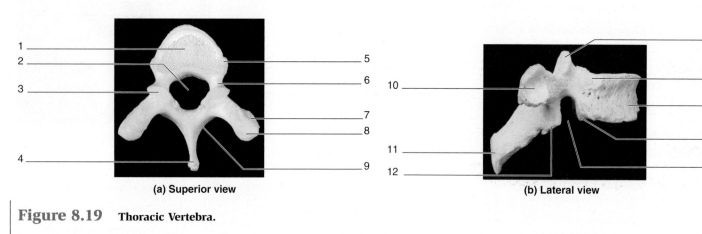

(a) Superior view   (b) Lateral view

Figure 8.19   **Thoracic Vertebra.**

☐ body

☐ costal demifacet

☐ costal facet

☐ inferior articular process

☐ lamina

☐ pedicle

☐ spinous process

☐ superior articular process

☐ superior costal facet

☐ transverse process

☐ vertebral (spinal) foramen

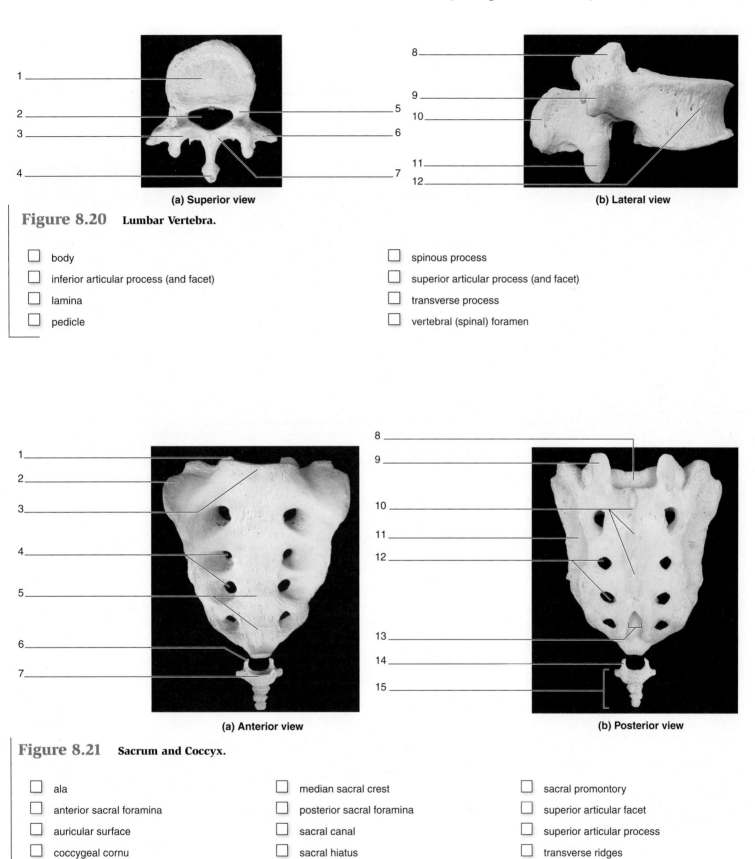

(a) Superior view

(b) Lateral view

Figure 8.20 **Lumbar Vertebra.**

☐ body

☐ inferior articular process (and facet)

☐ lamina

☐ pedicle

☐ spinous process

☐ superior articular process (and facet)

☐ transverse process

☐ vertebral (spinal) foramen

(a) Anterior view

(b) Posterior view

Figure 8.21 **Sacrum and Coccyx.**

☐ ala

☐ anterior sacral foramina

☐ auricular surface

☐ coccygeal cornu

☐ coccyx

☐ median sacral crest

☐ posterior sacral foramina

☐ sacral canal

☐ sacral hiatus

☐ sacral promontory

☐ superior articular facet

☐ superior articular process

☐ transverse ridges

The Thoracic Cage

The thoracic cage consists of the sternum, ribs, and thoracic vertebrae. Its main function is to protect vital organs such as the heart and lungs. However, the bones of the thoracic cage also serve as important attachment sites for muscles involved with respiratory movements and muscles involved with movements of the back, chest, and neck. **Table 8.4** summarizes the key features of the sternum and ribs. Refer to Table 8.3 for descriptions of the features of the thoracic vertebrae, which articulate with the ribs.

Table 8.4	The Axial Skeleton: Sternum and Ribs	
Bone	**Bone Feature**	**Description and Related Structures of Importance**
STERNUM		
Manubrium	Clavicular notch	Point of articulation with the clavicle.
	Suprasternal notch	Depression at the superior border.
	Notch for rib 1	Location of articulation with costal cartilage of rib 1.
	Sternal angle	Joint between the manubrium and the body; point of articulation with the costal cartilage of rib 2.
Body	Notches for ribs 2–7	Point of articulation for the costal cartilages of ribs 2–7. The notch for rib 2 is a partial notch.
	Xiphisternal joint	Joint between the body and the xiphoid; point of articulation with the superior part of the costal cartilage of rib 7.
Xiphoid	Partial notch for rib 7	Point of articulation for the inferior part of the costal cartilage of rib 7.
RIBS		
Typical Rib	Head	The part of a rib that articulates with the bodies of the thoracic vertebrae.
	Superior articular facet	A facet on the head of a rib that articulates with the inferior costal facet on the body of the vertebra that lies one level above it (i.e., superior articular facet of rib 6 with T5).
	Inferior articular facet	A facet on the head of a rib that articulates with the superior costal facet on the body of the numerically equivalent thoracic vertebra (i.e., inferior articular facet of rib 6 to T6).
	Shaft	The main part (body) of a rib, which begins at the angle of the rib and projects anteriorly.
	Neck	A narrow region where the head meets the tubercle of the rib.
	Tubercle	A projection at the junction between the shaft and neck; contains a facet for articulation with the transverse process of a thoracic vertebra.
	Angle	The location where the rib curves anteriorly.
	Costal groove	A groove on the inferior, deep border of the shaft; contains the intercostal artery, vein, and nerve.
	Cup	The point of articulation for a costal cartilage.
First Rib	Scalene tubercle	Attachment for the anterior scalene muscle.
	Groove for subclavian artery	A depression indicating the location where the subclavian artery passes out of the thoracic cavity.
	Groove for subclavian vein	A depression indicating the location where the subclavian vein passes into the thoracic cavity.
	Articular facet	A singular facet on the head of the rib (a typical rib has two facets).
Second Rib	All markings of a typical rib	Unique features of rib 2 are a rough tuberosity and a shallow costal groove.
11th and 12th Ribs	Articular facet	A singular facet on the head of the rib (a typical rib has two facets).
	Tubercle	Absent.
	Neck	Absent.

WHAT DO YOU THINK?

5 What part of the vertebral column is removed when a laminectomy is performed? For what purpose(s) do you think a surgeon would perform this procedure?

The Sternum

1. Observe the thoracic cage on an articulated skeleton and locate the sternum (**figure 8.22**). The sternum has three sections: the **manubrium**, the **body**, and the **xiphoid process** (*xiphos*, sword). The depression on the superior part of the manubrium is the **suprasternal notch.**

2. Palpate the sternal notch on your own body. Keeping your fingers on your manubrium, move your fingers inferiorly until you feel a rough ridge. This is the **sternal angle**. The sternal angle is located where the manubrium meets the body of the sternum. It is an important clinical landmark, because this is where the second rib articulates with the sternum.

3. Using Table 8.4 and your textbook as guides, identify the structures listed in figure 8.22 on a sternum. Then label them in figure 8.22.

Figure 8.22 **The Sternum.** Anterior view.

☐ body
☐ manubrium   1 _____
☐ second rib   2 _____
☐ sternal angle 3 _____
☐ suprasternal notch
☐ xiphoid process
                4 _____

                5 _____

                6 _____

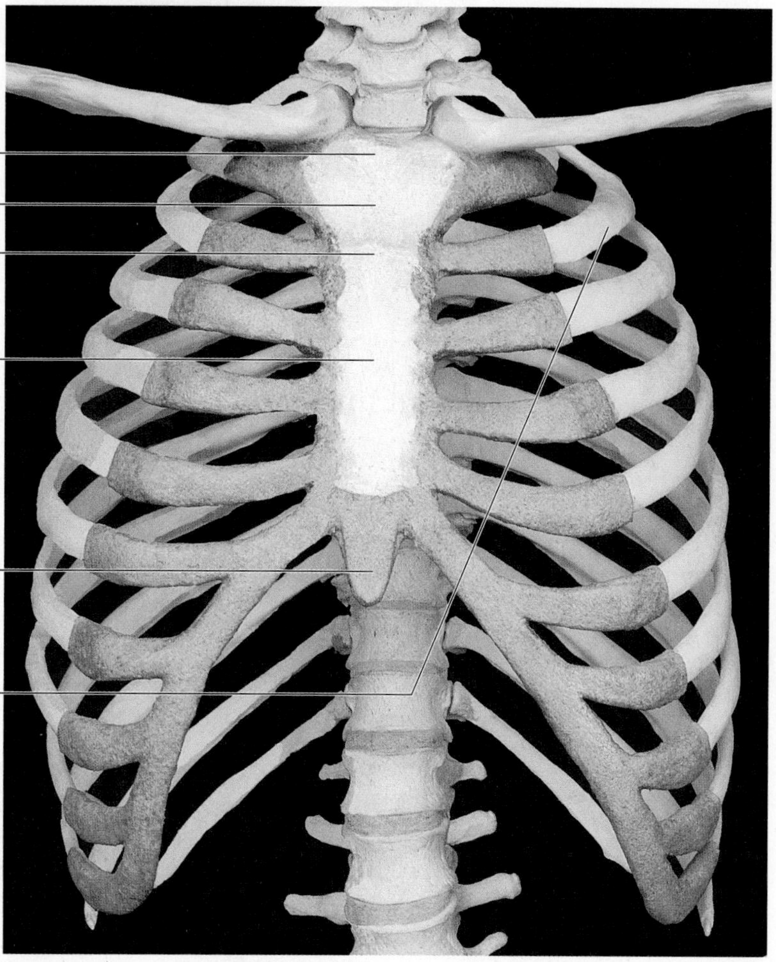

The Ribs

EXERCISE 8.10A: Typical Ribs

1. There are twelve pairs of **ribs**, one pair for each thoracic vertebra. Obtain a typical rib (any rib other than ribs 1, 2, 11, or 12) (**figure 8.23**).

2. As you observe the features of a typical rib, pay particular attention to the surfaces of the rib that articulate with the thoracic vertebrae.

3. Observe the articulations between the ribs and the thoracic vertebrae on an articulated skeleton, and review the unique features of thoracic vertebrae that allow them to form articulations with the ribs.

4. Using Table 8.4 and your textbook as guides, identify the structures listed in figure 8.23 on a typical rib. Then label them in figure 8.23.

WHAT DO YOU THINK?

6 Needles inserted into the thoracic cavity must always be placed along the superior border of a rib so as not to injure important structures. What important structures could be damaged by insertion of a needle too close to the inferior border of a rib? (Hint: Refer to Table 8.4.)

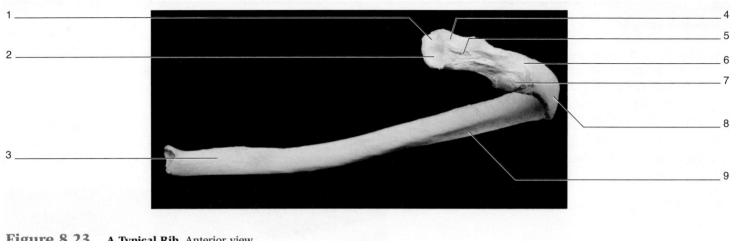

Figure 8.23   **A Typical Rib.** Anterior view.

☐ angle   ☐ head   ☐ shaft

☐ articular facet for transverse process   ☐ inferior articular facet   ☐ superior articular facet

☐ costal groove   ☐ neck   ☐ tubercle

EXERCISE 8.10B: The First Rib

1. Obtain a first rib (**figure 8.24**). Notice how it differs from a typical rib in size and shape.

2. The first rib is unusual in that it does not have all the features of the remaining eleven ribs (notable absence of an **angle**). It also has additional features such as the **scalene tubercle**, which is an attachment point for the anterior scalene muscle, a muscle that assists with elevation of the ribs during inhalation. On either side of the scalene tubercle you will find grooves for the passage of the subclavian artery and vein. These vessels convey blood between the thoracic cavity and the upper limb

3. Using Table 8.3 and your textbook as guides, identify the structures listed in figure 8.24 on the first rib. Then label them in figure 8.24.

4. *Optional Activity:* **AP|R Skeletal System**—Visit the Quiz area to test yourself on the bones of the axial skeleton and their prominent features.

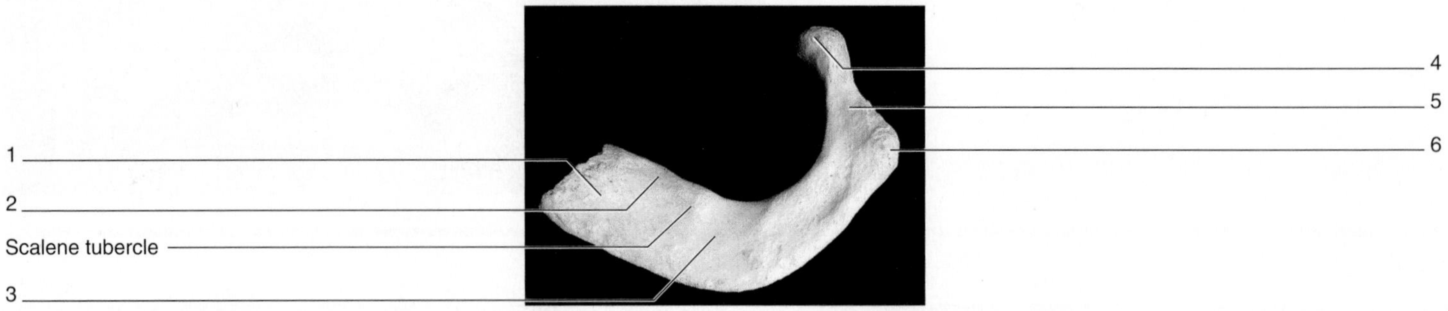

Scalene tubercle

Figure 8.24   **The First Rib.** Superior view.

☐ groove for subclavian artery   ☐ scalene tubercle

☐ groove for subclavian vein   ☐ shaft

☐ head   ☐ tubercle

☐ neck

Chapter 8: The Skeletal System: Axial Skeleton

Name: _____

Date: _____ Section: _____

POST-LABORATORY WORKSHEET

1. List all of the major bones that form the orbit of the eye. _____
   _____
   _____

2. List all of the major bones that form the wall of the nasal cavity. _____
   _____
   _____

3. List all of the major bones that form the nasal septum. _____
   _____
   _____

4. Which bones form the zygomatic arch? _____
   _____

5. Describe the three cranial fossae. Include a description of the major bony structures used to separate the fossae from each other. _____
   _____
   _____
   _____
   _____
   _____
   _____

6. To what does the term *sella turcica* refer? _____
   _____

7. What endocrine gland is found within the sella turcica in a living human? _____

8. What does the term *petrous* mean?_____
   Why is this term used to describe the petrous part of the temporal bone? _____
   _____
   What structures are found deep within the petrous part of the temporal bone? _____
   _____

9. What does the term *lacero* mean?_____
   Why do you think the term "lacero" is used to describe the foramen lacerum? _____
   _____
   What is unique about the foramen lacerum (as compared to other cranial foramina)? _____
   _____

10. Why do the fontanelles persist until well after birth?_____
_____

11. By what age do most of the fontanelles completely fill in? _____

12. What foramen is present in cervical vertebrae that is not present in other vertebrae? _____
    What structure runs through this foramen in a living human? _____

13. How do the superior and inferior articular processes of the atlas differ from the same processes on a "typical" vertebra? _____
_____
_____

14. What is the dens (odontoid process)? _____
    From what structure did the dens arise developmentally? _____

15. What are the two locations on a thoracic vertebra where the ribs articulate?

    a. _____

    b. _____

16. What is a functional consequence of the shape (and arrangement) of the superior and inferior articular processes of the lumbar vertebrae? (Hint: Put
    two of them together and see what movement is, or is not, allowed.) _____
_____

17. The anterior and posterior sacral foramina are the equivalent of the _____
    foramina in other regions of the vertebral column.

18. What is the clinical significance of the sternal angle? _____
_____

19. In what ways does the first rib differ from the remaining ribs? _____
_____

20. What structure(s) run in the costal groove? _____
_____

The Skeletal System: Appendicular Skeleton

INTRODUCTION

The term **appendicular** comes from the term *appendage*. A dictionary might define an appendage as *something that is added or attached to an item that is larger or more important, as an adjunct*. While our appendages may be described as simply "added" or "adjunct" structures, most of us would consider our arms and legs to be essential. In chapter 8 you looked at the bones that constitute the main structural support of the body: the bones of the axial skeleton. In this chapter you will study the bones composing the **pectoral girdle** and the **pelvic girdle**, which act as attachment points for muscles and ligaments that anchor the upper and lower limbs to the axial skeleton. In addition, you will study the bones of the **upper limb** and the **lower limb**, which act as attachment points for muscles and ligaments that allow movement and dexterity of the limbs. The more successfully you accomplish the task of learning bony features in this chapter, the better prepared you will be to understand muscle actions when you get to chapter 12. Bony features such as tuberosities, trochanters, and tubercles exist because of the pulling action of muscles that attach to them and stress

OUTLINE and OBJECTIVES

Module 5: SKELETAL SYSTEM

them during development. For example, without a sternocleidomastoid muscle pulling on the mastoid process of the temporal bone, a mastoid process would not exist! As you observe the features of the bones, remember that all bony features tell a story about the development and history of the entire musculoskeletal system.

As you observe the different bones and their bony features refer to the tables in this chapter for derivatives of the names of the features. If you understand the word origins, it will be easier for you to connect a structure with its name. For example, the *conoid tubercle* of the clavicle gets its name from its conical shape (*konoeides*, cone-shaped). The *coracoid process* of the scapula is named for its resemblance to a crow's beak (*karakodes*, like a crow's beak). The two names look very similar and can be easily confused if you do not pay close attention to their meanings. The process of looking up word origins will help you focus your attention and remember which process is which.

Chapter 9: The Appendicular Skeleton

Name: _____
Date: _____ Section: _____

PRE-LABORATORY WORKSHEET

1. List the two bones that make up the pectoral girdle.

   a. _____

   b. _____

2. List the three bones that make up the pelvic girdle.

   a. _____

   b. _____

   c. _____

3. In the anatomic position, the radius lies _medial / lateral_ to the ulna. (Circle the correct answer.)

4. In the anatomic position, the tibia lies _medial / lateral_ to the fibula. (Circle the correct answer.)

5. The carpal bones are located in the _____.

6. Tarsal bones are located in the _____.

7. Match the description given in column A with the appropriate bone listed in column B.

   Column A

   _____ 1. a bone that has two large tubercles on its proximal end

   _____ 2. a bone that has two large trochanters on its proximal end

   _____ 3. a bone that has an olecranon process

   _____ 4. a bone found in the wrist

   _____ 5. a sesamoid bone found in the knee

   _____ 6. the largest bone in the leg

   Column B

   a. tibia

   b. ulna

   c. humerus

   d. femur

   e. patella

   f. carpal

8. What region of the body is the calcaneus located in? _____.

9. A bone that has both an acromial and a coracoid process is the _____.

10. What region of the body is the pisiform bone located in? _____.

IN THE LABORATORY

In this laboratory session, you will study the bones that compose the appendicular skeleton on an articulated human skeleton, on disarticulated bones of the human skeleton, or on bone models.

You will identify the major landmarks and identifying features of each bone, and will associate the observed structures with their functions.

Gross Anatomy

The Pectoral Girdle

The **pectoral girdle** (*girdle*, a belt) consists of the paired clavicles and scapulae. The function of these bones is to act as the bony support for muscles that attach the upper limb to the axial skeleton and to attach each upper limb to the axial skeleton by one bony joint, the sternoclavicular joint. **Table 9.1** lists the bones of the pectoral girdle and describes their key features.

Table 9.1	Appendicular Skeleton: Pectoral Girdle		
Bone	**Bony Landmark**	**Description**	**Word Origins**
Clavicle clavicula, *a small key*	Acromial end (lateral)	The lateral end of the bone, which is flattened horizontally.	*akron*, tip, + *-omos*, shoulder
	Conoid tubercle	A small "cone-shaped" tubercle on the lateral, inferior end of the bone.	*konoeides*, cone-shaped
	Costal tuberosity	A rough impression on the inferior surface of the sternal end of the bone that serves as the attachment point for the costoclavicular ligament.	*costa*, rib
	Sternal end (medial)	The medial end of the bone, which is triangular in shape.	*sternon*, chest
Scapula scapula, *the shoulder blade*	Acromion	The large process at the lateral tip of the scapular spine, which projects laterally and slightly anteriorly.	*akron*, tip, + *-omos*, shoulder
	Coracoid process	The smaller of the two major scapular processes, which projects anteriorly.	*korakodes*, like a crow's beak
	Glenoid fossa	A shallow depression that forms the articulation between the scapula and the humerus.	*glenoeides*, resembling a socket
	Inferior angle	The angle between the medial and lateral borders.	*inferior*, lower
	Infraglenoid tubercle	A rough projection at the inferior border of the glenoid fossa; attachment point for the long head of the triceps brachii muscle.	*infra*, below, + *glenoeides*, resembling a socket
	Infraspinous fossa	A large depression inferior to the scapular spine; origin for the infraspinatus and teres minor muscles.	*infra*, below, + *spina*, spine
	Lateral (axillary) border	The border of the scapula that has the glenoid fossa on its superior part.	*axilla*, armpit
	Medial (vertebral) border	The longest border of the scapula; contains very few remarkable features.	*medialis*, middle
	Spine	A long "spiny" process on the posterior surface; attachment point for trapezius and deltoid muscles.	*spina*, spine
	Subscapular fossa	A large depression on the anterior surface of the bone; origin of the subscapularis muscle.	*sub*, under, + *spina*, spine
	Superior angle	The angle between the superior and medial borders.	*superus*, above
	Superior border	The border from which the coracoid and acromial processes project.	*superus*, above
	Supraglenoid tubercle	A rough projection at the superior border of the glenoid fossa; attachment point for the long head of the biceps brachii muscle.	*supra*, on the upper side, + *glenoeides*, resembling a socket
	Suprascapular notch	A small deep notch just medial to the coracoid process; the suprascapular nerve, artery, and vein pass through.	*supra*, on the upper side, + *scapula*, shoulder blade
	Supraspinous fossa	A large depression superior to the scapular spine; origin of the supraspinatus muscle.	*supra*, on the upper side, + *spina*, spine

EXERCISE 9.1 Bones of the Pectoral Girdle

EXERCISE 9.1A: The Clavicle

1. Obtain a clavicle or observe the clavicle on an articulated skeleton (**figure 9.1**).

2. The **clavicle** forms part of the only articulation between the axial skeleton and the upper limb, at the **sternoclavicular joint**. The clavicle has very few muscular attachments, compared to other bones. Rather than acting as a rigid attachment point for muscles, the clavicle functions more like a strut that pushes the shoulders laterally and keeps them from collapsing anteriorly toward the sternum. The superficial location of the clavicle allows most parts of the bone to be easily palpated.

3. Palpate your own clavicle. As you do this, pay attention to the two ends of the clavicle and name the bones that articulate with each end of the clavicle.

4. Using table 9.1 and your textbook as a guide, identify the structures listed in figure 9.1 on the clavicle. Then label them in figure 9.1.

5. Pick up the clavicle and hold it in front of you. Notice its S shape. How can you tell if it is a right clavicle or a left

   clavicle? _____

   _____

   _____

EXERCISE 9.1B: The Scapula

1. Obtain a scapula or observe the scapula on an articulated skeleton (**figure 9.2**).

2. The **scapula** (*scapula*, shoulder blade) is a large, irregular bone that is not directly attached to the axial skeleton. Recall that the only location where the bones of the pectoral girdle attach to the axial skeleton is where the sternal end of the clavicle articulates with the manubrium of the sternum at the sternoclavicular joint.

3. Using table 9.1 and your textbook as a guide, identify the structures listed in figure 9.2 on the scapula. Then label them in figure 9.2.

4. How can you tell whether the bone you are holding is a

   right scapula or a left scapula? _____

   _____

   _____

   _____

5. Pick up the scapula and hold it in front of you. Find the glenoid fossa. With what structure does this fossa articulate

   (articulate = form a joint)? _____

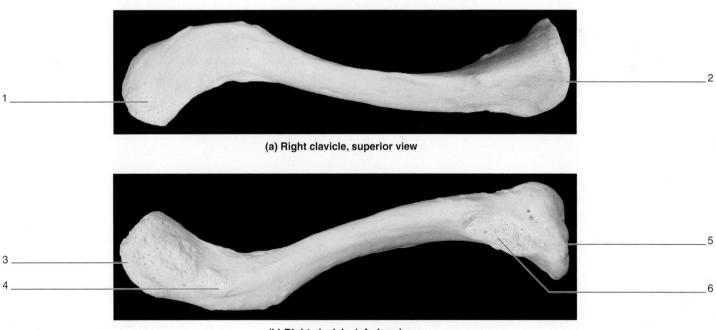

(a) Right clavicle, superior view

(b) Right clavicle, inferior view

Figure 9.1   **The Clavicle.**

☐ acromial end (lateral)   ☐ costal tuberosity
☐ conoid tubercle   ☐ sternal end (medial)

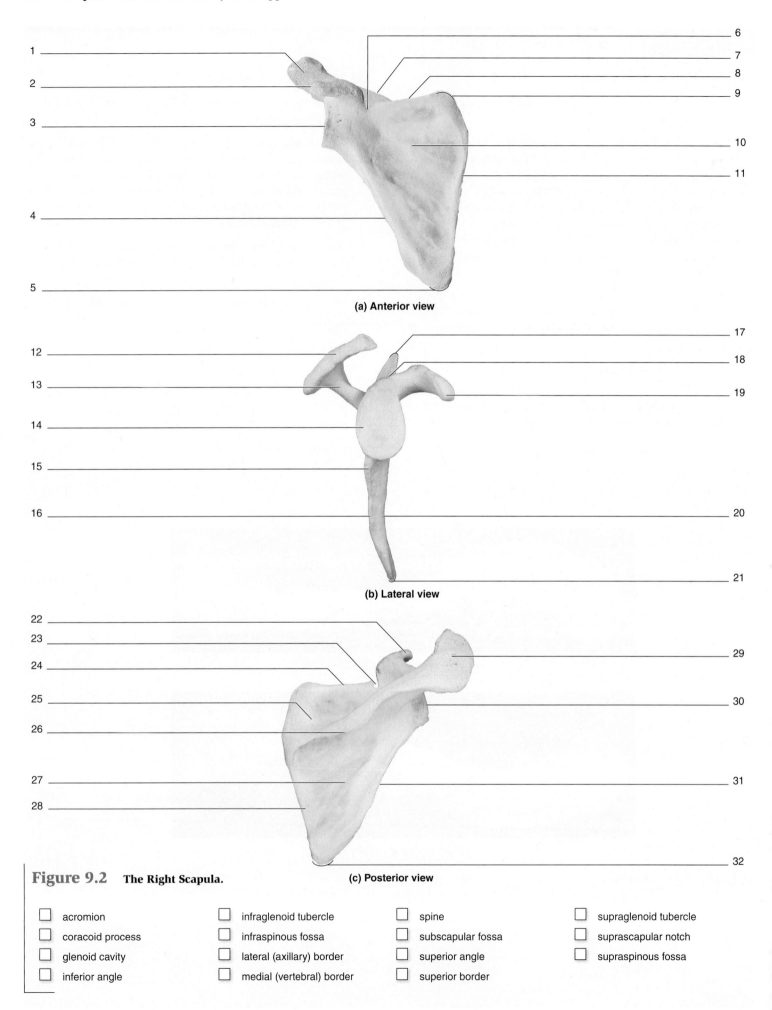

1
2
3
4
5

6
7
8
9
10
11

(a) Anterior view

12
13
14
15
16

17
18
19
20
21

(b) Lateral view

22
23
24
25
26
27
28

29
30
31
32

(c) Posterior view

Figure 9.2 **The Right Scapula.**

- [ ] acromion
- [ ] coracoid process
- [ ] glenoid cavity
- [ ] inferior angle
- [ ] infraglenoid tubercle
- [ ] infraspinous fossa
- [ ] lateral (axillary) border
- [ ] medial (vertebral) border
- [ ] spine
- [ ] subscapular fossa
- [ ] superior angle
- [ ] superior border
- [ ] supraglenoid tubercle
- [ ] suprascapular notch
- [ ] supraspinous fossa

6. Now that you have observed both the clavicle and the scapula, obtain a right clavicle and right scapula and articulate them with each other. In the box to the right, sketch the relationship between the clavicle and the scapula.

7. What is the name of the joint formed between the clavicle and the scapula? _____

_____

8. Palpate this joint on your own upper limb or your lab partner's upper limb. To do this, begin by palpating your clavicle. Then "walk" your fingers laterally until you reach the tip of your shoulder, where you will feel the acromion of the scapula. If you raise and lower your upper limb while keeping your fingers on the bones, you will be able to feel the joint between the clavicle and scapula.

The Upper Limb

The upper limb consists of the humerus, radius, ulna, carpals, metacarpals, and phalanges. Nearly all of the projections on these bones serve as attachment points for the muscles that move the upper limb. **Table 9.2** lists the bones of the upper limb and describes their key features.

Table 9.2	The Appendicular Skeleton: Upper Limb		
Bone	**Bony Landmark**	**Description**	**Word Origins**
Humerus humerus, *shoulder*	Anatomical neck	The narrow part between the head and the tubercles; location of the former epiphyseal (growth) plate.	
	Capitulum	The rounded surface (condyle) that articulates with the radius.	*caput*, head
	Coronoid fossa	A depression for the coronoid process of the ulna.	*corona*, crown, + *eidos*, resemblance, + *fossa*, trench
	Deltoid tuberosity	A rough projection on the proximal diaphysis.	*delta*, triangle, + *eidos*, resemblance
	Greater tubercle	A large lateral projection on the proximal epiphysis.	*tuber*, a knob
	Head	A rounded projection on the medial side of the proximal epiphysis; a facet that forms half of the glenohumeral joint.	*head*, the rounded extremity of a bone
	Intertubercular sulcus (groove)	A groove between the greater and lesser tubercles; sometimes referred to as the bicipital groove.	*inter-*, between, + *tuber*, knob
	Lateral epicondyle	A rough ridge proximal to the capitulum.	*epi-*, above, + *kondylos*, knuckle
	Lesser tubercle	A small, rough projection on the anterior side of the proximal epiphysis.	*tuber*, a knob
	Medial epicondyle	A rough ridge proximal to the trochlea.	*epi-*, above, + *kondylos*, knuckle
	Olecranon fossa	A depression for the olecranon process of the ulna.	*olecranon*, the head of the elbow, + *fossa*, trench
	Radial fossa	A depression for the head of the radius.	*radius*, spoke of a wheel, + *fossa*, trench
	Supracondylar ridges	Sloped ridges (medial and lateral) proximal to the epicondyles.	*supra-*, on the upper side, + *kondylos*, knuckle
	Surgical neck	The location where the head meets the shaft; the most common site of fracture.	
	Trochlea	The rounded surface (condyle) that articulates with the ulna.	*trochileia*, a pulley

(continued on next page)

Table 9.2	The Appendicular Skeleton: Upper Limb *(continued)*		
Bone	**Bony Landmark**	**Description**	**Word Origins**
Ulna ulna, *elbow*	Coronoid process	A projection on the anterior surface of the ulna that articulates with the humerus.	*corona*, crown, + *eidos*, resemblance
	Olecranon	A large projection on the proximal ulna, which forms the point of the "elbow" and serves as an attachment point for the triceps brachii muscle.	*olecranon*, the head of the elbow
	Radial notch	A depression on the lateral, proximal surface of the ulna that articulates with the head of the radius.	*radius*, spoke of a wheel
	Styloid process	A pointed process on the distal ulna that forms the medial aspect of the wrist.	*stylos*, pillar, + *eidos*, resemblance
	Trochlear notch	A ridge on the middle of the anterior surface of the ulna that separates the two depressions that articulate with the condyles of the humerus.	*trochileia*, a pulley
	Tuberosity of ulna	A process on the anterior surface of the proximal ulna that serves as an attachment point for the brachialis muscle.	*ulna*, elbow
Radius radius, *spoke of a wheel*	Head	The disc-shaped proximal end of the radius.	
	Neck	Location where the head of the radius meets the shaft of the bone.	
	Radial tuberosity	A large projection on the medial surface distal to the proximal epiphysis of the bone that serves as an attachment point for the biceps brachii muscle.	*radio-*, ray
	Styloid process of the radius	A small pointed projection on the distal radius.	*stylos*, pillar, + *eidos*, resemblance
	Ulnar notch	Location where the medial surface of the radius articulates with the distal end of the ulna.	
Carpals carpus, *wrist* *Proximal row*	Scaphoid	A large "boat-shaped" bone that articulates with the radius.	*skaphe*, boat, + *eidos*, resemblance
	Lunate	A "moon-shaped" bone that articulates with the radius.	*luna*, moon
	Triquetrum	A triangular bone on the medial, proximal aspect of the wrist; articulates with the ulna.	*triquetrus*, three-cornered
	Pisiform	A "pea-shaped" bone on the medial, palmar surface of the wrist.	*pisum*, pea, + *forma*, appearance
Distal row	Trapezium	A "table-shaped" bone that lies at the base of the first metacarpal (base of the thumb).	*trapezion*, a table
	Trapezoid	A "table-shaped" bone that lies at the base of the second metacarpal.	*trapezion*, a table, + *eidos*, resemblance
	Capitate	A "head-shaped" bone that lies in the center of the wrist, at the base of the third metacarpal.	*caput*, head
	Hamate	A "hook-shaped" bone that lies at the base of the fifth metacarpal.	*hamus*, a hook
Metacarpals meta-, *after,* + carpus, *wrist*	Base	The proximal epiphysis of the metacarpal.	
	Body	The diaphysis of the metacarpal.	
	Head	The distal epiphysis of the metacarpal.	
Phalanges	NA	Bones of the fingers and thumb.	*phalanx*, line of soldiers
II through V	Proximal	The phalanx closest to the palm of the hand.	*proximus*, nearest
	Middle	The middle phalanx.	
	Distal	The small cone-shaped distal bone of the digits.	*distalis*, away
Pollex pollex, *thumb*	Proximal	The phalanx closest to the palm of the hand.	*proximus*, nearest
	Distal	The small cone-shaped distal bone of the thumb.	*distalis*, away

EXERCISE 9.2 Bones of the Upper Limb

EXERCISE 9.2A: The Humerus

1. Obtain a **humerus** or observe the humerus on an articulated skeleton (**figure 9.3**).

2. Using table 9.2 and your textbook as a guide, identify the structures listed in figure 9.3 on the humerus, and then label them in figure 9.3.

3. How can you tell whether the bone you are holding is a right humerus or a left humerus?

_____

_____

_____

4. How can you determine which is the proximal end of the humerus?

_____

_____

_____

5. How can you distinguish the anterior surface of the humerus from the posterior surface?

_____

_____

_____

6. Obtain a scapula and a humerus and articulate them with each other. In the box below, sketch the relationship between the scapula and the humerus.

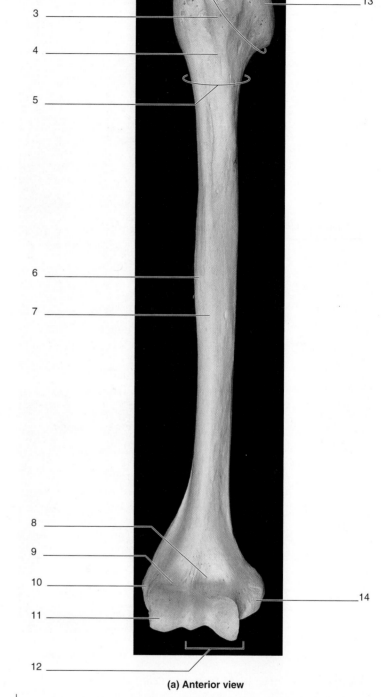

(a) Anterior view

Figure 9.3   **The Right Humerus.**

☐ anatomical neck
☐ capitulum
☐ coronoid fossa
☐ deltoid tuberosity
☐ greater tubercle
☐ head
☐ humerus

☐ intertubercular sulcus
☐ lateral epicondyle
☐ lesser tubercle
☐ medial epicondyle
☐ olecranon fossa
☐ radial fossa

☐ radius
☐ shaft
☐ supracondylar ridges
☐ surgical neck
☐ trochlea
☐ ulna

WHAT DO YOU THINK?

① Which of the two necks of the humerus do you think is more likely to fracture in an accident? What is the name of the joint formed between the scapula and the humerus?

(continued on next page)

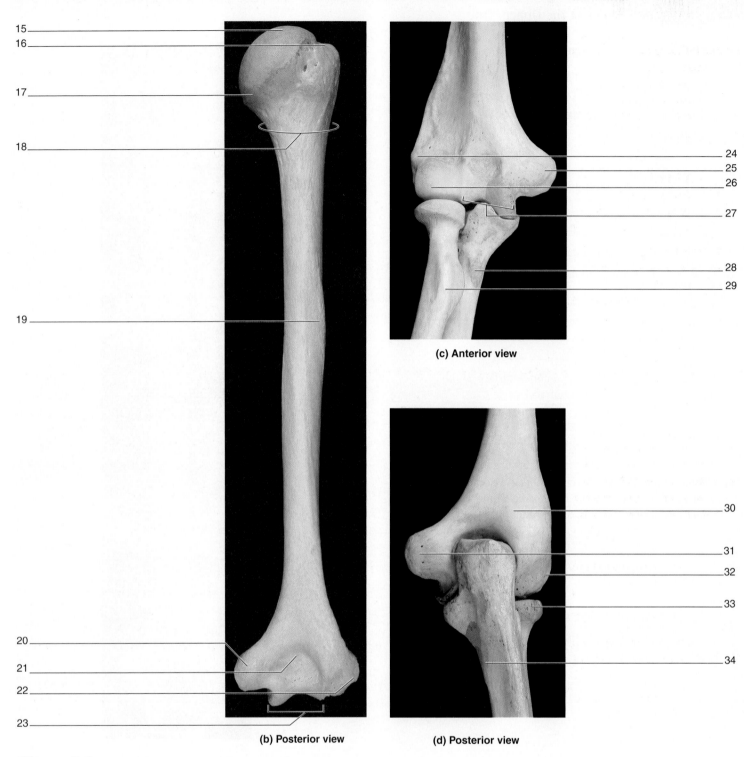

15
16
17
18
19
20
21
22
23

(b) Posterior view

24
25
26
27
28
29

(c) Anterior view

30
31
32
33
34

(d) Posterior view

Figure 9.3   **The Right Humerus.** *Continued.*

EXERCISE 9.2B: The Radius

1. Obtain a **radius** or observe the radius on an articulated skeleton (**figure 9.4**).

2. Using table 9.2 and your textbook as a guide, identify the structures listed in figure 9.4 on the radius, and then label them in figure 9.4.

3. How can you tell whether the bone you are holding is a right radius or a left radius?

_____

_____

_____

4. How can you determine which end of the radius is the proximal end?

_____

_____

_____

5. How can you distinguish the anterior surface of the radius from the posterior surface?

_____

_____

_____

EXERCISE 9.2C: The Ulna

1. Obtain an **ulna** or observe the ulna on an articulated skeleton (**figure 9.5**).

2. Using table 9.2 and your textbook as a guide, identify the structures listed in figure 9.5 on the ulna, and then label them in figure 9.5.

3. How can you tell whether the bone you are holding is a right ulna or a left ulna?

_____

_____

_____

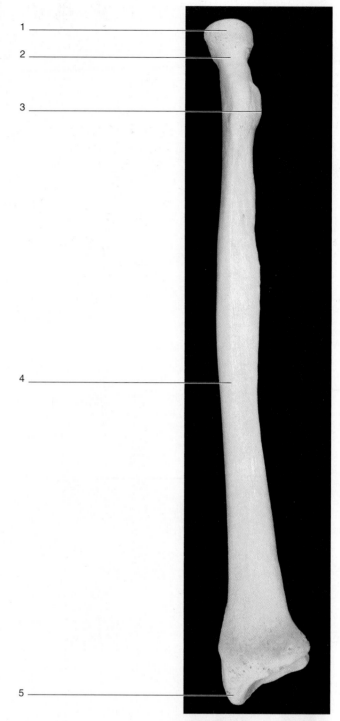

(a) Right radius, anterior view

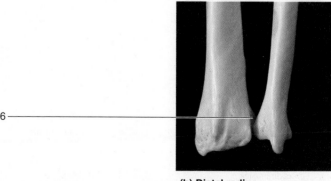

(b) Distal radius

Figure 9.4   **The Radius.**

☐ head

☐ neck

☐ radial tuberosity

☐ shaft

☐ styloid process of radius

☐ ulnar notch

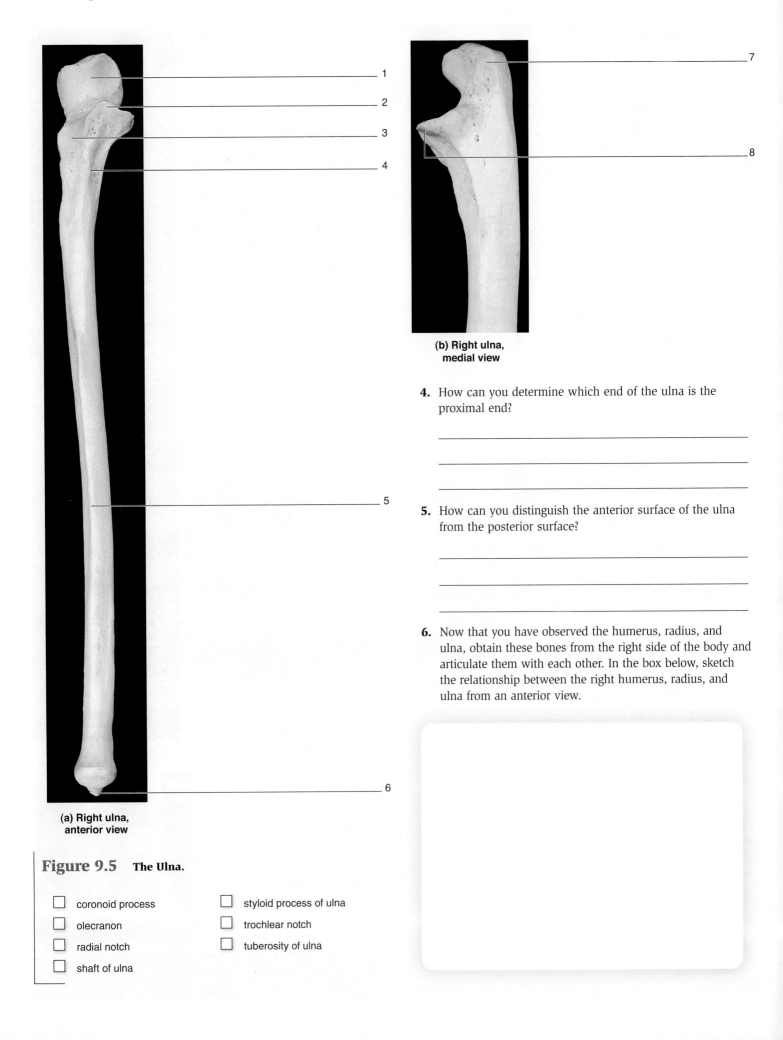

(b) **Right ulna,
medial view**

(a) **Right ulna,
anterior view**

Figure 9.5   **The Ulna.**

☐ coronoid process   ☐ styloid process of ulna
☐ olecranon   ☐ trochlear notch
☐ radial notch   ☐ tuberosity of ulna
☐ shaft of ulna

4. How can you determine which end of the ulna is the proximal end?

_____

_____

_____

5. How can you distinguish the anterior surface of the ulna from the posterior surface?

_____

_____

_____

6. Now that you have observed the humerus, radius, and ulna, obtain these bones from the right side of the body and articulate them with each other. In the box below, sketch the relationship between the right humerus, radius, and ulna from an anterior view.

EXERCISE 9.2D: The Carpals

1. Obtain articulated bones of the wrist (carpals), or observe the **carpal** bones on an articulated skeleton (**figure 9.6**). The term *carpus* means "wrist." Thus, the carpal bones are the bones of the wrist.

2. The eight carpal bones are arranged in two rows of four bones each. The proximal row, which is adjacent to the radius and ulna, contains the scaphoid, lunate, triquetrum, and pisiform. The bones of the distal row—the trapezium, trapezoid, capitate, and hamate—articulate with the metacarpals. Use the spaces below to draw the carpal bones of the proximal and distal rows as they would be positioned in an anterior view. Then label the bones in each row.

Proximal row

Distal row

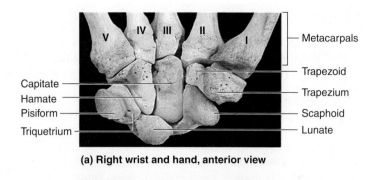

(a) **Right wrist and hand, anterior view**

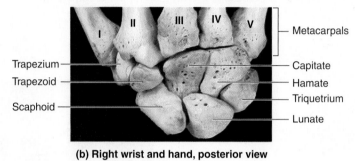

(b) **Right wrist and hand, posterior view**

Figure 9.6   **The Carpals.**

3. Using word origins to associate the shape of each carpal bone with its name will help you identify the bones individually. Refering to table 9.2 as a guide, complete the following chart to help you remember the carpal bones.

Carpal Bone	Word Origin	Bone Shape/ Appearance
Scaphoid		
Lunate		
Triquetrum		
Pisiform		
Trapezium		
Trapezoid		
Capitate		
Hamate		

4. *Optional Activity:* **AP|R Skeletal System**—Visit the Quiz area for focused drill and practice on the bones of the appendicular skeleton.

WHAT DO YOU THINK?

2 Observe the relationship between the carpal bones and the distal radius and ulna on an articulated skeleton. Which of the carpal bones do you think is most likely to fracture when someone falls on an outstretched hand?

EXERCISE 9.2E: The Metacarpals and Phalanges

1. Obtain articulated bones of the hand (metacarpals and phalanges), or observe bones of the hand on an articulated skeleton (**figure 9.7**).

2. **Metacarpals:** The prefix *meta-* means "after." Thus, the metacarpal bones are the bones that come after the carpus, or wrist, and they are located in the palm of the hand. There are five metacarpal bones, numbered I through V. The first metacarpal is on the lateral surface of the hand and forms the base of the thumb, or *pollex*. Palpate the palm of your hand to feel the metacarpal bones.

3. **Phalanges:** The term *phalanx* means "a line of soldiers." The next time you are typing on a keyboard writing a term paper or some other assignment, think about how your little "soldiers" are marching along doing great work for you.

4. The **pollex** is the thumb. How do the phalanges of the pollex (metacarpal I) differ from the phalanges of digits II–V?

   _____

   _____

   _____

5. Using table 9.2 and your textbook as a guide, identify the metacarpals and phalanges on an articulated skeleton. Then label them in figure 9.7.

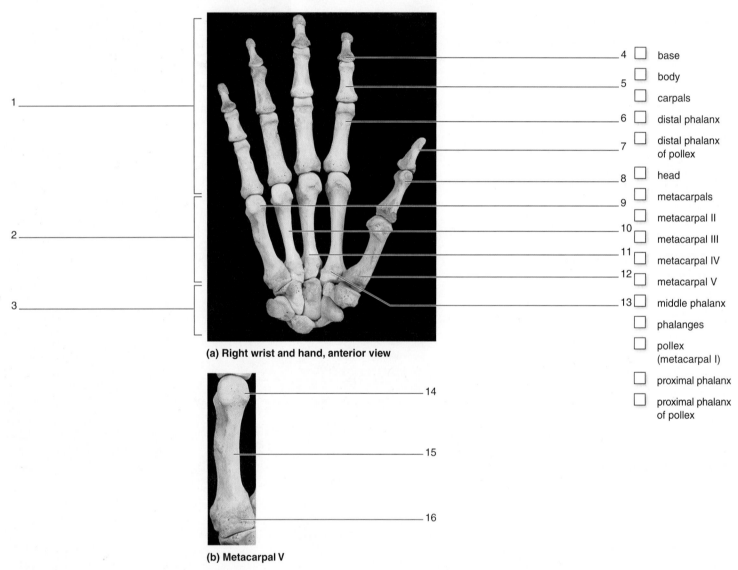

4 ☐ base
5 ☐ body
  ☐ carpals
6 ☐ distal phalanx
7 ☐ distal phalanx of pollex
8 ☐ head
9 ☐ metacarpals
10 ☐ metacarpal II
11 ☐ metacarpal III
12 ☐ metacarpal IV
  ☐ metacarpal V
13 ☐ middle phalanx
  ☐ phalanges
  ☐ pollex (metacarpal I)
  ☐ proximal phalanx
  ☐ proximal phalanx of pollex

(a) Right wrist and hand, anterior view

14
15
16

(b) Metacarpal V

Figure 9.7   **The Metacarpals and Phalanges.**

Study Tip!

Here is a mnemonic that may help you remember the names of the carpal bones, and the order in which they are found: **S**o **L**ong **T**op **P**art, **H**ere **C**omes **T**he **T**humb

So Long Top Part = Scaphoid, Lunate, Triquetrum, Pisiform
Here Comes The Thumb = Hamate, Capitate, Trapezium, Trapezoid

Proximal row first, moving from lateral to medial (anatomic position).
Distal row second, moving from lateral to medial (anatomic position).

EXERCISE 9.3 Surface Anatomy Review—Pectoral Girdle and Upper Limb

1. Palpate the manubrium and suprasternal (jugular) notch on yourself. If you move your fingers just lateral from the sternal notch, you will palpate the joint between the manubrium and the proximal end of the clavicle: the **sternoclavicular joint**. Recall that the only bony attachment between the pectoral girdle and the axial skeleton is at the sternoclavicular joint.

2. Palpate along the **clavicle** and make note of the curvatures of the clavicle as you move your fingers from medial to lateral. At the tip of the shoulder, you will feel the joint between the lateral aspect of the clavicle and the **acromial process** of the scapula: the **acromioclavicular joint**.

3. Continue to palpate along the acromial process as it curves posteriorly and becomes the **spine of the scapula**.

4. Palpate the inferior, lateral border of the deltoid muscle. Where the deltoid attaches to the humerus, at the **deltoid tuberosity**, you can often feel part of the diaphysis of the humerus because there is very little muscle between the bone and the skin at that point.

5. Moving down to your elbow, palpate the large **olecranon** of the ulna. This is the bony process that rests on a table when you lean on your elbows.

6. Just proximal from the olecranon on the medial aspect of the elbow, palpate the **medial epicondyle of the humerus**. If you place your thumb in the hollow between the olecranon of the ulna and the medial epicondyle of the humerus, you may be able to feel the cable-like **ulnar nerve**. This nerve is what causes the pain or tingly sensations that you feel when you hit your "funny bone."

7. Palpate the olecranon once again. Continue to palpate along the ulna distally until you come to the wrist joint. The bump you feel on the medial aspect of your wrist is the **styloid process of the ulna**. If you palpate the corresponding location on the lateral aspect of the wrist, you will feel the **styloid process of the radius**.

8. Finally, palpate the small metacarpal and phalangeal bones of the hand (see figure 9.7). As you do this, review the names of the bones.

9. Using your textbook as a guide, label the surface anatomy structures in **figure 9.8**.

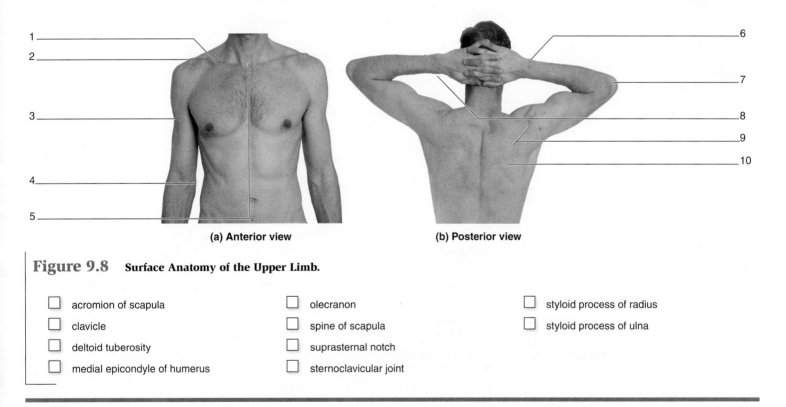

(a) Anterior view   (b) Posterior view

Figure 9.8 **Surface Anatomy of the Upper Limb.**

☐ acromion of scapula   ☐ olecranon   ☐ styloid process of radius

☐ clavicle   ☐ spine of scapula   ☐ styloid process of ulna

☐ deltoid tuberosity   ☐ suprasternal notch

☐ medial epicondyle of humerus ☐ sternoclavicular joint

The Pelvic Girdle

The pelvic girdle consists of the paired **ilium, ischium,** and **pubis** bones. Together, the three bones compose the *os coxae* (*os*, bone, + *coxa*, hip). **Table 9.3** lists the bones composing the os coxae and describes their key features. A complete pelvis is formed when right and left os coxae come together and articulate with the sacrum. Unlike the bones of the pectoral girdle, the bones of the pelvic girdle fuse together during development to become one solid structure.

In the following exercises you will first consider the features of the os coxae as a whole. You will then consider the features of the individual bones that compose the os coxae. Finally, you will observe the structural differences between the male and female pelvic girdles to discover the functional differences between them.

WHAT DO YOU THINK?

③ Why do you think it is functionally important that the bones of the os coxae fuse together rather than remain independent bones?

Table 9.3	The Appendicular Skeleton: Pelvic Girdle		
Bone	**Bony Landmark**	**Description**	**Word Origins**
Os Coxae os, *bone,* + coxa, *hip*	Acetabulum	A bony socket that forms the articulation with the head of the femur.	*acetabula*, a shallow cup
	Linea terminalis (pelvic brim)	An oblique ridge on the inner surface of the ilium and pubic bones that separates the true pelvis (below) from the false pelvis (above); consists of the pubic crest, pectineal line, and arcuate line.	*linea*, line, + *terminalis*, ending
	Lunate surface	The half-moon shaped (curved) smooth surface on the superior border of the acetabulum, which articulates with the head of the femur.	*luna*, moon
	Obturator foramen	A large, oval hole in the inferior part of the os coxae located anterior and medial.	*obturo*, to occlude
Ilium ilium, *flank*	Anterior gluteal line	A rough line running obliquely on the lateral surface of the ilium from the iliac crest to the greater sciatic notch.	*gloutos*, buttock
	Anterior inferior iliac spine	A process inferior to the anterior superior iliac spine; origin of rectus femoris muscle.	*spina*, a spine
	Anterior superior iliac spine	A projection at the anteriormost part of the iliac crest; lateral attachment point for the inguinal ligament; origin for sartorius muscle.	*spina*, a spine
	Arcuate line	An oblique line between the ilium and ischium that composes the iliac part of the linea terminalis (pelvic brim) of the bony pelvis.	*arcuatus*, bowed
	Auricular surface	An "earlike" rough surface on the medial aspect of the ilium; the point of articulation with the sacrum.	*auris*, ear
	Greater sciatic notch	A deep notch on the posterior surface of the ilium inferior to the posterior iliac spines.	*sciaticus*, the hip joint
	Iliac crest	The superior border of the ilium, beginning at the sacrum and ending on the lateral aspect of the hip.	*crista*, a ridge
	Iliac fossa	A large fossa on the anteromedial (internal) surface of the ilium inferior to the iliac crest.	*ilium*, flank, + *fossa*, a trench
	Iliac tuberosity	A large projection on the posterior, superior aspect of the ilium.	*ilium*, flank, + *tuber*, a knob
	Inferior gluteal line	A rough line running transversely on the lateral surface of the ilium just superior to the acetabulum.	*gloutos*, buttock
	Posterior gluteal line	A rough line running vertically on the lateral surface of the ilium from the iliac crest to the posterior rim of the greater sciatic notch.	*gloutos*, buttock
	Posterior inferior iliac spine	A small projection on the posterior inferior point of the ilium.	*spina*, a spine
	Posterior superior iliac spine	A projection on the posterior superior point of the ilium.	*spina*, a spine

Table 9.3	The Appendicular Skeleton: Pelvic Girdle *(continued)*		
Bone	**Bony Landmark**	**Description**	**Word Origins**
Ischium ischion, *hip*	Ischial spine	A small, sharp spine on the posterior aspect of the ischium.	*ischion*, hip
	Lesser sciatic notch	A notch located immediately inferior to the ischial spine on the posterior surface of the ilium.	
	Ramus of ischium	The inferior part of the ischium that connects to the pubis anteriorly and forms the inferior part of the obturator foramen.	*ramus*, branch
	Ischial tuberosity	A large, rough projection on the posterior, inferior surface of the ischium; attachment point for hamstring muscles.	*ischion*, hip, + *tuber*, a knob
Pubis pubis, *pubic bone*	Inferior pubic ramus	The inferior part of the pubis that joins with the ischium.	*ramus*, branch
	Pectineal line	Rough ridge on the medial surface of the superior ramus of the pubis.	*pectineal*, relating to the pubis
	Pubic crest	A ridge on the lateral part of the superior ramus of the pubis.	*crista*, crest
	Pubic tubercle	A projection composing the anterior most point of the bone; medial attachment point for the inguinal ligament.	*tuber*, a knob
	Superior pubic ramus	The superior part of the pubis that joins with the ilium.	*ramus*, branch

EXERCISE 9.4   Bones of the Pelvic Girdle

EXERCISE 9.4A: The Os Coxae

1. Obtain an **os coxae** or observe the os coxae on an articulated skeleton (**figure 9.9**).

2. Using table 9.3 and your textbook as a guide, identify the structures listed in figure 9.9 on the os coxae. Then label them in figure 9.9.

3. How can you tell whether the bone you are holding is a right os coxae or a left os coxae?

_____

_____

_____

4. How can you determine which end of the os coxae is the superior end?

_____

_____

_____

5. How can you distinguish the anterior surface of the os coxae from the posterior surface?

_____

_____

_____

6. With which part of the vertebral column does the os coxae articulate?

_____

EXERCISE 9.4B: Male and Female Pelves

1. Obtain a male pelvis and a female pelvis and lay them next to each other on your workspace with the anterior surfaces facing toward you. **Figure 9.10** demonstrates the features of male and female pelves.

2. There are numerous features that help distinguish a **male pelvis** from a **female pelvis**. For example, a female pelvis is generally wider and more flared, has a broader subpubic angle, a smaller, triangular obturator foramen, a wide, shallow greater sciatic notch, and ischial spines that rarely project into the pelvic outlet. All of these are adaptations that allow for childbirth. However, no method of sexing a pelvis is completely foolproof. At the very least, you can make a determination that a pelvis is more male-like than female-like, or vice versa. As you observe male and female pelves, develop your own method of differentiating the male pelvis from the female pelvis. If you have difficulty, consult your main textbook for assistance.

Study Tip!

One method of estimating female vs. male subpubic angles is to compare them to the angles formed between the digits on your hand when you spread them apart. For example, the angle between your thumb and index finger when you spread them apart approximates the wide subpubic angle of a female pelvis, whereas the angle formed between your index and middle fingers when you spread them apart approximates the narrower subpubic angle of a male pelvis.

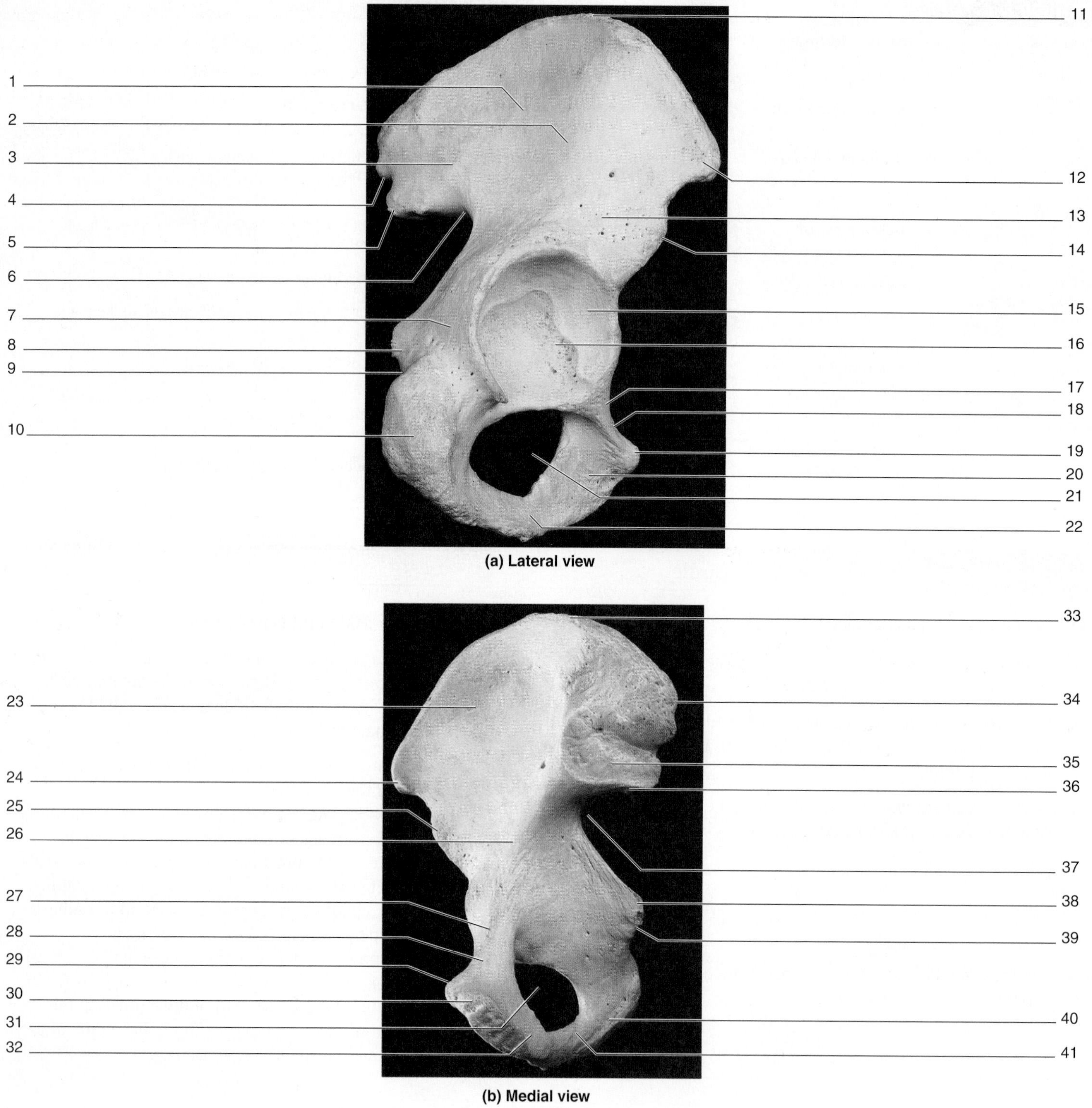

(a) Lateral view

(b) Medial view

Figure 9.9 **The Right Os Coxae.**

- [ ] acetabulum
- [ ] ala
- [ ] anterior gluteal line
- [ ] anterior inferior iliac spine
- [ ] anterior superior iliac spine
- [ ] arcuate line
- [ ] auricular surface
- [ ] greater sciatic notch

- [ ] iliac crest
- [ ] iliac fossa
- [ ] iliac tuberosity
- [ ] inferior gluteal line
- [ ] inferior pubic ramus
- [ ] ischial body
- [ ] ischial ramus
- [ ] ischial spine

- [ ] ischial tuberosity
- [ ] lesser sciatic notch
- [ ] lunate surface
- [ ] obturator foramen
- [ ] pectineal line
- [ ] posterior gluteal line
- [ ] posterior inferior iliac spine
- [ ] posterior superior iliac spine

- [ ] pubic crest
- [ ] pubic tubercle
- [ ] superior pubic ramus
- [ ] symphysial surface of pubic bone

3. In the spaces below, sketch the male and female pelves. Then list the features you used to distinguish the male pelvis from the female pelvis. Use the photos in figure 9.10 as a guide if you do not have samples of both a male and a female pelvis in your laboratory.

Male Pelvis

Female Pelvis

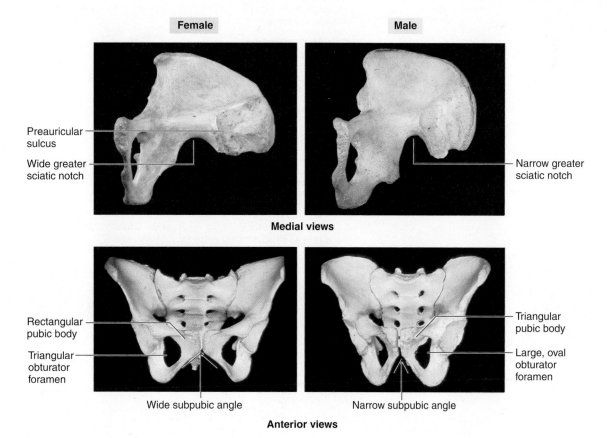

Figure 9.10   **Male and Female Pelves.**

The Lower Limb

The lower limb consists of the femur, patella, tibia, fibula, tarsals, metatarsals, and phalanges. Nearly all of the projections on these bones are attachment points for the muscles that move the limb. **Table 9.4** lists the bones of the lower limb and describes their key features.

Table 9.4	The Appendicular Skeleton: Lower Limb		
Bone	**Bony Landmark**	**Description**	**Word Origins**
Femur femur, *thigh*	Adductor tubercle	A small projection above the medial condyle.	*tuber*, a knob
	Fovea	A circular depression on the proximal end of the head.	*fovea*, a dimple
	Gluteal tuberosity	A projection on the proximal aspect of the linea aspera; gluteus maximus muscle attaches to it.	*gloutos*, buttock, + *tuber*, *a knob*
	Greater trochanter	A very large projection on the lateral surface of the proximal epiphysis.	*trochanter*, a runner
	Head	A very large, ball-shaped structure on the proximal epiphysis.	NA
	Intercondylar fossa	A depression on the distal end of the femur between the two condyles; serves as an attachment point for the cruciate ligaments of the knee.	*inter-*, between, + *kondylos*, knuckle
	Intertrochanteric crest	A large ridge that runs between the greater and lesser trochanters on the posterior surface.	*inter-*, between, + *trochanter*, a runner
	Intertrochanteric line	A shallow ridge that runs between greater and lesser trochanters on the anterior surface.	*inter-*, between, + *trochanter*, a runner
	Lateral condyle	A large rounded surface that articulates with the lateral condyle of the tibia.	*kondylos*, knuckle
	Lateral epicondyle	A rough surface superior to the lateral condyle.	*epi*, above, + *kondylos*, knuckle
	Lesser trochanter	A large projection on the medial surface of the proximal epiphysis.	*trochanter*, a runner
	Linea aspera	A "rough line" that runs along the posterior surface of the diaphysis.	*linea*, line, + *aspera*, rough
	Medial condyle	A rounded surface that articulates with the medial condyle of the tibia.	*kondylos*, knuckle
	Medial epicondyle	A rough surface superior to the medial condyle.	*epi*, above, + *kondylos*, knuckle
	Neck	The narrow portion where the head meets the shaft of the bone; this is the part of the bone that is fractured in a "broken hip".	
	Patellar surface	A smooth depression on the anterior surface of the distal epiphysis; location where the patella articulates with the femur.	(patella) *patina*, a shallow disk
	Pectineal line	A line on the posterior, superior aspect of the femur that serves as an attachment point for the pectineus muscle.	*pictineal*, ridged or comblike
	Popliteal surface	A triangular region on the posterior aspect of the distal femur.	*popliteal*, the back of the knee
	Shaft	The diaphysis of the bone.	NA

Table 9.4	The Appendicular Skeleton: Lower Limb *(continued)*		
Bone	**Bony Landmark**	**Description**	**Word Origins**
Tibia tibia, *the large shin bone*	Anterior border	A ridge on the anterior surface extending distally from the tibial tuberosity; commonly referred to as the "shin".	
	Intercondylar eminence	A prominent projection between the two condyles on the proximal epiphysis.	*eminentia*, a raised area on a bone
	Lateral condyle	A large, flat surface on the lateral aspect of the proximal epiphysis; the point of articulation with the lateral condyle of the femur.	*kondylos*, knuckle
	Medial condyle	A large, flat surface on the medial aspect of the proximal epiphysis; point of articulation with the medial condyle of the femur.	*kondylos*, knuckle
	Medial malleolus	A projection on the medial surface of the distal epiphysis.	*malleus*, hammer
	Shaft	The diaphysis of the bone.	
	Soleal line	A rough line running obliquely on the posterior surface of the proximal part of the bone; attachment for part of the soleus muscle.	*solea*, a sandal
	Tibial tuberosity	A projection on the anterior surface of the proximal epiphysis; attachment point for the patellar ligament (quadriceps femoris muscle attachment).	*tibia*, shin bone
Fibula fibula, *a clasp or buckle*	Head	The rounded proximal end of the bone.	
	Lateral malleolus	A projection on the lateral surface of the distal epiphysis.	*malleus*, hammer
	Neck	The narrow portion where the head meets the diaphysis.	
	Shaft	The diaphysis of the bone.	
Tarsals tarsus, *a flat surface*	Calcaneus	Bone that forms the heel of the foot; attachment point for the calcaneal tendon.	*calcaneus*, the heel
	Cuboid	A cube-shaped bone located at the base of the fourth metatarsal.	*kybos*, cube, + *eidos*, resemblance
	Intermediate cuneiform	A "wedge-shaped" bone located at the base of the second metatarsal.	*cuneus*, wedge, + *forma*, shape
	Lateral cuneiform	A "wedge-shaped" bone located at the base of the third metatarsal.	*cuneus*, wedge, + *forma*, shape
	Medial cuneiform	A "wedge-shaped" bone located at the base of the first metatarsal.	*cuneus*, wedge, + *forma*, shape
	Navicular	A bone shaped like a boat ("ship") located just anterior to the talus.	*navis*, ship
	Talus	The major weight-bearing bone of the ankle; articulates with the tibia and fibula.	*talus*, ankle
Metatarsals meta-, *after*, + tarsus, *a flat surface*	Base	The proximal epiphysis of the bone.	
	Head	The distal epiphysis of the bone.	
	Shaft	The diaphysis of the bone.	
Phalanges phalanx, *line of soldiers*			
II through V	Proximal	The phalanx closest to the metatarsal bones.	*proximus*, nearest
	Middle	The middle phalanx.	
	Distal	The small cone-shaped distal bone of the digits.	*distalis*, away
Hallux hallux, *the big toe*	Proximal	The phalanx closest to the sole of the foot.	*proximus*, nearest
	Distal	The distal phalanx.	*distalis*, away

Bones of the Lower Limb

EXERCISE 9.5A: The Femur

1. Obtain a **femur** or observe the femur on an articulated skeleton (**figure 9.11**).

2. Using table 9.4 and your textbook as a guide, identify the structures listed in figure 9.11 on the femur. Then label them in figure 9.11.

3. How can you tell whether the bone you are holding is a right femur or a left femur?

   _____

   _____

   _____

4. How can you determine which end of the femur is the proximal end?

   _____

   _____

   _____

5. How can you distinguish the anterior surface of the femur from the posterior surface?

   _____

   _____

   _____

6. Obtain a right os coxae and a right femur and articulate them with each other. In the box below, sketch the relationship between the bones of the os coxae and the femur.

   [ ]

7. What is the name of the joint formed between the bones of the os coxae and the femur?

   _____

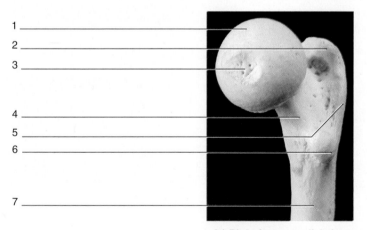

1 _____
2 _____
3 _____
4 _____
5 _____
6 _____
7 _____

(a) Right femur, medial view

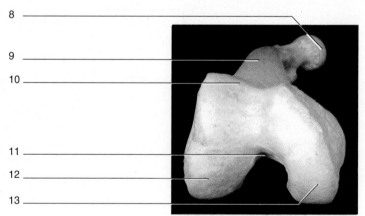

8 _____
9 _____
10 _____
11 _____
12 _____
13 _____

(b) Right femur, inferior view

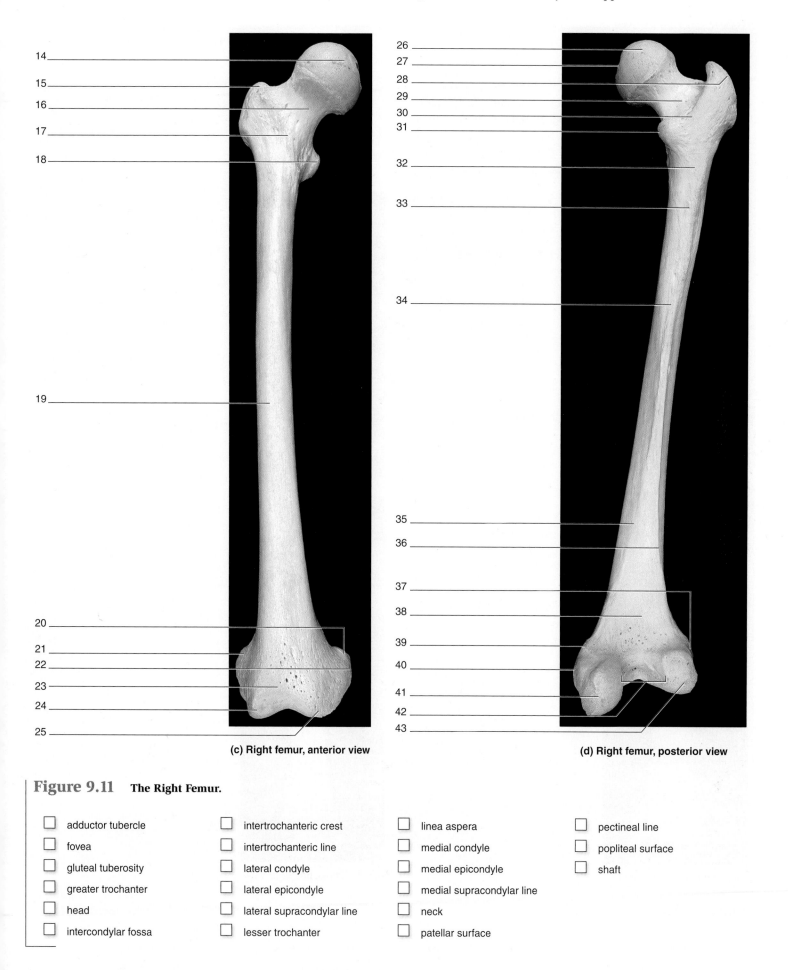

14 _____
15 _____
16 _____
17 _____
18 _____
19 _____
20 _____
21 _____
22 _____
23 _____
24 _____
25 _____

(c) Right femur, anterior view

26 _____
27 _____
28 _____
29 _____
30 _____
31 _____
32 _____
33 _____
34 _____
35 _____
36 _____
37 _____
38 _____
39 _____
40 _____
41 _____
42 _____
43 _____

(d) Right femur, posterior view

Figure 9.11 **The Right Femur.**

- ☐ adductor tubercle
- ☐ fovea
- ☐ gluteal tuberosity
- ☐ greater trochanter
- ☐ head
- ☐ intercondylar fossa
- ☐ intertrochanteric crest
- ☐ intertrochanteric line
- ☐ lateral condyle
- ☐ lateral epicondyle
- ☐ lateral supracondylar line
- ☐ lesser trochanter
- ☐ linea aspera
- ☐ medial condyle
- ☐ medial epicondyle
- ☐ medial supracondylar line
- ☐ neck
- ☐ patellar surface
- ☐ pectineal line
- ☐ popliteal surface
- ☐ shaft

EXERCISE 9.5B: The Tibia

1. Obtain a **tibia** or observe the tibia on an articulated skeleton. (**figure 9.12**).

2. Using table 9.4 and your textbook as a guide, identify the structures listed in figure 9.12 on the tibia. Then label them in figure 9.12.

3. How can you tell whether the bone you are holding is a right tibia or a left tibia?

_____

_____

_____

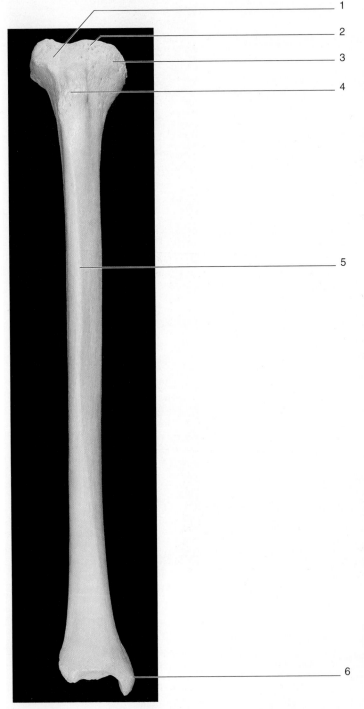

(a) Right tibia, anterior view

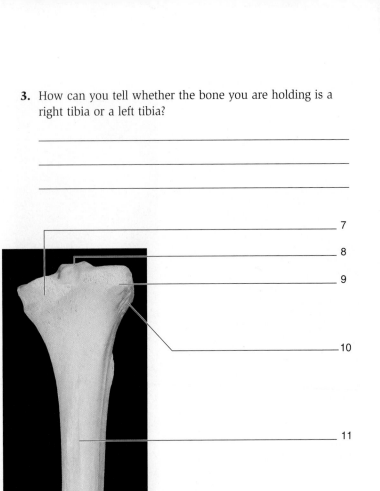

(b) Right tibia, posterior view

Figure 9.12   **The Tibia.**

☐ anterior border

☐ fibular articular facet

☐ intercondylar eminence

☐ lateral condyle

☐ medial condyle

☐ medial malleolus

☐ soleal line

☐ tibial tuberosity

4. How can you determine which end of the tibia is the proximal end?

_____

_____

_____

5. How can you distinguish the anterior surface of the tibia from the posterior surface?

_____

_____

_____

EXERCISE 9.5C: The Fibula

1. Obtain a **fibula** or observe the fibula on an articulated skeleton (**figure 9.13**).

2. Using table 9.4 and your textbook as a guide, identify the structures listed in figure 9.13 on the fibula. Then label them in figure 9.13.

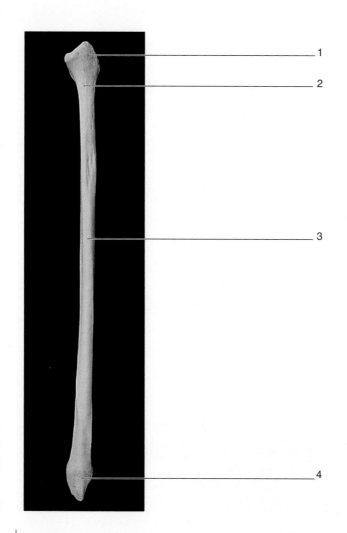

        1

        2

        3

        4

Figure 9.13   **The Right Fibula.** Lateral view.

☐ head   ☐ neck
☐ lateral malleolus   ☐ shaft

3. How can you tell whether the bone you are holding is a right fibula or a left fibula?

_____

_____

_____

4. How can you determine which end of the fibula is the proximal end?

_____

_____

_____

5. How can you distinguish the anterior surface of the fibula from the posterior surface?

_____

_____

_____

6. Now that you have observed the femur, tibia, and fibula, obtain these bones (all from the same side of the body) and articulate them with each other. In the box below, sketch the relationship between the right femur, tibia, and fibula from an anterior view. Include in your sketch the location of the patella.

7. Does the fibula form part of the knee joint?

_____

8. Does the fibula form part of the ankle joint?

_____

WHAT DO YOU THINK?

4 A fracture to which of the leg bones (tibia or fibula) would result in the greatest loss of function of the lower limb?

EXERCISE 9.5D: The Tarsals

1. Obtain bones of the ankle (tarsals) or observe the **tarsal** bones on an articulated skeleton (**figures 9.14** and **9.15**).

2. The term *tarsal* means "flat surface," as in the sole of the foot. The tarsal bones are much larger than their counterparts in the wrist (the carpal bones). The largest differences in structure are seen in the talus and calcaneus, two bones that form a major part of the weight-bearing ankle joint.

3. Using table 9.4 and your textbook as guides, identify the tarsal bones on an articulated skeleton. Then label them in figure 9.14.

4. The seven tarsal bones are arranged in two rows. The proximal row includes the talus, calcaneus, and navicular. The bones of the distal row—the medial cuneiform, intermediate cuneiform, lateral cuneiform, and cuboid— articulate with the metatarsal bones. Use the spaces to the right to diagram the tarsal bones of the proximal and distal rows as they would be positioned in a superior view. Then label the bones in each row.

Proximal Row

Distal Row

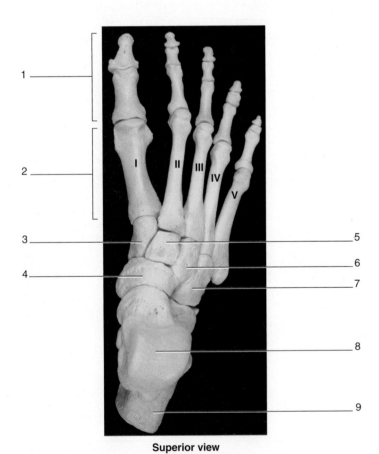

Superior view

Figure 9.14   **The Tarsals.**

- ☐ calcaneus
- ☐ cuboid
- ☐ intermediate cuneiform
- ☐ lateral cuneiform
- ☐ medial cuneiform
- ☐ metatarsals
- ☐ navicular
- ☐ phalanges
- ☐ talus

5. The names of the tarsal bones provide good clues for identifying each bone. Using table 9.4 as a guide, complete the following chart to help you remember the shapes of the tarsal bones.

Tarsal Bone	Word Origin	Bone Shape/ Appearance
Talus		
Calcaneus		
Navicular		
Medial Cuneiform		
Intermediate Cuneiform		
Lateral Cuneiform		
Cuboid		

6. Now that you have observed the tibia, fibula, calcaneus, and talus, obtain these bones (all from the same side of the body) and articulate them with each other. In the box on the right, sketch the relationship between the tibia, fibula, calcaneus, and talus.

7. Which bones or parts of bones compose the ankle joint?

_____

_____

_____

EXERCISE 9.5E: The Metatarsals and Phalanges

1. Obtain bones of the foot (metatarsals and phalanges) or observe the bones of the foot on an articulated skeleton (figures 9.14 and 9.15).

2. **Metatarsals:** The term *meta-* means "after." Thus, the metatarsal bones are the bones that come after the tarsus. These bones are found in the sole of your foot. Palpate the sole of your foot and feel for the metatarsal bones. There are five metatarsal bones, numbered I through V. The first metatarsal is on the medial surface of the foot and forms the base of the big toe, or *hallux*.

3. The **hallux** is the big toe. How do its phalangeal bones differ from the phalanges of digits II–V?

_____

4. Using table 9.4 and your textbook as a guide, identify the metatarsal bones and phalanges on an articulated skeleton. Then label them in figure 9.15.

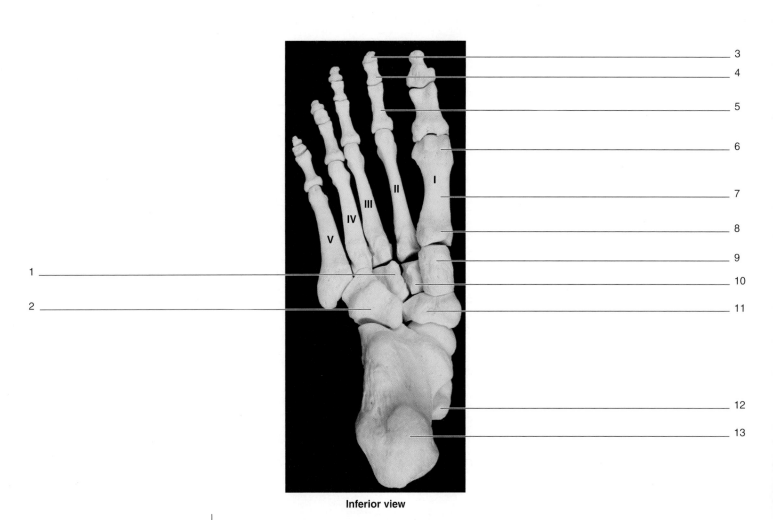

Inferior view

Figure 9.15   The Metatarsals and Phalanges.

Metatarsals	Phalanges	Tarsals
☐ hallux (metatarsal I)	☐ proximal phalanx	☐ calcaneus
☐ metatarsal II	☐ middle phalanx	☐ cuboid
☐ metatarsal III	☐ distal phalanx	☐ intermediate cuneiform
☐ metatarsal IV		☐ lateral cuneiform
☐ metatarsal V		☐ medial cuneiform
☐ base		☐ navicular
☐ shaft		☐ talus
☐ head		

EXERCISE 9.6 Surface Anatomy Review—Pelvic Girdle and Lower Limb

1. Place your hands on your hips. The bony ridge you feel under your skin is the **iliac crest**. The iliac crest is located at vertebral level L1/L2. Palpate anteriorly along the iliac crest until you come to the large bony projection on the anterior surface, the **anterior superior iliac spine**.

2. Palpate laterally along the iliac crest until your fingers are on the lateral aspect of the hip. Palpate the soft tissue of the gluteus medius muscle just inferior to the lateral portion of the iliac crest.

3. As you continue to palpate inferiorly, you will soon feel another bony projection. This is the **greater trochanter of the femur**. Depending upon the amount of fat and muscle in your gluteal region, you may also be able to palpate the **ischial tuberosities** deep within the posterior region of the buttocks.

4. Palpate the **patella** on the anterior aspect of the knee.

5. Place your thumb and pinky finger (fifth phalanx) on the medial and lateral aspects of the knee. Here you will feel the large **medial** and **lateral epicondyles of the femur**.

6. Palpate the lateral epicondyle of the femur. Now move your fingers distally until you feel the knoblike **head of the fibula**. As you continue to palpate distally, you will be unable to feel the shaft of the fibula because of the fibularis muscles that overlie it.

7. As you palpate distally along the fibula toward the ankle joint, you will eventually feel the **lateral malleolus** of the fibula in the ankle joint.

8. Palpate the patella once again. Distal to the patella on the anterior surface of the leg, you will feel the **tibial tuberosity**. Continue to palpate distally along the tibia. Notice that you can feel the entire shaft of the bone because it is only covered with skin and a little bit of fat. The subcutaneous part of the tibia here is commonly referred to as the shin.

9. As you palpate the most distal end of the tibia, you will feel the **medial malleolus** of the tibia in the ankle joint.

10. Palpate the large **calcaneus** in your heel. The talus and other tarsal bones are more difficult to palpate. However, if you wiggle your toes, you can see and palpate the **metatarsal bones** and the **phalanges**. As you palpate these bones, review their names. Recall that the first metatarsal is the metatarsal that lines up with the base of the big toe, or hallux, and the numbering continues, II through V, as you move toward the lateral aspect of the foot.

11. Using your textbook and atlas as guides, label the surface anatomy structures in **figure 9.16**.

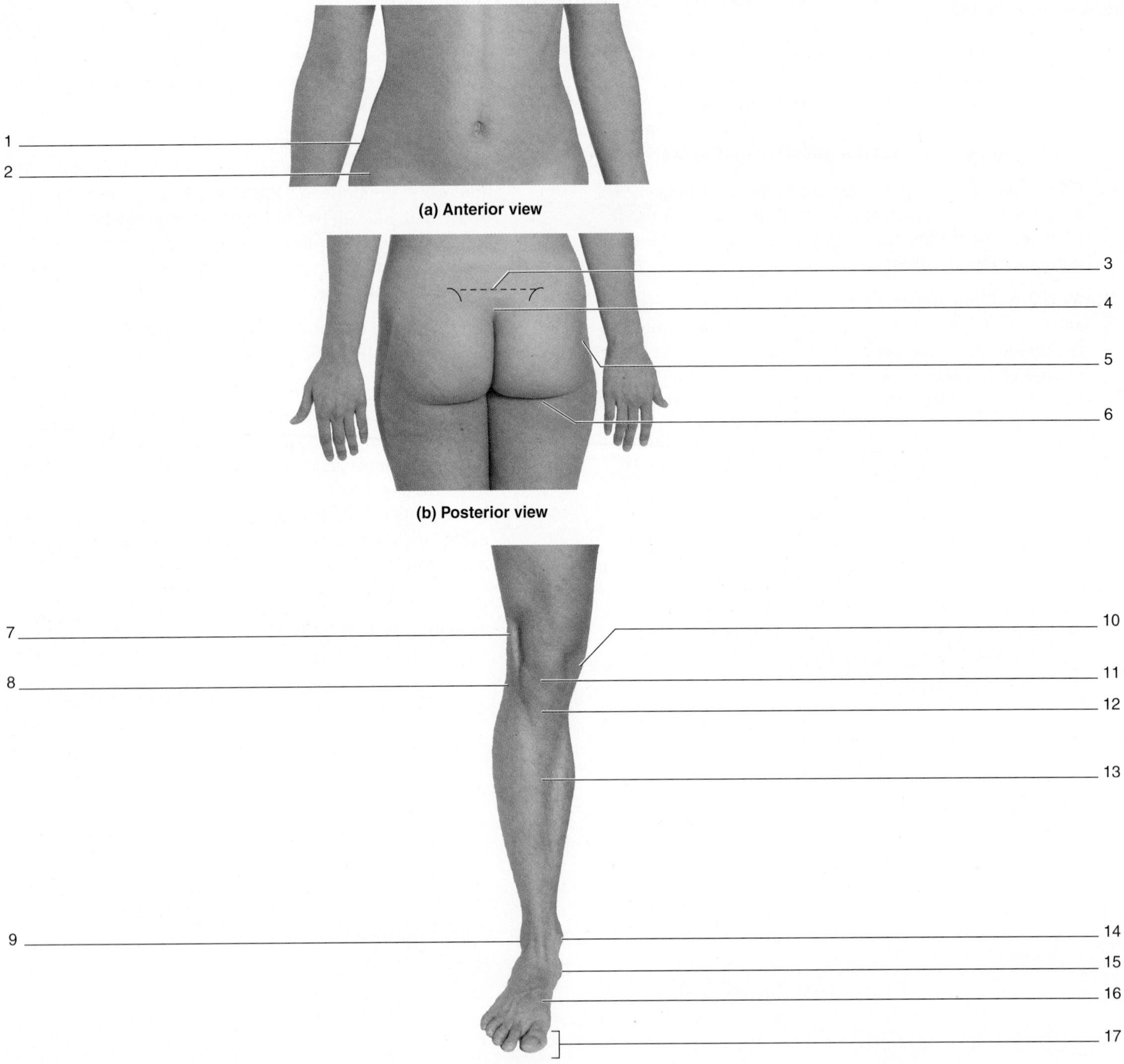

1 _____

2 _____

(a) Anterior view

3 _____

4 _____

5 _____

6 _____

(b) Posterior view

7 _____   10 _____

8 _____   11 _____

                                           12 _____

                                           13 _____

9 _____   14 _____

                                           15 _____

                                           16 _____

                                           17 _____

(c) Anterior view

Figure 9.16   Surface Anatomy of the Lower Limb.

- [ ] anterior superior iliac spine
- [ ] calcaneus
- [ ] greater trochanter of femur
- [ ] head of fibula
- [ ] iliac crest
- [ ] ischial tuberosity
- [ ] lateral malleolus
- [ ] lateral condyle of femur
- [ ] medial condyle of femur
- [ ] medial malleolus
- [ ] metatarsals
- [ ] patella
- [ ] phalanges
- [ ] sacrum
- [ ] shaft of tibia
- [ ] tibial tuberosity

Chapter 9: The Appendicular Skeleton

Name: _____
Date: _____ Section: _____

POST-LABORATORY WORKSHEET

1. Which process of the scapula can be palpated at the tip of the shoulder? _____

2. The head of the humerus articulates with the _____ of the scapula.

3. The only point of articulation between the pectoral girdle and the axial skeleton is the _____.

4. What are the structural and functional differences between the anatomical and surgical necks of the humerus? _____

_____

_____

_____

5. Describe how the humerus forms articulations with both the radius and ulna. _____

_____

_____

_____

6. Devise a mnemonic device to assist you with your recall of the carpal bones. Write your mnemonic here: _____

_____

_____

7. List three of the features that help you differentiate a male pelvis from a female pelvis.

Male Pelvis   *Female Pelvis*

a. _____   a. _____

b. _____   b. _____

c. _____   c. _____

8. Describe how the femur forms articulations with the tibia. In addition, describe how the fibula and patella relate to the tibia.

_____

_____

_____

_____

9. Devise a mnemonic device to assist you with your recall of the tarsal bones. Write your mnemonic here: _____

_____

_____

10. The anatomic name for the thumb is the _____.

11. The anatomic name for the big toe is the _____.

12. What bone or bony process of the upper limb serves the same function as the patella in the lower limb? (**Hint:** The patella's function is to act as a lever. It forces the tendons of the muscles on the anterior surface of the thigh farther away from the center of rotation of the joint in order to give the muscles a greater mechanical advantage—greater leverage.) _____

Articulations

10

OUTLINE and OBJECTIVES

Module 5: SKELETAL SYSTEM

INTRODUCTION

If someone asks whether you have ever injured a joint, you will probably think of your ankle, knee, hip, wrist, elbow, and shoulder joints. However, you may not have considered the fact that the sutures between your skull bones are also joints, as are the connections between your ribs and sternum. A joint, or **articulation** (*articulatio*, a forming of vines), is formed wherever one bone comes together with another bone. The study of joints is called **arthrology** (*arthron*, a joint, + *logos*, the study of). Joints are classified both by movement (synarthrotic, amphiarthrotic, or diarthrotic) and by structure (fibrous, cartilaginous, and synovial). Your main textbook defines these classifications and describes the joint movements (flexion, extension, etc.) that occur at diarthrotic synovial joints. Before engaging in the laboratory activities in this chapter, review these joint definitions and the descriptions of joint movements from your textbook so you will have a firm grip on the terminology, which will prepare you for the laboratory. For your reference, **table 10.1** summarizes the *functional* (movement) classifications of joints. The *structural* classifications of joints are presented in the exercises in this chapter. As you work through the structural classifications of joints, also practice using the functional terms from table 10.1 to describe the movement allowed at the joint.

Table 10.1	Functional (Movement) Classification of Joints		
Type of Joint	**Description**	**Examples**	**Word Origins**
Synarthrosis	An immobile joint.	Skull suture and tooth gomphosis	*syn*, together, + *arthron-*, a joint
Amphiarthrosis	A slightly mobile joint.	Syndesmosis, synchondrosis, and symphysis	*amphi*, on both sides, + *arthron*, a joint
Diarthrosis	A freely mobile joint.	All synovial joints	*di-*, two, + *arthron*, a joint

Chapter 10: Articulations

PRE-LABORATORY WORKSHEET

1. Define the following terms that relate to the classification of joints based on movement:

   a. *synarthrotic:* _____

   _____

   b. *amphiarthrotic:* _____

   _____

   c. *diarthrotic:* _____

   _____

2. List the three types of fibrous joints. For each type, give an example of a location in the body where that type of joint is found.

   Type of Fibrous Joint   *Location*

   a. _____   a. _____

   b. _____   b. _____

   c. _____   c. _____

3. List the two types of cartilaginous joints. For each type, give an example of a location in the body that type of joint is found.

   Type of Cartilaginous Joint   *Location*

   a. _____   a. _____

   b. _____   b. _____

4. Describe how synovial joints differ from fibrous and cartilaginous joints.

   _____

   _____

   _____

   _____

5. All synovial joints are classified as _____ when classified based on *movement*.

IN THE LABORATORY

In this laboratory session you will begin by briefly investigating the nature of fibrous and cartilaginous joints. The remainder of the laboratory session will be focused on an in-depth look at the structure and function of a representative synovial joint: the knee joint.

All of the exercises in this chapter can be performed either on cadaver specimens or on joint models, so the text is written simply to guide you through the process of finding the structures associated with each of the joints using whatever materials are available in the classroom. Under each heading you will find a brief description of the joint(s) in question and instructions regarding what you are to observe in the laboratory. Use the tables and figures provided in this manual, which contain detailed information about the joints you will be observing, to guide you in your observations.

Gross Anatomy

Fibrous Joints

Fibrous joints are characterized by having some amount of fibrous connective tissue connecting neighboring bones. Fibrous joints are classified as either synarthrotic (immobile) or amphiarthrotic (slightly mobile), based on the amount of movement allowed. Structurally, they are classified as sutures, syndesmoses, or gomphoses (**table 10.2**).

EXERCISE 10.1 Fibrous Joints

1. **Suture** – Observe a skull on an articulated skeleton or by itself. Observe the numerous **sutures** (*sutura*, a seam) between the cranial bones. Notice how tightly the bones fit together. In a preserved skeleton, the sutural joints may be somewhat loose because the membranous connective tissue that normally holds the bones together was destroyed when the skeleton was prepared. However, in a living adult the skull bones are held tightly together by membranous connective tissue that lies in the spaces between the bones, and also by the interlocking shapes of the articulating bones. Recall that the flat bones of the skull form by **intramembranous ossification**. The fibrous connective tissue within the sutures is a remnant of the original membrane that served as the structural framework for the developing bones. How would you classify a sutural joint based on the amount of *movement* allowed? _____

2. **Syndesmosis** – Observe the radius and ulna on an articulated skeleton. Even though the articulated skeleton no longer contains a **syndesmosis** (*syn-*, together, + *desmos*, a band) joint (it was destroyed during preparation of the skeleton), you can still observe the movement at the distal radioulnar joint. Take hold of the radius and ulna, and then move the bones to mimic the actions of supination and pronation

of the forearm (table 9.6). Notice how the radius pivots around the ulna during this motion. In a living human, these two bones are anchored to each other by an **interosseous membrane (figure 10.1)**, which composes the syndesmosis between these two bones. This membrane also contributes to the connective tissue that separates the forearm into anterior and posterior compartments. How would you classify a syndesmosis based on the amount of *movement* allowed?

_____

3. **Gomphosis** – Observe the maxilla and mandible on a skull or on an articulated skeleton. The joints between the teeth and their sockets are **gomphosis** joints (*gomphos*, nail, + *-osis*, condition). The teeth might be very loose (or absent) in the skeleton because the **periodontal membrane** that normally holds them tightly in place was destroyed when the skeleton was prepared. If the skull has some empty tooth sockets (alveolar processes), observe the shape of the inside of the socket that the cone-shaped root of the tooth fits into. How would you classify a gomphosis based on the amount of *movement* allowed? _____

4. Using table 10.2 and your textbook as a guide, label the types of fibrous joints in figure 10.1.

Table 10.2	Classification of Fibrous Joints		
Fibrous Joints	**Structure and Description**	**Examples**	**Word Origins**
Suture	Found exclusively between skull bones; consists of a small amount of connective tissue (the sutural ligament) holding the bone surfaces together.	Lambdoid suture, sagittal suture	*sutura*, a seam
Syndesmosis	Consists of large surfaces of bones that are anchored together by a connective tissue membrane called an interosseous membrane.	Distal radioulnar joint; distal tibiofibular joint	*syn-*, together, + *desmos*, a band
Gomphosis	Consists of a cone-shaped peg fitting into a socket and anchored by the periodontal membrane.	Teeth articulating with alveolar processes of mandible or maxilla	*gomphos*, nail, + *-osis*, condition

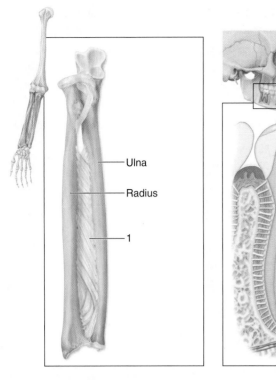

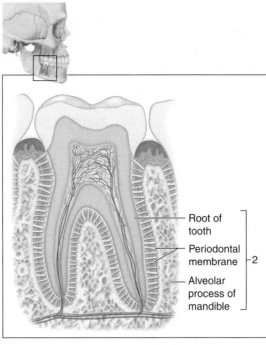

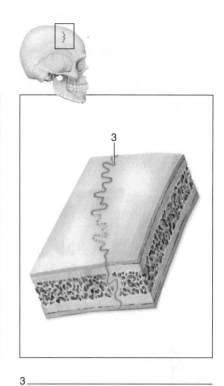

1 _____   2 _____   3 _____

Figure 10.1   Fibrous Joints.

☐ gomphosis   ☐ suture   ☐ syndesmosis

Cartilaginous Joints

Cartilaginous joints are characterized by having bone connected to bone with only cartilage between the bones. The joints are classified based on the type of cartilage found in the joint: either **hyaline cartilage** or **fibrocartilage**. Cartilaginous joints are classified as synarthroses or amphiarthroses, based on movement. Structurally, they are classified as synchondroses and symphyses (**table 10.3**).

EXERCISE 10.2   Cartilaginous Joints

1. *Synchondrosis* – A **synchondrosis** (*syn-*, together, + *chondrion*, cartilage) consists of bone connected to bone with **hyaline cartilage**. Examples are the sternocostal joints (the joints between the ribs and sternum) and the epiphyseal plates. Synchondroses should not be confused with synovial joints (or vice versa). Although synovial joints have hyaline cartilage as *part* of their structure (the articular cartilages), that cartilage is not the *only* thing found in between the bones.

2. Observe the articulations between the ribs and sternum on an articulated skeleton. On the articulated skeleton, the "cartilage" between the ribs and sternum is some sort of replacement material such as plastic or rubber. In a living human, this would be hyaline cartilage. However, observing the skeleton will give you an idea of the structure of a synchondrosis.

Table 10.3	Classification of Cartilaginous Joints		
Cartilaginous Joints	**Structure and Description**	**Examples**	**Word Origins**
Synchondrosis	Consists of bone connected to bone by hyaline cartilage.	Sternocostal joints; epiphyseal plates	*syn-*, together, + *chondrion*, cartilage
Symphysis	Consists of bone connected to bone by fibrocartilage.	Pubic symphysis; intervertebral discs	*symphysis*, a growing together

3. How would you classify the sternocostal joints based on the amount of *movement* allowed? _____

4. *Symphysis* – a **symphysis** (*symphysis*, a growing together) joint consists of bone connected to bone with **fibrocartilage**. Examples are the pubic symphysis and the intervertebral discs. The intervertebral discs have a more complex structure than the pubic symphysis. They consist of an outer band of fibrocartilage, the **anulus fibrosus**, which surrounds a gel-like interior, the **nucleus pulposus**. A "ruptured," "herniated," or "slipped" disc occurs when the annulus fibrosis tears and the nucleus pulposus leaks out.

The leaked nucleus pulposus can compress nerve fibers and cause neurological problems such as pain and numbness.

5. Observe the **pubic symphysis** and the **intervertebral discs** on the articulated skeleton. On the articulated skeleton, the "cartilage" is some sort of replacement material. In a living human this would be fibrocartilage.

6. How would you classify a symphysis joint based on the amount of *movement* allowed? _____

7. Using table 10.3 and your textbook as a guide, label the types of cartilaginous joints shown in **figure 10.2**.

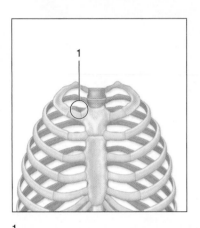

1 _____

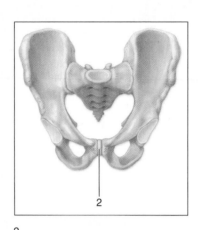

2 _____

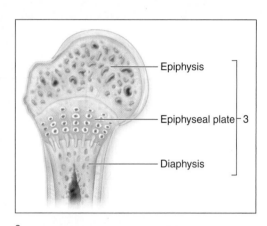

Epiphysis

Epiphyseal plate ⎱ 3

Diaphysis

3 _____

4 ⎡ Body of vertebra
⎣ Intervertebral disc

Nucleus pulposus
Anulus fibrosus

4 _____

Figure 10.2   Cartilaginous Joints.

☐ symphysis   ☐ synchondrosis

Synovial Joints

Synovial joints have a complex structure, which includes a joint cavity filled with fluid. The term *synovial* literally means, "together with egg" (*syn*, together, + *ovum*, egg). This term refers to the fluid inside the joint (the synovial fluid), which has the consistency and appearance of egg white. The joint cavity filled with synovial fluid allows the articulating bones to move easily past one another with very little friction in between the bones. The joint cavity allows synovial joints to have a great deal of movement. Thus, all synovial joints are classified as diarthroses based on movement. Structurally they are classified based on the shapes of the articulating bones (for example: ball-and-socket). Table 10.5 summarizes the structural classifications of synovial joints.

EXERCISE 10.3 General Structure of a Synovial Joint

1. Observe a model of a synovial joint, preferably a model of the knee joint.

2. Using **table 10.4** and your textbook as a guide, identify the features of a typical synovial joint listed in **figure 10.3** on the model. Then label them in figure 10.3.

3. *Optional Activity:* AP|R **Skeletal System**—Watch the "Synovial joint" animation for a summary of synovial joint structure and types.

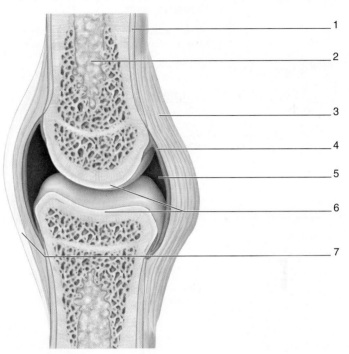

Figure 10.3 **General Structure of a Synovial Joint.**

☐ articular capsule
☐ articular cartilage
☐ fibrous layer of articular capsule
☐ ligament
☐ periosteum
☐ synovial (joint) cavity
☐ synovial fluid
☐ synovial membrane
☐ yellow bone marrow

Table 10.4	Components of Synovial Joints	
Structure	**Description**	**Word Origins**
Articular Capsule	Consists of two layers: an outer fibrous capsule and an inner synovial membrane.	*arthron*, a joint, + *capsa*, a box
Articular Cartilages	Hyaline cartilage found on the epiphyses of the articulating bones.	*arthron*, a joint
Bursae and Tendon Sheaths (most joints)	Either small, round sacs (bursae) or elongated structures that wrap around tendons (tendon sheaths); lined with synovial membrane and filled with synovial fluid; function to reduce friction between joint structures.	*bursa*, a purse
Fibrous Layer of Articular Capsule	A dense irregular connective tissue that anchors the two articulating bones to each other; anchors to the periosteum of the articulating bones; thickenings of the fibrous capsule form several joint ligaments.	*fibra*, fiber
Menisci (some joints)	Crescent-shaped pads of fibrocartilage found within the joint that provide cushioning between the articulating bones.	*meniskos*, crescent
Synovial Cavity	A cavity within the joint that is lined with synovial membrane and filled with synovial fluid.	*syn*, together, + *ovum*, egg, + *cavus*, hollow
Synovial Fluid	A very slippery fluid consisting of hyaluronic acid and other glycoproteins; primary function is to reduce friction in the joint.	*syn*, together, + *ovum*, egg, + *fluidus*, to flow
Synovial Membrane	A thin connective tissue membrane that lines all structures within the joint, including intra-articular ligaments; responsible for the formation of synovial fluid.	*syn*, together, + *ovum*, egg, + *membrana*, a skin

EXERCISE 10.4 Structural Classifications of Synovial Joints

Synovial joints are classified by structure using several different methods, most of which involve a description of the shape of the bones and the movement allowed at the joint (**table 10.5**). For example, in a ball-and-socket joint the "ball" (rounded) part of one bone fits into the "socket" (concave) part of another bone.

1. Observe an articulated skeleton to see examples of synovial joints.

2. Using table 10.5 and your textbook as a guide, identify the joints in **figure 10.4** on the articulated skeleton. Then label each type of joint on figure 10.4.

Table 10.5	Structural Classification of Synovial Joints		
Surface Shape	**Structure and Description**	**Examples**	**Word Origins**
Ball-and-Socket	Formed when a spherical head fits into a concave socket.	Hip joint; shoulder joint	*ball*, a round mass, + *soccus* a shoe or sock
Condylar (ellipsoid)	Formed when a convex oval surface fits into an elliptical concavity.	Radiocarpal joints; metacarpophalangeal joints	*kondylos*, knuckle, or *ellips*, oval, + *eidos*, form
Hinge (gynglymoid)	Formed when a convex surface fits into a concave surface and only allows movement along a single plane.	Knee joint; elbow joint	*gynglymos*, a hinge joint
Pivot	Formed when a round surface fits into a ring formed by a ligament and a depression in another bone.	Atlantoaxial joint	*pivot*, a post upon which something turns
Plane (gliding)	Formed when two flat surfaces come together.	Intermetatarsal joints; some intercarpal joints	*planus*, flat
Saddle (sellar) Joints	Formed when bones having both concave and convex surfaces come together such that the concave surfaces are at right angles to each other.	Carpometacarpal joint of the thumb; ankle joint; calcaneocuboid joint	*sella*, saddle

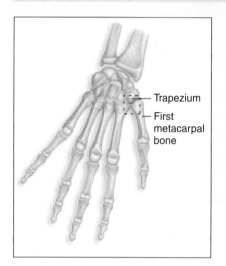

1 _____

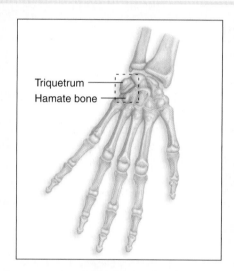

2 _____

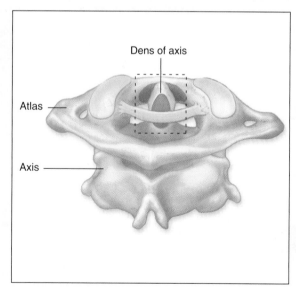

3 _____

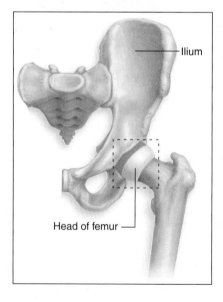

4 _____

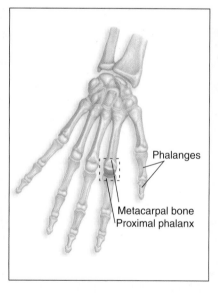

5 _____

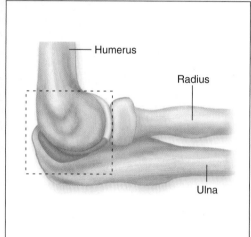

6 _____

Figure 10.4   Structural Classifications of Synovial Joints.

☐ ball-and-socket   ☐ hinge   ☐ plane

☐ condylar   ☐ pivot   ☐ saddle

EXERCISE 10.5 Practicing Synovial Joint Movements

1. All synovial joints are classified as **diarthrotic** (*di-*, two, + *arthron*, a joint) based on movement. There are several types of movement possible with synovial joints. **Table 10.6** summarizes the types of movements possible at synovial joints.

2. Practice all of the movements listed in table 10.6 with your laboratory partner until you feel confident that you can demonstrate or describe them all. This knowledge is critically important for your success in the later chapters that cover muscle actions. Muscle actions are described in terms of joint movements, so if you are able to describe the movement of the joint, you will also be able to describe the action of a muscle that acts about the joint.

3. *Optional Activity:* **AP|R** **Skeletal System**—Review the series of joint movement animations to see examples of multiple movements at each joint.

Table 10.6	Movements of Synovial Joints		
Movement	**Description**	**Opposite Movement**	**Word Origins**
Abduction	Movement of a body part away from the midline.	Adduction	*ab*, from, + *duco*, to lead
Adduction	Movement of a body part toward the midline.	Abduction	*ad-*, toward, + *duco*, to lead
Circumduction	Movement of the distal part of an extremity in a circle.	NA	*circum-*, around, + *duco*, to lead
Depression	Movement of a body part inferiorly.	Elevation	*depressio*, to press down
Dorsiflexion	Movement of the ankle such that the foot moves toward the dorsum (back).	Plantar flexion	*dorsum*, the back, + *flexus*, to bend
Elevation	Movement of a body part superiorly.	Depression	*e-levo-atus*, to lift up
Eversion	Movement of the ankle such that the plantar surface of the foot faces laterally.	Inversion	*e-*, out, + *versus*, to turn
Extension	An increase in joint angle.	Flexion	*extensio*, a stretching out
Flexion	A decrease in joint angle.	Extension	*flexus*, to bend
Inversion	Movement of the ankle such that the plantar surface of the foot faces medially.	Eversion	*in-*, inside, + *versus*, to turn
Opposition	Placement of the thumb (pollex) such that it crosses the palm of the hand and can touch all of the remaining digits.	Reposition	*op -*, against, + *positio*, a placing
Plantar Flexion	Movement of the ankle such that the foot moves toward the plantar surface.	Dorsiflexion	*plantaris*, the sole of the foot, + *flexus*, to bend
Pronation	Movement of the palm from anterior to posterior.	Supination	*pronatus*, to bend forward
Protraction	Movement of a body part anteriorly (especially: mandible and scapula).	Retraction	*pro-*, before + *tractio*, to draw
Reposition	Movement of the thumb to anatomic position after opposition.	Opposition	*re-*, backward, + *positio*, a placing
Retraction	Movement of a body part posteriorly (especially, mandible and scapula).	Protraction	*re-*, backward, + *tractio*, to draw
Rotation	Movement of a body part around its axis.	NA	*rotatio*, to rotate
Supination	Movement of the palm from posterior to anterior.	Pronation	*supinatus*, to bend backward

EXERCISE 10.6 The Knee Joint

In this exercise we will observe the structure of the knee joint as an example of a synovial joint. The knee joint is complex, as are most synovial joints, and it contains several modifications such as bursae, tendon sheaths, and menisci that are often associated with synovial joints. The knee joint also contains several strong ligaments, which help to stabilize the joint. **Table 10.7** summarizes the structures composing the knee joint.

Many of us have some peripheral knowledge of the knee joint, having known individuals who have suffered a ruptured ACL, torn meniscus, or other knee injury, even if we have no idea what an ACL is. The ACL (anterior cruciate ligament) is one of two **cruciate ligaments** (*cruciatus*, resembling a cross) found in the knee joint, and a ruptured ACL is a knee injury that is common in football players, downhill skiers, or others involved in contact sports.

1. Observe a model of the knee joint or the knee joint of a cadaver.

2. The knee joint is actually two joints: the **tibiofemoral joint**, which is classified as a synovial hinge joint that acts in flexion and extension (though it allows for some rotational movement as well), and the **patellofemoral joint**, which is classified as a planar (gliding) joint. The bony structure of the tibiofemoral joint includes the medial and lateral femoral condyles, which sit on top of the medial and lateral tibial condyles. The bony structure of the patellofemoral joint includes the patellar surface of the femur, which consists of the smooth anterior surface between the femoral condyles, and the medial and lateral facets of the patella.

3. Using table 10.7 and your textbook as a guide, identify the structures listed in **figure 10.5** on the model of the knee joint. Then label them in figure 10.5.

Table 10.7	Structures of the Knee Joint	
	Description	**Word Roots**
LIGAMENTS		
Anterior Cruciate Ligament	Connects the anterior intercondylar eminence of the tibia to the medial surface of the lateral condyle of the femur.	*ante-*, in front of, + *cruciatus*, shaped like a cross
Fibular (lateral) Collateral Ligament	Connects the lateral epicondyle of the femur to the head of the fibula.	*co*, together, + *latus*, side
Patellar Ligament	Connects the patella to the tibial tuberosity.	*patina*, a shallow disk
Posterior Cruciate Ligament	Connects the posterior intercondylar eminence of the tibia to the anterior part of the lateral surface of the medial condyle of the femur.	*post-*, behind, + *cruciatus*, shaped like a cross
Tibial (medial) Collateral Ligament	Connects the medial epidondyle of the femur to the medial surface of the tibia; its deep surface is anchored to the medial meniscus.	*co*, together, + *latus*, side
MENISCI		
Lateral Meniscus	A crescent-shaped pad of fibrocartilage located between the lateral condyle of the femur and the lateral condyle of the tibia.	*meniskos*, crescent
Medial Meniscus	A crescent-shaped pad of fibrocartilage located between the medial condyle of the femur and the medial condyle of the tibia.	*meniskos*, crescent
BURSAE		
Infrapatellar Bursa	Located between the proximal tibia and the patellar ligament.	*infra*, below, + *patina*, a shallow disc, + *bursa*, a purse
Prepatellar Bursa	Located between the patella and the overlying skin.	*pre-*, before, + *patina*, a shallow disc, + *bursa*, a purse
Suprapatellar Bursa	Located between the distal femur and the quadriceps femoris tendon; communicates with the synovial cavity of the knee joint.	*supra*, on the upper side, *patina*, a shallow disc, + *bursa*, a purse

WHAT DO YOU THINK?

1. Why do you think the anterior cruciate ligament is so often injured during contact sports?

WHAT DO YOU THINK?

2. Injuries to the tibial collateral ligament are often accompanied by a torn medial meniscus. After observing the fibrous tissues that form the joint capsule of the knee, why do you think this pattern of injury is so common?

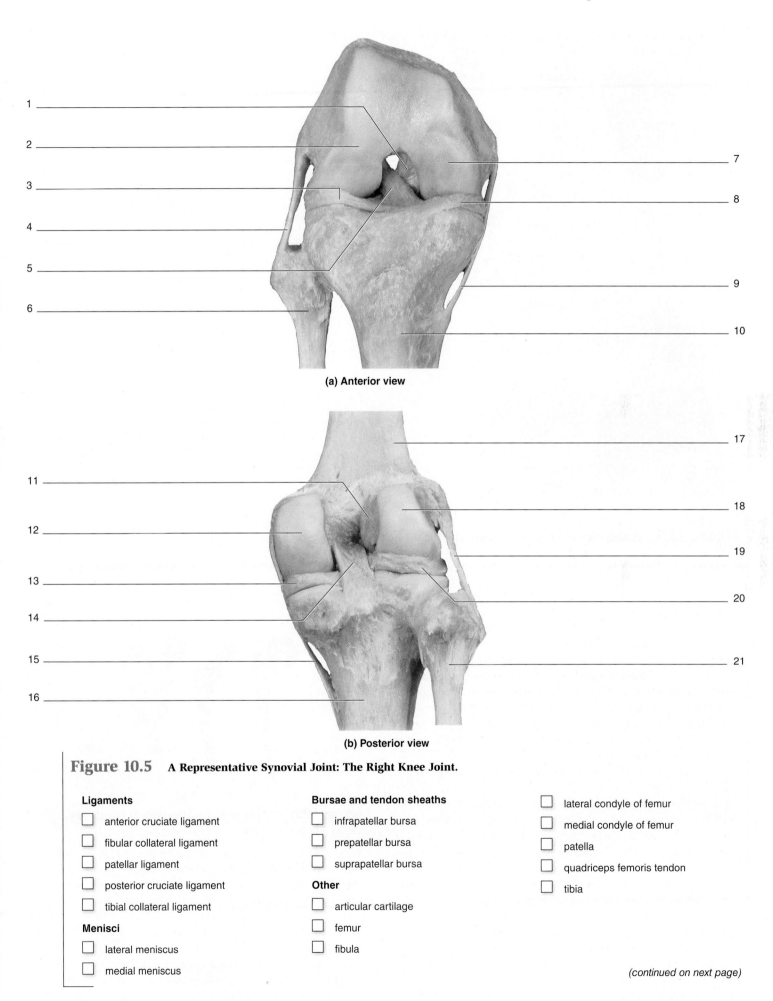

(a) Anterior view

(b) Posterior view

Figure 10.5   **A Representative Synovial Joint: The Right Knee Joint.**

Ligaments
- ☐ anterior cruciate ligament
- ☐ fibular collateral ligament
- ☐ patellar ligament
- ☐ posterior cruciate ligament
- ☐ tibial collateral ligament

Menisci
- ☐ lateral meniscus
- ☐ medial meniscus

Bursae and tendon sheaths
- ☐ infrapatellar bursa
- ☐ prepatellar bursa
- ☐ suprapatellar bursa

Other
- ☐ articular cartilage
- ☐ femur
- ☐ fibula

- ☐ lateral condyle of femur
- ☐ medial condyle of femur
- ☐ patella
- ☐ quadriceps femoris tendon
- ☐ tibia

(continued on next page)

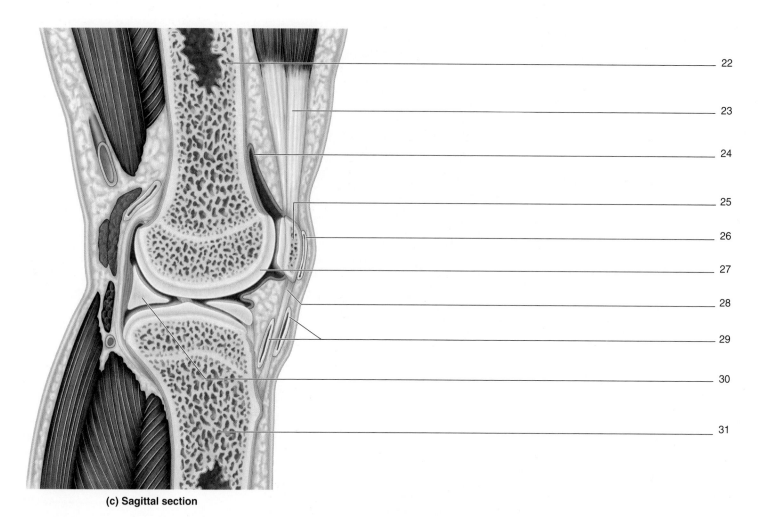

22

23

24

25

26

27

28

29

30

31

(c) Sagittal section

Figure 10.5   A Representative Synovial Joint: The Right Knee Joint. *Continued.*

Focus | Hip Replacement Surgery

What do you think of when you think of someone's "hip"? Many people think this is the part of the body that lies superficial to the iliac crest. In fact, the term *ilium* literally means "hip." When you "put your hands on your hips," you are putting them on your iliac crests. However, when someone falls and suffers a fractured hip, it has little to do with the iliac bones at all. Instead, it concerns a fracture of another bone that composes part of the hip joint: the femur. Specifically, a fractured hip refers to a fracture of the neck of the femur.

The hip joint is the joint formed between the acetabulum of the os coxae and the head of the femur. This joint suffers wear and tear with age, and is a common site of osteoarthritis. In the elderly, particularly elderly women, the bones that compose the hip joint may become brittle over time. These individuals are at increased risk of developing a femoral or "hip" fracture if they fall. Unfortunately, hip fractures do not heal particularly well, because the fracture often disrupts the blood supply to the head of the femur. Therefore, instead of repairing the fractured bones, a surgeon might opt to completely replace the hip joint.

Hip replacement surgery involves cleaning out the acetabulum. In many cases it must be reconstructed with bone grafts to create a more efficient and complete socket that will be a better fit for the two parts of the prosthetic hip. Within the acetabulum, the surgeon places an artificial cup, which will compose the "socket" of this "ball and socket" joint. Next, the fractured head and neck of the femur are removed, and a prosthetic is placed on the

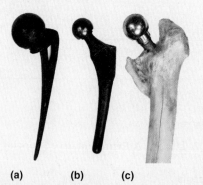

(a)   (b)   (c)

Figure 10.6   Three different examples of prostheses used for hip replacements.

proximal end of the femur, thus composing the new "ball" of the "ball and socket" joint. Figure 10.6 shows a few different types of prosthetics. Figure 10.6*a* is an example of a very old hip prosthetic. Notice how large the ball is. Figure 10.6*b* is an example of a prosthetic with a much smaller head, and figure 10.6*c* shows a similar prosthetic that is still within the bone. Notice in figure 10.6*c* that the greater and lesser trochanters of the femur are still intact. This is necessary to maintain the connections between the bone and the gluteal muscles and other muscles that act about the hip. Such muscles are separated during the surgery so the surgeon can access the hip joint. However, they must be reconnected to the bone after the surgery.

1. Fully classify the glenohumeral joint, using both functional (movement) and structural terms. _____

_____

_____

2. Explain why the knee joint is *not* classified as a synchondrosis even though it has hyaline cartilage. _____

_____

_____

3. The fibrous connective tissue component that holds together a gomphosis (holds a tooth into a socket) is called the_____.

4. How does a symphysis joint differ from a synchondrosis? _____

_____

_____

5. Give an example of a synovial hinge joint. _____

6. Give an example of a synovial ball-and-socket joint. _____

7. How are synovial joints classified based on *movement*? _____

8. How is a syndesmosis classified based on *movement*? _____

9. The atlantoaxial joint is classified as a(n)_____ joint.

10. Define *bursitis*. _____

_____

11. Where do the articular cartilages arise from developmentally? _____

_____

12. What type of tissue is the synovial membrane composed of? _____

_____

13. What are the main components of synovial fluid? _____

_____

Muscle Tissue and Introduction to the Muscular System

11

OUTLINE and OBJECTIVES

Anatomy & Physiology REVEALED®
aprevealed.com

Module 6: MUSCULAR SYSTEM

INTRODUCTION

As you are reading the text on this page, skeletal muscles connected to your eyeballs (the *extrinsic* eye muscles) are contracting to produce the very fine movements necessary for your eyes to track the words on the paper. At the same time, smooth muscle within the ciliary bodies of the eyes are contracting (or relaxing) to alter the shape of the lens so the image on the page is focused clearly upon your retina. And, as always, contraction of cardiac muscle in your heart is creating the force necessary to propel blood through your arteries to deliver oxygen and glucose to the working tissues within your eyes, brain, and the rest of your body. Clearly, you require the use of muscle tissue to read the words on this page. Indeed, properly functioning muscle tissue is essential for your very survival.

Muscle tissue is one of the most metabolically active tissues in the body and is the one tissue capable of creating movement of the body or body organs. All types of muscle tissue are characterized by the following properties: *excitability, contractility, elasticity,* and *extensibility*. **Excitable** tissues are able to generate and propagate special electrical signals called *action potentials* (APs). **Contractile** tissues are able to actively shorten themselves and produce force. **Elastic** tissues are able to be stretched and will spring back to their original shape once the stretching force is released. **Extensible** tissues are able to be lengthened by the pull of an external force (such as an external weight or the action of an opposing muscle).

There are three types of muscle tissue: skeletal muscle, smooth muscle, and cardiac muscle. **Skeletal muscle** composes the voluntary muscles that move the skin of the face and skeleton, **smooth muscle** is found mainly in the walls of the viscera (such as the blood vessels, stomach, urinary bladder, intestines, and uterus), and **cardiac muscle** is found in the heart. These three types of muscle tissue are distinguished from each other based on location, neural control (voluntary vs. involuntary), the presence or absence of visible striations, the shape of the cells, and the number of nuclei per cell. In this laboratory session, you will explore each of these features to gain an appreciation for the similarities and differences between the three types of muscle tissue. You will also begin your exploration of the human musculoskeletal system.

Study Tip!

Terminology related to the action of muscle fibers can be very confusing. Recall from chapter 10 that the term *flexion* refers to any joint movement in which the angle between two bones is decreased. However, people often use the term *flex* (as in, "flex your bicep for me") when they are asking you to *contract* a muscle. The term *contract* is the term scientists prefer to use to describe the action of a muscle fiber producing force. Even so, the term *contract* can also be misleading at times! Literally, to contract means to shorten. All muscle fibers have the ability to shorten when they produce force. However, striated muscle fibers can also produce force while remaining at a fixed length, or even while being forcibly lengthened.

Chapter 11: Muscle Tissue and Introduction to the Muscular System

PRE-LABORATORY WORKSHEET

1. List the three types of muscle tissue, and for each type give a location in the body where that type of tissue is found.

   Muscle Tissue   **Location in Body**

   a. _____   a. _____

   b. _____   b. _____

   c. _____   c. _____

2. Muscle is *excitable* tissue. What does it mean for a tissue to be excitable?

   _____

   _____

3. Muscle is *contractile* tissue. What does it mean for a tissue to be contractile?

   _____

4. The structural and functional unit of skeletal muscle is called a _____.

5. Define *tendon*: _____

   _____

6. Intercalated discs are characteristic of _____ muscle tissue.

7. Cylindrical bundles of contractile proteins located inside skeletal muscle fibers are called: _____

8. Skeletal muscles are given names that reflect location, shape, attachments, or other features related to the muscles. These names are based on Latin and Greek word roots. Match the word root given in column A with its corresponding meaning in column B.

   Column A   **Column B**

   ____ 1. *brevis*   a. around

   ____ 2. *capitis*   b. at an angle

   ____ 3. *endo-*   c. belly

   ____ 4. *epi-*   d. between

   ____ 5. *gastro-*   e. head

   ____ 6. *inter-*   f. large

   ____ 7. *magnus*   g. short

   ____ 8. *oblique*   h. straight

   ____ 9. *peri-*   i. upon

   ____ 10. *rectus*   j. within

IN THE LABORATORY

In this laboratory session you will observe the detailed structure of skeletal, smooth, and cardiac muscle tissues as seen under the microscope. In addition, you will observe common morphologies of skeletal muscles and begin to develop an understanding of the logic behind the naming of skeletal muscles. Finally, you will begin your study of the muscles of the body: often an overwhelming experience for the beginning student of anatomy. The exercises in this chapter will give you an overview of the organization of the human musculoskeletal system without overwhelming you with details. If you take to this introductory task and learn the information well, you will be prepared to learn the detailed names, bony attachments, and actions of the individual muscles of the body that will be introduced in chapters 12

and 13. Over time you will forget much of the detailed information you will learn in chapters 12 and 13 (until you have to know it again; after you've learned it a second time, the information will stay with you longer). However, even though this memory loss is inevitable, you will always be able to fall back on what you have learned in this chapter regarding the overall organization of the human musculoskeletal system and the major actions of groups of muscles as a starting point for review. If you start with an understanding of the overall organization of the human musculoskeletal system and the functions of specific muscle groups, the task of remembering the minutiae for each individual muscle will become much easier and will seem less overwhelming.

Histology

In the exercises that follow you will observe the histology of skeletal, smooth, and cardiac muscle tissues. **Table 11.1** lists the characteristics of the three types of muscle tissue, which you can use as a reference as you work through the exercises.

Skeletal Muscle Tissue

Skeletal muscle cells are some of the largest cells in the body. Their enormous size results from the fusion of hundreds of *myoblasts* during development into a single muscle fiber. A mature muscle fiber is multinucleate, containing 200 to 300 nuclei per millimeter of fiber length. The nuclei of normal skeletal muscle fibers are located peripherally, just under the *sarcolemma* (plasma membrane).

A whole muscle, such as the biceps brachii muscle of the arm, consists of several bundles, or **fascicles** (*fascis*, bundle), of muscle fibers (**figure 11.1a**). Each fascicle contains hundreds of long, cylindrical **muscle fibers** (figure 11.1b). Each muscle fiber contains within it a number of cylindrical bundles of contractile proteins called **myofibrils** (figure 11.1c). The myofibrils contain the **myofilaments** actin and myosin (the main contractile proteins of muscle), which are arranged into **sarcomeres** (the structural and functional unit of skeletal muscle, figure 11.1d). The regular arrangement of actin and myosin into sarcomeres gives each myofibril a striated or banded appearance, with visible *A bands* (dark bands), *I bands* (light bands), and *Z discs* (a dark line in the middle of an I band) (figure 11.1d,e). An adult skeletal

muscle fiber typically contains about 2000 myofibrils per cell. Because intermediate filaments within the muscle anchor and align the Z discs of adjacent myofibrils to each other, the entire muscle fiber takes on the same regular striated appearance of A bands and I bands found in the myofibrils when viewed under the light microscope.

When viewed under the light microscope, the A band appears dark in color. The dark color is due to the presence of thick filaments composed of **myosin**. Because the filaments are very thick, they absorb light, which is what causes the A band to appear dark. On the other hand, the I band contains thin filaments composed primarily of **actin**. Because the filaments are so thin, light passes easily through them, causing the I band to appear light. Finally, there are anchoring proteins found at the **Z discs**, which makes them dense like the A bands and causes light to be absorbed in that location as well. Thus, Z discs appear dark and are visible in the middle of each I band when the slide is viewed at high power.

Figure 11.1 demonstrates the relationships between the gross structure of a skeletal muscle and the microstructure of a skeletal muscle fiber. Though you will not be able to see myofibrils or myofilaments when you observe the muscle tissue under the microscope, you should be able to visualize A bands, I bands, and Z discs, and relate these bands to the corresponding arrangement of myofilaments into sarcomeres. That is, you should be able to relate what you see under the microscope to the structures shown in the drawing of a sarcomere in figure 11.1.

Table 11.1	Muscle Tissues		
Type of Muscle	**Description**	**Functions**	**Identifying Characteristics**
Skeletal	Elongate, cylindrical cells with multiple nuclei that are peripherally located. Tissue appears striated (light and dark bands along the length of the cell).	Produces voluntary movement of the skin and the skeleton.	Length of cells (extremely long); striations; multiple, peripheral nuclei.
Smooth	Elongate, spindle-shaped cells (fatter in the center, narrowing at the tips) with single, "cigar-shaped" nuclei that are centrally located. No striations.	Produces movement within visceral organs such as intestines, bladder, uterus, and stomach. Propels blood through blood vessels; food through the intestines, etc.	Spindle-shape of the cells; lack of striations; cigar-shaped or spiral nuclei.
Cardiac	Short, branched cells with single nuclei that are centrally located. Dark lines (intercalated discs) are seen where two cells come together. Tissue appears striated (light and dark bands along the length of each cell).	Performs the contractile work of the heart. Responsible for creating the pumping action of the heart.	Branched, uninucleate cells; striations; intercalated discs.

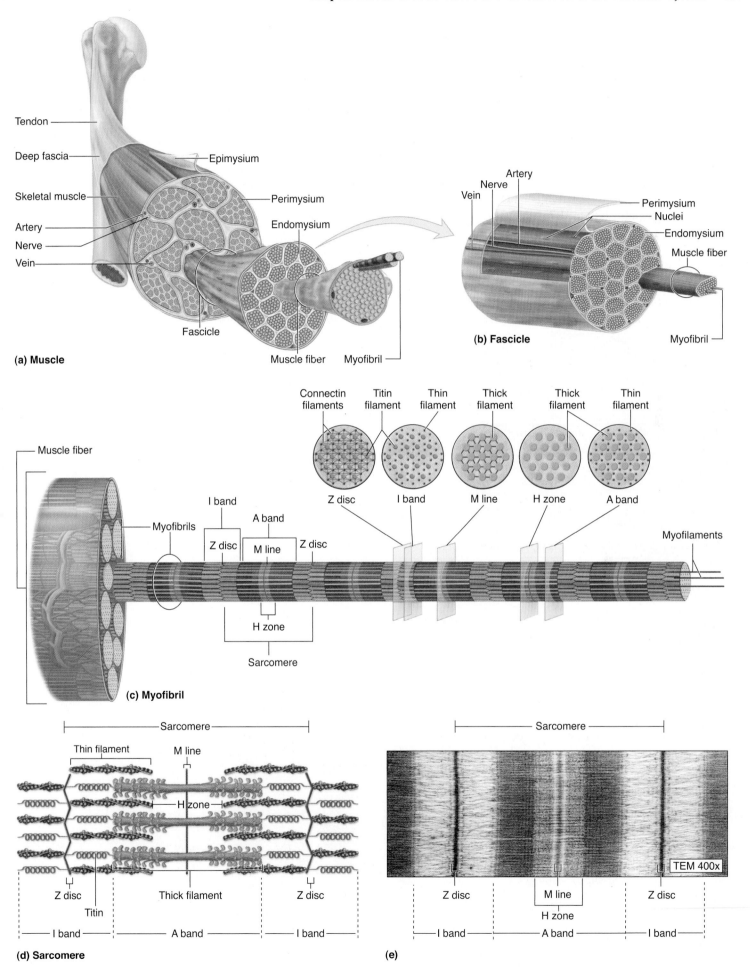

(a) Muscle

APR Figure 11.1 **Levels of Structural Organization of Skeletal Muscle.** (a) A whole muscle, (b) a muscle fascicle, (c) a muscle fiber, (d) a sarcomere, (e) an electron micrograph of a sarcomere.

EXERCISE 11.1   Histology of Skeletal Muscle Fibers

1. Obtain a slide of skeletal muscle tissue and place it on the microscope stage.

2. Bring the tissue into focus using the scanning objective. Switch to low power, bring the tissue sample into focus once again, and then switch to high power. Many slides of skeletal muscle contain muscle fibers shown in both longitudinal section and cross section. Scan the slide and identify muscle fibers shown in both longitudinal section and cross section (**figures 11.2** and **11.3**).

3. Focus on muscle fibers shown in longitudinal section (figure 11.2) and observe them at high power. Identify an individual muscle fiber. Notice the numerous peripherally located nuclei. Most of the nuclei that you see on the slide belong to the muscle fibers. However, about 5% to 15% of

the visible nuclei are those of **satellite cells**, myoblast-like cells located between the muscle fibers. Satellite cells give skeletal muscle a limited ability to repair itself after injury. Under the light microscope you will not be able to tell which nuclei belong to satellite cells.

4. Using table 11.1 and figure 11.2 as guides, identify the following structures on the slide of skeletal muscle tissue:

☐ A band   ☐ sarcomere(if visible)

☐ I band   ☐ Z disc (if visible)

☐ nucleus

5. In the space below, compose a brief sketch of skeletal muscle fibers as seen under the microscope. Be sure to label the structures listed above in your drawing.

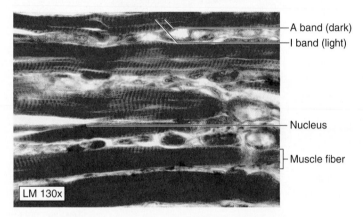

A band (dark)
I band (light)

Nucleus

Muscle fiber

LM 130x

Figure 11.2   **Skeletal Muscle in Longitudinal View.** Z discs and sarcomeres are not visible in this micrograph.

──────── ✕

EXERCISE 11.2   Connective Tissue Coverings of Skeletal Muscle

Now that you have observed the structure of individual muscle fibers, it is time to direct your attention to the relationship between individual muscle fibers and whole muscles. A whole skeletal muscle, such as the biceps brachii muscle, consists of many individual muscle fibers bundled together with connective tissue (figure 11.3). Each individual muscle fiber is covered by a layer of connective tissue called the **endomysium** (*endo*, within, + *mys*, muscle). Several muscle cells are bundled together into **fascicles** (*fascis*, bundle) by a surrounding layer of connective tissue called the **perimysium** (*peri*, around, + *mys*, muscle). Finally, the entire skeletal muscle is surrounded by a layer of connective tissue called the **epimysium** (*epi*, upon, + *mys*, muscle). The epimysium is an extension of the **deep fascia**, which will be discussed shortly.

1. You should still have the slide of skeletal muscle on the microscope stage. View the slide at low magnification and move the stage until you can see muscle fibers in cross section. In this view you will be able to identify the

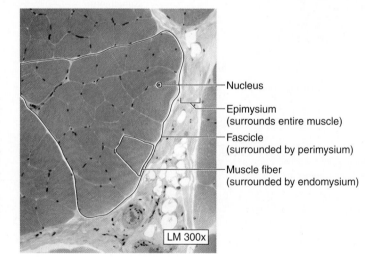

Nucleus

Epimysium (surrounds entire muscle)

Fascicle (surrounded by perimysium)

Muscle fiber (surrounded by endomysium)

LM 300x

Figure 11.3   **Connective Tissue Coverings of Skeletal Muscle.**   Muscle fibers are seen in a cross sectional view.

connective tissues that bundle the muscle fibers together to form a whole skeletal muscle.

2. Using figure 11.3 as a guide, identify the following structures:*

☐ endomysium   ☐ fascicle   ☐ nucleus

☐ epimysium   ☐ muscle fiber   ☐ perimysium

3. In the space below, compose a brief sketch of the connective tissue coverings of skeletal muscle as viewed under the microscope. Be sure to label the structures listed in step 2 in your drawing.

─────

*If the slide is skeletal muscle from a small mammal such as a mouse, you should be able to identify all three layers of connective tissue. However, if it is from a larger mammal, you will most likely be able to identify only endomysium and perimysium.

────────── ✕

Smooth Muscle Tissue

Spindle-shaped cells with cigar-shaped or spiral nuclei and an absence of striations are characteristic of smooth muscle tissue. Individual cells have tapered ends, and there is only one centrally located nucleus per cell. Most of the nuclei will appear to be somewhat cigar-shaped. However, in cells that have contracted, the nuclei take on a corkscrew or spiral appearance (figure 11.4a), which can be a key identifying feature.

Smooth muscle is generally found in two layers around tubular organs such as the small intestine. The most common arrangement is an *inner circular layer* and *outer longitudinal layer* (figure 11.4b).

EXERCISE 11.3 Smooth Muscle Tissue

1. Obtain a slide containing smooth muscle tissue (**figure 11.4**) and place it on the microscope stage.

2. Bring the tissue into focus using the scanning objective. Switch to low power, bring the tissue sample into focus once again, and then switch to high power. If you are viewing a slide of the small intestine, identify smooth muscle cells that have been cut in both longitudinal section and cross section. Note that the cells do not appear to be of uniform diameter when viewed in cross section. This is because some cells are sliced through the tapered ends, while others are sectioned through the thickest part of the fiber, which contains the nucleus.

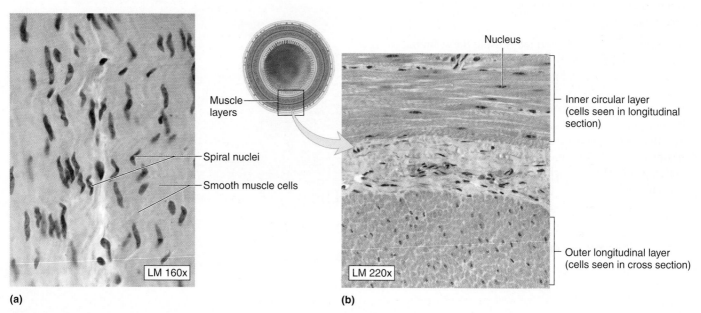

(a)   (b)

Figure 11.4   **Smooth Muscle Tissue.** (a) Close-up view of smooth muscle demonstrating spiral nuclei, (b) circular and longitudinal layers of smooth muscle tissue.

3. Using table 11.1 and figure 11.4 as guides, identify the following structures on the slide of smooth muscle tissue:

☐ inner circular layer   ☐ nucleus

☐ muscle cell in cross section   ☐ outer longitudinal layer

☐ muscle cell in longitudinal section

Study Tip!

Smooth muscle tissue and dense regular connective tissue can often be confused. A few key features will help you to distinguish them from each other. First, fibroblast nuclei (found in dense regular connective tissue) appear flattened and are located between fibers, whereas smooth muscle nuclei are plumper and can be seen *within* the cells. In addition, there will be relatively fewer nuclei per unit area in dense regular connective tissue than in smooth muscle tissue. Finally, the appearance of spiral or corkscrew nuclei is a good clue that the tissue is smooth muscle, not dense regular connective tissue. See if you can find a spiral nucleus on the slide of smooth muscle tissue to make sure you understand what it is.

4. In the space below, compose a brief sketch of smooth muscle cells as viewed under the microscope. Be sure to label the structures listed in step 3 in your drawing.

——————— ×

Cardiac Muscle Tissue

Branching cells with centrally located nuclei are characteristic of **cardiac muscle tissue** (**figure 11.5**). Like skeletal muscle fibers, cardiac muscle cells are striated. The striations appear because cardiac muscle cells also contain numerous myofibrils. Unlike skeletal muscle cells, the myofibrils in cardiac muscle cells are not anchored to each other with intermediate filaments at the Z discs. The resulting slight offset of myofibrils within the muscle fiber means that the banding pattern isn't always as clear in cardiac muscle as it is in skeletal muscle.

EXERCISE 11.4   Identification of Cardiac Muscle Tissue

1. Obtain a slide of cardiac muscle tissue and place it on the microscope stage.

2. Bring the tissue sample into focus using the scanning objective. Switch to low power, bring the tissue sample into focus once again, and then switch to high power. Notice that the cells contain only one or two nuclei, and that the cells are short and branched. Where two cells come together, you will notice a darkly stained line. This is an **intercalated disc**. Intercalated discs contain numerous **desmosomes**, which function to hold the cells together, and **gap junctions**, which allow electrical signals to be transmitted very rapidly from one cell to the next.

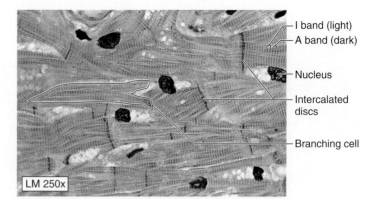

I band (light)
A band (dark)
Nucleus
Intercalated discs
Branching cell

LM 250x

Figure 11.5   **Cardiac Muscle Tissue.**

3. Using table 11.1 and figure 11.5 as guides, identify the following structures on the slide of cardiac muscle tissue:

☐ A band   ☐ intercalated disc

☐ branching cells   ☐ nucleus

☐ I band   ☐ Z disc (if visible)

4. In the space below, compose a brief sketch of cardiac muscle cells as viewed under the microscope. Be sure to label the structures listed in step 3.

Gross Anatomy

Gross Anatomy of Skeletal Muscles

The names of skeletal muscles often seem overly complex. However, if you begin to understand the basis of their names, you will find them far less mysterious. Over time you will begin to discover that the name of a muscle often gives you a clue to its location, size, shape, action, or attachment points. The efforts you have made to learn the Latin and Greek word roots of anatomical terms thus far will become even more valuable as you work through the next three chapters.

EXERCISE 11.5 Naming Skeletal Muscles

Table 11.2 summarizes some of the common ways skeletal muscles are named and gives word origins for the muscle names. Plan to spend some time on your own mastering these word origins so you will be better prepared to handle the material to come in chapters 12 and 13. To help you in this effort, the post-laboratory worksheet has a number of questions related to the naming of skeletal muscles.

WHAT DO YOU THINK?

1. You have discovered a new muscle in the body. The muscle is a long muscle that contains four heads. Based on your knowledge of word origins, suggest a logical name for this muscle.

Table 11.2	Common Methods for Naming Skeletal Muscles		
Name	**Meaning**	**Word Origins**	**Example**
NAMING SKELETAL MUSCLES BASED ON SHAPE			
Deltoid	Triangular	*delta*, the Greek letter delta (a triangle), + *eidos*, resemblance	Deltoid
Gracilis	Slender	*gracilis*, slender	Gracilis
Lumbrical	Wormlike	*lumbricus*, earthworm	Lumbricals
Rectus	Straight	*rectus*, straight	Rectus abdominis
Rhomboid	Diamond-shaped	*rhombo-*, an oblique parallelogram with unequal sides, + *eidos*, resemblance	Rhomboid major
Teres	Round	*teres*, round	Teres major
Trapezius	A four-sided geometrical figure having no two sides parallel	*trapezion*, a table	Trapezius

(continued on next page)

Table 11.2	Common Methods for Naming Skeletal Muscles *(continued)*		
Name	**Meaning**	**Word Origins**	**Example**
NAMING SKELETAL MUSCLES BASED ON SIZE			
Brevis	Short	*brevis*, short	Adductor brevis
Latissimus	Broadest	*latissimus*, widest	Latissimus dorsi
Longissimus	Longest	*longissimus*, longest	Longissimus capitis
Longus	Long	*longus*, long	Adductor longus
Major	Bigger	*magnus*, great	Teres major
Minor	Smaller	*minor*, smaller	Teres minor
NAMING SKELETAL MUSCLES BASED ON THE NUMBER OF HEADS AND/OR BELLIES			
Biceps	2 heads	*bi*, two, + *caput*, head	Biceps brachii
Digastric	2 bellies	*bi*, two, + *gastro*, belly	Digastric
Quadriceps	4 heads	*quad*, four, + *caput*, head	Quadriceps femoris
Triceps	3 heads	*tri*, three, + *caput*, head	Triceps brachii
NAMING SKELETAL MUSCLES BASED ON POSITION			
Abdominis	Abdomen	*abdomen*, the greater part of the abdominal cavity	Rectus abdominis
Anterior	On the front surface of the body	*ante-*, before, in front of	Serratus anterior
Brachii	Arm	*brachium*, arm	Biceps brachii
Dorsi	Back	*dorsum*, back	Latissimus dorsi
Femoris	Thigh	*femur*, thigh	Rectus femoris
Infraspinatus	Below the scapular spine	*infra-*, below, + *spina*, spine	Infraspinatus
Interosseous	In between bones	*inter*, between + *osseus*, bone	Interossei
Oris	Mouth	*oris*, mouth	Orbicularis oris
Pectoralis	Chest	*pectus*, chest	Pectoralis major
Posterior	On the back surface of the body	*posterus*, following	Serratus posterior
Supraspinatus	Above the scapular spine	*supra-*, on the upper side, + *spina*, spine	Supraspinatus
NAMING SKELETAL MUSCLES BASED ON DEPTH			
Externus	External	*external*, on the outside	Obturator externus
Internus	Internal	*internal*, away from the surface	Obturator internus
Profundus	Deep	*pro*, before, + *fundus*, bottom	Flexor digitorum profundus
Superficialis	Superficial	*super*, above, + *facies*, face	Flexor digitorum superficialis
NAMING SKELETAL MUSCLES BASED ON ACTION			
Abductor	Moves a body part away from the midline	*ab*, from, + *ductus*, to bring toward	Abductor pollicis brevis
Adductor	Moves a body part toward the midline	*ad*, toward, + *ductus*, to bring toward	Adductor pollicis
Constrictor	Acts as a sphincter and closes an orifice	*cum*, together, + *stringo*, to draw tight	Superior pharyngeal constrictor
Depressor	Flattens or lowers a body part	*de-*, away, + *pressus*, to press	Depressor anguli oris
Dilator	Causes an orifice to open, or dilate	*dilato*, to spread out	Dilator pupillae

Table 11.2	Common Methods for Naming Skeletal Muscles *(continued)*		
Name	**Meaning**	**Word Origins**	**Example**
Extensor	Causes an increase in joint angle	*ex-*, out of, + *-tensus*, to stretch	Extensor carpi ulnaris
Flexor	Causes a decrease in joint angle	*flectus*, to bend	Flexor carpi ulnaris
Levator	Raises a body part superiorly	*levo* + *atus*, a lifter	Levator scapulae
Pronator	Turns the palm of the hand from anterior to posterior	*pronatus*, to bend forward	Pronator teres
Supinator	Turns the palm of the hand from posterior to anterior	*supino* + *atus*, to bend backward	Supinator

EXERCISE 11.6 Architecture of Skeletal Muscles

The overall architecture of a skeletal muscle affects how the muscle functions. When you observe a whole muscle, the individual fibers and fascicles are visible, making it relatively easy to see how the fascicles are arranged within the muscle. Recall that when skeletal muscle contracts, it generally gets shorter and brings the two attachment points closer to each other. Thus, the orientation of the muscle fascicles compared to the attachment points of the muscle will directly affect the force produced by the muscle and the complexity of the muscle's actions. For instance, **pennate** architecture (*penna*, feather) allows a muscle to produce greater force per distance shortened than **parallel** architecture. In addition, muscles with more than two bony attachments (for example, biceps and triceps) produce more complex movements than muscles with only two attachments (one proximal attachment and one distal attachment). **Table 11.3** summarizes the common patterns of fascicle arrangement that contribute to skeletal muscle architecture.

1. Using classroom models of skeletal muscles or a prosected human cadaver (if available), observe the arrangement of fascicles in several different skeletal muscles of the body.

2. Locate the following muscles on the classroom models or the human cadaver. Then identify them on **figure 11.6**.

- ☐ deltoid
- ☐ extensor digitorum
- ☐ gastrocnemius
- ☐ orbicularis oculi
- ☐ pectoralis major
- ☐ rectus femoris
- ☐ sartorius
- ☐ trapezius
- ☐ triceps brachii

Table 11.3	Common Architectures of Skeletal Muscles					
Diagram						
Name	**Unipennate**	**Bipennate**	**Multipennate**	**Circular**	**Convergent**	**Parallel**
Word Origins	*uni*, one, + *penna*, feather	*bi*, two, + *penna*, feather	*multi*, many, + *penna*, feather	*circum*, around	*cum*, together, + *vergo*, to incline	*para*, alongside

3. Using table 11.3 and your textbook or atlas as guides, fill in the table below with the names of the muscles numbered in figure 11.6. Then, indicate the type of architecture each muscle represents.

Muscle Number	Muscle Name	Architecture
1		
2		
3		
4		
5		
6		
7		
8		
9		

Organization of the Human Musculoskeletal System

Table 11.4 lists the major muscle groups of the body and the common actions of each group of muscles. Obviously, each individual muscle within a group of muscles has its own specific actions. However, skeletal muscles do not act alone, so it makes functional sense to learn the muscles as groups and to learn the major actions of each group. Muscles contained within a group commonly act as synergists and often have a common nerve and blood supply. Thus, damage to the nerve and blood supply often affects a whole group of muscles, not just an individual muscle.

In the limbs, muscles are not only *functionally* organized into groups; they are also *physically* organized into groups. Muscle groups are separated from each other by extensions of the deep fascia that form **compartments** (**table 11.5**). Why is it important to learn this relationship? As an example, consider the muscles of the arm. In this laboratory session you will learn that the muscles located in the anterior compartment of the arm act primarily in flexion of the shoulder and elbow joints. In chapter 17 you will learn that this entire group of muscles is innervated by a single bundle of nerve fibers, the musculocutaneous nerve. Even if you forget the names of the individual muscles in the anterior compartment of the arm, you will be able to predict the loss of function that would result if the musculocutaneous nerve was damaged, based on your knowledge of compartments and muscle groups.

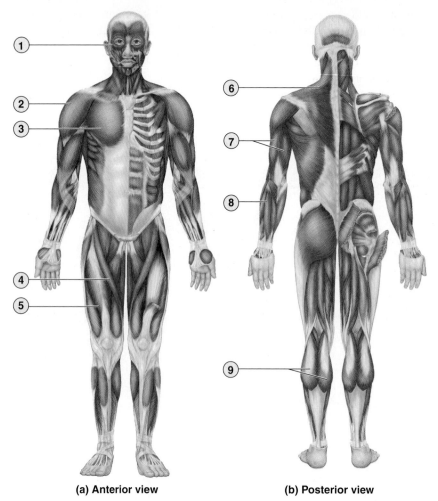

(a) Anterior view   (b) Posterior view

Figure 11.6   **Muscles of the Human Body.** (a) Anterior view, (b) posterior view.

Table 11.4	Major Muscle Groups and Their Actions
Major Muscle Group	**General Description of Muscle Actions**
Muscles of facial expression	Create facial expressions.
Muscles of mastication	Used for chewing.
Suprahyoid and infrahyoid muscles	Move the tongue and pharynx.
Muscles of the neck	Rotate, flex, and extend the head.
Muscles of respiration	Involved in breathing movements.
Muscles of the abdomen	Flex, bend, and rotate the spine.
Muscles of the back and spine	Extend, bend, and rotate the spine.
Muscles of the pelvic floor	Support pelvic contents and form sphincters around structures such as the urethra and the anus.
Muscles that act about the pectoral girdle	Stabilize the pectoral girdle and anchor the upper limb to the pectoral girdle.
Muscles that act about the pelvic girdle	Stabilize the pelvic girdle and anchor the lower limb to the pelvic girdle.

Table 11.5	Fascial Compartments of the Limbs and Their General Muscle Actions
Compartment	**General Description of Muscle Actions**
COMPARTMENTS OF THE ARM	
Anterior	Flexion of shoulder and elbow.
Posterior	Extension of shoulder and elbow.
COMPARTMENTS OF THE FOREARM	
Anterior	Flexion of the wrist and digits (fingers).
Posterior	Extension of the wrist and digits (fingers).
COMPARTMENTS OF THE THIGH	
Anterior	Extension of the knee, flexion of the hip.
Posterior	Extension of the hip, flexion of the knee.
Medial	Adduction of the thigh.
COMPARTMENTS OF THE LEG	
Anterior	Dorsiflexion and inversion of the ankle, extension of the digits (toes).
Posterior	Plantarflexion and inversion of the ankle, flexion of the digits (toes).
Lateral	Eversion of the ankle.

EXERCISE 11.7 Major Muscle Groups and Fascial Compartments of the Limbs

1. Using tables 11.4 and 11.5 as guides, identify each of the major muscle groups and compartments of the body.

2. As you identify each muscle group or compartment, practice performing the actions listed in the tables. Try to feel the contraction of the muscles by palpating the skin overlying the muscle groups. If you have difficulty remembering joint actions, refer back to chapter 10 for descriptions.

3. Figures 11.7 and 11.8 represent cross sections of the arm (brachium) and thigh, respectively. The details in cross-sectional views such as these are difficult to negotiate the first time through. Thus, right now just focus on the organization of the muscles into compartments. Visualize how the cross-sectional diagram relates to your own arm or thigh. As you perform the actions listed for each compartment, correlate the muscles used to perform each action with a specific compartment of the arm or thigh.

WHAT DO YOU THINK?

2 The proximal joint of the front limb of a dog contains two compartments of muscles, an anterior and a posterior compartment. If the muscles in the anterior compartment are all flexors, what is the common action of all the muscles in the posterior compartment?

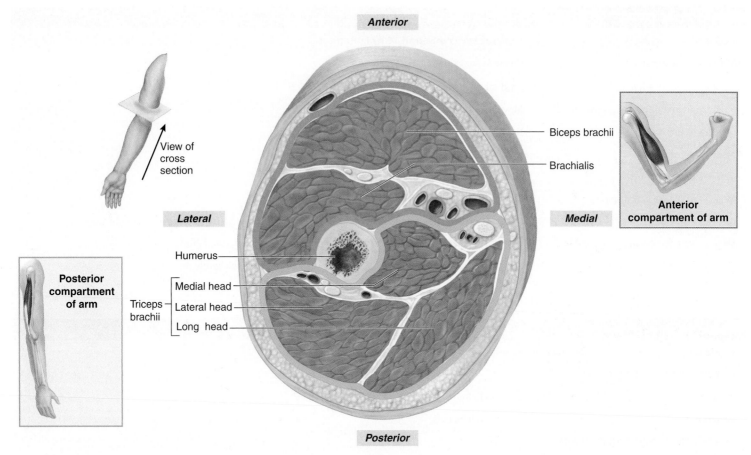

Figure 11.7   **Fascial Compartments of the Right Arm (Brachium).**

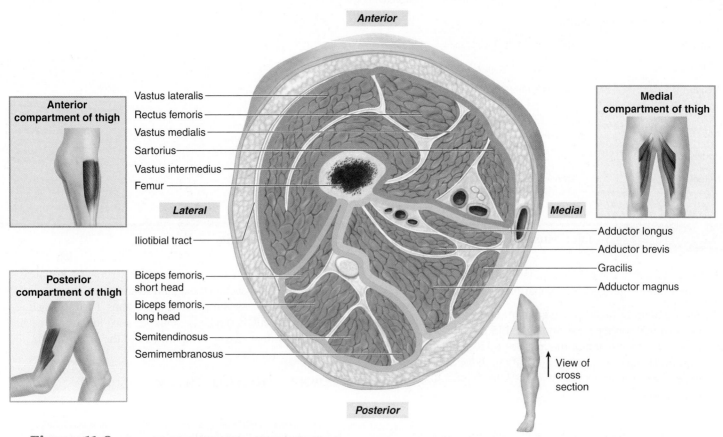

Figure 11.8   **Fascial Compartments of the Right Thigh.**

Chapter 11: Muscle Tissue and Introduction to the Muscular System

Name: _____

Date: _____ Section: _____

POST-LABORATORY WORKSHEET

1. Match the appropriate muscle type in column B with a characteristic listed in column A. You may use an answer choice more than once and you may list more than one type of muscle tissue for each characteristic.

Column A

_____ 1. branched fibers

_____ 2. central nuclei

_____ 3. cylindrical fibers

_____ 4. inner circular and outer longitudinal layers

_____ 5. intercalated discs

_____ 6. multinucleate

_____ 7. satellite cells

_____ 8. spindle-shaped cells

_____ 9. spiral nuclei

_____ 10. striated

Column B

a. skeletal muscle

b. cardiac muscle

c. smooth muscle

2. Which type(s) of muscle tissue are under involuntary (autonomic) control?

_____

3. Which type(s) of muscle tissue have striations?

_____

4. Which type(s) of muscle tissue have only one or two nuclei per cell?

_____

5. Match the appropriate word origin in column B with its meaning listed in column A.

Column A

_____ 1. smaller

_____ 2. round

_____ 3. a lifter

_____ 4. around

_____ 5. abductor

_____ 6. widest

_____ 7. straight

_____ 8. two-headed

_____ 9. slender

_____ 10. internal

Column B

a. peri-

b. rectus

c. latissimus

d. biceps

e. minor

f. internus

g. levator

h. to move away

i. teres

j. gracilis

6. In column B, write the letter of the structures labeled in the arm cross section that matches the description given in column A.

Column A

anterior (flexor) compartment of the arm

humerus

posterior (extensor) compartment of the arm

Column B

1. _____

2. _____

3. _____

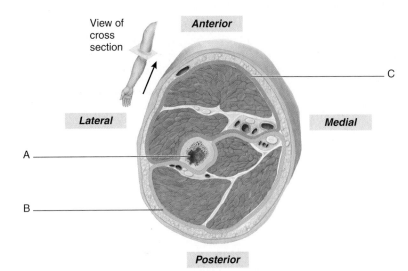

Fascial compartments of the arm (brachium)

7. In column B, write the letter of the structure labeled in the thigh cross section that matches the description given in column A.

Column A

femur

medial compartment of the thigh

anterior compartment of the thigh

posterior compartment of the thigh

Column B

1. _____

2. _____

3. _____

4. _____

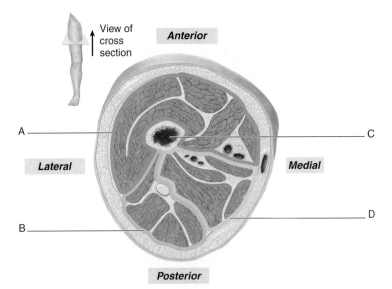

Fascial compartments of the thigh

Axial Muscles

12

OUTLINE and OBJECTIVES

INTRODUCTION

Imagine a time when you slept on a horrible mattress and woke up the next morning to find the muscles in your back and neck incredibly stiff and sore. For the next few days, every action seemed to cause pain. The classic comment on such an experience is, "I discovered muscles I never knew I had!" It seems that the only time we really contemplate the muscles of our back and neck is when they are injured. At such a time it is impossible *not* to contemplate these muscles—they constantly hurt. This example is a reminder of the importance of the muscles of the back, neck, and thorax for creating the most basic motions of our bodies. In this chapter you will explore the structure and function of muscles of the back, neck, and thorax. In addition, you will explore the structure and function of muscles important for chewing, creating facial expressions, and controlling such vital bodily functions as breathing, coughing, defecating, and urinating.

Chapter 12: Axial Muscles

PRE-LABORATORY WORKSHEET

1. Briefly state the difference between a muscle of *facial expression* and a muscle of *mastication*.

   _____

   _____

   _____

2. Define the following terms:

   aponeurosis:_____

   inguinal:_____

   extension:_____

   zygomatic:_____

   extrinsic:_____

3. The largest and most important respiratory muscle is a dome-shaped muscle that separates the thoracic cavity from the abdominal cavity. This muscle is

   called the _____.

4. List the four muscles that compose the anterior abdominal wall.

   a. _____

   b. _____

   c. _____

   d. _____

5. The three attachment points for the sternocleidomastoid muscle are:

   a. _____

   b. _____

   c. _____

6. List the three groups of muscles that collectively form the *erector spinae* group of muscles.

   a. _____

   b. _____

   c. _____

7. The pelvic floor is divided clinically into two triangular regions: the _____ triangle and the _____ triangle.

IN THE LABORATORY

In this laboratory session, you will identify, name, and explore the structure and function of muscles that move the axial skeleton. Before you begin your studies, find out from your laboratory instructor exactly which muscles you will be responsible for on your practical exams. Then, highlight or star those muscles on the tables throughout this chapter so you focus only on the muscles you are required to know for your course. In addition, before you begin the more detailed study of the axial musculature in this chapter, make sure you have successfully completed the Gross Anatomy section in chapter 11. Those exercises introduced you to the major muscle groups and the common actions of those muscle groups (table 11.4).

Study Tip!

Follow the procedure below as you study each major group of muscles. This will help you build your knowledge and understanding in logical steps. Once you have mastered the information from task 1, you will be ready to move on to task 2, task 3, and so on. You will find that each successive task requires more detailed knowledge. However, you will also find yourself prepared to deal with these increasing levels of knowledge if you follow the steps.

STEPWISE APPROACH TO LEARNING MUSCLES OF THE BODY

1. Describe the general location of the muscle group.

2. Describe the general actions that all muscles of the group have in common.

3. List the names of all muscles belonging to that group (or just the ones your instructor requires of you).

4. Identify the muscles on a model or a cadaver (this can be done concurrently with step 3).

5. Learn the outliers—the muscles that DO NOT share the common actions of the group.

6. Learn specific attachment points and actions of the muscles your instructor wants you to know in this amount of detail.

Follow these suggestions, study one muscle group at a time, take a lot of breaks, study using frequent, short time intervals—and the tasks will no longer feel overwhelming. You will be amazed at how well you are able to absorb the knowledge, and you might be pleasantly surprised at your results.

Gross Anatomy

Muscles of the Head and Neck

Muscles of the head and neck include the muscles of the face (muscles of facial expression and muscles of mastication), extrinsic eye muscles, muscles that move the tongue, and muscles that move the neck.

EXERCISE 12.1 Muscles of Facial Expression

1. Observe muscles of the face on a human cadaver or a classroom model.

2. Muscles of the face fall into two groups based on function and innervation: muscles of facial expression and muscles of mastication (chewing). **Muscles of facial expression (table 12.1)** allow you to express emotions such as fright, delight, confusion, surprise, and the like. These muscles are unique in that they have distal attachments on skin instead of bone, so when the muscles contract, they pull on the skin. This movement is easily seen on the surface of the face as a facial expression. Over time, the pulling of these muscles on the face causes a characteristic wrinkling of the skin. For this reason, it's a good idea to practice using your "smiling" muscles much more often than using your "frowning" muscles—you'll look happier in old age! All of the muscles of facial expression are innervated by the **facial nerves**, which are the paired seventh cranial nerves (CN VII). A person with damage to the facial nerve will be unable to demonstrate facial expressions on the affected side.

3. Look into a mirror (you may have to perform this part of the exercise at home, if there isn't a mirror in your laboratory classroom) and practice the following facial expressions. As you perform each expression, name the muscles used to create the expression (use table 12.1 as a guide).

   ☐ anger or doubt   ☐ surprise or delight

   ☐ sadness (frowning)   ☐ kissing (close mouth, purse cheeks, close eyes)

   ☐ happiness (smiling or laughter)

4. Using table 12.1 and your textbook as a guide, identify all of the muscles of facial expression listed in **figure 12.1** on the cadaver or model of the face. Then label them in figure 12.1.

5. *Optional Activity:* AP|R **Muscular System**—Watch the muscle action animations to review the actions of many of the muscles mentioned in chapters 12 and 13. You can also try the action, origin, and insertion questions found in the quiz area for challenging drill and practice.

Table 12.1	Muscles of Facial Expression*				
Muscle	**Origin**	**Insertion**	**Action**	**Innervation**	**Word Origins**
Buccinator	Mandible, molar region of mandible and maxilla	Orbicularis oris (corners of the lips)	Presses cheek against molar teeth, as in chewing, whistling, or playing a wind instrument	Facial (CN VII)	*bucca*, cheek
Corrugator Supercilii	Superciliary arch	Skin of eyebrow	Creates vertical wrinkles in medial forehead, as in frowning	Facial (CN VII)	*corrugo*, to wrinkle, + *superus*, above, + *cilium*, eyelid
Depressor Anguli Oris	Mandible (anterolateral surface of the body)	Muscles and skin in the lower lip near the angle of the mouth	Pulls corners of the mouth inferior, as in frowning	Facial (CN VII)	*depressus*, to press down, + *angulus*, angle, + *oris*, mouth
Depressor Labii Inferioris	Mandible (between the midline and the mental foramen)	Oribicularis oris and skin of the lower lip	Depresses the lower lip, as in expressions of doubt and sadness	Facial (CN VII)	*depressus*, to press down, + *labia*, lip, + *inferior*, lower
*Epicranius (Occipitofrontalis)***	Epicranial aponeurosis	Skin of the forehead (frontalis); superior nuchal line (occipitalis)	Elevates the eyebrows and creates horizontal wrinkles in the forehead, as in expressions of surprise or delight	Facial (CN VII)	*occiput*, the back of the head, + *frontalis*, in front
Levator Anguli Oris	Maxilla (lateral portion)	Skin at the superior corner of the mouth	Elevates the corners of the mouth and pulls them laterally, as in smiling	Facial (CN VII)	*levatus*, to lift, + *labia*, lip, + *superus*, above
Levator Labii Superioris	Maxilla (inferior to infraorbital foramen)	Orbicularis oris and skin of the upper lip	Elevates the upper lip, as in expressions of sadness or seriousness	Facial (CN VII)	*levatus*, to lift, + *anguli*, angle, + *oris*, mouth
Mentalis	Mandible (incisive fossa)	Skin of the chin	Wrinkles the skin of the chin and elevates and protrudes the lower lip, as in expressions of doubt	Facial (CN VII)	*mentum*, the chin
Nasalis	Maxilla	Alar cartilages of the nose	Flares the nostrils, widens the anterior nasal aperture	Facial (CN VII)	*nasus*, nose
Orbicularis Oculi	Skin around the margin of the orbit of the eye	Skin surrounding the eyelids	Closes the eyelids as in blinking	Facial (CN VII)	*orbiculus*, a small disk, + *oculus*, eye
Orbicularis Oris	Deep surface of skin of maxilla and mandible	Mucous membrane of the lips	Purses and protrudes the lips, closes the mouth	Facial (CN VII)	*orbiculus*, a small disk, + *oris*, mouth
Platysma	Fascia superficial to the deltoid and pectoralis major muscles at 1 and 2 ribs	Mandible (lower border) and skin of the cheek	Stretches the skin of the anterior neck, depresses the lower lip, as in expressions of fright	Facial (CN VII)	*platys*, flat, broad
Procerus	Nasal bones and nasal cartilages	Aponeurosis at the bridge of the nose and the skin of the forehead	Depresses the eyebrows and elevates the nose producing wrinkles in the skin of the nose, as in frowning and squinting the eyes	Facial (CN VII)	*procerus*, long or stretched out
Risorius	Fascia overlying the masseter muscles	Orbicularis oris and skin of the corner of the mouth	Pulls the corners of the mouth laterally, as in expressions of laughter and/or smiling	Facial (CN VII)	*risus*, to laugh
Zygomaticus (Major and Minor)	Zygomatic bone	Skin and muscle at corner of the mouth	Pulls the corners of the mouth posteriorly and superiorly, as in smiling	Facial (CN VII)	*zygon*, yoke

*All muscles in this table are innervated by the facial nerve (CN VII).
**The epicranius consists of the epicranial aponeurosis and the occipitofrontalis muscle, which has two bellies: frontal belly of occipitofrontalis and occipital belly of occipitofrontalis.

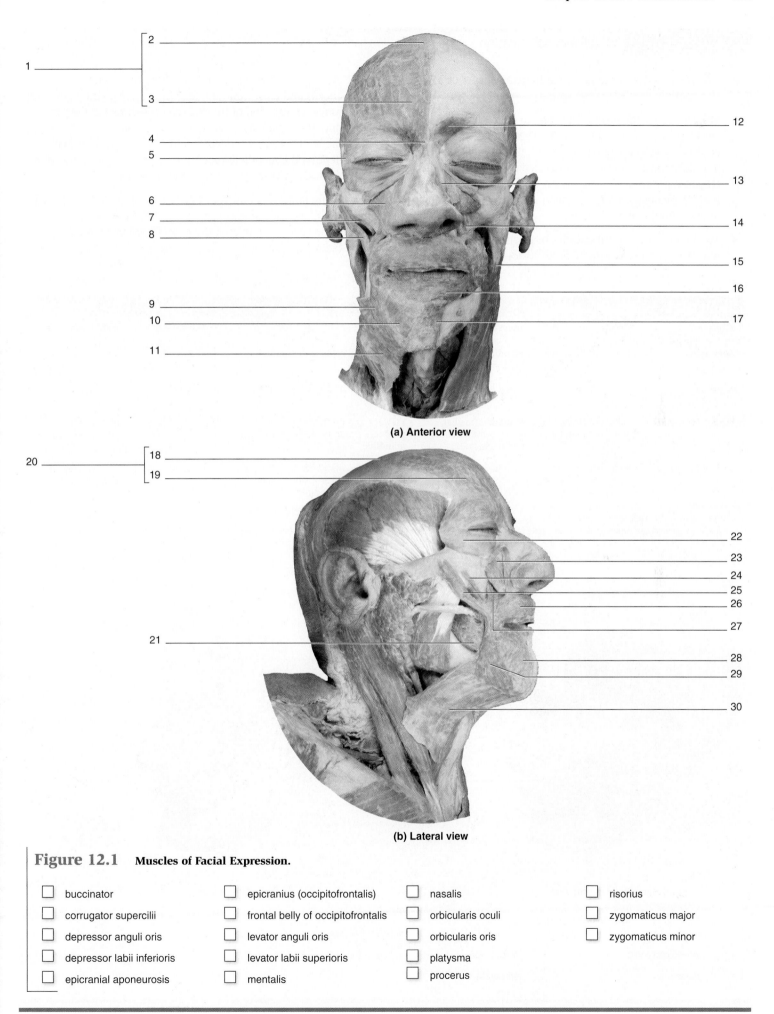

(a) Anterior view

(b) Lateral view

Figure 12.1 **Muscles of Facial Expression.**

- ☐ buccinator
- ☐ corrugator supercilii
- ☐ depressor anguli oris
- ☐ depressor labii inferioris
- ☐ epicranial aponeurosis

- ☐ epicranius (occipitofrontalis)
- ☐ frontal belly of occipitofrontalis
- ☐ levator anguli oris
- ☐ levator labii superioris
- ☐ mentalis

- ☐ nasalis
- ☐ orbicularis oculi
- ☐ orbicularis oris
- ☐ platysma
- ☐ procerus

- ☐ risorius
- ☐ zygomaticus major
- ☐ zygomaticus minor

EXERCISE 12.2 Muscles of Mastication

1. Observe muscles of the face on a human cadaver or a classroom model.

2. **Muscles of mastication** (*masticate*, to chew) are used in chewing movements (**table 12.2**). These muscles attach to the only mobile bone of the skull, the mandible. The muscles of mastication are innervated by branches of the **trigeminal nerves**, which are the paired fifth cranial nerves (CN V). Damage to the trigeminal nerve causes an inability to chew on the affected side.

3. The two most powerful muscles of mastication are the **masseter** and **temporalis** muscles. Place your fingers over the angle and ramus of the mandible (just below your cheek) and close your jaw forcefully (elevate the mandible) to feel contraction of the masseter. Then repeat the process, only this time place your fingers over your temples to feel the contraction of the temporalis muscle. Muscles that depress the mandible (open the mouth) are the infrahyoid muscles, which are covered in exercise 12.6.

4. Using table 12.2 and your textbook as a guide, identify all of the muscles of mastication listed in **figure 12.2** on a cadaver or the model of the face. Then label them in figure 12.2.

Table 12.2	Muscles of Mastication*				
Muscle	**Origin**	**Insertion**	**Action**	**Innervation**	**Word Origins**
Temporalis	Temporal fossa	Mandible (coronoid process)	Elevates and retracts the mandible	Trigeminal (CN V)	*tempus*, temple
Masseter	Zygomatic arch	Mandible (lateral surface of the ramus)	Elevates and protracts the mandible	Trigeminal (CN V)	*masétér*, chewer
*Medial Pterygoid***	Sphenoid (lateral pterygoid plate) and maxilla	Mandible (medial surface of the ramus and neck)	Elevates and protracts the mandible; produces a side-to-side motion of the mandible	Trigeminal (CN V)	*pteryx*, wing, + *eidos*, resemblance
*Lateral Pterygoid***	Sphenoid (greater wing and lateral pterygoid plate)	Mandible (neck)	Protracts the mandible and depresses the chin; produces a side-to-side motion of the mandible	Trigeminal (CN V)	*pteryx*, wing, + *eidos*, resemblance

*All muscles in this table are innervated by the trigeminal nerve (CN V).
**Lateral and medial pterygoids, when acting alone (alternating one side at a time), produce a side-to-side grinding motion.

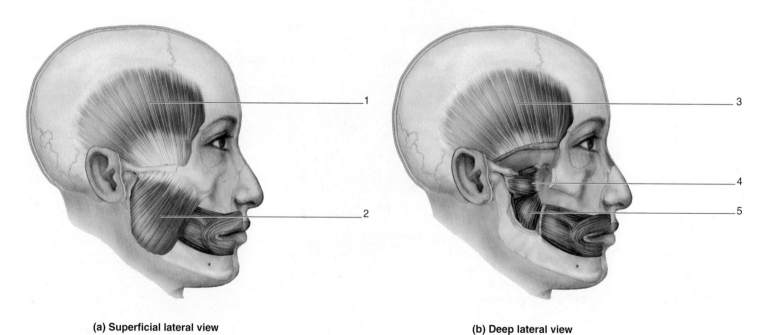

(a) Superficial lateral view   (b) Deep lateral view

Figure 12.2   **Muscles of Mastication.**

☐ lateral pterygoid   ☐ medial pterygoid
☐ masseter   ☐ temporalis

EXERCISE 12.3 Extrinsic Eye Muscles

1. Observe a model of the eye with **extrinsic** muscles.

2. The **extrinsic**, or extraocular (*extra-*, outside of, + *oculus*, eye), muscles of the eye (**table 12.3**) allow us to move our eyes up, down, side to side, and at an angle. These muscles originate on bone and insert onto the sclera of the eye. The **sclera** is the tough white connective tissue covering of the eyeball. They are named based on location, shape, or function, so they are quite easy to identify and remember. Our coverage of them in this exercise will be brief. However, we will return to these muscles when we cover the cranial nerves and their functions in chapter 16. If you understand the functions

of these muscles well now, you will find it much easier to understand the signs and symptoms of cranial nerve disorders when you get to the exercises in chapter 16.

3. Ask your laboratory partner to look in different directions and observe his or her eye movements. As his or her eyes move, name the muscles (in *both* eyes because they will be different!) used to create the movement (use table 12.3 as a guide).

4. Using table 12.3 and your textbook as a guide, identify the **extrinsic eye muscles** listed in **figure 12.3** on the model of the eye. Then label them in figure 12.3.

Table 12.3	Extrinsic Muscles of the Eye				
Muscle	**Origin**	**Insertion**	**Action**	**Innervation**	**Word Origins**
Inferior Oblique	Maxilla (anterior portion of orbit)	Sclera on the anterior, lateral surface of the eyeball, deep to the lateral rectus muscle	Elevates, abducts, and laterally rotates the eyeball	Oculomotor (CN III)	*inferior*, lower, + *obliquus*, slanting
Inferior Rectus	Sphenoid (tendinous ring around optic canal)	Sclera on the anterior, inferior surface of the eyeball	Depresses, adducts, and medially rotates the eyeball	Oculomotor (CN III)	*inferior*, lower, + *rectus*, straight
Lateral Rectus	Sphenoid (tendinous ring around optic canal)	Sclera on the anterior, lateral surface of the eyeball	Abducts the eyeball	Abducens (CN VI)	*lateralis*, lateral, + *rectus*, straight
Levator Palpebrae Superioris*	Sphenoid (lesser wing anterior and superior to the optic canal)	Skin of the upper eyelid	Elevates the upper eyelid	Oculomotor (CN III)	*levatus*, to lift, + *palpebra*, eyelid, + *superus*, above
Medial Rectus	Sphenoid (tendinous ring around optic canal)	Sclera on the anterior, medial surface of the eyeball	Adducts the eyeball	Oculomotor (CN III)	*medialis*, middle, + *rectus*, straight
Superior Oblique	Sphenoid (tendinous ring around optic canal)	Sclera on the posterior, superiolateral surface of the eyeball just deep to the belly of the superior rectus muscle	Depresses, abducts, and medially rotates the eyeball	Trochlear (CN IV)	*superus*, above, + *obliquus*, slanting
Superior Rectus	Sphenoid (tendinous ring around optic canal)	Sclera on the anterior, superior surface of the eyeball	Elevates, adducts, and medially rotates the eyeball	Oculomotor (CN III)	*superus*, above, + *rectus*, straight

*This muscle, while associated with the eye, does not attach to, or move, the eyeball itself.

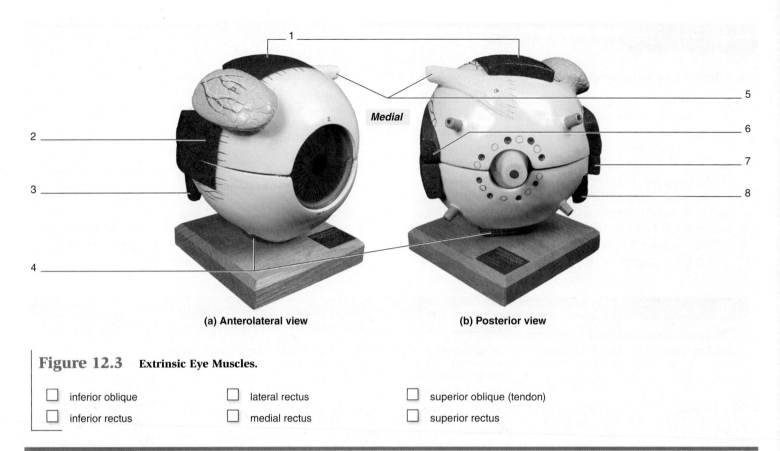

(a) Anterolateral view   **(b) Posterior view**

Figure 12.3 **Extrinsic Eye Muscles.**

☐ inferior oblique   ☐ lateral rectus   ☐ superior oblique (tendon)

☐ inferior rectus   ☐ medial rectus   ☐ superior rectus

EXERCISE 12.4 Muscles That Move the Tongue

1. Observe a model of a head that has been cut in a
   midsagittal plane so the muscles of the tongue are visible.

2. The tongue has both intrinsic and extrinsic muscles
   (**table 12.4**). The *intrinsic* muscles, within the tongue

Table 12.4	Muscles That Move the Tongue				
Muscle	**Origin**	**Insertion**	**Action**	**Innervation**	**Word Origins**
Genioglossus	Mandible (mental spine)	Hyoid bone (body) and inferior portion of the tongue	Depresses and protrudes the tongue	Hypoglossal (CN XII)	*geneion*, chin, + *glossa*, tongue
Hyoglossus	Hyoid bone (body and greater horn)	Inferior and lateral aspects of the tongue	Depresses and retracts the tongue	Hypoglossal (CN XII)	*Hyo-*, hyoid bone, + *glossa*, tongue
Palatoglossus	Soft palate (palatine aponeurosis)	Lateral aspect of the tongue	Depresses the soft palate and elevates the posterior aspect of the tongue	Vagus (CN X)	*palatum*, palate, + *glossa*, tongue
Styloglossus	Styloid process of the temporal bone	Inferior and lateral aspects of the tongue	Retracts and elevates the tongue during swallowing	Hypoglossal (CN XII)	*stylo*, styloid process, + *glossa*, tongue

itself, change the shape of the tongue. The *extrinsic* muscles, which connect the tongue to bony structures of the head and neck, create fine movements of the tongue necessary to form speech, manipulate food, and so on.

3. Look into a mirror as you stick your tongue out, (you may have to perform this part of the exercise at home if there isn't a mirror in your laboratory classroom) pull it in, alter its shape, and so on. As you perform these actions, decide if the muscles used are intrinsic or extrinsic muscles of the tongue.

4. Using table 12.4 and your textbook as a guide, identify the **muscles that move the tongue** listed in **figure 12.4** on a cadaver or on a model of the head and neck. (Note: These are all extrinsic muscles of the tongue.) Then label them in figure 12.4.

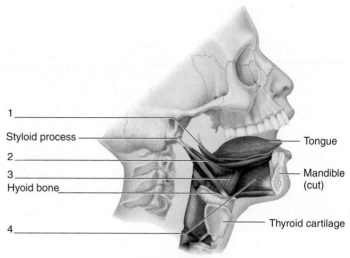

Right lateral view

Figure 12.4 **Muscles That Move the Tongue (and Associated Structures).**

☐ genioglossus   ☐ palatoglossus

☐ hyoglossus   ☐ styloglossus

EXERCISE 12.5 Muscles of the Pharynx

1. Observe a prosected cadaver, a model of the larynx (which will also demonstrate muscles of the pharynx), or a model of the head and neck demonstrating muscles of the pharynx.

2. The pharyngeal muscles (**table 12.5**) are used during the swallowing process. To get a feel for how these muscles function, swallow some saliva or a drink of water and pay

Table 12.5	Muscles of the Pharynx				
Muscle	**Origin**	**Insertion**	**Action**	**Innervation**	**Word Origins**
Levator Veli Palatini	Temporal bone (petrous portion) and cartilage of the auditory tube	Soft palate (palatine aponeurosis)	Elevates the soft palate, as in swallowing and yawning	Vagus (CN X)	*levatus*, to lift, + *velum*, veil, + *palatum*, palate
Palatopharyngeus	Soft palate (palatine aponeurosis)	Lateral wall of the pharynx	Elevates the larynx and pharynx	Vagus (CN X)	*palatum*, palate, + *pharyngo-*, pharynx
Stylopharyngeus	Styloid process of temporal bone	Larynx (thyroid cartilage)	Elevates the larynx and pharynx	Glossopharyngeal (CN IX)	*stylo-*, styloid process, + *pharyngo-*, pharynx
Tensor Veli Palatini	Sphenoid bone (pterygoid process)	Soft palate (palatine aponeurosis)	Elevates the soft palate	Trigeminal (CN V)	*tensus*, to stretch, + *velum*, veil, + *palatum*, palate
PHARYNGEAL CONSTRICTORS					
Inferior Constrictor	Larynx (thyroid and cricoid cartilages)	Posterior median raphe	Constricts the pharynx	Vagus (CN X)	*inferior*, lower, + *constringo*, to draw together
Middle Constrictor	Hyoid bone	Posterior median raphe	Constricts the pharynx	Vagus (CN X)	*middle*, middle, + *constringo*, to draw together
Superior Constrictor	Sphenoid bone (pterygoid process)	Posterior median raphe	Constricts the pharynx	Vagus (CN X)	*superus*, above, + *constringo*, to draw together

attention to the role these muscles play during swallowing. Notice that as you swallow, your larynx moves superiorly. (Your larynx is best felt by palpating your thyroid cartilage, or "Adam's apple." It is located just anterior to the pharynx.) The muscles that elevate the larynx *are not pharyngeal muscles*. Rather they are the suprahyoid muscles (see exercise 12.6). The muscles you are focusing on here are the ones you feel at the back of your throat at the very end of the swallowing process. The pharyngeal constrictors move the bolus of food, saliva, etc., into the esophagus.

3. Using table 12.5 as a guide, identify the following **muscles of the pharynx** listed in **figure 12.5** on a cadaver, on a model of the head and neck, or on a model of the larynx. Then label them in figure 12.5.

Figure 12.5   Muscles of the Pharynx.

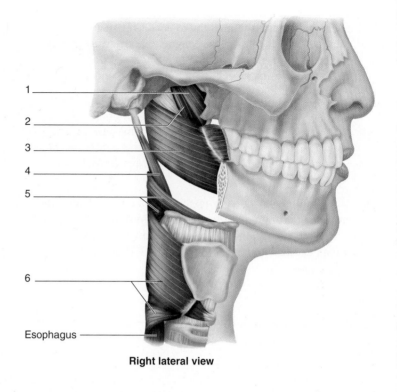

Right lateral view

☐ inferior constrictor

☐ levator veli palatini

☐ middle constrictor

☐ palatopharyngeus

☐ stylopharyngeus

☐ superior constrictor

☐ tensor veli palatini

EXERCISE 12.6   Muscles of the Neck

1. Observe a prosected cadaver or a classroom model demonstrating muscles of the neck.

2. The most prominent muscle in the anterior neck is the **sternocleidomastoid** muscle. The attachment of this muscle can be easily palpated under the skin. To do this, place your fingers just behind your ears to palpate the **mastoid process**, the insertion of the sternocleidomastoid.

Next, palpate your sternum, just lateral to the sternal notch as you rotate your neck from right to left, and feel for the tendons of the **sternal head** of the sternocleidomastoid. Finally, palpate just superior to the medial third of the clavicle while laterally flexing your neck to see if you can feel contraction of the **clavicular head** of the sternocleidomastoid. Understanding the locations of the

---

Focus | Understanding the Actions of Agonists, Synergists, and Antagonists

An **agonist** (*agon*, a contest), or **prime mover**, is a muscle used to create a given action about a joint. An **antagonist** (*anti*, against, + *agon*, a contest) is a muscle whose action opposes the action of the agonist. In contrast, a synergist (*syn*, together) is a muscle whose action assists the agonist. The sternocleidomastoid and splenius capitis are muscles that provide a good example of how muscles can act as either synergists or antagonists depending upon the movement required. These muscles can act as synergists for the action of lateral rotation of the neck, whereas they act as antagonists for the actions of flexion and extension of the neck. Palpate the sternal head of the sternocleidomastoid on your right side, and then laterally rotate your head to the right. Do you feel tension in the muscle? _____ Still palpating the muscle, laterally rotate your head to the left. Do you feel tension in the muscle? _____ Based on your observations, which direction does the right sternocleidomastoid rotate the neck?_____.

The most superficial muscle on the posterior side of the neck is the trapezius, a muscle that can move the neck, but is more of a prime mover of the scapula (see table 13.1). If you palpate the back of your neck as you rotate your neck, you can feel the trapezius muscle and a smaller muscle located deep to the trapezius, the splenius capitis, which is a prime mover of the neck (see table 12.6 for reference to its actions). Place your right hand over the back of the right side of your neck and rotate your head to the right. Keeping your hand in the same location, rotate your head to the left. In which direction of rotation do you feel more tension in the muscles on the back of the neck? _____ Based on your observations, which direction does the right splenius capitis muscle rotate the neck? _____.

In summary: to rotate your neck to the right, you use the sternal head of the sternocleidomastoid on the _____ side of the neck, and the splenius capitis muscle on the _____ side of the neck.

attachments, bellies, and borders of the two heads of this muscle is important because the sternocleidomastoid is a major clinical landmark in the neck.

3. Using **table 12.6** as a guide, identify the muscles of the neck listed in **figure 12.6** on a cadaver or a model of the head and neck. Then label them in figure 12.6.

Table 12.6	Muscles of the Neck				
Muscle	**Origin**	**Insertion**	**Action**	**Innervation**	**Word Origins**
MUSCLES THAT MOVE THE NECK					
Longissimus Capitis	T_1–T_4 (transverse processes) and C_4–C_7 (articular processes)	Mastoid process of temporal bone	Bilateral: extends the neck Unilateral: laterally rotates the neck to the same side	Cervical and thoracic spinal nerves	*longissimus*, longest, + *caput*, head
Semispinalis Capitis	T_1–T_5 (spinous processes), C_4–C_7 (articular processes)	Occipital bone (between superior and inferior nuchal lines)	Bilateral: extends the neck Unilateral: laterally flexes the neck to the same side	Dorsal rami of cervical spinal nerves	*semi*, half, + *spina*, spine, + *caput*, head
Splenius Capitis	Ligamentum nuchae and T_1–T_6 (spinous processes)	Superior nuchal line of occipital bone (lateral aspect) and mastoid process of temporal bone	Bilateral: extends the neck Unilateral: laterally rotates and laterally flexes the neck to the same side	Dorsal rami of spinal nerves	*splenion*, a bandage, + *caput*, head
Sternocleidomastoid	Sternal head: manubrium of the sternum Clavicular head: clavicle (medial third)	Mastoid process of temporal bone	Bilateral: flexes the neck Unilateral: laterally rotates the neck to the opposite side	Accessory (CN XI)	*sterno-*, sternum, + *cleido-*, clavicle, + *mastoid*, resembling a breast
SUPRAHYOID MUSCLES					
Digastric	Mastoid process (digastric groove on medial aspect)	Mandible (lower border near the midline)	When the mandible is fixed, it elevates the hyoid (posterior belly). When the hyoid is fixed, it depresses the mandible (anterior belly).	Posterior belly: facial (CN VII). Anterior belly: mandibular branch of the trigeminal (CN V)	*di-*, two, + *gastro-*, belly
Geniohyoid	Mandible (mental spine)	Hyoid bone	Protraction of hyoid. When hyoid is fixed, it depresses the mandible.	Hypoglossal (CN XII)	*geneion*, chin, + *hyoides*, shaped like the letter U
Mylohyoid	Mandible (mylohyoid line)	Hyoid bone	Elevates the floor of the mouth and tongue. When the hyoid is fixed it depresses the mandible.	Mandibular branch of the trigeminal (CN V)	*myle*, a mill, + *hyoides*, shaped like the letter U
Stylohyoid	Styloid process of temporal bone	Hyoid bone	Elevates the hyoid	Facial (CN VII)	*stylos*, pillar, + *hyoides*, shaped like the letter U
INFRAHYOID MUSCLES					
Omohyoid	Scapula (between the superior angle and the scapular notch)	Hyoid bone	Depresses the hyoid	Cervical spinal nerves C1–C3 through ansa cervicalis	*omos*, shoulder, + *hyoides*, shaped like the letter U
Sternohyoid	Manubrium of sternum (posterior surface) and first costal cartilage	Hyoid bone	Depresses the hyoid	Cervical spinal nerves C1–C3 through ansa cervicalis	*sternon*, chest, + *hyoides*, shaped like the letter U
Sternothyroid	Manubrium of sternum (posterior surface) and first costal cartilage	Thyroid cartilage	Depresses the larynx	Cervical spinal nerves C1–C3 through ansa cervicalis	*sternon*, chest, + *thyroid*, shaped like a shield
Thyrohyoid	Thyroid cartilage	Hyoid bone	Moves hyoid toward the larynx	First cervical spinal nerve C1 via hypoglossal (CN XII)	*thyro-*, thyroid, + *hyoides*, shaped like the letter U

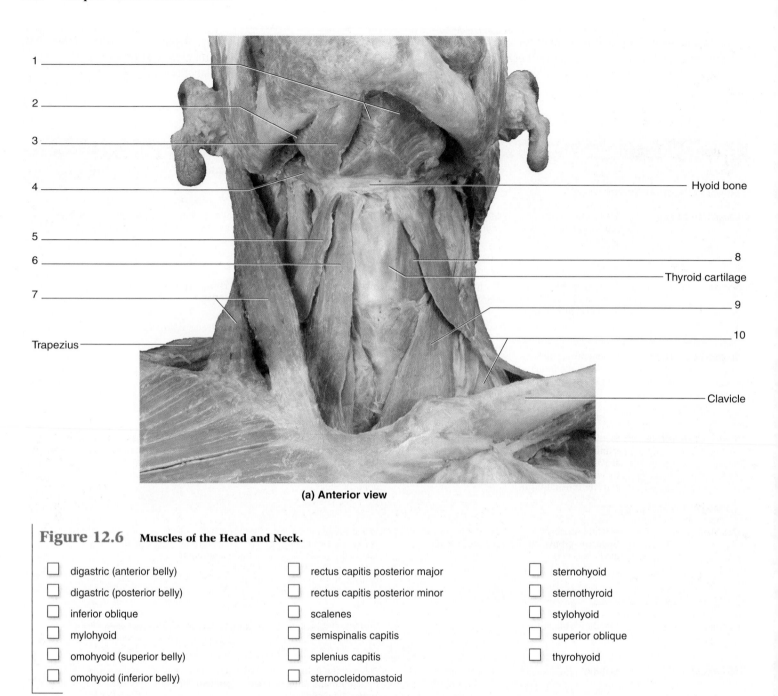

1

2

3

4 ———————————————— Hyoid bone

5

6 ———————————————————————————————— 8

———— Thyroid cartilage

7 ———————————————————————————— 9

———————————————————————— 10

Trapezius

———————— Clavicle

(a) Anterior view

Figure 12.6   Muscles of the Head and Neck.

☐ digastric (anterior belly)   ☐ rectus capitis posterior major   ☐ sternohyoid

☐ digastric (posterior belly)   ☐ rectus capitis posterior minor   ☐ sternothyroid

☐ inferior oblique   ☐ scalenes   ☐ stylohyoid

☐ mylohyoid   ☐ semispinalis capitis   ☐ superior oblique

☐ omohyoid (superior belly)   ☐ splenius capitis   ☐ thyrohyoid

☐ omohyoid (inferior belly)   ☐ sternocleidomastoid

Study Tip!

When studying muscles, or any other anatomical structure for that matter, remember that directional terms like "anterior/posterior" or "medial/lateral" are used in the naming of muscles only when necessary. That is, they are generally used only when there are two or more similar muscles with the same name. For example: if there is a "major" (e.g., zygomaticus major), then there must also be a "minor" (e.g., zyogmaticus minor); if there is a "superior" (e.g., superior oblique), there must also be an "inferior" (e.g., inferior oblique). This may seem simple, but it is not always obvious to the beginning student of anatomy.

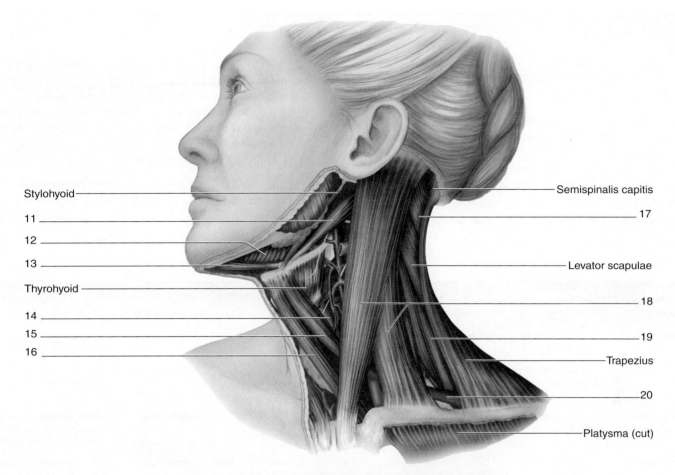

Stylohyoid

11

12

13

Thyrohyoid

14

15

16

Semispinalis capitis

17

Levator scapulae

18

19

Trapezius

20

Platysma (cut)

(b) Anterolateral view

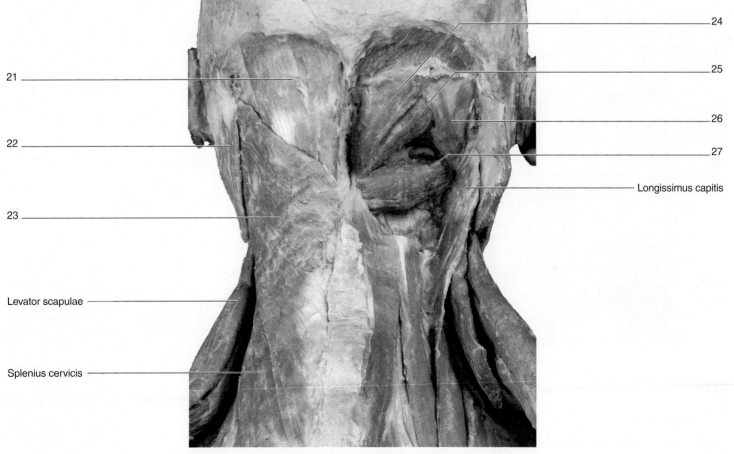

21

22

23

Levator scapulae

Splenius cervicis

24

25

26

27

Longissimus capitis

(c) Posterior view

EXERCISE 12.7 Suboccipital Muscles and the Suboccipital Triangle

1. Observe a prosected cadaver or a classroom model of the head and neck demonstrating the suboccipital muscles (**table 12.7**).

2. The suboccipital muscles are responsible for creating the small, fine movements of the head. They attach to the occipital bone, the atlas, and the axis. Before you study these muscles, recall the movements that occur at the **atlanto-occipital joint** and the **atlantoaxial joint**:

   Atlanto-occipital joint: flexion/extension, as in saying yes.
   Atlantoaxial joint: lateral rotation, as in saying no.

3. In addition to their role in creating fine neck movements, three of the muscles of the suboccipital region (the superior oblique, inferior oblique, and rectus capitis posterior major) compose a clinically relevant landmark called the **suboccipital triangle** (**figure 12.7**). The **vertebral artery** and **suboccipital nerve** are located within this triangle. Can you identify these structures on the cadaver or models?

4. Using table 12.7 as a guide, identify the **suboccipital muscles** listed in figure 12.7 on a cadaver or on a classroom model of the head and neck. Then label them in figure 12.7.

Table 12.7	Suboccipital Muscles				
Muscle	**Origin**	**Insertion**	**Action**	**Innervation**	**Word Origins**
Inferior Oblique	Axis (spinous process)	Atlas (transverse process)	Laterally rotates the neck (at the atlantoaxial joint)	Suboccipital	*inferior*, lower + *obliquus*, slanted,
Superior Oblique	Atlas (transverse process)	Occipital bone (lateral third of the inferior nuchal line)	Extends and laterally rotates the neck (at the atlanto-occipital joint)	Suboccipital	*superus*, above + *obliquus*, slanted,
Rectus Capitis Posterior Major	Axis (spinous process)	Occipital bone (middle portion of the inferior nuchal line)	Laterally rotates the neck	Suboccipital	*rectus*, straight, + *caput*, head, + *posterior*, the back, + *magnus*, great
Rectus Capitis Posterior Minor	Atlas (posterior tubercle)	Occipital bone (medial third of the inferior nuchal line)	Laterally rotates the neck (at the atlanto-occipital joint)	Suboccipital	*rectus*, straight, + *caput*, head, + *posterior*, the back, + *minor*, smaller

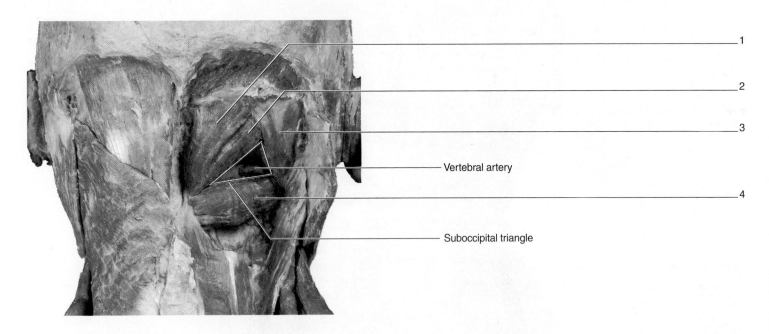

1
2
3
Vertebral artery
4
Suboccipital triangle

Figure 12.7 Suboccipital Muscles. Posterior view of the neck.

☐ inferior oblique   ☐ rectus capitis posterior minor
☐ rectus capitis posterior major   ☐ superior oblique

5. In the space to the right, draw a representative sketch of the suboccipital muscles. On top of your drawing, note the boundaries of the suboccipital triangle and the locations of the vertebral artery and the suboccipital nerve.

WHAT DO YOU THINK?

1 How does the vertebral artery get into the suboccipital triangle from the thoracic cavity? (Hint: Refer back to the special characteristics of cervical vertebrae described in table 8.5.)

Muscles of the Vertebral Column

Muscles that move the vertebral column are complex in both location and function. Your ability to identify all of the muscles listed here will depend on the degree to which the cadaver in your laboratory is dissected or the type(s) of models available to you in the laboratory. To simplify the process of learning these muscles, initially focus your attention on learning how the groups of muscles are arranged from superficial to deep, and from medial to lateral. Then turn your attention to specifically identifying the individual muscles belonging to each group.

EXERCISE 12.8 Muscles of the Vertebral Column

1. Observe a prosected cadaver or a classroom model of the thorax/abdomen that demonstrates muscles of the vertebral column (**table 12.8**).

2. Using table 12.8 as a guide, identify the **muscles of the vertebral column** listed in **figure 12.8** on a cadaver or on models of the thorax and abdomen. Then label them in figure 12.8.

3. The largest muscles of the back that move the vertebral column are collectively referred to as the **erector spinae** (*erector*, to make erect or straight, + *spina*, spine). The erector spinae consist of three muscle groups that form long columns along both sides of the vertebral column. The muscle groups, from medial to lateral are: spinalis, longissimus, and iliocostalis. Deep to the erector spinae are the **transversospinal** group of muscles, so named because they attach to transverse and spinous processes of adjacent vertebrae, and several smaller muscles that create fine movements of the vertebral column: interspinales and intertransversarii. The deeper muscles can be difficult to identify on cadavers or models. In general, the **semispinalis** muscles are the most superficial of the transversospinal muscles, span 5–6 vertebrae, and are most highly developed in the cervical and upper thoracic regions of the vertebral column. The **multifidus** lie deep to the semispinalis muscles, span 3–4 vertebrae, and are most highly developed in the lumbar region of the vertebral column. Finally, the **rotatores** are the deepest muscles of the transversospinal group, span 1–2 vertebrae, and are most highly developed in the lower thoracic region of the vertebral column.

Table 12.8	Muscles of the Vertebral Column					
Muscle Group	**Individual Muscles**	**Origin**	**Insertion**	**Action**	**Innervation**	**Word Origins**
SUPERFICIAL LAYER — SPLENIUS MUSCLES						
Splenius Muscles	The splenius muscles are thick, flat muscles on the lateral and posterior aspect of the neck.	Midline	Cervical vertebrae and skull	Holds the deep neck muscles in position; extends, laterally flexes, and laterally rotates the neck	Dorsal rami of spinal nerves	*splenius*, bandage
REGIONS	Capitis (splenius capitis)	Ligamentum nuchae and T_1–T_6 vertebrae (spinous processes)	Superior nuchal line (lateral aspect)	Bilateral: extends the neck Unilateral: laterally rotates and laterally flexes the neck to the same side Bilateral: extends the neck	Dorsal rami of spinal nerves	*splenion*, a bandage, + *caput*, head
	Cervicis (splenius cervicis)	Nuchal ligament and C_7–T_4 (spinous processes)	C_1–C_3 vertebrae (posterior tubercles)	Unilateral: laterally rotates and laterally flexes the neck to the same side	Dorsal rami of spinal nerves	*cervix*, neck
INTERMEDIATE LAYER — ERECTOR SPINAE (SACROSPINALIS) MUSCLES						
Erector Spinae	The erector spinae muscles compose the intermediate layer of back muscles. They are arranged into groups.	Broad tendon covering the posterior iliac crest, the lumbar vertebrae, and the sacrum	Vertebrae and ribs	Bilateral: extends the vertebral column and the head/neck Unilateral: laterally bends the vertebral column	Dorsal rami of spinal nerves	*erector*, to make erect, + *spina*, spine
GROUPS	Iliocostalis (lateral group)	Broad tendon covering the posterior iliac crest, the lumbar vertebrae, and the sacrum	Ribs (angles of lower ribs) and cervical vertebrae (transverse processes)	Extends the vertebral column	Dorsal rami of spinal nerves	*ilium*, groin, + *costal*, rib
	Longissimus (intermediate group)	Broad tendon covering the posterior iliac crest, the lumbar vertebrae, and the sacrum	Ribs (between tubercles and angles), cervical and thoracic vertebrae (transverse processes), and mastoid process of temporal bone	Extends the neck and vertebral column and laterally rotates the head	Dorsal rami of spinal nerves	*longissimus*, longest
	Spinalis (medial group)	Broad tendon covering the posterior iliac crest, the lumbar vertebrae, and the sacrum	Vertebrae (spinous processes of upper thoracic), and skull	Extends the neck and vertebral column and laterally rotates the head	Dorsal rami of spinal nerves	*spina*, spine
SPINAL FLEXORS — QUADRATUS LUMBORUM						
	Quadratus lumborum	Iliac crest and transverse processes of lower lumbar vertebrae	Rib 12, transverse processes of upper lumbar vertebrae	Abducts the trunk	Ventral rami of lumbar spinal nerves	*quadratus*, square, + *lumbus*, loin

Table 12.8	Muscles of the Vertebral Column *(continued)*					
Muscle Group	**Individual Muscles**	**Origin**	**Insertion**	**Action**	**Innervation**	**Word Origins**
DEEP LAYER — TRANSVERSOSPINALIS MUSCLES						
Transversospinal Group	The transversospinal muscles are the deepest muscles of the back and lie between transverse and spinous processes of adjacent vertebrae.	Transverse processes of inferior vertebrae	Spinous process of cervical and thoracic vertebra above and/or the posterior aspect of the occipital bone	Extends and rotates the vertebral column; stabilizes the vertebrae during local movements of the vertebral column	Dorsal rami of spinal nerves	*transverse*, across, + *spina*, spine
Multifidus	NA	T_1–T_3 vertebrae (transverse processes), C_4–C_7 (articular processes), ilium and sacrum	Spinous process of vertebra located 2–4 segments superior to vertebra of origin	Assists with local extension and rotation of the vertebral column	Dorsal rami of spinal nerves	*multus*, much, + *findo*, to cleave
Rotatores	NA	Transverse processes of all vertebrae (most developed in the thoracic region)	Vertebral arch (between lamina and transverse process) of vertebra superior to the vertebra of origin	Assists with local extension and rotation of the vertebral column	Dorsal rami of spinal nerves	*rotatus*, to rotate
Semispinalis Group	The semispinalis muscles are the deepest muscles of the back and lie between transverse and spinous processes of adjacent vertebrae.	Transverse processes of inferior vertebrae	Spinous process of vertebra above and/or the posterior aspect of the occipital bone	Extends and rotates the vertebral column; stabilizes the vertebrae during local movements of the vertebral column	Dorsal rami of spinal nerves	*semis*, half, + *spina*, spine
Regions	Capitis (semispinalis capitis)	Inferior cervical and superior thoracic vertebrae (spinous and transverse processes)	Occipital bone (between superior and inferior nuchal lines)	Extends and rotates the vertebral column; stabilizes the vertebrae during local movements of the vertebral column	Dorsal rami of spinal nerves	*caput*, head
	Cervicis (semispinalis cervicis)	T_1–T_6 vertebrae (transverse processes)	C_2–C_3 vertebrae (spinous processes)	Extends and rotates the vertebral column; stabilizes the vertebrae during local movements of the vertebral column	Dorsal rami of spinal nerves	*cervix*, neck
	Thoracis (semispinalis thoracis)	T_6–T_{10} vertebrae (transverse processes)	C_5–T_4 vertebrae (spinous processes)	Extends and rotates the vertebral column; stabilizes the vertebrae during local movements of the vertebral column	Dorsal rami of spinal nerves	*thoracis*, thorax
MINOR DEEP BACK MUSCLES						
	Interspinales	Cervical and lumbar vertebrae (superior surfaces of spinous processes)	Spinous process of the vertebra superior to the vertebra of origin (inferior surface)	Extends and rotates the vertebral column	Dorsal rami of spinal nerves	*inter*, between, + *spina*, spine
	Intertransversarii	Cervical and lumbar vertebrae (transverse processes)	Transverse processes of vertebra above or below vertebra of origin	Laterally flexes the vertebral column	Dorsal rami of spinal nerves	*inter*, between, + *transversarii*, relating to the transverse process

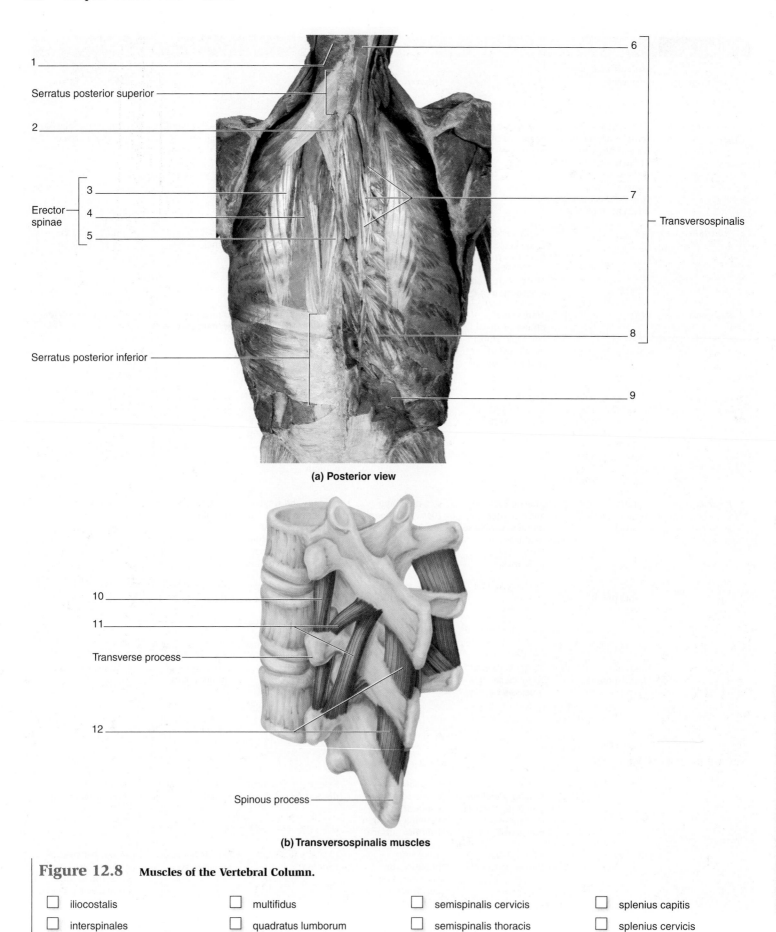

1 _____

Serratus posterior superior _____

2 _____

6

Erector spinae
3 _____
4 _____
5 _____

7

Transversospinalis

8

Serratus posterior inferior _____

9

(a) Posterior view

10 _____

11 _____

Transverse process _____

12 _____

Spinous process _____

(b) Transversospinalis muscles

Figure 12.8   **Muscles of the Vertebral Column.**

☐ iliocostalis

☐ interspinales

☐ intertransversarii

☐ longissimus

☐ multifidus

☐ quadratus lumborum

☐ rotatores

☐ semispinalis capitis

☐ semispinalis cervicis

☐ semispinalis thoracis

☐ spinalis

☐ splenius capitis

☐ splenius cervicis

Muscles of Respiration

The muscles of the thoracic cage and the diaphragm are the primary **muscles of respiration** (**table 12.9**). These muscles include the external and internal intercostals, the transversus thoracis, and the diaphragm.

Table 12.9	Muscles of Respiration				
Muscle	**Origin**	**Insertion**	**Action**	**Innervation**	**Word Roots**
Diaphragm	Inferior borders of rib 12, sternum, and the xiphoid process; costal cartilages of ribs 6–12, and lumbar vertebrae	Central tendon	Prime mover for inspiration; flattens when contracted, and increases intra-abdominal pressure and the size of the thoracic cavity	Phrenic nerves	*diaphragma*, a partition wall
External Intercostals	Inferior border of superior rib	Superior border of the rib below	Elevates the ribs	Intercostal nerves	*externus*, on the outside, + *inter*, between, + *costal*, rib
Internal Intercostals	Superior border of inferior rib	Inferior border of the rib above	Depresses the ribs	Intercostal nerves	*internus*, away from the surface, + *inter*, between, + *costal*, rib
Transversus Thoracis	Posterior surface of the lower half of the body of the sternum	Costal cartilages of ribs 2–6 (posterior surface)	Depresses the ribs	Intercostal nerves	*transversus*, crosswise, + *thoracis*, thorax
Anterior Scalene	C_3–C_6 (transverse processes)	First rib (scalene tubercle)	Elevates the first rib	Cervical plexus	*skalenos*, uneven
Middle Scalene	C_2–C_6 (transverse processes)	First rib (posterior to groove for subclavian artery)	Elevates the first rib	Cervical plexus	*skalenos*, uneven
Posterior Scalene	C_4–C_6 (transverse processes)	Second rib (lateral surface)	Elevates the second rib	Cervical and brachial plexuses	*skalenos*, uneven

EXERCISE 12.9 Muscles of Respiration

1. Observe a prosected cadaver or a classroom model of the thorax/abdomen that demonstrates muscles of the thoracic cage.

2. Using table 12.9 and your textbook as a guide, identify the muscles of respiration listed in **figure 12.9** on a cadaver or on a model of the thorax. Then label them in figure 12.9.

3. *Intercostals*–The majority of the muscle mass of the **external intercostals** is located on the posterior and lateral thorax, extending from the vertebral column to the **midclavicular line** (a vertical line that passes through the middle of the clavicle; see **figure 12.10**) on the anterior surface of the thorax. The **external intercostal membrane** lies in place of the external intercostal muscles in the space between the midclavicular line and the sternum. Notice that the muscle fibers of the external intercostals are arranged obliquely, pointing in an inferomedial direction.

4. On a cadaver you can identify the external intercostal membrane because the connective tissue fibers parallel the direction of the muscle fibers of the external intercostals. In fact, if you look closely, you will see muscle adjacent to the sternum in the intercostal spaces, but that muscle is an *internal* intercostal muscle. If you are observing classroom models you will most likely not be able to identify this membrane.

5. In contrast to the external intercostals, the majority of the muscle mass of the **internal intercostals** is located on the anterior surface of the thorax. These muscles extend from the sternum to the **scapular line** (a vertical line that passes through the inferior angle of the scapula; figure 12.10) on the posterior thorax. The **internal intercostal membrane** lies in place of the internal intercostal muscles on the posterior thorax between the scapular line and the vertebral column. The muscle fibers of the internal intercostals are arranged obliquely, pointing in an inferolateral direction, at right angles to the fibers of the external intercostals.

6. If possible, remove the breast plate on the cadaver (or on a model of the thorax) and observe its interior surface. Here you will see, lying adjacent to the inferior part of the sternum, the **transversus thoracis** muscle, which consists of several muscle bellies running obliquely. The transversus thoracis assists in depression of the ribs during forced expiration.

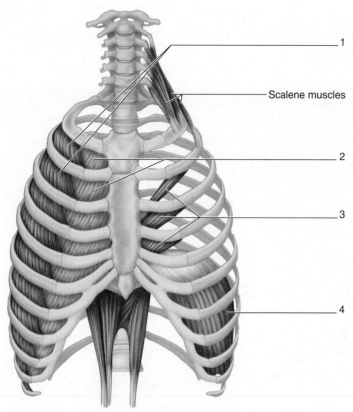

(a) Anterior view

1 — Scalene muscles

2

3

4

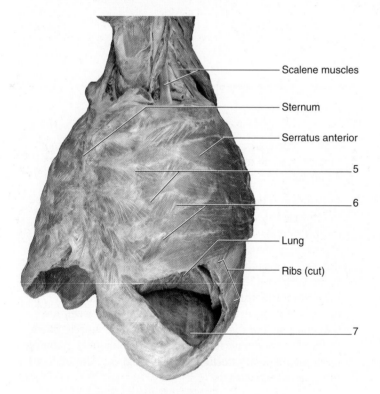

Scalene muscles

Sternum

Serratus anterior

5

6

Lung

Ribs (cut)

7

(b) Anterolateral view

Figure 12.9   Muscles of Respiration.

☐ diaphragm   ☐ internal intercostals

☐ external intercostals   ☐ transverse thoracis

7. In the space below, make a representative drawing of the intercostal muscles, noting the fiber orientation for each of them.

8. *Diaphragm*–The **diaphragm** has a broad origin along the bones that constitute the lower border of the thoracic cage. The muscle has a unique central attachment, the **central tendon of the diaphragm**. Several important structures pass through the diaphragm, including the aorta, inferior vena cava, and esophagus. The aorta and inferior vena cava have passages through the central tendon, whereas the passage for the esophagus is surrounded by the muscle fibers of the diaphragm. Very practical consequences result from this arrangement. As the diaphragm contracts during inspiration, its muscle fibers squeeze the esophagus and act as a sphincter to prevent stomach contents from being pushed back into the esophagus. In contrast, because the aorta and inferior vena cava pass through the central tendon, they are not constricted during inspiration. Remove the breast plate from the cadaver or model and observe the diaphragm. The structures within the diaphragm are best seen from an inferior view. If possible, try and observe the diaphragm from both superior and inferior points of view.

9. In the space below, make a representative drawing of the diaphragm, noting the locations of the passages for the aorta, inferior vena cava, and esophagus.

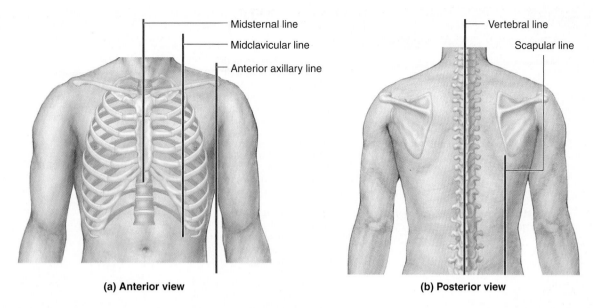

Figure 12.10 Points of Reference on the Anterior and Posterior Thorax.

Muscles of the Abdominal Wall

Muscles of the abdominal wall include the external obliques, internal obliques, transversus abdominis, and rectus abdominis.

EXERCISE 12.10 Muscles of the Abdominal Wall

1. Observe a prosected cadaver or a model of the thorax/ abdomen that demonstrates muscles of the abdominal wall (**table 12.10** and **figures 12.11–12.13**).

2. The muscle fiber orientation of the abdominal muscles parallels the muscle fiber orientation of the intercostal muscles. Like the external intercostals, the **external**

Table 12.10	**Muscles of the Abdominal Wall**				
Muscle	**Origin**	**Insertion**	**Action**	**Innervation**	**Word Origins**
External Oblique	Anterior surface of inferior 8 ribs	Linea alba and the anterior iliac crest	Flexes and rotates the trunk; compresses the abdominal viscera	Spinal nerves T8–T12, L1	*externus*, on the outside, + *obliquus*, slanting
Internal Oblique	Thoracolumbar fascia, lateral half of inguinal ligament, and iliac crest	Linea alba, iliac crest, pubic tubercle, and the inferior border of last 4 ribs, costal cartilages of ribs 8–10	Flexes and rotates the trunk; compresses the abdominal viscera	Spinal nerves T8–T12, L1	*internus*, away from the surface, + *obliquus*, slanting
Rectus Abdominis	Pubic symphysis and crest	Xiphoid process and the costal cartilages of ribs 5–7	Flexes and rotates the trunk; compresses the abdominal viscera	Spinal nerves T7–T12	*rectus*, straight, + *abdominis*, the abdomen
Transversus Abdominis	Lateral third of inguinal ligament, iliac crest, costal cartilages of inferior 6 ribs	Linea alba and pubic crest	Compresses the abdominal viscera	Spinal nerves T8–T12, L1	*transversus*, crosswise, + *obliquus*, slanting

STRUCTURES RELATED TO ABDOMINAL MUSCULATURE

Structure	**Attachment 1**	**Attachment 2**	**Description**	**Innervation**	**Word Origins**
Aponeurosis	NA	NA	A broad, flat tendon, such as those connecting abdominal muscles to the linea alba	NA	*apo-*, from, + *neuron*, sinew

(continued on next page)

Table 12.10	Muscles of the Abdominal Wall (continued)				
Structure	**Attachment 1**	**Attachment 2**	**Description**	**Innervation**	**Word Origins**
Iguinal Canal	NA	NA	An oblique passage in the anterior abdominal wall located superior to the inguinal ligament; it is formed from the aponeuroses of the external and internal oblique muscles	NA	*inguen*, groin
Inguinal Ligament	Anterior superior iliac spine	Pubic tubercle	A structure formed from the aponeurosis of the external oblique muscle; an important anatomical landmark in the inguinal region	NA	*inguen*, groin
Linea Alba	Xiphoid process of the sternum	Pubic symphysis	Literally, the "white line"; a tendinous structure that acts as the insertion point for the oblique and transversus abdominis muscles	NA	*linea*, line, + *alba*, white
Rectus Sheath	NA	NA	A connective tissue sheath that surrounds the rectus abdominis muscle and is formed from the aponeuroses of the external oblique, internal oblique, and transversus abdominis muscles	NA	*rectus*, referring to the rectus abdominis muscle
Tendinous Intersections	NA	NA	Tendinous bands that run across a muscle; in this case, the tendinous intersections are the structures that separate the parts of the rectus abdominis muscle	NA	*tendo-*, to stretch out, + *inscriptio*, to write on

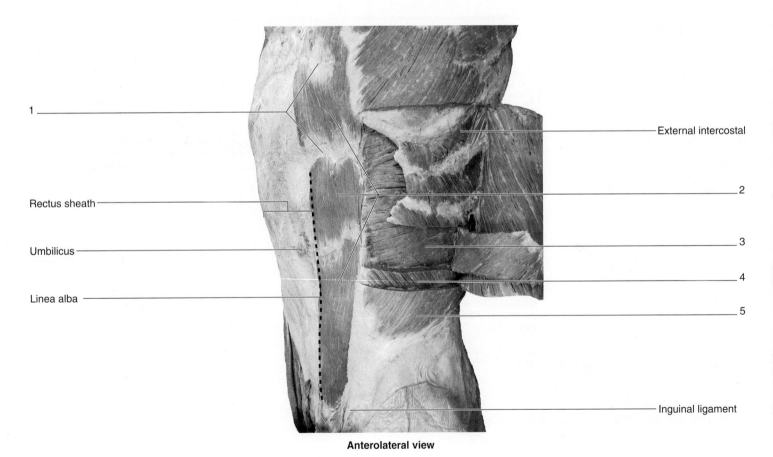

1

External intercostal

2

Rectus sheath

3

Umbilicus

4

Linea alba

5

Inguinal ligament

Anterolateral view

Figure 12.11   **Muscles of the Abdominal Wall.**

☐ external oblique   ☐ rectus abdominis   ☐ transversus abdominis

☐ internal oblique   ☐ tendinous intersections

obliques have muscle fibers that point in an inferomedial direction. Like the internal intercostals, the **internal obliques** have muscle fibers that point in an inferolateral direction. Deep to the internal obliques, the **transversus abdominis** has muscle fibers that run in a transverse, or horizontal, direction. All three of these muscles have broad, flat tendons called **aponeuroses** (*apo-*, from, + *neuron*, sinew). The aponeuroses of these muscles begin at the midclavicular line and extend to a central, tendonlike structure called the **linea alba**. In addition, the aponeuroses of these muscles form the **rectus sheath**, which surrounds the fourth abdominal muscle, the **rectus abdominis**.

3. Using table 12.10 as a guide, identify the **abdominal muscles** listed in figure 12.11 on a cadaver or on a classroom model of the abdomen. Then label them in figure 12.11.

EXERCISE 12.11 The Rectus Sheath, Inguinal Ligament, and Inguinal Canal

1. Observe a prosected cadaver or a classroom model of the thorax/abdomen that demonstrates muscles of the abdominal wall. In this exercise you will focus on important structures associated with the abdominal musculature.

2. The **rectus sheath (figure 12.12**; see table 12.10) is a structure formed from the aponeuroses of the external and internal obliques and the transversus abdominis muscles (this relationship is true only of the sheath superior to the umbilicus, so your observations should be made in that location). The sheath has an *anterior border* formed by the aponeurosis of the external oblique and part of the aponeurosis of the internal oblique. It has a *posterior border* formed by the remaining part of the aponeurosis of the internal oblique and the aponeurosis of the transversus abdominis. The rectus abdominis muscle is located within the rectus sheath, and is divided into four sections by connective tissue partitions called **tendinous intersections**. The tendinous intersections effectively divide one very long muscle into four smaller muscles, arranged in a series. This allows the muscle as a whole to make a greater overall change in length, in addition to increasing its force of contraction.

3. The **inguinal ligament (figure 12.13**; table 12.10) is formed by the *aponeurosis of the external oblique* muscle. This ligament extends from the anterior superior iliac spine (ASIS) laterally, to the pubic tubercle medially. Instead of being a straight ligament, the ligament folds back upon itself, forming a trough. The inguinal ligament is an important landmark of the abdomen and thigh. In addition,

the trough formed by the aponeurosis of the external oblique forms part of the *inguinal canal*.

4. The **inguinal canal** (figure 12.13; table 12.10) is an oblique passageway in the inferior abdominal wall. Its floor is formed by the trough of the **aponeurosis of the external oblique**. Its roof is formed by fibers of the **aponeurosis of the internal oblique**. Within this canal, structures pass from the abdomen into the subcutaneous tissues of the **perineum** (*perneon*, the area between the thighs below the pelvic diaphragm). In males, structures that pass through compose the **spermatic cord**, which consists of the testicular artery, vein, and nerve; lymphatic vessels; and the ductus deferens. In females, the **round ligament of the uterus**, a suspensory ligament, passes through. The **superficial inguinal ring**, is located lateral to the pubic symphysis. It is composed of fibers of the aponeuroses of the external oblique aponeurosis, and it is the path of exit for structures that pass through the inguinal canal. On a male cadaver or on a classroom model of the abdomen, identify the spermatic cord as it passes through the superficial inguinal ring. Attempt to identify the round ligament of the uterus on a female cadaver, keeping in mind that identification of the round ligament of the uterus is often quite difficult.

WHAT DO YOU THINK?

2 In an **indirect inguinal hernia**, abdominal contents, such as a loop of small intestine, pass into the inguinal canal. Why do you think such hernias are more common in males than in females?

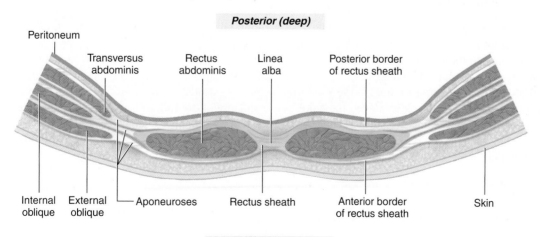

Posterior (deep)

Peritoneum
Transversus abdominis
Rectus abdominis
Linea alba
Posterior border of rectus sheath

Internal oblique
External oblique
Aponeuroses
Rectus sheath
Anterior border of rectus sheath
Skin

Anterior (superficial)

Figure 12.12 **The Rectus Sheath.** A transverse section through the abdominal wall demonstrates components of the rectus sheath.

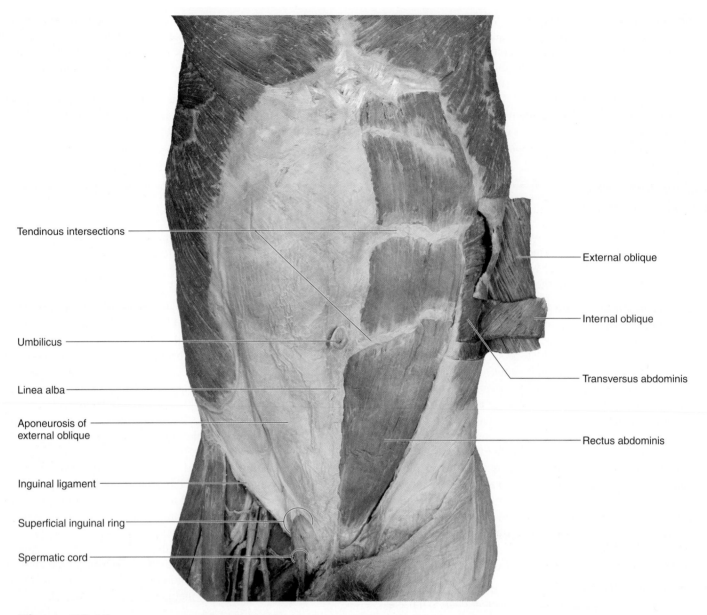

Tendinous intersections

External oblique

Umbilicus

Internal oblique

Linea alba

Transversus abdominis

Aponeurosis of external oblique

Rectus abdominis

Inguinal ligament

Superficial inguinal ring

Spermatic cord

Figure 12.13   **The Anterior Abdominal Wall.**

Muscles of the Pelvic Floor

Muscles of the pelvic floor are extremely important for holding pelvic contents in, and for forming sphincters around key openings, such as the external openings to the urethra and anus. In addition, they surround the female vaginal orifice. When we are young, we don't often think of these muscles in terms of muscles we want to make "bigger and stronger." However, the older we get, the more we understand the importance of these muscles, and the more we understand why keeping them strong is as important as—arguably even *more* important than—having strong abdominal or limb muscles. Weak pelvic floor muscles are a contributing factor in a number of very uncomfortable and potentially embarrassing disorders, such as **incontinence** (the inability to control urination) and **prolapse** (*prolapsus*, a falling) of the uterus (where the uterus extrudes itself inferiorly through the vaginal orifice) or prolapse of the rectum (where a portion of the rectum extrudes inferiorly through the anal orifice). Clinically, two regions are described in the pelvic floor: the **urogenital triangle**, and the **anal triangle** (**figure 12.14**).

Figure 12.14   **The Urogenital and Anal Triangles.**
The division between urogenital and anal triangles is a line
that runs horizontally between the ischial tuberosities.

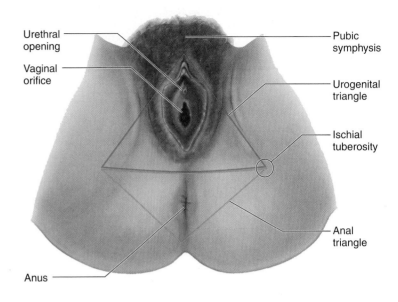

EXERCISE 12.12   Muscles of the Pelvic Floor

1. Observe a classroom model demonstrating muscles of the
   pelvic floor (**table 12.11**).

2. Using table 12.11 as a guide, identify the **muscles of the
   pelvic floor** listed in **figure 12.15** on classroom models.
   Then label them in figure 12.15.

Table 12.11	Muscles of the Pelvic Floor					
Muscle Group	**Individual Muscles**	**Origin**	**Insertion**	**Action**	**Innervation**	**Word Origins**
MUSCLES OF THE ANAL TRIANGLE						
	Coccygeus	Ischial spine	Sacrum (inferior end)	Supports the pelvic viscera and flexes the coccyx	Sacral (S4–S5)	*coccyx*, a cuckoo (relating to the bony coccyx)
Levator Ani	The muscles in this group form the posterior aspect of the *urogenital diaphragm*. As such, they support the pelvic floor. In addition, the levator ani muscles form the external anal sphincter.	Pubic bone and ischial spine	Coccyx and median raphe	Resists increases in intra-abdominal pressure; constricts (and closes) the opening to the anus	Pudendal (S2–S4)	*levatus*, to lift up, + *anus*, inferior opening of GI tract
	Iliococcygeus	Pubic bone and ischial spine	Coccyx and median raphe	Supports the pelvic viscera and flexes the coccyx	Pudendal (S2–S4)	*ilium*, hip, + *coccyx*, a cuckoo (relating to the bony coccyx)
	Pubococcygeus	Pubic bone (posterior aspect of the body)	Coccyx and median raphe	Supports the pelvic viscera and flexes the coccyx	Pudendal (S2–S4)	*pubo-*, pubic (bone), + *coccyx*, a cuckoo (relating to the bony coccyx)
	Puborectalis	Pubic bone and ischial spine	Coccyx and median raphe	Forms a sphincter around the anal opening; must relax for defecation to occur	Pudendal (S2–S4)	*pubo-*, pubic (bone), + *rectus*, straight (relating to the rectum)

(continued on next page)

Table 12.11	Muscles of the Pelvic Floor *(continued)*					
Muscle Group	**Individual Muscles**	**Origin**	**Insertion**	**Action**	**Innervation**	**Word Origins**
MUSCLES OF THE UROGENITAL TRIANGLE						
Superficial Layer	Bulbospongiosus (female)	Connective tissue at base of clitoris	Perineal body	Narrows the vaginal opening and assists with erection of the clitoris	Pudendal (S2–S4)	*bulbus*, a bulb, + *spongiosus*, relating to the corpus spongiosum
	Bulbospongiosus (male)	Connective tissue at base of penis	Perineal body	Compresses erectile tissues in the penis to assist with erection and ejects urine or semen	Pudendal (S2–S4)	*bulbus*, a bulb, + *spongiosus*, relating to the corpus spongiosum
	Ischiocavernosus	Ischial tuberosities and ischial ramus	Pubic symphysis	Compresses erectile tissues at base of clitoris or penis to assist with erection	Pudendal (S2–S4)	*ischion*, hip joint, + *cavernosum*, relating to the corpus cavernosum
	Superficial transverse perineal muscle	Ischial ramus	Perineal body	Supports the pelvic viscera and flexes the coccyx	Pudendal (S2–S4)	*pubo-*, pubic (bone), + *coccyx*, a cuckoo (relating to the bony coccyx)
Deep Layer	Deep transverse perineal muscle	Ischial ramus	Median raphe of urogenital diaphragm	Supports pelvic viscera	Pudendal (S2–S4)	*perineon*, the area between the thighs between the coccyx and pubis
	External urethral sphincter	Ischial ramus and pubic ramus	Median raphe of urogenital diaphragm	Forms a sphincter around the membranous urethra; must relax for micturition (urination) to occur	Pudendal (S2–S4)	*urethro-*, urethra, + *sphincter*, a band or lace

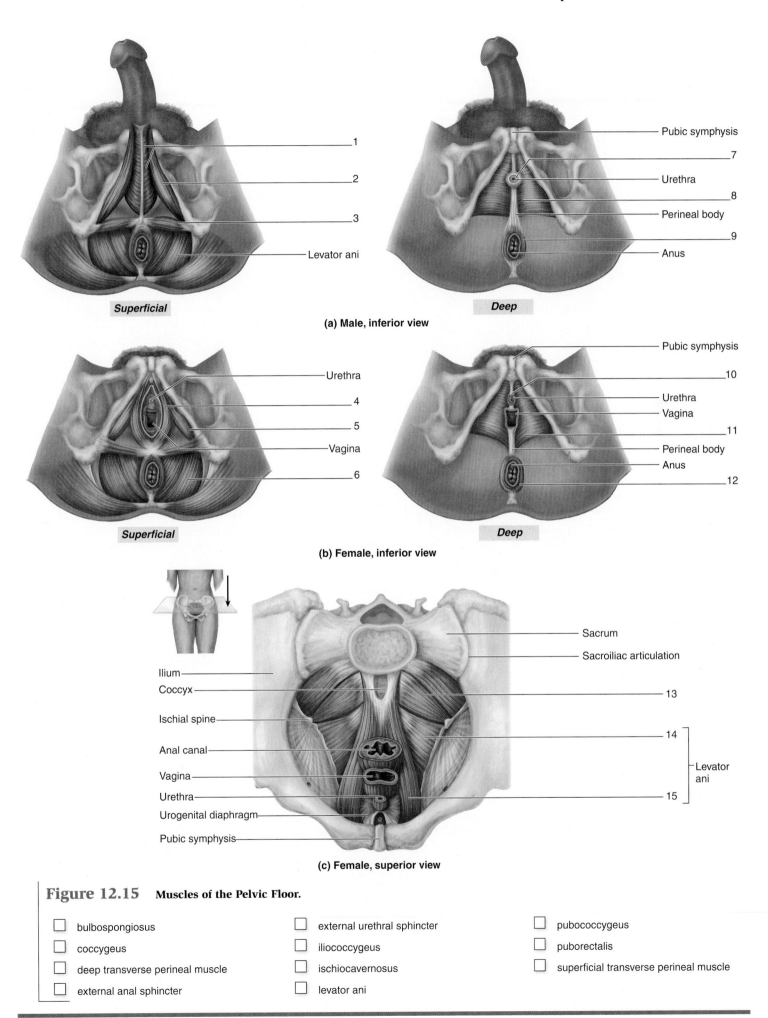

1

2

3

Levator ani

Superficial

Pubic symphysis

7

Urethra

8

Perineal body

9

Anus

Deep

(a) Male, inferior view

Urethra

4

5

Vagina

6

Superficial

Pubic symphysis

10

Urethra

Vagina

11

Perineal body

Anus

12

Deep

(b) Female, inferior view

Sacrum

Sacroiliac articulation

Ilium

Coccyx

Ischial spine

Anal canal

Vagina

Urethra

Urogenital diaphragm

Pubic symphysis

13

14

15

Levator ani

(c) Female, superior view

Figure 12.15 Muscles of the Pelvic Floor.

- [ ] bulbospongiosus
- [ ] coccygeus
- [ ] deep transverse perineal muscle
- [ ] external anal sphincter
- [ ] external urethral sphincter
- [ ] iliococcygeus
- [ ] ischiocavernosus
- [ ] levator ani
- [ ] pubococcygeus
- [ ] puborectalis
- [ ] superficial transverse perineal muscle

Chapter 12: Axial Muscles

Name: _____

Date: _____ Section: _____

POST-LABORATORY WORKSHEET

1. If the facial nerve is severed, which of the following muscles will *not* be paralyzed?

   a. buccinator

   b. masseter

   c. frontalis

   d. zygomaticus

   e. risorius

2. List the muscles used for normal inspiration:

   _____

   _____

3. List the muscles used for normal expiration:

   _____

   _____

4. List the muscles used for forced expiration (as in coughing):

   _____

   _____

5. Explain how the external obliques, internal obliques, and transversus abdominis muscles relate to the rectus sheath.

   _____

   _____

6. The inguinal ligament is formed from the aponeurosis of which abdominal muscle?

   _____

7. What structures run through the inguinal canal in males and females?

   Males: _____

   Females: _____

8. Explain how the sternocleidomastoid and splenius muscles can act as either synergists or antagonists for the actions of neck flexion, extension, and lateral rotation.

   _____

   _____

   _____

   _____

   _____

   _____

9. List the three structures that form the borders of the suboccipital triangle:

   a. _____

   b. _____

   c. _____

10. Explain the clinical significance of the suboccipital triangle (that is, what structures are located within the triangle?).

   _____

   _____

   _____

   _____

   _____

11. For each of the eye movements listed below, list the extrinsic eye muscle that creates that action (consider only movement of the right eye).

   Look up: _____

   Look down: _____

   Look lateral: _____

   Look medial: _____

   Look down and out: _____

   Look up and out: _____

   Look down and medial: _____

   Look down and lateral: _____

12. In the space below, describe the boundaries of the urogenital and anal triangles.

   _____

   _____

   _____

   _____

Appendicular Muscles

13

OUTLINE and OBJECTIVES

Gross Anatomy 245

Muscles That Move the Pectoral Girdle and Glenohumeral Joint 245

EXERCISE 13.1: MUSCLES THAT MOVE THE PECTORAL GIRDLE AND GLENOHUMERAL JOINT 246
- Identify the muscles that connect the pectoral girdle to the thorax
- Describe the locations and actions of muscles that connect the pectoral girdle to the thorax
- Demonstrate the actions of the trapezius, and name a muscle that acts as a synergist for the actions of elevation and adduction of the scapula
- Describe the consequences of damage to the serratus anterior
- List the four muscles that make up the rotator cuff, and explain the functional importance of these muscles

Upper Limb Musculature 248

EXERCISE 13.2: COMPARTMENTS OF THE ARM 248
- Identify the muscles of the anterior and posterior compartments of the arm
- Describe the locations and actions of muscles of the anterior and posterior compartments of the arm
- Explain the role of the biceps brachii in the actions of elbow flexion and forearm supination

EXERCISE 13.3: COMPARTMENTS OF THE FOREARM 251
- Identify the muscles of the anterior and posterior compartments of the forearm
- Describe the locations and actions of muscles of the anterior and posterior compartment of the forearm
- Demonstrate the location of the palmaris longus on yourself
- Describe the relationship between the tendons of the flexor digitorum superficialis and the flexor digitorum profundus as they attach to the phalanges
- List the three structures forming the boundaries of the anatomic snuffbox, and explain why the anatomic snuffbox is a relevant clinical landmark

EXERCISE 13.4: INTRINSIC MUSCLES OF THE HAND 256
- Identify the intrinsic muscles of the hand
- Describe the locations of the thenar eminence and the hypothenar eminence
- Describe the actions of the dorsal and palmar interossei

Muscles That Move the Hip Joint 258

EXERCISE 13.5: MUSCLES THAT MOVE THE HIP 258
- Identify the muscles that act about the hip, and describe their actions
- Explain the roles of the gluteus medius and gluteus minimus in locomotion
- Explain the importance of the piriformis as a clinically relevant landmark
- Describe the composition, location, and function of the iliopsoas

Lower Limb Musculature 260

EXERCISE 13.6: COMPARTMENTS OF THE THIGH 261
- Identify the muscles of the anterior, medial, and posterior compartments of the thigh
- Describe the locations and actions of muscles of the anterior, medial, and posterior compartments of the thigh
- Name the two muscles of the anterior compartment of the thigh that flex the hip joint
- Identify the borders of the femoral triangle, and explain the clinical significance of the femoral triangle
- Explain how the gracilis is an exception to the rule for muscles of the medial compartment of the thigh

EXERCISE 13.7: COMPARTMENTS OF THE LEG 265
- Identify the muscles of the anterior, lateral, and posterior compartments of the leg
- Describe the locations and actions of muscles of the anterior, lateral, and posterior compartments of the leg
- Name two muscles that act as antagonists to the fibularis muscles for the function of everting the ankle
- Name the muscles that compose the triceps surae
- Explain how the tibialis anterior and tibialis posterior muscles can act as either synergists or antagonists of each other

EXERCISE 13.8: INTRINSIC MUSCLES OF THE FOOT 268
- Identify the intrinsic muscles of the foot
- Describe the locations and actions of the intrinsic muscles of the foot
- Identify similarities and differences between the intrinsic muscles of the hand and the intrinsic muscles of the foot

Module 6: MUSCULAR SYSTEM

INTRODUCTION

The **appendicular muscles** are the muscles that most of us tend to be most familiar with. Anyone who has gone to a gym and lifted weights, or who has admired the muscular body of a basketball player or gymnast, has at least some familiarity with these muscles. They are the muscles we use to perform active tasks such as typing, walking, running, and lifting. They are the bulging muscles we see in the arms and legs of elite athletes. When lifting a heavy weight, we can feel the muscles in our upper limbs working, so it is fairly easy to determine which muscles we are using to perform the particular action. If we cannot figure this out right away from the fatigue or pain we experience immediately in the working muscle, we most definitely will figure it out in the next two days as delayed-onset muscle soreness (DOMS) sets in.

All of these activities give us a fascination and curiosity about the muscles responsible for creating movement. However, once we are required to view these muscles on a model, photo, or cadaver, and are asked to know their names, attachments, and actions in the context of a human anatomy class, we often forget to think about them in practical terms (such as, "What muscles am I using to perform this action?"). Instead, they become the bane of the anatomy student, their study dreaded as one of the most difficult tasks in the course.

However, you need not fear this task as long as you approach it with the right frame of mind. As you work through the task of learning names, attachments, and actions for the appendicular muscles, try to identify the muscles on your own body. Practice using them. An excellent way to do this is to go to a gym that has weight machines. Observe the illustrations on the weight machines that demonstrate what muscle or muscles the exercise is meant to work on. Then practice the movement *without using any weights* (especially if you have never lifted weights before). As you perform the stated action of the muscle(s), feel the muscle(s) produce tension under your skin. Once you can make the connection between the muscles in your own body and the muscles you see on the cadaver, on models, or in photographs, you will begin to truly appreciate their functions. In addition, the one "cheat sheet" you *can* bring into the laboratory with you on exam day is yourself!

Chapter 13: Appendicular Muscles

Name: _____

Date: _____ Section: _____

PRE-LABORATORY WORKSHEET

1. Match the compartment listed in column A with the action(s) described in column B. Actions are those performed by the majority of muscles within the listed compartment.

   Column A

   _____ 1. anterior arm

   _____ 2. posterior arm

   _____ 3. anterior forearm

   _____ 4. posterior forearm

   _____ 5. anterior thigh

   _____ 6. posterior thigh

   _____ 7. medial thigh

   _____ 8. anterior leg

   _____ 9. posterior leg

   _____ 10. lateral leg

   Column B

   a. adduction of the thigh

   b. dorsiflexion and inversion of the ankle, extension of the digits

   c. extension of the hip, flexion of the knee

   d. extension of shoulder and elbow

   e. extension of the knee, flexion of the hip

   f. flexion of shoulder and elbow

   g. extension of the wrist and digits

   h. eversion of the ankle

   i. flexion of the wrist and digits

   j. plantarflexion and inversion of the ankle, flexion of the digits

2. Match the bony process (or small bone) listed in column A with the appropriate bone (or appendage) listed in column B. Not all answer choices will be used, but each answer choice can be used only once.

   Column A

   _____ 1. greater trochanter

   _____ 2. calcaneus

   _____ 3. lesser tubercle

   _____ 4. olecranon process

   _____ 5. pisiform

   _____ 6. cuneiform

   _____ 7. lateral malleolus

   _____ 8. coracoid process

   Column B

   a. tarsal bone

   b. tibia

   c. clavicle

   d. humerus

   e. femur

   f. scapula

   g. ulna

   h. carpal bone

   i. radius

   j. fibula

3. Define the following terms:

   prime mover (agonist): _____

   _____

   synergist: _____

   _____

   antagonist: _____

   _____

4. Describe the region of the body that each of the following terms refers to.

upper limb: _____

_____

arm: _____

_____

forearm: _____

_____

lower limb: _____

_____

thigh: _____

_____

leg: _____

_____

IN THE LABORATORY

In this laboratory session, you will identify, name, and explore the structure and function of muscles that move the appendicular skeleton. Before you begin, find out from your laboratory instructor exactly which muscles you will be responsible for on your practical exams. By doing this first, you can cross out or edit the summary tables in this chapter so you focus only on those muscles you are required to know for your course. In addition, before you begin the more detailed study of the appendicular musculature in this chapter, make sure you have completed the exercises related to appendicular musculature in chapter 11. Those exercises introduced you to the major muscle groups and the common actions of those groups. Refer to tables 11.4 and 11.5 to review the muscle compartments of the upper and lower limbs, and the major actions of the muscles in each compartment.

Gross Anatomy

Muscles That Move the Pectoral Girdle and Glenohumeral Joint

The **pectoral girdle** consists of the clavicle and scapula. As you learned in chapter 10, the shoulder joint (glenohumeral joint) is a highly movable joint. This is largely because there is only minimal bony attachment between the pectoral girdle and the axial skeleton (the only point of bony attachment is the sternoclavicular joint).

Because of this arrangement, there must be very strong muscular attachments between the pectoral girdle and the axial skeleton. **Table 13.1** lists the characteristics of muscles whose primary action is about the pectoral girdle, and **table 13.2** lists muscles that move the glenohumeral Joint.

Table 13.1	Muscles That Move the Pectoral Girdle				
Muscle	**Origin**	**Insertion**	**Action***	**Innervation**	**Word Roots**
Pectoralis Minor	Ribs 3–5 (anterior surface)	Coracoid process of scapula	Protracts and depresses the scapula	Medial pectoral nerve	*pectus*, chest, + *minor*, smaller
Serratus Anterior	Ribs 1–8 (outer surface)	Scapula (medial border)	Protracts the scapula and rotates it superiorly; most important for holding the scapula flat against the ribcage; damage causes a "winging" of the scapula	Long thoracic nerve	*serratus*, a saw, + *anterior*, the front surface
Trapezius (Upper Portion)	Occipital bone, ligamentum nuchae, spine of C_7	Clavicle (lateral third) and scapula (acromial process)	Elevates the scapula and rotates it superiorly	Accessory nerve (CN XI)	*trapeza*, a table
Trapezius (Middle Portion)	Spines of T_1–T_5	Spine of scapula	Retracts, adducts, and stabilizes the scapula	Accessory nerve (CN XI)	*trapeza*, a table
Trapezius (Lower Portion)	Spines of T_1–T_{12}	Spine of scapula	Depresses the scapula and rotates it superiorly	Accessory nerve (CN XI)	*trapeza*, a table
Levator Scapulae	Transverse processes of C_1–C_4	Scapula (superior vertebral border)	Elevates the scapula and tilts the glenoid inferiorly	Dorsal scapular nerve	*levatus*, to lift, + *scapula*, the shoulder blade
Rhomboid Minor	Spines of C_7–T_1	Scapula (medial border)	Retracts, adducts, and stabilizes the scapula, tilts the glenoid inferiorly	Dorsal scapular nerve	*rhomboid*, resembling an oblique parallelogram, + *minor*, smaller
Rhomboid Major	Spines of T_2–T_5	Scapula (medial border)	Retracts, adducts, and stabilizes the scapula, tilts the glenoid inferiorly	Dorsal scapular nerve	*rhomboid*, resembling an oblique parallelogram, + *major*, larger

*Only actions that apply to movement of the pectoral girdle are listed.

Table 13.2	Muscles That Move the Glenohumeral Joint				
Muscle	**Origin**	**Insertion**	**Action***	**Innervation**	**Word Roots**
Pectoralis Major	Sternum, medial clavicle, and costal cartilages 2–6	Lateral part of intertubercular groove of humerus	Flexes and adducts arm	Medial pectoral nerve	*pectus*, chest, + *major*, larger
Pectoralis Minor	Anterior surfaces of ribs 3–5	Coracoid process of scapula	Flexes, adducts, medially rotates arm	Lateral pectoral nerve	*pectus*, chest, + *minor*, smaller
Coracobrachialis	Coracoid process of scapula	Midhumerus (medial surface)	Flexes and adducts arm	Musculocutaneous nerve	*coraco*, referring to the coracoid process, + *brachium*, the arm
Biceps Brachii (Long Head)	Supraglenoid tubercle	Radial tuberosity	Flexes arm (weak)	Musculocutaneous nerve	*bi*, two, + *caput*, head, + *brachium*, arm
Biceps Brachii (Short Head)	Coracoid process of scapula	Radial tuberosity	Flexes arm (weak)	Musculocutaneous nerve	*bi*, two, + *caput*, head, + *brachium*, arm
Deltoid (Anterior Fibers)	Clavicle (lateral third)	Humerus (deltoid tuberosity)	Flexes, adducts, medially rotates arm	Axillary nerve	*deltoid*, resembling the Greek letter delta (a triangle)
Deltoid (Middle Fibers)	Acromion, clavicle (lateral part), and scapula (lateral spine)	Humerus (deltoid tuberosity)	Abducts arm	Axillary nerve	*deltoid*, resembling the Greek letter delta (a triangle)
Deltoid (Posterior Fibers)	Spine of scapula	Humerus (deltoid tuberosity)	Extends, adducts, laterally rotates arm	Axillary nerve	*deltoid*, resembling the Greek letter delta (a triangle)
Supraspinatus (RC)	Supraspinous fossa	Humerus (greater tubercle)	Stabilizes arm, assists in abduction	Suprascapular nerve	*supra*, above, + *spina*, referring to the spine of the scapula
Latissimus Dorsi	Spinous process of T$_6$–L$_5$, iliac crest, and ribs 10–12	Humerus (intertubercular groove)	Extends, adducts, medially rotates arm	Thoracodorsal nerve	*latissimus*, widest, + *dorsum*, the back
Teres Major	Posterior surface of scapula at inferior angle	Crest of lesser tubercle on anterior humerus	Extends, adducts, medially rotates arm	Lower subscapular nerve	*teres*, round, + *major*, larger
Triceps Brachii (Long Head)	Infraglenoid tubercle of scapula	Olecranon process of ulna	Extends, adducts arm	Radial nerve	*tri*, three, + *caput*, head, + *brachium*, arm
Infraspinatus (RC)	Infraspinous fossa of scapula	Greater tubercle of humerus	Adducts and laterally rotates arm	Suprascapular nerve	*infra*, below, + *spina*, referring to the spine of the scapula
Teres Minor (RC)	Inferior angle of scapula	Greater tubercle of humerus	Adducts and laterally rotates arm	Axillary nerve	*teres*, round, + *minor*, smaller
Subscapularis (RC)	Subscapular fossa	Lesser tubercle of humerus	Medially rotates arm	Upper and lower scapular nerves	*sub*, under, + *scapula*, the shoulder blade

*Only actions that apply to movement of the glenohumeral joint are listed.
RC = a rotator cuff muscle, which is important in stabilization of glenohumeral joint, in addition to functions above.

EXERCISE 13.1   Muscles That Move the Pectoral Girdle and Glenohumeral Joint

1. Observe a prosected human cadaver or a classroom model demonstrating muscles of the thorax and upper limb.

2. Using tables 13.1 and 13.2 and your textbook as a guide, identify the **muscles that move the pectoral girdle and glenohumeral joint** listed in **figure 13.1** on the cadaver or on a classroom model. Then label them in figure 13.1.

3. *Trapezius*: The largest muscle that moves the pectoral girdle is the superficial **trapezius** muscle, which acts to anchor the scapula to the entire superior two-thirds of the vertebral column and to the back of the head. As noted in table 13.1, this muscle performs multiple actions on the scapula, depending upon which part of the muscle contracts at a given time. Deep to the trapezius are

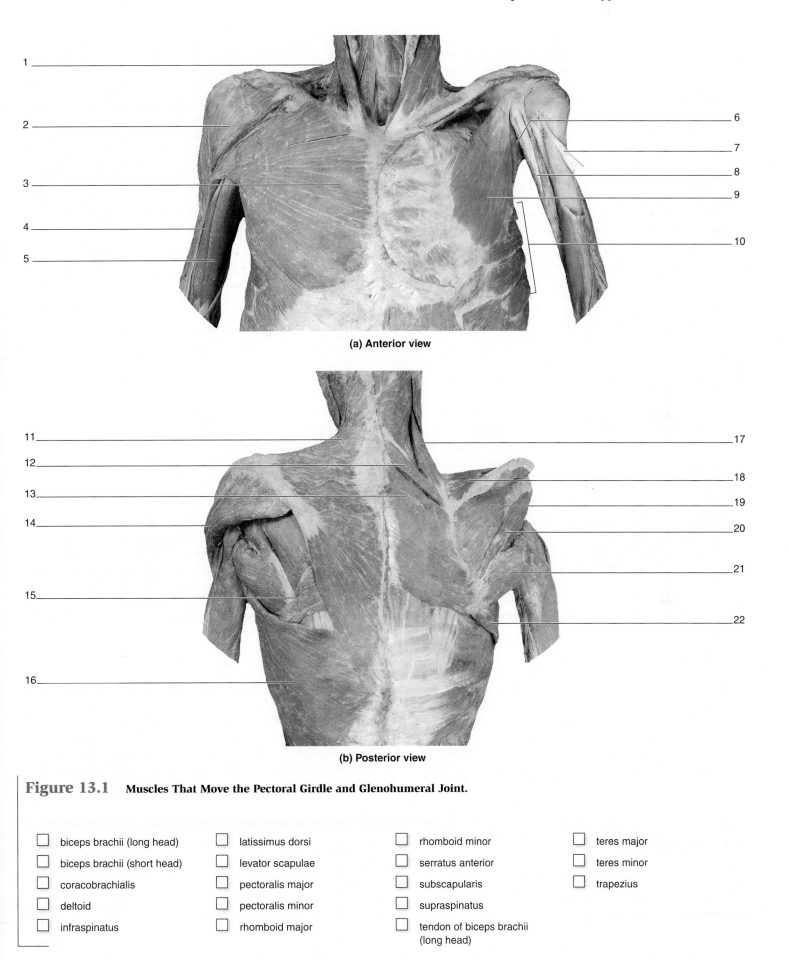

(a) Anterior view

(b) Posterior view

Figure 13.1 Muscles That Move the Pectoral Girdle and Glenohumeral Joint.

- [ ] biceps brachii (long head)
- [ ] biceps brachii (short head)
- [ ] coracobrachialis
- [ ] deltoid
- [ ] infraspinatus
- [ ] latissimus dorsi
- [ ] levator scapulae
- [ ] pectoralis major
- [ ] pectoralis minor
- [ ] rhomboid major
- [ ] rhomboid minor
- [ ] serratus anterior
- [ ] subscapularis
- [ ] supraspinatus
- [ ] tendon of biceps brachii (long head)
- [ ] teres major
- [ ] teres minor
- [ ] trapezius

smaller, more numerous muscles that act as **synergists** of the trapezius muscle. These include the **rhomboids** (major and minor) and the **levator scapulae**. As you identify these muscles on the cadaver or models, think about the actions they share with the trapezius. Practice performing the actions listed in table 13.1 and make sure that they make sense to you in light of the location and fiber orientation of the muscles themselves.

4. *Serratus Anterior*: One of the most important muscles for stabilization of the scapula is the **serratus anterior**. This muscle's main action is to prevent the scapula from pulling away from the rib cage ("winging"). To test the action of the serratus anterior muscle, stand and face a wall. With your arms held straight out horizontally in front of you, place your palms flat against the wall in front of you and then push against the wall. If the serratus anterior is working properly, your scapula will remain flat against your rib cage. If not, your scapula will "wing" and its medial border will be pushed away from your rib cage.

5. *Pectoralis Minor*: The **pectoralis minor** is a small muscle that can have multiple actions, depending upon what other muscles of the thorax and shoulder are contracting at the same time. If the scapula is fixed, the pectoralis minor assists in elevation of the rib cage (as in inspiration). If the scapula is not fixed, the pectoralis minor acts to pull the scapula anteriorly.

6. *Rotator Cuff Muscles*: The glenohumeral (shoulder) joint is the most mobile joint in the body, but this mobility comes at the expense of stability. Because of this, the muscles that connect the upper limb to the pectoral girdle perform two critical functions: stabilization of the glenohumeral joint, and movement of the glenohumeral joint. The **supraspinatus**, **infraspinatus**, **subscapularis**, and **teres minor** compose a musculotendinous cuff,

the **rotator cuff**, whose main function is to stabilize the glenohumeral joint. More than any other factor, the strength of these muscles determines how stable the joint is. Weaknesses in these muscles contribute to many musculoskeletal problems in the glenohumeral joint. Rotator cuff injuries are common in baseball players and in older adults who fall on an outstretched arm. The term *rotator cuff* comes from the fact that these muscles act to medially or laterally rotate the humerus, in addition to stabilizing the glenohumeral joint.

7. In the space below, make a representative drawing of the rotator cuff muscles, as seen in a posterior view.

Study Tip!

The following mnemonic can help you remember the names of the rotator cuff muscles. "A baseball pitcher who tears his rotator cuff **SITS** out the season." (**SITS**: S = supraspinatus, **I** = infraspinatus, **T** = teres minor, **S** = subscapularis). To remember it is the teres **minor** (not the teres major) that forms part of the rotator cuff, you can say that this injured pitcher will then be relegated to the **minor** leagues.

Upper Limb Musculature

The upper limb consists of the **arm** and the **forearm**. Each is divided into anterior and posterior compartments. The **anterior compartment** is collectively referred to as a **flexor compartment** because its muscles act to flex the arm, elbow, or wrist and fingers. The **posterior compartment** is collectively referred to as an **extensor compartment** because its muscles act to extend the arm, elbow, or wrist and fingers. Understanding the compartmental nature of the muscles not only is important for simplifying the task of learning the muscles within each compartment—it becomes even more important when you learn the distribution of peripheral nerves. In most cases, one nerve innervates one compartment. If you understand the common function of the muscles in an entire compartment, you will easily be able to determine the deficits an individual will suffer when a particular nerve is damaged.

EXERCISE 13.2 Compartments of the Arm

EXERCISE 13.2A: Anterior Compartment of the Arm

1. Observe a prosected human cadaver or a classroom model demonstrating muscles of the upper limb.

2. Using **table 13.3** and your textbook as a guide, identify the **muscles of the anterior compartment of the arm** listed in **figure 13.2** on the cadaver or on a classroom model. Then

label them in figure 13.2. As you identify these muscles, make note of the following:

Common Actions: Muscles in this compartment flex the arm or forearm.

Exceptions: The coracobrachialis muscle acts only about the glenohumeral joint.

The brachialis muscle acts only about the elbow joint.

Table 13.3	Anterior (Flexor) Compartment of the Arm				
Muscle	**Origin**	**Insertion**	**Action**	**Innervation**	**Word Roots**
Biceps Brachii (Long Head)	Supraglenoid tubercle of scapula	Radial tuberosity	Supinates forearm and flexes forearm (weak)	Musculocutaneous nerve	*bi*, two, + *caput*, head, + *brachium*, arm
Biceps Brachii (Short Head)	Coracoid process of scapula	Radial tuberosity	Supinates forearm and flexes forearm (weak)	Musculocutaneous nerve	*bi*, two, + *caput*, head, + *brachium*, arm
Coracobrachialis	Coracoid process of scapula	Midhumerus (medial surface)	Flexes arm and adducts (weak)	Musculocutaneous nerve	*coraco*, referring to the coracoid process, + *brachium*, the arm
Brachialis	Humerus (lower half of anterior surface)	Ulna (coronoid process and ulnar tuberosity)	Flexes forearm	Musculocutaneous nerve	*brachium*, the arm

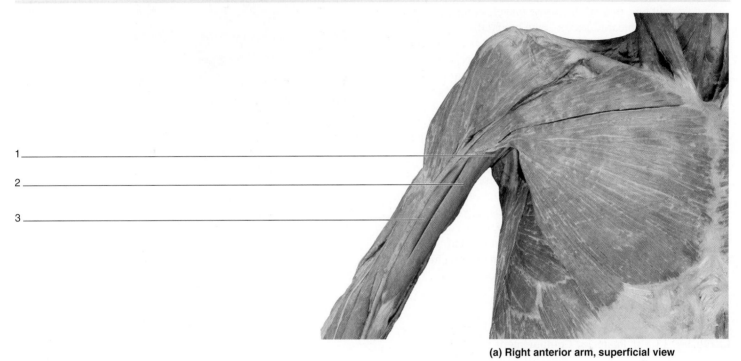

1 _____

2 _____

3 _____

(a) Right anterior arm, superficial view

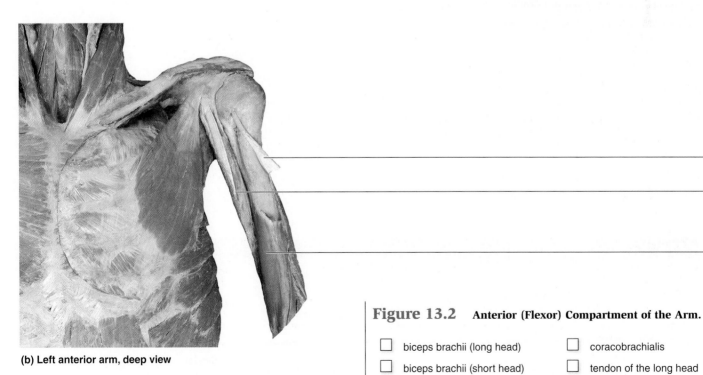

4

5

6

(b) Left anterior arm, deep view

Figure 13.2 Anterior (Flexor) Compartment of the Arm.

☐ biceps brachii (long head)   ☐ coracobrachialis

☐ biceps brachii (short head)   ☐ tendon of the long head of biceps brachii

☐ brachialis

3. *Brachialis and Biceps Brachii*: A common misconception is that the biceps brachii muscle is the most powerful flexor of the forearm (elbow). It is not. You can test this by performing the following activity. With your right wrist pronated, place the palm of your left hand over the belly of the biceps brachii on your right arm. Flex your elbow and note the degree to which the biceps brachii muscle produces tension. Next, supinate your wrist and perform the flexion action again. Did the biceps brachii produce more tension when the wrist was pronated or supinated? _____. You should have noticed a large difference in the amount of tension produced under the two conditions. In fact, the **prime mover**, or **agonist**, for the action of forearm flexion is the **brachialis** muscle, not the biceps brachii muscle. The biceps brachii is a **synergist** for this action. The biceps brachii itself is a very powerful **supinator** of the wrist and forearm. Thus, when we perform actions that require both flexion of the elbow and supination of the wrist and forearm, this is when we use our biceps brachii to its full capacity. Think about this the next time you are trying to twist off a stubborn bottle lid.

4. When identifying the biceps brachii muscle on models or cadavers, do not be confused by what you see. Because the **long head of the biceps brachii** disappears into the intertubercular groove of the humerus on its way to its attachment on the supraglenoid tubercle, only a small portion of its tendon is visible. On the other hand, the entire tendon of the **short head of the biceps brachii** is visible attaching to the coracoid process of the scapula. Thus, on models and cadavers, the short head of the biceps brachii will generally appear to be longer.

EXERCISE 13.2B: Posterior Compartment of the Arm

1. Observe a prosected human cadaver or a classroom model demonstrating muscles of the upper limb.

2. The posterior compartment of the arm consists of one muscle, the triceps brachii, with three heads (long, lateral, and medial).

3. Using **table 13.4** and your textbook as a guide, identify the **muscles of the posterior compartment of the arm** listed in **figure 13.3** on the cadaver or on a classroom model. Then label them in figure 13.3. As you identify these muscles, make note of the following:

Common Actions: Extend the forearm (elbow).

Exception: The long head also extends the arm.

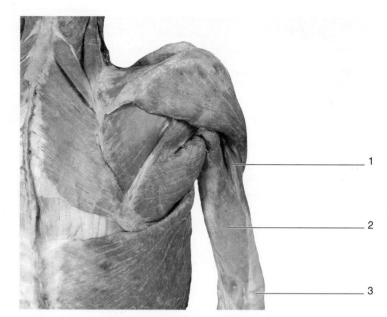

Right posterior arm, superficial view

Figure 13.3   **Posterior (Extensor) Compartment of the Arm.**

- ☐ long head of triceps brachii
- ☐ lateral head of triceps brachii
- ☐ olecranon process of ulna

Study Tip!

The *long* head of the *triceps brachii* attaches to the *infraglenoid tubercle* (of the scapula), while the *long* head of the *biceps brachii* attaches to the *supraglenoid tubercle* (of the scapula). Also, the medial head of the triceps brachii can be visualized only if the lateral head is bisected. It is a deep muscle, and it attaches directly to the humerus.

Table 13.4	Posterior (Extensor) Compartment of the Arm				
Muscle	**Origin**	**Insertion**	**Action**	**Innervation**	**Word Roots**
TRICEPS BRACHII					*tri*, three, + *caput*, head, + *brachium*, arm
Long Head	Infraglenoid tubercle of scapula	Olecranon of ulna	Extends forearm, extends arm	Radial nerve	
Lateral Head	Proximal half of posterior humerus	Olecranon of ulna	Extends forearm	Radial nerve	
Medial Head	Distal half of posterior humerus	Olecranon of ulna	Extends forearm	Radial nerve	

EXERCISE 13.3 Compartments of the Forearm

EXERCISE 13.3A: Anterior Compartment of the Forearm

1. Observe a prosected human cadaver or a classroom model demonstrating muscles of the upper limb.

2. The muscles of the **anterior compartment of the forearm** cause flexion of the wrist and the digits (fingers). They are arranged into two layers—a superficial group and a deep group. In most cases, the name of the muscle tells exactly what the muscle does or where the muscle is located. Thus, although the names of the muscles seem long, they can be very useful if you understand the meanings of the names.

3. Using **table 13.5** and your textbook as a guide, identify the muscles of the anterior compartment of the forearm listed in **figure 13.4** on the cadaver or on a classroom model. Then label them in figure 13.4.

4. *Identification of Superficial Muscles and Tendons*: Place your left palm on the medial epicondyle of your right humerus (**figure 13.5**). In this position, the order of the muscles, on your right forearm. From lateral to medial is:

Index finger—pronator teres (PT)

Middle finger—flexor carpi radialis (FCR)

Ring finger—palmaris longus (PL)

Pinky finger—flexor carpi ulnaris (FCU)

As you perform this exercise, flex your wrist and digits to identify the tendons, from lateral to medial, of the flexor carpi radialis, palmaris longus (if present, see below), and flexor carpi ulnaris.

5. *Palmaris Longus*: Approximately 15% to 20% of humans do not have a palmaris longus muscle. The palmaris longus passes *superficial* to the **flexor retinaculum** (a band of connective tissue that wraps around the wrist) and inserts onto the **palmar aponeurosis**. Perform the following exercise to determine whether you have a palmaris longus muscle. With your forearm supinated, touch the tips of your first and fifth digits (thumb and pinky finger) together and flex the wrist *slightly*. If palmaris longus is present, you will see its tendon passing longitudinally in the middle of your wrist. If no tendon is visible, the muscle is probably absent.

6. *Flexor Digitorum*: There are two muscles that flex the digits, the **flexor digitorum superficialis** (superficial) and the **flexor digitorum profundus** (deep) (**figure 13.6**). These muscles, as their names suggest, flex the fingers. They do so by attaching to the phalanges. These muscles take somewhat unique routes to get to their respective attachments on the phalanges. The flexor digitorum profundus tendon attaches to the **distal phalanx**. Thus, it must pass through the tendon of the more superficial

Table 13.5	Anterior (Flexor) Compartment of the Forearm				
Muscle	Origin	Insertion	Action	Innervation	Word Roots
SUPERFICIAL GROUP					
Pronator Teres	Humerus (medial epicondyle)	Radius (lateral shaft)	Pronates forearm and flexes elbow (weak)	Median nerve	*pronatus*, to bend forward, + *teres*, round
Flexor Carpi Radialis	Humerus (medial epicondyle)	Metacarpals II and III (base)	Flexes and abducts wrist	Median nerve	*flex*, to bend, + *carpus*, wrist, + *radialis*, radius
Palmaris Longus	Humerus (medial epicondyle)	Palmar aponeurosis	Flexes wrist	Median nerve	*palmaris*, the palm, + *longus*, long
Flexor Digitorum Superficialis	Humerus (medial epicondyle)	Middle phalanx of digits 2–5	Flexes the metacarpophalangeal and proximal interphalangeal joints of digits 2–5	Median nerve	*flex*, to bend, + *digit*, finger, + *superficialis*, surface
Flexor Carpi Ulnaris	Humerus (medial epicondyle)	Pisiform, hamate, and the metacarpal V (base)	Flexes and adducts wrist	Ulnar nerve	*flex*, to bend, + *carpus*, wrist, + *ulnaris*, ulna
DEEP GROUP					
Flexor Pollicis Longus	Radius (anterior surface) and interosseous membrane	Distal phalanx of the thumb	Flexes the distal phalanx of the thumb	Median nerve	*flexus*, to bend, + *pollex*, the thumb, + *longus*, long
Pronator Quadratus	Ulna (distal anterior shaft)	Radius (distal, anterior shaft)	Pronates the forearm	Median nerve	*pronatus*, to bend forward, + *quadratus*, square
Flexor Digitorum Profundus	Ulna (anteromedial surface) and interosseous membrane	Distal phalanx of digits 2–5	Flexes the distal phalanx of digits 2–5	Ulnar and median nerves	*flexus*, to bend, + *digitorum*, finger, + *profundus*, deep

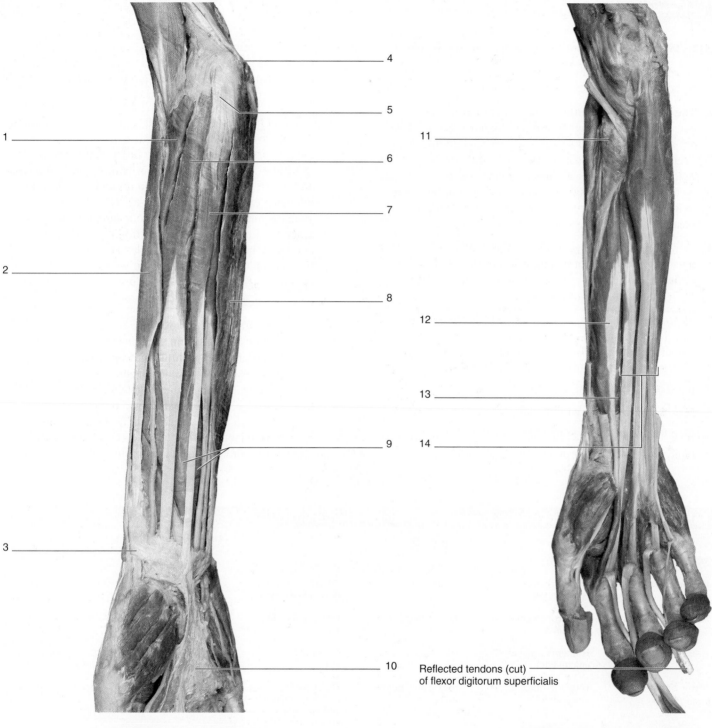

(a) Right anterior forearm, superficial view

(b) Right anterior forearm, deep view

Reflected tendons (cut)
of flexor digitorum superficialis

Figure 13.4   **Anterior (Flexor) Compartment of the Forearm.**

☐ brachioradialis

☐ common flexor tendon

☐ flexor carpi radialis

☐ flexor carpi ulnaris

☐ flexor digitorum profundus

☐ flexor digitorum superficialis

☐ flexor pollicis longus

☐ flexor retinaculum

☐ medial epicondyle of humerus

☐ palmar aponeurosis

☐ palmaris longus

☐ pronator quadratus

☐ pronator teres

☐ supinator

(Left hand covers
medial epicondyle)

Figure 13.5 **Locating Superficial Anterior Forearm Muscles.** Locate the superficial anterior forearm muscles by placing the left hand on the right forearm and noting the position of the digits, which correspond to the locations of the superficial forearm muscles.

muscle (the tendon of the flexor digitorum superficialis) as it travels to the distal phalanx. Observe a cadaver or a classroom model of the hand, and locate the middle and distal phalanges of one digit. The tendon of flexor digitorum superficialis inserts onto the **middle phalanx** of the digit. Before it reaches its insertion point, it has a slit through which the tendon of flexor digitorum profundus passes. Thus, though both muscles flex the digits, they do so at different joints. Perform an experiment to test the function of these two muscles. With your right hand, pull the tip of the third digit (middle finger) of your left hand posteriorly. This stretches the flexor digitorum profundus muscle beyond the point at which it can contract. Now try to flex the other digits of your hand. Notice that you can flex the joint between your proximal and middle phalanges, but you cannot flex the joint between your middle and distal phalanges. Once you release your hold on the third digit, you will be able to flex all the interphalangeal joints together.

7. *Optional Activity:* **AP|R** **Muscular System**—Visit the quiz area for drill and practice on muscles of the upper limb.

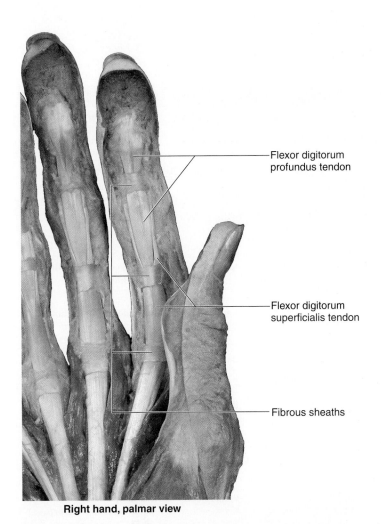

Flexor digitorum profundus tendon

Flexor digitorum superficialis tendon

Fibrous sheaths

Right hand, palmar view

Figure 13.6 **Flexor Tendons Within the Hand.** The tendon of the flexor digitorum profundus pierces the tendon of the flexor digitorum superficialis as it travels to its attachment point on the distal phalanx.

EXERCISE 13.3B: Posterior Compartment of the Forearm

1. Observe a prosected human cadaver or a classroom model demonstrating muscles of the upper limb.

2. Using **table 13.6** and your textbook as a guide, identify the **muscles of the posterior compartment of the forearm** listed in **figure 13.7** on the cadaver or on a classroom model. Then label them in figure 13.7.

3. *The Anatomic Snuffbox*: The tendons of the **abductor pollicis longus**, **extensor pollicis longus,** and **extensor pollicis brevis** muscles form the borders of a triangular area called the **anatomic snuffbox.** To see the anatomic snuffbox, extend your thumb as far as it will go (as if you were signaling to hitch a ride). The tendons of the bordering muscles will pop up, and a concavity can be seen between them. The anatomic snuffbox is clinically important because the radial artery is located in its floor and it is an area of palpation to evaluate fractures of the scaphoid bone. With your thumb relaxed, place a finger on the snuffbox and see if you can feel the pulse of the radial artery.

4. In the space below, make a representative drawing of the anatomic snuffbox, noting the locations of the three muscles that form its boundaries.

Table 13.6	Posterior (Extensor) Compartment of the Forearm				
Muscle	**Origin**	**Insertion**	**Action**	**Innervation**	**Word Roots**
Brachioradialis	Humerus (lateral supracondylar ridge)	Radius (styloid process)	Flexes the forearm	Radial nerve	*brachium*, arm, + *radialis*, the radius bone
Extensor Carpi Radialis Longus	Humerus (lateral supracondylar ridge)	Metacarpal II (base)	Extends and abducts the wrist	Radial nerve	*extendo*, to stretch out, + *carpus*, wrist, + *radialis*, radius, + *longus*, long
Extensor Carpi Radialis Brevis	Humerus (lateral epicondyle)	Metacarpal III (base)	Extends and abducts the wrist	Radial nerve	*extendo*, to stretch out, + *carpus*, wrist, + *radialis*, radius, + *brevis*, short
Extensor Digitorum	Humerus (lateral epicondyle)	Distal and middle phalanges of digits 2–5	Extends digits 2–5 at the metacarpophalangeal joint, extends the wrist	Radial nerve	*extendo*, to stretch out, + *digit*, finger
Extensor Digiti Minimi	Humerus (lateral epicondyle)	Proximal phalanx of digit 5 (little finger)	Extends the little finger at the metacarpophalangeal and interphalangeal joints	Radial nerve	*extendo*, to stretch out, + *digit*, finger, + *minimi*, smallest
Extensor Carpi Ulnaris	Humerus (lateral epicondyle) and proximal ulna	Metacarpal V (base)	Extends and adducts the wrist	Radial nerve	*extendo*, to stretch out, + *carpus*, wrist, + *ulnaris*, ulna
Supinator	Humerus (lateral epicondyle) and proximal ulna	Radius (proximal third)	Supinates the forearm	Radial nerve	*supinatus*, to move backward
Abductor Pollicis Longus	Ulna and radius (posterior surface) and interosseous membrane	Metacarpal I (base)	Abducts the thumb	Radial nerve	*abduct*, to move away from the midline, + *pollex*, thumb, + *longus*, long
Extensor Pollicis Brevis	Radius (posterior surface) and interosseous membrane	Proximal phalanx of the thumb	Extends the proximal phalanx of the thumb at the metacarpophalangeal joint	Radial nerve	*extendo*, to stretch out, + *pollex*, thumb, + *brevis*, short
Extensor Pollicis Longus	Ulna (middle third, posterior surface) and interosseous membrane	Distal phalanx of the thumb	Extends the distal phalanx of the thumb at the metacarpophalangeal and interphalangeal joints	Radial nerve	*extendo*, to stretch out, + *pollex*, thumb, + *longus*, long
Extensor Indicis	Ulna (posterior surface) and interosseous membrane	Tendon of extensor digitorum of digit 2 (index finger)	Extends the index finger	Radial nerve	*extendo*, to stretch out, + *indicis*, the forefinger

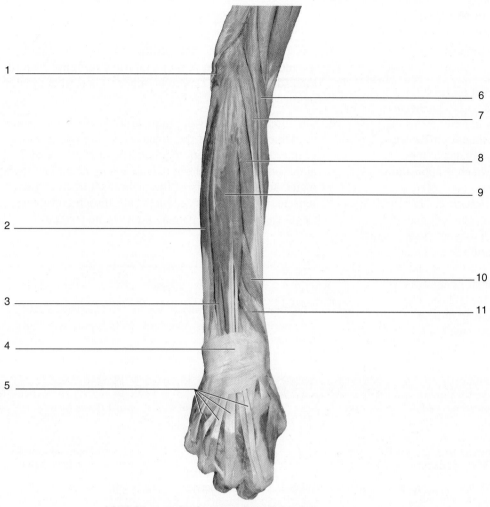

(a) Right posterior forearm, superficial view

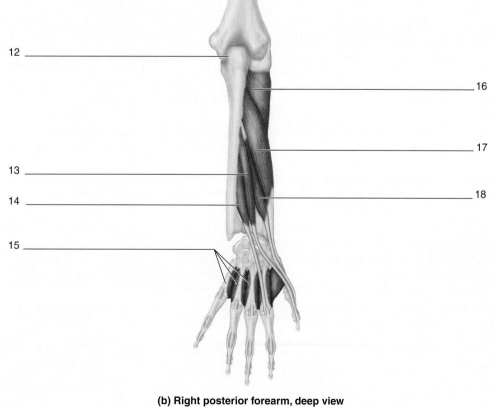

(b) Right posterior forearm, deep view

Figure 13.7 **Posterior (Extensor) Compartment of the Forearm.**

- ☐ abductor pollicis brevis
- ☐ abductor pollicis longus
- ☐ anconeus
- ☐ brachioradialis
- ☐ dorsal interossei
- ☐ extensor carpi radialis brevis
- ☐ extensor carpi radialis longus
- ☐ extensor carpi ulnaris
- ☐ extensor digiti minimi
- ☐ extensor digitorum
- ☐ extensor digitorum tendons
- ☐ extensor indicis
- ☐ extensor pollicis brevis
- ☐ extensor pollicis longus
- ☐ extensor retinaculum
- ☐ olecranon process of ulna
- ☐ supinator

EXERCISE 13.4 Intrinsic Muscles of the Hand

1. Observe a prosected human cadaver or a model demonstrating muscles of the hand.

2. The **intrinsic muscles of the hand** consist of the lumbricals, the interossei, and the thenar and hypothenar groups of muscles. The **thenar** group of muscles forms the large pad at the base of the thumb, the **thenar eminence**. Muscles in this group create special movements of the pollex (thumb), the most important of which is **opposition**. The **hypothenar** group of muscles forms a pad at the base of the fifth digit, the **hypothenar eminence**. Muscles in this group create special movements of the fifth digit (little finger). Observe both the thenar and hypothenar eminences on one of your hands. The **midpalmar** group is a group of muscles that lies between the thenar and hypothenar muscle groups. The midpalmar group consists of the lumbricals, interossei, and adductor pollicis. The **lumbricals** and **interossei** muscles are small muscles, located very deep in the hand, that create small, intricate movements of the fingers.

3. Using **table 13.7** and your textbook as a guide, identify the intrinsic muscles of the hand listed in **figure 13.8** on the cadaver or on a classroom model. Then label them in figure 13.8.

4. *The Interossei*: The two groups of interossei muscles act to abduct and adduct the digits. There are two groups of interossei, dorsal and palmar. The *dorsal interossei* abduct the digits, and the *palmar interossei* adduct the digits. To feel the belly of the first dorsal interosseous muscle, forcibly press (adduct) your thumb and index finger together. Then palpate between metacarpals

> ## Study Tip!
> The **D**orsal interossei **AB**duct the digits (mnemonic is **DAB**), whereas the **P**almar interossei **AD**duct the digits (mnemonic is **PAD**).

Table 13.7	Intrinsic Muscles of the Hand				
Muscle	**Origin**	**Insertion**	**Action**	**Innervation**	**Word Roots**
MIDPALMAR GROUP					
Adductor Pollicis	Capitate bone, metacarpals II–III	Medial side of proximal phalanx of thumb	Adducts thumb	Ulnar nerve	*adduct*, to move toward the midline, + *pollex*, thumb
Dorsal Interossei	Metacarpals II–IV (medial and lateral surface)	Tubercle of the proximal phalanx and dorsal aponeurosis	Abducts digits 2–4	Ulnar nerve	*dorsal*, the back, + *inter-*, between, + *os*, bone
Palmar Interossei	Metacarpals (medial and lateral surface)	Tubercle of the proximal phalanx and the dorsal aponeurosis	Adducts digits 2–5	Ulnar nerve	*palmar*, the palm of the hand, + *inter-*, between, + *os*, bone
Lumbricals	Flexor digitorum profundus tendon	Lateral side of the dorsal expansion of digits 2–5	Flexes metacarpal-phalangeal joint	Ulnar and median nerves	*lumbricus*, earthworm
THENAR GROUP					*thenar*, the palm of the hand
Abductor Pollicis Brevis	Flexor retinaculum, scaphoid and trapezium (tubercles)	Base of the proximal phalange of the thumb (lateral side)	Abducts the thumb	Recurrent branch of the median nerve	*abduct*, to move away from the midline, + *pollex*, thumb, + *brevis*, short
Flexor Pollicis Brevis	Flexor retinaculum and trapezium (tubercles)	Base of the proximal phalange of the thumb (lateral side)	Flexes the thumb	Recurrent branch of the median nerve	*flexus*, to bend, + *pollex*, thumb, + *brevis*, short
Opponens Pollicis	Flexor retinaculum and trapezium (tubercles)	Metacarpal I (lateral side)	Assists in opposition and medial rotation of the thumb	Recurrent branch of the median nerve	*oppono*, to place against, + *pollex*, thumb
HYPOTHENAR GROUP					*hypo-*, under, + *thenar*, the palm of the hand
Abductor Digiti Minimi	Pisiform bone and tendon of flexor carpi ulnaris	Base of the proximal phalanx of digit 5 (medial side)	Abducts digit 5	Ulnar nerve	*abduct*, to move away from the median plane, + *digitus*, a finger, + *minimi*, smallest
Flexor Digiti Minimi Brevis	Hamate (hook) and flexor retinaculum	Base of the proximal phalanx of digit 5 (medial side)	Flexes the proximal phalanx of digit 5	Ulnar nerve	*flex*, to bend, + *digitus*, a finger, + *minimi*, smallest
Opponens Digiti Minimi	Hamate (hook) and flexor retinaculum	Metacarpal V (medial border)	Medially rotates and opposes digit 5 toward the thumb	Ulnar nerve	*oppono*, to place against + *digitus*, a finger, + *minimi*, smallest

I and II using your other hand to feel the first dorsal interosseous muscle. Try this out on yourself now.

5. *Lumbricals*: The lumbricals are relatively easy to identify because they appear slim and "wormlike," as their name implies. There is a lumbrical muscle located between every metacarpal bone. The lumbricals flex the metacarpophalangeal joints and extend the interphalangeal joints.

6. In the space to the right, make a representative drawing of the intrinsic muscles of the hand.

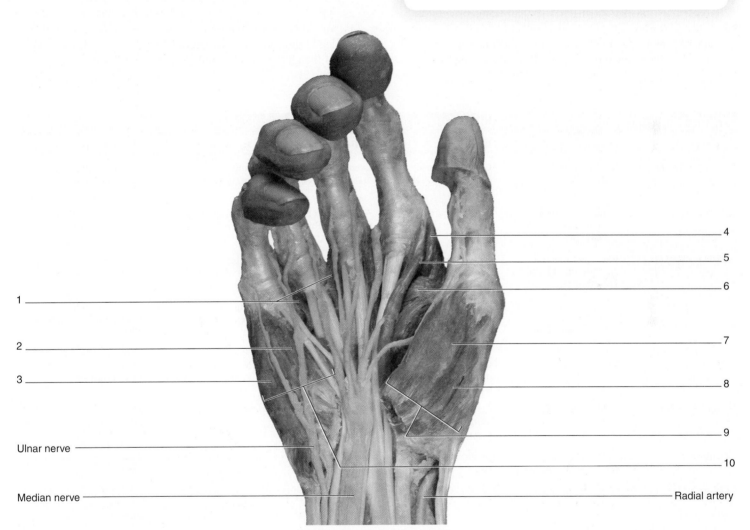

Intrinsic muscles of the hand

Figure 13.8   Intrinsic Muscles of the Hand.

- ☐ abductor digiti minimi
- ☐ abductor pollicis brevis
- ☐ adductor pollicis
- ☐ first dorsal interosseous
- ☐ flexor digiti minimi brevis

- ☐ flexor pollicis brevis
- ☐ hypothenar group
- ☐ lateral lumbricals
- ☐ medial lumbrical
- ☐ thenar group

Muscles That Move the Hip Joint

The **muscles that move the hip** include the gluteal muscles, the deep lateral rotators, and the iliopsoas muscle. The gluteal muscles and the iliopsoas are very large and powerful muscles. We use them for standing, walking, running, cycling, and just about every other locomotor activity. Failure of these muscles to work properly leads to gait disturbances.

EXERCISE 13.5 Muscles That Move the Hip

1. Observe a prosected human cadaver or a model demonstrating muscles of the hip.

2. Using **table 13.8** and your textbook as a guide, identify the muscles that move the hip listed in **figure 13.9** on the cadaver or on a classroom model. Then label them in figure 13.9.

3. *Gluteal Muscles*: The gluteus maximus extends the hip and *laterally* rotates the thigh. The gluteus medius and minimus muscles abduct the hip and *medially* rotate the thigh.

4. Consider the action of **hip abduction**. It may not seem that we perform this action often, except perhaps in a kickboxing class. In fact, we rarely use the gluteus medius and minimus muscles for extreme abduction of the hip, as in kicking the lower limb out to the side. Instead, when we attempt to abduct the hip against resistance, as when standing on one leg, these muscles prevent the hip from *adducting*, which would cause the hip on the side of the stance (standing) leg to protrude laterally (**figure 13.10**). To test the function of the gluteus medius and minimus muscles, first take a

Table 13.8	Muscles That Act About the Hip				
Muscle	**Origin**	**Insertion**	**Action**	**Innervation**	**Word Roots**
Gluteus Maximus	Dorsal ilium, sacrum, and coccyx	Gluteal tuberosity of the femur, iliotibial tract	Extends the thigh, assist in lateral rotation	Inferior gluteal nerve (L5–S2)	*gloutos*, buttock, + *maximus*, greatest
Gluteus Medius	Ilium (between the anterior and posterior gluteal lines)	Greater trochanter of femur	Abducts and medially rotates the thigh	Superior gluteal nerve (L4–S1)	*gloutos*, buttock, + *medius*, middle
Gluteus Minimus	Ilium (between the anterior and inferior gluteal lines)	Greater trochanter of femur	Abducts and medially rotates the thigh	Superior gluteal nerve (L4–S1)	*gloutos*, buttock, + *minimus*, smallest
Tensor Fascia Latae	Iliac crest (anterior aspect), anterior superior iliac spine (ASIS)	Iliotibial tract	Abducts the thigh	Superior gluteal nerve (L4–S1)	*tensus*, to stretch, + *fascia*, a band, + *latus*, side
Superior Gemellus	Ischial spine	Greater trochanter of the femur (medial surface)	Laterally rotates the thigh	Nerve to obturator internus (L5 and S1)	*geminus*, twin, + *superus*, above
Inferior Gemellus	Ischial tuberosity	Greater trochanter of the femur (medial surface)	Laterally rotates the thigh	Nerve to quadratus femoris (L5 and S1)	*geminus*, twin, + *inferior*, lower
Piriformis	Anterior sacrum, sacrotuberous ligament	Greater trochanter of femur	Laterally rotates the thigh	Nerve to piriformis (S1–S2)	*pirum*, pear, + *forma*, form
Obturator Internus	Obturator membrane (posterior surface)	Margins of obturator foramen	Laterally rotates the thigh	Nerve to obturator internus (L5 and S1)	*obturo*, to occlude, + *internus*, away from the surface
Quadratus Femoris	Ischial tuberosity (lateral border)	Intertrochanteric crest of femur	Laterally rotates the thigh	Nerve to quadratus femoris (L5 and S1)	*quad*, four (sided), + *femur*, thigh
ILIOPSOAS					
Psoas Major	T_{12}–L_5 (bodies and transverse processes)	Femur (lesser trochanter)	Flexes the hip joint	Lumbar plexus (L2–L3)	*psoa*, the muscles of the loins
Iliacus	Iliac bone (iliac fossa)	Femur (lesser trochanter)	Flexes the hip joint	Femoral nerve (L2–L3)	*ilium*, groin

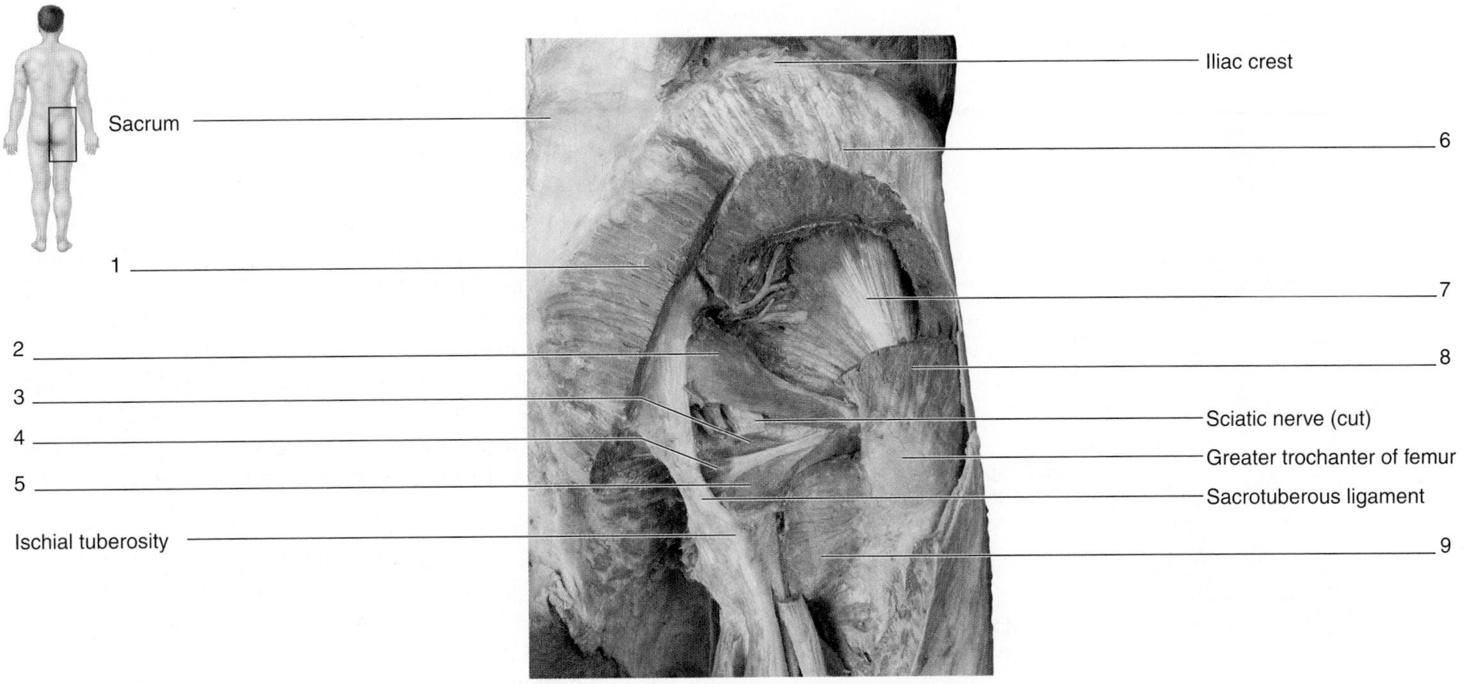

Iliac crest

Sacrum

6

1

7

2

8

3

Sciatic nerve (cut)

4

Greater trochanter of femur

5

Sacrotuberous ligament

Ischial tuberosity

9

(a) Right thigh and hip, deep posterior view

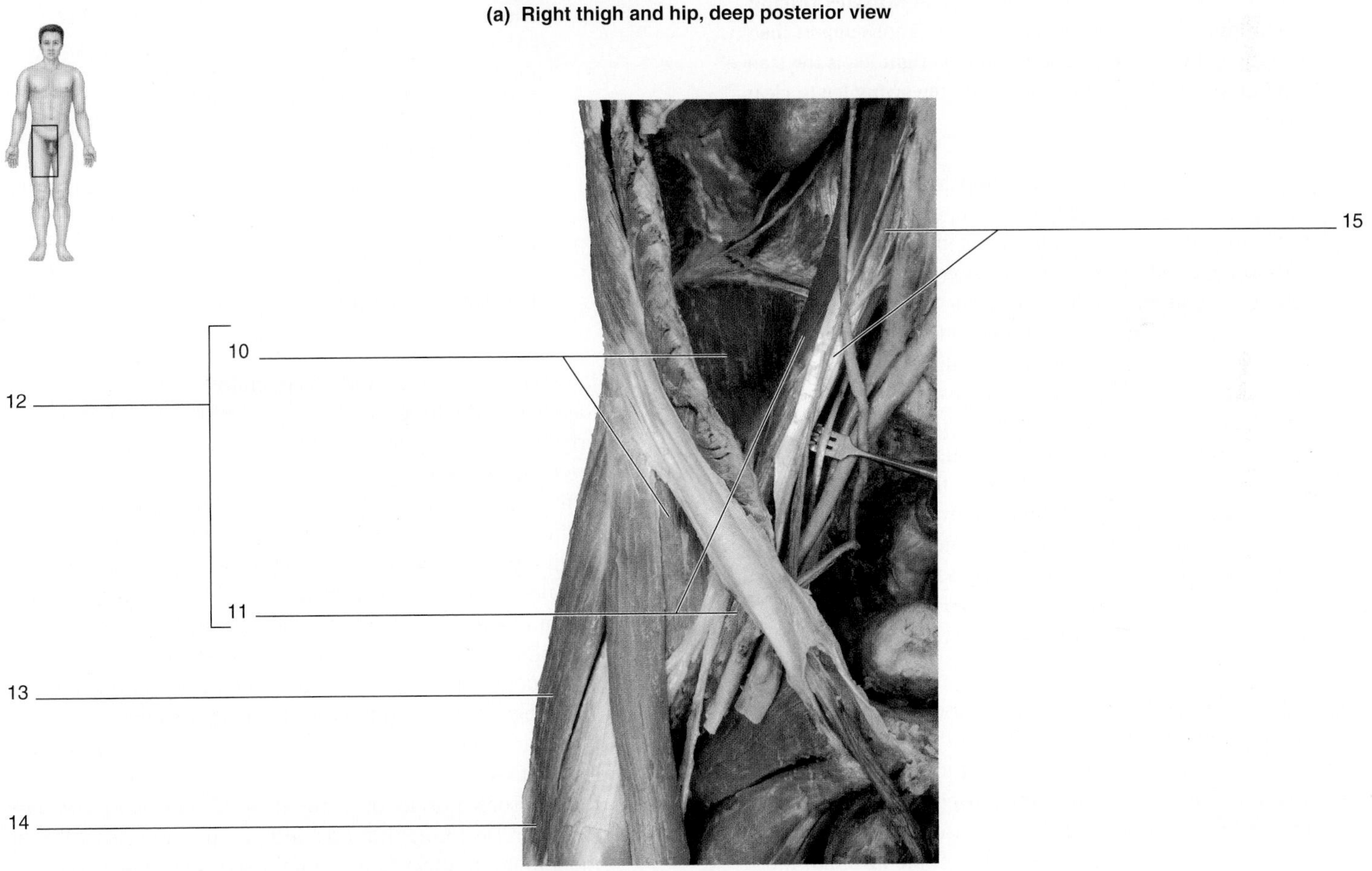

15

10

12

11

13

14

(b) Right thigh and hip, deep anterior view

Figure 13.9 Muscles That Move the Pelvic Girdle.

☐ gluteus maximus   ☐ iliacus   ☐ inferior gemellus   ☐ psoas major   ☐ superior gemellus

☐ gluteus medius   ☐ iliopsoas   ☐ obturator internus   ☐ psoas minor   ☐ tensor fascia latae

☐ gluteus minimus   ☐ iliotibial tract   ☐ piriformis   ☐ quadratus femoris

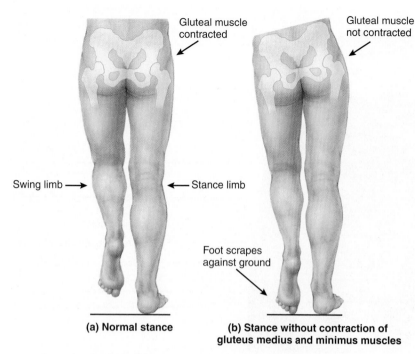

Gluteal muscle contracted

Gluteal muscle not contracted

Swing limb →   ← Stance limb

Foot scrapes against ground

(a) Normal stance

(b) Stance without contraction of gluteus medius and minimus muscles

Figure 13.10   Actions of the Gluteal Muscles During Locomotion. The gluteus medius and minimus support the hip on the right side of the body when the right leg is the stance (supporting) leg. This allows the foot on the swing leg to clear the ground.

standing position. Next, place the palm of your right hand flat against your right lateral hip (superficial to the location of the gluteus medius and minimus muscles). Now flex the knee on your left lower limb bringing your left foot off the ground so you are balancing on your right limb only. Do you feel tension in the muscles deep to your palm? If you are holding your hip vertical, you will. If you then relax the gluteus medius and minimus muscles, you will find that your hip protrudes laterally and the hip tilts to the unsupported side. The gluteus medius and minimus hold the hip in the vertical position every time the limb is in its stance position. By holding the hip in such a position, they allow the foot on the swing limb to move forward without scraping against the ground. It may seem like a minor action to you, but it can be quite problematic for an individual whose muscles are not functioning properly due to nerve injury or other disease.

5. *Iliotibial Tract*: The iliotibial tract (also called the IT band) is a thickening of the deep fascia of the thigh, which extends from the anterior iliac crest to the lateral tibia. Two muscles insert onto it: the **gluteus maximus** and the **tensor fascia latae**. When these muscles become stiff from overuse, they put excessive tension on the IT band. Most commonly this causes pain in the knee joint because the

tight IT band compresses structures in the lateral knee and the underlying vastus lateralis muscle.

6. *Lateral Rotators*: Several small muscles extend between the margins of the obturator foramen of the pelvis and insert onto the greater trochanter of the femur. They are called lateral rotators because lateral rotation of the thigh is their primary action. However, they are also very important in stabilizing the hip joint, in much the same way that the rotator cuff muscles stabilize the shoulder joint. A very important lateral rotator is the **piriformis** muscle because it is a major landmark in the gluteal region. For example, the sciatic nerve exits inferior to this muscle, and the superior and inferior gluteal arteries and nerves are named for their passages above and below the piriformis, respectively.

7. Observe a prosected cadaver or a model in which the gluteus maximus muscle has been reflected or cut away. First identify the gluteus medius and minimus muscles. Inferior to the gluteus medius is the pear-shaped piriformis muscle. Note the tough **sacrotuberous ligament** that overlies the medial attachment of the piriformis, and the large **sciatic nerve**, which exits inferior to the piriformis. Finally, locate the following lateral rotators, from superior to inferior:

☐ superior gemellus

☐ obturator internus (in this location, you will only see its tendon)

☐ inferior gemellus

☐ quadratus femoris

8. *Iliopsoas*: The **iliacus** and **psoas major** are considered together as the **iliopsoas** muscle. They have separate origins, but they come together and insert together on the **lesser trochanter** of the femur. These muscles are difficult to identify on the cadaver and models if you look at the anterior thigh to find them, because only a small portion of the insertion is visible there. The bellies of these muscles are located deep within the pelvis on the posterior abdominopelvic wall. Thus, when identifying them on the cadaver or on models, be sure to view the inside of the abdominopelvic cavity where the muscles originate. The longer of the two muscles is the psoas major. It is an important landmark in the pelvis because structures such as the ureters course superficial to it. In the anterior thigh, the iliopsoas muscle forms the floor of the *femoral triangle* (see exercise 13.6). You must look deep to the structures in the femoral triangle (the femoral nerve, artery, and vein) in order to see the iliopsoas muscle here.

Lower Limb Musculature

The lower limb is composed of the **thigh** (from hip to knee), the **leg** (from knee to ankle), and the **foot**. The thigh and the leg each consist of three compartments. As with the upper limb, each compartment consists of muscles that perform similar actions, and each compartment is served by one peripheral nerve branch.

EXERCISE 13.6 Compartments of the Thigh

EXERCISE 13.6A: Anterior Compartment of the Thigh

1. Observe a prosected human cadaver or a classroom model demonstrating muscles of the thigh.

2. Using **table 13.9** and your textbook as a guide, identify the **muscles of the anterior compartment of the thigh** listed in **figure 13.11** on the cadaver or on a classroom model.

Table 13.9	Anterior Compartment of the Thigh				
Muscle	**Origin**	**Insertion**	**Action**	**Innervation**	**Word Roots**
Sartorius	ASIS (anterior superior iliac spine)	Medial tibia	Crosses legs: flexes the hip and knee, abducts and laterally rotates the thigh	Femoral nerve (L2–L4)	*sartor*, a tailor
QUADRICEPS FEMORIS GROUP					
Rectus Femoris	Anterior inferior iliac spine	Tibial tuberosity	Extends the knee, flexes the hip	Femoral nerve (L2–L4)	*rectus*, straight, + *femur*, thigh
Vastus Lateralis	Femur (medial and posterior)	Tibial tuberosity	Extends the knee	Femoral nerve (L2–L4)	*vastus*, great, + *lateralis*, to the side
Vastus Intermedius	Femur (medial and lateral)	Tibial tuberosity	Extends the knee	Femoral nerve (L2–L4)	*vastus*, great, + *intermedius*, in between
Vastus Medialis	Femur (inferior and posterior)	Tibial tuberosity	Extends the knee	Femoral nerve (L2–L4)	*vastus*, great, + *medialis*, medial

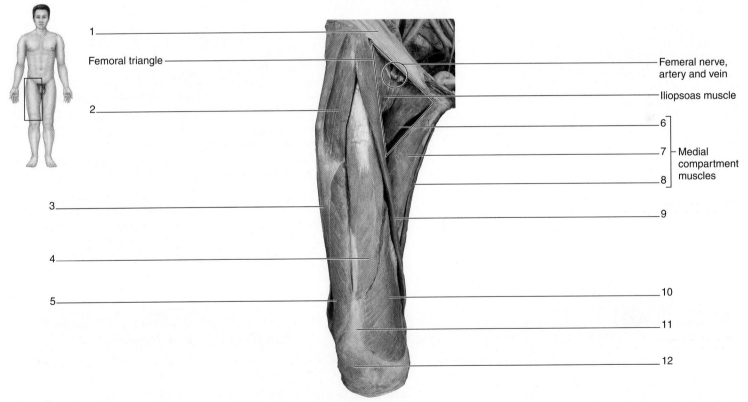

Right thigh, anterior view

Figure 13.11   Anterior Compartment of the Thigh.

- ☐ adductor longus
- ☐ gracilis
- ☐ iliotibial tract
- ☐ inguinal ligament
- ☐ patella
- ☐ pectineus
- ☐ quadriceps tendon
- ☐ rectus femoris
- ☐ sartorius
- ☐ tensor fasciae latae
- ☐ vastus lateralis
- ☐ vastus medialis

Then label them in figure 13.11. As you identify these muscles, make note of the following:

Common Actions: Extend the knee.

Exceptions: The rectus femoris and sartorius muscles flex the hip joint. Sartorius also flexes the knee joint.

3. *Sartorius*: The word *sartorius* literally means "tailor." If you remember this and think about the typical cross-legged position tailors take to mend something by hand, it will be easy to remember the actions of the sartorius muscle. Try this out for yourself. Sit in a chair and cross your right leg over your left so that your right ankle is resting on your left knee. Next, note the positions of your right hip and knee joints. The hip is flexed and laterally rotated, and the knee is also flexed. These are the actions of the sartorius muscle.

4. *Femoral Triangle*: The **femoral triangle** is a triangular space in the upper, anterior thigh. Its borders are the **sartorius**, the **adductor longus**, and the **inguinal ligament**. It is clinically relevant because of the structures located within the space, which include the femoral nerve, artery, and vein. The only structures superficial to these are fat and skin, so it is a location where the vascular system can be accessed with relative ease.

5. In the space below, make a representative drawing of the femoral triangle. Be sure to draw and label the structures that form its boundaries, and draw and label the contents found within the triangle.

Study Tip!

The femoral triangle contains some very important structures. The mnemonic for remembering the contents of the femoral triangle is **NAVEL**, which lists the contents of the femoral triangle from lateral to medial:

N—femoral **N**erve

A—femoral **A**rtery

V—femoral **V**ein

E—**E**mpty space

L—**L**ymphatic vessel

EXERCISE 13.6B: Medial Compartment of the Thigh

1. Observe a prosected human cadaver or a classroom model demonstrating muscles of the thigh.

2. Using **table 13.10** and your textbook as a guide, identify the **muscles of the medial compartment of the thigh** listed in **figure 13.12** on the cadaver or on a classroom model. Then label them in figure 13.12. As you identify these muscles, make note of the following:

Common Action: Adduct the thigh.

Exceptions: Although also an adductor of the thigh, the gracilis muscle does not attach to the femur. It is also the only muscle of the group to cross and flex the knee joint.

Study Tip!

- All the muscles in the medial compartment of the thigh attach to the pubic bone and, except for the gracilis, to the linea aspera of the femur.

- Think of these muscles as "the short one" (adductor brevis), "the long one" (adductor longus), "the really big one" (adductor magnus), and "the graceful one" (gracilis) to make them easier to identify relative to each other.

- As you view the anterior thigh, you will see these muscles medial to the sartorius. From superior to inferior, the order of muscles is pectineus, adductor longus, and adductor magnus. The adductor brevis is located deep to the pectineus and adductor longus muscles, so you will have to move them aside to see it clearly.

Table 13.10	Medial Compartment of the Thigh					
Muscle	**Origin**	**Insertion**	**Action**	**Innervation**	**Word Roots**
Pectineus	Pubis	Femur (pectineal line)	Adducts and laterally rotates the thigh	Obturator or femoral nerve (L2–L4)	*pectineal*, a ridged or comblike structure
Adductor Longus	Pubis	Femur (linea aspera)	Adducts and laterally rotates the thigh	Obturator nerve (L2–L4)	*adduct*, to bring toward the median plane, + *longus*, long
Adductor Brevis	Pubis	Femur (linea aspera)	Adducts and laterally rotates the thigh	Obturator nerve (L2–L3)	*adduct*, to bring toward the median plane, + *brevis*, short
Adductor Magnus	Pubis	Femur (linea aspera)	Adducts and laterally rotates the thigh	Obturator nerve (L2–L4)	*adduct*, to bring toward the median plane, + *magnus*, large
Gracilis	Pubis	Tibia (medial condyle)	Adducts the thigh and flexes the leg	Obturator nerve (L2–L4)	*gracilis*, slender

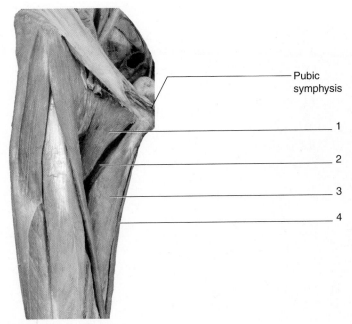

Right thigh, anterior view

Pubic symphysis

1

2

3

4

Figure 13.12 **Medial Compartment of the Thigh.** The adductor magnus is not visible in this photo.

- ☐ adductor brevis
- ☐ adductor longus
- ☐ adductor magnus
- ☐ gracilis
- ☐ pectineus

3. *Gracilis:* In addition to being the outlier in terms of its attachments, this muscle is a weak adductor of the thigh—so weak, in fact, that it can be removed without a patient experiencing great loss in function. When a patient needs a muscle graft, a common muscle used for this purpose is the **gracilis**. The superficial location and limited function of this muscle make it a good candidate for grafting procedures.

4. In the space below, make a representative drawing of the muscles of the medial compartment of the thigh.

5. *Optional Activity:* AP|R **Muscular System**—Visit the quiz area for drill and practice on muscles of the lower limb.

EXERCISE 13.6C: Posterior Compartment of the Thigh

1. Observe a prosected human cadaver or a classroom model demonstrating muscles of the thigh.

2. The **posterior compartment of the thigh** contains only three muscles: the semimembranosus, the semitendinosus, and the biceps femoris. These muscles are collectively referred to as the **hamstring** muscles. They received this name because of their association with the analogous muscles in pigs. The meat we call ham comes from these posterior thigh muscles of a pig. In a slaughterhouse, pig thighs are hung by the long tendons of these muscles, hence the term *hamstrung.* In fact, when we say that we feel "hamstrung," we say that we feel (figuratively) crippled. If our hamstring muscles were severed, we literally would be crippled.

3. Using **table 13.11** and your textbook as a guide, identify the **muscles of the posterior compartment of the thigh** listed in **figure 13.13** on the cadaver or on a classroom model.

Table 13.11	Posterior Compartment of the Thigh				
Muscle	**Origin**	**Insertion**	**Action**	**Innervation**	**Word Roots**
Biceps Femoris					*bi*, two, + *caput*, head, + *femur*, thigh
Long Head	Ischial tuberosity	Head of fibula	Flexes the leg extends the hip	Tibial nerve (L4–S1)	
Short Head	Linea aspera of femur	Head of fibula	Flexes the leg	Fibular nerve (L5–S1)	
Semimembranosus	Ischial tuberosity	Medial tibia	Flexes the leg extends the hip	Tibial nerve (L4–S1)	*semi*, half, + *membrana*, a membrane
Semitendinosus	Ischial tuberosity	Medial tibia	Flexes the leg extends the hip	Tibial nerve (L4–S1)	*semi*, half, + *tendinosus*, tendon

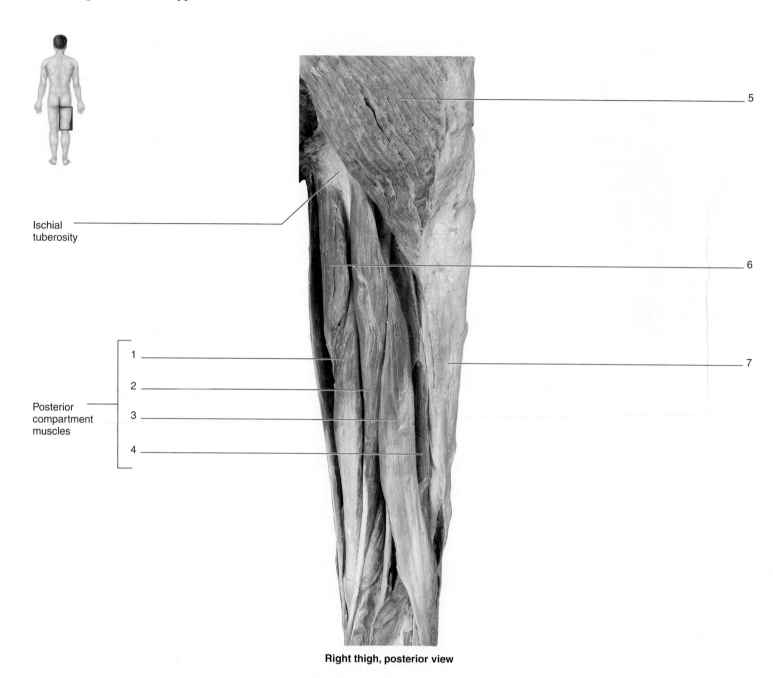

Ischial tuberosity

Posterior compartment muscles

1
2
3
4

5

6

7

Right thigh, posterior view

Figure 13.13   Posterior Compartment of the Thigh.

☐ adductor magnus

☐ biceps femoris, long head

☐ biceps femoris, short head

☐ gluteus maximus

☐ iliotibial tract

☐ semimembranosus

☐ semitendinosus

Then label them in figure 13.13. As you identify these muscles, make note of the following:

Common Actions: Extend the hip and flex the leg.

Exceptions: The short head of the biceps femoris does not cross the hip joint. Thus, it does not extend the hip.

4. *Semitendinosus and Semimembranosus*: These two muscles are named for the appearance of their connective tissue components. The **semitendinosus** is the more superficial of the two muscles. At its *distal* end it forms a long, slender tendon, hence the name ending "tendinosus." The **semimembranosus** muscle is the deeper of the two muscles. At its *proximal* end, it forms a broad, flat tendon that looks like a membrane, hence the name ending "membranosus."

EXERCISE 13.7 Compartments of the Leg

EXERCISE 13.7A: Anterior Compartment of the Leg

1. Observe a prosected human cadaver or a classroom model demonstrating muscles of the leg.

2. Using **table 13.12** and your textbook as a guide, identify the **muscles of the anterior compartment of the leg** listed in **figure 13.14** on the cadaver or on a classroom model. Then label them in figure 13.14. As you identify these muscles, make note of the following:

   Common Actions: Dorsiflex the ankle and extend the digits.

   Exception: The tibialis anterior does not act on the digits. In addition, it inverts the ankle as it dorsiflexes.

EXERCISE 13.7B: Lateral Compartment of the Leg

1. Observe a prosected human cadaver or a classroom model demonstrating muscles of the leg.

2. Using **table 13.13** and your textbook as a guide, identify the **muscles of the lateral compartment of the leg** listed in **figure 13.15** on the cadaver or on classroom models. Then label them in figure 13.15.

3. *Fibularis Longus and Brevis*: The fibularis muscles were formerly referred to as peroneus muscles. Thus, you should be aware that the terms relate to the same muscles. The **fibularis (peroneus) longus** has a very long tendon that lies superficial to the belly of the **fibularis (peroneus) brevis** muscle. The tendons of both muscles pass immediately posterior to the lateral malleolus en route to their attachments on the plantar surface of the foot. Their major action is to evert the ankle.

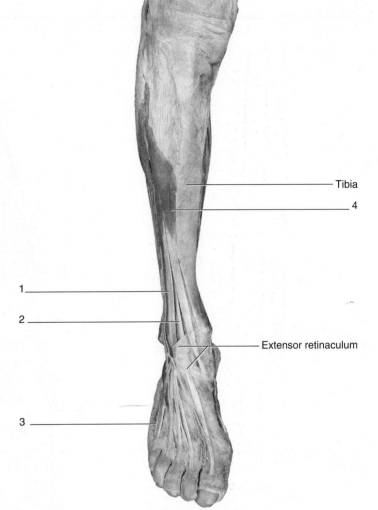

Right leg, anterior view

Figure 13.14 **Anterior Compartment of the Leg.**

☐ extensor digitorum longus   ☐ fibularis tertius tendon

☐ extensor hallucis longus   ☐ tibialis anterior

Study Tip!

pEroneus (fibularis) muscles always Evert, whereas tIbialis muscles always Invert.

Table 13.12	Anterior Compartment of the Leg				
Muscle	**Origin**	**Insertion**	**Action**	**Innervation**	**Word Roots**
Tibialis Anterior	Tibia (lateral condyle and upper two-thirds)	Middle cuneiform (inferior surface) and metatarsal I	Dorsiflexes and inverts the foot	Deep fibular nerve	*tibia*, the shinbone, + *anterior*, the front surface
Extensor Hallucis Longus	Fibula (anteromedial surface) and interosseous membrane	Distal phalanx of big toe	Extends the great toe and dorsiflexes the ankle	Deep fibular nerve	*extendo*, to stretch out, + *hallux*, the great toe, + *longus*, long
Extensor Digitorum Longus	Tibia (lateral condyle and proximal three-fourths)	2nd and 3rd phalanges of digits 2–5	Extends digits 2–5 and dorsiflexes the ankle	Deep fibular nerve	*extendo*, to stretch out, + *digit*, a toe, + *longus*, long
Fibularis Tertius	Fibula (anterior, distal surface) and interosseous membrane	Base of metatarsal V	Dorsiflexes and weakly everts foot	Deep fibular nerve	*fibularis*, fibula, + *tertius*, third

Table 13.13	Lateral Compartment of the Leg				
Muscle	**Origin**	**Insertion**	**Action**	**Innervation**	**Word Roots**
Fibularis Longus	Head of fibula	Metatarsal I and middle cuneiform	Everts the foot and plantarflexes the ankle	Superficial fibular nerve	*fibularis*, relating to the fibula, + *longus*, long
Fibularis Brevis	Distal fibula	Metatarsal V	Everts the foot and plantarflexes the ankle	Superficial fibular nerve	*fibularis*, relating to the fibula, + *brevis*, short

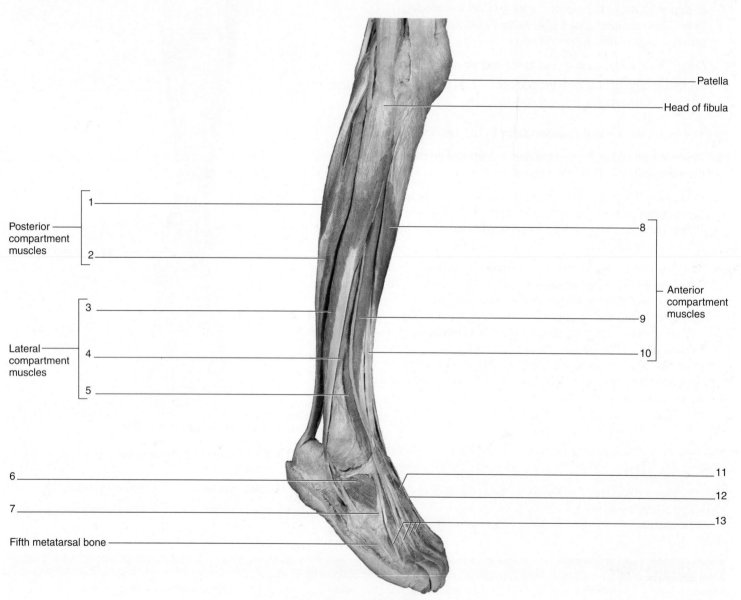

Right leg, lateral view

Figure 13.15   Lateral View of the Leg. In a lateral view of the leg, muscles from all three leg compartments can be seen.

☐ extensor digitorum brevis

☐ extensor digitorum longus

☐ extensor digitorum longus tendons

☐ extensor hallucis brevis

☐ extensor hallucis longus

☐ extensor hallucis longus tendon

☐ fibularis brevis

☐ fibularis longus

☐ fibularis tertius

☐ fibularis tertius tendon

☐ gastrocnemius

☐ soleus

☐ tibialis anterior

EXERCISE 13.7C: Posterior Compartment of the Leg

1. Observe a prosected human cadaver or a classroom model demonstrating muscles of the leg.

2. Using **table 13.14** and your textbook as a guide, identify the **muscles of the posterior compartment of the leg** listed in **figure 13.16** on the cadaver or on a classroom model. Then label them in figure 13.16.

3. *Triceps Surae*: The **triceps surae** includes the **soleus** and the two-headed **gastrocnemius** muscle. The tendons of these muscles come together to insert onto the calcaneus via the calcaneal (Achilles) tendon. The most superficial of the muscles is the gastrocnemius, though the lateral edges of the soleus can be seen as they poke out beneath the inferior portion of the gastrocnemius. The **plantaris** is a small muscle that can be seen only by reflecting or removing the gastrocnemius and soleus muscles. The plantaris has a very small belly, located just below the popliteal fossa in the back of the knee. It has a very long, flat tendon that often resembles ribbon used to decorate presents. The muscle itself is a very weak flexor of the leg. Because its actions are minimal and its tendon is long, the tendon is often removed and used for tendon grafts. If you are viewing a cadaver and cannot find the plantaris, this may be because it isn't there. The plantaris is absent in approximately 6% to 8% of humans, and it is more commonly absent in the left leg than in the right leg.

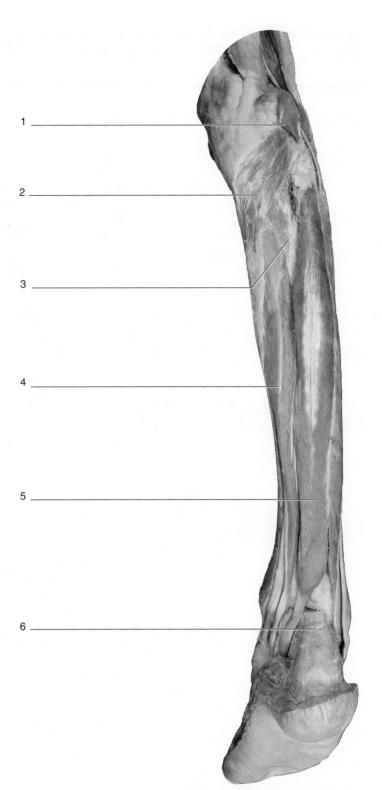

Right leg, deep posterior view

Figure 13.16   **Posterior Compartment of the Leg.**

☐ calcaneal tendon   ☐ plantaris

☐ flexor digitorum longus   ☐ popliteus

☐ flexor hallucis longus   ☐ tibialis posterior

Table 13.14	Posterior Compartment of the Leg				
Muscle	**Origin**	**Insertion**	**Action**	**Innervation**	**Word Roots**
Gastrocnemius	Femoral condyles	Calcaneus	Plantarflexes the ankle and flexes the leg (weak)	Tibial nerve (L4–S1)	*gaster*, belly, + *kneme*, leg
Soleus	Proximal tibia, fibula	Calcaneus	Plantarflexes the ankle	Tibial nerve (L4–S1)	*solea*, a sandal, sole of the foot
Plantaris	Femur (supracondylar ridge)	Calcaneus	Plantarflexes the ankle and flexes the leg (weak)	Tibial nerve (L4–S1)	*plantar*, relating to the sole of the foot
Popliteus	Lateral condyle of femur	Posterior, proximal surface of tibia	Flexes leg, unlocks the knee joint	Tibial nerve	*poplit*, back of the knee
Tibialis Posterior	Proximal tibia and fibula	Medial cuneiform and navicular	Inverts the foot and plantarflexes the ankle	Tibial nerve (L5–S1)	*tibia*, the shinbone, + *posterior*, the back surface
Flexor Digitorum Longus	Posterior tibia	Distal phalanx of digits 2–5	Flexes digits 2–5	Tibial nerve (L5–S1)	*flexus*, to bend, + *digit*, a toe, + *longus*, long
Flexor Hallucis Longus	Inferior two-thirds of fibula	Distal phalanx of the great toe	Flexes the great toe	Tibial nerve (L5–S1)	*flexus*, to bend, + *hallux*, the great toe, + *longus*, long

---

EXERCISE 13.8 | Intrinsic Muscles of the Foot

1. Observe a prosected human cadaver or a classroom model demonstrating muscles of the foot.

2. The intrinsic muscles of the plantar surface of the foot are arranged in four layers, named layer 1 to layer 4 from superficial to deep. The foot also contains two neurovascular planes (a neurovascular plane is a region where the nerves and blood vessels that serve the region are located). The first is located between muscle layers 1 and 2, and the second is located between layers 3 and 4.

3. Using **table 13.15** and your textbook as a guide, identify the **intrinsic muscles of the foot** listed in **figure 13.17** on the cadaver or on a classroom model. Then label them in figure 13.17.

4. *Adductor Hallucis*: The **adductor hallucis** muscle is located in the third plantar muscle layer. It is a good landmark for the identification of other muscles in the foot. In addition, its two heads (oblique and transverse) appear like the

> ### Study Tip!
> Conveniently, the mnemonic for remembering the functions of the interossei in the foot is the same as in the hand: **DAB** and **PAD**. The **D**orsal interossei **AB**duct the digits (**DAB**), whereas the **P**lantar interossei **AD**duct the digits (**PAD**).

number 7 when viewed together. To identify this muscle and others in the deeper layers of the foot, the tough **plantar aponeurosis** must first be cut and reflected.

5. *Interossei*: Just like in the hand, there are two groups of interosseous muscles in the foot, though they are called dorsal and *plantar*, instead of dorsal and *palmar*. The functions of the interossei in the foot are the same as in the hand: the dorsal interossei abduct the digits, whereas the plantar interossei adduct the digits.

Table 13.15	Intrinsic Muscles of the Foot				
Muscle	**Origin**	**Insertion**	**Action**	**Innervation**	**Word Roots**
DORSAL MUSCULATURE					
Extensor Digitorum Brevis	Calcaneus and extensor retinaculum (inferior surface)	Digits 2–4 (dorsal surfaces)	Extends the proximal phalanx of digits 2–4	Deep fibular nerve	*dorsal*, the back, + *inter-*, between, + *os*, bone
Extensor Hallucis Brevis	Calcaneus and extensor retinaculum (inferior surface)	Dorsal surface of the great toe	Extends the proximal phalanx of the great toe	Deep fibular nerve	*palmar*, the palm of the hand (foot), + *inter-*, between, + *os*, bone
PLANTAR MUSCULATURE—LAYER 1					
Abductor Digiti Minimi	Calcaneus and plantar aponeurosis	Base of the proximal phalanx of digit 5 (lateral surface)	Abducts and flexes digit 5 (the little toe)	Lateral plantar nerve	*abduct*, to move away from the median plane, + *digitus*, toe, + *minimi*, smallest
Abductor Hallucis	Calcaneus and plantar aponeurosis	Base of the proximal phalanx of digit 1 (medial side)	Abducts the great toe	Medial plantar nerve	*abduct*, to move away from the median plane, + *hallux*, the great toe
Flexor Digitorum Brevis	Calcaneus and plantar aponeurosis	Middle phalanx of digits 2–5	Flexes digits 2–5	Medial plantar nerve	*flex*, to bend, + *digitus*, toe, + *brevis*, short
PLANTAR MUSCULATURE—LAYER 2					
Lumbricals	Tendons of the flexor digitorum longus	Tendons of the extensor digitorum longus	Flexes the proximal phalanges and extends the distal phalanges of digits 2–5	Medial and lateral plantar nerve	*lumbricus*, earthworm
Quadratus Plantae	Calcaneus (medial surface and lateral margin)	Tendon of the flexor digitorum longus (lateral side)	Flexes digits 2–5	Lateral plantar nerve	*quadratus*, having four sides, + *plantae*, sole of the foot
PLANTAR MUSCULATURE—LAYER 3					
Adductor Hallucis (Oblique Head)	Base of metatarsals II–V	Base of the proximal phalanx of the great toe	Adducts the great toe and helps maintain the transverse arch of the foot	Lateral plantar nerve	*adduct*, to move toward the median plane, + *hallux*, the great toe
Adductor Hallucis (Transverse Head)	Capsules of the metatarsophalangeal joints III–V	Base of the proximal phalanx of the great toe	Adducts the great toe and helps maintain the transverse arch of the foot	Lateral plantar nerve	*adduct*, to move toward the median plane, + *hallux*, the great toe
Flexor Digiti Minimi Brevis	Base of metatarsal V	Proximal phalanx of digit 5	Flexes the proximal phalanx of digit 5 (the little toe)	Lateral plantar nerve	*flex*, to bend, + *digitus*, a toe, + *minimi*, smallest, + *brevis*, short
Flexor Hallucis Brevis	Cuboid and lateral cuneiform (plantar surfaces)	Proximal phalanx of the great toe	Flexes the proximal phalanx of the great toe	Medial plantar nerve	*flex*, to bend, + *hallux*, the great toe, + *brevis*, short
PLANTAR MUSCULATURE—LAYER 4					
Dorsal Interossei	Metatarsals I–V (adjacent surfaces)	Proximal phalanges of digits 2–4 (sides)	Abducts digits 2–4 and flexes the metatarsophalangeal joints	Lateral plantar nerve	*dorsal*, the back, + *inter-*, between, + *os*, bone
Plantar Interossei	Bases of metatarsals III–V	Bases of proximal phalanges of digits 3–5 (medial side)	Adducts digits 2–4 and flexes the metatarsophalangeal joints	Lateral plantar nerve	*plantae*, the sole of the foot, + *inter-*, between, + *os*, bone

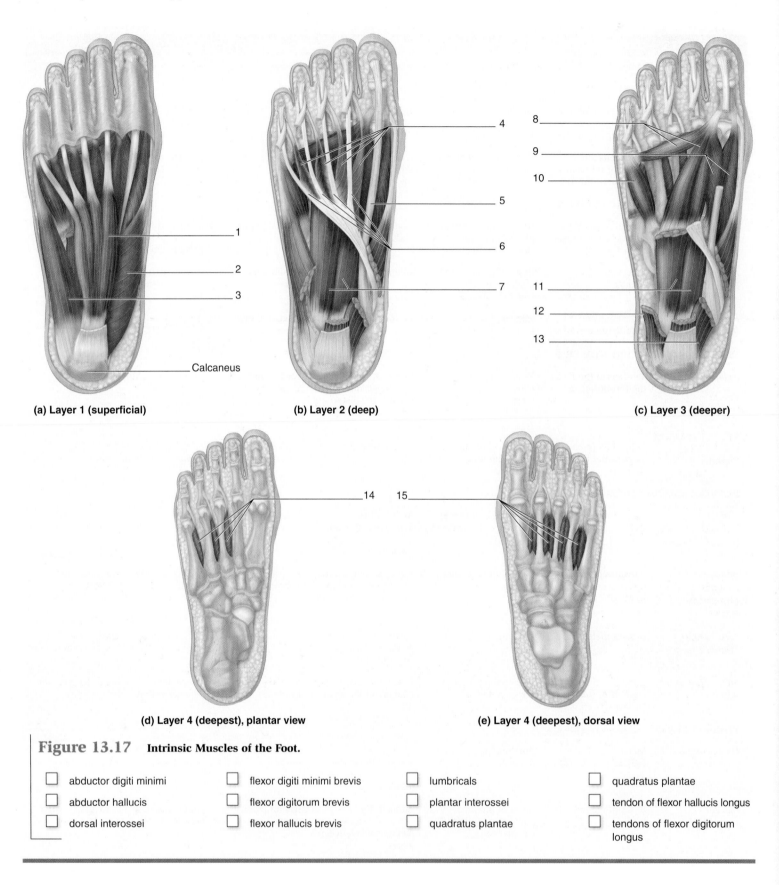

(a) Layer 1 (superficial)

1
2
3
Calcaneus

(b) Layer 2 (deep)

4
5
6
7

(c) Layer 3 (deeper)

8
9
10
11
12
13

(d) Layer 4 (deepest), plantar view

14

(e) Layer 4 (deepest), dorsal view

15

Figure 13.17 **Intrinsic Muscles of the Foot.**

☐ abductor digiti minimi   ☐ flexor digiti minimi brevis   ☐ lumbricals   ☐ quadratus plantae

☐ abductor hallucis   ☐ flexor digitorum brevis   ☐ plantar interossei   ☐ tendon of flexor hallucis longus

☐ dorsal interossei   ☐ flexor hallucis brevis   ☐ quadratus plantae   ☐ tendons of flexor digitorum longus

Chapter 13: Appendicular Muscles

Name: _____

Date: _____ Section: _____

POST-LABORATORY WORKSHEET

1. Name three muscles that attach to the coracoid process of the scapula.

   a. _____

   b. _____

   c. _____

2. For each of the actions of the trapezius listed below, name a muscle that acts as a synergist for that action.

   a. Elevation of the scapula: _____

   b. Adduction of the scapula: _____

3. Describe the involvement of the biceps brachii in the following movements:

   a. Flexion of the forearm: _____

   _____

   b. Supination of the forearm/wrist: _____

   _____

4. True or False: The flexor digitorum superficialis muscle of the anterior compartment of the forearm flexes the distal phalanges. _____.

   If you answered "false," the name of what muscle would make the statement true? _____

5. The gluteus maximus muscle is a major extensor of the hip. What muscle acts as the major antagonist to the gluteus maximus muscle? (Hint: The muscle you name would be a prime mover for hip *flexion*.)

   _____

6. List the three structures that form the borders of the femoral triangle.

   a. _____

   b. _____

   c. _____

7. List three structures that are found *within* the femoral triangle.

   a. _____

   b. _____

   c. _____

8. What two muscles insert onto the iliotibial tract (IT band)?

   a. _____

   b. _____

9. List three similarities between the muscles of the hand and the muscles of the foot, regarding their structure, orientation, functions, and/or names.

   a. _____

   b. _____

   c _____

10. How can you distinguish between the semitendinosus muscle and the semimembranosus muscle?

_____

_____

_____

_____

11. Compare and contrast the functions of the tibialis anterior and tibialis posterior muscles by answering the following three questions:

a. For what action do the tibialis anterior and posterior act as *antagonists*?

_____

b. For what action do the tibialis anterior and posterior act as *synergists*?

_____

c. What two muscles act as antagonists to the tibialis anterior and tibialis posterior for the action you gave in part (b) of this question?

i. _____

ii. _____

12. The action of the dorsal interosseous muscles of the hand is to _____ the digits, whereas the action of the palmar interosseous muscles of the hand is to _____ the digits.

13. What three muscles insert on the calcaneal tendon?

a. _____

b. _____

c. _____

14. The muscles in question 13 are collectively referred to as the _____.

15. One action of the gluteal muscles (gluteus maximus, medius, and minimus) is to rotate the hip.

a. Which gluteal muscle(s) laterally rotate the hip? _____.

b. Which gluteal muscle(s) medially rotate the hip? _____.

16. Which three muscles comprise the rotator cuff?

a. _____

b. _____

c. _____

17. Which muscle of the medial compartment of the thigh does not insert onto the linea aspera of the femur? _____

18. What muscle of the quadriceps muscle group is the only muscle to cross the hip joint? _____.

19. What is the *origin* (proximal attachment) of the muscle you listed in your answer to question 18? _____ _____. What is the *insertion* (distal attachment)? _____

20. Describe the difference between the patellar *tendon* and the patellar *ligament*, and describe the location of each. (Hint: Think about the specific definitions of *tendon* and *ligament*.)

_____

_____

_____

_____

Nervous Tissues

14

OUTLINE and OBJECTIVES

Module 7: NERVOUS SYSTEM

INTRODUCTION

Structures of the nervous system are not particularly impressive in the absence of any knowledge of what they do. Imagine how unimpressed early anatomists must have been with the brain. All they saw was a very dense, oatmeal-colored mass of tissue with no known function. Their observations of brain tissue could hardly have been as fascinating or romantic as their observations of the heart. At first, the most inspiring names they could find for brain tissues were "gray matter" and "white matter." Today we know that the nervous system is fascinating, and that scientists are barely beginning to grasp its complexity. Information about the complexity of the nervous system has been gained through dynamic studies of the brain using magnetic resonance imaging (MRI), not through static, gross observation.

As a student of anatomy learning about the gross structures of the nervous system, you will have to trust what others tell you about the functions of those structures because these functions cannot be seen at the gross level. However, when you observe nervous tissue under the microscope, it becomes much more interesting and you will begin to understand some of the relationships between structure and function.

In this laboratory session you will observe the histological structure of nervous tissues. As you proceed to subsequent chapters that cover the gross structures of the nervous system, it will be important for you to return to this chapter and relate what you observe microscopically with what you observe grossly.

Chapter 14: Nervous Tissues

PRE-LABORATORY WORKSHEET

1. For each name, write the letter for the structure as labeled in the diagram below.

_____ 1. axon

_____ 2. axon hillock

_____ 3. dendrite

_____ 4. myelin sheath/neurolemmocyte (Schwann cell)

_____ 5. chromatophilic substance

_____ 6. neurofibril node (node of Ranvier)

_____ 7. nucleolus

_____ 8. nucleus

_____ 9. cell body (soma)

_____ 10. synaptic knobs

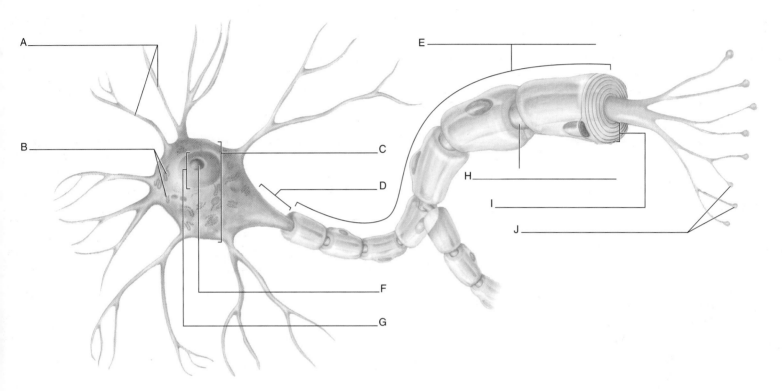

2. Match the cell type in column B with its description in column A.

Column A

_____ 1. cells that myelinate axons in the central nervous system

_____ 2. cells that help reinforce the blood-brain barrier

_____ 3. glial cells found in peripheral nerve ganglia

_____ 4. cells that myelinate axons in the peripheral nervous system

_____ 5. cells that become phagocytes in response to tissue injury

_____ 6. cells that line the ventricles of the brain

Column B

a. astrocytes

b. ependymal cells

c. microglia

d. oligodendrocytes

e. satellite cells

f. neurolemmocytes (Schwann cells)

3. Match the connective tissue structure in column B with its description in column A.

Column A

_____ 1. surrounds a fascicle of axons

_____ 2. surrounds an individual axon

_____ 3. surrounds the entire nerve

Column B

a. endoneurium

b. epineurium

c. perineurium

IN THE LABORATORY

Nervous tissue is characterized by its excitability. Recall that excitability is the ability to generate and propagate electrical signals called action potentials. Nervous tissue is composed of two basic cell types. *Neurons* are the excitable cells. They have a limited ability to divide and multiply in the adult. *Neuroglia* (glial cells) are supporting cells whose functions are to support and protect neurons. Glial cells retain the ability to divide and multiply, and they constitute more than half of the matter found in neural tissue. In this laboratory session you will observe the histological structure of both types of cells.

Histology

At the gross level, nervous tissue is classified as either white matter or gray matter. **White matter** consists mainly of myelinated axons. Bundles of white matter in the central nervous system are **tracts**, and bundles of white matter in the peripheral nervous system are **nerves**. White matter is white because of the presence of myelin, a fatty substance produced by glial cells that is used to insulate axons to increase the speed of conducting action potentials along the axon. **Gray matter** consists mainly of neuron cell bodies, dendrites, and unmyelinated axons. Collections of gray matter within the central nervous system are **nuclei**, and collections of gray matter within the peripheral nervous system are **ganglia**. Gray matter is gray because of an absence of myelin. In this laboratory session you will begin by observing the histological appearance of white matter and gray matter and then will proceed to observe specific cell types found within nervous tissues.

EXERCISE 14.1 Gray and White Matter

1. Obtain a slide of a cross section of the spinal cord.

2. Observe the slide with your naked eye before placing it on the microscope stage. Notice that the spinal cord consists of an inner, butterfly-shaped core of **gray matter** surrounded by an outer region of **white matter** (**figure 14.1**). What neuronal structures do you expect to find within the gray

   matter? _____. What neuronal structures to you expect to find within the white matter?

   _____.

3. In the space below, make a simple line drawing of your microscopic observations. Label the following in your drawing:

   ☐ gray matter   ☐ white matter

   _____ ×

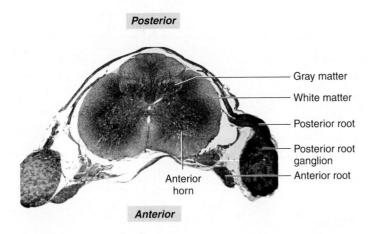

Posterior

— Gray matter

— White matter

— Posterior root

— Posterior root ganglion

Anterior horn — Anterior root

Anterior

Figure 14.1   Gray Matter and White Matter.   A stained, histological cross section through the spinal cord showing gray matter (pink) and white matter (purple).

WHAT DO YOU THINK?

1 Which type of neural tissue is more involved with the integration of information (rather than the transmission or conducting of information)—white matter or gray matter?

Neurons

Neurons are typically the largest cells seen in any slide of nervous tissue. They have very large, light-colored nuclei with prominent nucleoli. The **cell body** (soma) of the neuron contains a large amount of rough endoplasmic reticulum that stains darkly and is called **chromatophilic substance** (Nissl bodies). Generally **dendrites** are not easily seen, even at high magnification. On the other hand, one can often identify a single large **axon** leaving the cell body. There will be an absence of chromatophilic substance at the location where the axon leaves the cell body, an area of the neuron called the **axon hillock**. Table 14.1 summarizes the parts of a typical neuron and their functions.

Neurons are classified based on the number of cellular processes that come off of the cell body. **Multipolar** neurons (**table 14.2**) have many dendrites and a single axon. **Bipolar** neurons, found only in special sensory organs such as the retina, have one dendrite and one axon. Finally, **unipolar** neurons have a single process (*uni*, one). Unipolar neurons are found only in the posterior (dorsal) root ganglia, which are peripheral nerve ganglia.

Table 14.1	Parts of a Neuron	
Part of the Neuron	**Description and Function**	**Word Origins**
Axon	A single large process of a neuron that conducts information away from the cell body of a neuron.	*axon*, axis
Axon Hillock	A light-staining region where the axon leaves the cell body of a neuron; devoid of chromatophilic substance; the location where action potentials are generated by a neuron.	*hillock*, a small elevation
Cell Body (Soma)	The part of the neuron that contains the nucleus and cellular organelles.	*soma*, body
Chromatophilic (Nissl) Substance	Dark-staining material found within the soma of a neuron, but absent in the axon hillock region; rough endoplasmic reticulum that functions in the production of membrane-associated proteins.	*chroma*, color, + *phileo*, to love
Dendrite	Multiple branching processes of a neuron that bring information into the cell body.	*dendron*, tree
Neurofibril Node (Node of Ranvier)	A bare region on a myelinated axon where there is an absence of myelin and where action potentials are generated.	*neuron*, nerve, + *fibrilla*, fiber; *Louis Ranvier*, French pathologist
Synaptic Knobs	Swellings on the ends of an axon that form synapses with effector organs such as glands and muscle fibers.	*syn-*, together, + *hapto*, to clasp, + *knob*, a protuberance
Telodendria (axon terminals)	Branches at the end of an axon, with each process containing a synaptic knob at its end.	*telos*, end, + *dendron*, tree

Table 14.2	Multipolar Neurons in the Cerebrum and Cerebellum	
Neuron Type	**Location**	**Features**
Anterior Horn Cell	Anterior horn of the spinal cord	Very large neuron with prominent chromatophilic substance. Cell body is irregularly shaped with multiple dendrites extending from the soma.
Pyramidal Cell	Cerebral cortex	Cell body has a triangular shape and multiple dendrites extend from the cell body.
Purkinje Cell	Cerebellum	Cell bodies appear "basket-like" with a rounded area facing the granular layer of the cerebellum, and a tuft of dendrites extending into the molecular layer of the cerebellum.

EXERCISE 14.2 General Multipolar Neurons—Anterior Horn Cells

1. Obtain a slide of the *spinal cord in cross section*.

2. Place the slide on the microscope stage and bring the tissue sample into focus on low power (**figure 14.2a**). Again, note the inner core of gray matter surrounded by an outer region of white matter. Move the microscope stage so the inner gray matter is in the center of the field of view and then switch to high power (figure 14.2b). Look for very large, multipolar neurons found in the anterior (ventral) horn of the gray matter. These cells are called **anterior horn cells**, for their location. They are large motor neurons whose axons exit the spinal cord and travel through peripheral nerves to skeletal muscles in the body. Because their axons are so long, the cell bodies of these neurons are quite large and easy to see.

3. Using tables 14.1 and 14.2 and figure 14.3 as guides, identify the following structures on the spinal cord slide:

☐ gray matter   ☐ nucleus

☐ chromatophilic   ☐ cell body of anterior
   substance   horn cell

☐ nucleolus of anterior   ☐ white matter
   horn cell

4. In the space below, make a brief sketch of anterior horn cells as seen through the microscope.

_____ ×

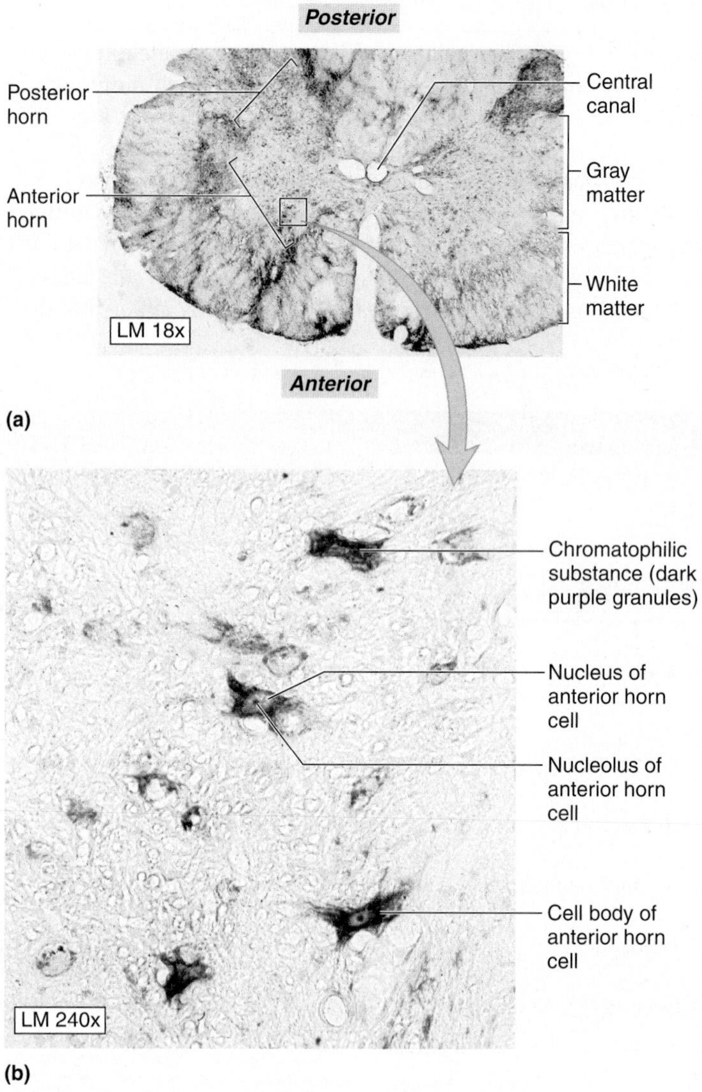

(a)

(b)

Figure 14.2   Gray Matter of the Spinal Cord. (a) Cross section of spinal cord. (b) Close up showing large motor neurons (anterior horn cells).

EXERCISE 14.3 Cerebrum—Pyramidal Cells

1. Obtain a slide of the cerebrum that has been stained with Nissl stain. Nissl stain colors the **rough endoplasmic reticulum** (the chromatophilic substance) of the neurons an intense blue color.

2. Place the slide on the microscope stage and bring the tissue sample into focus on low power. Identify areas of gray matter and white matter on the slide (**figure 14.3**).

3. Bring an area of gray matter to the center of the field of view and switch to high power. Note the very large cells located within the gray matter. These cells are **neurons**. Note the very large nuclei of the *neurons*, which are surrounded by many cells with much smaller nuclei, the *glial cells*. Locate large neurons whose cell bodies appear triangular in shape. These cells are special neurons of the cerebrum called **pyramidal cells** (figure 14.3, table 14.2).

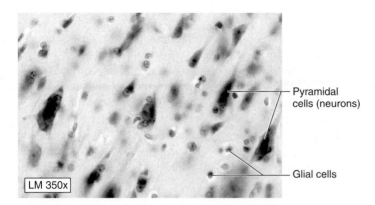

LM 350x

Figure 14.3 **Pyramidal Cells in the Cerebral Cortex.** The pyramidal cells (neurons) are the large, triangular cells. The smaller nuclei visible in the slide are mainly those of glial cells.

Pyramidal cells (neurons)

Glial cells

4. In the space below, sketch what you observed under the microscope. Label the following in your drawing:

☐ glial cells   ☐ pyramidal cells

_____ ×

EXERCISE 14.4 Cerebellum—Purkinje Cells

1. Obtain a slide of the **cerebellum**. Place it on the microscope stage and bring the tissue sample into focus on low power (**figure 14.4a**).

2. Note the folds of tissue with an outer region, the *molecular layer*, and an inner region, the *granular layer* (figure 14.4a). Deep to the folds is the **white matter** of the cerebellum. Move the microscope stage to bring the junction between the molecular and granular layers to the center of the field of view. Then switch to high power (figure 14.4b). Identify the very large cells located at this juncture. These cells are **Purkinje cells** (table 14.2), also known as basket cells, which are large multipolar neurons of the cerebellum (the layer where they are located is the *Purkinje cell layer*).

3. In the space below, sketch what you observed at high power. Label the following in your drawing:

☐ granular layer   ☐ Purkinje cells
☐ molecular layer   ☐ white matter

_____ ×

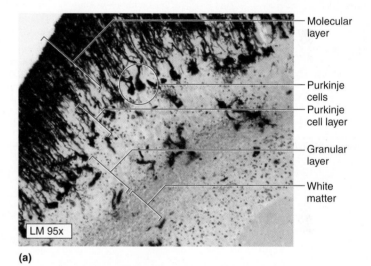

Molecular layer

Purkinje cells

Purkinje cell layer

Granular layer

White matter

LM 95x

(a)

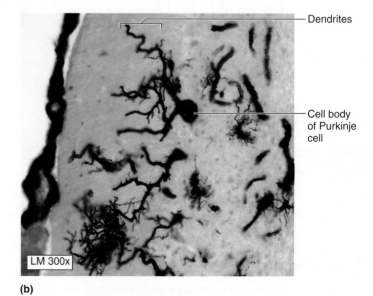

Dendrites

Cell body of Purkinje cell

LM 300x

(b)

Figure 14.4 **Cerebellum—Purkinje Cells.** The cerebellum, demonstrating Purkinje cells between the granular and molecular layers. (a) Low power. (b) High power.

Glial Cells

Glial cells, or **neuroglia** (literally, "nerve glue"), are much more abundant than neurons in the nervous system, and they retain the capacity for cell division. Thus, most brain tumors arise from glial cells and are referred to as *gliomas*. The two most abundant glial cells in the central nervous system are **astrocytes** and **oligodendrocytes**. Because of their large size, these cells are collectively referred to as *macroglia*. The central nervous system also contains resident macrophages called **microglia**. As their name implies, microglia are very tiny cells. Their name is a bit of a misnomer, however, because these cells become very

large phagocytic cells when tissue injury occurs. You will not see microglial cells in the laboratory. Special epithelial cells found lining the fluid-filled ventricles of the brain are **ependymal cells**.

The peripheral nervous system contains only one type of glial cell, but it is named differently—and functions differently—depending upon its location. Thus, we typically refer to the peripheral nervous system as having two glial cell types: *neurolemmocytes (Schwann cells)* and *satellite cells*. **Table 14.3** summarizes the characteristics of each type of glial cell.

Table 14.3	Glial Cells			
Cell Name	**Location**	**Description**	**Function(s)**	**Word Origins**
Astrocytes	CNS	Star-shaped cells that are very abundant in the central nervous system.	General supporting cells in the CNS. Transfer nutrients to neurons from the blood. Reinforce the blood-brain barrier. Maintain the extracellular environment around neurons.	*astron*, star, + *kytos*, cell
Oligodendrocytes	CNS	Cells with a few long processes that wrap around axons in the CNS.	Myelinate axons in the central nervous system. Each oligodendrocyte can myelinate multiple axons.	*oligos*, few, + *dendro-*, like a tree, + *kytos*, cell
Microglia	CNS	Small cells with oval nuclei and multiple branching processes.	Microglia are derived from blood monocytes, and they are the resident macrophages in the CNS. They are normally very small (hence the name), but transform into very large, phagocytic cells (macrophages) when tissues are injured. As macrophages they engulf dead tissue and/or pathogens and remove them from the CNS.	*mikros*, small, + *glia*, glue
Ependymal Cells	CNS	Cuboidal to columnar shaped cells with microvilli and cilia on their apical surfaces.	Ependymal cells are epithelial cells that line the ventricles (fluid-filled spaces) of the brain and the central canal of the spinal cord. Unlike other epithelia, they DO NOT have a basement membrane.	*ependyma*, an upper garment
Neurolemmocytes (Schwann Cells)	PNS	Large cells that wrap their plasma membrane around peripheral nerve axons.	Myelinate axons in the peripheral nervous system. Each neurolemmocyte can myelinate only one axon.	*neuron*, nerve, + *lemma*, husk, + *kytos*, cell; *Theodor Schwann*, German histologist and physiologist
Satellite Cells	PNS	Small glial cells found surrounding the cell bodies of somatic sensory cells in the posterior (dorsal) root ganglion.	Satellite cells sit right outside the cell bodies of somatic sensory neurons, hence the appearance of "satellites" around those neurons. They provide general support for the neurons and are analagous in function to astrocytes in the CNS.	*satelles*, attendant

Glial Cells of the Central Nervous System

EXERCISE 14.5 Astrocytes

1. Obtain a slide of the cerebrum or cerebellum that has been stained with silver stain. Silver stain makes the general supporting cells of the central nervous system, the **astrocytes**, stain very dark so they are visible under the microscope (**figure 14.5**).

2. Place the slide on the microscope stage and bring the tissue sample into focus on low power. Then switch to high power and bring the tissue sample into focus once again.

3. The two most prominent cell types you will see are the large neurons (can you identify pyramidal cells or Purkinje cells?) and the smaller astrocytes. Astrocytes, as their name implies, are shaped like stars (*astron-*, star). They have multiple long cellular processes that wrap themselves around neurons and around blood vessels in the central nervous system. These processes allow them to perform one of their main functions, which is to transport nutrients from the blood to the neurons.

4. In the space below, make a sketch of an astrocyte as seen through the microscope.

_____ ×

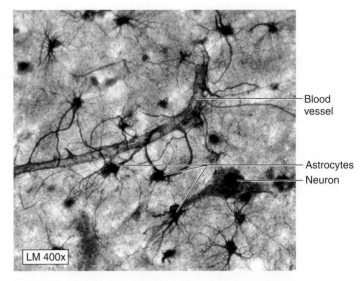

Blood vessel

Astrocytes
Neuron

LM 400x

Figure 14.5 **Astrocytes.** Two astrocytes in the center of the slide have processes that can be seen touching a blood vessel.

EXERCISE 14.6 Ependymal Cells

1. Obtain a slide of the brain or of choroid plexus and place it on the microscope stage. Bring the tissue sample into focus on low power. Then switch to high power and bring the tissue sample into focus once again. Look for the "empty" spaces on the slide, because those spaces will likely be the ventricular spaces that are lined with ependymal cells.

2. **Ependymal cells (figure 14.6)** are cuboidal to columnar-shaped epithelial cells that line the ventricles of the brain and the central canal of the spinal cord. They also form the outer layer of the choroid plexus. They have both *cilia* and

microvilli on their apical surfaces, but they are unique as epithelia in that they do not have a basement membrane.

3. Using figure 14.6 as a guide, locate ependymal cells on the slide.

4. In the space below, make a simple line drawing of ependymal cells as seen through the microscope. Label the following in your drawing:

☐ choroid plexus (if visible)   ☐ ventricular space

☐ ependymal cells

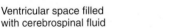

Ventricular space filled with cerebrospinal fluid

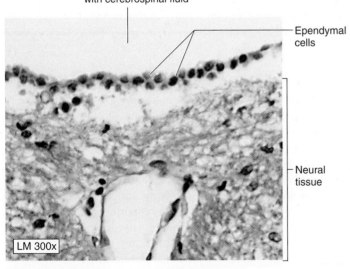

Ependymal cells

Neural tissue

LM 300x

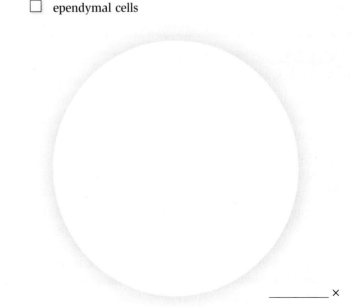

_____ ×

Figure 14.6 **Ependymal Cells.** Cuboidal ependymal cells lining a ventricular surface of the brain. Choroid plexus is not visible in this slide.

Glial Cells of the Peripheral Nervous System

Neurolemmocytes (Schwann Cells)

1. Obtain a slide of a longitudinal section of a peripheral nerve and place it on the microscope stage (**figure 14.7**). A nerve is a bundle of myelinated axons, so you should expect to observe axons, myelin sheaths, and connective tissue on this slide.

2. Bring the tissue sample into focus on low power and note the wavy appearance of the axons. The axons appear wavy because the nerve is not stretched. When a nerve is stretched, the axons also stretch out. The fact that peripheral nerves have elasticity is important because it allows the nerves to stretch during movement without being damaged. This elasticity is imparted to the nerves by the connective tissue coverings that surround the axons.

3. Observe the slide on high power. Note the very dark lines, which are the axons. Each axon is surrounded by light material, which is the **myelin sheath**. Note the elongated nuclei seen throughout the slide. These nuclei belong to **neurolemmocytes** (**Schwann cells**), the glial cells responsible for myelinating axons in the peripheral nervous system. Each individual neurolemmocyte myelinates only a single axon and a single axon has hundreds of neurolemmocytes along its length. The bare areas of axon found between myelin sheaths are called **neurofibril nodes**, or **nodes of Ranvier**.

4. Scan the slide to see if you can locate a neurofibril node.

5. In the space below, make a simple line drawing of a peripheral nerve as seen through the microscope. Label the following in your drawing:

☐ axon ☐ neurofibril node
☐ myelin sheath ☐ neurolemmocyte nucleus

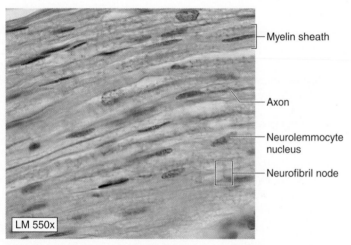

———— ×

Figure 14.7 **Neurofibril Nodes.** Longitudinal section of a nerve demonstrating neurofibril nodes.

Labels on figure: Myelin sheath, Axon, Neurolemmocyte nucleus, Neurofibril node, LM 550x

---

Satellite Cells

1. Obtain a slide of a spinal ganglion (the slide may be labelled dorsal root ganglion, posterior root ganglion, or peripheral nerve ganglion) and place it on the microscope stage.

2. Bring the tissue sample into focus on low power and locate both the nerve *root* and the *ganglion* (**figure 14.8a**).

3. Recall that the peripheral nervous system is mainly composed of nerves, which are bundles of myelinated axons. Occasionally collections of neuron cell bodies are found along these nerves. A collection of neuron cell bodies in the peripheral nervous system is called a **ganglion** (*ganglion*, swelling). The name comes from the fact that the area where the neuron cell bodies aggregate appears as a "swelling" on the cordlike nerve. The posterior roots of the spinal cord contain swellings called **posterior (dorsal) root ganglia** (peripheral nerve ganglia). These structures contain the neuron cell bodies of somatic sensory neurons, which are classified as unipolar neurons. Unipolar neurons have only a single process extending from the cell body of the neuron. The absence of multiple dendrites coming right off the cell bodies of the neurons allows the glial cells (which are also found in the ganglion) to lie directly adjacent to the cell bodies of the neurons. They look like small "satellites" surrounding the neuron cell body, and are called **satellite cells**. Satellite cells are general supporting cells for neurons within the posterior root ganglia. They have the same embryonic origin as neurolemmocytes, but not the same function, which is why they are given a different name.

4. Observe the ganglion at high power. The large nuclei you see that contain prominent nucleoli are the cell bodies of unipolar neurons. The small nuclei found surrounding the neurons are the nuclei of the satellite cells (figure 14.8b).

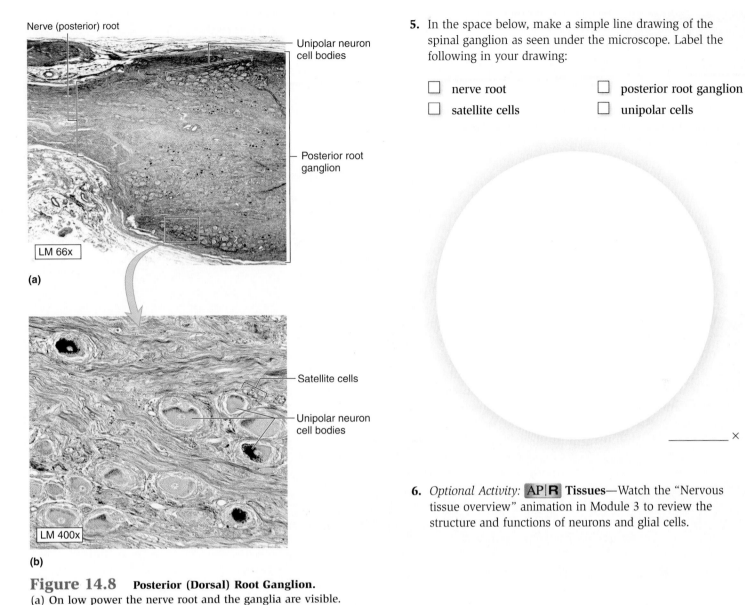

Nerve (posterior) root

Unipolar neuron cell bodies

Posterior root ganglion

LM 66x

(a)

Satellite cells

Unipolar neuron cell bodies

LM 400x

(b)

Figure 14.8 **Posterior (Dorsal) Root Ganglion.**
(a) On low power the nerve root and the ganglia are visible.
(b) On high power the unipolar neurons surrounded by satellite cells are visible.

5. In the space below, make a simple line drawing of the spinal ganglion as seen under the microscope. Label the following in your drawing:

☐ nerve root   ☐ posterior root ganglion
☐ satellite cells   ☐ unipolar cells

———— ×

6. *Optional Activity:* AP|R **Tissues**—Watch the "Nervous tissue overview" animation in Module 3 to review the structure and functions of neurons and glial cells.

Peripheral Nerves

Both cranial nerves and spinal nerves are collections of myelinated axons that are bundled together with connective tissue. The bundling pattern is exactly like the bundling pattern seen with skeletal muscle, and the prefixes used for the names of the connective tissue coverings are also the same. An entire peripheral nerve is bundled by connective tissue called **epineurium**. Within an entire nerve, the axons are grouped into bundles called **fascicles**. Each fascicle is surrounded by a connective tissue covering called the **perineurium**. Within each fascicle are individual axons. Each individual axon is surrounded by a connective tissue covering called the **endoneurium**. Between the fascicles are blood vessels that supply nutrients to the nerve.

EXERCISE 14.9 Coverings of a Peripheral Nerve

1. Obtain a slide of a cross section of a peripheral nerve and place it on the microscope stage.

2. Bring the tissue sample into focus on low power. Using **figure 14.9** as a guide, identify epineurium, fascicles, and perineurium.

3. Observe the slide at high power so you can see the individual *axons* in cross section. Each axon is surrounded by a myelin sheath, which is then surrounded by a connective tissue covering of endoneurium. In between the fascicles, locate the many small blood vessels.

4. In the space below, sketch a simple line drawing of a cross section of a peripheral nerve as seen through the microscope. Label the following in your drawing:

☐ axon   ☐ fascicle

☐ blood vessel   ☐ myelin sheath

☐ endoneurium   ☐ perineurium

☐ epineurium

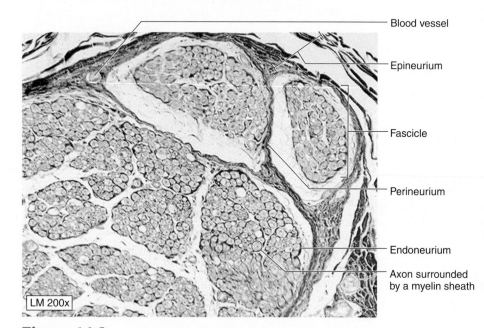

Blood vessel

Epineurium

Fascicle

Perineurium

Endoneurium

Axon surrounded by a myelin sheath

LM 200x

Figure 14.9   **Cross Section Through a Peripheral Nerve.**

——————— ×

WHAT DO YOU THINK?

2 When a peripheral nerve is compressed, blood vessels that run between fascicles of axons are also compressed. What might a consequence of this be?

Chapter 14: Nervous Tissues

Name: _____

Date: _____ Section: _____

POST-LABORATORY WORKSHEET

1. The counterpart of a neurolemmocyte (Schwann cell) in the CNS is a(n) _____.

2. Is it more common for brain tumors to arise from neurons or from glial cells? _____. Explain your answer:

   _____

   _____

   _____

3. A collection of neuron cell bodies in the CNS is called a(n) _____.

4. A collection of neuron cell bodies in the PNS is called a(n) _____.

5. A collection of myelinated axons in the CNS is called a(n) _____.

6. A collection of myelinated axons in the PNS is called a(n) _____.

7. Multiple sclerosis (MS) is a disease in which the immune system attacks myelin sheaths. Often a patient suffering from MS will experience a bout of illness, which involves muscle weakness and sensory alterations, followed by a period of recovery, in which normal function partially or completely returns. Using your knowledge of glial cells, explain the biological basis for the partial or complete recovery of nerve function in an MS patient.

   _____

   _____

   _____

   _____

   _____

   _____

   _____

   _____

   _____

8. Ependymal cells are unique as epithelia in that they do not have a _____.

9. Glial cells in the CNS derived from circulating monocytes are _____.

10. List three functions of astrocytes.

    1. _____

    2. _____

    3. _____

11. A fascicle of axons is surrounded by a connective tissue sheath called the _____.

12. A multipolar neuron has many _____ and a single _____.

13. a. What is chromatophilic substance? _____

    b. Where is chromatophilic substance found? _____

    c. The part of a neuron distinguished by its notable *absence* of chromatophilic substance is called the _____.

14. The branches at the end of an axon are called _____; and they contain swellings called _____

    _____.

15. Describe the structure and function of microglia:

    _____

    _____

    _____

16. Describe the structure, function, and specific location of satellite cells:

    _____

    _____

    _____

17. A patient with an *astrocytoma* (*-oma,* tumor) has a tumor that is derived from _____. This tumor would be located in
    what part of the nervous system?

    _____

18. Compare and contrast the structure, location, and function of neurolemmocytes and oligodendrocytes:

    _____

    _____

    _____

    _____

    _____

19. a. What is a neurofibril node (node of Ranvier)?

    _____

    b. What is the function of a neurofibril node?

    _____

20. Give the location where each of the neuron types listed below is found:

    a. Purkinje cell _____

    b. Anterior horn cell _____

    a. Pyramidal cell _____

Central Nervous System: The Brain

15

OUTLINE and OBJECTIVES

Module 7: NERVOUS SYSTEM

INTRODUCTION

Recall a time when you were unable to figure out a word problem in your math class. At the time you or someone else may have said, "Come on, use your gray matter!"—meaning "Use your brain! Think!" Having completed the exercises in chapter 14, you now know that your brain consists not only of gray matter (neuron cell bodies) but also of white matter (fiber tracts). However, it still looks like a mass of "matter" or "stuff" that is difficult to understand functionally merely from observing its gross anatomical structure. Because of this, the exercises in this laboratory session may prove challenging because they might seem to be simply exercises in memorization.

To help you make sense of the structures you will view in this laboratory session, the tables in this chapter list the names of the structures you will see along with brief summaries of their functions and information about the derivation of their names. Keep in mind that brain structures were named well before their functions were known. Most of the names do not relate to function; instead they derive from the general appearance of the structure. For instance, "mammillary bodies" have little or nothing to do with the female reproductive system, but to early anatomists they appeared to be "little breasts" on the inferior surface of the brain. The area known as the thalamus apparently looked like a "bed or bedroom." As you will see, the thalamus has very little, if anything, to do with the bedroom or sleep.

As you learn the structures of the brain, try to imagine how and why they were given their names. This will make it easier for you to recall the names when you are taking an exam. In addition, pay attention to the function(s) of each structure you identify. Knowledge of function is even more important than knowledge of structure, because when an area of the brain is damaged you will expect to see deficiencies directly related to the functions of that area of the brain.

Chapter 15: CNS—The Brain

PRE-LABORATORY WORKSHEET

Name: _____

Date: _____ Section: _____

1. Define *gyrus*: _____

_____

2. Define *sulcus*: _____

_____

3. Brain structures such as the mammillary bodies, which are involved with the suckling reflex and emotions, are part of the _____ system.

4. The three meninges that cover the brain and spinal cord are the:

   a. _____

   b. _____

   c. _____

5. Cerebrospinal fluid circulates between the _____ mater and the _____ mater.

6. The ventricles of the brain that are located within the cerebral hemispheres are the _____.

7. A dural septum that lies between the two cerebral hemispheres is the _____.

8. All dural venous sinuses eventually drain into the _____ vein.

9. Dural venous sinuses carry _____ _____ and _____.

10. The dura mater is composed of _____ _____ connective tissue.

11. The _____ _____ separates the frontal lobe from the parietal lobe of the brain.

12. The _____ _____ is a fiber tract that connects the left cerebral hemisphere to the right cerebral hemisphere.

13. The _____ lobe of the brain is involved with the sense of vision.

14. The _____ _____ separates the cerebral hemispheres of the brain from the cerebellar hemispheres of the brain.

IN THE LABORATORY

In this laboratory session you will first observe the protective coverings of the brain, the meninges. You will then observe the ventricular system of the human brain and relate its adult structure to the development of the brain. This will be followed up with observations of preserved human brains or models of human brains. Finally, you will dissect a representative mammalian brain, the sheep brain.

Gross Anatomy

The Meninges

The **meninges** are connective tissue coverings of the brain and spinal cord and consist of the dura mater, arachnoid mater, and pia mater. These coverings perform many functions and their structure varies slightly depending on whether they are covering the brain or the spinal cord. In exercise 15.1 we will consider meningeal structures as they apply to the brain, and in chapter 17 we will consider meningeal structures as they apply to the spinal cord.

EXERCISE 15.1 Cranial Meninges

1. Obtain a model of the dura mater of the brain or a cadaveric specimen of the head with intact dural structures.

2. Observe the dura mater (**figure 15.1**). The **dura mater** (*dura*, hard, + *mater*, mother) is composed of two layers: the outer layer is the **periosteal dura**, which is simply the inner periosteum that lines the cranial bones, and the inner layer is the **meningeal dura**, which forms structures such as cranial dural septa and dural venous sinuses. **Cranial dural septa** are infoldings of the meningeal dura that separate the cranial cavity into compartments. The **dural venous sinuses** transmit venous blood and cerebrospinal fluid (CSF) draining from the brain to the internal jugular vein. The largest of the dural venous sinuses is the **superior sagittal sinus**. Eventually all sinuses drain into the **sigmoid** (*sigmoid*, s-shaped) **sinus**, which drains into the **internal jugular vein**. **Table 15.1** includes a list of the dural venous sinuses and gives their general locations, and **figure 15.2** describes the flow of CSF through the dural venous sinuses. Identify as many of the sinuses as you can on the specimens available to you in the laboratory.

3. If you have a cadaveric specimen available, locate the superior sagittal sinus and note the small granular

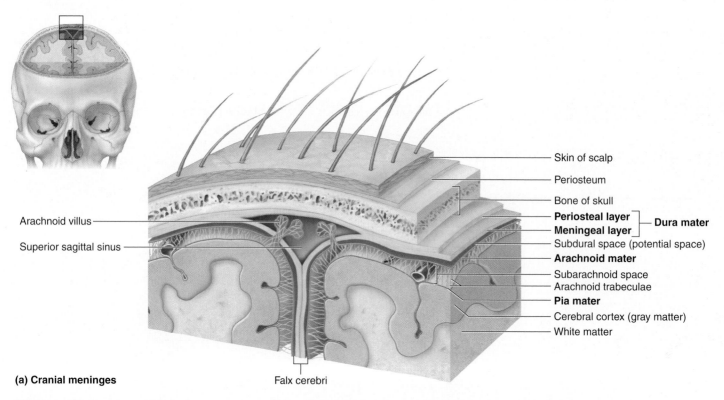

Arachnoid villus

Superior sagittal sinus

Skin of scalp
Periosteum
Bone of skull
Periosteal layer ⎱ **Dura mater**
Meningeal layer ⎰
Subdural space (potential space)
Arachnoid mater
Subarachnoid space
Arachnoid trabeculae
Pia mater
Cerebral cortex (gray matter)
White matter

(a) Cranial meninges

Falx cerebri

Figure 15.1 **The Meninges** (a) Coronal section through the superior sagittal sinus. (b) Dural venous sinuses and cranial dural septa.

Table 15.1	Meninges, Dural Septa, and Dural Venous Sinuses	
Structure	**Description and Function**	**Word Origin**
MENINGEAL LAYER		
Dura Mater	Very tough, durable membrane composed of dense irregular connective tissue that protects CNS structures within the cranial cavity and vertebral canal.	*durus*, hard, + *mater*, mother
Periosteal Dura	Outer layer of dura mater that composes the inner periosteum of the cranial bones, so anchors the dura tightly to the cranial bones (not within the vertebral canal). In most places within the cranial cavity it is anchored to, or continuous with, the meningeal dura.	*periosteal*, relating to the periosteum, + *dura*, relating to the dura mater
Meningeal Dura	Inner layer of dura mater that forms cranial dural septa and dural venous sinuses within the cranial cavity; continuous with the dura mater of the vertebral canal.	*meninx*, membrane, + *dura*, relating to the dura mater
Arachnoid Mater	Thin, loose connective tissue membrane that lies between the dura mater and the pia mater; contains numerous web-like extensions of connective tissue that anchor it to the pia mater and creates a space (the subarachnoid space) for cerebrospinal fluid (CSF) to flow around the brain and spinal cord.	*arachno*, spider cobweb, + *eidos*, resemblance, + *mater*, mother
Pia Mater	A thin, highly vascular connective tissue membrane in direct contact with the brain and spinal cord. It follows all the convolutions (gyri) of the brain and spinal cord and is generally inseparable from brain tissue.	*pia*, soft or tender, + *mater*, mother
CRANIAL DURAL SEPTA	Folds of the meningeal dura that reflect away from the bones of the cranial cavity to create partitions within the cranial cavity that separate cerebral and cerebellar hemispheres.	*durus*, hard, + *septa*, fold
Diaphragma Sellae	Forms a "roof" over the sella turcica of the sphenoid bone and contains a hole for the passage of the infundibulum.	*diaphragm*, diaphragm, + *sella*, saddle
Falx Cerebelli	Located between the two cerebellar hemispheres.	*falx*, sickle, + *cerebelli*, relating to the cerebellum
Falx Cerebri	Located between the two cerebral hemispheres, within the longitudinal fissure of the brain; anchored anteriorly to the crista galli of the ethmoid bone.	*falx*, sickle, + *cerebelli*, relating to the cerebrum
Tentorium Cerebelli	Drapes across the cerebellar hemispheres horizontally within the transverse fissure of the brain between the cerebellum and the cerebrum.	*tentorium*, a tent, + *cerebelli*, relating to the cerebellum

(continued on next page)

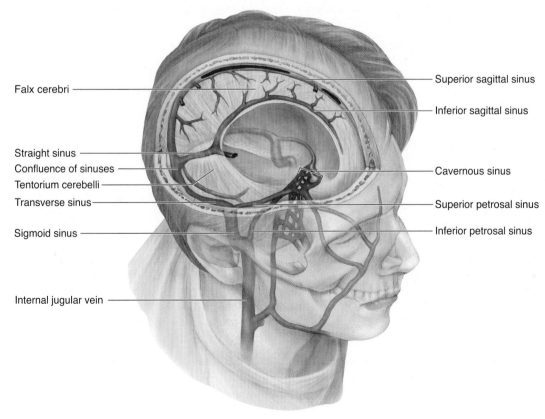

(b) Dural venous sinuses

Table 15.1	Meninges, Dural Septa, and Dural Venous Sinuses *(continued)*	
Structure	**Description and Function**	**Word Origin**
DURAL VENOUS SINUSES	Spaces formed between layers of meningeal and periosteal dura within the cranial cavity that transport CSF and venous blood from the subarachnoid space and the brain to the internal jugular vein.	
Cavernous Sinuses	Located on either side of the sella turcica of the sphenoid bone. Transmit CSF and venous blood from the anterior cranial fossa to the petrosal sinuses.	*cavernous*, relating to a cavity, + *sinus*, a channel
Inferior Sagittal Sinus	Located inferior to the falx cerebri. Transmits CSF and venous blood from the subarachnoid space and the brain (respectively) to the straight sinus.	*inferior*, below, + *sagitta*, an arrow (relating to the sagittal plane), + *sinus*, a channel
Petrosal Sinuses	Dural sinuses located on the superior ridge of the petrous part of the temporal bone (superior petrosal sinus) or in the groove between the petrous part of the temporal bone and the occipital bone (inferior petrosal sinus). Transmits CSF and venous blood from the cavernous sinuses to the sigmoid sinuses (or the terminal portions of the transverse sinuses).	*petrosal*, relating to the petrous part of the temporal bone, + *sinus*, a channel
Sigmoid Sinuses	Located in the posterior cranial fossa just posterior to the petrous part of the temporal bone and extending into the jugular foramen. Transmit CSF and venous blood from the transverse sinuses to the internal jugular vein.	*sinus*, a channel, + *sigmoid*, shaped like an S
Confluence of Sinuses	Located within the posterior cranial cavity deep to the external occipital protruberance. Transmit CSF and venous blood from the superior sagittal sinus and the straight sinus to the transverse sinuses.	*confluens*, to flow together, + *sinus*, a channel
Straight Sinus	Located at the junction between the falx cerebri, falx cerebelli, and tentorium cerebelli. Transmit CSF and venous blood from the inferior sagittal sinus to the confluence of sinuses.	*sinus*, a channel, + straight
Superior Sagittal Sinus	Largest sinus within the cranial cavity; located superior to the falx cerebri. Transmits CSF and venous blood from the subarachnoid space and the brain (respectively) to the confluence of sinuses.	*superior*, above, + *sagitta*, an arrow (relating to the sagittal plane), + *sinus*, a channel
Transverse Sinuses	Located posterior to the tentorium cerebelli. Runs from the confluence of sinuses, along the posterior aspect of the occipital bone, to the posterior cranial fossa just posterior to the petrous part of the temporal bone. Transmits CSF and venous blood from the confluence of sinuses to the sigmoid sinuses.	*transversus*, across, + *sinus*, a cavity
MENINGEAL SPACES		
Epidural Space	Space between the dura mater and the walls of the vertebral canal (there is no epidural space within the cranial cavity). Clinically relevant for administration of anesthetic (an "epidural"). Anesthetic within this space numbs the roots of the nerves that exit the spinal cord. Thus, areas located below the spinal cord level where anesthetic is administered become completely devoid of sensation.	*epi*, above, + *dura*, relating to the dura mater
Subarachnoid Space	Space between the arachnoid mater and the pia mater where CSF flows as it circulates around the brain and spinal cord. Blood vessels are located within this space.	*sub*, under, + *arachnoid*, relating to the arachnoid mater
Subdural Space	Potential space between the dura mater and the arachnoid mater. In a healthy individual, this space does not exist; but traumatic injury may cause bleeding into the subdural space (subdural hematoma).	*sub*, under, + *dural*, relating to the dura mater
OTHER MENINGEAL STRUCTURES		
Denticulate Ligaments	"Tooth"-shaped extensions of the pia mater within the vertebral canal that anchor the pia mater to the arachnoid mater. Located between the posterior and anterior roots of the spinal cord.	*denticulus*, a small tooth
Filum Terminale	A long, thin extension of the pia mater at the caudal end of the spinal cord that anchors it to the end of the sacral canal.	*filum*, thread, + *terminus*, an end

structures located in and around the sinus. These structures are called **arachnoid granulations** (villi) and are responsible for transporting cerebrospinal fluid from the subarachnoid space into the dural venous sinuses.

4. Obtain a human brain with the arachnoid mater and pia mater intact. If you do not have one available, this activity will be completed when you perform the sheep brain dissection. On the human brain, note the thin, transparent covering that lies on top of the brain and does not dip down into the grooves (sulci). This is the **arachnoid mater** (*arachnoid*, shaped like a spider web). Deep to the arachnoid mater is the finest, thinnest meningeal layer, the **pia mater** (*pia*, soft or tender). The pia mater is in direct contact with the neural tissue of the brain and follows all of the convolutions (gyri) and grooves (sulci) on its surface. The space between the arachnoid mater and the pia mater is the **subarachnoid space**. What fluid is normally found in the subarachnoid space? _____

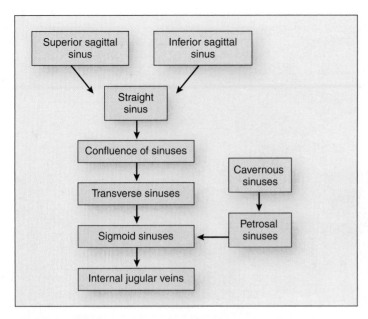

Figure 15.2   **Flow of Fluid Through the Dural Venous Sinuses.**

5. Using table 15.1 and figure 15.1 as guides, identify the **meningeal structures** listed in **figure 15.3** on a model or cadaveric specimen. Then label them in figure 15.3.

6. *Optional Activity:* AP|R **Nervous System**—Watch the "Meninges" and "Dural Sinus Blood Flow" animations to reinforce your understanding of these structures and their relationships.

WHAT DO YOU THINK?

1 If the passage of fluid is blocked at the confluence of sinuses, into which sinuses will fluid back up?

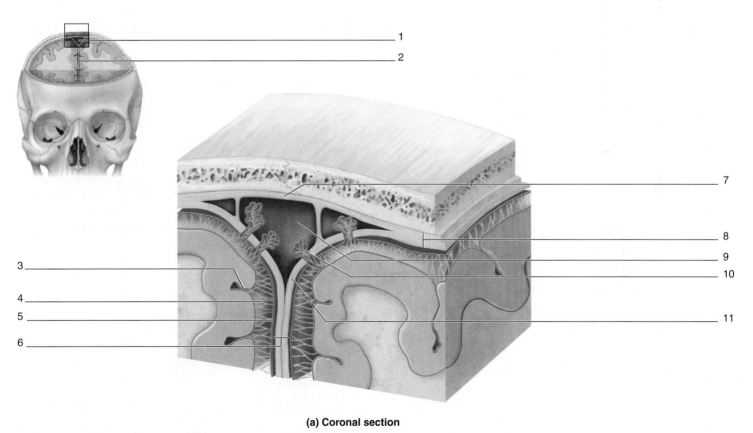

(a) Coronal section

Figure 15.3   **Meningeal Structures.**

- [ ] arachnoid mater
- [ ] arachnoid villi
- [ ] dura mater
- [ ] falx cerebri
- [ ] meningeal dura
- [ ] periosteal dura
- [ ] pia mater
- [ ] subarachnoid space
- [ ] superior sagittal sinus

(continued on next page)

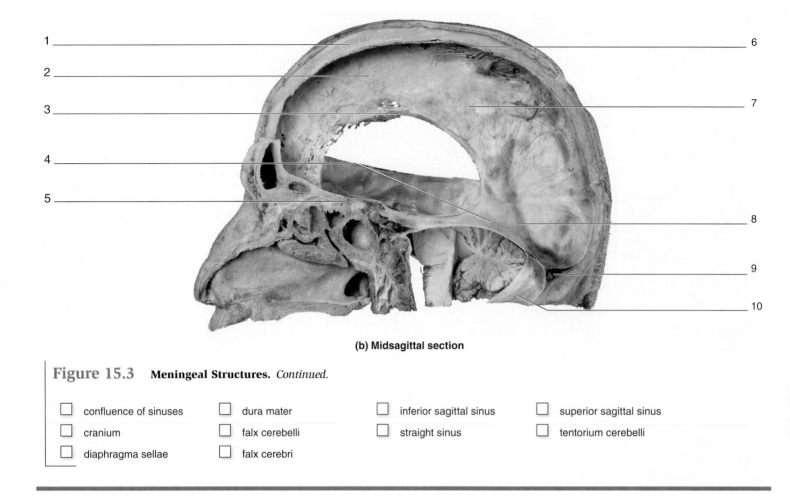

1 _____

2 _____

3 _____

4 _____

5 _____

6 _____

7 _____

8 _____

9 _____

10 _____

(b) Midsagittal section

Figure 15.3   **Meningeal Structures.** *Continued.*

☐ confluence of sinuses

☐ cranium

☐ diaphragma sellae

☐ dura mater

☐ falx cerebelli

☐ falx cerebri

☐ inferior sagittal sinus

☐ straight sinus

☐ superior sagittal sinus

☐ tentorium cerebelli

Ventricles of the Brain

The **ventricles** are fluid-filled spaces within the central nervous system that are complex in shape. In exercise 15.2 you will view a *cast* of the ventricles, which is created by filling the ventricular spaces with plastic, allowing the plastic to harden, and then removing the brain tissue so only the cast is left. A cast allows you to visualize the three-dimensional structure of the ventricles without the brain literally "getting in the way." If your laboratory does not have casts of the brain ventricles, this exercise can be performed using charts or figures.

Recall that the central nervous system initially develops as a neural tube. As it grows, the neural tube begins to change size and shape. The cephalic end develops into the brain, while the rest develops into the spinal cord. Both the brain and the spinal cord contain fluid-filled spaces inside. The pattern of growth of the neural tissue surrounding the neural tube changes the size and shape of the fluid-filled spaces within. Thus, because the spinal cord remains mostly a tubular structure as it grows, the fluid-filled space inside, the **central canal**, remains tubular. On the other hand, because the cephalic (brain) end of the neural tube undergoes extensive folding as it grows, the fluid-filled spaces within develop into irregular shapes. These shapes tell a story about how the parts of the brain developed.

The cephalic end of the neural tube first develops into three **primary vesicles** (prosencephalon, mesencephalon, and rhombencephalon) and then into five **secondary vesicles** (telencephalon, diencephalon, mesencephalon, metencephalon, and myelencephalon). **Figure 15.4** lists the secondary vesicles of the brain and the parts of the ventricular system that develop from

each of them. For instance, the telencephalon undergoes extensive growth as it develops into the cerebral hemispheres. This growth creates the horseshoe-shaped structure of the **lateral ventricles** in the adult brain. On the other hand, the mesencephalon does not undergo such extensive growth as it develops into the midbrain. Hence the fluid-filled space inside, the **cerebral aqueduct**, remains tubular in shape within the adult brain.

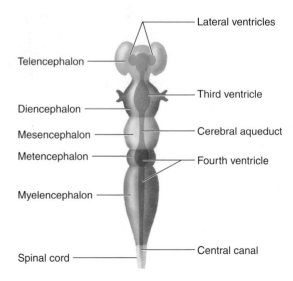

Figure 15.4   **Secondary Brain Vesicles and Associated Ventricular Structures of the Brain.**

EXERCISE 15.2 Brain Ventricles

1. Obtain a cast of **the ventricles of the brain** (**figure 15.5**).

2. Identify the ventricles of the brain listed in **table 15.2** on the cast of the ventricles. As you identify each of the ventricles, relate each ventricle to the **secondary brain vesicle** from which it developed (table 15.2 and figure 15.4).

3. Using table 15.2 and your textbook as guides, identify the ventricular structures listed in figure 15.5 on a **cast of the brain ventricles**. Then label them in figure 15.5.

4. *Optional Activity*: AP|R **Nervous System**—Watch the "Brain Ventricles" and "CSF flow" animations to solidify your understanding of how cerebrospinal fluid flows through the brain ventricles.

Study Tip!

As you observe whole brains or brain models in the laboratory, always begin by locating the ventricular spaces and associating each ventricular space with a secondary brain vesicle (table 15.2). Next, identify the adult brain structures that surround the ventricular spaces. Finally, correlate each adult brain structure to the secondary brain vesicle from which it formed. For example, the lateral ventricles (ventricular space), which are part of the telencephalon (secondary brain vesicle), are surrounded by the cerebral hemispheres (adult brain structures). Thus, the cerebral hemispheres (brain structure) are derived from the telencephalon (secondary brain vesicle).

(a) Anterolateral view

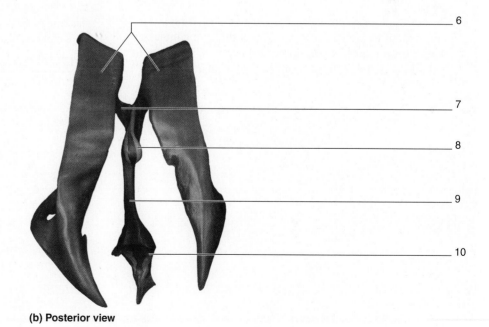

(b) Posterior view

Figure 15.5   **Cast of the Ventricles of the Brain.**

- ☐ cerebral aqueduct
- ☐ fourth ventricle
- ☐ interventricular foramen
- ☐ lateral ventricles
- ☐ third ventricle

Table 15.2	Ventricles of the Brain		
Structure	**Location and Description**	**Secondary Brain Vesicle Derivative**	**Word Origin**
Lateral Ventricles	Horseshoe-shaped ventricles within the cerebral hemispheres containing anterior and posterior horns whose shape follows the developmental shape of the cerebral hemispheres.	Telencephalon	*latus*, to the side, + *ventriculus*, belly
Third Ventricle	Narrow, quadrilateral-shaped ventricle located in the midsagittal plane inferior to the corpus callosum and medial to the thalamic nuclei; surrounded by structures of the diencephalon.	Diencephalon	*ventriculus*, belly
Cerebral Aqueduct	Narrow channel that lies in the midbrain between the cerebral peduncles and the corpora quadrigemina (tectal plate).	Mesencephalon	*aquaeductus*, a canal
Fourth Ventricle	Diamond-shaped ventricle located anterior to the cerebellum and posterior to the pons.	Metencephalon	*ventriculus*, belly
Central Canal	Narrow channel that lies in the center of the medulla oblongata and spinal cord.	Myelencephalon	*centrum*, center, + *canalis*, a channel

The Human Brain

In the next series of exercises you will identify structures that are visible in four views of the human brain: superior, lateral, inferior, and midsagittal. **Table 15.3** lists the main brain structures that are visible in these views, and lists the views in which each structure is visible. Next to each structure is a description of its function.

As you identify parts of the brain, follow these steps to make things easier:

■ Name the structures in a logical order, such as the order in which the structures appear from anterior to posterior. If you concentrate on learning the structures in an orderly fashion, you will be more able to recall the information later on.

■ Do not think of brain regions as isolated structures. The structures are easier to identify within the context of their surroundings. If you do not believe this, think about it the next time you are traveling to your anatomy class—would you know how to get to the anatomy classroom if all of the buildings on campus (with the exception of the one you are traveling to) were suddenly moved around?

■ Always associate a function with each structure you identify. Use table 15.3 as a reference.

Study Tip!

Because structure/function relationships are not easily visualized when it comes to the brain, consider making flashcards using table 15.3 as a reference. List the brain structure on one side of the card and its function on the other side. Once you have mastered the names and locations of the listed structures, quiz your lab partners on the structure and function of each item.

Table 15.3	Brain Structures Visible in Superficial Views of Whole or Sagitally Sectioned Brains			
Brain Structure	**Description**	**Function(s)**	**Word Origin**	**Views Where Visible**
Central sulcus	A deep groove that runs along the coronal plane.	Separates the frontal lobe from the parietal lobe.	*central*, in the center, + *sulcus*, a furrow	Superior, lateral, and midsagittal
Cerebellum	The second largest part of the brain after the cerebral hemispheres.	Regulation of muscle tone (a low-level muscle contraction), coordination of motor activity, and maintenance of balance and equilibrium.	*cerebellum*, little brain	Lateral, inferior, and midsagittal
Cerebral Aqueduct	The part of the ventricular system of the brain that lies in the mesencephalon.	Carries cerebrospinal fluid from the third ventricle to the fourth ventricle.	*aquaeductus*, a canal	Midsagittal
Cerebral Peduncles	Fiber tracts located between the mesencephalon and the pons.	The fibers connect the forebrain (cerebral hemispheres and diencephalon) to the hindbrain (medulla oblongata, pons, and cerebellum).	*cerebrum*, brain, + *pedunculus*, a little foot	Inferior, midsagittal
Cingulate Gyrus	A gyrus located just superior to the corpus callosum.	This area of the brain is not well understood. It is predominantly motor and may play a role in the limbic system (such as controlling motor functions with a strong emotional component).	*cingo*, to surround, + *gyros*, circle	Midsagittal
Corpora Quadrigemina (Tectal Plate)	Consists of four twin bodies, the superior and inferior colliculi.	Play a role in visual and auditory reflexes.	*corpus*, body, + *quad*, four, + *geminus*, twin	Midsagittal
Corpus Callosum	A fiber tract located superior to the lateral ventricles.	Contains axons from association neurons that connect the lobes of the two cerebral hemispheres to each other (except for the temporal lobes).	*corpus*, body, + *callosus*, thick-skinned	Midsagittal
Epithalamus	A small projection extending posteriorly from the superior/caudal portion of the third ventricle.	Contains the pineal body (pineal gland) along with other structures.	*epi*, above, + *thalamos*, a bed or bedroom	Midsagittal
Fornix	An arching fiber tract located inferior to the septum pellucidum.	Connects limbic system structures to each other.	*fornix*, arch	Midsagittal
Fourth Ventricle	Ventricle of the brain that lies in the metencephalon and contains choroid plexus.	CSF from the cerebral aqueduct flows into this ventricle before moving on to the central canal of the spinal cord and exiting into the subarachnoid space through the median and lateral apertures.	*ventriculus*, belly	Midsagittal
Frontal Lobe	Lies deep to the frontal bone.	Somatic motor functions. Contains Broca's area, which controls motor speech. Controls conjugate eye movement (the ability to move the eyes together). Higher level functions include judgment and foresight (the ability to think before acting).	*frontal*, in the front, + *lobos*, lobe	Superior, lateral, inferior, and midsagittal
Hypothalamus	Located deep to the walls of the inferior/rostral part of the third ventricle.	Regulates body temperature, metabolism (hunger/thirst), sleep, sex, and emotional control (limbic system functions). It is also a "master" endocrine gland, controlling hormone secretion from the pituitary gland.	*thalamos*, a bed or bedroom	Midsagittal

(continued on next page)

Table 15.3	Brain Structures Visible in Superficial Views of Whole or Sagitally Sectioned Brains *(continued)*			
Brain Structure	**Description**	**Function(s)**	**Word Origin**	**Views Where Visible**
Inferior Colliculus	A pair of oval projections that make up inferior part of the corpora quadrigemina (tectal plate).	Controls *auditory reflexes*, such as the sudden turning of the head toward the source of a very loud sound.	*inferior*, lower, + *colliculus*, a mound or hill	Midsagittal
Infundibulum	A funnel-shaped inferior extension of the brain located immedately posterior to the optic chiasm.	Consists of fiber tracts that connect the hypothalamus to the posterior pituitary (pars nervosa).	*infundibulum*, a funnel	Inferior and midsagittal
Intermediate Mass (Interthalamic Adhesion)	A fiber tract that crosses the third ventricle. In a midsagittal section of the brain you will see the cut end of this structure in the middle of the third ventricle.	A fiber tract that connects the two thalamic nuclei to each other. It is absent in about 20% of human brains.	*intermediate*, in the middle, + *mass*, a mass	Midsagittal
Lateral Fissure	A deep groove that runs along a horizontal plane between the frontal/parietal lobes and the temporal lobe.	Separates the frontal and parietal lobes from the temporal lobe.	*latus*, the side, + fissure, a deep furrow	Lateral
Longitudinal Fissure	A deep fissure between the two cerebral hemispheres.	Separates the two cerebral hemispheres from each other. The falx cerebri lies in this fissure in a living human.	*longus*, long, + *fissure*, a deep furrow	Superior and inferior
Mammillary Bodies	Two small bump-like ("breast-shaped") structures located immediately posterior to the infundibulum.	Involved in short-term memory processing; part of the limbic system (the emotional brain). Also involved with suckling and chewing reflexes.	*mammillary*, shaped like a breast	Inferior and midsagittal
Medulla Oblongata	The most caudal aspect of the brainstem, forming the transition zone between the brain and spinal cord, located just inferior to the pons and anterior to the cerebellar hemispheres.	Contains the centers for regulation of respiration and cardiac function, as well as the *reticular formation*, which is a group of nuclie that are important in regulating wakefulness and selective attention.	*medius*, middle, + *oblongus*, rather long	Lateral, inferior, and midsagittal
Occipital Lobe	Lies deep to the occipital bone.	Primary visual area (the first area of the cerebral cortex where visual information synapses, after the thalamus). Visual association area (the ability to make sense of visual information).	*occiput*, the back of the head, + *lobos*, lobe	Superior, lateral, and midsagittal
Olfactory Bulbs	Swellings connected to the anterior end of the olfactory tracts that lie on the inferior surface of the frontal lobes of the brain lateral to the longitudinal fissure.	Location where cranial nerve I (CN I) the olfactory nerves, first synapse after passing through the cribriform plate of the ethmoid bone.	*olfactus*, to smell, + *bulbus*, a globular structure	Inferior
Olfactory Tracts	Nerve fibers that extend from the olfactory bulbs posteriorly to the junction where the frontal lobes meet the optic chiasm.	Carry the axons of neurons from the olfactory bulbs toward structures in other areas of the brain involved with olfaction.	*olfactus*, to smell, + *tractus*, a drawing out	Inferior
Optic Chiasm	The X-shaped structure formed where the two optic nerves come together just anterior to the infundibulum.	Location where fibers from both optic nerves cross over and travel in the optic tract on the opposite side. Not all fibers from the optic nerves cross over.	*optikos*, relating to the eye or vision, + *chiasma*, two crossing lines	Inferior and midsagittal
Parietal Lobe	Lies deep to the parietal bone.	Somatic sensory functions including stereognosis (the ability to recognize by touch). Controls some motor functions. Higher level functions include logical reasoning (math, problem solving).	*parietal*, a wall, + *lobos*, lobe	Superior, lateral, and midsagittal

(continued on next page)

Table 15.3	Brain Structures Visible in Superficial Views of Whole or Sagitally Sectioned Brains *(continued)*			
Brain Structure	**Description**	**Function(s)**	**Word Origin**	**Views Where Visible**
Parieto-occipital Sulcus	Groove that runs along a coronal plane.	Separates the parietal lobe from the occipital lobe.	*parieto-occipital*, between the parietal and occipital lobes, + *sulcus*, a furrow	Superior, lateral, and midsagittal
Pineal Body (Gland)	Small gland found within the epithalamus. You will not be able to establish the difference between the epithalamus and the pineal gland on gross observation alone.	Secretes the hormone melatonin from its precursor molecule, serotonin, in response to *decreased* light levels. Melatonin has an effect on circadian rhythms. May also play a role in establishing the onset of puberty as pineal tumors delay the onset of puberty and loss of cells within the pineal gland lead to an early onset of puberty.	*pineal*, shaped like a pine cone	Midsagittal
Pons	Collection of white matter that lies anterior to the cerebellum and the fourth ventricle; appears as a large mass just superior to the medulla oblongata.	A "bridge" of nerve tracts that connect the cerebral hemispheres to the cerebellar hemispheres. Contains centers for control of respiration.	*pons*, bridge	Lateral, inferior, and midsagittal
Postcentral Gyrus	Fold of brain tissue located immediately posterior to the central sulcus.	Primary somatic sensory area of the brain. Sensory information that comes in from the periphery will first synapse in the thalamus, but the first part of the cerebral cortex where this information is received is the postcentral gyrus.	*post*, after, + *central*, relating to the central sulcus, + *gyros*, circle	Superior, lateral, and midsagittal
Precentral Gyrus	Fold of brain tissue located immediately anterior to the central sulcus.	Primary somatic motor area of the brain. Neurons from this gyrus are somatic motor neurons that send their signals down to the spinal cord to control voluntary muscle activity.	*pre*, before, + *central*, relating to the central sulcus, + *gyros*, circle	Superior, lateral, and midsagittal
Septum Pellucidum	A thin membrane located between the corpus callosum (above) and fornix (below).	Contains neurons and glial cells and forms a thin connection between the corpus callosum above and the fornix below; also forms a thin wall between the two anterior horns of the lateral ventricles.	*saeptum*, a partition, + *pellucidus*, allowing the passage of light	Midsagittal
Superior Colliculus	A pair of rounded projections that make up superior part of the corpora quadrigemina (tectal plate).	Controls visual reflexes, such as the sudden turning of the head toward the source of a flashing light.	*superus*, above, + *colliculus*, a mound or hill	Midsagittal
Temporal Lobe	Lies deep to the temporal bone.	Primary auditory area of the brain (the first area where auditory nerve fibers synapse within the cerebral cortex—after the thalamus).	*tempus*, time	Lateral
Thalamus	Paired nuclei located deep to the lateral walls of the third ventricle. A pin pierced through the lateral wall of the third ventricle adjacent to the intermediate mass will pass into the thalamic nuclei.	Primary relay center for all sensory information coming into the brain (except olfaction).	*thalamos*, a bed or bedroom	Midsagittal
Third ventricle	The part of the ventricular system of the brain that lies in the diencephalon.	CSF from the lateral ventricles flows into this ventricle before moving on to the cerebral aqueduct.	*ventriculus*, belly	Midsagittal
Transverse Fissure	A deep fissure between the cerebrum and the cerebellum.	Separates the cerebral hemispheres from the cerebellar hemispheres. The tentorium cerebelli lies in this fissure in a living human.	*transversus*, across, + *fissure*, a deep furrow	Lateral

EXERCISE 15.3 Superior View of the Human Brain

1. Obtain a human brain or models of a human brain.

2. Observe the superior surface of the brain (**figure 15.6**). The most prominent feature in this view is the **longitudinal fissure**, which separates the two cerebral hemispheres from each other. A major sulcus you must identify is the **central sulcus**. Identification of the central sulcus is difficult, though not impossible, on a real human brain. The following are two features to look for:

   ■ The pre- and postcentral gyri should become continuous with each other on the lateral aspect of the central sulcus just above the lateral sulcus. This means the central sulcus will not enter the lateral sulcus.
   ■ The central sulcus will dip down into the longitudinal fissure.

3. Once you have identified the central sulcus, you will be able to identify the associated lobes and gyri located nearby.

4. Using table 15.3 and your textbook as guides, identify the structures listed in figure 15.6 on the superior view of the brain. Then label them in figure 15.6.

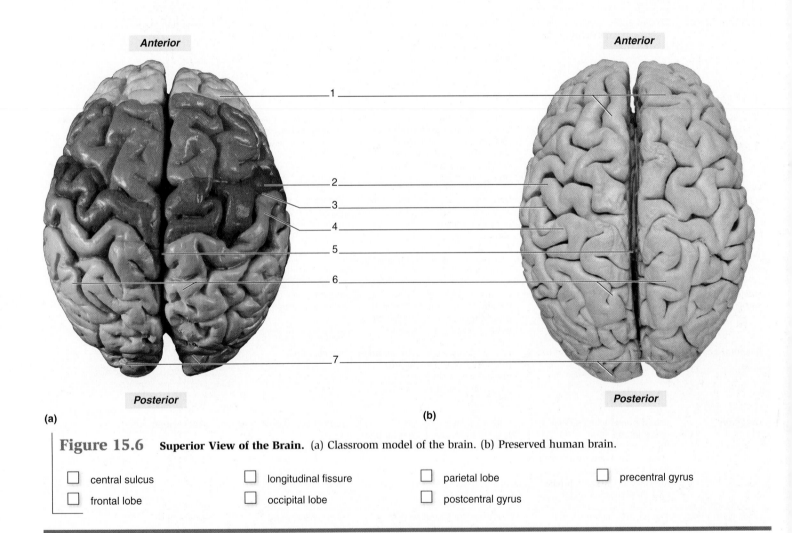

(a)   (b)

Figure 15.6 **Superior View of the Brain.** (a) Classroom model of the brain. (b) Preserved human brain.

☐ central sulcus   ☐ longitudinal fissure   ☐ parietal lobe   ☐ precentral gyrus

☐ frontal lobe   ☐ occipital lobe   ☐ postcentral gyrus

EXERCISE 15.4 Lateral View of the Human Brain

1. Obtain a human brain or models of a human brain.

2. Observe the lateral surface of the brain (**figure 15.7**). As with the superior view, you should be able to see the **central sulcus** by identifying the location where the pre- and postcentral gyri become continuous with each other just above the **lateral sulcus**.

3. Using table 15.3 and your textbook as guides, identify the structures listed in figure 15.7 on the lateral view of the brain. Then label them in figure 15.7.

4. *Optional Activity:* **AP|R** Nervous System—Watch the "Divisions of brain" animation for an overview of the regions of the brain and their general functions.

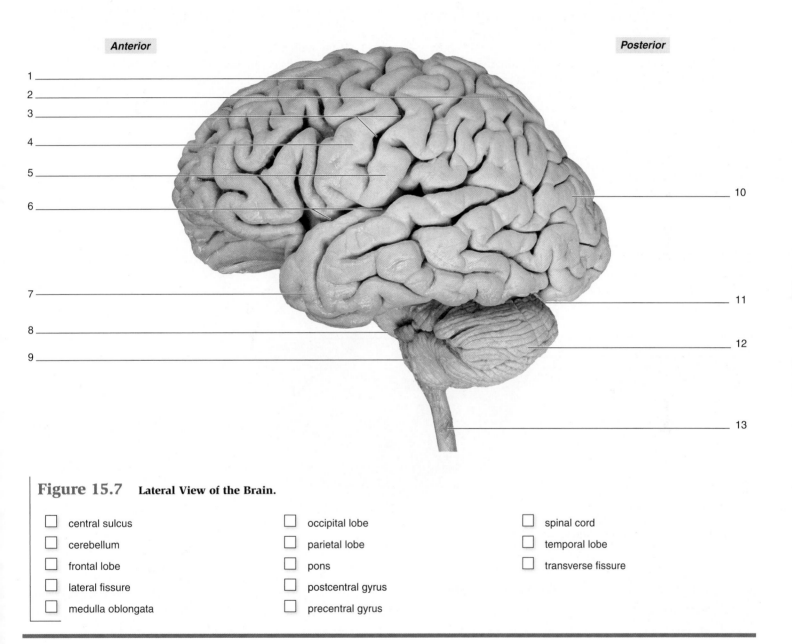

Figure 15.7   **Lateral View of the Brain.**

- ☐ central sulcus
- ☐ cerebellum
- ☐ frontal lobe
- ☐ lateral fissure
- ☐ medulla oblongata
- ☐ occipital lobe
- ☐ parietal lobe
- ☐ pons
- ☐ postcentral gyrus
- ☐ precentral gyrus
- ☐ spinal cord
- ☐ temporal lobe
- ☐ transverse fissure

EXERCISE 15.5 Inferior View of the Human Brain

1. Obtain a human brain or models of a human brain.

2. Observe the inferior surface of the brain (**figure 15.8**). This view is considerably more complicated than the superior or lateral views because of the cranial nerves that come off of the diencephalon and hindbrain. Do not concern yourself with the cranial nerves at this time. You will come back to them in chapter 16.

   One of the more problematic structures to identify in this view on a real brain is the infundibulum. When a brain is removed from the cranium, the pituitary gland almost always gets left behind within the sella turcica of the sphenoid bone because of a tough dural septa called the

diaphragma sellae, which lies between the pituitary gland and the rest of the brain. The only structure left connected to the brain is the stalk of tissue that connects the pituitary gland to the hypothalamus, which is the **infundibulum**, or **pituitary stalk**. If you are observing a model of the brain, the pituitary gland may be shown intact. If not, the infundibulum can be identified as a small strand of tissue that is located directly posterior to the **optic chiasm** and directly anterior to the **mammillary bodies**.

3. Using table 15.3 and your textbook as guides, identify the structures listed in figure 15.8 on the inferior view of the brain. Then label them in figure 15.8.

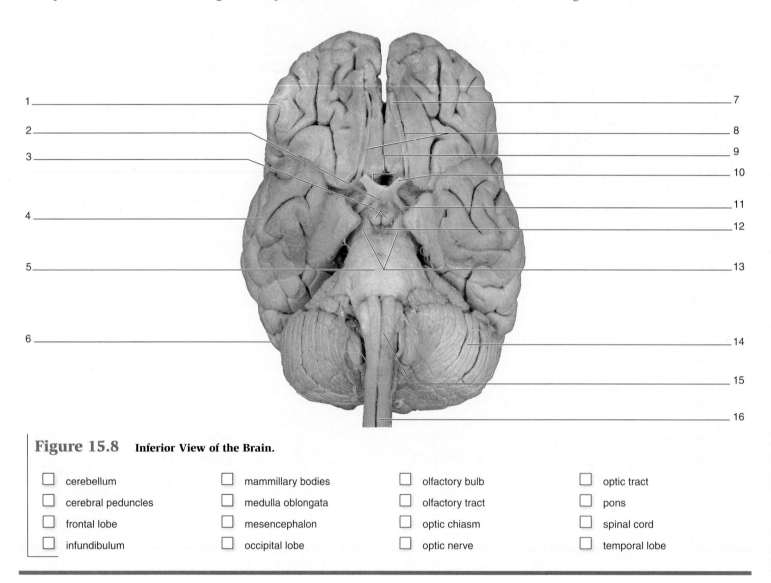

Figure 15.8 **Inferior View of the Brain.**

☐ cerebellum	☐ mammillary bodies	☐ olfactory bulb	☐ optic tract
☐ cerebral peduncles	☐ medulla oblongata	☐ olfactory tract	☐ pons
☐ frontal lobe	☐ mesencephalon	☐ optic chiasm	☐ spinal cord
☐ infundibulum	☐ occipital lobe	☐ optic nerve	☐ temporal lobe

EXERCISE 15.6 Midsagittal View of the Human Brain

1. Obtain a human brain or brain model that has been sectioned along the midsagittal plane and observe its medial surface (**figure 15.9**).

2. In the very center of your view, you will see the **third ventricle**. The third ventricle is the central depressed area that appears to have a cut nerve in the center. The "cut nerve" isn't really a nerve, but it is similar. It is a fiber tract called the **interthalamic adhesion** (or **intermediate mass**), which connects the two thalamic nuclei to each other. Use the interthalamic adhesion, the thalamus, and the third ventricle as reference points for identification of

other structures in this view. Many of the structures that are located around the third ventricle of the brain belong to a system called the **limbic system** (*limbus*, border), so named because structures of the limbic system are located at the border of the third ventricle and the brainstem. The limbic system is referred to as the "emotional brain" because its structures play a role in our emotions.

3. Using table 15.3 and your textbook as a guide, identify the structures listed in figure 15.9 on the midsagittal view of the brain. Then label them in figure 15.9.

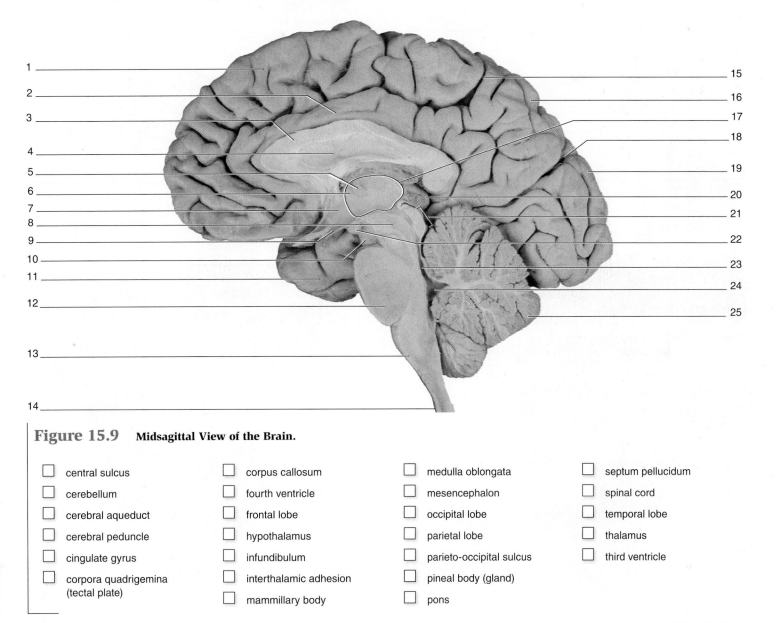

Figure 15.9 **Midsagittal View of the Brain.**

☐ central sulcus
☐ cerebellum
☐ cerebral aqueduct
☐ cerebral peduncle
☐ cingulate gyrus
☐ corpora quadrigemina (tectal plate)

☐ corpus callosum
☐ fourth ventricle
☐ frontal lobe
☐ hypothalamus
☐ infundibulum
☐ interthalamic adhesion
☐ mammillary body

☐ medulla oblongata
☐ mesencephalon
☐ occipital lobe
☐ parietal lobe
☐ parieto-occipital sulcus
☐ pineal body (gland)
☐ pons

☐ septum pellucidum
☐ spinal cord
☐ temporal lobe
☐ thalamus
☐ third ventricle

The Sheep Brain

In the following exercises you will identify most of the structures you identified on a human brain on a sheep brain. Sheep brains share many similarities with human brains, and are readily available for laboratory studies. The experience of dissecting a real brain (as opposed to observing a plastic model of a brain) will give you an appreciation for the true appearance and texture of the brain and

its associated structures. You may notice that the brain tissue itself is relatively delicate. Even so, the tissue is much more solid than it would be if you were observing a fresh brain because it has been fixed with chemicals. Living brain tissues are *extremely* delicate, and have the consistency of firm gelatin. In these dissection exercises you will identify both brain structures *and* cranial nerves on the sheep brain. In chapter 16 you will observe the structure, function, and identification of cranial nerves on a human brain.

EXERCISE 15.7   Sheep Brain Dissection

EXERCISE 15.7A: Dura Mater

1. Obtain a sheep brain, dissecting tray, and dissecting tools (forceps, scissors, a scalpel, and gloves). Place the sheep brain in the dissecting tray and take turns with your laboratory partner(s) observing its gross structure. This exercise requires a sheep brain with dura mater intact. If the dura mater is missing, proceed to the section "Inferior View of the Sheep Brain" (p. 305). Otherwise begin here.

2. Using your hands or blunt forceps, feel the toughness of the dura mater. Notice how it surrounds the entire brain. What type of tissue is the dura mater composed of?

_____

3. Observe the dura mater on the superior surface of the brain (**figure 15.10b**), and locate the following structures:

   ☐ confluence of sinuses   ☐ transverse sinuses
   ☐ superior sagittal sinus

4. Rotate the brain so it is resting on its superior surface and you are viewing its inferior surface (figure 15.10a). Notice the relatively large **pituitary gland** projecting beneath the dura mater. Your goal in this part of the dissection will be to cut away the dura mater without disconnecting the pituitary from the rest of the brain. Notice the tufts of capillaries found just posterior and lateral to the pituitary gland. Just lateral to these capillaries on both sides you

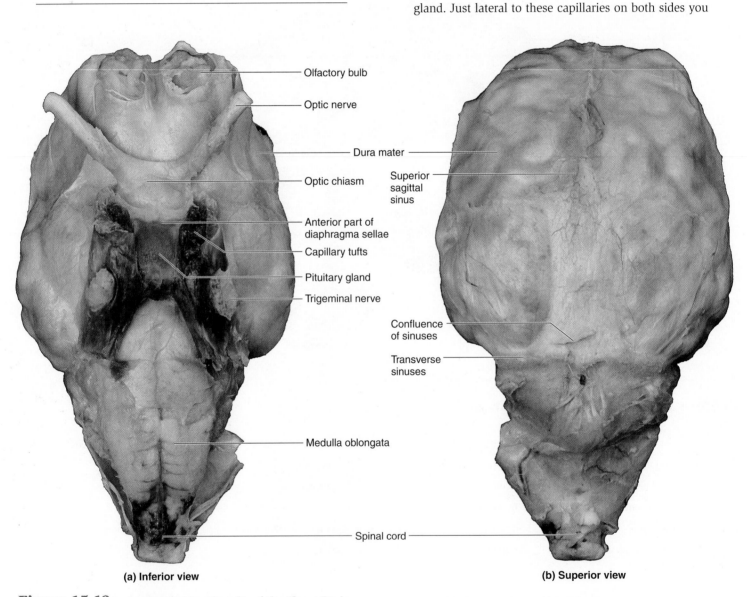

Olfactory bulb

Optic nerve

Dura mater

Optic chiasm

Superior sagittal sinus

Anterior part of diaphragma sellae

Capillary tufts

Pituitary gland

Trigeminal nerve

Confluence of sinuses

Transverse sinuses

Medulla oblongata

Spinal cord

(a) Inferior view   (b) Superior view

Figure 15.10   Anatomical Landmarks of the Sheep Brain.

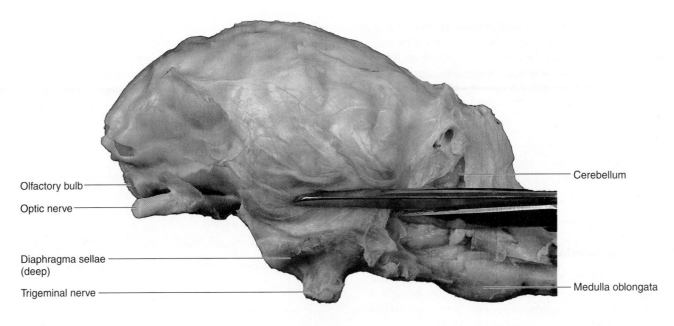

(a) Lateral view

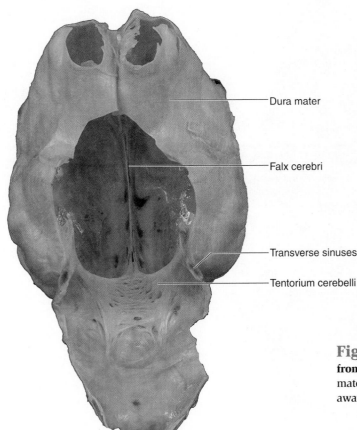

(b) Inferior view

Figure 15.11 **Incisions to Remove the Dura Mater from the Sheep Brain.** (a) Initial incision in the dura mater. (b) Appearance of the dura mater after it has been cut away from the brain.

will see the large **trigeminal nerves** (CN V). Using a blunt probe, feel the dura mater surrounding the base of the pituitary gland. This dural membrane is the **diaphragma sellae**, which lies between the pituitary and the rest of the brain (except where the pituitary stalk exits the sella turcica of the sphenoid bone).

5. To free the connections between the dura and the rest of the brain without breaking off the pituitary gland, you will first cut around the trigeminal nerves and capillary tufts.

Figure 15.11*a* shows where to make the initial incision. Cut *around* (lateral to) the optic chiasm, diaphragma sellae, pituitary gland, and trigeminal nerves to make a complete circle, which will free the dura mater from its attachments.

6. Next, make an anterior cut in the dura mater along the midsagittal plane between the olfactory bulbs and olfactory tracts (figure 15.11*b*). Once you have freed the dura mater from its connections to the diaphragma sellae, gently pull the dura away from the brain. Pull in a posterior, superior

direction so the falx cerebri and tentorium cerebelli slip out of their respective fissures without damaging the delicate brain tissues. As you pull the dura mater away from the brain, gently tease away any remaining connections. Figure 15.11*b* shows what the dura mater should look like after you have cut it away and removed it from the brain.

7. Once you have completely freed the dura from the brain, observe it closely and compare the dural septa and sinuses in the sheep brain to those you identified in the human brain. What dural septa is missing in a sheep brain that is present in a human brain?

EXERCISE 15.7B: Inferior View of the Sheep Brain

1. Obtain a sheep brain without the dura mater intact, or use the brain from which you just removed the dura mater. Place it in the dissecting pan on its superior surface so you are looking at its inferior surface (**figure 15.12**).

2. When sheep brains are collected by a commercial vendor for use in the laboratory, the dura mater is first separated from the cranial bones. In such specimens, most of the dura has been dissected away from the cranium and the only part remaining is the diaphragma sellae, a membrane between the sella turcica and the rest of the brain (see figure 15.11*a*). Surrounding the diaphragma sellae are some capillary tufts and large cranial nerves, the trigeminal nerves (CN V).

3. Using figure 15.12 as a guide, identify the following on the sheep brain you are dissecting:

☐ capillary tufts   ☐ optic chiasm

☐ diaphragma sellae   ☐ pituitary gland

☐ olfactory bulb   ☐ trigeminal nerves (CN V)

☐ olfactory tract

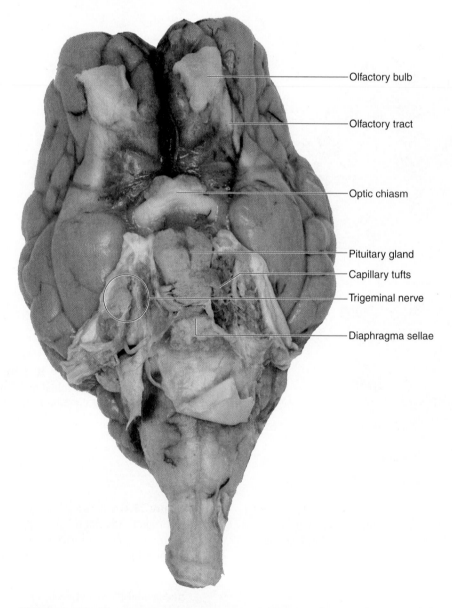

Olfactory bulb

Olfactory tract

Optic chiasm

Pituitary gland
Capillary tufts
Trigeminal nerve

Diaphragma sellae

Figure 15.12 **Inferior View of the Sheep Brain.** The dura mater surrounding the pituitary gland has not yet been removed.

4. Next you will dissect the diaphragma sellae and the capillary tufts away from the pituitary gland without damaging the cranial nerves, without detaching the pituitary from the infundibulum, and without detaching the trigeminal nerves from the brain. You will need to dissect carefully, because it is very easy to accidentally detach these structures from the brain if you pull too hard on the diaphragma sellae while you are trying to remove it.

5. Gently lift the dura mater posterior to the pituitary gland to see the small nerves that enter the dura mater on its deep surface (**figure 15.13**).

6. Using scissors or a scalpel, detach the nerves where they enter the dura mater. Cut the nerves where they attach to the dura (not where they attach to the brain!) and then cut the dura and bony material away, removing as much of it as possible while keeping the pituitary intact. You will need to be careful as you do this, because the connection between the pituitary and the rest of the brain is delicate.

7. Using **figure 15.14***a* as a guide, identify the following structures in the inferior view of the sheep brain you are dissecting:

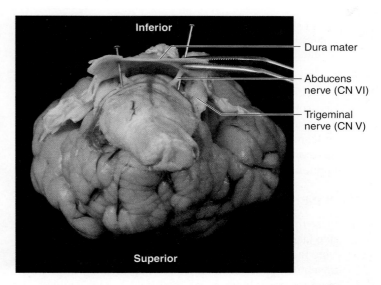

Figure 15.13 **Cranial Nerves Entering the Dura Mater.** The abducens and trigeminal nerves can be seen exiting the inferior surface of the sheep brain and piercing the dura mater.

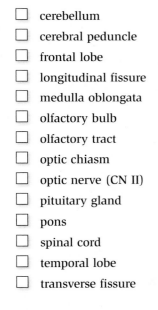

☐ cerebellum
☐ cerebral peduncle
☐ frontal lobe
☐ longitudinal fissure
☐ medulla oblongata
☐ olfactory bulb
☐ olfactory tract
☐ optic chiasm
☐ optic nerve (CN II)
☐ pituitary gland
☐ pons
☐ spinal cord
☐ temporal lobe
☐ transverse fissure

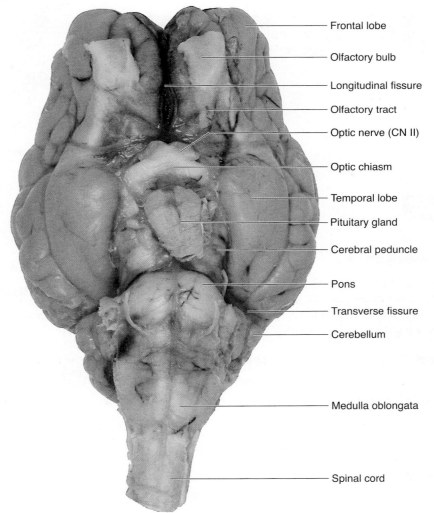

(a) Inferior view, pituitary gland intact

Figure 15.14 **Inferior Views of the Sheep Brain.** *(continued on next page)*

8. Gently lift the pituitary to observe the **mammillary body** (figure 15.14b). Note that the sheep brain has only a single mammillary body, whereas the human brain has two.

9. Finally, attempt to identify the cranial nerves listed in figure 15.14b. Identification of cranial nerves IX (glossopharyngeal nerve) through XII (hypoglossal nerve) may not be possible because these nerves are very small and may have been damaged or torn off the brain as the dura mater was removed from the brain, or as the brain was removed from the cranium.

EXERCISE 15.7C: Superior View of the Sheep Brain

1. Place the brain in the dissecting tray with the inferior side facing down (away from you) (**figure 15.15a**). Note the thin, transparent **arachnoid mater** that covers the entire surface of the brain without dipping into the **sulci** (grooves) between the **gyri** (folds) of the brain. Note the numerous **blood vessels** that lie between the arachnoid mater and the **pia mater**. The space occupied by the blood vessels is also a space where cerebrospinal fluid flows in the living animal. What is the name of this space?

_____

2. Using figure 15.15a as a guide, identify the following structures in the superior view of the sheep brain you are dissecting:

☐ arachnoid mater
☐ blood vessels
☐ cerebellum
☐ cerebrum
☐ gyrus
☐ longitudinal fissure
☐ spinal cord
☐ sulcus
☐ transverse fissure

3. Pick up the brain and gently pull the cerebral hemispheres away from the cerebellum so you can see into the transverse fissure. Using figure 15.15b as a guide, identify the following structures:

☐ cerebellum   ☐ pineal gland
☐ cerebrum   ☐ superior colliculus
☐ inferior colliculus

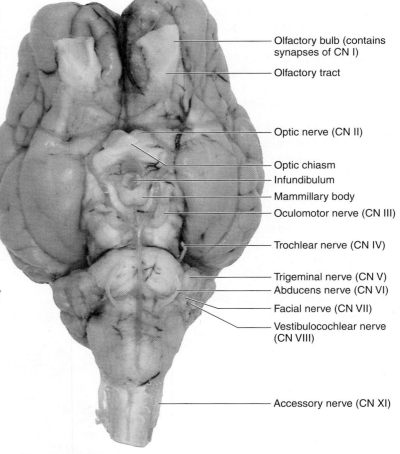

(b) Inferior view, pituitary gland removed

Figure 15.14 **Inferior views of the sheep brain.** *Continued.*

WHAT DO YOU THINK?

② Now that you have identified the pineal gland on both a human brain and a sheep brain, compare the relative size of the pineal gland to the size of the entire brain in each. Why do you think the pineal gland is so much larger (relative to brain size) in sheep than in humans? What does this tell you about the relative importance of the pineal gland in the functioning of the entire organism?

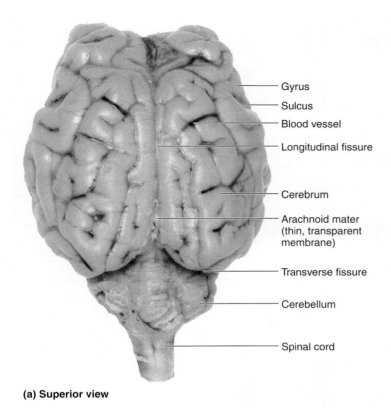

Gyrus

Sulcus

Blood vessel

Longitudinal fissure

Cerebrum

Arachnoid mater (thin, transparent membrane)

Transverse fissure

Cerebellum

Spinal cord

(a) Superior view

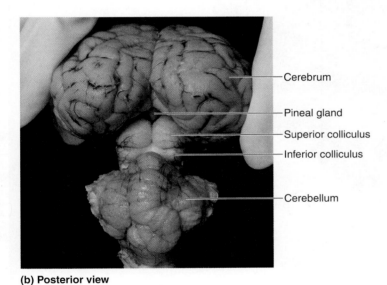

Cerebrum

Pineal gland

Superior colliculus

Inferior colliculus

Cerebellum

(b) Posterior view

Figure 15.15　Superior and Posterior Views of the Sheep Brain. (a) Superior view of the sheep brain. (b) Posterior view, cerebral hemispheres are pulled away from the cerebellum to reveal deeper structures.

EXERCISE 15.7D: Midsagittal and Coronal Sections of the Sheep Brain

1. Some of you will perform a midsagittal section of the sheep brain; others will perform a coronal section. Ask your instructor which section to make before you begin cutting. Although you will perform only one of these dissections, be sure to observe brains that have been sectioned along both planes.

2. *Midsagittal Section*: Place the sheep brain in your dissecting tray with its superior surface facing up (facing you). Using a scalpel, cut the brain in half along the midsagittal plane. Start your cut on the anterior end of the brain by placing the scalpel blade within the longitudinal fissure. What is the first structure the scalpel blade will cut through?

   _____.

3. Once you have cut the brain in half, observe its medial surface. Using **figure 15.16** as a guide, identify the following structures on the sheep brain you are dissecting:

☐ central canal	☐ medulla oblongata
☐ cerebellum	☐ optic chiasm
☐ cerebral aqueduct	☐ pineal gland
☐ cerebral peduncle	☐ pituitary gland
☐ cerebrum	☐ pons
☐ corpus callosum	☐ spinal cord
☐ fornix	☐ superior colliculus
☐ fourth ventricle	☐ thalamus
☐ mammillary body	

4. *Coronal Section*: Place the sheep brain in your dissecting tray with the superior surface down (facing away from you), and identify the pituitary gland (or the pituitary stalk, if the pituitary gland has been removed). Using a scalpel, cut the brain in half along a coronal plane that travels through the pituitary gland and continues toward the cerebral hemispheres.

5. Once you have cut the brain in half, observe the cut surface. Using **figure 15.17** as a guide, identify the following structures on the sheep brain you are dissecting:

☐ cerebral peduncle	☐ lateral ventricle
☐ cerebrum	☐ longitudinal fissure
☐ corpus callosum	☐ pons
☐ fornix	☐ thalamus
☐ hypothalamus	☐ third ventricle
☐ internal capsule	

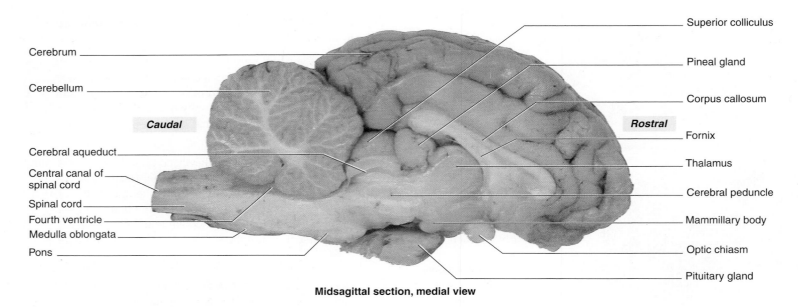

Cerebrum

Cerebellum

Caudal

Cerebral aqueduct

Central canal of spinal cord

Spinal cord

Fourth ventricle

Medulla oblongata

Pons

Superior colliculus

Pineal gland

Corpus callosum

Rostral

Fornix

Thalamus

Cerebral peduncle

Mammillary body

Optic chiasm

Pituitary gland

Midsagittal section, medial view

Figure 15.16   **Midsagittal View of the Sheep Brain.**

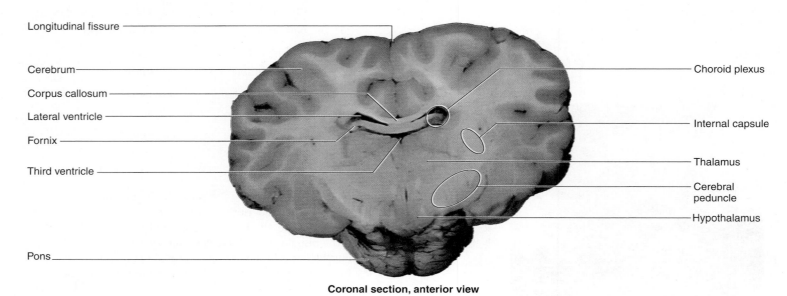

Longitudinal fissure

Cerebrum

Corpus callosum

Lateral ventricle

Fornix

Third ventricle

Pons

Choroid plexus

Internal capsule

Thalamus

Cerebral peduncle

Hypothalamus

Coronal section, anterior view

Figure 15.17   **Coronal Section Through the Sheep Brain.**

6. The **corpus callosum, fornix, internal capsule,** and **cerebral peduncles** are all fiber tracts. Recall from chapter 14 that a fiber tract is a bundle of myelinated axons. What is the function of these fiber tracts? _____

_____

7. Using forceps, open the lateral ventricles a bit and see if you can identify **choroid plexus** in the wall of the ventricle. Upon gross observation, the choroid plexus kind of looks like a bunch of "junk" inside the ventricle, but it is really a tuft of capillaries covered by glial cells. The histology of the choroid plexus was covered in chapter 14 (exercise 14.6).

What two structures make up the choroid plexus?

_____

_____

What is the function of the choroid plexus?

_____

_____

8. When you have finished your dissection, collect all the organic material from your dissecting pan and dispose of it in the proper containers (ask your laboratory instructor what these are). Dispose of the scalpel blades in the sharps container, and throw used paper towels and gloves into the garbage. Clean your dissecting tools and dissecting pan and return them to the proper storage area, and disinfect your laboratory workstation.

Chapter 15: CNS—The Brain

POST-LABORATORY WORKSHEET

1. You should have noticed that the corpora quadrigemina (tectal plate) are much larger in the sheep brain, relative to total brain size, than in the human brain. Using this information, answer the following questions:

   a. What is the function of the superior colliculus? _____

   _____

   b. What is the function of the inferior colliculus? _____

   _____

   c. What does the difference in size between the superior and inferior colliculi tell you about the influence this region of the brain has on the overall functioning of a sheep vs. a human (that is, compare how much influence this area of the brain has on control over body functions).

   _____

   _____

   _____

2. Continuing with the line of reasoning from question 1, compare and contrast structures of the human brain and the sheep brain by filling in the following table. That is, write down information about the relative size of the structure compared to the size of the entire brain. Then, based on function, explain why the structure might be more important for survival of the human or the sheep.

Brain Structure	Human Brain	Sheep Brain
Frontal Lobe		
Inferior Colliculus		
Mammillary Bodies		
Medulla Oblongata		
Olfactory Bulbs		
Pineal Body (Gland)		
Superior Colliculus		

3. What would be the effect of severing the corpus callosum?

_____

_____

_____

4. Label the following figure of the inferior view of a sheep brain.

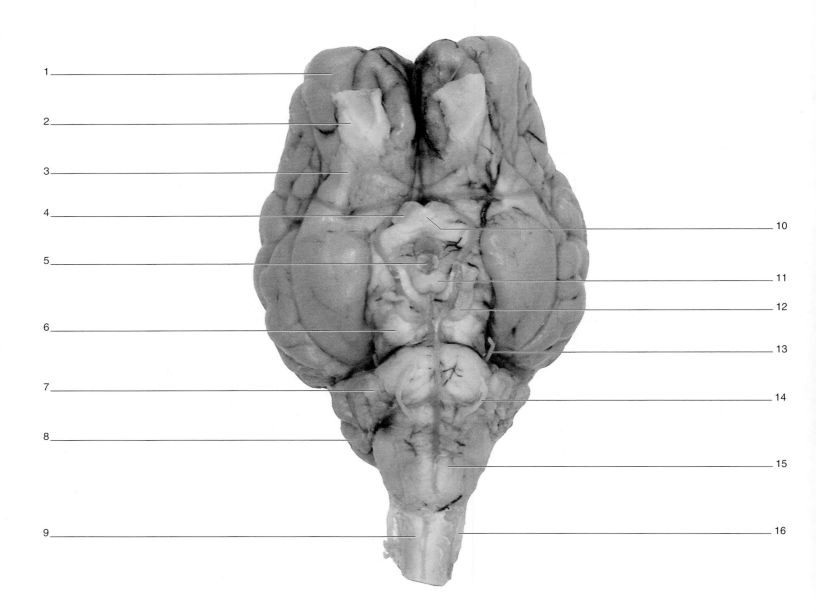

1 _____

2 _____

3 _____

4 _____   10 _____

5 _____

                                                                             11 _____

                                                                             12 _____

6 _____

                                                                             13 _____

7 _____

                                                                             14 _____

8 _____

                                                                             15 _____

9 _____   16 _____

5. Label the following figure of a midsagittal section of a sheep brain.

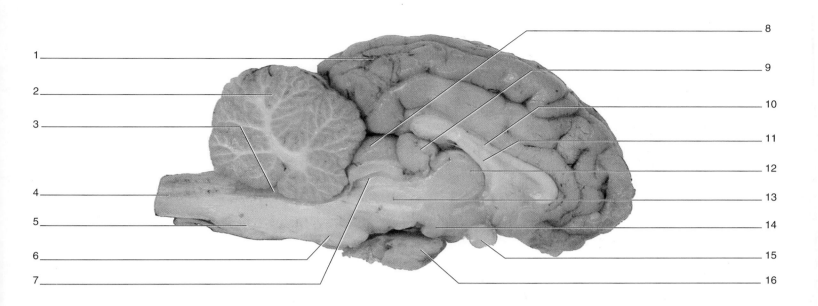

1 _____

2 _____

3 _____

4 _____

5 _____

6 _____

7 _____

8 _____

9 _____

10 _____

11 _____

12 _____

13 _____

14 _____

15 _____

16 _____

6. Label the following figure of a coronal section of a sheep brain.

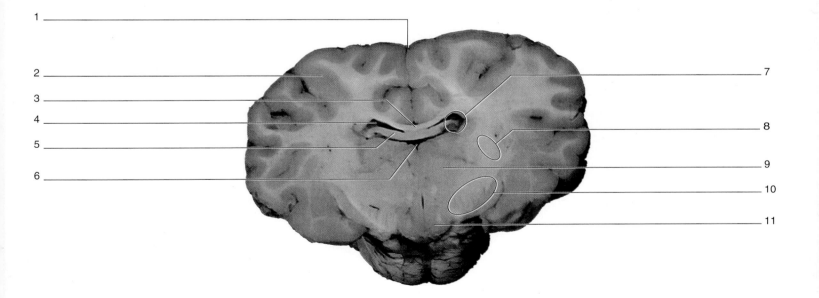

1 _____

2 _____

3 _____

4 _____

5 _____

6 _____

7 _____

8 _____

9 _____

10 _____

11 _____

Cranial Nerves

16

OUTLINE and OBJECTIVES

INTRODUCTION

Imagine waking up one morning to find that the muscles on the left side of your face no longer worked! You look into the mirror, try to smile, and find that your smile is very crooked. The expressions on the right side of your face are fine, but the left side of your face is stone cold and expressionless. You rush to the doctor's office hoping for a cure, only to discover that there is nothing modern medicine can do for you. Alas! The doctor informs you that you have a condition called Bell palsy, a disorder of the seventh cranial nerve, the facial nerve. The doctor informs you that authorities on the disorder think it has something to do with a viral infection in the nerve root. The only assurance she has for you is that you will eventually recover, but it will take time. You return home, both sides of your face droopy now (though the right side of your face is a little droopier than the left). At home, after pondering your condition for awhile, it suddenly occurs to you why you had to learn those pesky names, numbers, and functions of the 12 cranial nerves in your anatomy class. Though you are dismayed at your condition, you take comfort in the fact that you have some knowledge of what is going on in your body. A smile gently spreads across the right side of your face.

Module 7: NERVOUS SYSTEM

Chapter 16: Cranial Nerves

Name: _____

Date: _____ Section: _____

PRE-LABORATORY WORKSHEET

1. The three parts of the brainstem are the _____, _____, and _____.

2. The pons is associated with which ventricular structure? _____.

3. The medulla oblongata is associated with which ventricular structure? _____.

4. A cranial nerve is a nerve that originates from the _____.

5. Somatic motor neurons innervate _____ muscle tissue.

6. Visceral motor neurons innervate _____ or _____ muscle tissue, or

_____.

7. An example of a special sensory organ is _____.

8. Define the following terms:

   a. *somatic nervous system:*_____

   _____

   b. *autonomic nervous system:*_____

   _____

   c. *ganglion:*_____

   _____

   d. *tract:*_____

   _____

   e. *sensory:*_____

   _____

   f. *motor:*_____

   _____

IN THE LABORATORY

The peripheral nervous system consists of cranial nerves, which come off of the brain, and spinal nerves, which come off of the spinal cord. There are 12 pairs of cranial nerves and 31 pairs of spinal nerves. Although there are fewer cranial nerves than spinal nerves, learning the names and functions of the cranial nerves is generally more challenging, because each nerve is given both a number and a name, whereas spinal nerves are only given numbers (T1, T2, etc.). In addition, most of the cranial nerves take complicated pathways through the skull as they travel to their target organs. Remember all of the foramina of the skull that you learned in chapter 8? Most of those foramina are passageways for cranial nerves to exit the cranial cavity.

The exercises in this laboratory session are organized so that if you complete them in the proper sequence, your knowledge of the cranial nerves will start out basic and then become more detailed. The first exercises will help you establish a base knowledge about the locations and names of the nerves. Subsequent exercises will build on that base knowledge and help you learn the detailed functions and locations of the cranial nerves.

Gross Anatomy

Cranial Nerves

An inferior view of the brain allows you to see all of the cranial nerves as they come off of the brain. Cranial nerves are numbered, starting from the anterior (rostral) part of the brain and moving posterior (caudal), using Roman numerals I through XII. **Figure 16.1** demonstrates the inferior surface of the brain and the cranial nerves. The olfactory bulbs (where CN I synapses) and the optic nerves (CN II) are very large, easily identifiable structures located on the inferior surface of the frontal lobes of the brain. The remainder of the cranial nerves (CN III through CN XII) are generally smaller and are located closer together in the region of the midbrain, pons, and medulla, which makes their identification a little more challenging. In exercise 16.1 you will identify the cranial nerves on a brain or on a model of the brain.

EXERCISE 16.1 Identification of Cranial Nerves on a Brain or Brainstem Model

1. Obtain a human brain or a model of a human brain or brainstem.

2. Turn the brain over and observe its inferior surface (figure 16.1). Note all of the small nerves exiting the brain from various locations. These are the cranial nerves **(table 16.1)**.

3. Using table 16.1 and figure 16.1 as guides, identify the 12 cranial nerves on the inferior surface of the brain.

Table 16.1	Names of Cranial Nerves	
Nerve Number	**Name**	**Word Origins**
I	Olfactory	*olfacio*, to smell
II	Optic	*optikos*, relating to the eye or vision
III	Oculomotor	*oculo-*, the eye, + *motorius*, moving
IV	Trochlear	*trochileia*, a pulley
V	Trigeminal	*tri-*, three, + *geminus*, twins
VI	Abducens	*abductio-*, to move away from the median plane
VII	Facial	*facialis*, relating to the face
VIII	Vestibulocochlear	*vestibulum*, entrance, + *cochlea*, snail shell
IX	Glossopharyngeal	*glossus*, tongue, + *pharyngeus*, pharynx
X	Vagus	*vagus*, wanderer
XI	Accessory	*spina*, spine, + *accessory*, an extra structure
XII	Hypoglossal	*hypo*, beneath, + *glossus*, tongue

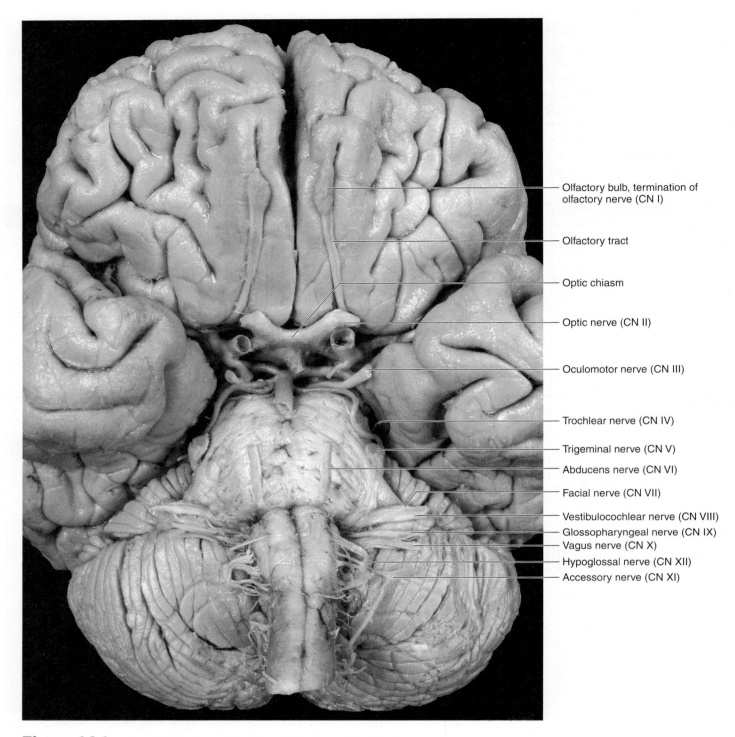

Olfactory bulb, termination of olfactory nerve (CN I)

Olfactory tract

Optic chiasm

Optic nerve (CN II)

Oculomotor nerve (CN III)

Trochlear nerve (CN IV)

Trigeminal nerve (CN V)

Abducens nerve (CN VI)

Facial nerve (CN VII)

Vestibulocochlear nerve (CN VIII)

Glossopharyngeal nerve (CN IX)

Vagus nerve (CN X)

Hypoglossal nerve (CN XII)

Accessory nerve (CN XI)

Figure 16.1   **Cranial Nerves on the Inferior Surface of the Brain.**

4. As you continue observing the inferior surface of the brain, notice where each nerve exits the brain or brainstem. Knowledge of the general area where a specific nerve emerges can be helpful in identifying the nerves, even if by process of elimination. Complete the following chart for cranial nerves III through XII to organize your thoughts. Use figure 16.1 as a guide.

Point of Exit	Cranial Nerve
Midbrain	
Pons	
Medulla Oblongata	

5. Mnemonic devices, simple phrases or plays on words that aid in memory recall, make remembering long strings of anatomical names easier. One popular technique is to associate a word with the first letter of the corresponding word you want to remember. A mnemonic that will help you recall the names of the cranial nerves in numerical order is:

Oh **O**nce **O**ne **T**akes **T**he **A**natomy **F**inal, **V**ery **G**ood **V**acations **A**re **H**eavenly.

Mnemonics work best when they mean something to you. Develop your own mnemonic for remembering the cranial

nerves. _____

_____

_____

_____

6. *Optional Activity*: AP|R **Nervous System**—Visit the quiz area to test yourself on cranial nerve location, composition, and related foramina.

Study Tip!

If you are observing a real brain instead of a model, you might not see the trochlear nerve (CN IV). It is very small, and often detaches from the brain when the brain is removed from the skull. If this is the case, make sure you identify the trochlear nerve on a model of the human brain. This nerve is unique because it is the only cranial nerve that originates from the posterior part of the brainstem, rather than the anterior or medial aspect.

Foramina of the Skull Associated with Cranial Nerves

All of the cranial nerves must eventually exit the cranial cavity to get to their target organs. They do so by traveling through various foramina (sing. foramen) of the skull, which you learned in chapter 8. In exercise 16.2 you will review the foramina of the skull that are relevant to the cranial nerves. **Table 16.2** lists the foramen that each cranial nerve uses to exit from the cranial cavity. It is useful to first identify each nerve and its foramen of exit. Then go back and consider each foramen individually and list all of the cranial nerves that travel through each one, because it's common for several cranial nerves to travel through each foramen.

Study Tip!

The trigeminal nerve (CN V) has three branches. Hence the name: *tri* = three, *geminus* = twins. Each branch is given its own name and designation as follows:

V$_1$ = Ophthalmic branch

V$_2$ = Maxillary branch

V$_3$ = Mandibular branch

Table 16.2	Cranial Nerves and Foramina of Exit	
Nerve Number	**Name**	**Foramen of Exit from the Cranial Cavity**
I	Olfactory	Olfactory foramina in the cribriform plate of the ethmoid bone
II	Optic	Optic canal
III	Oculomotor	Superior orbital fissure
IV	Trochlear	Superior orbital fissure
V	Trigeminal	Superior orbital fissure (V$_1$—ophthalmic) Foramen rotundum (V$_2$—maxillary) Foramen ovale (V$_3$—mandibular)
VI	Abducens	Superior orbital fissure
VII	Facial	Internal acoustic meatus (exits via the stylomastoid foramen)
VIII	Vestibulocochlear	Internal acoustic meatus
IX	Glossopharyngeal	Jugular foramen
X	Vagus	Jugular foramen
XI	Accessory	Jugular foramen (accessory division) Foramen magnum (spinal division)
XII	Hypoglossal	Hypoglossal canal

EXERCISE 16.2 Identification of Cranial Nerve Foramina of Exit on a Skull

1. Obtain a human skull or a model of a human skull.

2. Remove the skullcap and observe the interior surface of the base of the cranium (**figure 16.2**).

3. Using table 16.2 and your textbook or atlas as guides, identify the structures listed in figure 16.2 on the interior surface of the base of the cranium.

4. Complete the following chart by writing the names of the cranial nerves that travel through each of the listed foramina of the skull (use the last column in table 16.2 as a guide).

Foramen	Nerves Traveling Through
Olfactory Foramina	
Optic Canal	
Superior Orbital Fissure	
Foramen Rotundum	
Foramen Ovale	
Internal Acoustic Meatus	
Jugular Foramen	
Hypoglossal Canal	

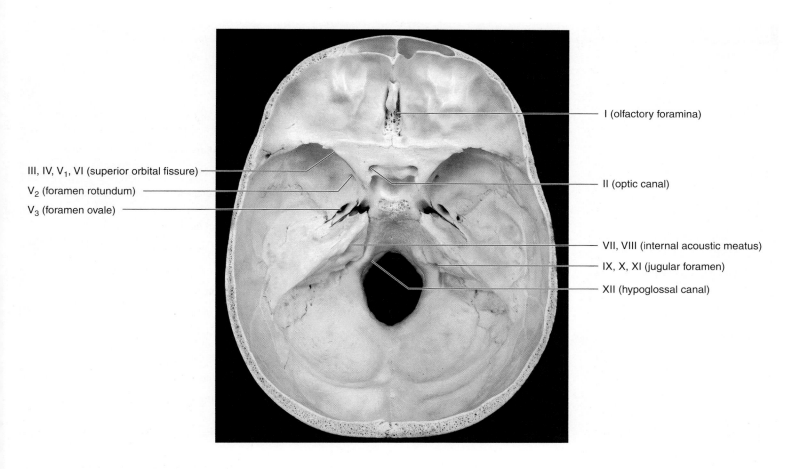

Figure 16.2   **Cranial Nerve Passages Through Skull Foramina.**

- ☐ olfactory nerves (CN I)
- ☐ optic nerve (CN II)
- ☐ oculomotor nerve (CN III)
- ☐ trochlear nerve (CN IV)
- ☐ V₁ - ophthalmic branch of trigeminal nerve (CN V)

- ☐ V₂ - maxillary branch of trigeminal nerve (CN V)
- ☐ V₃ - mandibular branch of trigeminal nerve (CN V)
- ☐ abducens nerve (CN VI)
- ☐ facial nerve (CN VII)

- ☐ vestibulocochlear nerve (CN VIII)
- ☐ glossopharyngeal nerve (CN IX)
- ☐ vagus nerve (CN X)
- ☐ spinal accessory nerve (CN XI)
- ☐ hypoglossal nerve (CN XII)

WHAT DO YOU THINK?

① An **acoustic neuroma** is a tumor that arises from neurolemmocytes (Schwann cells) surrounding the vestibular portion of the vestibulocochlear nerve (CN VIII). The tumor is benign, but generally grows within the confined space of the petrous part of the temporal bone, thus compressing the nerve and creating problems with balance and hearing loss. What nerve other than CN VIII would you expect to be affected by this tumor (due to its close proximity)?

Testing Cranial Nerve Functions

Cranial nerves carry different types of information from the brain to or from the target organ(s) for each nerve. **Sensory** (afferent) information travels from the target organ to the brain, whereas **motor** (efferent) information travels from the brain to the target organ.

The types of information carried by cranial nerves are classified in the chart on the right.

Table 16.3 lists the type of information carried by each cranial nerve and lists the specific functions of each cranial nerve.

Somatic sensory	Sensation from cutaneous touch, pain, temperature, position, and pressure receptors.
Somatic motor	Motor output to skeletal muscle.
Special sensory	Sensation from special sensory organs such as the eye and ear.
Visceral sensory	Sensation from sensory receptors within visceral organs.
Visceral motor	Motor output to cardiac muscle, smooth muscle, and glands.

Table 16.3	Detailed Functions and Divisions of the Cranial Nerves		
Nerve Number	**Name**	**Type of Nerve**	**Function(s)**
I	Olfactory	Special sensory	Sensory nerve of smell.
II	Optic	Special sensory	Sensory nerve of vision.
III	Oculomotor	Somatic motor	Motor to levator palpebrae superioris (raises the upper eyelid) and all extrinsic eye muscles except for the superior oblique and the lateral rectus.
		Visceral motor (parasympathetic)	Motor to the ciliary body and pupillary sphincter muscles of the eye.
IV	Trochlear	Somatic motor	Motor to superior oblique muscle of the eye.
V	Trigeminal	Mixed	Sensation from the cornea, scalp, forehead, face, and teeth; motor to the muscles of mastication.
V_1	Ophthalmic branch	Somatic sensory	Sensory from the cornea, scalp, and forehead.
V_2	Maxillary branch	Somatic sensory	Sensory from the face, cheeks, and maxillary (upper) teeth.
V_3	Mandibular branch	Somatic sensory	Sensory from the chin, mandibular (lower) teeth, and tongue.
		Somatic motor	Motor to the muscles of mastication.
VI	Abducens	Somatic motor	Motor to the lateral rectus muscle of the eye.
VII	Facial	Mixed	Sensory fibers for taste from the anterior two-thirds of the tongue.
		Somatic motor	Motor to the muscles of facial expression.
		Special sensory	Sensory fibers for taste from the anterior two-thirds of the tongue.
		Visceral motor (parasympathetic)	Motor to the lacrimal glands, and the submandibular and sublingual salivary glands.
VIII	Vestibulocochlear	Special sensory	Sensory nerve of hearing and balance.
IX	Glossopharyngeal	Mixed	Sensory fibers for taste from the posterior one-third of the tongue, sensation from the ear and pharynx, motor to the stylopharyngeus muscle and the parotid salivary gland.
		Somatic sensory	Cutaneous sensation from the external ear; general and taste sensation from posterior 1/3 of tongue.
		Somatic motor	Motor to the stylopharyneus muscle.
		Special sensory	Sensory fibers for taste from the posterior one-third of the tongue.
		Visceral sensory	Sensory from the carotid sinus and carotid body, mastoid air cells, pharynx, middle ear, and tympanic cavity.
		Visceral motor (parasympathetic)	Motor to the parotid salivary gland.
X	Vagus	Mixed	Motor to the pharynx and thoracic and abdominal viscera, sensation from the pharynx and ear.
		Somatic sensory	Sensory from the external acoustic meatus, tympanic membrane, dura mater of the posterior cranial fossa, and auricle of the ear.
		Somatic motor	Motor to muscles of the pharynx, larynx, and palate (except stylopharyngeus and tensor veli palatini).
		Special sensory	Sensory fibers for taste from the palate and epiglottis.
		Visceral motor (parasympathetic)	Motor to glands of the pharynx and larynx, and smooth muscle of the heart, lungs, and abdominal viscera.
		Visceral sensory	Sensory from the pharynx, larynx, bronchi, aorta, and abdominal viscera.
XI	Accessory	Somatic motor	Motor to the trapezius and sternocleidomastoid muscles.
XII	Hypoglossal	Somatic motor	Motor to the intrinsic and extrinsic muscles of the tongue.

EXERCISE 16.3   Testing Specific Functions of the Cranial Nerves

Neurological tests similar to some of those described in this exercise are performed by physicians when testing for damage to one or more cranial nerves. These tests can indicate if a cranial nerve is damaged. However, they are not infallible. For instance, an inability to hear could indicate damage to the vestibulocochlear nerve. However, the damage could also reside in the auditory cortex of the brain. Each of the tests described here were chosen because they are both easy and quick to perform. **Table 16.4** lists the common disorders of the cranial nerves along with potential signs and symptoms, and causes of the disorders.

Table 16.4	**Common Disorders of the Cranial Nerves**		
Nerve Number	**Name**	**Signs and Symptoms of Damage**	**Potential Cause of the Disorder**
I	Olfactory	Inability to smell (anosmia).	A fracture of the cribriform plate of the ethmoid can damage the olfactory nerves.
II	Optic	Blindness on the affected side (hemianopia).	Intracranial tumor or stroke that damages the nerve or nerve tract prevents visual information from reaching the brain.
III	Oculomotor	Pupil dilation (mydriasis). Eye deviates down and out (strabismus) from muscle paralysis resulting in double vision (diplopia). Eyelid droops (ptosis).	Increased intracranial pressure is a common cause of compression of the nerve. The parasympathetic fibers that innervate the pupillary sphincter muscle are located on the surface of the nerve, so pupil dilation is often the first sign of nerve damage or increased intracranial pressure. Nerve damage results in paralysis of all extraocular muscles except the superior oblique and lateral rectus muscles. Paralysis of the levator palpebrae superioris muscle causes ptosis.
IV	Trochlear	Difficulty turning the eye inferior and lateral, which leads to double vision (diplopia).	Nerve damage results in paralysis of the superior oblique muscle.
V	Trigeminal	Trigeminal neuralgia (tic douloureux), a sudden, intense pain along the course of one of the divisions of the nerve.	Pressure on the nerve from the artery that courses alongside it stimulates sensory fibers within the nerve. Pain is often triggered by touching structures inside the mouth.
VI	Abducens	Eye deviates medially (adducts) causing double vision (diplopia).	Any disorder that increases intracranial pressure (for example, a stroke) can cause this nerve to be crushed against the clivus (sloped portion) of the sphenoid bone.
VII	Facial	Bell palsy—paralysis of the muscles of facial expression on the side of the face with the affected nerve. Loss of taste sensation on the anterior two-thirds of the tongue (ageusia). Decreased salivation (hypoptyalism).	A viral infection that causes inflammation of the facial nerve is the most likely source. This problem often resolves itself within a couple of months.
VIII	Vestibulocochlear	Loss of balance and equilibrium, nausea, vomiting, and dizziness or inability to hear (anacusis).	Acoustic neuroma—a tumor originating in neurolemmocytes within the internal acoustic meatus causes compression of the nerve.
IX	Glossopharyngeal	Difficulty swallowing (dysphagia), loss of taste sensation on the posterior one-third of tongue (ageusia). Decreased salivation (hypoptyalism).	Nerve damage interrupts the sensory component of the swallowing reflex.
X	Vagus	Difficulty swallowing (dysphagia) or hoarseness (dysphonia).	Nerve damage interrupts the motor component of the swallowing reflex. Hoarseness results from paralysis of the muscles of the larynx.
XI	Accessory	Difficulty elevating the scapula or rotating the head.	Nerve damage results in paralysis of the sternocleidomastoid and/or trapezius muscles.
XII	Hypoglossal	When sticking out the tongue, it moves in the direction of the damaged nerve.	Compression of nerve from increased intracranial pressure.

EXERCISE 16.3A: Olfactory (CN I)

The olfactory nerves (**figure 16.3**) are unique as cranial nerves in that there are more than two of them (the rest of the cranial nerves are paired—a right and left for each), and they are constantly being replaced. The nerves lie within the nasal epithelium, and their axons project through the **olfactory foramina** within the cribriform plate of the ethmoid bone (table 16.2). The olfactory neurons then synapse with neurons within the **olfactory bulbs** and the signals are sent to the brain via the **olfactory tracts**.

1. Obtain vials of peppermint, lemon, and vanilla oils.

2. Have your laboratory partner close his eyes. Open the vial of peppermint oil and pass it just under your laboratory partner's nose, then ask him to breathe in and identify the smell.

3. Repeat this process with the vials of lemon and vanilla oils. Allow some time between applications of the different oils so the olfactory nerves won't confuse the odors. Damage to the olfactory nerves results in an inability to identify odors. Excessive smoking or inflammation of the nasal mucosa as a result of a viral infection can inhibit the sense of smell, and the sense of smell also declines with age.

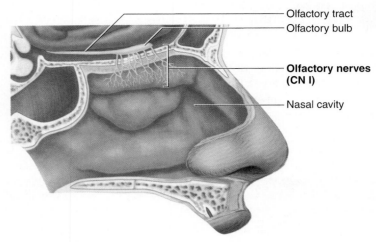

Figure 16.3 Location of the Olfactory Nerves (CN I).

EXERCISE 16.3B: Optic (CN II)

Each of the optic nerves (**figure 16.4**) begins as a number of ganglion cells located within the **retina** of the eye. Different parts of the retina correspond to different portions of the visual field. The axons of these nerves exit the eye at the optic disc, or "blind spot" of the eye, at which point they become the **optic nerve**, which then travels posteriorly toward the diencephalon. Anterior to the pituitary stalk, several of the fibers cross over at the prominent **optic chiasm** (*chiasma*, a crossing of two lines). The fibers then continue to travel posteriorly to reach the visual cortex in the occipital lobe of the brain. Figure 16.4 demonstrates the pattern of flow of visual information from the retina to the brain. When damage to the retina or optic nerve is suspected, visual field tests are performed to discover the location of the damage. These tests of visual function are beyond the scope of this course and will not be performed in this laboratory session.

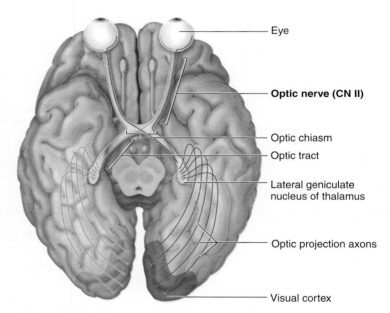

Figure 16.4 The Optic Nerve (CN II).

EXERCISE 16.3C: Oculomotor (CN III)

The oculomotor nerves (**figure 16.5**) send motor fibers to the majority of the extraocular eye muscles as well as to the muscles that control pupil diameter (tables 16.3 and 16.4).

1. Obtain a small flashlight. Look into one of your laboratory partner's eyes and observe the size of the pupil. While looking into her eye, gently shine the light into the eye (if it is a bright light, just bring it near the eye so that more light enters the eye—the goal here is not to blind your laboratory partner with the light).

   Did you observe a change in pupil diameter?_____

   If so, what happened?_____

2. Repeat the above activity, but this time observe the pupil of the eye that you are NOT shining the light into. Did

you observe a change in pupil diameter?_____

If so, what happened?_____

WHAT DO YOU THINK?

2. When a light is shined into a patient's right eye, an examiner expects to see a change in pupil diameter in both eyes. The response, called the *consensual light reflex*, is used to test the function of two cranial nerves. The reflex involves one cranial nerve sending the afferent (sensory) signal toward the brain, and another cranial nerve sending the efferent (motor) signal out to the pupil. Which cranial nerve carries the afferent (sensory) signal to the brain?

EXERCISE 16.3D: Trochlear (CN IV)

The trochlear nerve (**figure 16.6**) controls only one extraocular eye muscle—the superior oblique (tables 16.3 and 16.4). Ask your laboratory partner to look down and out (inferior and lateral). Weakness or an inability to perform this action indicates a weak superior oblique muscle or damage to the trochlear nerve.

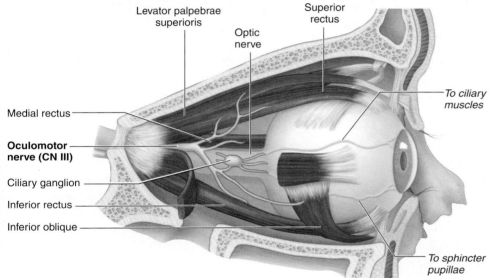

Levator palpebrae superioris
Optic nerve
Superior rectus
To ciliary muscles
Medial rectus
Oculomotor nerve (CN III)
Ciliary ganglion
Inferior rectus
Inferior oblique
To sphincter pupillae

Figure 16.5 The Oculomotor Nerve (CN III).

Optic nerve (CN II)
Superior oblique
Trochlear nerve (CN IV)

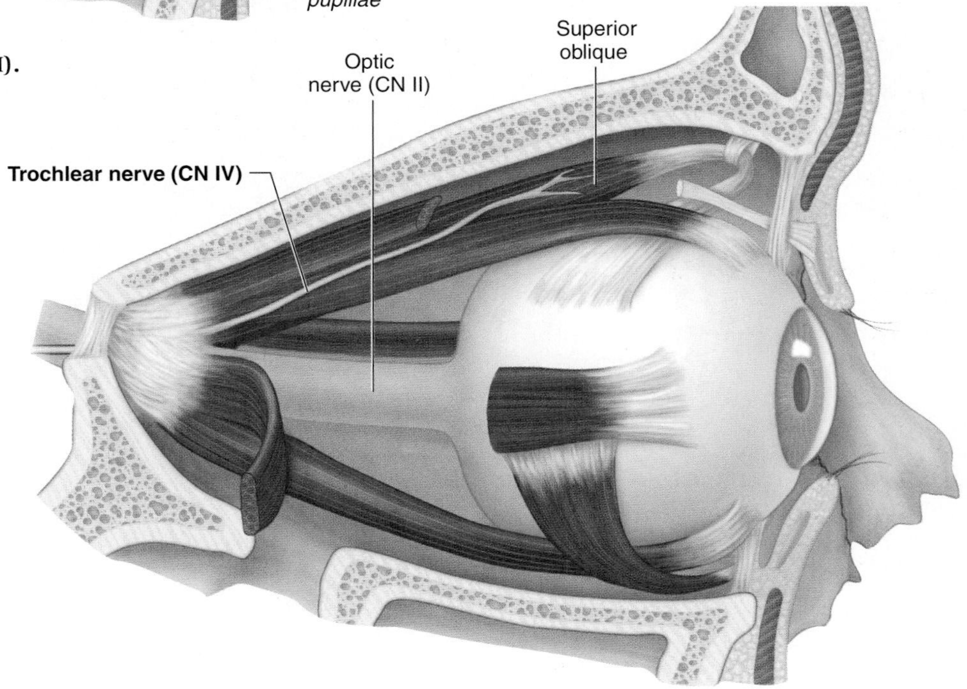

Figure 16.6 The Trochlear Nerve (CN IV).

EXERCISE 16.3E: Trigeminal (CN V)

The trigeminal nerve (**figure 16.7**) is the largest and most complex of the cranial nerves. It has three branches: the ophthalmic, maxillary, and mandibular branches, which are named V_1, V_2, and V_3, respectively. It is the predominant nerve carrying sensory information from the face, but it also carries motor output to the muscles of mastication.

1. Obtain a feather and a cotton ball.

2. Have your laboratory partner close his eyes, and then proceed to gently touch his face with the feather in the sensory distribution areas of the trigeminal nerve shown in figure 16.7*b*. An inability to feel this sensation in one or more locations indicates damage to a branch of the trigeminal nerve.

3. Have your laboratory partner keep his eyes open and look up and away from you. *Very gently and lightly* touch a few strands of the fibers from the cotton ball to his cornea. This is a test for the *corneal reflex*, whose sensory component is carried by a branch of the trigeminal nerve. Touching the cornea with the cotton should cause him to blink. An absent corneal reflex indicates damage to the ophthalmic branch (V_1) of the trigeminal nerve (contact lens wearers may also have a diminished or absent corneal reflex).

EXERCISE 16.3F: Abducens (CN VI)

The abducens nerve (**figure 16.8**) controls only one extraocular eye muscle—the lateral rectus (tables 16.2 and 16.3). Ask your laboratory partner to look laterally to the right. Weakness or an inability to do so indicates a weak lateral rectus muscle in the right eye or damage to the right abducens nerve.

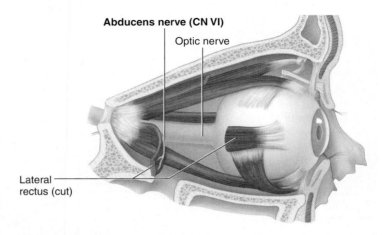

Figure 16.8   **The Abducens Nerve (CN VI).**

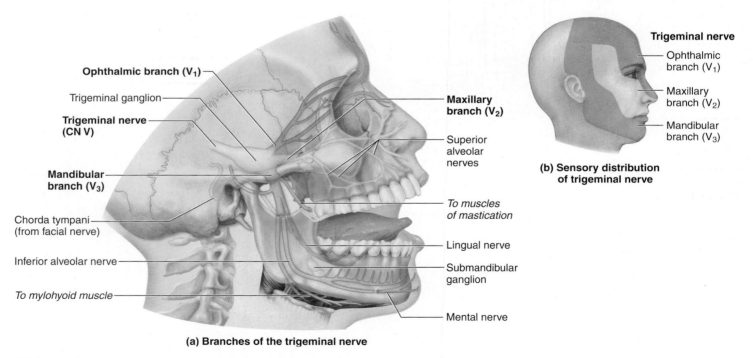

(a) Branches of the trigeminal nerve

Figure 16.7   **The Trigeminal Nerve (CN V).**

EXERCISE 16.3G: Facial (CN VII)

The facial nerve (**figure 16.9**) has several functions. Two major functions are to carry motor output to the muscles of facial expression and carry sensory information to the brain from taste buds on the anterior two-thirds of the tongue (**figure 16.10**).

1. Obtain vials of salt and sugar and a cup of drinking water.

2. Ask your laboratory partner to close her eyes, open her mouth, and stick out her tongue. Place a few grains of salt on the anterolateral surface of the right side of her tongue, and see if she can positively identify the taste as salty.

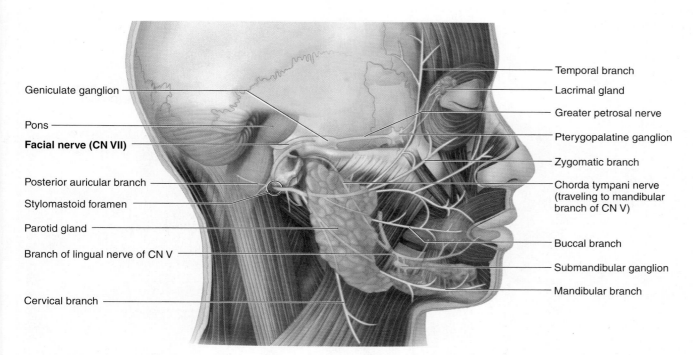

Figure 16.9 **The Facial Nerve (CN VII).**

3. Have your laboratory partner take a drink of water to refresh her taste buds before performing the next test. Once again, ask her to close her eyes, open her mouth, and stick out her tongue. Place a few grains of sugar on the tip of her tongue and see if she can positively identify the taste as sweet. An inability to identify these tastes may indicate damage to the facial nerve.

4. Ask your laboratory partner to demonstrate facial expressions, such as surprise, happiness, sadness, and confusion. An inability to express these emotions facially may indicate paralysis of the muscles of facial expression, a common consequence of damage to the facial nerve.

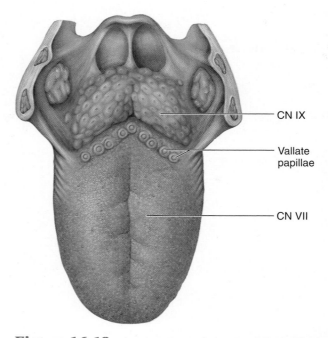

Figure 16.10 **Innervation of the Taste Buds of the Tongue.**

EXERCISE 16.3H: Vestibulocochlear (CN VIII)

The vestibulocochlear nerve (**figure 16.11**) is two nerves: the **vestibular nerve**, which carries sensory information about balance and equilibrium to the brain from the vestibule, and the **cochlear nerve**, which carries sensory information about sound to the brain from the cochlea. Both nerves enter the petrous part of the temporal bone through the **internal auditory canal**. Figure 16.11 demonstrates the special sensory structures within this bone that are targets of the vestibulocochlear nerve.

1. Obtain a tuning fork. Ask your laboratory partner to close his eyes.

2. Holding the tuning fork by its base, gently strike it on the table and then hold it near your laboratory partner's ear and ask him if he hears anything. An inability to hear the vibrations caused by the tuning fork can indicate damage to the vestibulocochlear nerve.

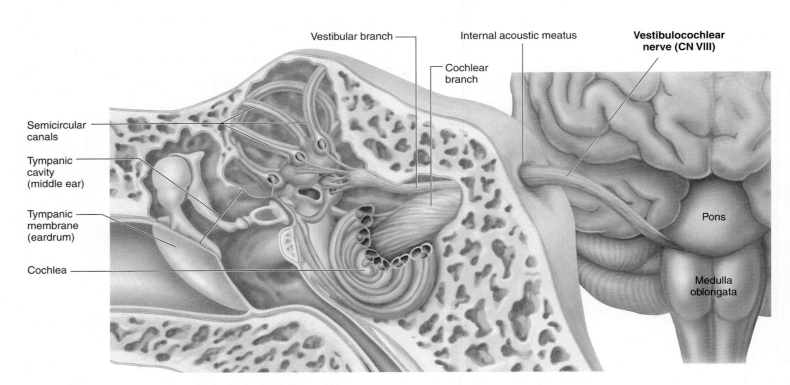

Figure 16.11   **The Vestibulocochlear Nerve (CN VIII).**

EXERCISE 16.3I: Glossopharyngeal (CN IX) and Vagus (CN X)

The glossopharyngeal and vagus nerves (**figures 16.12** and **16.13**, tables 16.3 and 16.4) both carry information to and from the soft palate, pharynx, and larynx to control reflexes such as the coughing and gagging reflexes. In addition, the glossopharyngeal nerve carries sensory information from the taste buds on the posterior one-third of the tongue (see figure 16.10). The vagus nerve is unique as a cranial nerve because it innervates many structures within the thoracic and abdominal cavities. It is the predominant pathway for parasympathetic information to travel from the brain to the visceral organs of the body. Though tests for the functioning of the glossopharyngeal and vagus nerves are not easy to perform, you will try to observe at least some of the functions of these nerves with this exercise.

1. Ask your laboratory partner to open her mouth and say "Ah" while you observe her soft palate and uvula.

2. Unilateral drooping of the soft palate or deviation of the uvula to one side may indicate damage to either the glossopharyngeal or vagus nerves. The glossopharyngeal nerve carries sensory information from the pharynx to the brain, while the vagus nerve carries motor information back out to the muscles that raise the palate and that are used in swallowing.

3. Another test for the functioning of these two nerves is to test for the gag reflex. This will not be attempted in the lab because inexperienced testing of this reflex could cause your laboratory partner to choke or, at the very least, become very uncomfortable.

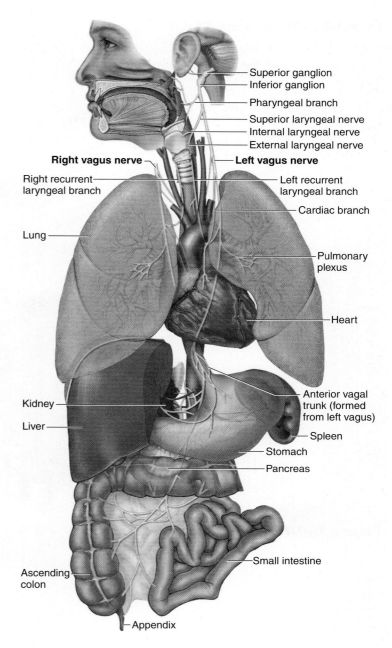

Figure 16.13   **The Vagus Nerve (CN X).**

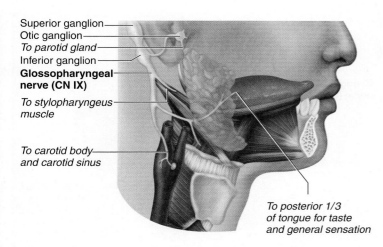

Figure 16.12   **The Glossopharyngeal Nerve (CN IX).**

EXERCISE 16.3J: Accessory (CN XI)

The accessory nerve (**figure 16.14**) carries motor information to the trapezius and sternocleidomastoid muscles (tables 16.3 and 16.4).

1. Ask your laboratory partner to elevate her scapula ("shrug" her shoulders) to test the function of the trapezius muscle.

2. Next, ask her rotate her head first to the right and then to the left to test the function of the sternocleidomastoid muscle. Damage to the accessory nerve would cause both of these actions to be weak or impossible due to paralysis of the muscles.

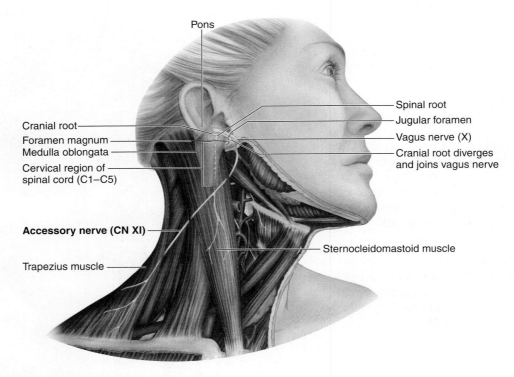

Figure 16.14   **The Accessory Nerve (CN XI).**

EXERCISE 16.3K: Hypoglossal (CN XII)

The hypoglossal nerve (**figure 16.15**) innervates intrinsic and extrinsic muscles of the tongue (tables 16.3 and 16.4). Ask your laboratory partner to stick out his tongue. He should be able to stick out his tongue without it deviating to one side. If the hypoglossal nerve is damaged, the tongue will deviate to the side of the damaged nerve.

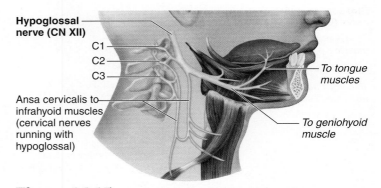

Figure 16.15   **The Hypoglossal Nerve (CN XII).**

Chapter 16: Cranial Nerves

POST-LABORATORY WORKSHEET

1. Write the names of the numbered structures shown on the illustration of the inferior brain in the spaces provided.

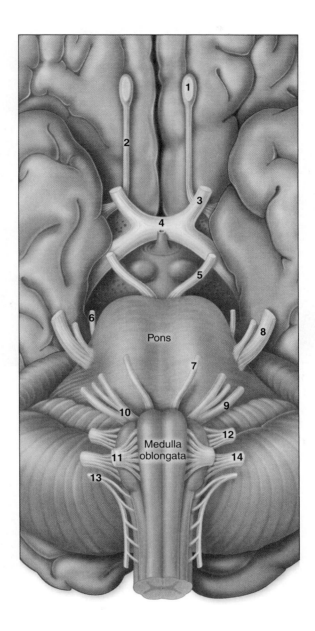

1. _____
2. _____
3. _____
4. _____
5. _____
6. _____
7. _____

8. _____
9. _____
10. _____
11. _____
12. _____
13. _____
14. _____

2. In the chart below, name each of the numbered foramina (shown in the photo of the superior view of the skull) and then list the cranial nerve(s) that travel through each foramen.

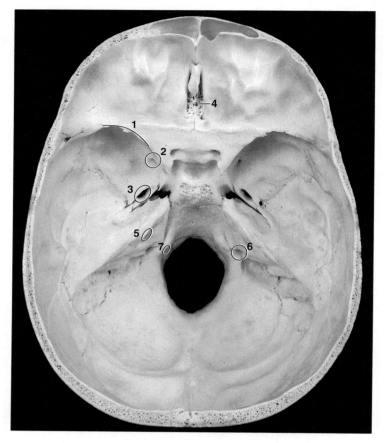

Foramen Number	Foramen Name	Cranial Nerve(s) That Travel(s) Through
1		
2		
3		
4		
5		
6		
7		

3. Four of the 12 cranial nerves carry parasympathetic motor output. In the chart below, list the nerves that carry parasympathetic motor output, and then give the parasympathetic function(s) of each nerve.

Cranial Nerve	Parasympathetic Function(s) of the Nerve

4. Match the disorder listed in column A with the cranial nerve associated with that disorder listed in column B.

Column A

_____ 1. tongue deviates to one side when it is stuck out of the mouth

_____ 2. soft palate droops on one side

_____ 3. weakness in elevation of the scapula

_____ 4. inability to smell

_____ 5. pupillary reflexes are absent

_____ 6. inability to taste bitter (sensed by posterior taste buds of the tongue)

_____ 7. inability to pucker the lips

_____ 8. blindness

_____ 9. inability to laterally rotate the eye

_____ 10. corneal reflex is absent

_____ 11. difficulty turning the eye inferior and lateral

_____ 12. inability to maintain balance and equilibrium

Column B

a. olfactory

b. optic

c. oculomotor

d. trochlear

e. trigeminal

f. abducens

g. facial

h. vestibulocochlear

i. glossopharyngeal

j. vagus

k. spinal accessory

l. hypoglossal

5. Three of the 12 cranial nerves carry somatic motor fibers to extraocular muscles of the eye. In the chart below, list the nerves that carry somatic motor fibers to the extraocular muscles, and then name the muscle(s) innervated by each nerve.

Cranial Nerve	Muscles Innervated

6. Several cranial nerves innervate structures of the tongue. In the chart below, list the nerves that innervate structures of the tongue, and then list the functions of each nerve.

Cranial Nerve	Tongue Structures Innervated by Nerve

The Spinal Cord and Spinal Nerves

17

OUTLINE and OBJECTIVES

Module 7: NERVOUS SYSTEM

INTRODUCTION

When we hear that someone has fractured a vertebra in an accident, we immediately think the worst: "They are going to be paralyzed!" Though paralysis is common when the spinal cord is severed from damage at the fracture site itself, the amount of paralysis and subsequent loss of function is highly dependent upon what part of the spinal cord is injured.

We know that the spinal cord is important for carrying information to and from the periphery and the brain. An understanding of where this information enters and leaves the spinal cord is important for understanding the degree of paralysis that might result from accidents such as this. For instance, the most common vertebral fractures occur in the lower lumbar region (L_3–L_5). Did you know that a fracture here cannot sever the spinal cord and rarely results in paralysis? At the end of this laboratory exercise you will understand why. In contrast, a fracture high in the vertebral column, such as between the atlas (C_1) and axis (C_2), is commonly fatal. Not because of paralysis or loss of sensation from the limbs, but because the nerve that controls the diaphragm (the phrenic nerve) can no longer stimulate the muscle to contract, and breathing ceases.

In the peripheral nervous system injuries most often are the result of peripheral nerve compression or irritation. Some familiar examples are *carpal tunnel syndrome*, which causes pain, weakness, or numbness in the hand, and *sciatica*, a condition in which pain or numbness occurs in the lower back, buttock, posterior thigh, and/or leg and foot. Understanding the mechanisms behind all of these disorders requires an understanding of the organization of the spinal cord and spinal nerves.

In this laboratory session we will investigate the structure and function of the spinal cord and spinal nerves, with a focus on major nerves of the body and the structures they serve.

Chapter 17: The Spinal Cord and Spinal Nerves

Name: _____

Date: _____ Section: _____

PRE-LABORATORY WORKSHEET

1. Which of the three meningeal layers forms the dural venous sinuses?

_____

2. Cerebrospinal fluid flows between

   a. the dura mater and the arachnoid mater.

   b. the dura mater and the pia mater.

   c. the arachnoid mater and the pia mater.

   d. the cranial bones and the dura mater.

3. The spinal cord lies within the _____ canal of the vertebral column.

4. True or False: A nerve is a bundle of myelinated axons. _____

5. True or False: Both cranial and spinal nerves are part of the peripheral nervous system. _____

6. Which of the following sections of the spinal cord does not form a nerve plexus?

   a. cervical

   b. thoracic

   c. lumbar

   d. sacral

7. The muscles in the anterior compartment of the arm share what common action(s)?

_____

8. The muscles in the posterior compartment of the thigh share what common action(s)?

_____

9. The muscles in the anterior compartment of the leg share what common action(s)?

_____

10. Which plexus (cervical, brachial, lumbar, or sacral) is composed of rami, trunks, divisions, and cords?_____

11. Define *tract*:_____

12. Define *ganglion*:_____

13. Somatic sensory neurons are structurally classified as _____ neurons.

14. Somatic motor neurons are structurally classified as _____ neurons.

IN THE LABORATORY

The materials available for laboratory study of the spinal cord and spinal nerves are often somewhat limited. However, this does not mean you cannot develop an appreciation for the anatomy of these organs, even if you must rely on models to do so. Anything that is described on a cadaver in this chapter can also be observed on a classroom model. However, you may need to look at more than one model to see everything. For instance, to see all the parts of the brachial plexus, you may first need to observe a model of the head, neck, and thorax and then view a model of the upper limb. As far as the spinal cord is concerned, understanding the cross-sectional anatomy of the spinal cord is critical for understanding the links between the central and peripheral nervous systems.

Histology

Spinal Cord Organization

Recall that the brain consists of an outer cortex of gray matter and an inner core of white matter (with some nuclei of gray matter as well). The spinal cord is organized just the opposite, with an outer cortex of white matter surrounding an inner core of gray matter. The gray matter of the spinal cord is organized into horns, so named because of their appearance. **Posterior horns** contain axons from somatic sensory neurons that enter the spinal cord through the posterior roots, and also contain the cell bodies of interneurons (association neurons). **Anterior horns** contain cell bodies of somatic motor neurons, whose axons exit the spinal cord through the anterior roots. **Lateral horns**, located only in the thoracic region and first 2 lumbar segments of the spinal cord, contain the cell bodies of sympathetic neurons, whose axons exit the spinal cord through the anterior roots along with axons of somatic motor neurons.

The spinal cord varies in diameter from part to part, and is the largest in the cervical and lumbar parts because of the information flow to and from the upper and lower limbs in these parts. In addition, there is a difference in the relative percentage of gray matter vs. white matter in the different parts of the spinal cord. In general, there is more white matter at the cranial end of the spinal cord and more gray matter at the caudal end of the spinal cord. As you observe histological cross-sections of the spinal cord, make note of the similarities and differences between the different parts of the spinal cord and try to associate these with functions. Also make note of similarities and differences between the organization of gray matter and white matter in the spinal cord as compared to the brain.

EXERCISE 17.1   Histological Cross Sections of the Spinal Cord

1. Obtain slides containing cross sections of the spinal cord (**figure 17.1**).

2. As you observe each slide, first observe it with the naked eye and attempt to identify the different parts of the spinal cord (cervical, thoracic, lumbar, and sacral) based on cross-sectional area and relative amounts of gray vs. white matter. **Table 17.1** compares the general characteristics of tissue sections from each spinal cord part, and **table 17.2** summarizes the features you will observe in a typical spinal cord cross section.

3. Place the slide on the microscope stage and bring the tissue sample into focus on low power.

4. Using figure 17.1, tables 17.1 and 17.2 and your textbook as guides, identify the following structures on the histology slide of the spinal cord.

   ☐ anterior horn   ☐ posterior horn
   ☐ anterior median fissure   ☐ posterior median sulcus
   ☐ anterior root   ☐ posterior root
   ☐ central canal   ☐ posterior root ganglion
   ☐ gray matter   ☐ white matter
   ☐ lateral horn
   ☐ nerve roots of cauda
     equina

5. In the spaces below, draw a simple cross section of each spinal cord part, making note of the major differences between the parts.

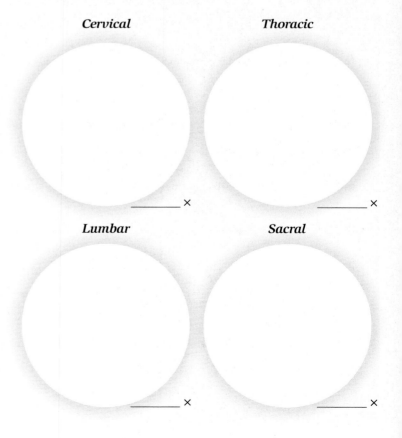

Cervical   *Thoracic*

_____ ✕   _____ ✕

Lumbar   *Sacral*

_____ ✕   _____ ✕

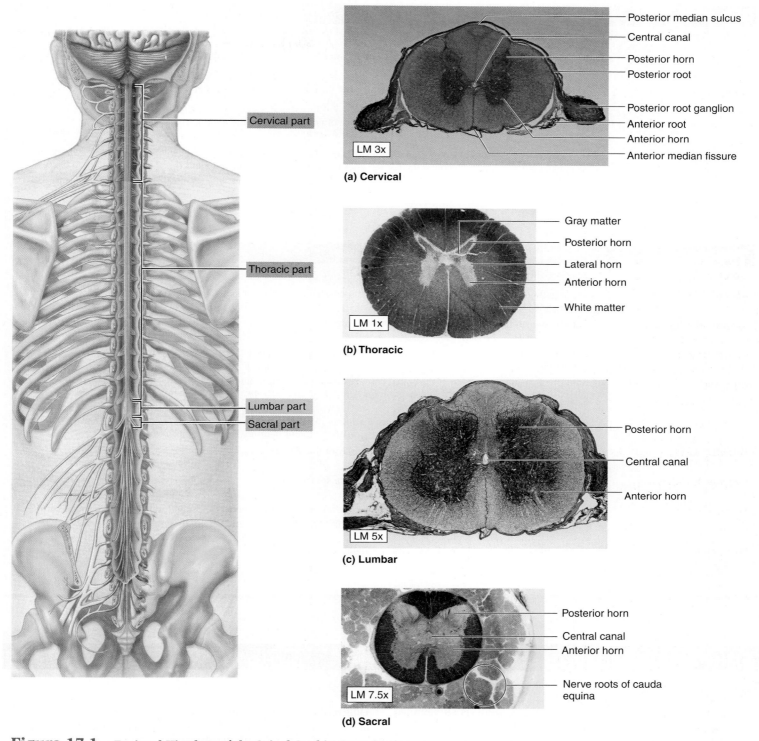

(a) Cervical

- Posterior median sulcus
- Central canal
- Posterior horn
- Posterior root
- Posterior root ganglion
- Anterior root
- Anterior horn
- Anterior median fissure

LM 3x

(b) Thoracic

- Gray matter
- Posterior horn
- Lateral horn
- Anterior horn
- White matter

LM 1x

(c) Lumbar

- Posterior horn
- Central canal
- Anterior horn

LM 5x

(d) Sacral

- Posterior horn
- Central canal
- Anterior horn
- Nerve roots of cauda equina

LM 7.5x

Cervical part
Thoracic part
Lumbar part
Sacral part

Figure 17.1 **Regional Histology of the Spinal Cord in Cross Section.**

Study Tip!

One way to correctly distinguish anterior from posterior horns is to remember that *posterior* horns reach all the way to the *back*. That is, in the posterior horns the gray matter extends all the way to the edge of the spinal cord, whereas in the anterior horns the gray matter does not extend to the edge of the spinal cord.

Table 17.1	Regional Characteristics of the Spinal Cord	
Region of the Spinal Cord	**Relative Size**	**Predominant Tissue**
Cervical	Relatively large cross-sectional area due to an abundance of nerves entering and exiting from the cervical and brachial plexuses.	White matter
Thoracic	Relatively small cross-sectional area. A distinguishing feature is the presence of lateral horns, which contain the cell bodies of sympathetic motor neurons.	White matter
Lumbar	The largest cross-sectional area, due to an abundance of nerves entering and exiting from the lumbar and sacral plexuses. The anterior horns are large.	Gray matter
Sacral	The smallest cross-sectional area. Cross-sections of numerous nerve roots surrounding the spinal cord are usually visible if the section goes through both the spinal cord and the roots of the cauda equina, which run adjacent to the spinal cord.	Gray matter

Table 17.2	Histology of the Spinal Cord in Cross Section	
Structure	**Description**	**Word Origins**
Anterior Horns	Gray matter on the anterolateral part of the spinal cord that carries only motor information.	*anterior*, the front, + *horn*, resembling a horn in shape
Anterior Median Fissure	Deep groove on the anterior surface of the spinal cord.	*anterior*, the front, + *median*, the middle, + *fissure*, a deep furrow
Central Canal	Small hole in the center of the spinal cord that is continuous with the ventricles of the brain.	*central*, in the center, + *canalis*, a duct or channel
Gray Matter	Butterfly-shaped inner region of the spinal cord, consisting of anterior and posterior horns. It contains cell bodies of motor neurons or interneurons, depending upon the location.	literally, a substance that appears gray in color
Lateral Horns	Small lateral extensions of gray matter found mainly in the thoracic region of the spinal cord, and containing cell bodies of sympathetic motor neurons.	*latus*, to the side, + *horn*, resembling a horn in shape
Posterior Horns	Gray matter on the posterolateral part of the spinal cord that carries only sensory information.	*posterior*, the back, + *horn*, resembling a horn in shape
Posterior Median Sulcus	Shallow groove located on the posterior surface of the spinal cord.	*posterior*, the back, + *median*, the middle, + *sulcus*, a furrow
Spinal Nerves	Nerves that form when anterior and posterior roots converge. A spinal nerve exits the vertebral canal through an intervertebral foramina, which forms between adjacent vertebrae.	*spinal*, relating to the spinal cord, + *nevus*, a white, cordlike structure
White Matter	Outer region of the spinal cord consisting of bundles of myelinated axons organized into tracts.	literally, a substance that appears white in color

Gross Anatomy

The Spinal Cord

The spinal cord, part of the central nervous system, is surrounded by the same three meninges that surround the brain: the dura mater, arachnoid mater, and pia mater (figures 17.2 and 17.3). The dura mater surrounding the spinal cord consists of only a single layer of tissue, whereas the dura mater surrounding the brain consists of two layers (periosteal dura and meningeal dura). In addition, there are modifications of the pia mater around the spinal cord: the denticulate ligaments and filum terminale.

Denticulate ligaments are extensions of the pia mater that anchor the spinal cord to the arachnoid mater and dura mater at intervals all along the length of the spinal cord. The **filum terminale** is an extension of the pia mater at the caudal end of the spinal cord that anchors it inferiorly to the coccyx. **Table 17.3** lists some of the key features of the spinal cord that you will identify upon gross observation, and describes their functions.

Table 17.3	Gross Anatomy of the Spinal Cord	
Structure	**Description**	**Word Origins**
*Anterior (Motor) Root**	A nerve root exiting the ventrolateral surface of the spinal cord that contains axons of somatic motor neurons and visceral motor neurons.	*anterior*, the front, + *root*, the beginning part
Cauda Equina	A collection of anterior and posterior roots that extend inferiorly from the lumbar and sacral parts of the spinal cord and lie within the vertebral canal. It is named for the fact that the bundle of nerve roots resembles a horse's tail.	*cauda*, tail, + *equinus*, horse
Conus Medullaris	The cone-shaped distal tip of the spinal cord.	*konos*, cone, + *medius*, middle
Denticulate Ligaments	Extensions of the pia mater located between anterior and posterior roots. These "ligaments" anchor the spinal cord to the arachnoid and dura mater at intervals between the locations where anterior and posterior roots pierce the dura mater.	*denticulus*, a small tooth, + *ligamentum*, a bandage
Filum Terminale	An extension of pia mater that extends beyond the distal end of the spinal cord. It begins at the conus medullaris and attaches to the distal end of the dura mater at the coccyx.	*filum*, thread, + *terminatio*, ending
*Posterior (Sensory) Root**	A nerve root exiting the posterolateral surface of the spinal cord that contains axons of somatic sensory neurons.	*posterior*, the back, + *root*, the beginning part
*Posterior Root Ganglion**	A swelling on the posterior root that contains cell bodies of somatic sensory neurons and satellite cells (glial cells).	*posterior*, the back, + *root*, the beginning part, + *ganglion*, a swelling
Rootlets (Radicular Fila)	Small branches of nerve fibers coming off of the spinal cord that come together to form the anterior and posterior roots.	*radicula*, a spinal nerve root, + *filum*, thread

*The anterior root is also known as the ventral root. The posterior root is also known as the dorsal root.

EXERCISE 17.2 Gross Anatomy of the Spinal Cord

1. Observe a whole spinal cord from a cadaver (meninges intact) or a model of the spinal cord.

2. Using table 17.3 and your textbook as guides, identify the structures listed in **figure 17.2** on a human spinal cord or a model of the spinal cord. Then label them in figure 17.2.

3. Observe a model of a cross section of the spinal cord.

4. Using tables 17.2 and 17.3 and your textbook as guides, identify the structures listed in **figure 17.3** on the spinal cord model. Then label them in figure 17.3.

5. Describe key features on the anterior and posterior surface of the spinal cord (as seen in cross section) that will enable you to distinguish the anterior surface from the posterior surface on laboratory practical exams.

_____

_____

_____

_____

_____

6. *Optional Activity*: AP|R **Nervous System**—Watch the "Typical Spinal Nerve" animation to help you visualize how spinal nerves relate to the spinal cord.

WHAT DO YOU THINK?

1 If a posterior root is severed, what loss of function will result?

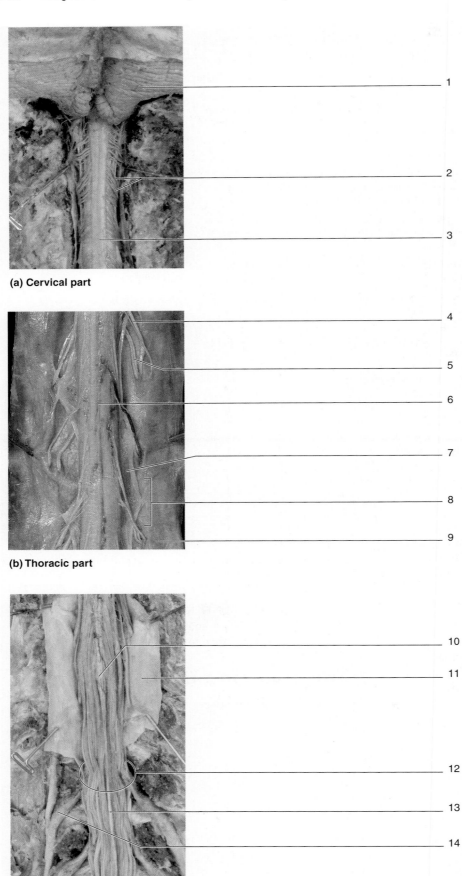

(a) Cervical part

(b) Thoracic part

(c) Lumbar part

Figure 17.2   **Regional Gross Anatomy of the Spinal Cord.**

- ☐ anterior rootlets
- ☐ anterior roots
- ☐ brain (cerebellum)
- ☐ cauda equina
- ☐ conus medullaris
- ☐ denticulate ligament
- ☐ dura mater
- ☐ filum terminale
- ☐ pia mater
- ☐ posterior median sulcus
- ☐ posterior root ganglion
- ☐ posterior rootlets
- ☐ posterior roots

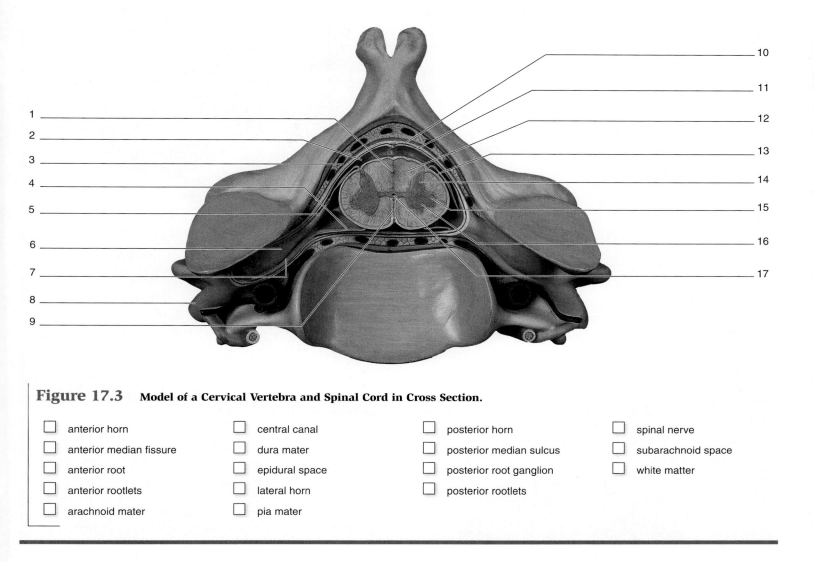

Figure 17.3   **Model of a Cervical Vertebra and Spinal Cord in Cross Section.**

☐ anterior horn

☐ anterior median fissure

☐ anterior root

☐ anterior rootlets

☐ arachnoid mater

☐ central canal

☐ dura mater

☐ epidural space

☐ lateral horn

☐ pia mater

☐ posterior horn

☐ posterior median sulcus

☐ posterior root ganglion

☐ posterior rootlets

☐ spinal nerve

☐ subarachnoid space

☐ white matter

Peripheral Nerves

The peripheral nervous system consists of cranial nerves and spinal nerves. The structure and function of the cranial nerves was covered in chapter 16. In this laboratory exercise we will investigate the structure and function of the **spinal nerves**. After spinal nerves exit the vertebral canal through intervertebral foramina, they immediately branch into posterior and anterior **rami**. Throughout the length of the vertebral column, the **posterior rami** innervate skin and muscles of the back that move the vertebral column (this excludes muscles located on the back that move the pectoral girdle and upper limb). In the thoracic region of the vertebral column, the **anterior rami** become

intercostal nerves that supply skin, bone, and muscle of the thoracic cage. In the cervical, lumbar, and sacral regions of the vertebral column, the anterior rami form complex networks called **plexuses** (*plexus*, a braid), which give off peripheral nerves that supply skin, muscle, and bones of the limbs. There are four major nerve plexuses: cervical, brachial, lumbar, and sacral. In this laboratory exercise we will investigate each of these plexuses and explore the major peripheral nerves that arise from each plexus. Because the lumbar and sacral plexuses share some anterior rami, we will consider them together as the **lumbosacral plexus**.

EXERCISE 17.3   The Cervical Plexus

1. Observe a prosected human cadaver or a model of the head, neck, and thorax demonstrating nerves of the cervical plexus.

2. The cervical plexus arises from the anterior rami of spinal nerves C1–C4. Most of the nerves arising from the cervical plexus carry sensory information from skin

on the lateral side of the neck and motor information to muscles that attach to the hyoid bone. Perhaps the single most important nerve arising from the cervical plexus is the **phrenic nerve**, which innervates the diaphragm (*phrenic*, relating to the diaphragm or mind—historically the diaphragm was thought to be the seat of the mind).

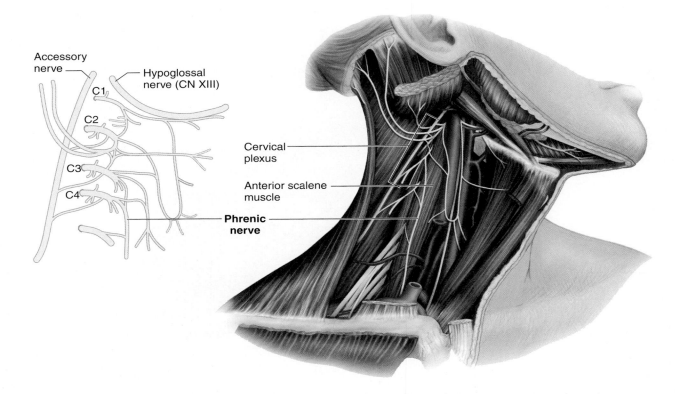

Figure 17.4   **The Cervical Plexus and Phrenic Nerve in the Posterior Triangle of the Neck.**

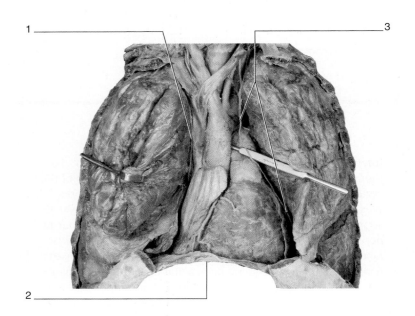

Figure 17.5   **The Phrenic Nerves in the Thoracic Cavity.**

☐ diaphragm   ☐ right phrenic nerve
☐ left phrenic nerve

In the neck, the phrenic nerves can be seen on the inferior part of the anterior scalene muscle, but they are difficult to identify in this location (**figure 17.4**). They are most easily identified within the thoracic cavity (**figure 17.5**). Here they travel within the mediastinum, between the pleural and pericardial cavities. If the heart or lungs have been removed from the cadaver, the phrenic nerves will be easy to identify (if they are still intact). If the heart and lungs remain (or if you are looking at a model of the thorax), you will need to look at the location where the pleural and pericardial cavities meet to locate the phrenic nerves.

3. Using your textbook as a guide, identify the structures listed in figure 17.5 on a human cadaver or a human torso model. Then label them in figure 17.5.

WHAT DO YOU THINK?

❷ As relates to spinal cord injuries, what is an advantage of having the phrenic nerve arise from the cervical plexus instead of arising from the thoracic part of the spinal cord (even though the thoracic spinal cord is closer in physical location to the diaphragm)?

EXERCISE 17.4 The Brachial Plexus

1. Observe a prosected human cadaver or a model of the axilla and upper limb demonstrating nerves of the brachial plexus.

2. The **brachial plexus** arises from the anterior rami of spinal nerves C5–T1. The overall organization of the brachial plexus is complex, but follows an organized pattern. We will investigate the branching pattern of the brachial plexus in detail in this laboratory exercise. However, please note that the branching pattern of the brachial plexus is unique to the brachial plexus. Other plexuses (cervical, lumbar, and sacral) have their own branching patterns, which we will not explore in detail.

3. Table 17.4 describes the organization of the brachial plexus. In this laboratory session, you will focus on identification of the **trunks**, **cords** and **branches** of the brachial plexus. There are three **trunks**, which pass between the anterior and middle scalene muscles of the neck. The **superior trunk** forms from the anterior rami of C5 and C6, the **middle trunk** is a continuation of the anterior ramus of C7, and the **inferior trunk** forms from the anterior rami of C8 and T1. Each trunk divides into two **divisions**, anterior and posterior. All posterior divisions come together to form the **posterior cord**, and all the anterior divisions come together to form the **medial cord** and **lateral cord**. The cords are named for their location relative to the axillary artery. The best way to identify the terminal branches that arise from the medial and lateral cords of the brachial plexus (the musculocutaneous, median, and ulnar nerves) is to first locate the medial and

lateral cords around the axillary artery and then use your fingers to spread apart the terminal branches that arise from the cords. Notice that the connections between the cords and branches appear to form a letter 'M' (**figure 17.6**).

4. *The Posterior Cord*: There are only two major nerves (**table 17.5**) that arise from the posterior cord, the **axillary nerve**, which remains in the axillary region, and the **radial nerve,** a large nerve that continues to travel posteriorly along the arm and forearm and innervate skin and muscle along the way (the *posterior* cord innervates all structures in the *posterior* compartments of the upper limb).

5. *The Medial and Lateral Cords*: The medial and lateral cords form three main terminal nerves, the musculocutaneous, median, and ulnar nerves. The **musculocutaneous nerve** forms from the lateral cord and innervates muscles of the anterior compartment of the arm (biceps brachii, coracobrachialis, and brachialis). It is most easily identified where it pierces through the coracobrachialis muscle. The **ulnar nerve** forms from the medial cord and innervates muscles on the ulnar surface of the forearm along with most

Table 17.4	Organization of the Brachial Plexus		
Structure	**Description**	**Number**	**Names**
Rami	These are the anterior rami of cervical spinal nerves. The rami combine to form trunks as they pass between the anterior and medial scalene muscles of the neck.	5	C5, C6, C7, C8, T1
Trunks	Located between the anterior and middle scalene muscles of the neck.	3	Superior, middle, inferior
Divisions	Each trunk divides into an anterior and posterior division. The posterior divisions of each trunk come together to form the posterior cord. The anterior divisions of each trunk come together to form the medial and lateral cords.	6	Anterior and posterior division (for each trunk)
Cords	Located in the axilla and named for their location relative to the axillary artery.	3	Medial, lateral, and posterior
Branches (Terminal Nerves)	Each terminal nerve innervates a compartment of the arm or forearm, with the exception of the median and ulnar nerves, which both innervate the anterior compartment of the forearm.	5	Axillary, radial, musculocutaneous, median, ulnar

Figure 17.6   Organizational Scheme of the Brachial Plexus.

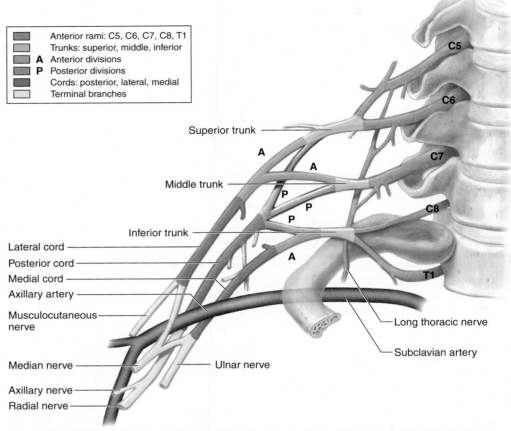

▨	Anterior rami: C5, C6, C7, C8, T1
▨	Trunks: superior, middle, inferior
▨ **A**	Anterior divisions
▨ **P**	Posterior divisions
▨	Cords: posterior, lateral, medial
▨	Terminal branches

Table 17.5	Major Nerves of the Brachial Plexus			
Nerve	**Description**	**Motor Innervation**	**Sensory Innervation**	**Word Origins**
Axillary *(C5–C6)*	Arises from the posterior cord and runs around the surgical neck of the humerus en route to the shoulder. It can be easily injured when the surgical neck of the humerus is fractured.	Deltoid and teres minor	Skin on the dorsolateral aspect of the shoulder	*axilla*, the armpit
Long thoracic *(C5–C7)*	Arises from the roots of C5, C6, and C7 and runs superficial to the serratus anterior muscle, which predisposes it to injury.	Serratus anterior	None	*longus*, long, + *thoracis*, relating to the thorax
Median *(C5–T1)*	Arises from both the medial and lateral cords. Runs along the anteriomedial aspect of the arm, passes deep to the bicipital aponeurosis in the cubital fossa, then enters the forearm. After passing through the forearm, it courses deep to the flexor retinaculum through the carpal tunnel and into the hand. The tight passage through the carpal tunnel predisposes the nerve to injury (carpal tunnel syndrome).	Anterior compartment of the forearm except for the flexor carpi ulnaris and ulnar half of the flexor digitorum profundus	Skin of the lateral three and one-half digits on the palmar surface of the hand (and the distal aspect of the same digits on the dorsal surface), skin of the medial palm proximal to its innervation of the digits	*median*, in the middle
Musculocutaneous *(C5–C7)*	Arises from both the medial and lateral cords of the brachial plexus. It pierces through the coracobrachialis muscle and then runs between the brachialis and biceps brachii muscles.	Anterior compartment of the arm; biceps brachii, brachialis, and coracobrachialis	Skin overlying the lateral surface of the forearm	*muscus*, a mouse (muscle), + *cutaneous*, relating to the skin
Radial *(C5–T1)*	Arises from the posterior cord and runs deep within the posterior compartment of the arm adjacent to the humerus.	Posterior (extensor) compartments of the arm and forearm	Skin overlying the posterior compartments of the arm and forearm and the dorsum of the hand (except the tips of the fingers and the entire fifth digit)	*radialis*, relating to the radius
Ulnar *(C8–T1)*	Arises from the medial cord. Runs along the medial aspect of the arm before becoming superficial immediately posterior to the medial epicondyle of the humerus (where it is easily injured).	Flexor carpi ulnaris and ulnar half of flexor digitorum profundus, as well as most intrinsic hand muscles	Skin of the medial one and one-half digits on the palmar surface of the hand, skin of the lateral surface of the hand on both palmar and dorsal surfaces	*ulnar*, relating to the ulna

intrinsic muscles of the hand. It is most easily identified where it passes superficially behind the medial epicondyle of the humerus. In this location it is vulnerable to injury. When it is struck, as when you bang your elbow against a counter, you feel a tingly sensation along the course of the nerve. For this reason, this region of the elbow supplied by the ulnar nerve is commonly referred to as the "funny bone." The **median nerve** forms from branches of both medial and lateral cords. The median nerve is most easily identified in the distal, anterior compartment of the wrist, where it passes through the **carpal tunnel** into the hand. If your laboratory has a cadaver, see if you can pass a blunt probe through the carpal tunnel alongside the median nerve. This will allow you to better understand why the median nerve is so easily irritated by repetitive motions of the wrist. Its passage through the carpal tunnel is very narrow. As the flexor muscles of the forearm contract, their tendons rub against the median nerve, causing inflammation that results in **carpal tunnel syndrome.**

6. *Additional Nerves of the Brachial Plexus:* The **long thoracic nerve** arises from the anterior rami of C5–C7 and innervates the **serratus anterior muscle.** Unlike most nerves, which lie deep to the muscles they innervate, the long thoracic nerve lies superficial to the muscle, which makes it susceptible to injury. Injury can result in paralysis of the serratus anterior muscle. Remembering the discussion in chapter 13, describe how to diagnose a paralyzed or nonfunctional serratus anterior muscle.

_____

_____

7. Using tables 17.4 and 17.5 and your textbook as a guide, identify the structures of the brachial plexus listed in **figure 17.7** on a human cadaver or on models of the upper limb. Then label them in figure 17.7.

8. In the space below, make a simple line drawing of the brachial plexus and label the structures using the list in figure 17.7.

Study Tip!

The *cord* of the brachial plexus is *medial,* as in located medial to the axillary artery; the *nerve* is *median,* as in located in the middle (of the branches that form from medial and lateral cords).

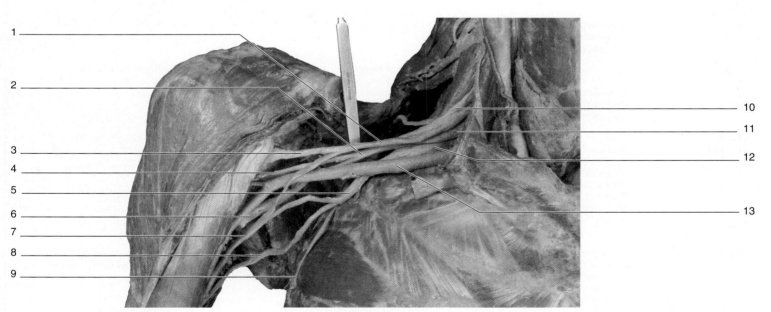

Right axilla, anterior view

Figure 17.7   **Major Nerves of the Brachial Plexus.**

- ☐ axillary artery
- ☐ axillary nerve
- ☐ inferior trunk
- ☐ lateral cord

- ☐ long thoracic nerve
- ☐ medial cord
- ☐ median nerve
- ☐ middle trunk

- ☐ musculocutaneous nerve
- ☐ posterior cord
- ☐ radial nerve

- ☐ superior trunk
- ☐ ulnar nerve

EXERCISE 17.5   The Lumbosacral Plexus

1. Observe a prosected human cadaver or a model of the abdomen and lower limb demonstrating nerves of the lumbosacral plexus.

2. The lumbosacral plexus is really two plexuses, the **lumbar plexus** (arising from the anterior rami of L1–L4) and the

sacral plexus (arising from the anterior rami of L4–S4). The branching pattern of the nerves is unique, as with the other plexuses in the body. However, we will not concern ourselves with the branching pattern in this laboratory exercise. We will focus on the major nerves that arise from the lumbosacral plexus (**table 17.6**).

Table 17.6	Major Nerves of the Lumbosacral Plexus			
Nerve	**Description**	**Motor Innervation**	**Sensory Innervation**	**Word Origins**
Common Fibular (S) (L4–S2)	Begins at the bifurcation of the sciatic nerve proximal to the popliteal fossa, passes superficially near the head of the fibula, wraps around the neck of the fibula, and then divides into superfcial and deep branches.	Biceps femoris muscle, short head	Skin on the proximal posterolateral surface of the leg	*fibular*, relating to the fibula
Deep Fibular (S) (L4–S1)	Arises from the common fibular nerve at the neck of the fibula, passes through the extensor digitorum longus muscle into the anterior compartment of the leg.	Anterior compartment of the leg, dorsal musculature of the foot	Skin on the first interdigital cleft of the foot	*deep*, situated at a deeper level than a corresponding structure, + *fibular*, relating to the fibula
Femoral (L) (L2–L4)	Runs along the lateral border of the psoas major muscle, travels under the inguinal ligament into the femoral triangle to innervate the anterior compartment of the thigh.	Anterior compartment of the thigh	Skin of the anterior thigh and leg	*femoral*, relating to the femur or thigh
Inferior Gluteal (S) (L5–S2)	Exits the pelvis through the greater sciatic foramen inferior to the piriformis.	Gluteus maximus	NA	*inferior*, lower, + *gloutos*, the buttock
Obturator (L) (L2–L4)	Runs along the medial border of the psoas major muscle, travels through the obturator foramen into the medial compartment of the thigh.	Medial compartment of the thigh	Skin on the medial surface of the thigh	*obturatus*, to occlude or stop up
Pudendal (S) (S2–S4)	Exits the pelvis through the greater sciatic foramen inferior to the piriformis muscle, then travels through the lesser sciatic foramen to enter the perineum.	Perineal muscies, external anal sphincter, and external urethral sphincter	External genitalia in both males and females	*pudendal*, that which is shameful
Sciatic (S) (L4–S3)	Exits the pelvis through the greater sciatic foramen inferior to the piriformis muscle, then enters the posterior compartment of the thigh through a groove between the ischial tuberosity and greater trochanter of the femur.	Posterior compartment of the thigh (the tibial division of the sciatic is responsible for innervation of all posterior compartment muscles except for the short head of biceps femoris)	NA	*sciaticus*, the hip joint
Superficial Fibular (S) (L5–S2)	Arises from the common fibular nerve at the neck of the fibula and descends within the lateral compartment of the leg.	Lateral compartment of the leg	Skin on the distal, lateral surface of the leg and the dorsal surface of the foot	*superficialis*, the surface, + *fibular*, relating to the fibula
Superior Gluteal (S) (L4–S1)	Exits the pelvis through the greater sciatic foramen superior to the piriformis.	Gluteus medius, gluteus minimus, and tensor fascia lata muscles	NA	*superus*, above, + *gloutos*, the buttock
Tibial (S) (L4–S3)	Begins at the bifurcation of the sciatic nerve proximal to the popliteal fossa, runs along the tibialis posterior muscle, and then branches into two plantar nerves at the ankle.	Posterior compartment of the leg, plantar musculature of the foot	NA	*tibial*, relating to the tibia

S = Sacral; L = Lumbar

3. *Nerves Within the Gluteal Region*: The **gluteal nerves** (superior and inferior) are named for their exit location relative to the piriformis muscle of the deep buttock (**figure 17.8**). The **superior gluteal nerve** arises superior to the piriformis, while the **inferior gluteal nerve** arises inferior to the piriformis. A small, but very important, nerve that also arises in this region is the **pudendal nerve** (*pudendal*, that which is shameful).

> ## Study Tip!
>
> — If you remember that the word *pudendal* means "that which is shameful," it will be easy to remember the structures it innervates. It innervates all of our "shameful" structures—namely, the external genitalia. In addition, its fibers control the contraction of the external sphincters surrounding the anus and urethra. Wouldn't it be "shameful" if this nerve were damaged? A very minimal anesthetic given during childbirth is a **pudendal nerve block**. It blocks sensation from the birth canal (vagina), but the mother still receives sensations from the contracting uterus.

4. Using table 17.6 and your textbook as a guide, identify the nerves and muscles in the gluteal region listed in **figure 17.8** on a human cadaver or a model of the gluteal region. Then label them in figure 17.8.

5. The **sciatic nerve** deserves special attention. The sciatic nerve arises inferior to the **piriformis** muscle and travels into the posterior compartment of the thigh. Along the way, it passes close to the ischial tuberosity. If you sit on a hard surface for a long period of time, the areas served by the nerve can lose sensation due to a blockage of blood supply to the fibers within the nerve. Once pressure is relieved and blood flow is restored, the sensations return as a painful "pins and needles" sensation. The next time this happens to you, note that the pins-and-needles sensation occurs on the posterior thigh and in the entire leg. The anterior and medial thigh are spared, because they receive innervation from the femoral and obturator nerves, respectively. The sciatic is the *largest nerve in the body*. It is very large because it is actually two nerves, the **common fibular nerve** and the **tibial nerve**. These two nerves are bundled together in a common connective tissue sheath. The section where the two are bundled together is what we refer to as the sciatic nerve proper. Most commonly the two nerves separate from each other proximal to the popliteal fossa of the knee. However, it is not uncommon for them to separate much higher up in the posterior thigh. It is the **tibial division** of the sciatic that is actually responsible for innervating the posterior compartment of the thigh. However, because the nerve remains bundled as part of the sciatic, we usually say it is the **sciatic nerve** that innervates the posterior compartment of the thigh.

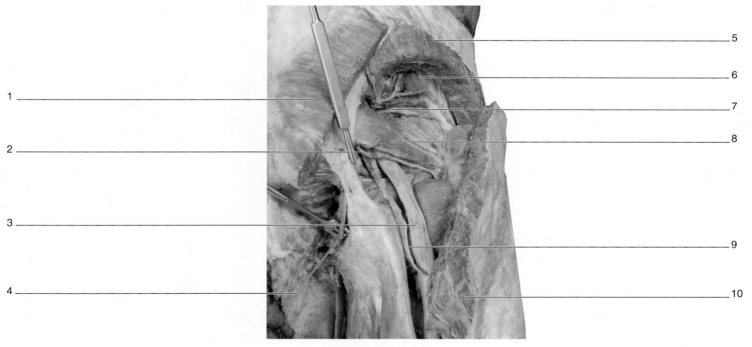

Right gluteal region

Figure 17.8   **Nerves of the Lumbosacral Plexus Within the Gluteal Region.**

- ☐ gluteus maximus muscle
- ☐ gluteus medius muscle
- ☐ gluteus minimus muscle
- ☐ inferior gluteal nerve
- ☐ piriformis muscle
- ☐ posterior femoral cutaneous nerve
- ☐ pudendal nerve
- ☐ sciatic nerve
- ☐ superior gluteal nerve

6. *Nerves of the Lower Limb*: It will be easiest to learn the nerves of the lower limb if you can relate each nerve to a single compartment. Here is where your knowledge of limb compartments from chapter 13 becomes very important. Recall that there are three compartments of the thigh (anterior, posterior, and medial) and three compartments of the leg (anterior, posterior, and lateral). Each compartment, for the most part, receives innervation from a single nerve of the lumbosacral plexus. If you first associate one nerve with one compartment, you will then be prepared to learn the "exceptions" to the rules. The general rules are:

Compartment	Nerve
Anterior thigh	Femoral nerve
Posterior thigh	Sciatic nerve (tibial division)
Medial thigh	Obturator nerve
Anterior leg	Deep fibular nerve
Posterior leg	Tibial nerve
Lateral leg	Superficial fibular nerve

7. Using table 17.6 and your textbook as a guide, identify the nerves and muscles listed in **figure 17.9** in the lower limb of a human cadaver or a model of the lower limb. Then label them in figure 17.9.

Figure 17.9 **Nerves of the Lumbosacral Plexus.**
(a) Pelvis and anterior thigh. (b) Posterior view of lower limb showing popliteal region.

- ☐ common fibular nerve
- ☐ femoral nerve
- ☐ genitofemoral nerve
- ☐ iliohypogastric nerve
- ☐ ilioinguinal nerve
- ☐ lateral femoral cutaneous nerve
- ☐ lateral sural cutaneous nerve
- ☐ medial sural cutaneous nerve
- ☐ obturator nerve
- ☐ subcostal nerve
- ☐ tibial nerve

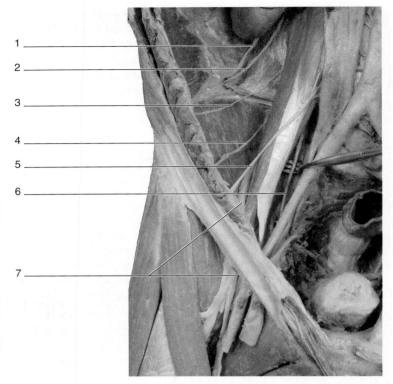

(a) Right pelvic region, anterior view

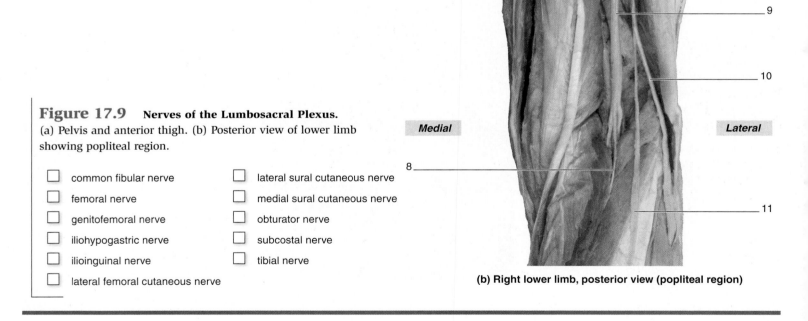

Medial   **Lateral**

(b) Right lower limb, posterior view (popliteal region)

Chapter 17: The Spinal Cord and Spinal Nerves

POST-LABORATORY WORKSHEET

Name: _____

Date: _____ Section: _____

1. Compare and contrast the features of the brain and spinal cord by filling in the following table.

Structure	Unique Features Within the Brain	Unique Features Within the Spinal Cord
Dura mater		
Arachnoid mater		
Pia mater		
Central space (for example, ventricles)		
Location of gray matter		
Location of white matter		

2. Explain, in your own words, why the cervical region of the spinal cord has more white matter than the other regions of the spinal cord.

_____

_____

_____

_____

_____

_____

3. Which parts of the spinal cord contain lateral horns? _____

What structures are located within the lateral horns? _____

_____

4. Fill in the table below with features of the main nerve plexuses.

Plexus	Formed from Anterior Rami of These Spinal Nerves	Major Nerves Formed from the Plexus
Cervical		
Brachial		
Lumbosacral		

5. In your own words, describe the organization (branching scheme) of the brachial plexus.

_____

_____

_____

_____

6. Fill in the following table with the nerve that innervates each compartment or listed structure.

Location and/or Muscles	Nerve
Anterior compartment of the arm	
Anterior compartment of the forearm	
Anterior compartment of the leg	
Anterior compartment of the thigh	
Deltoid and teres minor muscles	
External genitalia, external anal and urethral sphincter muscles	
Gluteus maximus muscle	
Gluteus medius, gluteus minimus, and tensor fascia lata muscles	
Lateral compartment of the leg	
Medial compartment of the thigh	
Medial forearm and most intrinsic hand muscles	
Posterior compartment of the arm	
Posterior compartment of the forearm	
Posterior compartment of the leg	
Posterior compartment of the thigh	
Serratus anterior muscle	

7. An individual's spinal cord is severed between the first and second cervical vertebrae. This is a life-threatening situation. Given what you know about the peripheral nerves that exit the spinal cord in this region, explain why the individual will die without immediate medical intervention. (Hint: What spinal nerve exits at this location, and how does it relate to the major nerve plexuses of the body?)

_____

_____

_____

_____

_____

_____

8. Using colored pencils, color in the segments of the brachial plexus in the illustration below as per the colors in the key. Then label the nerves indicated by the leader lines.

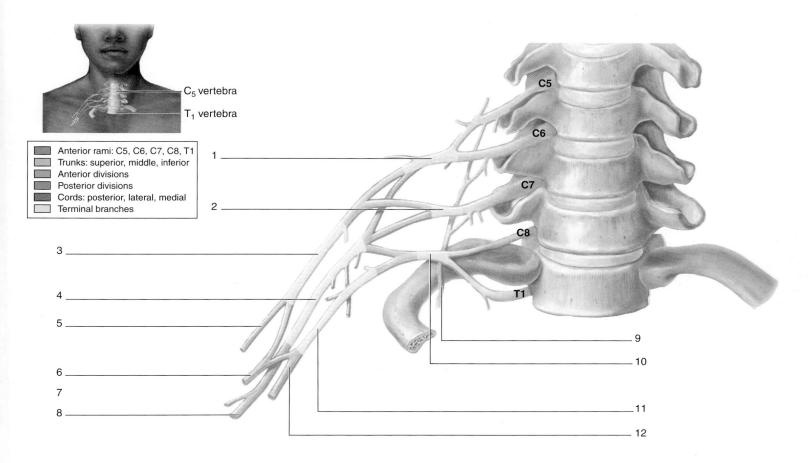

C₅ vertebra

T₁ vertebra

Anterior rami: C5, C6, C7, C8, T1
Trunks: superior, middle, inferior
Anterior divisions
Posterior divisions
Cords: posterior, lateral, medial
Terminal branches

C5

C6

C7

C8

T1

1 _____

2 _____

3 _____

4 _____

5 _____

6 _____

7 _____

8 _____

9 _____

10 _____

11 _____

12 _____

18

The Nervous System: General and Special Senses

OUTLINE and OBJECTIVES

Module 7: NERVOUS SYSTEM

INTRODUCTION

Your ability to read this chapter and perform the laboratory activities within is absolutely dependent upon the proper functioning of your sensory systems. You rely on the special sense of vision to read the words typed on the page. In the laboratory classroom, you rely on the special sense of hearing to hear the instructions delivered to you and to discuss the material with your classmates. You rely on the special sense of equilibrium to maintain your body's three-dimensional position in space so you don't fall over as you move about the classroom. In addition, sensory receptors in your skin constantly perceive things such as touch, pain, pressure, and the temperature of the environment around you, thus allowing you to orient yourself in the world and protect your body from harm.

Indeed, if you are reading this chapter introduction at home, you may be enjoying a cup of coffee or other food or drink as you read. Your enjoyment of such things is highly dependent upon the special senses of olfaction (smell) and gustation (taste). In fact, our enjoyment of things like music, warm baths, food, drink, and a good book are all completely dependent on the multitude of sensory receptors located throughout our bodies and their ability to receive signals from our environment. Of course, it is our brain that actually *interprets* such signals. However, without the appropriate input, there would be nothing at all for our brains to interpret. In this laboratory session we will explore the intricate structure and function of many of the amazing, beautiful, and incredibly *functional* sensory receptors and organs found throughout our bodies.

Chapter 18: General and Special Senses

Name: _____

Date:_____ Section:_____

PRE-LABORATORY WORKSHEET

1. Explain the difference between *general senses* and *special senses*.

_____

_____

_____

_____

_____

2. List the five special senses.

_____

_____

_____

_____

_____

3. Of the five special senses you listed in question 2, which two senses have receptor organs located within the petrous part of the temporal bone?

_____

_____

4. Tactile corpuscles are located within the _____ layer of the dermis.

5. Lamellated corpuscles are located within the _____ layer of the dermis.

6. The optic nerve carries the special sense of _____.

7. The olfactory nerves are located within the _____ of the ethmoid bone.

8. The vestibulocochlear nerve carries the special senses of _____ and _____.

9. Taste receptors on the posterior third of the tongue are innervated by the _____ nerve, while taste receptors on the anterior two-thirds of the tongue are innervated by the _____ nerve.

IN THE LABORATORY

In this laboratory exercise we will explore the histology of a select few somatic sensory receptors and special sensory organs of the body. We will then explore the gross structure of two special sensory organs (the eye and ear) through observations of classroom models and by dissection of a cow eye.

Histology

General Senses

All of our senses are classified as either **general senses** or **special senses**. **General senses** are transmitted through receptors that sense modalities such as temperature, pain, pressure, light touch, and the concentration of chemicals like oxygen, carbon dioxide, and hydrogen ion. General sensory receptors transmit either autonomic (visceral) or somatic sensation. **Visceral sensory receptors** respond to such things as changes in blood pressure, pH, and carbon dioxide levels in the blood. **Somatic sensory**

receptors respond to such things as light touch, temperature, pain, and pressure. In general, somatic sensory receptors sense things we are consciously aware of. Because we are consciously aware of the input coming from these sensory receptors, we can begin to understand the relationship between the structure of the sensory receptor and its associated function more easily than we can with autonomic senses. For this reason, we will focus only on general **somatic** senses the next two laboratory exercises. Specifically, we will focus on understanding the structure and function of sensory receptors found in the skin (**table 18.1**).

Table 18.1	Sensory Receptors in Thick Skin			
Receptor	**Location**	**Structure**	**Senses**	**Word Origins**
Free Nerve Ending	In the dermis, ending in glands and hair follicles, with some extending into the epidermis.	An unmodified, unencapsulated nerve ending.	Sustained touch, pain, temperature, itching, and hair movement.	*free*, referring to the fact there are no connective tissue coverings
Lamellated (Pacinian) Corpuscle	Deep within the reticular layer of the dermis.	Concentric layers of connective tissue surrounding a nerve ending at the core.	Deep pressure and high frequency vibration.	*lamina*, plate, + *corpus*, body
Tactile (Meissner) Corpuscle	Within dermal papillae.	An oval structure consisting of a few layers of connective tissue encapsulating a nerve ending.	Fine touch.	*tactus*, to touch, + *corpus*, body
Tactile Disc (Merkel Disc)	At the junction between the dermis and the epidermis.	An association between a modified keratinocyte in the epidermis, called a tactile cell, with a specialized nerve ending in the dermis, called a tactile disc.	Detect fine, light touch and texture.	*tactus*, to touch

EXERCISE 18.1 Tactile (Meissner) Corpuscles

1. Obtain a histology slide of thick skin and place it on the microscope stage. Bring the tissue sample into focus using the scanning objective.

2. Observe the slide on low power and review the layers of the skin (**figures 6.1 and 18.1**).

   Epidermis: stratum basale, stratum spinosum, stratum granulosum, stratum lucidum, and stratum corneum
   Dermis: papillary layer and reticular layer

3. Move the microscope stage so the junction between the dermis and epidermis is in the center of the field of view, and locate the **dermal papillae** (figure 18.1). Next, move the stage so a single papilla is in the center of the field of view, then change to high power.

4. Sensory receptors called **tactile corpuscles** are located within the dermal papillae of thick skin. These receptors are oval in shape with a surrounding capsule of connective tissue (figure 18.1 and table 18.1), and they function in sensing light touch. Look for tactile corpuscles within the dermal papilla. If you don't see one, scan the slide for other papillae that may have tactile corpuscles within.

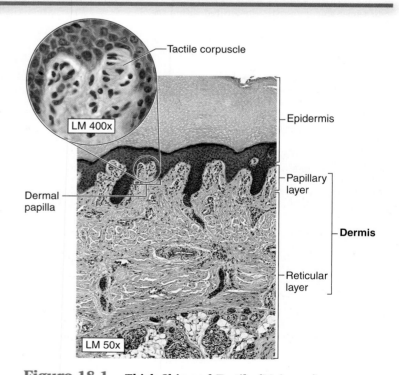

Figure 18.1   Thick Skin and Tactile (Meissner) Corpuscles.   Tactile corpuscles are found within the dermal papillae of the skin and are sensory receptors for fine touch.

5. In the space below, make a simple drawing of a tactile corpuscle within a dermal papilla.

_____ ×

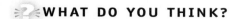
WHAT DO YOU THINK?

1 Why do you think tactile (Meissner) corpuscles are located relatively close to the surface of the skin rather than deep within the dermis? (What advantage does this serve?)

EXERCISE 18.2 Lamellated (Pacinian) Corpuscles

1. Obtain a histology slide of thick skin and place it on the microscope stage.

2. Observe the slide on low power and identify the dermis and epidermis (figure 18.1).

3. Move the microscope stage so the deepest part of the reticular layer of the dermis is in the center of the field of view. Within this portion of the dermis you will see cross sections through numerous sweat glands and blood vessels. You will also see cross sections of **lamellated corpuscles**. Lamellated corpuscles resemble onions in cross section because they have several layers of cells forming circular layers around a central sensory receptor (**figure 18.2 and table 18.1**).

4. Locate a lamellated corpuscle and then move the microscope stage so the lamellated corpuscle is in the center of the field of view. Change to a higher power to view its structure in more detail. Lamellated corpuscles function in the sensation of deep pressure. When enough pressure is applied to the surface of the skin, the layers of connective tissue surrounding the central sensory receptor are compressed, and the receptor sends action potentials to the brain.

5. In the space below, make a simple drawing of a lamellated corpuscle, making note of its location in the skin.

_____ ×

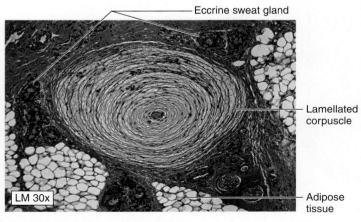

Figure 18.2 **Lamellated Corpuscles.** Lamellated corpuscles are located deep within the reticular layer of the dermis and are sensory receptors for deep pressure.

WHAT DO YOU THINK?

2 Why do you think lamellated corpuscles are located deep within the dermis instead of relatively close to the surface of the skin?

Special Senses

Special senses are transmitted through specialized organs and cells in the body that respond specifically to the modalities of olfaction, taste, vision, hearing, and equilibrium. Each special sensory organ has a relatively complex structure that allows it to function in its specific capacity. In the next set of laboratory exercises we will explore the structure and function of several of these special sensory organs.

EXERCISE 18.3 Gustation (Taste)

Gustation (taste) is one of the most pleasurable sensations humans can experience. However, to have a complete sense of gustation, our olfactory sense must also be intact. In fact, without the ability to smell, the ability to taste suffers tremendously. Perhaps one reason gustation is so pleasurable is that the gustatory and olfactory pathways are tied into the limbic system, our emotional brain.

The sensory receptor specialized for gustatory sensation is a **taste bud** (gustatory bud) (**table 18.2**). Taste buds are located throughout the oral cavity and pharynx, but are particularly concentrated on the tongue, and are associated with the raised epithelial ridges on the tongue (papillae). In this exercise we will explore the types of tongue papillae and the location and function of taste buds associated with the papillae.

There are four types of papillae located on the tongue: filiform, fungiform, vallate (circumvallate), and foliate. The detailed structure, function, and locations of the tongue papillae are shown in **figure 18.3** and summarized in **table 18.3**. The largest of the papillae are the **foliate** and **vallate (circumvallate)** papillae. In addition to being the largest papillae, the vallate papillae house more than half of our taste buds in the walls of the crypts that surround each papilla. Thus, we will focus our histological observations on the taste buds associated with the vallate papillae.

1. Obtain a histology slide of the tongue or a histology slide demonstrating mammalian *vallate papillae* (figure 18.3*d*). Scan the slide at low power and identify the papillae on the surface of the tongue. Move the microscope stage so one or two papillae are in the center of the field of view. Then increase the magnification, first to medium and then to high power. Locate a taste bud at the edge of the papilla (**figure 18.4**).

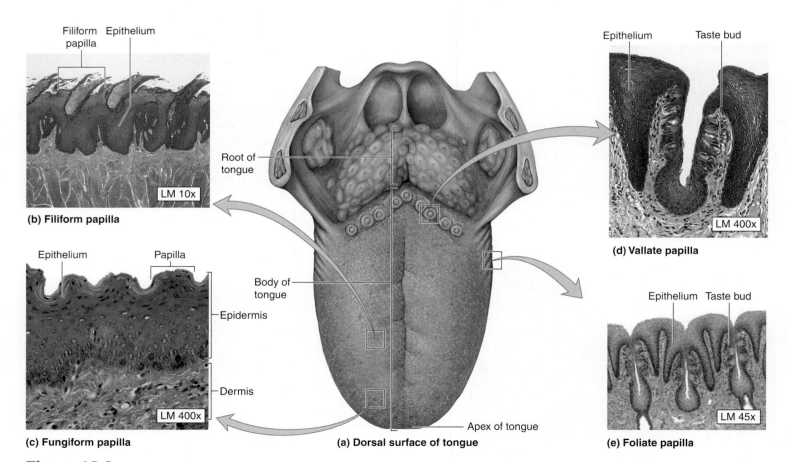

Figure 18.3 **Taste Buds.** Gustation (taste) requires taste buds, which are associated with tongue papillae. (a) Dorsal surface of tongue, (b) Filiform papilla, (c) Fungiform papilla, (d) Vallate papilla, (e) Foliate papilla.

Table 18.2	Cells Associated with Taste Buds	
Structure	**Description**	**Function**
Supporting Cells (Sustentacular Cells)	Located between the gustatory cells, these cells have more oval-shaped nuclei and stain darkly.	Support the gustatory cells by producing a glycoprotein; may also function in taste sensation.
Gustatory Cells	These cells have round nuclei and are very light in color. They are epithelial cells with a taste hair on the apical surface, which is composed of microvilli twisted around each other.	Receive information in the form of chemicals dissolved in solution; transmit taste information to the CNS by forming synapses with gustatory nerve endings.
Basal Cells	These are small stem cells found at the base of the taste bud.	Precursor cells to the supporting cells and gustatory cells.

Table 18.3	Tongue Papillae			
Type of Papilla	**Location**	**Description**	**Function**	**Word Origins**
Filiform	Anterior two-thirds of the tongue.	The most numerous of the papillae. They are conical in shape and contain sensory nerve endings.	General sensation.	*filum*, thread, + *forma*, shape
Fungiform	Anterior part of the tongue; most numerous at the tip.	Less numerous than filiform papillae. They are shaped like mushrooms and contain some taste buds.	Taste sensation.	*fungus*, a mushroom, + *forma*, shape
Foliate	Sides of the tongue.	Large papillae with deep crypts that house taste buds in their walls in infants and children.	Taste sensation.	*foliate*, resembling a leaf
Vallate (Circumvallate)	Posterior part of the tongue along the sulcus terminalis.	The largest papillae. They contain deep crypts that house taste buds in their lateral walls.	Taste sensation.	*circum-*, around, + *vallum*, wall

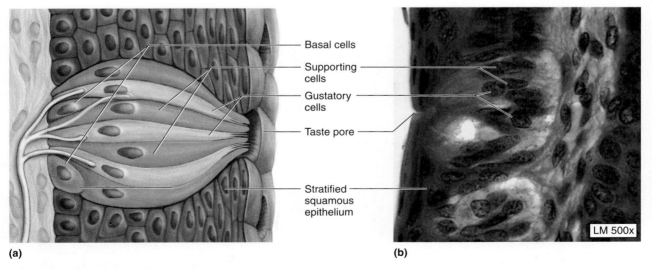

(a)   (b)

Figure 18.4   **Detailed Structure of a Taste Bud.** (a) Illustration. (b) Photomicrograph.

2. Using figures 18.3 and 18.4, and tables 18.2 and 18.3 as guides, identify the following structures (note: you may not be able to identify all of the types of papillae on a single slide):

☐ basal cells
☐ filiform papillae
☐ foliate papillae
☐ fungiform papillae
☐ gustatory cells

☐ supporting cells
☐ taste pores
☐ vallate (circumvallate) papillae

3. Locate a taste bud at the edge of the papilla in the crevice (crypt) between two papillae (figure 18.4).

4. In the space below, make a simple line drawing of the histology of a vallate papilla, noting the location of the taste buds.

_____ ×

Olfaction is the sense of smell. It is an important sensory modality, not just for smell, but also for gustation (taste). Olfactory sensation is received by special sensory cells found within the epithelium lining the roof of the nasal cavity—**olfactory epithelium** (**figure 18.5***a* and **table 18.4**). The **olfactory receptor cells** are neurons, but they are unique because they are continuously replaced. The olfactory receptor cells compose the **olfactory nerves (CN I)**. The axons of the olfactory nerves travel through the **cribriform foramina** within the cribriform plate of the ethmoid bone to synapse with neurons within the **olfactory bulbs**, which subsequently transmit the information to the brain via the **olfactory tract**.

1. Obtain a histology slide of olfactory epithelium (figure 18.5*b*) and place it on the microscope stage.

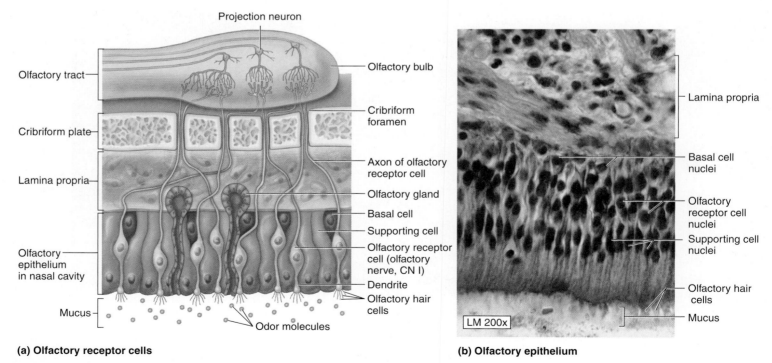

(a) **Olfactory receptor cells**

(b) **Olfactory epithelium**

Figure 18.5   **Olfactory Epithelium.**   (a) Cell types within the olfactory epithelium. (b) Histology of the olfactory epithelium.

Table 18.4	Olfactory Epithelium		
Structure	**Description**	**Function**	**Word Origins**
Basal Cells	Cuboidal (triangular) cells whose nuclei are located near the basal surface of the olfactory epithelium.	Thought to be the precursor cells to the olfactory receptor cells and supporting cells.	*basalis*, situated near the base
Cilia	Surface modifications of epithelial cells that contain microtubules; specialized cilia in that they are not motile.	Apical ends are the site of interaction between dissolved odiferous substances and the olfactory receptor cells.	*cilium*, eyelash
Cribriform Plate of the Ethmoid Bone	The superior portion of the ethmoid bone containing olfactory foramina.	Forms the roof of the nasal cavity.	*cribriform*, shaped like a sieve, + *ethmoid*, resembling a sieve
Cribriform Foramina	Numerous small holes in the cribriform plate of the ethmoid.	Allow for the passage of axons of the olfactory receptor cells from the olfactory epithelium to the olfactory bulb, where they will synapse with other neurons.	*olfactus*, to smell, + *foramen*, to pierce
Olfactory Bulb	A white swelling on the superior and anterior aspect of the cribriform plate of the ethmoid bone.	The location where axons from olfactory receptor cells synapse with other neurons that will carry the information to the brain via the olfactory tract.	*olfactus*, to smell, + *bulb*, a globular structure
Olfactory Receptor Cells	Columnar cells whose large nuclei are located in the middle of the olfactory epithelium (between apical and basal surfaces) and have cilia on their apical surfaces.	Bipolar neurons that function as sensory receptors for olfaction (smell).	*olfactus*, to smell, + *recipio*, to receive, + *cella*, a chamber
Olfactory Tract	A white extension of the olfactory bulb that lies superior to the frontal bone and inferior to the frontal lobe of the brain.	Consists of myelinated axons of neurons that originate in the olfactory bulb and carry olfactory information to the brain.	*olfactus*, to smell, + *tractus*, a drawing out
Supporting (Sustentacular) Cells	Columnar cells whose nuclei are located near the apical surface of the olfactory epithelium.	Surround and support the specialized olfactory receptor cells.	*supporto*, to carry, + *cella*, a chamber

2. Bring the tissue sample into focus on low power. Increase magnification to medium power, and then to high power so you can clearly see the cells composing the olfactory epithelium. It will be difficult to distinguish between the three major cell types of the olfactory epithelium: basal cells, olfactory receptor cells, and supporting cells. In general, the nuclei closest to the basement membrane of the epithelium are the nuclei of *basal cells* (the cells appear triangular in shape), the nuclei in the middle of the epithelium are the nuclei of *olfactory receptor cells*, and the nuclei closest to the apical surface of the epithelium are the nuclei of *supporting cells*.

3. Using figure 18.5 and table 18.4, as guides, identify the following structures on the histology slide of olfactory epithelium:

☐ basal cells   ☐ olfactory receptor cells
☐ cilia   ☐ supporting cells

4. In the space below, make a simple drawing of the olfactory epithelium, labeling all of the structures listed in step 3.

_____ ✕

EXERCISE 18.5 Vision (The Retina)

The **retina** (**figure 18.6**) of the eye is also referred to as the **neural tunic**, because this layer is composed of neural tissue. In fact, the retina develops as a direct outgrowth of the brain. Thus, the retina is the only part of the brain visible to us without surgical intervention (though you do need an ophthalmoscope). Axons from neurons within the retina travel to the brain through the **optic nerve (CN II)**. The retina is responsible for translating information that comes into the eye as light rays into electrical signals (action potentials) that the brain can "read." This information is relayed by the optic nerves to the **thalamus** and then on to the **occipital lobe** of the brain, where the image is truly processed and "visualized." The retina is a very complex yet beautifully organized structure. In this laboratory exercise we will explore its structure and function at a very basic level by observing the cells that are visible histologically. In the gross anatomy section of this laboratory exercise you will place the retina in context of other structures of the eye.

1. Obtain a histology slide of the retina and place it on the microscope stage.

2. Bring the tissue sample into focus on low power, then move the microscope stage so the retina (figure 18.6) is in the center of the field of view. Switch to medium power and bring the tissue sample into focus once again.

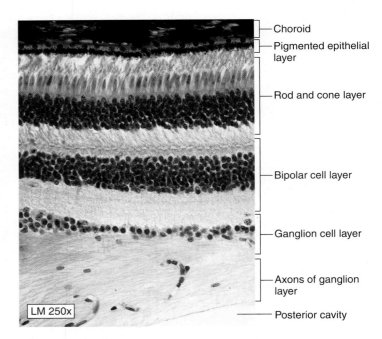

Choroid
Pigmented epithelial layer
Rod and cone layer
Bipolar cell layer
Ganglion cell layer
Axons of ganglion layer
Posterior cavity

LM 250x

Figure 18.6   **Histology of the Retina.**

Table 18.5	The Retina		
Structure	**Description**	**Function**	**Word Origins**
Bipolar Cell Layer	Middle layer of the retina, composed of cells with intermediate-sized nuclei; as the name suggests, these neurons are bipolar neurons.	Receives signals from rods and cones and transmits them to ganglion cells.	*bipolar*, relating to bipolar neurons
Choroid	The vascular and pigmented layer of the eye located between the retina (internal) and the sclera (external). It is recognized histologically by numerous blood vessels and by dark staining characteristics of the melanin.	Blood vessels of the choroid supply nutrients to the tissues of the retina and sclera, and the pigment absorbs excess light waves.	*choroideus*, like a membrane
Cones	Photoreceptor cells with a light-transducing portion located in the outermost layer of the retina and with nuclei located in the layer just internal to that. You will not be able to distinguish rods from cones using a light microscope.	Photoreceptor cell specializing in color vision.	*conus*, shaped like a cone
Fovea Centralis	Histologically, an area of the retina devoid of bipolar and ganglion cell layers. Its photoreceptor layer is composed exclusively of cones.	Area of highest visual acuity in the eye. When you focus on something, you move your eyes so the light entering the eye is focused on the fovea.	*fovea*, a pit, + *centralis*, in the center
Ganglion Cell Layer	Innermost layer of the retina, composed of cells with very large nuclei.	Receives information from bipolar cells and sends that information to the brain.	*ganglion*, a swelling or knot
Retina	Referred to as the "neural tunic" of the eye; consists of numerous layers of neurons involved in phototransduction.	Phototransduction: transduction of information that enters the eye as light waves into signals (action potentials) that can be interpreted by the brain.	*rete*, a net
Rods	Photoreceptor cells with a light-transducing portion located in the outermost layer of the retina and with nuclei located in the layer just internal to that; not distinguishable from cones using a light microscope.	Photoreceptor cells specializing in black-and-white vision; very sensitive, most useful when light is dim.	*rod*, shaped like a rod
Rod and Cone Layer	Outermost layer of the retina (closest to the choroid and sclera), containing the light-transducing portions of photoreceptor cells (rods and cones). The layer immediately internal to this layer contains the nuclei of rods and cones; the smallest and most numerous nuclei of the retina.	Layer of the retina where light waves are initially transduced into neuronal action potentials; the cells in this layer synapse with neurons in the bipolar cell layer.	NA
Sclera	Dense irregular connective tissue that surrounds the entire eye except for its anterior aspect where the cornea is located; histologically, the most external layer, composed of collagen fibers and fibroblasts.	Protects the eye, serves as an attachment point for extraocular eye muscles, and helps maintain the round shape of the eye.	*skleros*, hard

3. Using **table 18.5** and figure 18.6, as guides, identify the following structures on the slide of the retina (you may need to change to high power to view all of the structures or to see them in greater detail):

- ☐ bipolar cell layer
- ☐ choroid
- ☐ ganglion cell layer
- ☐ pigmented layer
- ☐ rod and cone layer
- ☐ sclera

4. With the medium power objective in place, scan the slide and locate the **fovea centralis** (**figure 18.7a**). The fovea is a thinner than normal area of the retina, which is devoid of bipolar and ganglion cell layers and has the highest concentration of cones of the entire retina. When light focuses on the fovea, visual acuity is at its highest. When we focus our gaze on something, we move our eyes so the light entering the eye hits the fovea and the image becomes most clear.

5. Scan the slide and locate the **optic disc**, the location where the optic nerve leaves the eye (figure 18.7b). The **optic disc** (blind spot) is easily identifiable because all retinal layers are absent at this location. Notice that the optic disc and optic nerve are approximately the same color as the cells in the ganglion cell layer of the retina. This should help you understand that the ganglion cell layer, the optic disc, and the optic nerve all contain parts of the same cells—namely, the axons of ganglion cells. The axons of the ganglion cell layer leave the eye at the optic disc. Once they enter the optic nerve, the axons become myelinated.

6. In the space below, make a simple drawing of the retina and label the following layers: rod and cone layer, bipolar cell layer, and ganglion cell layer.

_____ ×

7. _Optional Activity:_ **AP|R** **Nervous System**—Watch the "Vision" animation to learn the sequence of events involved in vision and the functions of cells in the retina.

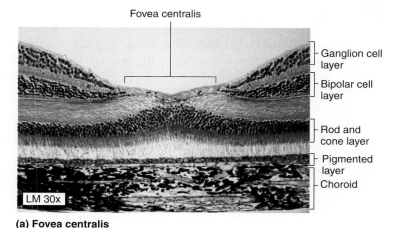

(a) Fovea centralis

Fovea centralis

- Ganglion cell layer
- Bipolar cell layer
- Rod and cone layer
- Pigmented layer
- Choroid

LM 30x

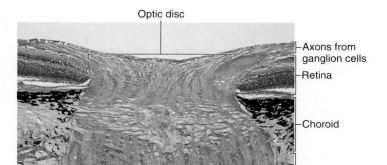

(b) Optic disc

Optic disc

- Axons from ganglion cells
- Retina
- Choroid
- Sclera
- Optic nerve (CN II)

LM 25x

Figure 18.7 **Specialized Areas of the Neural Tunic of the Eye.** (a) The fovea is the area of the retina where visual acuity is the highest. Ganglion and bipolar cell layers are absent, and there is an abundance of cones in the rod and cone layer. (b) The optic disc is the area of the retina where the axons of ganglion cells exit the retina to become the optic nerve. There are no rods or cones in this area, which is why the optic disc is also referred to as the "blind spot" of the retina.

WHAT DO YOU THINK?

3 The optic disc is referred to as the "blind spot" of the eye. Based on your histological observation of the optic disc, explain why this is the case.

EXERCISE 18.6 Hearing

Equilibrium (the sense of balance) and hearing are functions of the **semicircular ducts** and the **cochlea**, respectively. These special sensory organs are located within the petrous part of the temporal bone. We will explore the location, gross structure, and function of these organs in the gross anatomy section of this chapter. In this section we will focus on the histological features of the highly specialized epithelium that lines the cochlea, the **spiral organ** (of Corti), which will provide us with insight into how this organ performs its function: transformation of sound waves into electrical signals that can be interpreted by the brain.

1. The **cochlea** (**figure 18.8**) is the organ responsible for transforming fluid vibrations received at the oval window into electrical signals that are sent to the thalamus and then on to the temporal lobe of the brain where they are processed into sounds.

2. Within the cochlea, the **spiral organ** (organ of Corti) sits upon the **basilar membrane**, within the scala media (cochlear duct). The scala media is surrounded by the **scala tympani** and **scala vestibuli** (table 18.6). As the basilar membrane vibrates due to movement of the fluid (perilymph) in the scala tympani, the sensory cells attached to the basilar membrane also vibrate. Cilia of the **hair cells** of the cochlea are embedded in the immobile **tectorial membrane** that lies above them. Thus, vibrations of the basilar membrane cause the cilia to bend. As the

cilia bend, electrical signals (action potentials) are generated in the hair cells. These action potentials are transmitted to the brain by the **cochlear branch** of the **vestibulocochlear nerve (CN VIII)**. The fluid vibrations are eventually dampened when vibrations of the perilymph reach the **round window** at the end of the **scala tympani**.

3. Obtain a slide of the *cochlea* (**figure 18.9**) and place it on the microscope stage. Bring the tissue sample into focus on low power and then increase the magnification. Move the microscope stage until a single cross section through the cochlea is in the center of the field of view.

4. Using figures 18.8 and 18.9 and **table 18.6** as guides, identify the three chambers within the cochlea and the membranes that separate the chambers from each other:

☐ basilar membrane   ☐ scala vestibuli
☐ scala media   ☐ vestibular membrane
   (cochlear duct)
☐ scala tympani

5. Once you have identified the *scala media* (cochlear duct), move the microscope stage so the scala media is in the center of the field of view. Increase the magnification to high power and focus in on the *spiral organ* (organ of Corti) (figure 18.9).

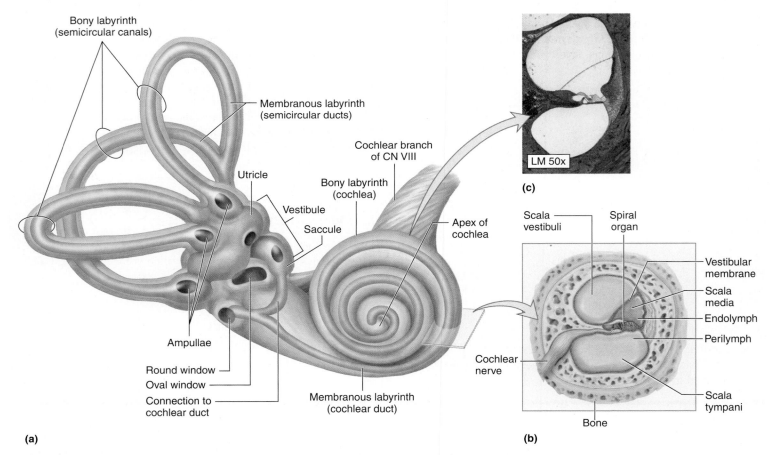

Figure 18.8 **The Cochlea.** (a) The semicircular canals and cochlea are part of the inner ear. (b) The cochlea houses the spiral organ, which contains specialized cells that translate sound waves into sensory impulses. (c) Light micrograph demonstrating a cross-section through the cochlea.

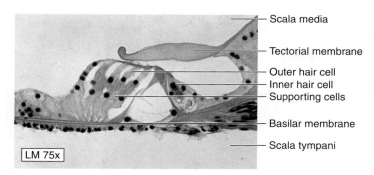

Scala media
Tectorial membrane
Outer hair cell
Inner hair cell
Supporting cells
Basilar membrane
Scala tympani

LM 75x

Figure 18.9 Histology of the Spiral Organ.

6. Using table 18.6 and figure 18.9 as guides, identify the following structures on the slide of the cochlea:

☐ basilar membrane   ☐ scala vestibuli
☐ cochlear nerve   ☐ sensory cells
☐ endolymph   ☐ spiral ganglion
☐ perilymph   ☐ tectorial membrane
☐ scala media   ☐ vestibular membrane
☐ scala tympani

Table 18.6	The Cochlea		
Structure	**Description**	**Function**	**Word Origins**
Basilar Membrane	Forms the floor of the scala media and supports the cells of the spiral organ.	Vibrates in response to fluid vibrations of the perilymph in the scala tympani.	*basalis*, situated near the base, + *membrana*, a membrane
Endolymph	Fluid within the scala media (cochlear duct) that has a composition similar to that of intracellular fluid (high in potassium).	Nourishes the epithelial cells of the spiral organ.	*endon*, within, + *lympha*, a clear fluid
Hair Cells (Sensory Cells)	Highly specialized epithelial cells of the spiral organ that contain stereocilia, which embed in the tectorial membrane.	Bending of the stereocilia generates action potentials in the hair cells; axons of sensory cells project into the spiral ganglion and synapse with neurons that relay the signals to the brain.	*sensus*, to sense
Helicotrema	An opening at the end of the cochlea that is shaped like a half moon.	Allows perilymph from the scala tympani to communicate with perilymph from the scala vestibuli.	*helix*, a spiral, + *trema*, a hole
Perilymph	Fluid contained within the scala vestibuli and scala tympani that has a composition similar to that of extracellular fluid (high in sodium).	Transmits pressure waves through the scala tympani and scala vestibuli; vibrations of the stapes at the oval window create the vibrations, and they are dampened when they reach the round window.	*peri-*, around, + *lympha*, a clear fluid
Scala Media (Cochlear Duct)	Middle chamber of the cochlea, containing the spiral organ and filled with endolymph.	Contains the special sensory organ of sound, the spiral organ.	*scala*, a stairway, + *medialis*, middle
Scala Tympani	Chamber inferior to the scala media; filled with perilymph.	Transmits pressure waves of the perilymph from the helicotrema to the round window.	*scala*, a stairway, + *tympani*, relating to the tympanic membrane
Scala Vestibuli	Chamber superior to the scala media; filled with perilymph.	Transmits pressure waves of the perilymph from the oval window to the helicotrema.	*scala*, a stairway, + *vestibulum*, entrance court
Spiral Ganglion	Nerve ganglion located on the cochlear part of the auditory nerve.	Contains the cell bodies of bipolar neurons, which receive input from sensory cells of the spiral organ and then relay those signals to the brain.	*spiralis*, a coil, + *ganglion*, a swelling or knot
Spiral Organ (Formerly Organ of Corti)	Composed of specialized epithelium that sits on the basilar membrane; the epithelium is composed of sensory cells and supporting cells.	Special inner and outer sensory cells have stereocilia embedded in the tectorial membrane; when the basilar membrane vibrates, the stereocilia bend and the sensory cells send action potentials to the brain.	*spiralis*, a coil, + *organum*, a tool, instrument
Tectorial Membrane	Gelatinous membrane in which the cilia of hair cells of the spiral organ are embedded.	Does not vibrate itself, but when the basilar membrane vibrates, the cilia of the hair cells embedded in the tectorial membrane bend, causing the hair cells to generate action potentials.	*tectus*, to cover, + *membrana*, a membrane
Vestibular Membrane	Thin membrane between the scala vestibuli and the scala media.	Forms a partition separating endolymph within the scala media from perilymph within the scala vestibuli.	*vestibulum*, relating to the vestibule, + *membrana*, a membrane
Vestibulocochlear Nerve (CN VIII)	Cranial nerve arising from axons of sensory cells of the spiral organ and from the vestibular apparatus.	Transmits sound information to the brain.	*audio*, to hear

7. In the space below, make a simple drawing of a cross section through the cochlea and label the locations of the structures listed in step 6.

_____ ×

Gross Anatomy

General Senses

The definitions of general senses and special senses are described at the beginning of the histology section of this chapter. If you are starting your laboratory observations with the gross anatomy exercises, read the introduction to general senses on p. 358 before you proceed.

Most sensory receptors responsible for general sensation (things such as: touch, pain, pressure, temperature) are located in the skin. Thus, in exercise 18.7 you will observe a classroom model of the skin, with special emphasis on sensory receptors located within the skin.

EXERCISE 18.7 Sensory Receptors in the Skin

1. Observe a classroom model of thick skin (**figure 18.10**). Some somatic sensory receptors in the skin, such as tactile menisci and free nerve endings, are too small to view under the microscope, so you will identify them on models instead.

2. Locate a tactile disc. **Tactile discs** (Merkel discs) are located at the dermal/epidermal junction of thick skin, although their location is not restricted to dermal papillae, as tactile corpuscles are. Their function is the sensation of light touch.

3. Find some free nerve endings on the skin model. **Free nerve endings** are just that—nerve endings with no specialized cells surrounding them. The ends of these neurons are located near the dermal/epidermal junction, with many of them extending into the epidermis. They also have endings in hair follicles and glands. Free

nerve endings function in the sensation of sustained touch, temperature, itching, and pain. Imagine a time when you had a blister that ripped open. A blister is created when the dermis separates from the epidermis, and fluid accumulates between the layers. When the epidermis is removed from a blister, it is very painful! Why? The free nerve endings in the dermis are now exposed to the environment, and this causes them to fire action potentials. This is why large superficial wounds ("scrapes") on the skin are much more painful than deep wounds. We perceive a deep wound as a more severe wound (as it typically is), but are often confused that it causes us less pain than a more superficial wound. The greater the surface area exposed, the greater the number of nerve endings stimulated. Hence the common exclamation, "It's only a scrape, but it hurts like crazy!"

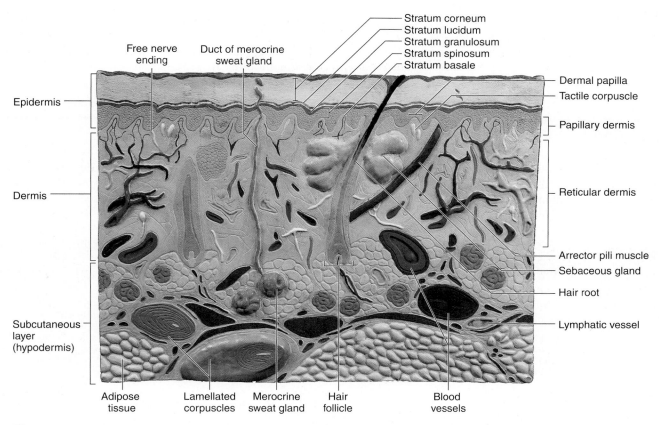

Figure 18.10   Skin. Model of the skin demonstrating sensory receptors such as tactile discs, free nerve endings, and lamellated corpuscles. This model is misleading in that it demonstrates hair follicles *and* thick skin (epidermis). Thick skin does not contain hair follicles.

4. Using figure 18.10, table 18.1, and your textbook as guides, identify the following on a model of the skin:

☐ dermal papillae   ☐ reticular dermis

☐ free nerve endings   ☐ stratum basale

☐ tactile corpuscle   ☐ stratum corneum

☐ tactile disc   ☐ stratum granulosum

☐ lamellated corpuscle   ☐ stratum lucidum

☐ papillary dermis   ☐ stratum spinosum

5. In the space below, make a simple drawing of thick skin and label the locations of the sensory receptors listed in table 18.1.

Special Senses

The definitions of general senses and special senses are described at the beginning of the histology section of this chapter. If you are starting your laboratory observations with the gross anatomy exercises, read the introduction to special senses on p. 360 before you proceed. The gross anatomy exercises in this section focus on two special sensory organs: the eye and the ear.

EXERCISE 18.8 Gross Anatomy of the Eye

In exercise 18.9 we will explore the anatomy of the eye, using a cow eye as a model organ. Though there is some variance in structure between a cow eye and a human eye, cow eyes are much easier to obtain and they are larger, which greatly facilitates making internal observations of the eye. In this laboratory exercise you will identify all structures listed in **tables 18.7** and **18.8** on a classroom model or on yourself, because not all structures associated with the eye are visible in the dissected cow eye or in the histology slide of the retina.

Table 18.7	Internal Structures of the Eye		
Structure	**Description**	**Function**	**Word Origins**
Anterior Chamber	The space between the cornea and the iris/pupil.	Filled with aqueous humor, which is described elsewhere in this table.	*anterior*, the front surface, + *camera*, an enclosed space
Anterior cavity	The space anterior to the lens and posterior to the cornea. It is subdivided by the iris into the anterior and posterior chambers, which are described elsewhere in this table.	Filled with aqueous humor.	*anterior*, the front surface, + *cavus*, hollow
Aqueous Humor	Watery fluid, similar in composition to cerebrospinal fluid. It is secreted by the ciliary processes and circulates within the anterior and posterior chambers of the eye.	Provides nourishment to the structures within the anterior and posterior chambers in the eye, in particular the avascular lens and cornea.	*aqueous*, watery, + *humor*, fluid
Choroid Layer	The pigmented, vascular layer found in between the retina and the sclera.	Blood vessels of the choroid supply nutrients to the tissues of the retina and sclera, and its pigment absorbs excess light waves.	*choroideus*, like a membrane
Ciliary Body	The thickened extension of the vascular tunic, located between the choroid and the iris, that contains the ciliary muscle.	Produces aqueous humor; contraction of the ciliary muscle within alters the shape of the lens.	*cilium*, eyelid, + *bodig*, a thing or substance
Ciliary Muscle	Smooth muscle found within the ciliary body that is composed of both circular and radial fibers.	Contraction of this muscle relaxes the suspensory ligaments that attach it to the lens, which relaxes the lens and makes it become rounder to accommodate for near vision.	*cilium*, eyelid
Fovea Centralis	The depression ("central pit") in the macula lutea that contains only cones and lacks blood vessels.	The area of highest visual acuity in the eye.	*fovea*, a pit, + *centralis*, in the center
Iris	The colored portion of the eye, which makes up the anterior portion of the vascular tunic; the dilator pupillae and sphincter pupillae muscles are located within the iris.	Controls the amount of light entering the eye. Contraction of the radially arranged dilator pupillae muscle (under sympathetic stimulation) increases pupil diameter, whereas contraction of the circularly arranged sphincter pupillae muscle (under parasympathetic stimulation) decreases the diameter of the pupil.	*iris*, rainbow
Lens	A transparent, biconvex structure composed of highly specialized, modified epithelium.	Bends light waves (refraction) so that they hit the retina optimally for clear vision.	*lens*, a lentil
Macula Lutea	A "yellow spot" on the retina located superior to the optic disc on the posterior wall of the eye, which contains the fovea centralis within.	Contains the fovea centralis.	*macula*, a spot, + *luteus*, yellow
Optic Disc ("Blind Spot")	An area of the retina where there is an absence of photoreceptors because it is where the axons of ganglion cells leave the eye to travel in the optic nerve.	The location where axons exit the eye.	*optikos*, the eye, + *discus*, disc
Ora Serrata	Anteriormost portion of the choroid, which appears serrated (hence the name).	Demarcates the division of the visual retina from the nonvisual retina.	*ora*, an edge, + *serratus*, a saw
Posterior Chamber	The space between the iris/pupil and the lens and ciliary body.	Filled with aqueous humor, which is described elsewhere in this table.	*posterior*, the back surface, + *camera*, an enclosed space
Pupil	The space (opening) in the center of the iris.	The size of the pupil (which is controlled by the iris) regulates the amount of light entering the eye.	*pupilla*, pupil

Table 18.7	Internal Structures of the Eye *(continued)*		
Structure	**Description**	**Function**	**Word Origins**
Retina	Also called the neural tunic of the eye. It is the inner layer of the wall of the eye, composed of a pigmented layer, rods, cones, bipolar cells, and ganglion cells.	Transduces information that enters the eye as light waves into a nervous signal (action potential) that can be interpreted by the brain.	*rete*, a net
Suspensory Ligament	Ligament that extends between the ciliary body and ciliary muscles and the lens.	Attaches the lens to the ciliary body and ciliary muscles so that contraction and/or relaxation of ciliary muscles can alter the shape of the lens.	*suspensio*, to hang up, + *ligamentum*, a bandage
Tapetum Lucidum	Metallic-appearing, opalescent inner layer of the sclera; present in many animals (e.g., the cow eye) but not in humans.	Scatters light waves within the eye; allows for better vision in dim limited light (humans do not have this layer).	*tapeta*, a carpet, + *lucidus*, clear
Vascular Tunic	Middle layer of the wall of the eye; consists of the choroid, ciliary body, and the iris.	Provides nourishment to structures within the eye.	*vasculum*, a small vessel
Vitreous Chamber (Posterior Cavity)	A space posterior to the lens and anterior to the retina.	Occupied by the vitreous humor, which is described elsewhere in this table.	*vitreus*, glassy, + *camera*, an enclosed space
Vitreous Humor	Jellylike substance composed of a connective tissue stroma surrounded by liquid (vitreous humor) that fills up the interior of the eye.	Helps maintain the round shape of the eye and is critical in holding the retina against the wall of the eye.	*vitreus*, glassy, + *humor*, fluid

Table 18.8	External Structures of the Eye		
Structure	**Description**	**Function**	**Word Origins**
Cornea	Transparent tissue on the anterior surface of the eye consisting of an external layer of stratified squamous epithelium, a middle layer of regularly arranged collagen fibers, and an inner layer of endothelium.	The primary structure used to refract (bend) light waves in the eye.	*corneus*, horny
Extraocular Muscles	Muscles that originate from a common tendinous ring and insert on the sclera of the eye.	Creates the very fine eye movements that are necessary for focusing light upon the retina.	*extra*, outside of, + *oculus*, the eye
Fibrous tunic	Tough outer connective tissue covering of the eye, composed of the sclera and cornea.	As shown elsewhere in this table, for sclera and cornea.	*fibro-*, fiber + *tunic*, a coat
Lacrimal Caruncle	Small, fleshy mound of tissue at the medial aspect of the eye containing ciliary glands (modified sweat glands) and tarsal glands.	Ciliary glands are modified sweat glands that keep the eyelashes lubricated. Tarsal glands are sebaceous glands that secrete a lipid secretion that prevents tears from overflowing and prevents the eyelids from sticking together.	*caruncula*, a small fleshy mass
Lacrimal Gland	Almond-shaped serous gland located in the superior and lateral aspect of the orbit.	Secretes tears.	*lacrima*, a tear
Lacrimal Puncta	Two small openings in the lacrimal caruncle; the tiny holes on the "bump" at the inferomedial aspect of the eye.	Facilitates the drainage of tears into the lacrimal sac.	*lacrima*, a tear, + *punctum*, a prick or point
Lacrimal Sac	A swelling at the superior part of the nasolacrimal duct, medial to the lacrimal bone, lateral to the nasal bone, and deep to the maxilla.	Receives tears from the lacrimal canals and transmits them to the nasolacrimal duct.	*lacrima*, a tear
Nasolacrimal Duct	A duct that runs from the lacrimal sac into the nasal cavity.	Conducts tears from the lacrimal sac into the nasal cavity.	*nasal*, relating to the nose, + *lacrima*, a tear, + *ductus*, to lead
Optic Nerve	CN II; a large nerve exiting the posteromedial aspect of the eye that exits the orbit through the optic foramen; consists of myelinated axons of ganglion cells.	Carries visual information from the eye to the lateral geniculate nucleus of the thalamus.	*optikos*, relating to the eye or vision
Orbital Fat Pad	A thick capsule of adipose connective tissue that fills in all the spaces between the eye, extraocular eye muscles, nerves, and orbit.	Cushions the eye and helps support it and hold it in place.	*orbital*, relating to the orbit of the eye
Sclera	Dense irregular connective tissue that surrounds the entire eye except for its anterior aspect, where the cornea is located.	Protects the eye, serves as an attachment point for extraocular eye muscles, and helps maintain the round shape of the eye.	*skleros*, hard

1. Observe a classroom model of the eye (**figure 18.11**).

2. Using figure 18.11, tables 18.7 and 18.8, and your textbook as guides, identify the following structures on a model of the eye:

   ☐ anterior cavity
   ☐ anterior chamber
   ☐ choroid
   ☐ ciliary body
   ☐ cornea
   ☐ fovea centralis
   ☐ inferior oblique muscle
   ☐ inferior rectus muscle
   ☐ iris
   ☐ lacrimal gland
   ☐ lateral rectus muscle
   ☐ lens
   ☐ medial rectus muscle
   ☐ nasolacrimal duct
   ☐ optic disc ("blind spot")
   ☐ optic nerve
   ☐ ora serrata
   ☐ posterior chamber
   ☐ pupil
   ☐ retina
   ☐ sclera
   ☐ superior oblique muscle
   ☐ superior rectus muscle
   ☐ suspensory ligament
   ☐ vascular tunic
   ☐ vitreous chamber (posterior cavity)

   Note: You should have identified the extraocular eye muscles in chapter 12. If necessary, refer to table 12.3 on page 217 for guidance on the structure and function of these muscles because they are not listed in the tables in this chapter.

3. Obtain a mirror and observe the externally visible structures of your eye. If you have no mirror, perform this observation on your lab partner. Using table 18.8 and your textbook as guides, identify all of the structures listed in **figure 18.12** on your eye (or your lab partner's eye), and then label them in figure 18.12.

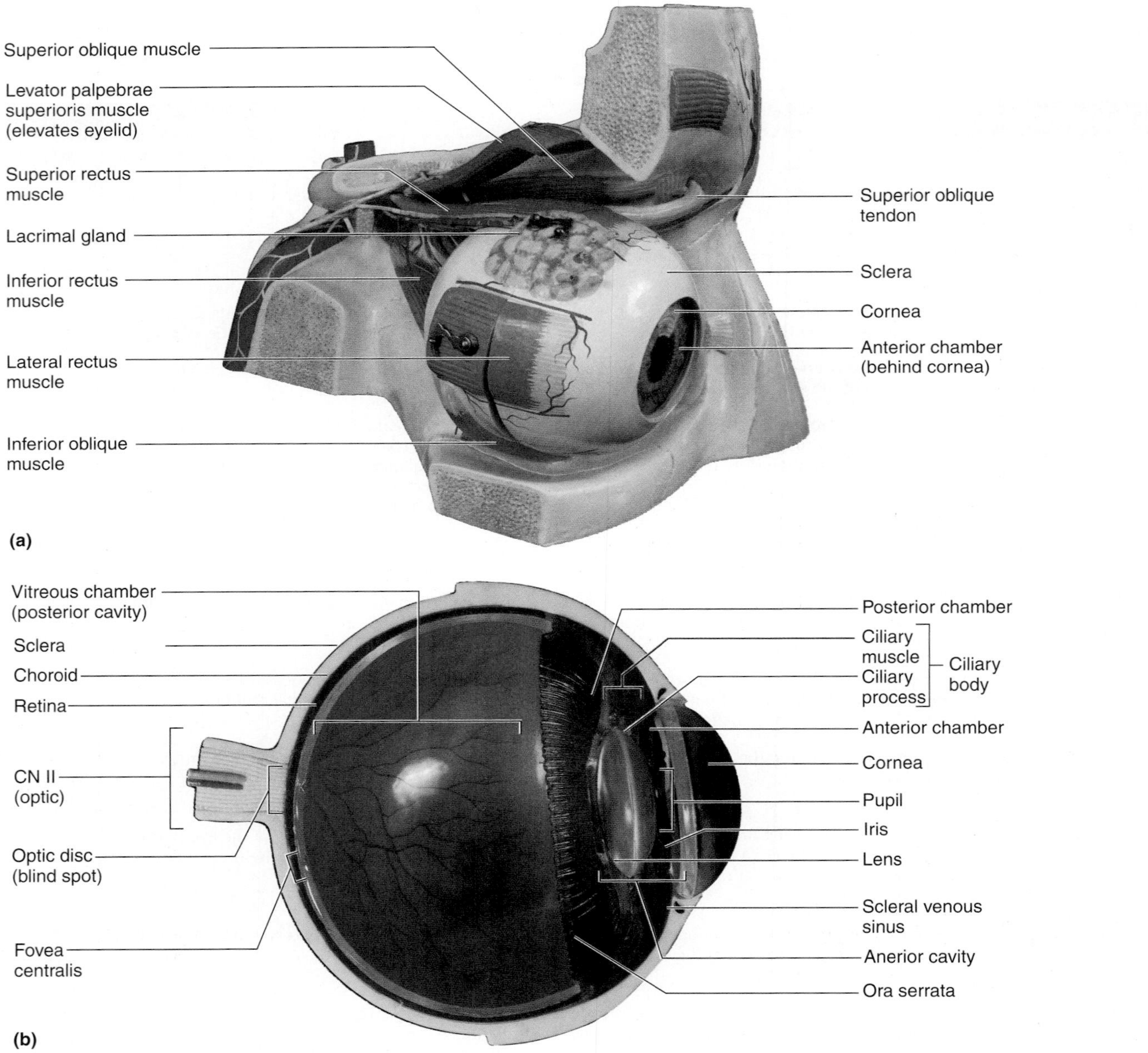

(a)

(b)

Figure 18.11   Classroom Model of the Eye.   (a) Anterior view. (b) Internal view.

Figure 18.12   **External Eye Structures.**

- ☐ eyebrow
- ☐ eyelashes
- ☐ inferior eyelid
- ☐ iris
- ☐ lacrimal caruncle
- ☐ lateral palpebral commissure
- ☐ medial palpebral commissure
- ☐ palpebral fissure
- ☐ pupil
- ☐ sclera
- ☐ superior eyelid

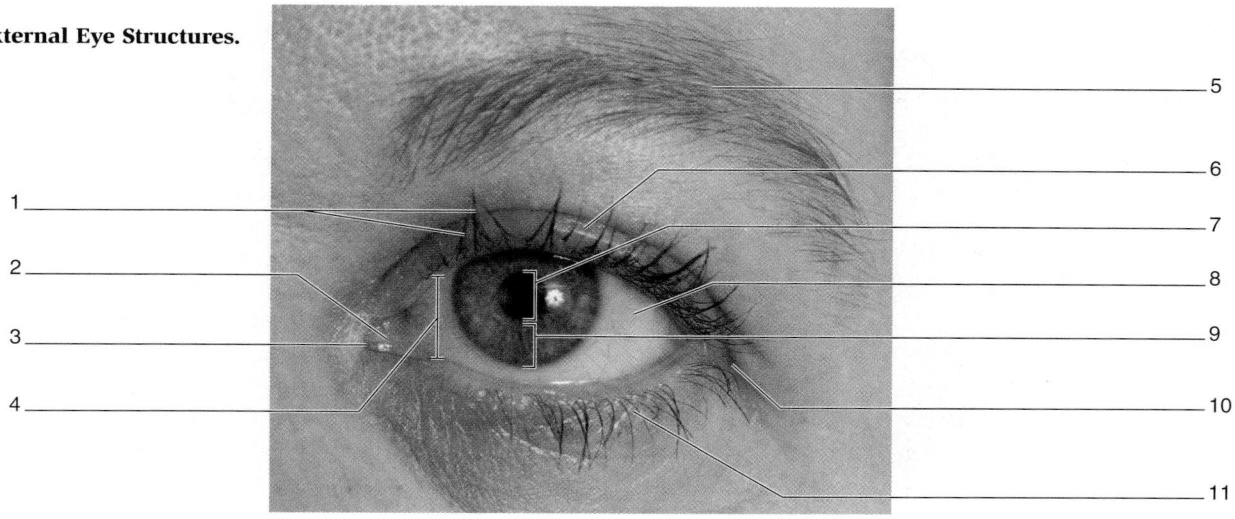

EXERCISE 18.9 Cow Eye Dissection

This exercise is designed to be done using fresh cow eyes, although preserved pig eyes may be substituted if necessary. If you are dissecting a preserved pig eye, the tissues will be tougher and the cornea will be opaque instead of transparent.

1. Obtain a dissecting pan, dissecting tools, and a fresh cow eye. Observe the gross structure of the eye before making any cuts. Using tables 18.7 and 18.8 and **figure 18.13a** as guides, identify the following structures:

   - ☐ cornea
   - ☐ extraocular muscles
   - ☐ optic nerve
   - ☐ orbital fat pad
   - ☐ sclera

2. Using scissors and forceps (the tissue will be slippery!), remove the orbital fat pad and extraocular muscles, leaving the optic nerve intact (figure 18.13b). Once you have cleaned this off, you will be able to view the entire eye and get oriented more easily. Notice how tough the outer covering or **sclera** of the eye is. This is the layer of tissue you will need to cut through in order to see structures within the eye.

3. Using scissors and forceps, cut the eye open by making a coronal incision through the sclera that completely encircles the eye. Once you do this, you will notice that a jelly-like fluid oozes out of the posterior chamber of the eye. This fluid is the **vitreous humor** which fills the **vitreous chamber** (**figure 18.14b**), and whose functions include holding the **retina** against the posterolateral walls of the eye. In the cow eye, much of the **choroid**, which contains black pigment, becomes mixed into the vitreous humor once the eye is cut open. Thus, you may find that you have a gooey, black mess on your hands almost immediately. Before things get too "mixed up," identify structures in the

posterior half of the dissected eye. Look for a yellowish, thin membrane that is connected to the posterior of the eye at only one spot (figure 18.14a). This is the **retina**, which contains neurons responsible for transmitting

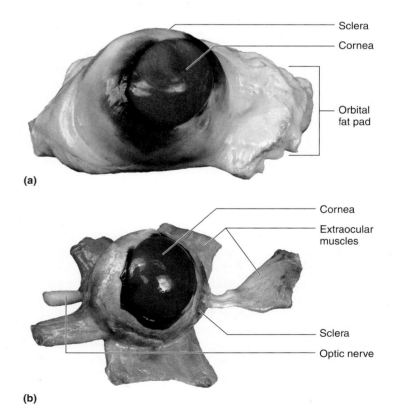

(a)

(b)

Figure 18.13   **Fresh Cow Eye.**   (a) Before dissection with orbital fat pad intact. (b) After dissecting away the orbital fat pad to expose the extraocular muscles and optic nerve.

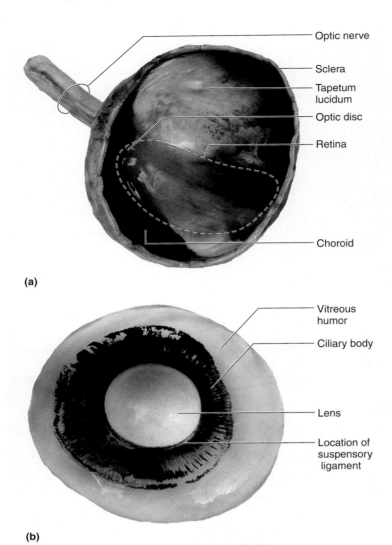

Optic nerve

Sclera

Tapetum lucidum

Optic disc

Retina

Choroid

(a)

Vitreous humor

Ciliary body

Lens

Location of suspensory ligament

(b)

Figure 18.14 **Coronal Sections of Cow Eye.** (a) Posterior part demonstrating the retinal, optic disc, and tapetum lucidum. (b) Anterior part demonstrating the choroid, lens, and iris.

visual information to the brain histology of (the retina was covered in detail in exercise 18.5). The retina is very delicate and easily falls away from the posterior wall of the eye when the vitreous humor is not there to hold it in place.

4. Find the location where the retina attaches to the posterior wall of the eye (it will "pucker" in this area). If you have difficulty finding this spot, first find the optic nerve on the outside of the eye and then look inside the eye for the location where the optic nerve leaves the eye. This spot within the eye is the **optic disc** (blind spot). It is called a "blind" spot because it is devoid of photoreceptors. This is where axons from the ganglion cell layer of the retina (see exercise 18.5) leave the eye and travel through the brain via the optic nerve (CN II).

5. Observe the inner walls of the posterior half of the eye. Notice the very colorful, iridescent **tapetum lucidum** (figure 18.14*a*). This structure is not present in humans, but it is present in animals that must be able see well in dim light. The tapetum lucidum reflects light. Thus, when it is dark outside and very little light is entering the eye, the tapetum lucidum causes light waves to bounce around within the eye and increases the frequency with which light rays stimulate the retina. This makes things more visible, but the image does not become sharper. It is the reflection of light waves from the tapetum lucidum that causes a cat's eyes (or those of other animals) to "glow" when a light shines on them at night. The next time you are out at night, notice that this does not happen in humans. In humans, the inside of the eye is completely coated with a black choroid, which absorbs excess light. This makes it more difficult for us to see things in the dark. On the plus side, the images that we do see are sharper.

6. Now focus your attention on the anterior portion of the eye (figure 18.14*b*). Again, it will be somewhat difficult to see many structures because the choroid covers everything, making the structures very dark. Notice the semitransparent **lens**, which is suspended in place by a ring of black-colored tissue. This tissue is the **ciliary body**, whose function is to suspend the lens. The cavity anterior to the lens and posterior to the cornea is the **anterior cavity** of the eye. The anterior cavity is further subdivided by the iris into an **anterior chamber** (between the cornea and the iris) and a **posterior chamber** (between the iris and the lens). In a living organism, the anterior cavity is filled with a clear, watery fluid called **aqueous humor**. Try to find the fine, delicate structures composing the **suspensory ligament**, which runs between the ciliary body and the lens.

7. Carefully remove the lens from the eye. You will notice that although it is not completely transparent, you can still see through it. Place it on a piece of paper containing text (your lab manual will work fine if you don't mind getting it a little bit wet). As shown in **figure 18.15** place the lens over a letter or two of text and make note of the change in appearance of the text, if any, as seen through the lens. What happens?

8. Identify the following structures on the interior of the dissected cow eye (use tables 18.7 and 18.8 and figures 18.13 through 18.15 as guides):

- ☐ anterior cavity
- ☐ anterior chamber
- ☐ choroid
- ☐ ciliary body
- ☐ lens
- ☐ optic disc
- ☐ posterior chamber
- ☐ retina
- ☐ suspensory ligament
- ☐ tapetum lucidum
- ☐ vitreous humor

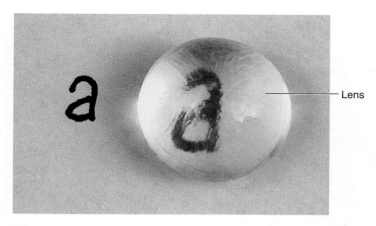

Figure 18.15   Lens. After removing the lens from the cow eye, place it over some text to see how it changes the image of the text.

9. When you have finished your dissection, clean up your workspace: Dispose of the cow eye debris in the organic waste. Dispose of used scalpel blades in the sharps container. Dispose of used paper towels and other paper waste in the wastebasket. Rinse off your dissecting tray and dissection instruments, and lay them out to dry. Finally, wipe down your laboratory workstation with disinfectant so it is clean for the next person who comes into the laboratory.

WHAT DO YOU THINK?

4 An individual who suffers a strong blow to the head may end up with a detached retina that causes visual problems. Why do you think the retina easily detaches from the posterior wall of the eye? What structure normally holds the retina in place?

EXERCISE 18.10 Gross Anatomy of the Ear

The ear is responsible for two important special sensory modal–ities: equilibrium (balance) and hearing. The organs responsible for both of these are located deep within the *petrous part of the temporal bone*, and therefore they are nearly impossible to iden-tify on a cadaver. Thus, our exploration of the gross anatomy of the ear will be accomplished using models of the ear.

1. Obtain a model of the ear (**figure 18.16**). On the model, first distinguish between the external-ear, middle-ear, and inner-ear cavities, and the structures that link the cavities to each other (**table 18.9**). The **tympanic membrane** is the link between the external-ear and the middle-ear cavities, while the **oval window** is the link between the middle-and inner-ear cavities.

2. Using figure 18.16, and table 18.9, as guides, identify the following structures on the model:

- ☐ auditory tube
- ☐ auricle
- ☐ cochlea
- ☐ ear ossicles
- ☐ endolymph
- ☐ external acoustic meatus
- ☐ incus
- ☐ malleus
- ☐ oval window
- ☐ perilymph

- ☐ round window
- ☐ saccule
- ☐ semicircular canals
- ☐ spiral organ
- ☐ stapedius muscle
- ☐ stapes
- ☐ tectorial membrane
- ☐ tensor tympani muscle
- ☐ tympanic membrane
- ☐ utricle

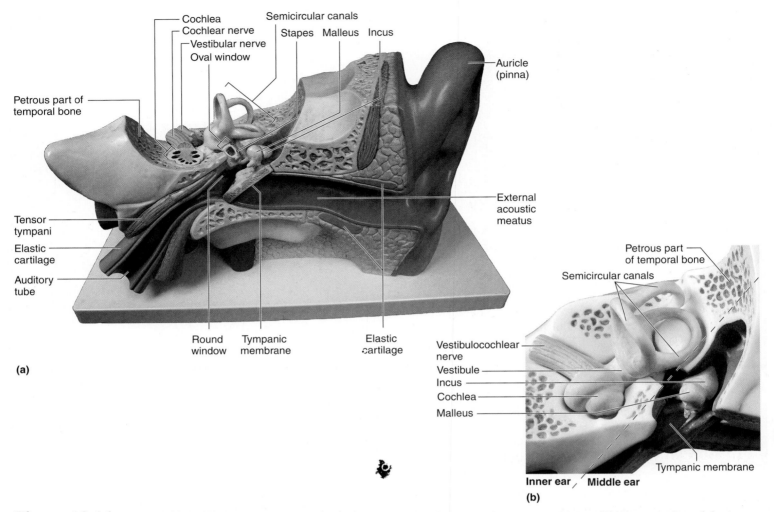

Figure 18.16 **Classroom Model of the Ear.** (a) Anterior view of the petrous part of the temporal bone. (b) Close-up view of the inner ear on the model.

Table 18.9	Structures of the External, Middle, and Internal Ear		
Structure	**Description**	**Function**	**Word Origins**
EXTERNAL EAR			
Auricle (Pinna)	External ear, composed of an elastic cartilage skeleton that is covered with skin.	Collects sound waves from the environment and funnels them into the external auditory meatus.	*pinna*, a wing
External Acoustic Meatus	Canal leading from the auricle of the ear to the tympanic membrane.	Transmits sound vibrations that arrive at the auricle to the tympanic membrane, where they cause it to vibrate.	*externa*, outside, + *acoustic*, relating to sound, + *meatus*, a passage
MIDDLE EAR	Cavity that begins at the tympanic membrane and ends at the oval window.	Contains the ear ossicles.	
Auditory Ossicles	Three tiny bones (malleus, incus, and stapes) found within the middle ear.	Transmits pressure vibrations from the tympanic membrane to the oval window of the cochlea, creating fluid vibrations in the perilymph of the scala vestibuli.	*ossiculum*, a bone
Incus	Tiny bone located within the middle-ear cavity that is shaped like an anvil.	Transmits pressure vibrations from the malleus to the stapes, thus participating in the amplification of vibrations of the tympanic membrane.	*incus*, an anvil

Table 18.9	Structures of the External, Middle, and Internal Ear *(continued)*		
Structure	**Description**	**Function**	**Word Origins**
Malleus	Tiny bone located within the middle-ear cavity that is shaped like a hammer.	Transmits pressure vibrations from the tympanic membrane to the incus thus participating in the transmission and amplification of the vibrations that arrive as sound waves on the tympanic membrane.	*malleus,* a hammer
Stapes	Tiny bone located within the middle-ear cavity that is shaped like a stirrup.	Transmits pressure vibrations from the incus to the oval window of the cochlea, creating fluid vibrations in the perilymph of the scala vestibuli, also participates in amplification of the vibrations of the tympanic membrane.	*stapes,* a stirrup
Auditory (Pharyngotympanic or Eustachian) Tube	Tube lined with elastic cartilage that connects the middle-ear cavity to the nasal cavity.	Opening of this channel allows air to enter or leave the middle ear cavity such that the pressure in the middle ear equilibrates with the environmental pressure, this allows the tympanic membrane to vibrate freely.	*audio,* to hear, + *tubus,* a canal
Oval Window	Opening into the scala vestibuli that is covered by the foot of the stapes.	Vibrations of the stapes at the oval window create fluid vibrations in the perilymph of the scala vestibuli.	*oval,* egg-shaped, + *window,* an opening
Stapedius Muscle	Small muscle connecting the neck of the stapes to the temporal bone.	Contraction of this muscle acts to dampen vibrations of the stapedius as a protective measure against excessive vibration on the oval window from very loud noises.	*stapedius,* relating to the stapes
Tensor Tympani Muscle	Small muscle connecting the handle of the malleus to the cartilage of the auditory tube.	Contraction of this muscle pulls the malleus medially and tenses the tympanic membrane as a protective measure against excessive vibration from very loud noises.	*tensus,* to stretch, + *tympani,* relating to the tympanic membrane
Tympanic Membrane	Drumlike, tight, thin membrane that separates the middle-ear cavity from the external-ear cavity.	Vibrates in response to sound waves that strike it as they reach the end of the external acoustic meatus, these vibrations cause vibrations in the ossicles of the middle ear cavity (malleus, incus, and stapes).	*tympanon,* drum, + *membrana,* a membrane
INNER EAR	Chamber located within the petrous part of the temporal bone that contains the cochlea and the vestibule.	Holds the organs responsible for the sensation of hearing (cochlea) and balance and equilibrium (vestibule).	
Cochlea	Spiral-shaped organ found within the inner ear.	Contains the spiral organ and associated structures that allow for the special sense of hearing.	*cochlea,* a snail shell
Spiral Organ (Organ of Corti)	Organ composed of specialized epithelium that is found within the scala media (cochlear duct) of the cochlea.	Special sensory organ for hearing.	*spiralis,* a coil, + *organon,* a tool or instrument
Semicircular Canals	Three ringlike canals that are oriented at right angles to each other and communicate with the vestibule.	Sense angular acceleration.	*semicircular,* shaped like a half circle, + *canalis,* a duct or channel
Vestibule	Located between the cochlea and the semicircular canals; contains the saccule and utricle.	Detect acceleration and deceleration movements.	*vestibulum,* entrance court
Saccule	Smallest membranous sac in the vestibule; connects with the cochlear duct.	Contains receptors that sense linear vertical acceleration.	*saccus,* a sac
Utricle	The largest membranous sac in the vestibule; semicircular canals arise from it.	Contains receptors that sense linear horizontal acceleration.	*uter,* leather bag
Vestibulocochlear Nerve	CN VIII; travels through the internal acoustic meatus.	Cranial nerve carrying information on balance, equilibrium, and hearing to the brain.	*vestibulo-,* referring to the vestibule, + *cochlea,* referring to the cochlea

3. In the space below, make a simple drawing of the ear and label the locations of the structures listed in step 2.

4. After you have identified all of the gross structures of the ear at least once, review the sequence of events required for the transmission of sound waves from the environment to the cochlea. As you do this, name all of the structures involved in the sequence. The sequence is as follows:

- Sound waves are focused on the **external acoustic meatus** by the contours of the outer ear (**auricle**) and create vibrations of the **tympanic membrane**. The **auditory tube** ensures that air pressure in the middle ear is the same as air pressure in the environment so the tympanic membrane can vibrate freely. It is lined with elastic cartilage and remains collapsed unless there is a large difference in pressure between the environment and the middle-ear cavity. When a difference in pressure exists, the auditory tube opens briefly and air moves to equalize the pressure in the middle ear with the pressure in the environment.

- Vibrations of the tympanic membrane create vibrations of the **auditory ossicles** (malleus, incus, and stapes). Excessive vibrations (from a loud noise) cause a reflexive contraction of the **tensor tympani** and **stapedius** muscles to dampen the vibrations of the ear ossicles and help protect the delicate cells of the inner ear.

- Vibration of the foot of the stapes against the **oval window** creates vibrations of the **perilymph**.

- Vibrations of the perilymph cause the **basilar membrane** to vibrate (see exercise 18.6). Vibrations of the basilar membrane cause stereocilia of the hair cells of the spiral organ to bend, sending action potentials to the brain.

- Remaining pressure waves exit the inner ear by way of the round window.

5. *Optional Activity*: AP|R **Nervous System**—Watch the "Hearing" animation to review the sequence of events involved in hearing.

Chapter 18: General and Special Senses

Name: _____

Date: _____ Section: _____

POST-LABORATORY WORKSHEET

1. Match the sensory receptor listed in column A with its appropriate location listed in column B.

   Column A

   _____ 1. tactile disc

   _____ 2. free nerve ending

   _____ 3. tactile corpuscle

   _____ 4. lamellated corpuscle

   Column B

   a. located at dermal/epidermal junction

   b. located deep in the reticular layer of the dermis

   c. located within the dermal papillae

   d. located throughout the dermis

2. Describe the anatomic relationship between lingual papillae and taste buds. That is, describe where a taste bud is located relative to a lingual papilla.

   _____

   _____

3. The following table lists the cell types associated with taste buds. Next to each cell type, give a brief description of the location of the cell within the taste bud and the function of the cell.

Cell Type	Location	Function
Supporting Cells		
Gustatory Cells		
Basal Cells		

4. The following table lists the cell types associated with olfactory epithelium. Next to each cell type, give a brief description of the location of the cell within the olfactory epithelium and the function of the cell.

Cell Type	Location	Function
Basal Cells		
Olfactory Receptor Cells		
Supporting Cells		

5. A fracture of the cribriform plate of the ethmoid bone can result in a loss of the sense of smell. Given your knowledge of the location of olfactory receptor cells and the pathway taken by their axons to reach the brain, explain why this can happen.

   _____

   _____

   _____

   _____

6. Of the cell layers that compose the retina, which layer contains cells whose axons leave the eye at the optic disc to become the optic nerve?

   _____

   _____

7. The spiral organ is located within the _____ _____ of the cochlea. This chamber is separated from the scala tympani by the _____, and separated from the scala vestibuli by the _____, and is filled with a fluid called _____.

8. Match the part of the eye listed in column A with its description listed in column B.

Column A

_____ 1. Vascular tunic

_____ 2. Suspensory ligament

_____ 3. Retina

_____ 4. Anterior chamber

_____ 5. Ora serrata

_____ 6. Choroid layer

_____ 7. Vitreous humor

_____ 8. Ciliary muscle

_____ 9. Fovea centralis

_____ 10. Lens

_____ 11. Optic disc ("blind spot")

_____ 12. Posterior chamber

_____ 13. Aqueous humor

_____ 14. Iris

_____ 15. Tapetum lucidum

_____ 16. Ciliary body

_____ 17. Macula lutea

Column B

a. space between the cornea and the iris/pupil

b. watery fluid that circulates within the anterior and posterior chambers of the eye

c. pigmented, vascular layer found in between the retina and the sclera

d. thickened extension of the vascular tunic containing the ciliary muscle

e. smooth muscle within the ciliary body composed of both circular and radial fibers

f. depression in the macula lutea that contains only cones

g. colored part of the eye

h. transparent, biconvex structure composed of highly specialized, modified epithelium

i. "yellow spot" on the retina located superior to the optic disc on the posterior wall of the eye

j. location where the axons of ganglion cells leave the eye to travel in the optic nerve

k. anteriormost part of the choroid, which appears serrated

l. space between the iris/pupil and the lens and ciliary body

m. neural tunic of the eye; composed of several layers of neurons involved with transducing light waves into nervous signals

n. ligament extending between the ciliary body and ciliary muscles and the lens

o. jellylike substance composed of a connective tissue stroma surrounded by liquid that fills up the interior of the eye

p. middle layer of the wall of the eye, consisting of blood vessels and pigment molecules

q. metallic-appearing, opalescent inner layer of the sclera; it is present in many animals (i.e., the cow eye), but not humans

9. Label the following diagram of the eye.

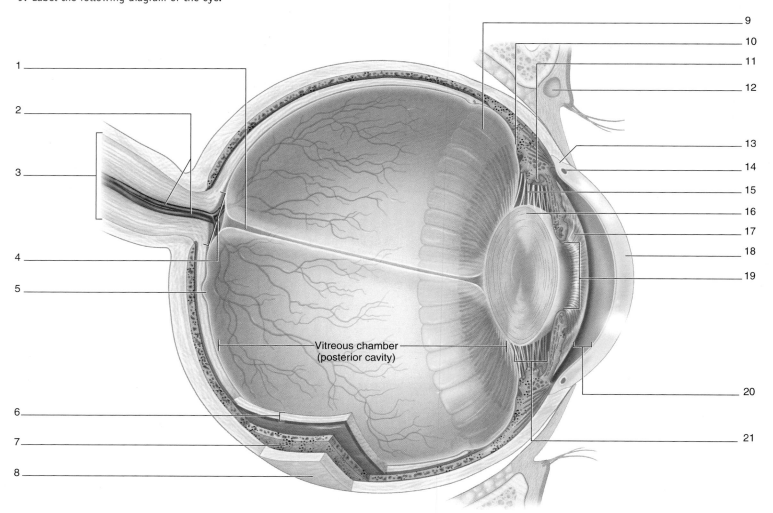

Vitreous chamber (posterior cavity)

10. Match the part of the ear listed in column A with its description listed in column B.

Column A

_____ 1. Semicircular canals

_____ 2. Vestibulocochlear nerve

_____ 3. Utricle

_____ 4. Incus

_____ 5. Cochlea

_____ 6. Stapes

_____ 7. External acoustic meatus

_____ 8. Middle ear

_____ 9. Saccule

_____ 10. Auricle

_____ 11. Stapedius muscle

_____ 12. Round window

_____ 13. Tensor tympani muscle

_____ 14. Internal acoustic meatus

_____ 15. Spiral organ

_____ 16. Inner ear

_____ 17. Tectorial membrane

_____ 18. Oval window

_____ 19. Endolymph

_____ 20. Tympanic membrane

_____ 21. Malleus

_____ 22. Ear ossicles

_____ 23. Perilymph

_____ 24. Auditory (pharyngotympanic) tube

Column B

a. Fibrocartilage-lined tube that connects the middle ear cavity to the nasal cavity

b. Spiral-shaped organ found within the inner ear

c. Three tiny bones (malleus, incus, and stapes) found within the middle ear

d. Fluid contained within the scala media

e. Canal leading from the auricle of the ear to the tympanic membrane

f. Tiny bone located within the middle ear cavity that is shaped like an anvil

g. Chamber located within the petrous part of the temporal bone that contains the cochlea and the vestibule

h. Canal on the medial, posterior aspect of the petrous part of the temporal bone

i. Tiny bone located within the middle ear cavity that is shaped like a hammer

j. Cavity that begins at the tympanic membrane and ends at the oval window

k. Opening into the scala vestibuli that has the foot of the stapes covering it

l. Fluid contained within the scala vestibuli and scala tympani

m. External ear, composed of an elastic cartilage skeleton that is covered with skin

n. Small hole at the end of the scala tympani that is covered by a membrane

o. Smallest membranous sac in the vestibule; contains receptors for sensing vertical acceleration

p. Three ring-like canals that are oriented at right angles to each other and communicate with the vestibule

q. Organ composed of specialized epithelium that is found within the scala media (cochlear duct) of the cochlea

r. Small muscle connecting the neck of the stapes to the temporal bone

s. Tiny bone located within the middle ear cavity that is shaped like a stirrup

t. Gelatinous membrane in which the hair cells of the spiral organ are imbedded

u. Small muscle connecting the handle of the malleus to the cartilage of the auditory tube

v. Drumlike, tight, thin membrane that separates the middle-ear cavity from the external-ear cavity.

w. Largest membranous sac in the vestibule; contains receptors for sensing horizontal acceleration.

x. Cranial nerve VIII, which travels through the internal acoustic meatus.

11. Label the following diagram of the ear.

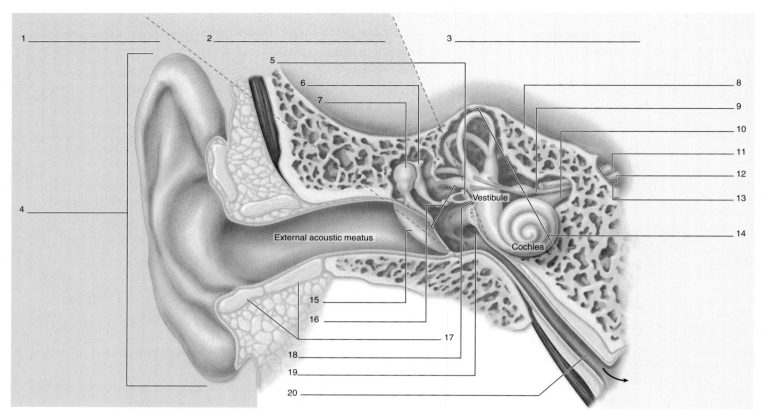

1 _____

2 _____

3 _____

4 _____

5 _____

6 _____

7 _____

8 _____

9 _____

10 _____

11 _____

12 _____

13 _____

14 _____

15 _____

16 _____

17 _____

18 _____

19 _____

20 _____

External acoustic meatus

Vestibule

Cochlea

The Endocrine System

19

OUTLINE and OBJECTIVES

Anatomy & Physiology REVEALED®
aprevealed.com

Module 8: ENDOCRINE SYSTEM

INTRODUCTION

How long has it been since you last ate? One hour? Two? Five? Even if it has been several hours since your last meal, your blood glucose and blood calcium levels remain remarkably stable, fluctuating only minor amounts around your body's normal physiological level (unless, of course, you have a disease such as diabetes). Maintenance of blood glucose and blood calcium levels are physiological imperatives, for if their levels are too high or too low, severe impairment of nervous and muscular activity will occur. However, it is not often that our blood glucose or calcium levels move drastically out of the normal range. This is because they are under tight regulation by the endocrine system. The hormones insulin and glucagon, produced by the pancreas, regulate blood glucose levels, and the hormones calcitonin and parathyroid hormone, produced by the thyroid and parathyroid glands, respectively, regulate blood calcium levels.

This brief description of these hormones and the variables they regulate provides only a glimpse at the functioning of the endocrine system, a system of chemical messengers that travel in the bloodstream and act on distant target cells. The endocrine system (**figure 19.1**) consists of a number of "classical" endocrine organs, such as the pituitary gland, adrenal glands, pancreatic islets,

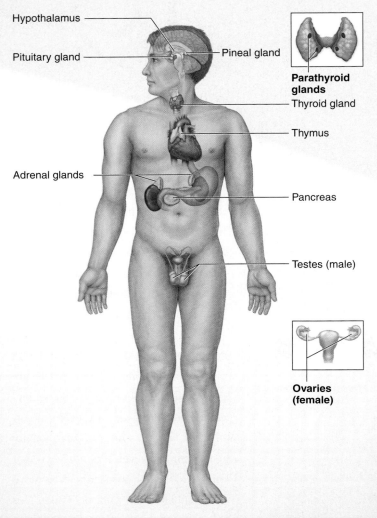

Figure 19.1   **Classical Endocrine Glands of the Human Body.**

Hypothalamus

Pituitary gland

Pineal gland

Parathyroid glands

Thyroid gland

Thymus

Adrenal glands

Pancreas

Testes (male)

Ovaries (female)

and thyroid gland. However, nearly every organ in the body contains cells or tissues that produce and secrete hormones. For instance, cells in the walls of the stomach secrete hormones that regulate appetite, gastric motility, and acid secretion.

In this laboratory exercise, we will explore the structure and function of the classical endocrine organs. In subsequent exercises covering the digestive, urinary, cardiovascular, and reproductive systems, we will explore the structure and function of endocrine cells and tissues (and associated hormones), related to each system in particular.

Chapter 19: The Endocrine System

PRE-LABORATORY WORKSHEET

Name: _____

Date: _____ Section: _____

1. Describe the basic differences between endocrine and exocrine glands.

_____

_____

_____

_____

_____

_____

2. Describe the embryonic origins of the anterior and posterior lobes of the pituitary gland.

_____

_____

_____

_____

_____

_____

3. Where are parafollicular C cells located? _____

4. What endocrine gland consists of a cortex and medulla, with each part embryonically derived from different tissues? _____

5. What endocrine gland consists of follicles lined with simple cuboidal epithelium? _____

6. What endocrine gland secretes hormones that regulate the secretion of hormones by the anterior pituitary gland?

_____

7. What endocrine gland secretes the hormone melatonin? _____

8. The _____ _____ are four small endocrine glands on the posterior part of the thyriod gland that secrete a hormone that regulates blood calcium levels.

9. Name the major endocrine gland that secretes sex steroid hormones in the male and in the female.

a. male _____

b. female _____

I N   T H E   L A B O R A T O R Y

A unique feature of the endocrine system is that it is not entirely limited to large, grossly visible organs. Several organs in the body contain cells that secrete hormones, and thus the organ has an endocrine role, yet the entire organ isn't necessarily referred to as an endocrine gland. The few organs that *are* strictly endocrine in nature are generally quite small. Only the thyroid and adrenal glands can be viewed easily on a cadaver. Thus, our exploration of the endocrine system in the anatomy laboratory will be carried out predominantly at the microscopic level. However, in the gross anatomy section of this chapter you will be asked to locate the major (large) endocrine organs on the cadaver or on classroom models. As you work through each histology slide, make associations between the tissues you are viewing with the gross anatomic location where the tissue is found.

Histology

In the following exercises the microscopic anatomy of the various endocrine glands are described in detail. The tables in these exercises list the glands, the hormones they produce, and the functions of each of the listed hormones. Your focus in the anatomy laboratory will be to differentiate the various endocrine glands from each other when viewing them under the microscope. Learning the details regarding the actions of the hormones produced by each gland is a secondary objective, and is something that you will cover in much greater detail in a course in human physiology.

EXERCISE 19.1   The Hypothalamus and Pituitary Gland

The **pituitary** gland, or hypophysis (*hypophysis*, an undergrowth), is a remarkable organ, for this incredibly tiny organ plays a gigantic role in endocrine regulation of the body. It is about the size and shape of a pea, but it secretes hormones that regulate the growth and development of nearly every other endocrine organ in the body. The secretion of pituitary hormones is tightly controlled by cells within the **hypothalamus,** a part of the brain whose structure and function was discussed in chapter 15. Your primary textbook covers the structure, function, and relationships between the hypothalamus and pituitary gland in detail. Because it is not possible to visualize the details of these relationships in the laboratory, we will focus our studies here on the structure and function of the pituitary gland alone. **Table 19.1** summarizes the cells and structures that compose the pituitary gland and lists the hormones secreted by each cell type, along with the releasing or inhibiting hormones secreted by the hypothalamus that influence the secretion of hormones by the pituitary gland.

1. Obtain a histology slide of the pituitary gland (**figure 19.2**). Before placing it on the microscope stage, observe the slide with your naked eye. Notice there is a distinctive difference in color between the two parts, or lobes, of the pituitary gland. The darker area is the **anterior pituitary** (also called the anterior lobe, or **adenohypophysis** [*adeno-*, a gland, + *hypophysis*, pituitary]), whereas the lighter area is the **posterior pituitary** (also called the posterior lobe, or **neurohypophysis** [*neuro-*, relating to nervous tissue, + *hypophysis*, pituitary]).

2. Place the slide on the microscope stage and bring the tissue sample into focus on low power. Once again, identify the anterior and posterior lobes (figure 19.2*a*). Recall that the anterior pituitary is derived embryologically from an outpocketing of the roof of the mouth and consists of epithelial tissue. Thus, the cells have an appearance that is characteristic of glandular epithelial tissue (figure 19.2*b*). The posterior pituitary is derived embryologically from a downgrowth of the diencephalon of the brain and consists of nervous tissue. Thus, the cells have an appearance that is characteristic of nervous tissue (figure 19.2*b*).

3. *Anterior Pituitary*—Using figures 19.2 and 19.3 as guides, identify the anterior pituitary gland. Then move the microscope stage so the anterior pituitary is in the center of the field of view.

4. Increase the magnification so you can see the glandular nature of the cells (**figure 19.3**). There are three cell types within the anterior pituitary: acidophils, basophils, and chromophobes. The first two kinds of cells are named for their "love" (*-phil*, to love) of acidic or basic dyes. *Acidophils* attract acidic dyes, and appear red in color. *Basophils* attract basic dyes and appear blue in color. *Chromophobes* (*chroma*, color, + *phobos*, fear) are cells that attract neither acidic nor basic dyes, and are thought to be cells that have released their hormone(s). It can be difficult to identify the various cell types within the anterior lobe using standardly prepared slides. However, it is useful to know the cell types because this knowledge allows you to use a couple of handy mnemonic devices to remember the names of the hormones produced by each cell.

5. The mnemonics for remembering which hormones are produced by which cells of the anterior pituitary are as follows:

   Mnemonic for acidophils: **GPA** (as in Grade Point Average: If you do not remember this information, it could be harmful to your GPA.)

   G = growth hormone (GH)
   P = prolactin (PRL)
   A = acidophil

   Mnemonic for basophils: **B-FLAT** (as in the musical note: If you remember this information, it will be beneficial to your GPA and you will be happily singing to the tune of B-flat!)

   B = basophil
   F = follicle-stimulating hormone (FSH)
   L = luteinizing hormone (LH)
   A = adrenocorticotropic hormone (ACTH)
   T = thyroid-stimulating hormone (TSH)

Table 19.1	Histology of the Pituitary Gland				
Pituitary Gland	**Cells**	**Description**	**Hormones Produced**	**Action of Pituitary Hormone(s)**	**Hypothalamic Releasing or Inhibiting Hormone**	**Word Origins**
Anterior Pituitary (Adenohypophysis)	Acidophils	Appear red in color due to their attraction for acidic stains.	GH, PRL	NA	NA	*acidus*, sour (relating to acidic dyes), + *-phil*, to love
			Growth hormone (GH)	Stimulates the liver and other tissues to produce IGF-1 (insulin-like growth factor-1), which promotes bone and muscle growth.	Growth-hormone-releasing hormone (GHRH) stimulates release, whereas growth-hormone-inhibiting hormone (GHIH, somatostatin) inhibits release.	*grōth*, growth, + *hormon*, to set in motion
			Prolactin (PRL)	Stimulates the mammary glands to develop and produce milk.	Prolactin-inhibiting hormone (PIH) inhibits release.	*pro-*, before, + *lac*, milk
	Basophils	Appear blue in color due to their attraction for basic stains.	FSH, LH, ACTH, TSH	See below.	See below.	*baso-*, basic (relating to basic dyes), + *-phil*, to love
			Follicle-stimulating hormone (FSH)	Stimulates the growth and maturation of ovarian follicles (females); stimulates spermatogenesis (males).	Gonadotropin-releasing hormone (GnRH) stimulates release, whereas the hormone inhibin inhibits its release.	*folliculus*, a small sac (referring to the ovarian follicles)
			Luteinizing hormone (LH)	Induces ovulation, stimulates the production of estrogen and progesterone by cells of the corpus luteum (females); stimulates interstitial cells to produce testosterone (males).	Gonadotropin-releasing hormone (GnRH) stimulates release.	*luteus*, yellow (referring to the corpus luteum of the female ovary)
			Adrenocorticotropic hormone (ACTH)	Stimulates the growth, development, and secretion of steroid hormones by the adrenal cortex.	Corticotropin-releasing hormone (CRH) stimulates release.	*adrenocortico*, referring to the adrenal cortex, + *trophe*, nourishment
			Thyroid-stimulating hormone (TSH)	Stimulates the secretion of thyroid hormones by the thyroid gland.	Thyrotropin-releasing hormone (TRH) stimulates release.	*thyroid*, shaped like a shield
	Chromophobes	Appear very light in color due to a lack of staining.	Thought to be devoid of hormone, hence the lack of staining properties	NA	NA	*chroma*, color, + *phobos*, fear
Pituitary Gland	**Cells**	**Description**	**Hypothalamic Hormones Stored and Released**	**Action of Pituitary Hormone(s)**	**Hypothalamic Nucleus Containing Neuron Cell Bodies**	**Word Origins**
Posterior Pituitary (Neurohypophysis)	Axon terminals	Axon terminals store hormone that was produced in the cell bodies of the neurons within the hypothalamus.	Oxytocin	Stimulates uterine contractions and milk ejection by mammary glands.	Paraventricular nucleus and supraoptic nucleus.	*axon*, axis, + *terminus*, the limit; *para-*, next to, + *ventricular*, relating to the third ventricle of the brain, + *nucleus*, a collection of neuron cell bodies; *okytckos*, swift birth
	Axon terminals	Axon terminals store hormone that was produced in the cell bodies of the neurons within the hypothalamus.	Antidiuretic hormone (ADH, vasopressin)	Increases water retention by the kidneys (increases blood volume and blood pressure); vasoconstriction.	Paraventricular nucleus and supraoptic nucleus.	*axon*, axis, + *terminus*, the limit; *supra-*, above, + *optic*, relating to the optic tract, + *nucleus*, a collection of neuron cell bodies; *anti*, against, + *diuresis*, excretion of urine
	Pituicytes	Derived from glial cells; have processes that surround axon terminals of the hormone-secreting neurons.	NA	NA	NA	*pituita*, a phlegm (relating to the pituitary gland), + *-cyte*, cell

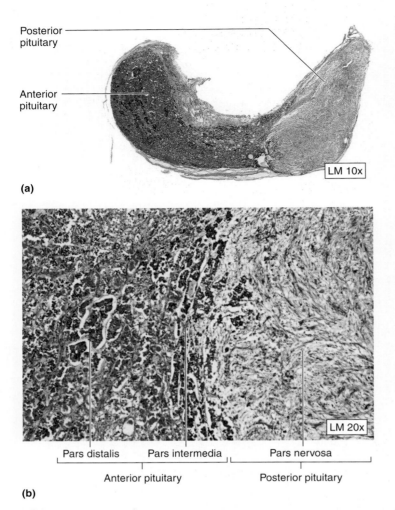

(a)

(b)

Figure 19.2 **Pituitary Gland.** (a) Low magnification view of both parts of the pituitary. (b) Medium magnification view of the anterior and posterior pituitary.

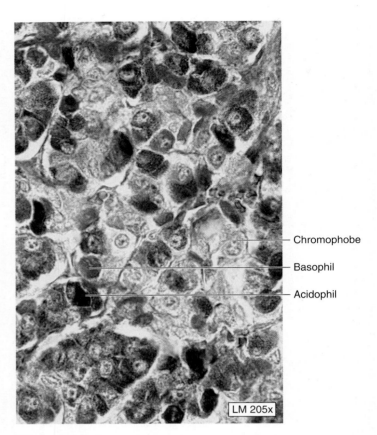

Figure 19.3 **Anterior Pituitary.** The three cell types in the anterior pituitary are basophils, which stain blue, acidophils, which stain red, and chromophobes, which don't take up biological stains.

6. *Posterior Pituitary*—Using figure 19.2 as a guide, identify the posterior pituitary. Then move the microscope stage so the posterior pituitary is in the center of the field of view.

7. Increase the magnification so you can see the cells clearly (**figure 19.4**). The majority of the nuclei visible within the slide are the nuclei of **pituicytes**, which are derived from glial cells. Pituicytes surround the axon terminals of neurons whose cell bodies are located in the paraventricular and supraoptic nuclei of the hypothalamus. These neurons secrete the hormones oxytocin and antidiuretic hormone (ADH, vasopressin).

8. Using figures 19.2 through 19.4, and table 19.1 as guides, identify the following structures on the slide of the pituitary gland.

☐ acidophils		☐ chromophobes
☐ anterior pituitary		☐ pituicytes
☐ basophils		☐ posterior pituitary

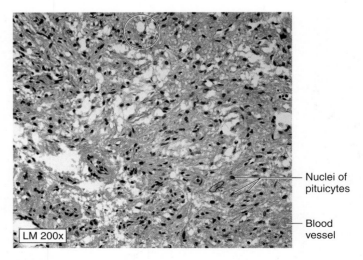

Figure 19.4 **Posterior Pituitary.** The majority of the nuclei seen in this micrograph are of pituicytes, which are glial cells.

9. In the space below, make a simple drawing of the pituitary gland, labeling the structures listed in step 8.

_____ ✕

WHAT DO YOU THINK?

① Tumors of the pituitary gland are not that uncommon. Symptoms of pituitary tumors often result from oversecretion of pituitary hormones. For instance, an increase in bone and muscle growth can occur if the tumor secretes excess growth hormone (GH). Based on your understanding of the location of the pituitary gland with respect to other brain structures, can you explain why individuals with pituitary gland tumors often also experience visual disorders? (Hint: none of the hormones secreted by the pituitary gland influence vision, so think about structures located next to the pituitary.)

EXERCISE 19.2 The Pineal Gland

The **pineal gland** (*pineus*, relating to a pine, shaped like a pine cone), also called the **pineal body,** is a small region in the epithalamus of the brain. Its primary cells, called **pinealocytes** (**figure 19.5**), secrete the hormone **melatonin.** Melatonin's effects in humans are unsubstantiated, but in other organisms melatonin is responsible for regulation of circadian rhythms. In humans, it may have a role in determining the onset of puberty. Pinealocytes are innervated by neurons from the sympathetic nervous system, and their secretion of hormone is affected by the amount of light received by the individual, which is relayed to the pineal through these neurons. Melatonin secretion increases when light levels are low (at night) and decreases when light levels are high (during the day). Pinealocytes appear in groups of cells within the pineal gland. They are surrounded by glial cells, whose function is similar to that of astrocytes in other parts of the brain. Clinically, one of the most important features of the pineal gland is the presence of calcium concretions, termed "**pineal sand**" (corpora arenacea). These concretions are easily visible in radiographs of the head, and provide radiologists with a landmark that is consistent and easy to identify. The number of concretions in the pineal gland increases with age.

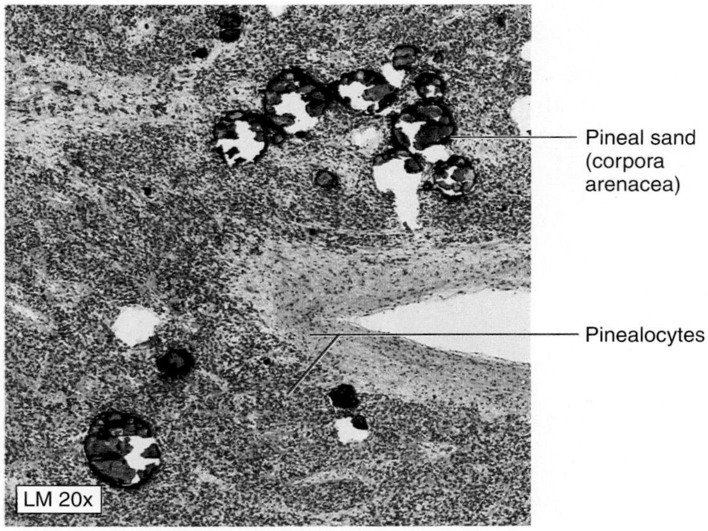

1. Obtain a histology slide of the pineal gland. Place the slide on the microscope stage and bring the tissue sample into focus on low power.

2. Using figure 19.5 as a guide, identify the following structures on the slide:

   ☐ pinealocytes   ☐ pineal sand

3. In the space below, make a simple drawing of the pineal gland as seen at medium or high magnification, labeling pinealocytes and pineal sand.

_____ ✕

Pineal sand (corpora arenacea)

Pinealocytes

LM 20x

Figure 19.5   **The Pineal Gland.**   Histology of the pineal gland.

EXERCISE 19.3 The Thyroid and Parathyroid Glands

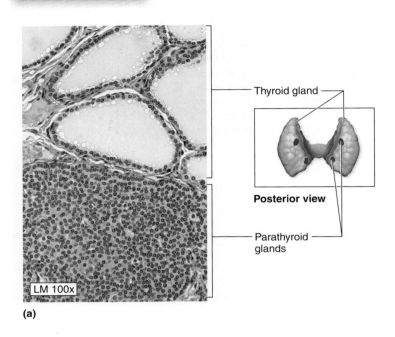

(a)

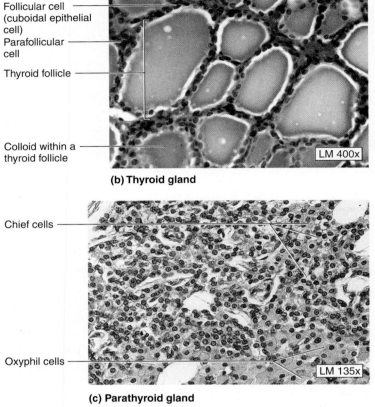

Follicular cell
(cuboidal epithelial
cell)
Parafollicular
cell
Thyroid follicle
Colloid within a
thyroid follicle

LM 400x

(b) Thyroid gland

Chief cells
Oxyphil cells

LM 135x

(c) Parathyroid gland

Figure 19.6 **Thyroid and Parathyroid Glands.** (a) The thyroid gland consists of large thyroid follicles and the parathyroid gland consists of tightly packed cells. (b) Thyroid follicles, which produce thyroid hormone, and parafollicular cells, which produce calcitonin. (c) High magnification view of the parathyroid gland demonstrating chief cells and oxyphil cells.

The **thyroid gland** (**figure 19.6**; **table 19.2**) is a butterfly-shaped gland located anterior to the trachea and inferior to the thyroid cartilage of the larynx. It consists of two main lobes connected to each other anteriorly by a narrow **isthmus** (*isthmus*, neck). The functional units of the thyroid gland are **thyroid follicles**, which are lined with a simple cuboidal epithelium. Inside each follicle is a mass of **colloid**, consisting largely of precursors to the thyroid hormones, called **thyroglobulins**. When combined with iodine, **thyroid hormone** breaks off from the colloid and is transported out of the follicles and into the blood. The spaces between the follicles contain another cell type, called **parafollicular cells**. These cells secrete the hormone **calcitonin**.

Closely related to the thyroid gland are a series of small glands (usually 4) called **parathyroid glands**. These glands consist of two cell types: **chief (principal) cells**, which are smaller, more abundant cells with relatively clear cytoplasm that and produce **parathyroid hormone** (parathormone), and **oxyphil cells**, which are larger, less abundant cells with granular pink cytoplasm, and whose function is unknown.

1. Obtain a histology slide of the thyroid and parathyroid glands and place it on the microscope stage. Bring the tissue sample into focus on low power and then change to high power.

2. Using figure 19.6 and table 19.2, as guides, identify the following structures on the slide:

☐ chief (principal) cells   ☐ oxyphil cells

☐ colloid   ☐ parafollicular cells

☐ follicular cells (cuboidal epithelial cells)   ☐ parathyroid gland

   ☐ thyroid follicles

3. In the space below, make a simple drawing of the thyroid and parathyroid glands, labeling all of the structures listed above.

×

Table 19.2	Histology of the Thyroid and Parathyroid Glands				
Cell Types	**Description**	**Hormones Produced**	**Action of Hormone**	**Mechanism of Action**	**Word Origins**
THYROID GLAND	A shield- or butterfly-shaped gland located anterior to the trachea and inferior to the thyroid cartilage. Consists of two lobes connected by a narrow isthmus anteriorly				thyroid, shaped like an oblong shield
Follicular Cells	Simple cuboidal epithelial cells that line the thyroid follicles; they have very dark nuclei.	Thyroid hormone	Increase basal metabolic rate (BMR); important in early development of the central nervous system.	Stimulates or inhibits transcription of certain genes in target cells.	folliculus, a small sac
Parafollicular Cells	Lighter-staining cells found in the interstitial spaces between the thyroid follicles; larger than the follicular cells, with nuclei that have an appearance similar to a clock face.	Calcitonin	Decrease blood calcium levels.	Inhibits the action of osteoclasts, increases urinary excretion of calcium, decreases digestive absorption of calcium.	para, next to, + folliculus, a small sac
PARATHYROID GLANDS	4–6 small glands located on the posterior surface of the thyroid gland				para, next to, + thyroid, shaped like an oblong shield
Chief (Principal) Cells	Relatively small cells that contain a centrally located, round nucleus with one or more nucleoli.	Parathyroid hormone (parathormone)	Increase blood calcium levels.	Indirectly increases the action of osteoclasts, decreases urinary excretion of calcium, and stimulates synthesis of vitamin D, which increases dietary absorption of calcium.	Principal, the predominant cell type of a gland
Oxyphil Cells	Larger than chief cells and more reddish in color.	Unknown	NA	NA	oxys, sour acid, + -phil, to love

EXERCISE 19.4 The Adrenal Glands

The adrenal glands are located directly superior to each kidney (*ad*, to, + *ren*, kidney). They are similar to the pituitary gland in that they are composed of two regions (**figure 19.7**), each with a separate embryological origin. The outer region, the **adrenal cortex**, is derived from mesoderm and has the appearance of typical glandular epithelium. The inner region, the **adrenal medulla**, is derived from modified postganglionic sympathetic neurons (which are derived from neural crest cells) and has the appearance of nervous tissue. The cells of the adrenal cortex synthesize steroid hormones (specifically, **corticosteroids**, *cortico*, relating to the adrenal cortex, + *steroid*, steroid hormone), whereas the cells of the adrenal medulla synthesize **catecholamine** hormones (that is, hormones derived from the amino acid tyrosine and that contain a catechol ring). The entire gland is surrounded by a dense irregular connective tissue **capsule**, which protects the gland and helps anchor it to the superior

border of the kidney. The adrenal cortex has three recognizable zones, and each zone has cells that predominantly secrete one category of corticosteroid hormones. Characteristics of the zones, and descriptions of the hormones secreted by cells within each zone, are summarized in **table 19.3**.

1. Obtain a histology slide of the adrenal gland (figure 19.7) and place it on the microscope stage. Bring the tissue sample into focus on low power and identify the two major regions, the cortex and the medulla (figure 19.7a).

2. Move the stage so the *adrenal cortex* (figure 19.7b) is in the center of the field of view. Then change to high power.

3. Using figure 19.7 and table 19.3 as guides, identify the following zones of the adrenal cortex: zona glomerulosa, zona fasciculata, and zona reticularis.

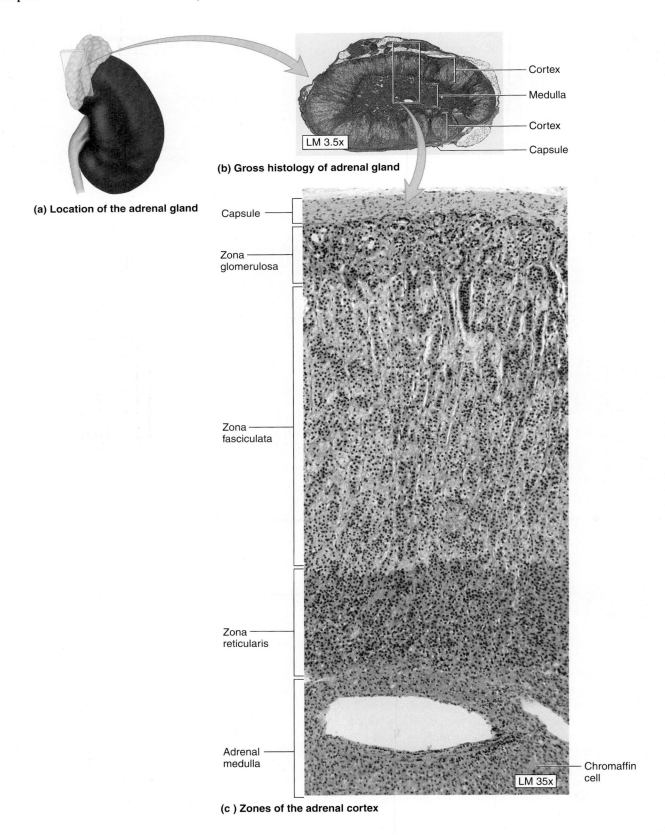

(a) Location of the adrenal gland

(b) Gross histology of adrenal gland

LM 3.5x

Cortex

Medulla

Cortex

Capsule

Capsule

Zona glomerulosa

Zona fasciculata

Zona reticularis

Adrenal medulla

Chromaffin cell

LM 35x

(c) Zones of the adrenal cortex

Figure 19.7   **Adrenal Glands.**   (a) The adrenal glands are located superior to each kidney. (b) Low magnification micrograph of the adrenal glands. (c) High magnification micrograph demonstrating the three layers of the adrenal cortex and part of the adrenal medulla.

4. Change back to the low power objective, move the microscope stage so the *adrenal medulla* is in the center of the field of view, and then change back to high power. The nuclei you see within the adrenal medulla are nuclei of *chromaffin cells*, which are modified post-ganglionic sympathetic neurons. What hormone(s) do these cells secrete? _____

5. Using figure 19.7 and table 19.3, as guides, identify the following structures on the slide of the adrenal glands:

☐ adrenal cortex   ☐ zona fasciculata

☐ adrenal medulla   ☐ zona glomerulosa

☐ capsule   ☐ zona reticularis

☐ chromaffin cells

6. In the space below, make a simple drawing of the adrenal gland, labeling all of the structures listed above.

7. *Optional Activity*: AP|R **Endocrine System**—Review the histology slides of the adrenal (suprarenal) gland, as well as the pituitary gland, thyroid gland, and endocrine pancreas.

Table 19.3	Histology of the Adrenal Glands					
Adrenal Gland Region	Zone and/or Cells	Description	Hormones Produced	Action of Hormone(s)	Word Origins	
Adrenal Cortex	Zona glomerulosa	The outermost region of the adrenal cortex, containing "balls" of cells, located immediately deep to the capsule of the adrenal gland.	Mineralcorticoids: aldosterone	Increases sodium and water retention by the kidneys (thus increasing blood volume and blood pressure); vasoconstriction.	*zona*, zone, + *glomus*, a ball of yarn	
	Zona fasciculata	The middle and largest region of the adrenal cortex, containing long cords, or "bundles," of cells.	Glucocorticoids: cortisol	Mobilization of glucose by stimulating protein breakdown and gluconeogenesis (the production of new glucose from amino acids), and lipolysis of adipose tissue.	*zona*, zone, + *fasciculus*, a bundle	
	Zona reticularis	The innermost region of the adrenal cortex, containing a "network" of cells, located between the zona fasciculata and the cells of the adrenal medulla.	Glucocorticoids and gonadocorticoids: androgens	Androgens are similar in structure and function to the male sex steroid hormone testosterone; in females they are responsible for sex drive, in males they have little function because they are secreted in very low amounts compared to testosterone secretion by the testes.	*zona*, zone, + *rete*, a net	
Adrenal Medulla	Chromaffin cells	Large, spherical cells that have a yellowish-brown tint when specially stained due to their reaction with chrome salts.	Catecholamines: epinephrine and norepinephrine	Epinephrine, secreted in the largest quantity (80–90% of all hormone secretion by the adrenal medulla), increases heart rate and contractility (force of heart muscle contraction); norepinephrine is a powerful vasoconstrictor.	*chroma*, color, + *affinis*, affinity, attraction for	

EXERCISE 19.5   The Endocrine Pancreas—Pancreatic Islets (of Langerhans)

The **pancreas** is largely an exocrine gland. **Exocrine** glands produce substances that are secreted into ducts. The majority of the cells within the exocrine pancreas produce digestive enzymes and bicarbonate, which are secreted into ducts that empty into the small intestine. The clusters of exocrine cells are **pancreatic acini** (singular: acinus). Interspersed between the acini are small islands of cells that have an endocrine function. **Endocrine** glands secrete hormones into the blood. The endocrine part of the pancreas consists of the **pancreatic islets** (islets of Langerhans). These islets contain hormone-secreting cells and a rich supply of blood capillaries. Four distinct cell types exist within the islets, although they cannot be distinguished in a normal histological preparation. Thus, when you observe the slide of the pancreas, your goal will be to simply identify the pancreatic islets, not the specific cell types within. Nonetheless, you should know which cell type secretes each hormone. The cell types and hormones secreted by each cell type are listed in **table 19.4.**

1. Obtain a histology slide of the pancreas, place it on the microscope stage, and bring the tissue sample into focus on low power. The majority of the cells you will see, which will look somewhat like cuboidal epithelial cells, are the exocrine cells of the pancreas (**figure 19.8**). As you scan the slide you will also see some small islands of cells, the **pancreatic islets,** which (typically) stain lighter in color than the exocrine cells.

2. Locate a pancreatic islet and move the microscope stage so the islet is in the center of the field of view.

3. Increase the magnification so you can see the islet in greater detail. Using figure 19.8 and table 19.4 as guides, identify the following on the slide:

☐ pancreatic acini (exocrine cells)   ☐ pancreatic islets

☐ secretory ducts

Table 19.4	Histology of the Pancreatic Islets (of Langerhans)				
Cell Types	**Description**	**Hormones Produced**	**Action of Hormone**	**Mechanism of Action**	**Word Origins**
Alpha Cells	Compose about 30% of islet cells; located on the periphery of the islet.	Glucagon	Increases blood glucose levels.	Stimulates glycogenolysis (breakdown of glycogen) in the liver, and gluconeogenesis (formation of new glucose from amino acids or fats).	*alpha,* the first letter of the Greek alphabet, + *glucose,* sugar, + *ago,* to lead
Beta Cells	Compose about 65% of islet cells; located in the center of the islet.	Insulin	Decreases blood glucose levels.	Stimulates glucose uptake by muscle cells, liver cells, and adipocytes, and stimulates glycogenesis (formation of glycogen from glucose) in the liver.	*beta,* the second letter of the Greek alphabet, + *insula,* an island
Delta Cells	Compose about 4% of islet cells; located on the periphery of the islet.	Somatostatin	Inhibits the release of glucagon and insulin by alpha and beta cells.	Inhibits the release of glucagon and insulin when nutrient levels in the bloodstream are high.	*delta,* the 4th letter of the Greek alphabet, + *soma,* body, + *stasis,* standing still
F Cells	All other (rare) cell types in the pancreatic islets (~1%) are grouped together and given this name.	Pancreatic polypeptide	Inhibits somatostatin release from delta cells.	Regulates secretion of somatostatin from delta cells.	NA

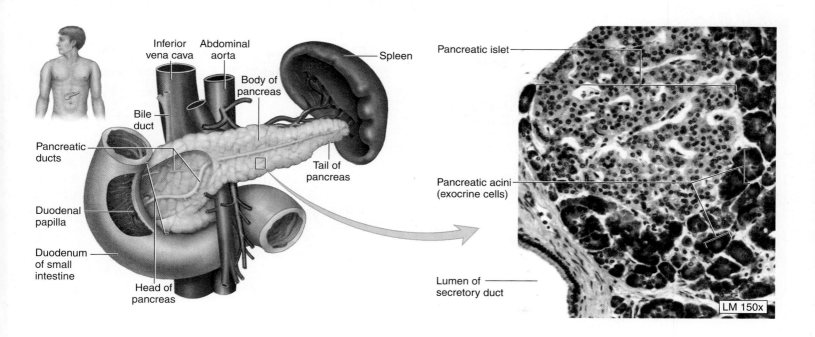

Figure 19.8   Endocrine Pancreas.   The endocrine part of the pancreas consists of the pancreatic islets, which are the lighter colored regions in this micrograph. The darker cells surrounding the islet are acinar cells, which compose the exocrine part of the pancreas.

4. In the space below, sketch a pancreatic islet and the exocrine cells that surround it. Be sure to label the structures listed in step 3.

_____ ×

Gross Anatomy

Some endocrine glands are difficult to identify on the cadaver. However, most classical glands and associated structures can be identified on the cadaver or on classroom models, so these will be the focus of this section.

EXERCISE 19.6   Gross Anatomy of Endocrine Organs

1. Observe a human cadaver or classroom models of the brain, thorax, abdomen, and skull.

2. Using figure 19.1 (p. 384) and your textbook as guides, identify the following organs or structures on the human cadaver or on classroom models, then label them in figure 19.9.

☐ adrenal gland   ☐ pineal gland

☐ hypothalamus   ☐ pituitary gland

☐ ovaries   ☐ testes

☐ pancreas   ☐ thyroid gland

3. *Optional Activity:* **AP|R** **Endocrine System**—Watch the endocrine system animations to review the structure and function of the hypothalamus and pituitary gland, pancreas, thyroid and parathyroid glands, and adrenal (suprarenal) glands.

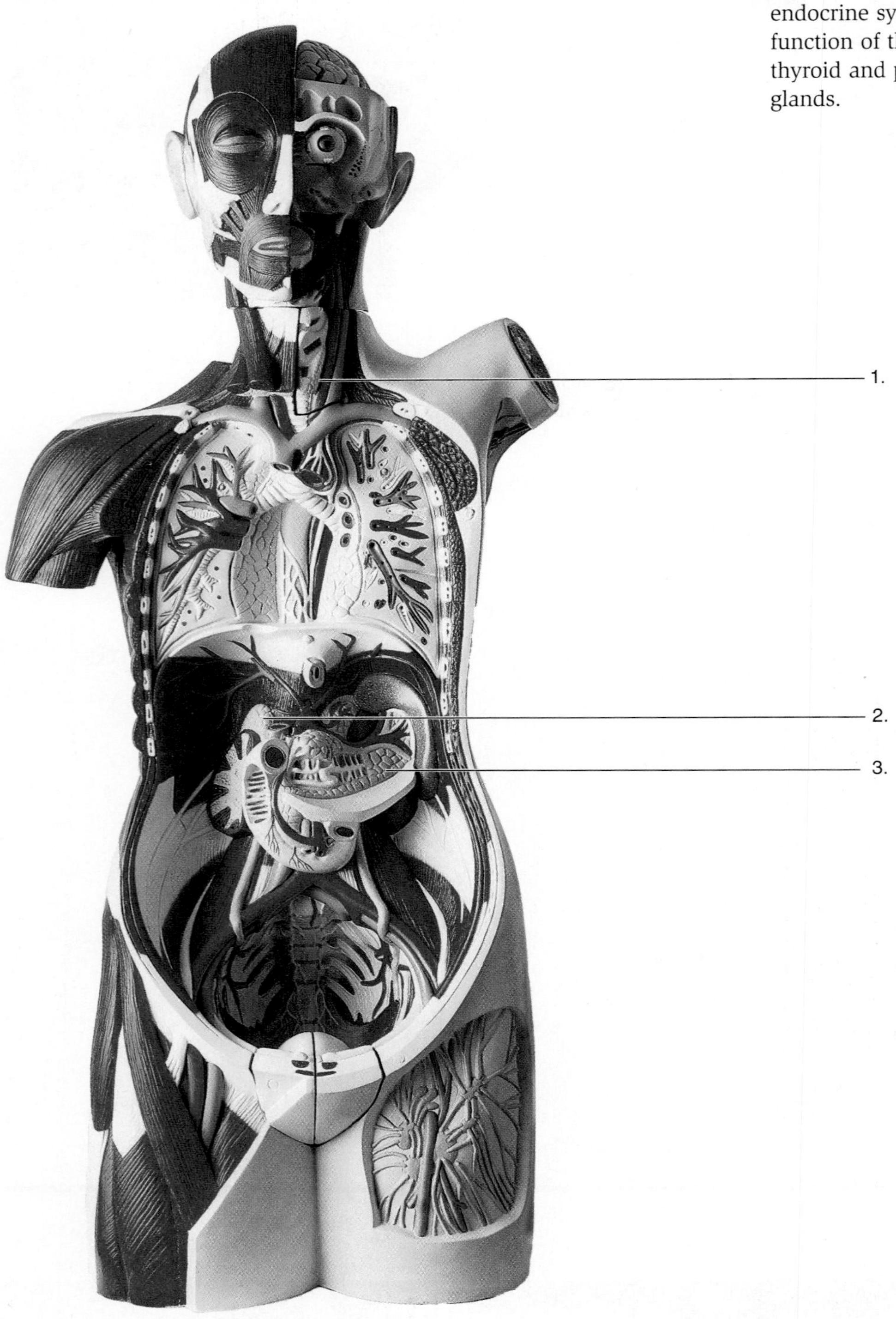

(a) Human torso model

Figure 19.9   **Classroom models demonstrating major endocrine glands of the body.**   (a) Human torso model. (b) Model of a sagittal section of the brain.

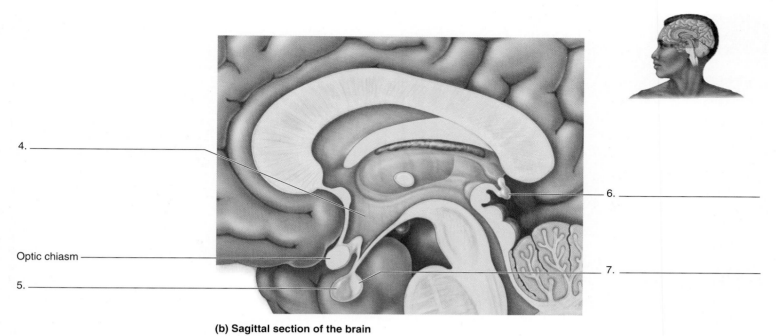

4. —————————

6. —————————————————

Optic chiasm —————

5. —————

7. —————————————————

(b) Sagittal section of the brain

Figure 19.9   **(Continued)**

Chapter 19: The Endocrine System

Name: _____

Date: _____ Section: _____

POST-LABORATORY WORKSHEET

1. Match the cells or structures listed in column A with the endocrine gland in which they are found in column B.

   Column A

   _____ 1. oxyphil cells

   _____ 2. pituicytes

   _____ 3. colloid

   _____ 4. parafollicular cells

   _____ 5. chromophobes

   _____ 6. chief (principal) cells

   _____ 7. chromaffin cells

   _____ 8. alpha cells

   _____ 9. zona fasciculata

   _____ 10. pinealocytes

   Column B

   a. adrenal gland

   b. anterior pituitary

   c. pancreas

   d. parathyroid gland

   e. pineal gland

   f. posterior pituitary

   g. thyroid gland

2. List the hormones produced by acidophils and basophils of the anterior pituitary.

   Acidophils   *Basophils*

   _____   _____

   _____   _____

                                                          _____

                                                          _____

3. What are chromophobes?

   _____

   _____

4. What structures are located within the paraventricular and supraoptic nuclei of the hypothalamus?

   _____

   _____

   _____

5. List the two hormones released in the posterior pituitary.

   _____

   _____

6. Describe the structure and function of the thyroid gland. In addition, list the two hormones produced by the thyroid gland and identify the cell types that secrete each hormone.

_____

_____

_____

_____

_____

_____

7. List the hormones secreted by each region of the adrenal gland.

zona glomerulosa _____

zona fasciculata _____

zona reticularis _____

medulla _____

8. List the hormones secreted by each cell type of the pancreatic islets (of Langerhans).

alpha cells _____ delta cells _____

beta cells _____ F cells _____

9. Describe why pineal sand is important to radiologists.

_____

_____

10. Identify the following glands:

a. _____ b. _____ c. _____ d. _____

The Cardiovascular System: Blood

20

OUTLINE and OBJECTIVES

INTRODUCTION

Blood as a bodily fluid has fascinated humans for centuries. Its rich red color adds to both its beauty and its mystery. For many of us, the sight of blood is quite disturbing because it usually indicates that something is very wrong! It's no wonder that for centuries humans have been fascinated by this most colorful bodily fluid. Early Greek physicians believed that blood was one of the four bodily "humors" (*humor*, a bodily fluid). The other bodily fluids were phlegm, black bile, and yellow bile (**table 20.1**). Each of these humors was associated with an element: air, water, earth, or fire, and had particular characteristics: hot, cold, wet, or dry. Before you put any confidence in this idea, know that modern science recognizes over 92 naturally occurring elements, and earth, air, fire, and water are not among them! Greek physicians believed that a balance of the four humors was required for optimal health of both the body and the mind, and that disease was the result of an imbalance between the bodily fluids. It was this concept of disease that made bloodletting a popular treatment for disease. By draining some blood from a patient, the physician supposedly was putting the patient's humors back into balance. In those days, studying anatomy or medicine wasn't nearly as complicated as it is today. On the

Table 20.1	The Four Greek Humors					
Bodily Fluid (Humor)	Element	Characteristics	Source	Mood	Mood Characteristics	Word Origins
Yellow Bile	Fire	Hot and dry	Liver	Choleric	Irritable	*chole*, bile
Black Bile	Earth	Dry and cold	Spleen	Melancholy	Sad, depressed	*melas*, black, + *chole*, bile
Phlegm	Water	Cold and wet	Brain	Phlegmatic	Calm and unexcitable	*phlegma*, inflammation
Blood	Air	Hot and wet	Heart	Sanguine	Lively and optimistic	*sanguis*, blood

other hand, we can all be thankful that we didn't have to be patients back then. To this day we continue to refer to this early concept of medicine, even though most of us are largely unaware we are doing so. We refer to an infant who cries all the time as being "colicky." We refer to someone who is sad as feeling "melancholy." The next time you are feeling sad, consider that if you had lived in the year 400 B.C., your doctor would have diagnosed you as having an excess of "black bile." Then have a good laugh and be thankful that you are living in the twenty-first century. If that isn't enough to give you a good laugh, consider what early Greeks thought to be the source of phlegm: the brain. That thought may be enough to transform you from melancholy to "sanguine," or lively and optimistic. Are you feeling lively and optimistic yet? Then you are ready to tackle the current "sanguine" topic: blood.

Focus | Human Blood Samples for Laboratory Investigations

Blood is an amazing fluid with a beautiful red color due to the presence of hemoglobin within red blood cells. Hemoglobin is essential for the transport of oxygen throughout the body. Blood transports much more than just oxygen, however. It also transports nutrients, waste products, carbon dioxide, hormones, and heat. Proper circulation of blood to the tissues is essential for their survival. In medicine, a sample of a patient's blood can yield important clues as to what diseases may be affecting the individual. For instance, a white blood cell count performed on a sample of the patient's blood can determine whether the patient has a bacterial infection. An excess of neutrophils will confirm the suspicion that the patient has such an infection, whereas a normal abundance of neutrophils might be a clue that an infection is viral, rather than bacterial, in nature.

Figure 20.1 summarizes the steps involved in obtaining a sample of human blood for laboratory analysis. A blood sample is collected by a *phlebotomist* (*phleps*, vein, + *tomē*, to cut) and placed in a test tube (figure 20.1, step 1). The test tube is then placed in a centrifuge, which spins the sample at high speed to separate the component parts (figure 20.1, step 2). Because the formed elements like erythrocytes and leukocytes are heavy, they fall to the bottom of the test tube, while the blood plasma remains floating on the top (figure 20.1, step 3). Leukocytes and platelets form a thin layer called a *buffy coat*, which sits on top of the layer of erythrocytes. The iron present in the hemoglobin in erythrocytes makes erythrocytes heavier than leukocytes, which explains why erythrocytes lie in the layer below leukocytes in the test tube.

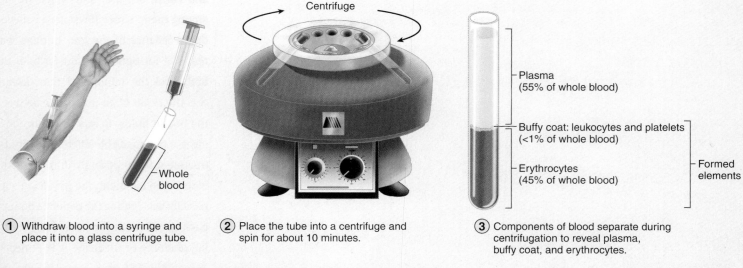

Centrifuge

Plasma (55% of whole blood)

Buffy coat: leukocytes and platelets (<1% of whole blood)

Erythrocytes (45% of whole blood)

Formed elements

① Withdraw blood into a syringe and place it into a glass centrifuge tube.

Whole blood

② Place the tube into a centrifuge and spin for about 10 minutes.

③ Components of blood separate during centrifugation to reveal plasma, buffy coat, and erythrocytes.

Figure 20.1 **Separation of a Whole Blood Sample by Centrifugation.** Using a centrifuge to separate whole blood into plasma (~55%) and formed elements (~45%) is the first step in determining the composition of whole blood.

Chapter 20: Cardiovascular System—Blood

Name: _____

Date: _____ Section: _____

PRE-LABORATORY WORKSHEET

1. List the three components of blood that separate from each other when a blood sample is centrifuged (see figure 20.1 for reference).

   _____

   _____

   _____

2. Explain the basis for the categorization and naming of the two major categories of white blood cells (granulocytes vs. agranulocytes).

   _____

   _____

   _____

   _____

3. Explain the basis for the categorization and naming of the three types of granulocytes (eosinophils, basophils, and neutrophils).

   _____

   _____

   _____

   _____

4. Define *hemopoiesis:*

   _____

   _____

   _____

5. Megakaryocytes give rise to formed elements called _____.

IN THE LABORATORY

In this chapter we will explore the normal constituents of human blood. Knowledge of the normal appearance and abundance of the various types of blood cells provides a basis for recognizing when cells or abundances have become abnormal, and what that means. Tables 20.2 and 20.3 summarize the characteristics of the formed elements of blood, which are the structures you will focus on iden-

tifying in this laboratory exercise. Considerably more detail on the structure and function of blood is covered in your main textbook for this course. Thus, it is important to make sure that you have read and understood the information in the chapter of your primary anatomy textbook before performing the laboratory exercises.

Table 20.2	Characteristics of the Formed Elements				
Formed Element	**Description**	**Function**	**Size**	**Percentage of Formed Elements**	**Word Origins**
Platelets	Small, purple-colored structures often mistaken for "junk" on the slide. They are cell fragments of megakaryocytes and are tiny compared to erythrocytes and leukocytes.	Play a role in hemostasis (blood clotting). When activated, the surfaces become spiny instead of smooth and they become very "sticky," forming a platelet plug that helps stop bleeding from the blood vessel wall.	~2 μm	1%	*platys*, flat
Erythrocytes (Red Blood Cells, RBCs)	Very uniform in size and shape. Shaped like biconcave discs, lack a nucleus, and are orange to red in color due to their attraction for eosinophilic stains.	Transport oxygen and carbon dioxide, and participate in regulation of the pH of blood.	~7.5 μm	99%	*erythros*, red, + *kytos*, a hollow (cell)
Leukocytes (White Blood Cells, WBCs)	Generally purple in color due to the basophilic staining properties of their nuclei. A summary of the characteristics of the various leukocytes is provided in table 20.3.	Protect the body from pathogens, fight infections, and remove dead or damaged tissues.	7–21 μm	<1%	*leukos*, white, + *kytos*, a hollow (cell)

Table 20.3	Leukocyte Characteristics							
Leukocyte	**Description**	**Function**	**Percentage of Leukocytes**	**Diameter Relative to That of an RBC**	**Nuclear Shape**	**Color of Cytoplasmic Granules**	**Common Conditions Causing an Increase in Abundance**	**Word Origins**
Eosinophils	Have orange to reddish cytoplasmic granules that are fairly light in color; the bluish-colored, bi-lobed nucleus can be seen easily.	Fight parasitic infections and mediate (neutralize) the effects of histamines. Phagocytize antigen–antibody complexes and allergens.	2–4%	~1.3x	Bi-lobed	Orange or red (eosinophilic)	Parasitic infections and allergic reactions.	*acidus*, sour, referring to acidic dyes such as eosin, + *philos*, to love
Basophils	Very rare and therefore difficult to locate on a blood smear. Cytoplasmic granules stain very dark blue, so the nucleus is generally not visible, though it is usually bilobed.	Release histamine and heparin, and are involved in the inflammatory response.	<1%	~1.3x	Bi-lobed	Blue or purple (basophilic)	Tissue injury and allergic reactions.	*baso*-, referring to basic dyes, + *philos*, to love
Neutrophils	Have very light, lavender-colored cytoplasmic granules; the multi-lobed nucleus can be easily seen.	Important in fighting bacterial infections. They migrate out of the blood toward the site of infection, where they phagocytize bacteria and damaged tissues.	60–70%	~1.3x	Multi-lobed (~5 lobes)	Lavender (pale blue-purple)	Bacterial infection.	*neutro*-, neutral, + *philos*, to love
Monocytes	Recognized by their enormous size, at least twice the size of an erythrocyte. Have a large, blue/purple nucleus that often has a small indentation in it, making it "horseshoe-shaped," which helps distinguish it from a lymphocyte.	Monocytes are circulating cells. When they migrate out of the bloodstream, they become large, phagocytic cells called macrophages. They have little to no function in circulating blood.	3–8%	~2–3x	Large and horseshoe-shaped	NA	Bacterial infection.	*monos*, single, + *kytos*, a hollow (cell)
Lymphocytes	Nearly the same size as an erythrocyte. Nuclei are very large and blue/purple in color, and are surrounded by a halo of pale blue cytoplasm.	Responsible for the specific immune response to infection. Each type of lymphocyte has a specific function in fighting pathogens.	20–25%	~1–2x	Large and round, nearly fills the entire cell	NA	Viral infections, autoimmune diseases.	*lympho*-, referring to the lymphatic system, + *kytos*, a hollow (cell)

Histology

EXERCISE 20.1 Identification of Formed Elements on a Prepared Blood Smear

1. Obtain a prepared slide of human blood or use the slide you prepared (see Focus: Making a Human Blood Smear). Place the slide on the microscope stage and observe at low power. You will not be able to see much at this magnification, but, as always, it is important to progress from low to high power so you will continue to keep the sample in focus before progressing to the oil immersion objective.

2. Change from low to medium, then to high power, making sure to bring the slide into focus at each power. After you have the slide in focus on high power, obtain a vial of immersion oil. Rotate the nosepiece (the part of the microscope that holds the objective lenses) so the high-power and oil immersion objectives lie on either side of the slide (**figure 20.3**). Place a drop of immersion oil over the center of the slide (where you see the light coming through the slide), and then carefully rotate the nosepiece to bring the immersion objective into place over the slide. Bring the slide into focus.

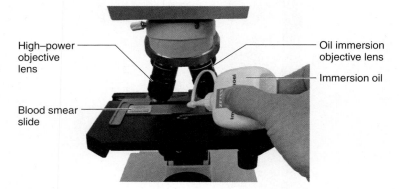

Figure 20.3   **Orienting the Objective Lenses for Placement of a Drop of Immersion Oil on the Prepared Blood Slide.**

Focus | Making a Human Blood Smear

In exercise 20.1 you will either use a prepared slide of human blood, or you will be asked to make your own slide using your own blood. To make a blood smear using your own blood, use the following procedure:

1. Obtain four glass slides, a lancet, cotton balls, alcohol prep pads, and a vial of Wright's stain. Clean the slides with soap and water and let them dry. It takes two slides to make a blood smear, and you may have to try several times before you get a good preparation, which is why you will most likely need to clean at least four slides.

2. Using the alcohol prep pad, clean the tip of the middle finger on the hand you use the least (that is, if you are right-handed, use the tip of the middle finger on your left hand).

3. Place the tip of the lancet on the clean finger, and pull the trigger so it pierces the skin.

4. Squeeze the finger until a small drop of blood appears on the surface. Blot away the first drop of blood using the cotton ball and then squeeze the finger again until another drop appears. Gently touch the drop of blood to the top surface of the microscope slide about 1.5–2 cm from the edge of the slide

(**figure 20.2**, step 1). You do not need a lot of blood to make the smear, so try not to squeeze the finger too hard.

5. Orient the slide so the end with the drop of blood on it is closest to you. Then place the edge of a second microscope slide on the drop of blood and hold it at a 45-degree angle to the slide with the drop of blood on it (figure 20.2, step 2).

6. Quickly push the second slide over the first slide (push away from you) to smear the blood (figure 20.2, step 3). You are aiming for a very thin, transparent layer of blood on the slide. If the layer is too thick, you will have difficulty seeing the cells because they will be too closely packed together. If you don't end up with an even smear, or good spread, start over with two new slides. It might take a couple of tries before you are able to get a good, even smear. Discard any slides that made contact with blood in a bowl containing a bleach solution. Allow the slide with the blood smear on it to air dry.

7. Place a couple of drops of Wright's stain on the blood smear slide so you end up with enough stain on the slide to cover the smear, but not so much that it overflows the slide. Keep track of the number of drops you use (number of drops = _____).

8. Let the slide stand for 3–4 minutes, until the smear begins to take on a blue/green color.

9. Add to the slide a volume of distilled water equal to the volume of Wright's stain you used. Generally this means you will use the same number of drops of distilled water as you used of Wright's stain. Move the slide around and blow on it a little bit to mix the distilled water and stain on the slide.

10. Let the slide stand for 5–10 minutes. During this time, blow on the slide occasionally to keep the distilled water and stain mixed.

11. Rinse the slide with distilled water for 2–3 minutes, and then tilt the slide to allow the excess water to drip off of the slide. Allow the slide to air dry. Once the slide is dry it will be ready to examine.

① Place a small drop of blood approximately 2 cm from the edge of the slide.   ② Place the edge of a second slide on the drop of blood and hold it at a 45° angle to the first slide.   ③ Push the second slide away from you to smear the blood on the first slide.

Figure 20.2   **Preparation of a Blood Smear.**

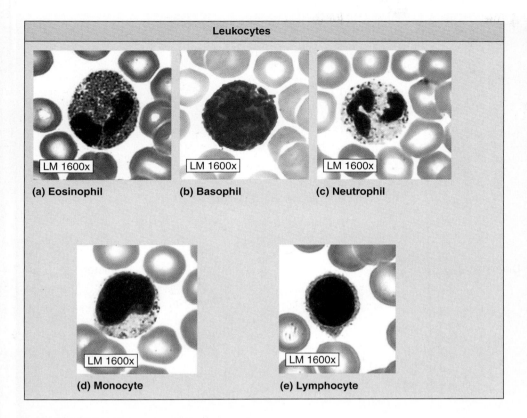

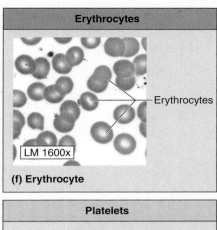

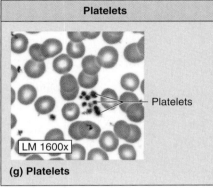

Figure 20.4   **Formed Elements in the Blood.**

3. Using tables 20.2 and 20.3 and **figure 20.4**, as guides, identify the following formed elements on the prepared blood slide. Note that some elements will be much easier to locate than others, due to their relative abundance in whole blood.

☐ basophils   ☐ monocytes

☐ eosinophils   ☐ neutrophils

☐ erythrocytes   ☐ platelets

☐ lymphocytes

4. After you locate each formed element, make a drawing of it in the appropriate box below. Use colored pencils to shade each structure the way it appears under the microscope. In addition, make note of the approximate diameter of each cell, using the diameter of an erythrocyte (7.5 µm) as a reference.

Erythrocyte

_____ ×

Diameter: _____ µm

Study Tip!

Although lymphocytes usually have circular nuclei and monocytes usually have indented, or "horseshoe-shaped" nuclei, sometimes monocytes can also have what look like circular nuclei. In these cases, they can appear very similar in structure to lymphocytes. To prevent yourself from misidentifying a monocyte as a lymphocyte, always consider the size of the cell in addition to the shape of the nucleus. A monocyte is going to be 2–3 times larger than a red blood cell, whereas a lymphocyte will be much closer to the same size as a red blood cell.

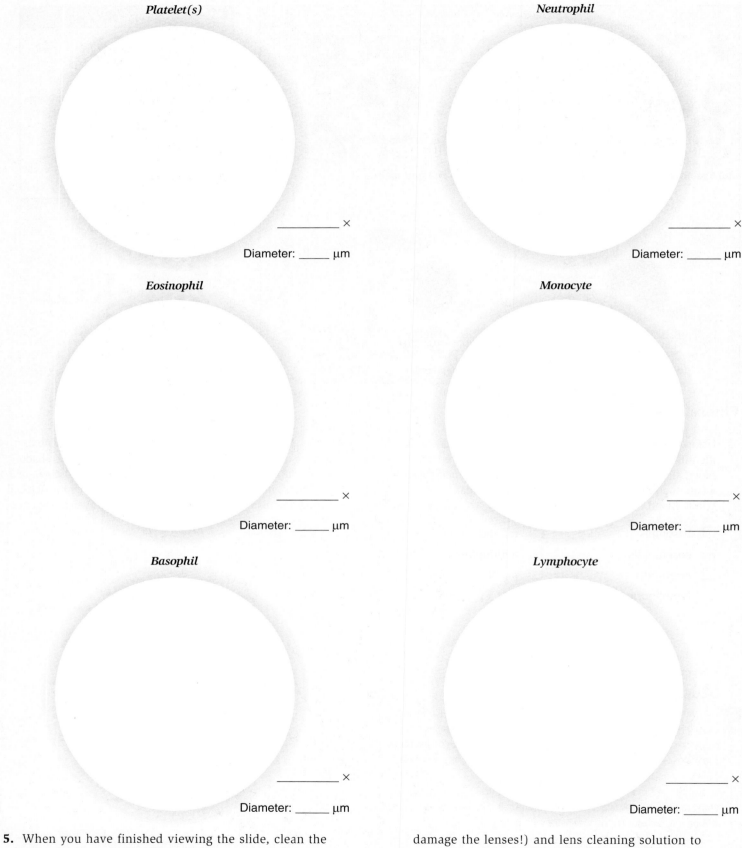

Platelet(s)

_____ ×

Diameter: _____ μm

Neutrophil

_____ ×

Diameter: _____ μm

Eosinophil

_____ ×

Diameter: _____ μm

Monocyte

_____ ×

Diameter: _____ μm

Basophil

_____ ×

Diameter: _____ μm

Lymphocyte

_____ ×

Diameter: _____ μm

5. When you have finished viewing the slide, clean the microscope and slide thoroughly to remove all traces of oil from the instruments. To do this, rotate the nosepiece so the high-power and oil immersion lenses are on either side of the slide once again (figure 20.3). Use *lens paper* (do not use anything else or it may damage the lenses!) and lens cleaning solution to carefully and thoroughly clean the oil from both the oil immersion objective and the blood smear slide. Blood smear slides prepared using your own blood should be disposed of by placing them in a bleach solution for disinfection.

EXERCISE 20.2 Identification of Megakaryocytes on a Bone Marrow Slide

All the formed elements of the blood are produced in the red bone marrow through the process of **hemopoiesis** (*hemo-*, blood, + *poiesis*, a making). Identifying the precursor cells of circulating blood cells is a complicated endeavor, which is generally undertaken only in upper-level histology courses. In this laboratory exercise you will not seek to identify all of the precursor cells for each lineage of blood cells. Instead, you will only locate the precursor cells of platelets (*megakaryocytes*) and investigate their structure and function. However, as you engage in the process of identifying megakaryocytes, try to also obtain an appreciation for the general appearance of the precursor cells of erythrocytes and leukocytes that you see on the slide.

1. Obtain a prepared slide of red bone marrow and place it on the microscope stage. Bring the slide into focus on high power. Using **figure 20.5** as a guide, identify the following on the slide:

   ☐ bone tissue   ☐ platelets
   ☐ megakaryocytes   ☐ red bone marrow

2. **Megakaryocytes** (*megas*, big, + *karyo*, kernel (referring to the nucleus) + *kytos*, a hollow, cell) are extremely large cells with enormous nuclei. They are easily identifiable on prepared slides of bone marrow. Megakaryocytes remain within the bone marrow. However, their products, **platelets,** are continuously delivered to the bloodstream. Megakaryocytes produce platelets in a process that involves

pinching off the cytoplasm of the megakaryocyte—a platelet is merely a fragment of the cytoplasm of a megakaryocyte enveloped by a portion of plasma membrane.

3. In the space below, make a simple line drawing of a megakaryocyte, making sure to indicate its relative size compared to the size of the developing blood cells that you observed on the bone marrow slide.

_____ ×

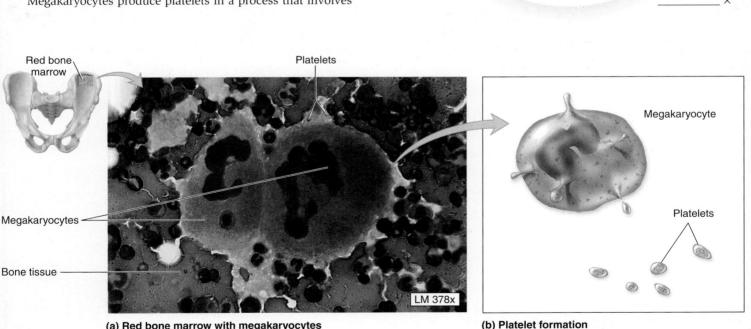

(a) Red bone marrow with megakaryocytes   (b) Platelet formation

Figure 20.5   Bone Marrow Slide. The megakaryocytes in red bone marrow give rise to platelets.

Gross Anatomy

EXERCISE 20.3 Identification of Formed Elements of the Blood on Classroom Models or Charts

1. Observe classroom models or charts demonstrating blood cells, and identify each of the following blood cell types on the models or charts.

   ☐ basophils   ☐ monocytes

   ☐ eosinophils   ☐ neutrophils

   ☐ erythrocytes   ☐ platelets

   ☐ lymphocytes

2. *Optional Activity:* **AP|R** **Cardiovascular System**—Watch the "Hemopoiesis" animation to review the formation and characterestics of each of the formed elements.

Chapter 20: Cardiovascular System—Blood

POST-LABORATORY WORKSHEET

1. The diameter of an erythrocyte is _____.

2. Label the formed elements of blood shown in the following figure.

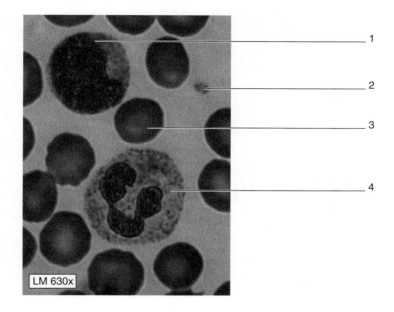

3. The rarest type of leukocyte is _____.

4. An agranulocyte with a large, spherical nucleus surrounded by pale blue cytoplasm is a(n) _____.

5. A leukocyte that contains reddish-colored granules and a bi-lobed nucleus is a(n) _____.

6. A significant increase in the number of circulating neutrophils may indicate a(n) _____ infection.

7. Small cellular fragments that are involved in the process of hemostasis are _____.

8. The most abundant type of leukocyte is _____.

9. Lymphocytes are responsible for initiating the _____ immune response to infection.

10. A significant increase in the number of circulating eosinophils may indicate a(n) _____ infection.

11. Platelets are derived from _____, which reside in the bone marrow.

12. Formed elements that compose less than 1% of whole blood and are involved in defense against disease are _____.

13. An inflammatory chemical secreted by basophils is _____.

14. A blood cell that is a precursor to tissue macrophages is a(n) _____.

The Cardiovascular System: The Heart

21

INTRODUCTION

The heart is an amazing organ that has been viewed with awe for ages. For many centuries physicians thought the heart, not the brain, was the control center and spiritual/emotional center of the body. Perhaps this is because structure and function relationships in the heart are relatively straightforward and easy to see, whereas such relationships are nearly impossible to discover through gross observation of the brain. As a student of anatomy who has studied the brain, you know that our emotions come not "from the heart" but from the brain. Yet the heart is still the organ that we associate with strong "life" forces, as well as that elusive force we call love. Even though scientists and lay-people alike recognize that the brain, not the heart, controls the functioning of the rest of the body, we also recognize that the heart *is* essential for the survival of all organs and tissues in the body: if the heart fails to pump blood, and that failure results in a lack of flow of oxygenated blood to the tissues, the tissues will die.

Have you ever considered how amazing it is that your heart continues to beat, day and night, day after day, year after year, without stopping? It is

Anatomy & Physiology REVEALED®
aprevealed.com

an enormous job. Failure of this amazing organ to perform its job results in the direst of consequences. Perhaps it is no surprise that heart disease is the number one cause of death for Americans (it accounts for approximately 32% of all deaths, while cancer, at number two, accounts for 23% of all deaths). A thorough understanding of the structure and function of the heart is critical for every one of us, whether we intend to go into health science fields or not. If your future career involves health care, you will be dealing with individuals suffering from heart disease on a daily basis. Even if your future career does not involve health care, your heart will be with you every moment, and how you take care of it now may very well determine the length and overall quality of your life.

Chapter 21: Cardiovascular System—The Heart

Name:_____

Date:_____ Section:_____

PRE-LABORATORY WORKSHEET

1. Fill in the following paragraph with the appropriate terms.

   Blood that enters the heart through the superior vena cava (SVC) and the inferior vena cava (IVC) enters the _____ (chamber).

   It then flows past the _____ valve and enters the _____ (chamber). Next, blood is pumped past

   the _____ valve and enters the _____ (vessel), which carries it to the _____. Blood

   returns to the heart through the _____ _____ (vessels), and enters the _____

   (chamber). From here, it flows past the _____ (valve) and enters the _____ _____

   (chamber). The blood is then pumped past the _____ _____ valve into the _____

   (vessel), where it travels through the systemic circulation and sends oxygenated blood to the body's tissues.

2. Which circuit pumps blood at higher pressure, pulmonary or systemic? _____

3. The term *coronary* means _____.

4. An artery is a vessel that carries blood _____ (direction) the heart.

5. A vein is a vessel that carries blood _____ (direction) the heart.

6. Most of the anterior surface of the heart receives oxygenated blood from the _____ coronary artery.

7. The muscles that attach to chordae tendineae are called _____ muscles.

8. The area where the great vessels (aorta, pulmonary trunk, etc.) leave the heart is called the _____ of the heart.

9. The wall of the _____ ventricle is much thicker than the wall of the _____ ventricle.

10. The layer of the pericardial sac that is in contact with the heart itself is the _____ layer.

IN THE LABORATORY

In this laboratory session you will review the structure and function of cardiac muscle and identify structures of the heart through observations of a preserved human heart or a model of the heart, and through dissection of a sheep heart. As you work through these exercises, be aware that most textbook figures of the heart are drawn to make identification of the chambers and vessels very straightforward. When you observe a real heart, you will find identification of the structures to be challenging because the chambers do not lie directly superior, inferior, or lateral to each other as they are often depicted in textbook drawings. In fact, most of the heart structures labeled "right" (such as the right atrium and right ventricle) lie not only on the right side of the heart, but also on the *anterior* surface of the heart. Likewise, most of the structures labeled "left" (such as the left atrium and left ventricle) lie not only on the left side of the heart, but also on the *posterior* surface of the heart.

Histology

EXERCISE 21.1 Cardiac Muscle

This activity is a review of observations of muscle tissue that were covered in chapter 11 of this laboratory manual. Depending on the time available in the laboratory, your instructor may want you to repeat your observations of cardiac muscle tissue, or may simply ask you to refer back to your notes on cardiac muscle tissue from chapter 11.

1. Obtain a slide of cardiac muscle and place it on the microscope stage. Bring the tissue sample into focus on low power, and then switch to high power. As you make your observations of cardiac muscle tissue, compare and contrast the structure and function of cardiac muscle with the structure and function of skeletal muscle by filling in **table 21.1**.

2. Using table 21.1 below as a guide, identify the structures listed in **figure 21.1** on the slide. Then label them in figure 21.1.
   What two types of cellular junctions are found in the intercalated discs?

   _____ and _____

   What is the purpose of the cellular junctions in the intercalated discs?

   _____

   _____

   _____

3. In the space below, make a drawing of cardiac muscle as seen through the microscope. Be sure to include and label all of the structures listed in figure 21.1.

   _____ ×

Table 21.1	Comparisons Between Cardiac and Skeletal Muscle Tissues	
Muscle Tissue	**Cardiac Muscle**	**Skeletal Muscle**
Nervous Control		
Appearance of Cells		
Number of Nuclei		
Location of Nuclei		

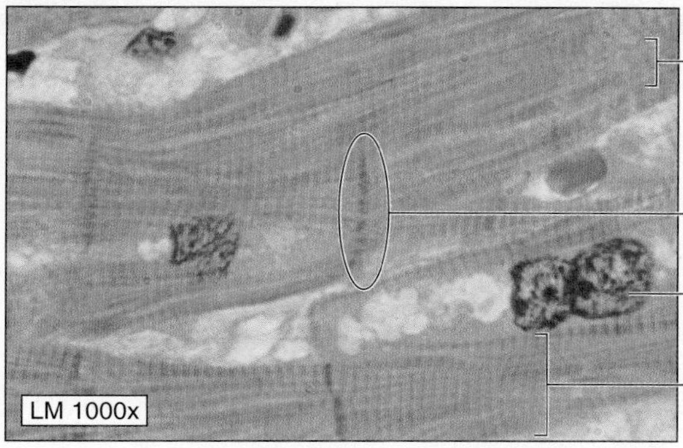

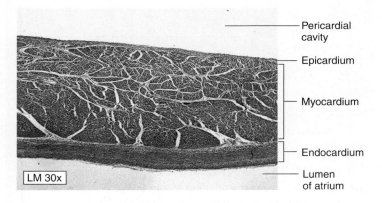

LM 1000x

Figure 21.1 **Cardiac Muscle Tissue.** Cardiac muscle tissue is characterized by short, branching cells with single, centrally-located nuclei. Intercalated discs are dark lines visible between the cells. A bands and I bands are also visible, indicating the presence of sarcomeres within the myofibers.

☐ cardiac muscle cell   ☐ intercalated disc   ☐ nucleus   ☐ striations

---

EXERCISE 21.2 Layers of the Heart Wall

1. Obtain a slide demonstrating the atria of the heart. Place it on the microscope stage and observe on low power. The wall of the heart is composed of three layers, the endocardium, myocardium, and epicardium. The **endocardium** consists of the simple squamous epithelium that lines the heart, called **endothelium**, plus an underlying layer of connective tissue (**figure 21.2**). The **myocardium** is composed of cardiac muscle and is by far the thickest layer of the heart wall. The **epicardium** is the same tissue that composes the visceral layer of the pericardial sac—a serous membrane referred to as the visceral layer of serous pericardium. What to call this layer of tissue depends on the context. When referring to it as part of the heart wall, the appropriate term is *epicardium*. When referring to it as part of the pericardial sac, the appropriate term is *visceral layer of serous pericardium*. You may see cross sections of the coronary vessels deep to the epicardium on the slide as well.

2. Using figure 21.2 as a guide, identify the following on the slide of the atrium:

☐ endocardium (endothelium)   ☐ myocardium (cardiac muscle)

☐ epicardium (visceral, serous pericardium)

3. In the space below, make a simple drawing of the layers of the heart wall as seen through the microscope. Be sure to include and label all of the structures listed above in your drawing.

Pericardial cavity

Epicardium

Myocardium

Endocardium

Lumen of atrium

LM 30x

Figure 21.2 **Histology of the Heart Wall.** This slide demonstrates the wall layers of the atrium of the heart. In the atrium, the endocardium is thick, consisting of endothelium plus a layer of underlying connective tissues. The myocardium is the thickest layer, consisting of cardiac muscle tissue, and the epicardium is the thinnest layer, consisting of the visceral pericardium (a serous membrane) with a thin layer of underlying connective tissue.

_____ ✕

Gross Anatomy

EXERCISE 21.3 The Pericardial Cavity

1. Observe a model of a human thorax or the thoracic cavity of a human cadaver (**figure 21.3**).

2. The heart is located within the **thoracic cavity**, a cavity that also houses the lungs, trachea, esophagus, and a variety of nerves and blood vessels. Within the thoracic cavity, the heart is located within a space called the **mediastinum**. The **mediastinum** is the space between the two pleural cavities, which contains the trachea, esophagus, thymus, nerves, and blood vessels, as well as the pericardial cavity. The **pericardial cavity** encases the heart.

3. Using your textbook as a guide, identify the structures listed in figure 21.3 on a classroom model of the thorax or on a human cadaver. Then label them in figure 21.3.

4. *Optional Activity*: **AP|R** **Cardiovascular System**—Explore the "Thorax" dissections to view the heart in the thoracic cavity and appreciate the surrounding structures.

5. The pericardial cavity has two layers to it, an outer, **parietal layer** (parietal, *wall*), and an inner, **visceral layer** (viscus, *internal organ*). The parietal layer is anchored to surrounding structures and helps keep the pericardial sac in place within the mediastinum. Observe the pericardial cavity within the thorax and name the structures the parietal pericardium is anchored to on each of the following surfaces:

   Superior:_____

   Inferior:_____

   Lateral:_____

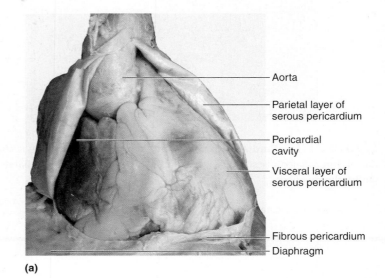

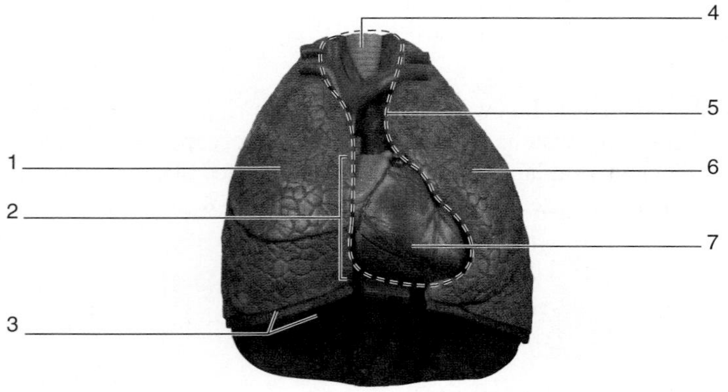

- Aorta
- Parietal layer of serous pericardium
- Pericardial cavity
- Visceral layer of serous pericardium
- Fibrous pericardium
- Diaphragm

(a)

(b)

Figure 21.4 **The Pericardial Sac.** (a) Location of the heart within the pericardial sac of a cadaver. The parietal pericardium has been cut and reflected to reveal the heart, which is covered by visceral pericardium. (b) Layers and tissues composing the pericardial sac.

- ☐ diaphragm
- ☐ endocardium
- ☐ fibrous pericardium
- ☐ myocardium
- ☐ parietal layer of serous pericardium
- ☐ pericardial cavity
- ☐ serous pericardium
- ☐ visceral layer of serous pericardium (epicardium)

Figure 21.3 **Location of the Heart Within the Thoracic Cavity.** Within the thoracic cavity the heart is located within the mediastinum, the space between the two pleural cavities, which house the lungs.

- ☐ diaphragm
- ☐ heart
- ☐ left lung/pleural cavity
- ☐ mediastinum
- ☐ pericardial cavity
- ☐ right lung/pleural cavity
- ☐ trachea

6. The **visceral layer** of the pericardium is in contact with the heart itself. To remove the heart from the pericardial sac, the parietal layer of the pericardial sac must be cut away from the visceral layer (**figure 21.4a**). Upon removal of the parietal layer, the visceral layer will remain with the heart itself. Note that the terms *parietal* and *visceral* refer only to the physical *layers* of the pericardial sac. The actual *tissue* that composes each layer is either a *serous tissue* or a *fibrous tissue* (serous pericardium or fibrous pericardium). **Serous pericardium** is composed of simple squamous epithelium called **mesothelium**. The serous tissue is a smooth, shiny layer of tissue that produces a small amount of **pericardial fluid**, which lubricates the inside of the pericardial sac. Which layers of the pericardium (parietal and/or visceral) contain serous pericardium? _____
Fibrous pericardium is composed of dense irregular connective tissue (collagen fibers and fibroblasts), which strengthens the sac and anchors it to surrounding structures. Which layers of the pericardium (parietal and/or visceral) contain fibrous pericardium?_____

7. Using figure 21.4a as a guide, identify the structures listed in figure 21.4b on the model of the thorax or on a human cadaver. Then label them in figure 21.4b.

EXERCISE 21.4   Gross Anatomy of the Human Heart

Obtain a preserved human heart from a cadaver or a classroom model of the heart. If you are using a preserved heart, place it in a dissecting pan and keep it moist while you make your observations. In addition, *use only a blunt probe* to point out structures on the heart so as not to damage the heart.

1. Note the size and shape of the heart. A normal heart is about the size and shape of a human fist. Based on this information, is the heart you are observing of normal size? _____ If the heart appears to be enlarged, make note of that observation, as it can be (though it is not necessarily) indicative of heart disease. One of the first things you may notice is that the heart you are holding in your hands looks very little like the drawings in your textbook. It is a twisted organ, so

identification of chambers can be challenging at first. Begin by identifying the **apex** (the pointed, inferior portion of the heart) and the **base** (the superior point where the great vessels enter and leave). In most instances, the term *base* refers to the bottom of an organ or tissue. Is this generalization true with the heart? _____ The heart has a relatively flat inferior surface, the **diaphragmatic surface**, which is the surface of the heart that lies on top of the diaphragm. Once you have identified these structures, place the heart in your right hand with the diaphragmatic surface in the palm of your hand, the apex directed toward your thumb and wrist, and the base directed toward the space between the tip of the thumb and the second digit (**figure 21.5**). You should now be viewing the anterior surface of the heart in

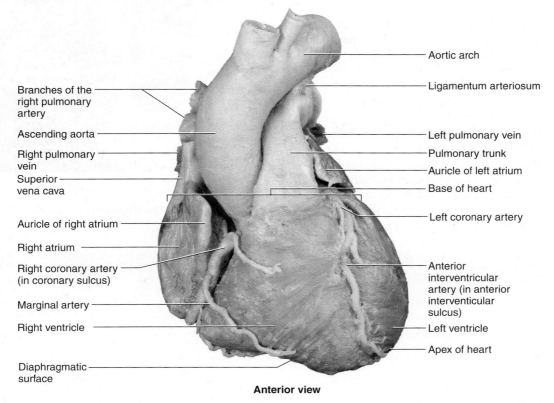

Branches of the right pulmonary artery

Ascending aorta

Right pulmonary vein

Superior vena cava

Auricle of right atrium

Right atrium

Right coronary artery (in coronary sulcus)

Marginal artery

Right ventricle

Diaphragmatic surface

Aortic arch

Ligamentum arteriosum

Left pulmonary vein

Pulmonary trunk

Auricle of left atrium

Base of heart

Left coronary artery

Anterior interventricular artery (in anterior interventicular sulcus)

Left ventricle

Apex of heart

Anterior view

Figure 21.5   Orientation of the Heart in the Right Hand of the Observer.   The heart shown here is oriented with the anterior surface facing the observer, which is how it should look when it is in the observer's right hand with the apex pointed toward the thumb and wrist, and the great vessels directed toward the space between the tip of the thumb and the second digit.

a position that very closely resembles the heart's orientation within the thorax. If you are unsure if you are viewing the anterior surface, check with your instructor before you proceed.

2. Identify the **right and left ventricles** (figure 21.5). Because you are viewing the anterior surface of the heart, the left ventricle will be on *your* right, and the right ventricle will be on *your* left. Notice that the mass of the left ventricle fills up nearly the entire palm of your hand because it has a much thicker myocardium than the right ventricle. Identify the **right atrium** superior and lateral to the right ventricle. The left atrium is not visible in this view. In the following sections you will first identify the layers of the heart wall, and then proceed to identify the heart chambers and major heart structures within each chamber.

3. *Layers of the Heart Wall*: Both atria and ventricles are composed of a three-layered wall. The layers of the heart wall, from outside to inside, are the epicardium, myocardium, and endocardium (these are described in greater detail in exercise 21.2). **Table 21.2** summarizes the characteristics of the wall layers of the heart.

4. Using table 21.2 and figure 21.4 as guides, identify the following components of the heart wall on a heart model or on a human heart.

   ☐ myocardium   ☐ epicardium

   ☐ endocardium

5. *The Right Atrium*: Holding the heart in your right hand once again, with the anterior surface directed toward you, identify the **superior and inferior vena cava** entering the right atrium (figure 21.5). Within the thorax, these vessels run vertically and meet at the right atrium. Next, look inside the right atrium. Notice the thin strands of **pectinate muscle** in the wall of the right atrium (**figure 21.6**). Pectinate muscle is found only in the wall of the right atrium (although some pectinate muscle is also found within the walls of both *auricles*, which are extensions of the atria), so its presence provides a good indicator that you are observing the right atrium and not the left atrium, if you are unsure. Locate the shallow depression covered with a thin membrane in the interatrial septum. This is the **fossa ovalis**, a remnant of a fetal shunt between the right and left atria called the **foramen ovale**. Is the hole completely closed off in your specimen? If not, what might some of the consequences be (this condition is called a *patent* foramen

ovale [*pateo*, to lie open])?_____ Just inferior to the fossa ovalis, look for the small opening of the **coronary sinus**, a vein that drains nearly all deoxygenated blood from the heart wall. Finally, observe the **right atrioventricular (AV) valve** and count the cusps. How many cusps are there? _____ Based on that information, is the right AV valve a tricuspid or bicuspid valve?_____

_____

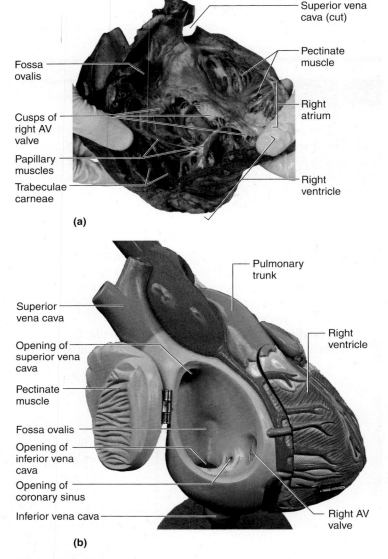

(a)

(b)

Figure 21.6 **The Right Atrium of a Human Heart.**
(a) Gross specimen. (b) Heart model.

Table 21.2	Layers of the Heart Wall		
Wall Layer	**Tissue**	**Description**	**Word Origins**
Endocardium	Endothelial cells plus some connective tissue and smooth muscle.	This layer is relatively thick in the atria, and thin in the ventricles.	*endon*, within, + *kardia*, heart
Myocardium	Composed of layers of cardiac muscle in addition to elastic fibers and loose connective tissue between the layers.	This is the thickest layer of the heart wall.	*mys*, muscle, + *kardia*, heart
Epicardium	Composed of a serous membrane with an underlying layer of elastic fibers and adipose connective tissue.	This is the visceral pericardium. It is relatively thicker in the ventricles than in the atria.	*epi-*, upon, + *kardia*, heart

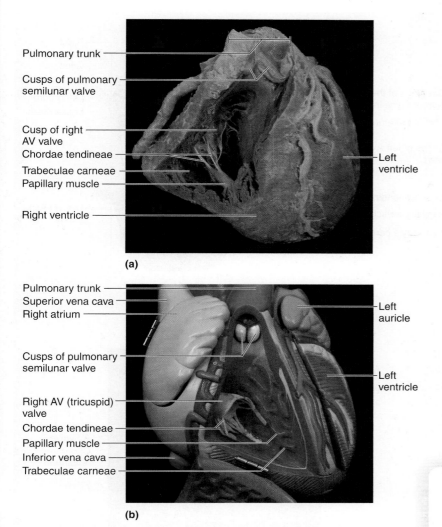

Pulmonary trunk

Cusps of pulmonary
semilunar valve

Cusp of right
AV valve

Chordae tendineae

Trabeculae carneae

Papillary muscle

Right ventricle

Left
ventricle

(a)

Pulmonary trunk

Superior vena cava

Right atrium

Left
auricle

Cusps of pulmonary
semilunar valve

Left
ventricle

Right AV (tricuspid)
valve

Chordae tendineae

Papillary muscle

Inferior vena cava

Trabeculae carneae

(b)

Figure 21.7   **The Right Ventricle of a Human Heart.**
(a) Gross specimen. (b) Heart model.

6. *The Right Ventricle*: Place a blunt probe in the right atrium,
and direct it into the right ventricle (**figure 21.7**). If the
right ventricle is not already cut open, ask your instructor
to cut it open to identify the structures inside. The most
prominent features in the walls of the right ventricles are
the strands of cardiac muscle in the walls of the ventricle
called **trabeculae carneae** (*trabs*, a beam, + *carneus*,
fleshy), and the nipple-like **papillary muscles** (*papilla*, a
nipple), which attach to the cusps of the right AV valve by
string-like structures called **chordae tendineae** (*chorda*,
cord, + *tendo*, to strech out). When the ventricles contract,
the papillary muscles also contract and pull down on the
cusps of the AV valve. Because blood is being pushed up
against the underside of the valve cusps as the ventricles
contract, the action of the papillary muscles pulling down
on the valve cusps keeps the valve closed. This prevents
blood from flowing back into the right atrium, and instead
forces the blood out through the **pulmonary trunk**.

7. Using figure 21.7 as a guide, identify the following structures
in the right ventricle of a human heart or heart model:

    ☐ chordae tendineae   ☐ papillary muscles

    ☐ cusps of the right AV valve   ☐ trabeculae carneae

8. Place the tip of a blunt probe in the right ventricle and
pass it out through the **pulmonary trunk**. To enter the
pulmonary trunk, the probe will have to pass through
the **pulmonary semilunar valve**. If the vessel has been
cut open, you will be able to see that the cusps of the
semilunar valve are shaped like "half moons," hence
their name. How many cusps are there?_____.

    Where will the blood travel after it leaves the
pulmonary trunk?_____.

9. In the space below, make a simple drawing of the
right atrium and right ventricle. Be sure to include
and label all of the structures listed below in your
drawing.

    ☐ chordae tendineae   ☐ pulmonary trunk

    ☐ fossa ovalis   ☐ right atrium

    ☐ inferior vena cava   ☐ right AV valve

    ☐ opening of coronary   ☐ right ventricle
       sinus
                                 ☐ superior vena cava
    ☐ papillary muscles
                                 ☐ trabeculae carneae
    ☐ pectinate muscle

    ☐ pulmonary semilunar
       valve

Figure 21.8 **The Left Atrium of a Human Heart.**
Posterior view of a human heart with left atrium and left
ventricle cut open.

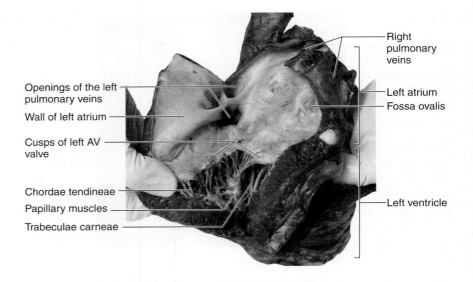

Openings of the left pulmonary veins

Wall of left atrium

Cusps of left AV valve

Chordae tendineae

Papillary muscles

Trabeculae carneae

Right pulmonary veins

Left atrium

Fossa ovalis

Left ventricle

10. *The Left Atrium*: Rotate the heart until you are viewing its
posterior surface (**figure 21.8**). Note the four **pulmonary
veins**, which collectively drain into the **left atrium**. Look
inside the left atrium. Notice that the wall of the left atrium
is thin and smooth and does *not* have pectinate muscles
(although the wall of the left *auricle* does). The left atrium
is little more than an expansion of the tissue where the
four pulmonary veins come together. Thus, its walls are
not always easy to identify. Now that you have identified
both right and left atria, place your index finger in the right
atrium and your thumb in the left atrium and again find the
fossa ovalis, which lies in the **interatrial septum**. What is
the name of the fetal structure of which the fossa ovalis is a

remnant?_____ Next, observe the **left
atrioventricular (AV) valve** and count the cusps. How

many cusps are there?_____ Based on that
information, is the left AV valve a tricuspid

or bicuspid valve?_____ What is another

name for this valve?_____

11. *The Left Ventricle*: Place a blunt probe in the left atrium
and pass it into the **left ventricle** (**figure 21.9**). If the left
ventricle is not already cut open, ask your instructor to
cut it open so you can identify the structures within. The
left ventricle contains all the same structures as the right
ventricle, and the **left atrioventricular (AV) valve** functions
the same way the right AV valve functions. The biggest
difference between the two chambers is the thickness of the
ventricular walls.

12. Using figure 21.9 as a guide, identify the following
structures in the left ventricle:

☐ chordae tendineae   ☐ papillary muscles
☐ left AV valve   ☐ trabeculae carneae

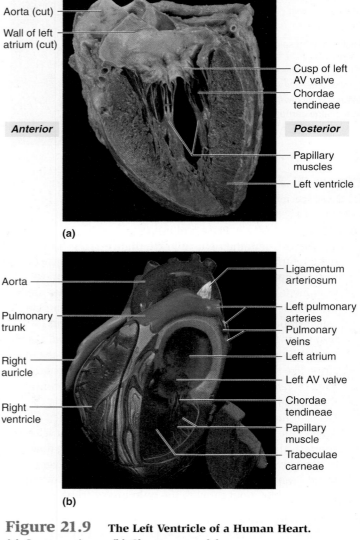

Aorta (cut)

Wall of left atrium (cut)

Anterior

Posterior

Cusp of left AV valve

Chordae tendineae

Papillary muscles

Left ventricle

(a)

Aorta

Pulmonary trunk

Right auricle

Right ventricle

Ligamentum arteriosum

Left pulmonary arteries

Pulmonary veins

Left atrium

Left AV valve

Chordae tendineae

Papillary muscle

Trabeculae carneae

(b)

Figure 21.9 **The Left Ventricle of a Human Heart.**
(a) Gross specimen. (b) Classroom model.

13. Note the difference in thickness between the myocardium in the wall of the left ventricle as compared to that of the right ventricle. What are the consequences of this difference? _____

_____

_____

Does the chamber size (volume) of the left ventricle appear to be greater than that of the right ventricle?_____

14. Place a probe in the left ventricle and pass it out through the **aorta**. To enter the aorta, the probe will have to pass through the **aortic semilunar valve**. If the vessel has been cut open, you will be able to see that the cusps of the semilunar valve are shaped like "half moons," hence their name. How many cusps are there?_____
Where will the blood travel after it leaves the
aorta?_____

15. Observe the outside of the arch of the aorta where it passes just superior to the pulmonary trunk. Look for a small ligament attaching the pulmonary trunk to the aorta (see figure 21.5 and figure 21.9b). This is the **ligamentum arteriosum**. Of what fetal structure is the ligamentum arteriosum a remnant?_____ This fetal structure shunts blood from the _____ to the _____, thereby shunting blood away from the _____.

16. In the space below, make a simple drawing of the left atrium and left ventricle. Be sure to include and label all of the structures listed below in your drawing.

☐ aorta   ☐ left atrium
☐ aortic semilunar valve   ☐ left ventricle
☐ left AV valve   ☐ papillary muscles
☐ chordae tendineae   ☐ pulmonary veins
☐ fossa ovalis   ☐ trabeculae carneae

17. *Optional activity*: **AP|R** **Cardiovascular System**—Watch the "Heart" animation to gain a 3D fly-through perspective of the internal heart.

WHAT DO YOU THINK?

❶ Do you think the volume of blood pumped by each ventricle should be different? Explain your answer.

EXERCISE 21.5 | The Coronary Circulation

1. Obtain a preserved human heart or a classroom model of the heart.

2. The coronary circulation is the circulation to the heart wall itself. Adequate blood flow through the coronary arteries is absolutely essential for the functioning of the heart. If any coronary vessel becomes blocked due to disease or other processes, the area served by the vessel may become **ischemic**, meaning it lacks blood flow (*ischio*, to keep back, + *chymos*, juice). Prolonged ischemia to heart muscle leads to **hypoxia** (*hypo-*, too little, + *oxia*, oxygen). When cardiac muscle lacks an oxygen supply for more than a few minutes, the tissue dies, or becomes **necrotic** (*nekrosis*, death). The area of dead tissue composes a **myocardial infarction** (*myocardial*, referring to the myocardium, + *in-farcio*, to stuff into), otherwise known as a "heart attack" or "MI." If a myocardial infarction results in vast tissue destruction, the heart may no longer be effective as a pump, which can lead to the death of the individual. If the individual survives the attack, the body will eventually repair the area of dead tissue and replace it with scar tissue. Scar tissue is mainly composed of a dense collection of collagen fibers. **Figure 21.10** demonstrates a human heart

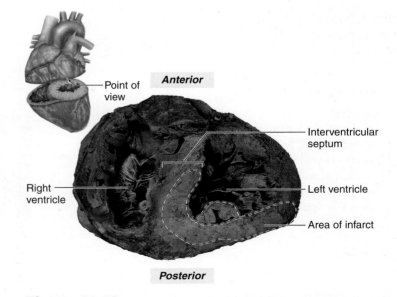

Figure 21.10   **Inferior View of a Transverse Section through a Human Heart.**   Scar tissue, which is evidence of a previous myocardial infarction, can be seen in the interventricular septum and posterior wall of the left ventricle.

Table 21.3	Arterial Supply to the Heart			
Vessel	**Description**	**Areas Served**	**Word Origins**	
Anterior Interventricular Artery	Located in the anterior interventricular sulcus (groove). Physicians typically refer to this as the "LAD" (left anterior descending).	Anterior parts of the right and left ventricles.	*inter*, between, + *ventricular*, referring to the ventricles of the heart (from *ventriculus*, belly)	
Circumflex Artery	Located in the coronary sulcus between the left atrium and left ventricle.	Left atrium and left ventricle (lateral part).	*circum*, around, + *flexus*, to bend	
Left Coronary Artery	Located posterior to the pulmonary artery; branches into the anterior interventricular and circumflex arteries just after it emerges from behind the pulmonary artery.	Anterior interventricular and circumflex arteries.	*corona*, a crown	
Marginal Artery	Branches off of the right coronary artery at the right margin of the heart and is located on the lateral part of the right ventricle.	Lateral part of the right ventricle.	*margo*, border or edge	
Posterior Interventricular Artery	Continuation of the right coronary artery located in the posterior interventricular sulcus.	Posterior parts of the right and left ventricles.	*inter*, between, + *ventricular*, referring to the ventricles of the heart (from *ventriculus*, belly)	
Right Coronary Artery	Located in the coronary sulcus between the right atrium and the right ventricle. Branches include the SA nodal artery, which supplies blood to the sinoatrial node within the right atrium.	Right atrium and marginal and posterior interventricular arteries.	*corona*, a crown	

with evidence of a healed myocardial infarction. If you are observing a human heart, does your specimen show any evidence of (scarring from) past myocardial infarctions (scar tissue is generally clear to whitish in appearance and is much tougher than muscle tissue)? _____

3. The blood supply to the heart arises from two main coronary vessels, the **right and left coronary arteries** (**table 21.3**). The openings into these arteries arise behind the cusps of the aortic semilunar valve. **Figure 21.11** shows the relationship between the cusps of the aortic semilunar valve and the openings to the right and left coronary arteries. Observe the aortic semilunar valve and identify the openings into the right and left coronary arteries (figure 21.11). When the left ventricle contracts and pushes blood into the aorta, the cusps of the semilunar valve fold over the openings to the coronary arteries. What consequence does this have in terms of blood flow to the heart during

ventricular contraction (systole)?_____

_____

_____

When the ventricles relax, the cusps of the semilunar valves fall shut as they fill with blood. Thus, they are no longer covering the openings to the coronary arteries. What consequence does this have in terms of blood flow to the heart during ventricular relaxation (diastole)?_____

_____

_____

4. *Cardiac Veins:* For the most part, venous drainage from the heart wall parallels the arterial supply. However, all venous blood draining the heart wall (with one exception, see table 21.4) eventually drains into one large vessel, the **coronary sinus**. The coronary sinus is located on the

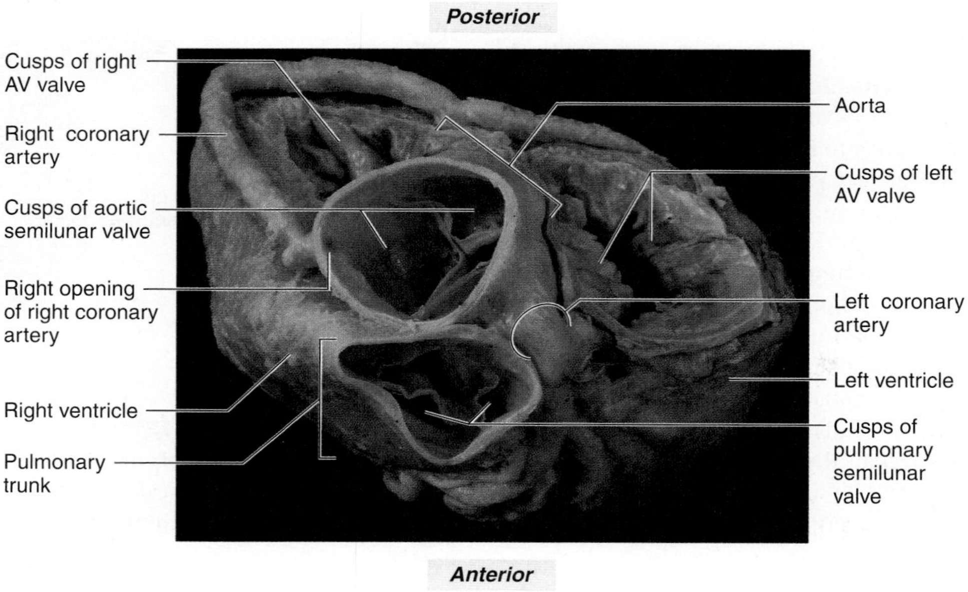

Figure 21.11 **The Coronary Circulation.** Superior view of the human heart with the atria removed. All four valves of the heart can be seen in this view. The openings to the right and left coronary arteries are located behind the cusps of the aortic semilunar valve. In this photo you can see where the right and left coronary arteries come off of the aorta and you can see the cusps of the aortic semilunar valve within the lumen of the aorta.

Posterior

Cusps of right AV valve
Right coronary artery
Cusps of aortic semilunar valve
Right opening of right coronary artery
Right ventricle
Pulmonary trunk

Aorta
Cusps of left AV valve
Left coronary artery
Left ventricle
Cusps of pulmonary semilunar valve

Anterior

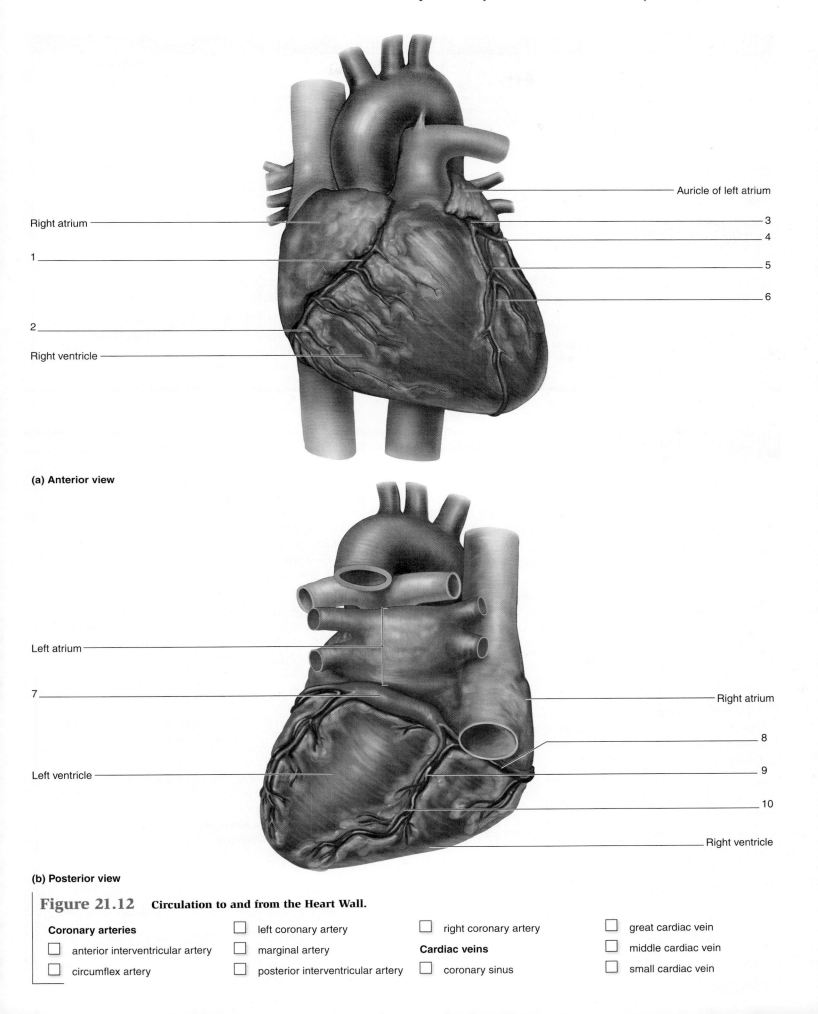

Auricle of left atrium

Right atrium

3

4

1

5

6

2

Right ventricle

(a) Anterior view

Left atrium

7

Right atrium

8

9

Left ventricle

10

Right ventricle

(b) Posterior view

Figure 21.12 **Circulation to and from the Heart Wall.**

Coronary arteries

☐ anterior interventricular artery

☐ circumflex artery

☐ left coronary artery

☐ marginal artery

☐ posterior interventricular artery

☐ right coronary artery

Cardiac veins

☐ coronary sinus

☐ great cardiac vein

☐ middle cardiac vein

☐ small cardiac vein

Table 21.4	Venous Drainage of the Heart		
Vessel	**Description**	**Areas Drained**	**Word Origins**
Coronary Sinus	Located in the coronary sinus on the posterior surface of the heart. The coronary sinus is the largest vein of the heart, and its opening can be located within the right atrium of the heart just inferior to the fossa ovalis.	Entire heart; all veins of the heart drain into the coronary sinus, with the exception of a few small veins of the right ventricle, which drain directly into the right atrium.	*corona*, a crown, + *sinus*, cavity
Great Cardiac Vein	Located in the anterior interventricular sulcus next to the anterior interventricular artery.	Anterior parts of the right and left ventricles.	*cardiacus*, heart, + *vena*, vein
Middle Cardiac Vein	Located in the posterior interventricular sulcus next to the posterior interventricular artery.	Posterior parts of the right and left ventricles.	*cardiacus*, heart, + *vena*, vein
Small Cardiac Vein	Located on the lateral part of the right ventricle, near the marginal artery.	Lateral part of the right ventricle.	*cardiacus*, heart, + *vena*, vein

posterior surface of the heart and runs in the coronary sulcus. What chamber does the coronary sinus empty into?

_____

5. Using tables 21.3 and **21.4**, and your textbook as guides, identify the vessels shown in **figure 21.12** on the cadaver

heart or on the classroom model of the heart. Then label them in figure 21.12.

6. *Optional Activity*: AP|R **Cardiovascular System**—Study the "Heart" dissections to review the vasculature and features of the heart, then use the Quiz feature to test yourself on these structures.

EXERCISE 21.6 Superficial Structures of the Sheep Heart

1. Obtain a dissecting pan, dissecting instruments, gloves, and a preserved sheep heart. Rinse the heart with water to remove any dried blood or other debris, and place it in the dissecting pan to begin your superficial observations.

2. **Figure 21.13** demonstrates superficial structures of the sheep heart from both anterior and posterior views. Begin your observations by distinguishing the anterior surface from the posterior surface. One way to know you are observing the anterior surface of the heart is that you will see the fairly distinctive ruffled borders of both the right and left **auricles**. The auricles are extensions of the right and left **atria**.

3. Observe the surface of the heart closely to locate the **visceral pericardium** (epicardium). Then observe the outer surfaces of the great vessels of the heart to see if you can find any remnants of the **parietal pericardium** where it attached to these vessels. Note the large amount of fatty tissue deep to the epicardium. One of the functions of this fatty tissue is to help cushion the heart within the pericardial cavity.

4. Using figure 21.13 as a guide, identify the following superficial features on the sheep heart:

☐ anterior interventricular sulcus
☐ apex
☐ coronary sulcus
☐ left atrium
☐ left auricle
☐ left ventricle
☐ posterior interventricular sulcus
☐ right atrium
☐ right auricle
☐ right ventricle

5. Next you will identify the great vessels: the aorta, pulmonary trunk, and venae cavae. These vessels are often cut very close to their attachments to the heart, which can make identification difficult. To make the task easier, carefully remove as much of the epicardial fat as you can from the superior aspect of the heart (leave the fat in place on the ventricles for now).

6. Once you have cleaned away as much fat as possible, proceed with identification of the great vessels of the heart. Begin by viewing the anterior surface of the heart (figure 21.13*a*). The two most prominent vessels coming off the heart are the pulmonary trunk and the aorta. Both vessels have thick, tough walls, which helps with identification. The pulmonary trunk is located the most anteriorly and points to the right (from your point of view). The aorta is directly posterior to the pulmonary trunk and points to the left (from your point of view).

7. Turn the heart over to view the posterior surface (figure 21.13*b*). Here you will find the pulmonary veins, which will be more to the left (from your point of view), and the superior and inferior venae cavae, which will be more to the right (from your point of view). Before moving on, make sure you have identified all of the following vessels:

☐ aorta
☐ inferior vena cava
☐ pulmonary trunk
☐ pulmonary veins
☐ superior vena cava

Instructors Sheet

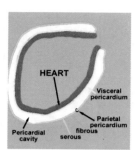

HEART

Visceral pericardium

Parietal pericardium

Pericardial cavity

fibrous serous

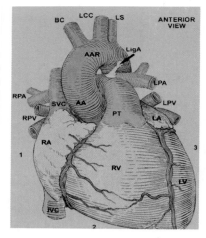

ANTERIOR VIEW

BC LCC LS

AAR LigA

RPA LPA

SVC AA LPV

RPV PT LA

RA 1 3

RV

LV

IVC 2

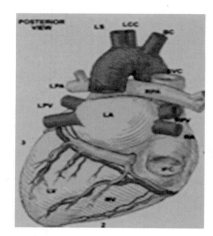

POSTERIOR VIEW

LS LCC BC

SVC

LPA RPA

LPV LA

LV

RV 2

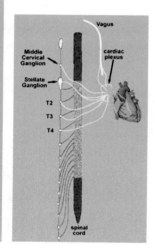

Vagus

Middle Cervical Ganglion

cardiac plexus

Stellate Ganglion

T2

T3

T4

spinal cord

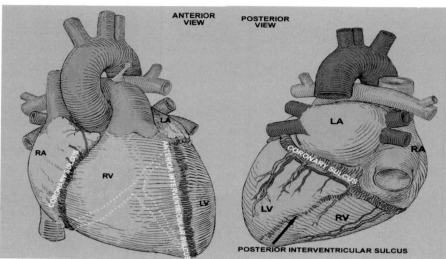

ANTERIOR VIEW   POSTERIOR VIEW

LA

RA RA

RV LA

CORONARY SULCUS

LV LV RV

ANTERIOR INTERVENTRICULAR SULCUS

CORONARY SULCUS

POSTERIOR INTERVENTRICULAR SULCUS

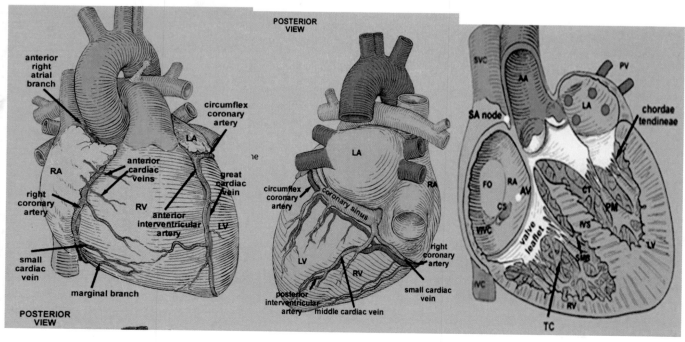

POSTERIOR VIEW

anterior right atrial branch

circumflex coronary artery

LA

anterior cardiac veins

great cardiac vein

RA

right coronary artery

RV

anterior interventricular artery

LV

small cardiac vein

marginal branch

POSTERIOR VIEW

LA

circumflex coronary artery

coronary sinus

RA

LV

RV

right coronary artery

small cardiac vein

posterior interventricular artery

middle cardiac vein

SVC

AA

PV

SA node

LA

chordae tendineae

FO RA AV

CT

CS

PM

IVC

IVS

valve leaflet

SMB

LV

TC

RV

13. Metatarsalphalangeal (MP)
14. Interphalangeal (IP)

15. Describe the effects of aging on the joints.

Required reading: <u>Human Anatomy</u>, 3rd <u>Edition</u>, chapter 8, 9, & 12

Master of Physician Assistant Practice

MPAP 501 – Anatomy Objectives

Topic: Appendicular Skeleton & Muscles (lower extremity) & Articulations

At the conclusion of this lecture and completion of required readings, the student will be able to:

Skeleton:
1. List the bones of the pelvic girdle and their prominent surface features.
2. Describe how each bone of the pelvic girdle contributes to its strength and function.
3. Compare and contrast a male vs. female pelvis.
4. Differentiate the stability of the pectoral vs. pelvic girdle.
5. Name the 3 bones that fuse to form each os coxae.
6. Name the 30 bones of the lower limb and the prominent markings on each bone:
   - 1 femur
   - 1 patella
   - 1 tibia, 1 fibula
   - 7 tarsal bones
   - 5 metatarsal bones
   - 14 phalanges

Muscles:
7. Describe the muscles and major movements of the pelvic girdle and lower limb.
8. Name the muscles that move the hip/thigh and their organization into movement groups.
9. Discuss the thigh muscles and their actions that move the knee/lower leg.
10. List the muscles of the lower leg and actions of these muscles that move the ankle, foot, and toes.
11. Name and state the function of the intrinsic muscles of the foot.

Articulations:
12. Describe the anatomical structure of the following joints of the lower extremity:
   - Pelvic girdle & Lower Limb
   - Hip
   - Knee
   - Ankle
   - Intertarsal joints
   - Tarsometatarsal joints

Master of Physician Assistant Practice

MPAP 501 – Anatomy LAB Objectives

Topic: Appendicular Skeleton & Muscles (lower extremity) & Articulations Lab

At the conclusion of this lab session and completion of required readings, the student will be able to:

Skeleton:
1. Identify the bony landmarks of the pelvic girdle.
2. List the three bones that compose the pelvic girdle.
3. Identify the landmarks of the bones of the lower limb and relate them to muscular attachments.
4. Associate the shapes of the tarsal bones with their respective names.
5. Compare and contrast the structure and function of the carpal vs. tarsal bones.

Muscles:
6. Identify the muscles that act about the hip and describe their actions.
7. Explain the roles of the gluteus medius and gluteus minimus in locomotion.
8. Explain the importance of the piriformis as a clinically relevant landmark.
9. Describe the composition, location, and function of the iliopsoas.
10. Identify the locations and actions of the following lower limb muscles:
    - Anterior, medial, and posterior compartments of the thigh
    - Anterior, lateral, and posterior compartments of the lower leg
11. Name the 2 muscles of the anterior compartment of the thigh that flex the hip joint.
12. Identify the borders of the femoral triangle.
13. Explain how the gracilis is an exception to the rule for muscles of the medial compartment of the thigh.
14. Name the 2 muscles that act as antagonists to the fibularis muscles for the function of everting the ankle.
15. Explain how the tibialis anterior and tibialis posterior muscles can act as either synergists or antagonists of each other.
16. Identify the location and action of the intrinsic muscles of the foot.

Articulations:
17. Identify and describe the structures that compose each of the following joints:
    - Hip
    - Knee
    - Ankle
    - Foot

Required reading: Human Anatomy Laboratory Manual, chapter 9, 10, & 13

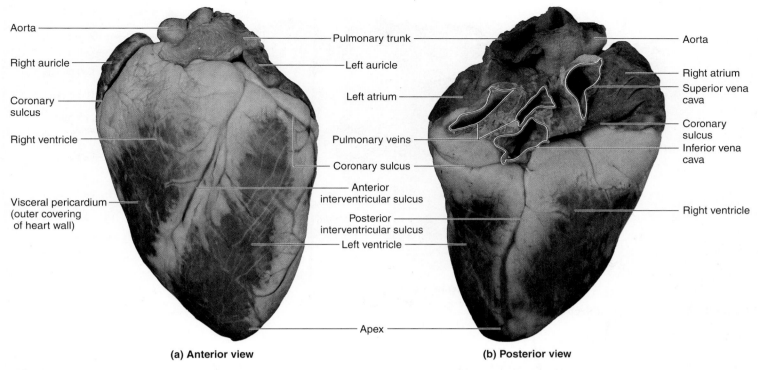

Aorta

Right auricle

Coronary
sulcus

Right ventricle

Visceral pericardium
(outer covering
of heart wall)

Pulmonary trunk

Left auricle

Left atrium

Pulmonary veins

Coronary sulcus

Anterior
interventricular sulcus

Posterior
interventricular sulcus

Left ventricle

Apex

(a) Anterior view

Aorta

Right atrium

Superior vena
cava

Coronary
sulcus

Inferior vena
cava

Right ventricle

(b) Posterior view

Figure 21.13 **Superficial Structures of the Sheep Heart.** (a) Anterior view, and (b) Posterior view.

8. To verify that you have identified the great vessels correctly, use a blunt probe or your fingers (or both) to see where each vessel comes from, or leads to, in the heart. Place the tip of the probe into the lumen of one of the vessels and see where it goes. If the vessel is large enough, put your index finger into the lumen of the vessel so you can *feel* where it goes. Now answer the following questions by giving the name of the heart chamber the probe will go into when placed into each vessel: A probe in the pulmonary trunk will lead into the _____; a probe in the aorta will lead into the _____; a probe in the pulmonary veins will lead into the _____; and a probe in the superior or inferior vena cava will lead into the _____.

In the next two exercises you will cut the entire heart in half to compare wall thicknesses of the right and left ventricles. Approximately two-thirds of the dissection groups in your laboratory should make coronal sections of the heart, and the remaining third will make transverse sections. The different sections yield different views of the chambers and the structures within each chamber. All students should observe hearts that have been sectioned both ways. Ask your laboratory instructor which type of section your group should make before you begin.

EXERCISE 21.7 Coronal Section of the Sheep Heart

1. Obtain a scalpel or a knife with a 6-inch blade and a plastic ruler with millimeter increments. Either the scalpel or the knife will work for this next task, although the knife will make a cleaner cut. Turn the heart upside down so the base is on the dissecting pan, the apex is pointed toward you, and the anterior surface of the heart is facing your body. Make a coronal section through the entire heart to separate it into anterior and posterior portions (**figure 21.14**).

2. Once you have completed the cut, identify the following structures within the ventricles:

   ☐ chordae tendineae   ☐ trabeculae carneae
   ☐ papillary muscles

3. Using a small ruler, measure the thickness of the right ventricular wall approximately 1 cm below the valve ring

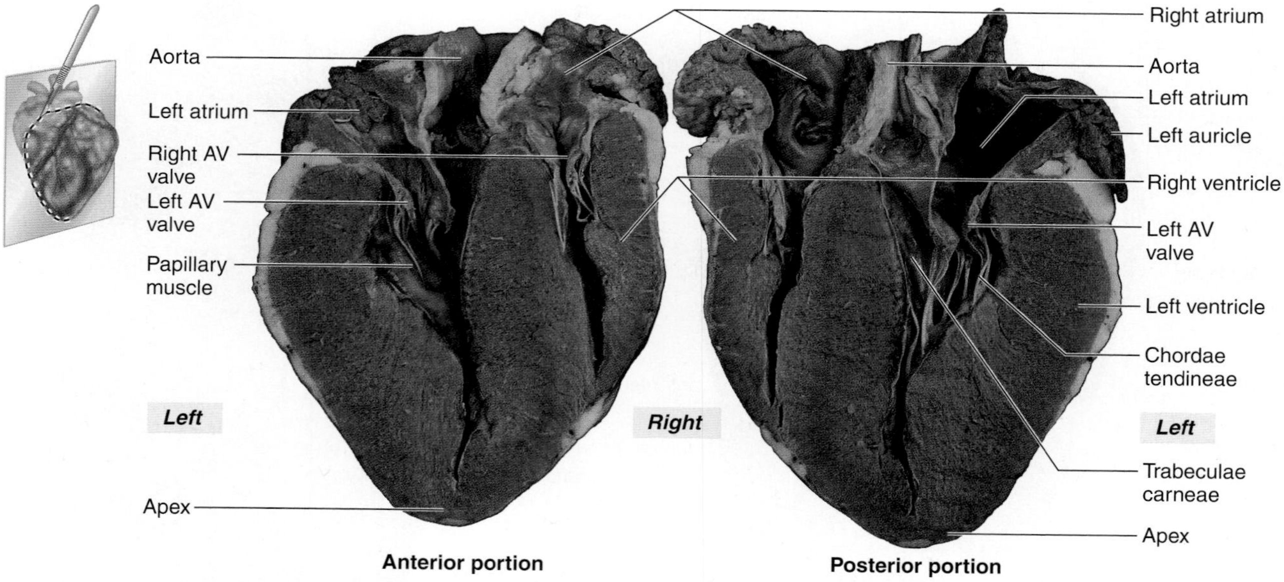

Figure 21.14 **Coronal Section of the Sheep Heart.**

(where the right atrioventricular valve is located) and record it here: _____ cm. Next measure the thickness of the left ventricular wall approximately 1 cm below the valve ring (where the left atrioventricular valve is located) and record it here: _____ cm. Approximately how much thicker is the wall of the left ventricle compared to the wall of the right ventricle?

_____ What is the *functional* consequence of this difference? _____

_____

_____

4. Do the chambers of the right and left ventricles appear to differ in *volume* (that is, the amount of blood each could hold)? _____ Explain the consequences of having each chamber pump a different volume of

blood. _____

_____

_____

_____

5. In the space below, draw a simple diagram of the coronal view of the ventricles. Be sure to include and label all the structures listed below in your drawing.

☐ aorta   ☐ left AV valve

☐ apex   ☐ papillary muscle

☐ chordae tendineae   ☐ right atrium

☐ left atrium   ☐ right AV valve

☐ left auricle   ☐ trabeculae carneae

6. When you have finished your observations, discard the scalpel blade in the sharps container, clean your dissection instruments with soap and water and let them air dry, and put the sheep heart back into the container it came in—or dispose of it according to your instructor's directions.

EXERCISE 21.8 **Transverse Section of the Sheep Heart**

1. Obtain an intact sheep heart, a scalpel or a knife with a 6-inch blade, and a plastic ruler with millimeter increments. Either the scalpel or the knife will work for this next task, although the knife will make a cleaner cut. To make this cut, slice the heart transversely approximately 1 cm inferior to the coronary sulcus (**figure 21.15**) so that you will have separated the entire heart into superior and inferior portions. Observe the cut ends of the heart. Using a small ruler, measure the thickness of right ventricular wall and record it here:

   _____ cm. Next measure the thickness of the

   left ventricular wall and record it here: _____ cm.

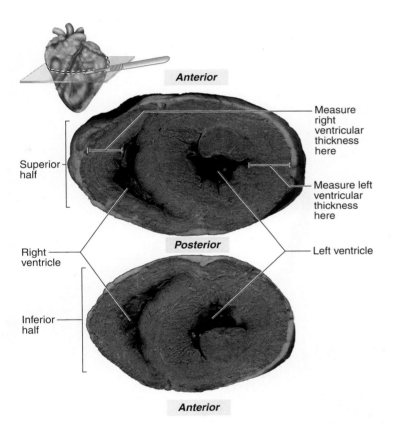

Figure 21.15 **Transverse Section of the Sheep Heart.**

Approximately how much thicker is the wall of the left ventricle compared to the wall of the right ventricle?

_____ What is the *functional* consequence of this

difference? _____

_____

_____

2. Do the chambers of the right and left ventricles appear to differ in *volume* (that is, the amount of blood each could hold)? _____ Explain the consequences of having each chamber pump a different volume of blood.

_____

_____

_____

3. In the space below, draw a simple diagram of the transverse view of the ventricles. Be sure to include and label the right and left ventricles in your drawing:

4. When you have finished your observations, discard the scalpel blade in the sharps container, clean your dissection instruments with soap and water and let them air dry, and put the sheep heart back into the container it came in—or dispose of it according to your instructor's directions.

Chapter 21: Cardiovascular System—The Heart

Name:_____

Date: _____ Section: _____

POST-LABORATORY WORKSHEET

1. Define *mediastinum*: _____

_____

2. The simple squamous epithelium that lines the heart and blood vessels is called _____.

3. A heart surgeon is about to perform a heart transplant. She has already cut through the sternum and entered the thoracic cavity and mediastinum. The heart, however, remains enclosed in the pericardial sac. The tissues the surgeon must cut through, from superficial to deep, to enter the pericardial cavity are:

   1. _____

   2. _____

   Will the surgeon need to cut through the visceral layer of serous pericardium to remove the heart from the pericardial sac? Why or why not?

   _____

   _____

   _____

4. Trace the route of a drop of blood through the heart beginning with the superior and inferior vena cavae, and ending with the descending aorta. Be sure to include the names of all of the valves the blood passes through along its journey.

   _____

   _____

   _____

   _____

   _____

   _____

   _____

   _____

   _____

   _____

   _____

   _____

   _____

   _____

   _____

   _____

5. Label the superficial structures of the heart in the following images of the anterior and posterior surfaces of the human heart.

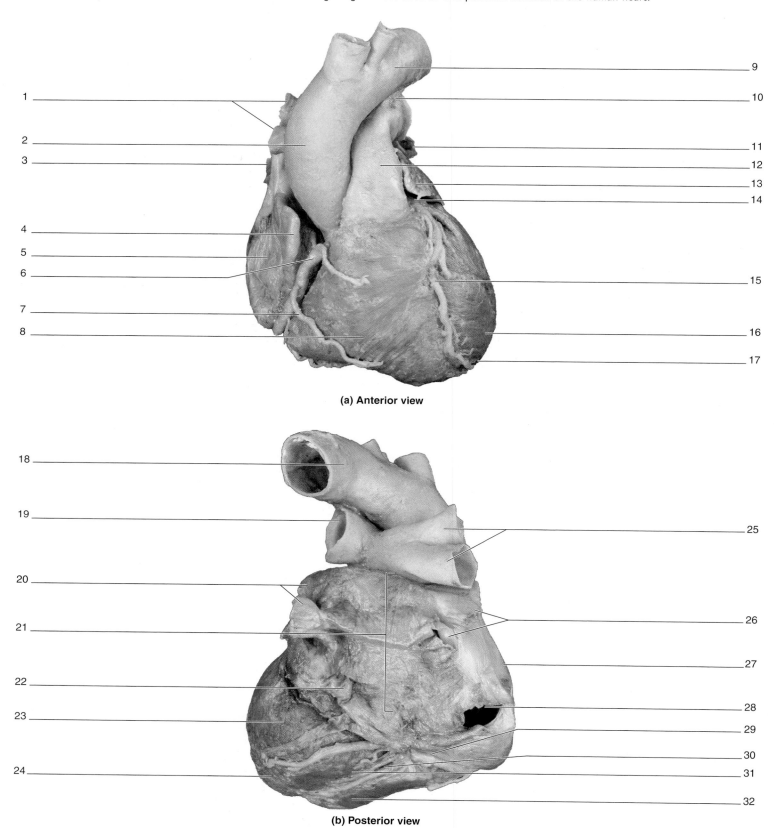

1 _____

2 _____
3 _____

4 _____
5 _____
6 _____

7 _____
8 _____

9 _____
10 _____

11 _____
12 _____
13 _____
14 _____

15 _____

16 _____
17 _____

(a) Anterior view

18 _____

19 _____

20 _____

21 _____

22 _____

23 _____

24 _____

25 _____

26 _____

27 _____

28 _____
29 _____

30 _____
31 _____

32 _____

(b) Posterior view

6. Which chamber of the heart contains pectinate muscles in its walls? _____

7. True or False: The right ventricle pumps blood at a lower pressure than the left ventricle. _____

8. True or False: The right ventricle pumps a lower volume of blood than the left ventricle. _____

9. Figure 21.10 on page 423 demonstrates a myocardial infarction in the interventricular septum and posterior wall of a human heart. This infarct occurred because one of the arteries to the heart wall was blocked and the tissue was starved of oxygen. Which artery was the most likely source of the blockage? (Hint: Think about the areas of the heart wall served by each vessel. Use table 21.3 as a guide if necessary). _____

_____

_____

10. Describe how papillary muscles and chordae tendineae function to keep the AV valves closed when the ventricles contract.

_____

_____

_____

_____

_____

_____

_____

_____

Vessels and Circulation

22

OUTLINE and OBJECTIVES

Anatomy & Physiology REVEALED®
aprevealed.com

**Module 9:
CARDIOVASCULAR SYSTEM**

INTRODUCTION

While proper functioning of the "heart" of the cardiovascular system is necessary for adequate blood flow, our vast system of blood vessels provides necessary conduits to direct the blood to locations where it is most needed by the tissues. **Arteries** are blood vessels that carry blood *away* from the heart, whereas **veins** are blood vessels that carry blood *toward* the heart. Arteries are high-pressure blood conduits (*conduit*, a channel) that carry blood at high pressure and high velocity toward the tissues. Arteries branch into smaller and smaller channels, ultimately forming **arterioles** (*arteriole*, a small artery). Arterioles have the special function of controlling the flow of blood into **capillary beds**. Capillaries are the site of exchange of gases, nutrients, wastes, and so forth between the blood and the tissues. Blood flows out of capillaries into small veins called **venules**, which come together to form the larger veins that return blood to the heart. **Veins** are low-pressure blood reservoirs (*reservoir*, a receptacle) that carry blood back to the heart. Generally, a vein travels next to each major artery and has the same name as the artery it travels with (the major exception to this rule is the hepatic portal system within the abdominal cavity). However, there are usually at least three times as many veins draining a structure as there are arteries supplying it. In the limbs, most of the veins that do not accompany an artery are superficial veins, located just under the skin. For example, the brachium (arm) is supplied by the brachial artery, which has the brachial vein traveling next to it. The brachial artery and vein travel fairly deep within the arm, where they are protected by the musculature of the arm. In addition to the brachial vein, two superficial veins also drain blood from the arm. These are the cephalic and basilic veins. Be aware that there is considerably more variation among individuals in the branching patterns and locations of veins than there is with arteries. Such variation often has clinical significance. For instance, when blood samples need to be collected from a patient, blood is commonly drawn from the median cubital vein. However, not all individuals have a median cubital vein. Lastly, when you are tracing blood flow through the venous system, remember that veins *drain* blood from an area of the body. This means that in your descriptions of blood flow through veins you should start by naming the most distal veins first, and then name the veins blood travels through as it proceeds toward the heart.

Chapter 22: Vessels and Circulation

PRE-LABORATORY WORKSHEET

Name: _____

Date: _____ Section: _____

1. List the three layers present in all blood vessel walls (except capillaries), starting with the outermost layer.

   a. _____

   b. _____

   c. _____

2. The wall of a capillary consists of only a tunica intima. The tunica intima of a capillary (and all vessels, for that matter) is composed of _____

   _____ and a _____.

3. The three types of capillaries are:

   a. _____

   b. _____

   c. _____

4. What specialization do veins have that no other type of blood vessel has?

   _____

5. The three arteries that branch off the aortic arch are:

   a. _____

   b. _____

   c. _____

6. The hepatic portal system is a system of veins that carry venous blood to the _____.

7. Which of the following is *not* a component of the hepatic portal system?

   a. Hepatic portal vein

   b. Superior mesenteric vein

   c. Hepatic veins

   d. Inferior mesenteric vein

   e. Splenic vein

8. Blood leaving the left ventricle of the heart enters the _____ circulation, while blood leaving the right ventricle

   of the heart enters the _____ circulation.

9. The major vein draining blood from the lower half of the body, which empties into the right atrium of the heart is the _____

   _____.

IN THE LABORATORY

In this laboratory session you will begin by investigating the histological characteristics of the different types of blood vessels (arteries, arterioles, capillaries, and veins). You will then identify the major circulatory routes blood uses to travel to and from the organs and tissues of the body by locating blood vessels on a human cadaver or on classroom models of the cardiovascular system. Once you have completed the exercises in this chapter, you should be able to describe the pathway a drop of blood takes as it travels from the heart to a target organ and back to the heart once again, which is called a *trace*.

Histology

Blood Vessel Wall Structure

All blood vessels except capillaries have three layers ("tunics") to their walls. The three layers of a blood vessel wall (tunica intima, tunica media, and tunica externa) are analogous in both structure and function to the three layers of the heart wall (endocardium, myocardium, and epicardium). The differences in structure and function between the different types of blood vessels (arteries, veins, and so on) come mainly from modifications of these three wall layers. In particular, the type of tissue that composes the tunica media greatly affects the function of the vessel. **Table 22.1** describes the general composition of the three layers of a blood vessel wall, and **table 22.2** summarizes unique features of each wall layer in the different types of blood vessels, such as arteries, veins, and capillaries.

Table 22.1	Layers of a Blood Vessel Wall		
Wall Layer	**Location**	**Components**	**Word Origins**
Tunica Intima	Innermost layer, in contact with the lumen of the vessel.	Endothelial cells and subendothelial areolar connective tissue.	*tunic*, a coat, + *intimus*, innermost
Tunica Media	Middle layer.	Varied amounts of collagen fibers, elastic fibers, and smooth muscle cells.	*tunic*, a coat, + *medius*, middle
Tunica Externa	Outermost layer.	Areolar connective tissue that anchors the vessel to surrounding structures.	*tunic*, a coat, + *externus*, on the outside

Table 22.2	Characteristics of Wall Layers in Specific Types of Blood Vessels				
Type of Vessel	**Tunica Intima**	**Tunica Media**	**Tunica Externa**	**Diameter**	**Characteristics and Special Functions**
Elastic Artery	An internal elastic lamina is present but not easily distinguished from the elastic tissue of the tunica media.	Contains numerous elastic fibers and concentric elastic laminae. Also contains smooth muscle cells and reticular fibers.	Underdeveloped in contrast to other vessels. Contains vasa vasorum, lymphatics, and nerves.	2.5 cm–1 cm	Expansion and contraction of elastic tissues smooths out the flow of blood.
Muscular Artery	Contains a very prominent internal elastic lamina.	Contains up to 40 layers of smooth muscle, and a prominent external elastic lamina. Also contains reticular and elastic fibers, which are secreted by the smooth muscle cells.	Contains vasa vasorum, lymphatics, and nerves.	1 cm–3 mm	Contraction of smooth muscle continues to push blood through the arterial system.
Arteriole	An internal elastic lamina is present only in the largest arterioles.	Contains only one or two layers of smooth muscle, with no external elastic lamina.	Very thin.	3 mm–10 μm	Smooth muscle forms sphincters, which control the flow of blood into capillary beds.
Capillary	Endothelium and a basement membrane only.	NA	NA	8–10 μm	See table 22.3.
Postcapillary Venule	Subendothelial layer is very thin, which assists with exchange.	Very thin with no smooth muscle cells.	Very thin.	10–50 μm	Drains blood from capillary beds. Site where leukocytes leave the circulation and enter the tissues via diapedesis.*
Venule	Endothelium and a thin subendothelial layer.	Very thin with very few smooth muscle cells.	Thickest layer of the wall.	50–100 μm	Venules are simply small veins, and are the counterpart to arterioles.
Vein	Infoldings form valves, which prevent the backflow of blood. Not all veins have valves.	Very thin with a small amount of smooth muscle.	Thickest layer of the wall. Contains vasa vasorum.	Greater than 100 μm	Thick tunica externa anchors the vessel to surrounding structures, which assists blood flow when skeletal muscles contract.

*diapedesis (*dia-*, through, + *pedesis*, a leaping) - the passage of leukocytes through the walls of blood vessels.

EXERCISE 22.1 Blood Vessel Wall Structure

1. Obtain a slide demonstrating an artery and a vein (they may both be on the same slide, or you may have two different slides).

2. Place the slide on the microscope stage and bring the tissue sample into focus on low power. Scan the slide and look for the circular or oval cross section of a vessel. If you see more than one vessel on the slide, you will need to determine which vessel is an artery and which is a vein. In general, arteries have relatively thick walls and small lumens, whereas veins have relatively thin walls and large lumens (**figure 22.1**). In addition, the lumens of veins are often collapsed because of the fragile, thin nature of the blood vessel wall.

3. Once you have identified an artery, and a vein, move the microscope stage so the wall of the *artery* is in the center of the field of view and increase the power on the microscope until you can see all the layers of the artery wall.

4. Using figure 22.1*b* and table 22.1 as guides, identify the structures listed below. Keep in mind that the innermost layer of the vessel (the tunica intima) will be incredibly thin and difficult to identify except on high power. Most likely you will see only the flattened nuclei of the endothelial cells, and very little of the rest of the cells.

☐ artery   ☐ tunica intima

☐ lumen   ☐ tunica media

☐ tunica externa   ☐ vein

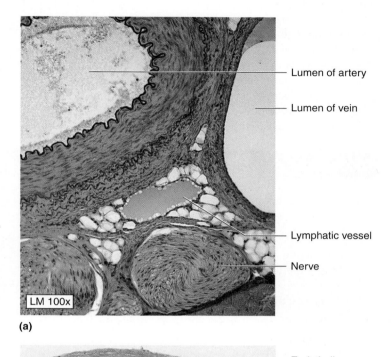

- Lumen of artery
- Lumen of vein
- Lymphatic vessel
- Nerve

LM 100x

(a)

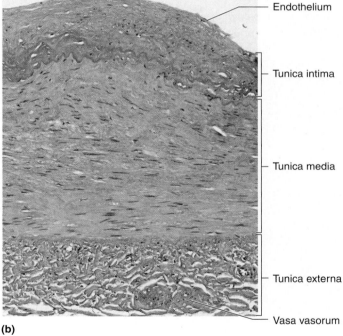

- Endothelium
- Tunica intima
- Tunica media
- Tunica externa
- Vasa vasorum

(b)

Figure 22.1   Blood Vessel Wall Structure.   (a) Cross-section through the center of a neurovascular bundle containing a nerve, artery, vein, and lymphatic vessel. (b) The three layers of the wall of a blood vessel: tunica intima, tunica media, and tunica externa. The tunica externa has its own blood supply, the vasa vasorum (literally the "vessels of the vessels").

Elastic Arteries

The aorta is an example of an **elastic artery**. Only the aorta and the pulmonary, brachiocephalic, common carotid, subclavian, and common iliac arteries are classified as elastic arteries (figure 22.1, table 22.2). These arteries, located very close to the heart, have walls that are thick enough to withstand the pressure of blood that is pumped into them from the ventricles of the heart. The ventricles create enough force to move blood through these vessels, so they need very little smooth muscle in their tunica media to assist with blood flow. Instead, elastic arteries have an abundance of collagen and elastic fibers in their tunica media, which makes them both tough (collagen fibers) and stretchy (elastic fibers). The ability of the vessel wall to stretch as it receives blood from the ventricles, and then recoil, greatly smooths out the flow of blood through these arteries.

Large vessels such as the aorta have tiny blood vessels called *vasa vasorum* (literally, "the vessels of the vessels," see figure 22.1*b*) in the tunica externa. The vasa vasorum are analogous in both structure and function to the coronary arteries in the outer layer of the heart wall (epicardium).

EXERCISE 22.2 Elastic Artery—The Aorta

1. Obtain a slide of the aorta (**figure 22.2**) and place it on the microscope stage. Bring the wall of the aorta into focus on low power.

2. Using figure 22.2 and tables 22.1 and 22.2 as guides, identify the following on the slide of the aorta:

   ☐ elastic fibers   ☐ tunica intima
   ☐ lumen   ☐ tunica media
   ☐ tunica externa   ☐ vasa vasorum

3. In the space below, draw a brief sketch of the wall of the aorta as seen under the microscope.

_____ ×

⚡ WHAT DO YOU THINK?

❶ Why do you think large blood vessels have their own blood vessels (vasa vasorum) in the tunica externa?

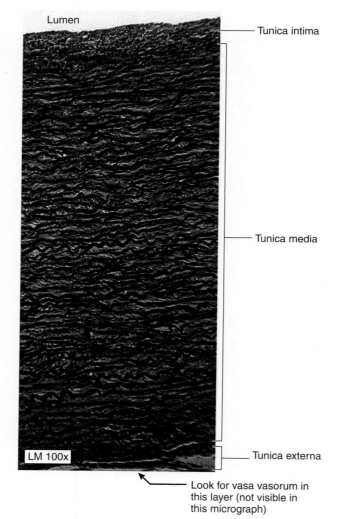

Figure 22.2 **Elastic Artery.** The wall of the aorta, an elastic artery, contains numerous elastic fibers (black) in the tunica media. No vasa vasorum are visible in the tunica externa in this micrograph.

Labels on figure: Lumen, Tunica intima, Tunica media, Tunica externa, Look for vasa vasorum in this layer (not visible in this micrograph), LM 100x

---

Muscular Arteries

As blood moves through the elastic arteries and travels farther away from the heart, the force of the heart is no longer enough to keep the blood moving through the vessels. Thus, the amount of elastic tissue in the tunica media of the arteries starts to decrease and the amount of smooth muscle in the tunica media starts to increase. Contraction of the smooth muscle keeps blood moving through the arteries as the blood gets farther away from the heart. **Muscular arteries** are easily distinguished from elastic arteries by the presence of two prominent bands of elastic fibers, the **internal elastic lamina** and the **external elastic lamina**, with several layers of smooth muscle sandwiched in between the two bands (see figure 22.3, table 22.2).

EXERCISE 22.3 Muscular Artery

1. Obtain a slide of a small, muscular artery, and place it on the microscope stage. Bring the tissue sample into focus on low power, then locate the wall of the vessel (**figure 22.3**).

2. Using figure 22.3 and tables 22.1 and 22.2 as guides, identify the following on the slide of the muscular artery:

   - ☐ external elastic lamina
   - ☐ tunica externa
   - ☐ internal elastic lamina
   - ☐ tunica intima
   - ☐ lumen
   - ☐ tunica media
   - ☐ smooth muscle

3. In the space below, draw a brief sketch of the wall of the muscular artery as seen under the microscope.

_____ ×

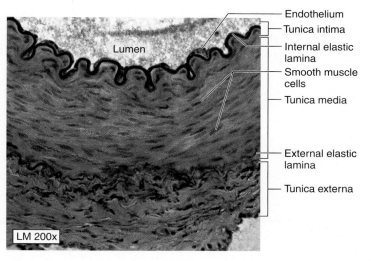

Figure 22.3 **Muscular Artery.** Muscular arteries contain distinct internal and external elastic laminae bordering the tunica media, which is predominantly smooth muscle.

Arterioles

Arterioles have a unique function in the cardiovascular system. Arterioles control the flow of blood into capillary beds (table 22.2). As such, the most prominent feature of the wall of an arteriole is a layer of circular smooth muscle in the tunica media. This smooth muscle acts as a sphincter to change the diameter of the vessel, thus regulating the flow of blood into the capillary beds.

EXERCISE 22.4 Arteriole

1. Obtain a slide of an arteriole and place it on the microscope stage. Bring the tissue sample into focus on low power. Scan the slide to locate an arteriole in cross-section (**figure 22.4**).

2. Using figure 22.4 and tables 22.1 and 22.2 as guides, identify the following on the slide of the arteriole:

   - ☐ lumen
   - ☐ tunica intima
   - ☐ smooth muscle
   - ☐ tunica media
   - ☐ tunica externa

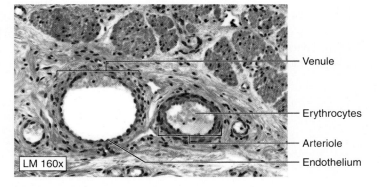

Figure 22.4 **Arteriole.** The tunica media of the arteriole contains circular smooth muscle that functions as a sphincter to alter the diameter of the vessel.

3. In the space below, draw a brief sketch of the arteriole as seen under the microscope.

_____ ×

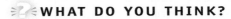
Veins

Veins are large vessels with thin walls that function to return blood to the heart at low pressure. They are characterized by having large lumens and thin walls (see figure 22.5, table 22.2). They may also contain valves, which are infoldings of the tunica intima. These valves prevent blood from flowing backward. Large veins, like large arteries, also contain vasa vasorum. **Venules** are simply small veins. The venules that come immediately after capillary beds, **postcapillary venules,** are the site where most white blood cells leave the circulation to enter the tissues via diapedesis (see table 22.2).

 EXERCISE 22.5 Vein

1. Place a slide of a large vein on the microscope stage and bring the tissue sample into focus on low power.

2. Using **figure 22.5** and tables 22.1 and 22.2 as guides, identify the following on the slide of the large vein:

☐ elastic fiber   ☐ tunica media

☐ smooth muscle   ☐ valve (may not be visible on the slide)

☐ tunica externa

☐ tunica intima   ☐ vasa vasorum

3. In the space below, draw a brief sketch of the large vein as seen under the microscope.

_____ ×

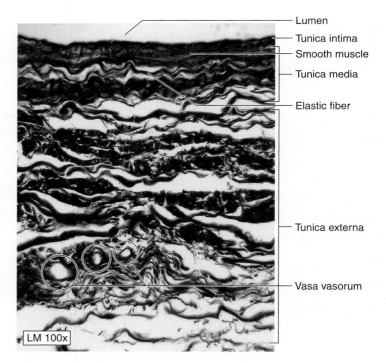

Figure 22.5   Vein. Cross-section through the wall of a vein. The thickest layer of the vessel wall is the tunica externa, which contains vasa vasorum.

Capillaries

Capillaries are unique blood vessels in that they consist *only* of a tunica intima (endothelium and basement membrane). This allows them to perform the important job of permitting exchange between the blood and the tissues. There are several different types of capillaries in the body, some of which allow more exchange between the blood and the tissues and some of which allow less. For instance, capillaries in the spleen and liver, organs where a lot of blood processing takes place, allow a great deal of exchange to occur. In contrast, capillaries in the brain, an organ that must be protected from harmful substances such as toxins and viruses that might be present in the blood, allow very little exchange.

EXERCISE 22.6 Observing Electron Micrographs of Capillaries

1. Because capillaries are so tiny, the only way to truly appreciate their structure is to view them with an electron microscope. Obviously you do not have access to an electron microscope in your anatomy laboratory. Thus, this portion of the exercise will be performed by viewing the electron micrographs in **table 22.3**.

2. Observe the electron micrograph of the **continuous capillary** in table 22.3*a*. Identify the endothelial cells in the micrograph and then look for the areas where two endothelial cells come in contact with each other. Notice that the endothelial cells overlap each other in these areas. Where the cells come together, there are numerous tight junctions that prevent substances from moving between the endothelial cells. The overlapping nature of the endothelial cells and the tight junctions between the cells together create a *continuous* barrier between the blood and the tissues. Thus, for exchange to occur between the blood and the tissues, substances must travel *through* the endothelial cells. This gives the endothelial cells control over what substances can and cannot be exchanged between the blood and the tissues.

3. Observe the electron micrograph of the **fenestrated capillary** in table 22.3*b*. Identify endothelial cells in the micrograph. In contrast to the smooth surface of the endothelial cells of the continuous capillaries, the endothelial cells of the fenestrated capillaries appear wavy, particularly in the area ajdacent to the nucleus of the endothelial cell. As with continuous capillaries, the endothelial cells of fenestrated capillaries form a continuous barrier between the blood and the tissues (that is, there are no spaces between endothelial cells). However, the endothelial cells themselves have pores, or **fenestrations**, in them that allow any substance that is smaller than the size of the pore to be exchanged easily between the blood and the tissues. Thus, fenestrated capillaries allow much greater exchange than continuous capillaries, and they are found in organs where a lot of exchange is necessary, such as within endocrine glands where hormones are secreted into the blood.

4. Observe the electron micrograph of the **sinusoidal capillary** in table 22.3*c*. Sinusoidal capillaries are located within organs that do a lot of processing of the blood, such as the liver and spleen. Sinusoidal capillaries are characterized by having *discontinuous* endothelial cells, which do not overlap each other. In fact, there are open spaces between endothelial cells. The endothelial cells themselves are also fenestrated. Thus, the open spaces and fenestrations together create a minimal barrier between the blood and the tissues, which allows for maximum exchange. The micrograph in table 22.3*c* demonstrates a sinusoidal capillary within the liver. The capillary runs from left to right in the top 1/3 of the micrograph. Inside the lumen of the capillary is a macrophage. The one large, prominent nucleus visible in the micrograph is that of a liver cell, or hepatocyte, which contains many small lipid droplets. There is another hepatocyte at the top of the micrograph as well. Observe the edges of the hepatocytes that face the lumen of the sinusoidal capillary. There you will see the thin endothelial cells that line the capillary. Now observe the area where there should be endothelium between the macrophage and the hepatocyte containing the lipid droplets and nucleus. Notice that there is no endothelial cell in between the two. Also, in the location between the macrophage and the hepatocyte above it, toward the left side of the micrograph, you can see fenestrations if you look very closely. The magnification of this micrograph is not quite high enough to see the fenestrations clearly, but they are there.

5. In the spaces to the right, draw a brief sketch of each of the three types of capillaries. Be sure to label the following on your drawings.

☐ basement membrane   ☐ fenestrations (if present)

☐ endothelial cells   ☐ lumen of capillary

WHAT DO YOU THINK?

3 Recall from chapter 14 (nervous tissues) that the blood vessels that supply nervous tissues of the brain are surrounded by glial cells called astrocytes, which collectively form the *blood-brain barrier*. What type of capillaries would you expect to find in this area of the brain, and why?

4 Also recall from chapter 14 that the blood vessels that form the choroid plexus within the brain ventricles are surrounded by glial cells called ependymal cells, which collectively form the *blood-cerebrospinal fluid (CSF) barrier*. What type of capillaries would you expect to find in this area of the brain and why?

5 Given your answers to questions 3 and 4, which of the two barriers (blood-brain or blood-CSF) represents a more complete barrier between the blood and the tissues of the brain?

Continuous

Table 22.3	Characteristics of the Three Types of Capillaries		

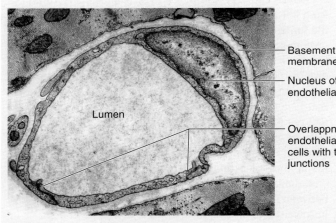

(a) Continuous Capillary

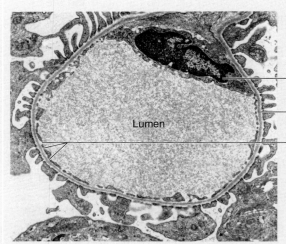

(b) Fenestrated Capillary

	(a) Continuous Capillary		(b) Fenestrated Capillary
Characteristics	Endothelial cells connected to each other by **tight junctions**.	*Characteristics*	Endothelial cells contain **fenestrations** (pores) covered by a thin diaphragm. Macromolecules pass through fenestrations to get into the tissues.
Endothelial Cells	Not fenestrated.	*Endothelial Cells*	Fenestrated.
Basement Membrane	Continuous.	*Basement Membrane*	Continuous.
Description and Function	Form a continuous barrier between blood and tissues for tight regulation of exchange.	*Description and Function*	Allow for increased exchange between blood and tissues.
Locations	Muscle tissues, skin, connective tissues, exocrine glands, and nervous tissues.	*Locations*	Glomerulus of kidney, lamina propria of intestine, choroid plexus of brain, ciliary body, and most endocrine glands.
Word Origin	*continuus*, continued.	*Word Origin*	*fenestra*, a window.

Fenestrated   *Sinusoidal*

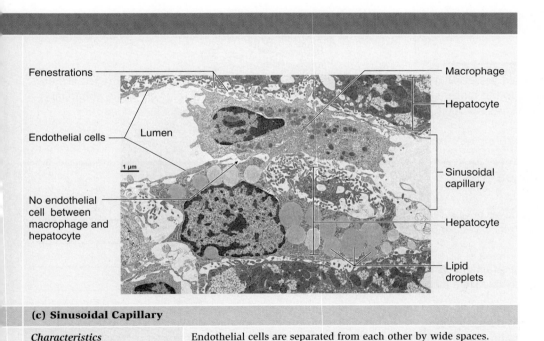

Fenestrations

Macrophage

Hepatocyte

Endothelial cells   Lumen

1 μm

No endothelial
cell between
macrophage and
hepatocyte

Sinusoidal
capillary

Hepatocyte

Lipid
droplets

(c) Sinusoidal Capillary

Characteristics	Endothelial cells are separated from each other by wide spaces. Endothelial cells are fenestrated without diaphragms. Macrophages are located among the endothelium.
Endothelial Cells	Fenestrated.
Basement Membrane	Discontinuous.
Description and Function	Allow blood and tissues to come into intimate contact with each other, which allows for maximum exchange between blood and tissues.
Locations	Liver, spleen, bone marrow, and some endocrine glands.
Word Origins	*sinus*, a channel, + *eidos*, resemblance.

Gross Anatomy

In the following laboratory exercises you will identify the major arteries and veins of the body on a prosected human cadaver or on classroom models or charts. If your laboratory has both prosected cadavers and models, be sure to identify the vessels on both, because it is likely that both will be used on laboratory practical exams. As you identify each vessel, consider the area of the body supplied by or drained by the vessel, and ask yourself what tissue(s) or organ(s) would suffer damage if the vessel were blocked or cut.

After identifying the major vessels of each region of the body, you will be asked to trace the flow of blood from the heart to an organ within each region, and back to the heart (see Study Tip! box on this page). Your initial traces will be done using figures and will include only vessels that are relatively close to the target organ. However, you will also be asked to write out the complete trace in words. This means you will need to complete the trace you started in the figure by extending the list of blood vessels to include the vessels close to the heart that are not shown in the figures (for example: the aorta). When asked to do a complete trace on a laboratory practical exam, you will also need to include blood flow through the heart and the pulmonary circulation, and name all of the heart valves blood travels through on its journey. A complete trace both begins and ends in the right atrium of the heart.

Study Tip! Blood Traces

The task of tracing the flow of blood through the blood vessels of the body is challenging for many students. If you find yourself having difficulty with this task, use the analogy of driving a car to your school. A drop of blood or a red blood cell represents your car, and the blood vessels represent the roads and highways. The organs represent your destination (school, the supermarket, and so on), and the heart represents your home. For example, to travel from your home to school, you must drive your car along a series of roadways that lead to the school. Each street has a name, which helps direct people driving their cars to their destinations. Similarly, as blood travels from the heart to a destination in the body (for example, the right hand), it must follow a given route. This route consists of several "streets," each with an identifying name. As you trace the flow of blood from your heart to your right hand, imagine yourself driving a car inside the blood vessels, and write down the names of the arteries you pass through as you travel to your destination. Of course, once at your destination, you must eventually return "home" to the heart. Thus, you also need to visualize the trip through the veins that you would take to get from the hand back to the heart.

Pulmonary Circulation

The **pulmonary circulation** is the system of blood vessels that carry blood from the right ventricle of the heart to the lungs and back to the left atrium of the heart. Blood leaves the right ventricle relatively deoxygenated, and returns to the left atrium highly oxygenated, having picked up oxygen in the pulmonary capillaries.

EXERCISE 22.7 Pulmonary Circulation

1. Observe the thoracic cavity of a human cadaver or observe classroom models or charts demonstrating blood vessels of the heart and lungs.

2. Using your textbook as a guide, identify the structures listed in **figure 22.6** on the cadaver, models, or charts. Then label them in figure 22.6.

3. *Optional Activity*: AP|R **Cardiovascular System**—Watch the "Pulmonary and Systemic Circulation" animation to review the differences between these two circuits.

4. In the spaces to the right, trace the flow of blood from the right ventricle of the heart through the pulmonary circuit to the left atrium of the heart in words. Be sure to indicate any valves encountered along the way.

_____

_____

_____

_____

_____

_____

_____

_____

_____

_____

_____

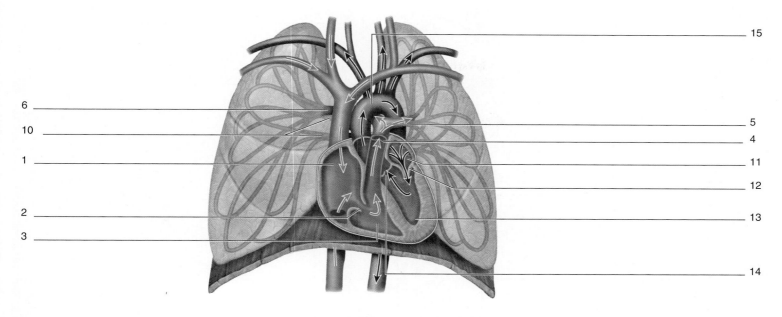

6
10
1
2
3
15
5
4
11
12
13
14

(a) Heart and lungs

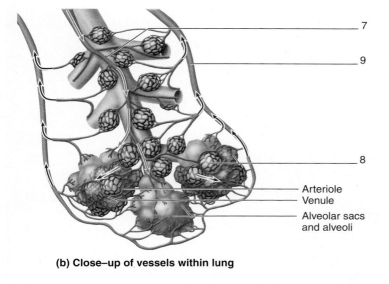

7
9
8

Arteriole
Venule
Alveolar sacs
and alveoli

(b) Close–up of vessels within lung

Figure 22.6 **Pulmonary Circulation.** Blue arrows indicate the path of deoxygenated blood. Red arrows indicate the path of oxygenated blood.

- [ ] aorta
- [ ] aortic semilunar valve
- [ ] branch of pulmonary artery
- [ ] branch of pulmonary vein
- [ ] left atrium
- [ ] left AV valve
- [ ] left ventricle
- [ ] pulmonary arteries
- [ ] pulmonary capillaries
- [ ] pulmonary semilunar valve
- [ ] pulmonary trunk
- [ ] pulmonary veins
- [ ] right atrium
- [ ] right AV valve
- [ ] right ventricle

Systemic Circulation

The **systemic circulation** is the system of blood vessels that carry blood from the left ventricle of the heart to body organs and back to the right atrium of the heart. Blood leaves the left ventricle highly oxygenated, and returns to the right atrium relatively deoxygenated, with oxygen diffused out of and carbon dioxide diffused into the tissue capillaries.

EXERCISE 22.8   Circulation to the Head and Neck

1. Observe the head and neck regions of a prosected human cadaver or observe classroom models or charts demonstrating blood vessels of the head and neck.

2. The major arteries carrying blood to structures of the head and neck are the external and internal carotid arteries (**figure 22.7**). The **external carotid arteries** supply most

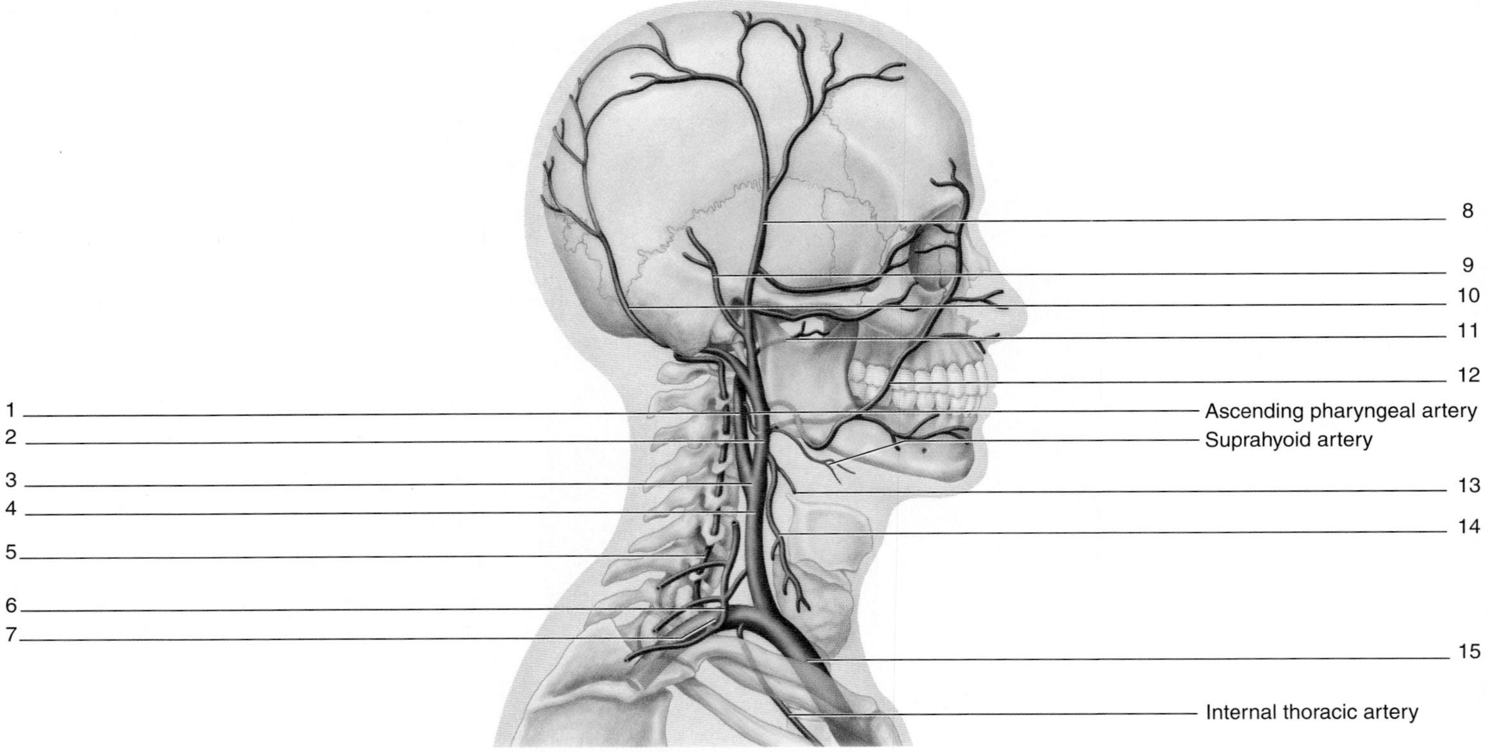

(a) Arteries, right lateral view

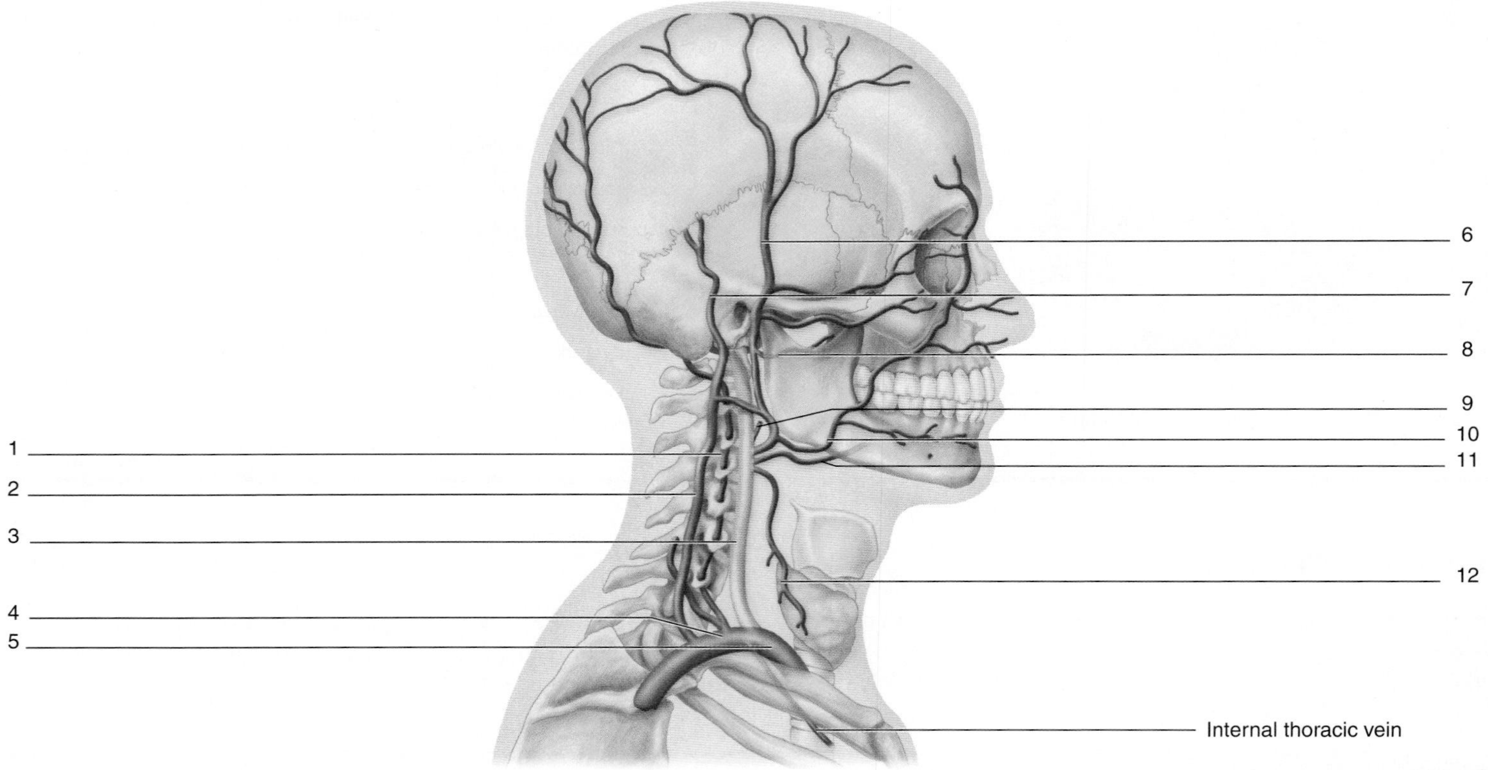

Ascending pharyngeal artery
Suprahyoid artery
Internal thoracic artery

(b) Veins, right lateral view

Internal thoracic vein

Figure 22.7 Circulation to the Head and Neck.

(a) Arterial Supply
- ☐ brachiocephalic trunk
- ☐ carotid sinus
- ☐ common carotid artery
- ☐ external carotid artery
- ☐ facial artery
- ☐ internal carotid artery
- ☐ maxillary artery

- ☐ occipital artery
- ☐ posterior auricular artery
- ☐ subclavian artery
- ☐ superficial temporal artery
- ☐ superior laryngeal artery
- ☐ superior thyroid artery
- ☐ thyrocervical trunk
- ☐ vertebral artery

(b) Venous Drainage
- ☐ external jugular vein
- ☐ facial vein
- ☐ internal jugular vein
- ☐ lingual vein
- ☐ maxillary vein
- ☐ pharyngeal vein

- ☐ posterior auricular vein
- ☐ right brachiocephalic vein
- ☐ subclavian vein
- ☐ superficial temporal vein
- ☐ superior thyroid vein
- ☐ vertebral vein

superficial structures of the head and neck. The **external jugular veins** drain most superficial areas of the scalp, face, and neck.

3. Using your textbook as a guide, identify the *arteries* listed in figure 22.7a that supply blood to the head and neck. Then label them in figure 22.7a.

4. Using your textbook as a guide, identify the *veins* listed in figure 22.7b that drain blood from the head and neck. Then label them in figure 22.7b.

5. The pathway a drop of blood takes to get from the aortic arch to the *skin overlying the anterior part of the right*

parietal bone of the skull and back to the superior vena cava is shaded in **figure 22.8.** Trace this flow of blood by writing in the names of the vessels in the figure. Label the vessels in order, starting at number 1, so you are figuratively tracing the pathway and naming the vessels encountered along the way.

6. *Optional Activity*: **AP|R** **Cardiovascular System—** *Anatomy & Physiology Revealed* includes numerous dissections showing vascular supply to all body regions; review these dissections and use the Quiz feature to test yourself on each region.

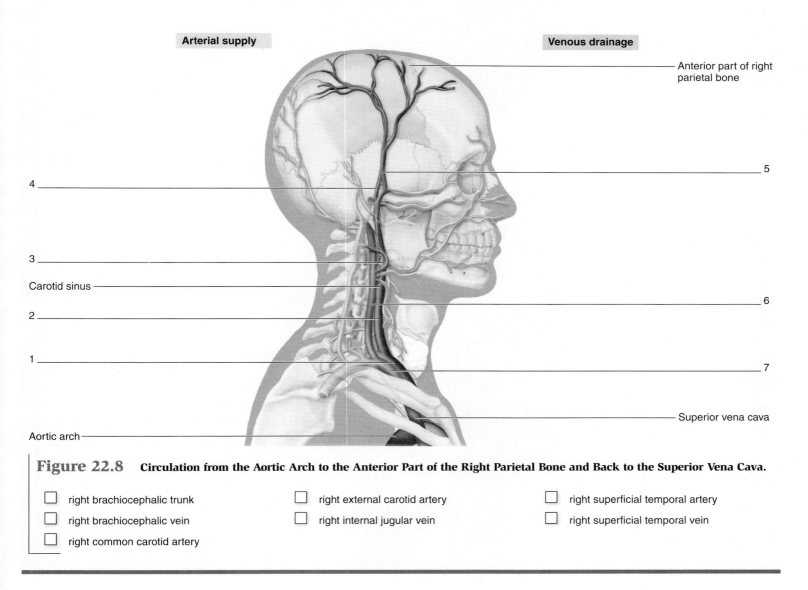

Figure 22.8 **Circulation from the Aortic Arch to the Anterior Part of the Right Parietal Bone and Back to the Superior Vena Cava.**

☐ right brachiocephalic trunk ☐ right external carotid artery ☐ right superficial temporal artery
☐ right brachiocephalic vein ☐ right internal jugular vein ☐ right superficial temporal vein
☐ right common carotid artery

EXERCISE 22.9 Circulation to the Brain

1. Observe the cranium and brain of a prosected human cadaver or observe classroom models or charts demonstrating blood vessels of the cranium and brain.

2. The major arteries carrying blood to structures of the brain are the internal carotid arteries and the vertebral

arteries (figures 22.7 and **22.9**). The **internal carotid arteries** supply 75% of the blood flow to the brain, while the **vertebral arteries** supply 25% of the blood flow to the brain. Both pairs of vessels supply blood to the **cerebral arterial circle** (figure 22.9a). The major

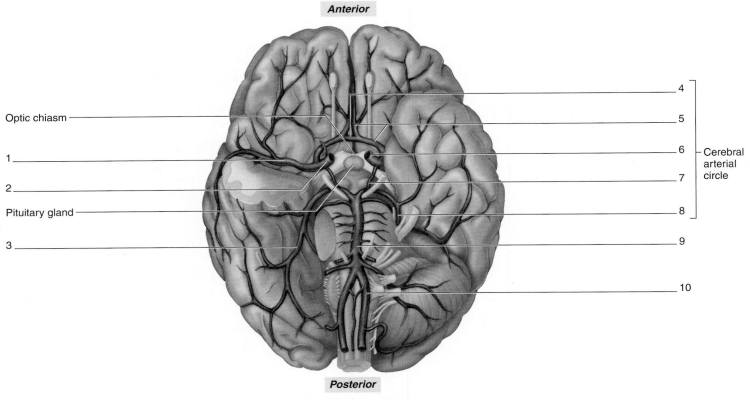

Anterior

Optic chiasm

4

5

6

7

Cerebral
arterial
circle

1

2

Pituitary gland

8

3

9

10

Posterior

(a) Arteries of the brain, inferior view

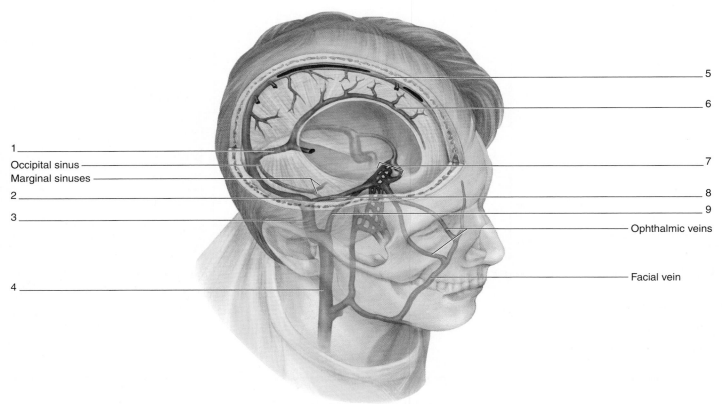

5

6

1

Occipital sinus

Marginal sinuses

7

2

8

3

9

Ophthalmic veins

Facial vein

4

(b) Cranial and facial veins, right superior anterolateral view

Figure 22.9 Circulation to the Brain.

(a) Arterial Supply

- [ ] anterior cerebral artery
- [ ] anterior communicating artery
- [ ] basilar artery
- [ ] internal carotid artery
- [ ] middle cerebral artery
- [ ] posterior cerebral artery
- [ ] posterior communicating artery
- [ ] vertebral artery

(b) Venous Drainage

- [ ] cavernous sinus
- [ ] inferior petrosal sinus
- [ ] inferior sagittal sinus
- [ ] internal jugular vein
- [ ] sigmoid sinus
- [ ] straight sinus
- [ ] superior petrosal sinus
- [ ] superior sagittal sinus
- [ ] transverse sinus

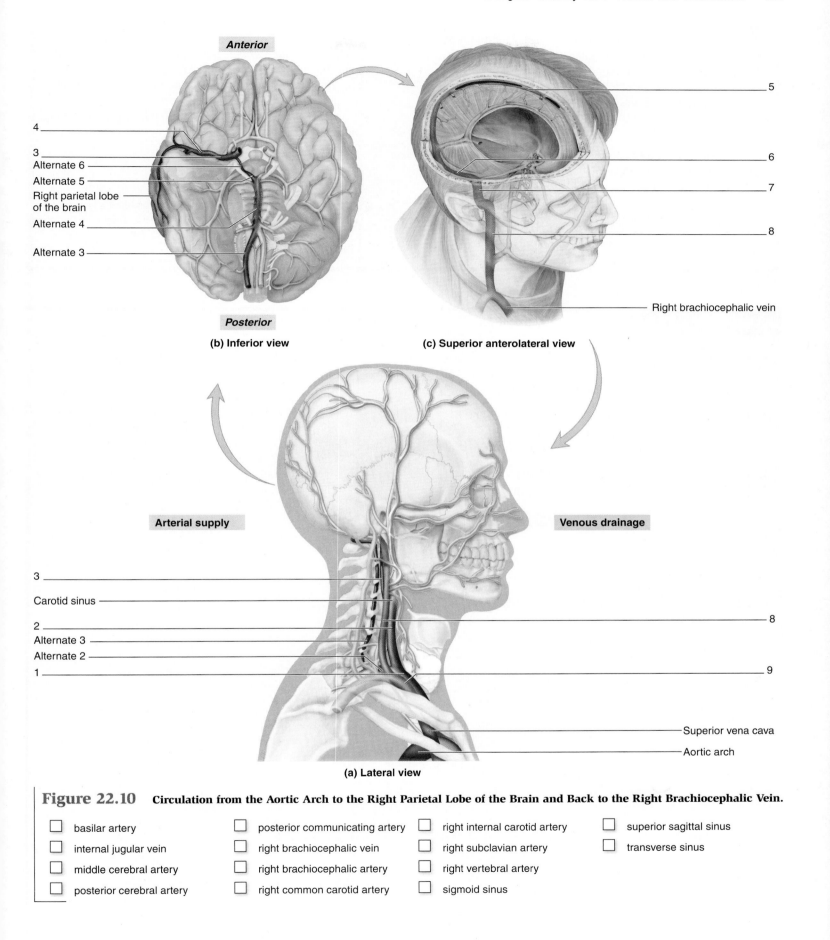

Anterior

4
3
Alternate 6
Alternate 5
Right parietal lobe
of the brain
Alternate 4

Alternate 3

Posterior

(b) Inferior view

5

6

7

8

Right brachiocephalic vein

(c) Superior anterolateral view

Arterial supply

Venous drainage

3

Carotid sinus

2
Alternate 3
Alternate 2
1

8

9

Superior vena cava

Aortic arch

(a) Lateral view

Figure 22.10 **Circulation from the Aortic Arch to the Right Parietal Lobe of the Brain and Back to the Right Brachiocephalic Vein.**

☐ basilar artery
☐ internal jugular vein
☐ middle cerebral artery
☐ posterior cerebral artery

☐ posterior communicating artery
☐ right brachiocephalic vein
☐ right brachiocephalic artery
☐ right common carotid artery

☐ right internal carotid artery
☐ right subclavian artery
☐ right vertebral artery
☐ sigmoid sinus

☐ superior sagittal sinus
☐ transverse sinus

veins draining blood from the head, neck, and brain are the external and internal jugular veins. The **internal jugular veins** drain all blood from inside the cranial cavity plus some superficial areas of the face. Blood draining from brain tissues first enters the **dural venous sinuses**, which collectively drain into the internal jugular vein (figure 22.9*b*).

3. Using your textbook as a guide, identify the *arteries* listed in figure 22.9*a* that supply blood to the brain. Then label them in figure 22.9*a*.

4. Using your textbook as a guide, identify the *veins* listed in figure 22.9*b* that drain blood from the brain. Then label them in figure 22.9*b*.

5. The pathway a drop of blood takes to get from the aortic arch to the *right parietal lobe of the brain* and back to the right brachiocephalic vein is shaded in **figure 22.10**. Trace this flow of blood by writing in the names of the vessels in the figure. Label the vessels in order, starting at number 1, so you are figuratively tracing the pathway and naming the vessels encountered along the way.

 WHAT DO YOU THINK?

6. A patient suffers a stroke of the middle cerebral artery that cuts off the blood supply to the right parietal lobe of the brain. What neurological deficits would result? (Hint: See table 15.3 on p. 297.)

EXERCISE 22.10   Circulation to the Thoracic and Abdominal Walls

1. Observe the thoracic cavity on a prosected human cadaver or observe classroom models or charts demonstrating blood vessels of the thoracic and abdominal cavities.

2. Using your textbook or atlas as guides, identify the *arteries* listed in **figure 22.11*a*** that supply blood to the thoracic and abdominal walls. Then label them in figure 22.11*a*.

3. Using your textbook as a guide, identify the *veins* listed in figure 22.11*b* that drain blood from

the thoracic and abdominal walls. Then label them in figure 22.11*b*.

4. The pathway a drop of blood takes to get from the left ventricle of the heart to the *right kidney* and back to the right atrium of the heart is shaded in **figure 22.12**. Trace this flow of blood by writing in the names of the vessels in the figure. Label the vessels in order, starting at number 1, so you are figuratively tracing the pathway and naming the vessels encountered along the way.

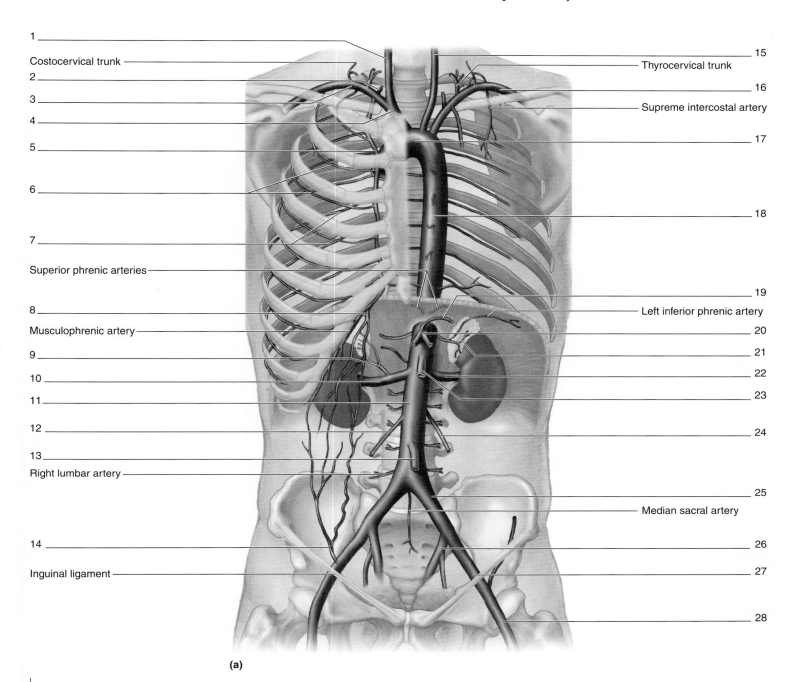

1 _____

Costocervical trunk _____

2 _____

3 _____

4 _____

5 _____

6 _____

7 _____

Superior phrenic arteries _____

8 _____

Musculophrenic artery _____

9 _____

10 _____

11 _____

12 _____

13 _____

Right lumbar artery _____

14 _____

Inguinal ligament _____

15 _____

Thyrocervical trunk _____

16 _____

Supreme intercostal artery _____

17 _____

18 _____

19 _____

Left inferior phrenic artery _____

20 _____

21 _____

22 _____

23 _____

24 _____

25 _____

Median sacral artery _____

26 _____

27 _____

28 _____

(a)

Figure 22.11   Circulation to the Thoracic and Abdominal Walls.

(a) Arterial Supply

☐ anterior intercostal arteries
☐ aortic arch
☐ brachiocephalic trunk
☐ celiac trunk
☐ descending abdominal aorta
☐ descending thoracic aorta
☐ inferior epigastric artery
☐ inferior mesenteric artery
☐ internal thoracic artery

☐ left common carotid artery
☐ left common iliac artery
☐ left external iliac artery
☐ left femoral artery
☐ left gastric artery
☐ left gonadal artery
☐ left internal iliac artery
☐ left superior suprarenal (adrenal) artery
☐ left renal artery
☐ left subclavian artery

☐ posterior intercostal arteries
☐ right common carotid artery
☐ right gonadal artery
☐ right inferior suprarenal (adrenal) artery
☐ right renal artery
☐ right subclavian artery
☐ superior epigastric artery
☐ superior mesenteric artery
☐ vertebral artery

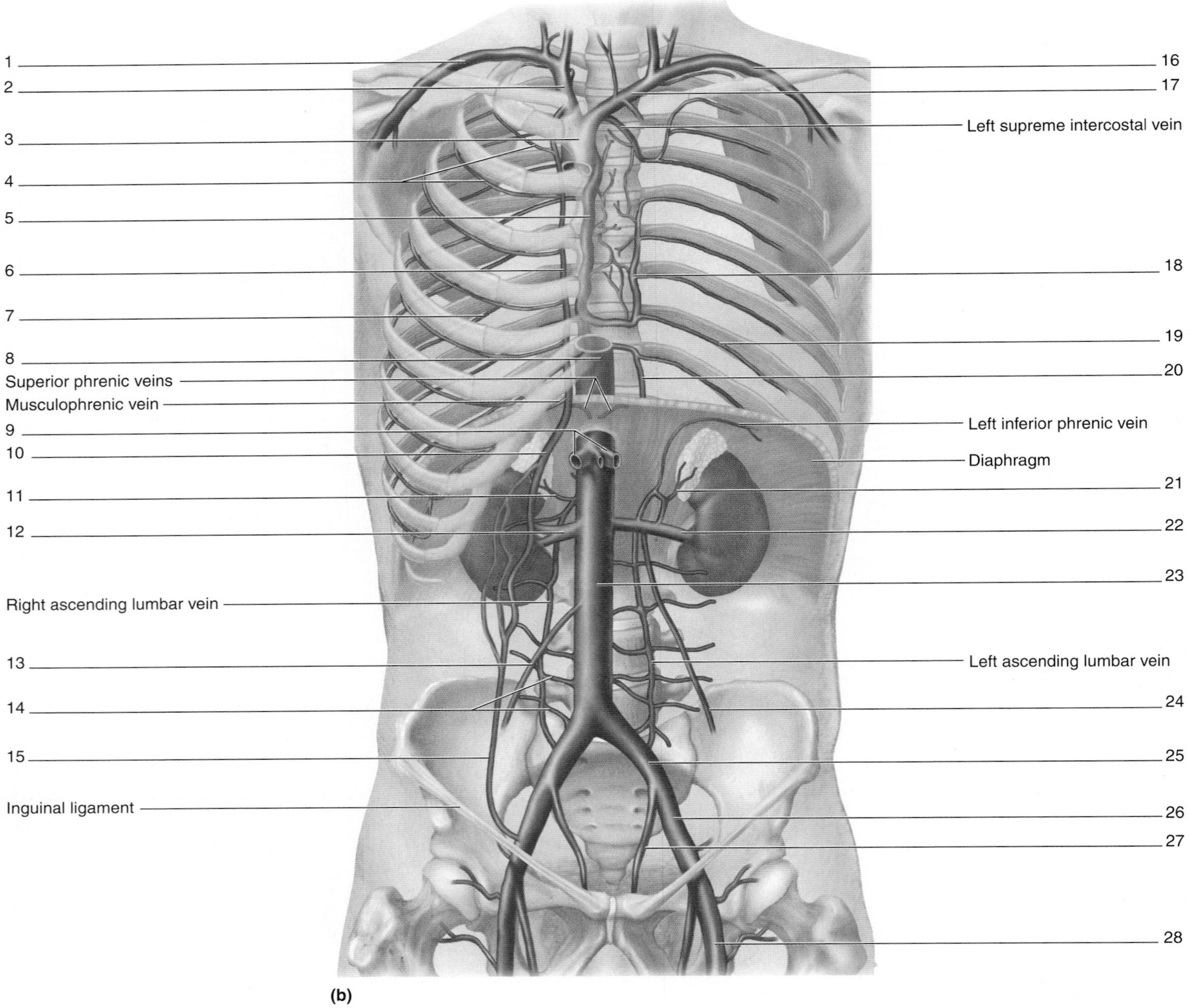

1 _____

2 _____

3 _____

4 _____

5 _____

6 _____

7 _____

8 _____

Superior phrenic veins _____

Musculophrenic vein _____

9 _____

10 _____

11 _____

12 _____

Right ascending lumbar vein _____

13 _____

14 _____

15 _____

Inguinal ligament _____

16 _____

17 _____

Left supreme intercostal vein _____

18 _____

19 _____

20 _____

Left inferior phrenic vein _____

Diaphragm _____

21 _____

22 _____

23 _____

Left ascending lumbar vein _____

24 _____

25 _____

26 _____

27 _____

28 _____

(b)

Figure 22.11   Circulation to the Thoracic and Abdominal Walls *(continued)*.

(b) Venous Drainage

☐ accessory hemiazygos vein
☐ anterior intercostal veins
☐ azygos vein
☐ hemiazygos vein
☐ hepatic veins
☐ inferior vena cava
☐ internal thoracic vein
☐ left brachiocephalic vein
☐ left common iliac vein

☐ left external iliac vein
☐ left femoral vein
☐ left gonadal vein
☐ left internal iliac vein
☐ left posterior intercostal vein
☐ left renal vein
☐ left subclavian vein
☐ left suprarenal vein
☐ right brachiocephalic vein
☐ right gonadal vein

☐ right inferior epigastric vein
☐ right lumbar veins
☐ right posterior intercostal vein
☐ right renal vein
☐ right subclavian vein
☐ right superior epigastric vein
☐ right suprarenal vein
☐ superior vena cava

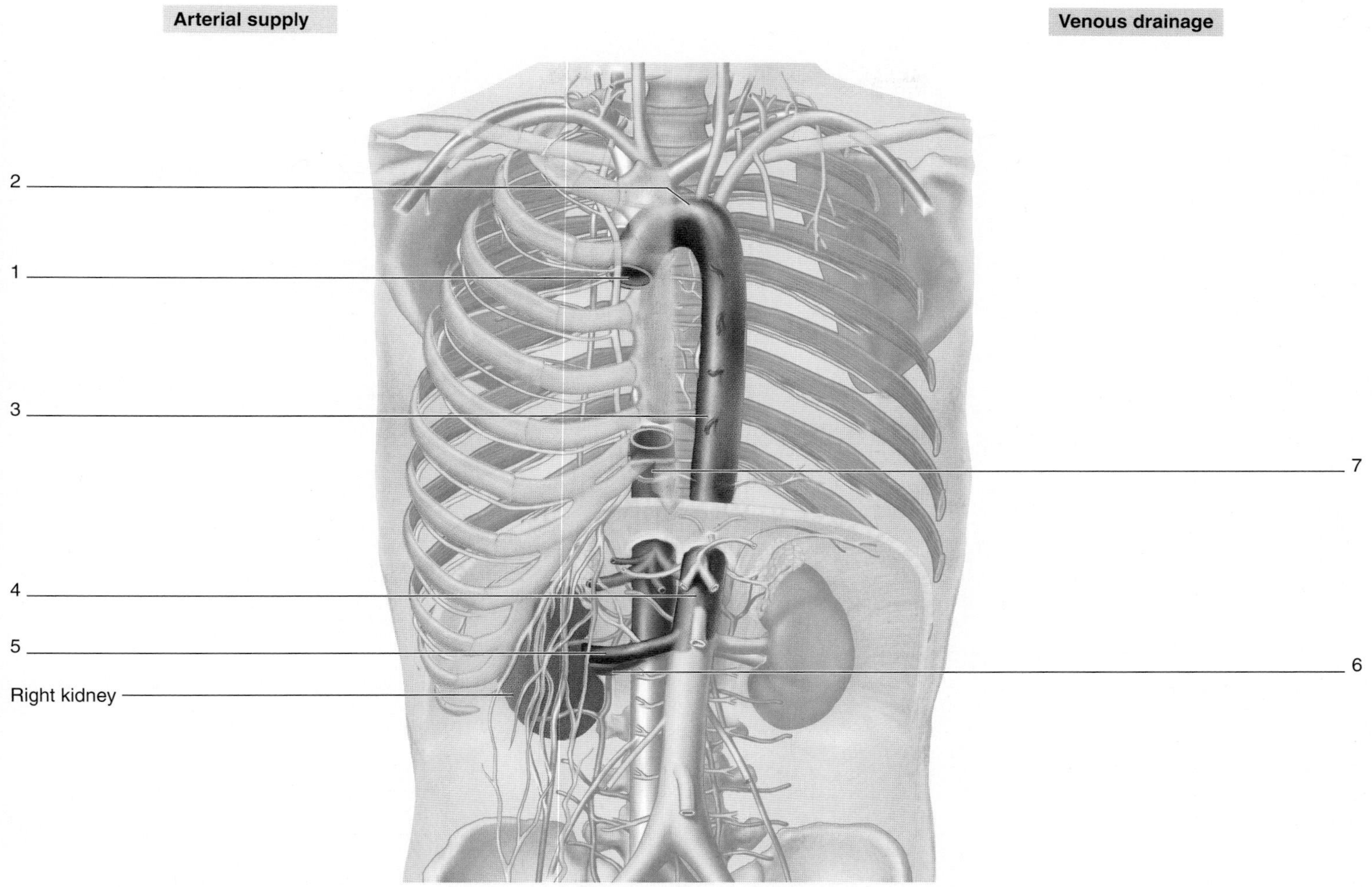

Arterial supply

Venous drainage

2

1

3

7

4

5

6

Right kidney

Figure 22.12   **Circulation from the Left Ventricle of the Heart to the Right Kidney and Back to the Right Atrium of the Heart.**

☐ abdominal aorta   ☐ ascending aorta   ☐ inferior vena cava   ☐ right renal vein

☐ aortic arch   ☐ descending thoracic aorta   ☐ right renal artery

---

EXERCISE 22.11   Circulation to the Abdominal Cavity

1. Observe the abdominal cavity on a prosected human cadaver and/or observe classroom models or charts demonstrating blood vessels of the abdominal cavity.

2. Using your textbook as a guide, identify the *arteries* listed in **figure 22.13** that supply blood to structures in the abdomen. Then label them in figure 22.13.

3. Venous drainage of abdominal organs is unique in that it is an example of a portal system, called the **hepatic portal system**. In this system there are two capillary beds—the first in an abdominal organ, and the second in the liver—connected to each other by a **portal vein**. An artery supplies blood to the first capillary bed, which is located in an abdominal organ such as the stomach, intestine, or spleen.

Blood drains from abdominal organs into three veins: the **splenic**, **inferior mesenteric**, and **superior mesenteric** veins. These veins then drain into the **hepatic portal vein**, which carries blood to the second capillary bed in the liver. This blood is high in nutrient content, but also high in toxins, bacteria, and other potentially dangerous substances. Because the blood flows from the abdominal organs directly to the liver, the liver is allowed to do its job of storing nutrients, detoxifying drugs, and removing bacteria from the blood before the blood enters the general circulation. Thus, the liver is said to have "first pass" at the blood that drains from the abdominal organs. Finally, venous blood drains from liver capillaries into **hepatic veins**, which carry it to the inferior vena cava and back to the heart.

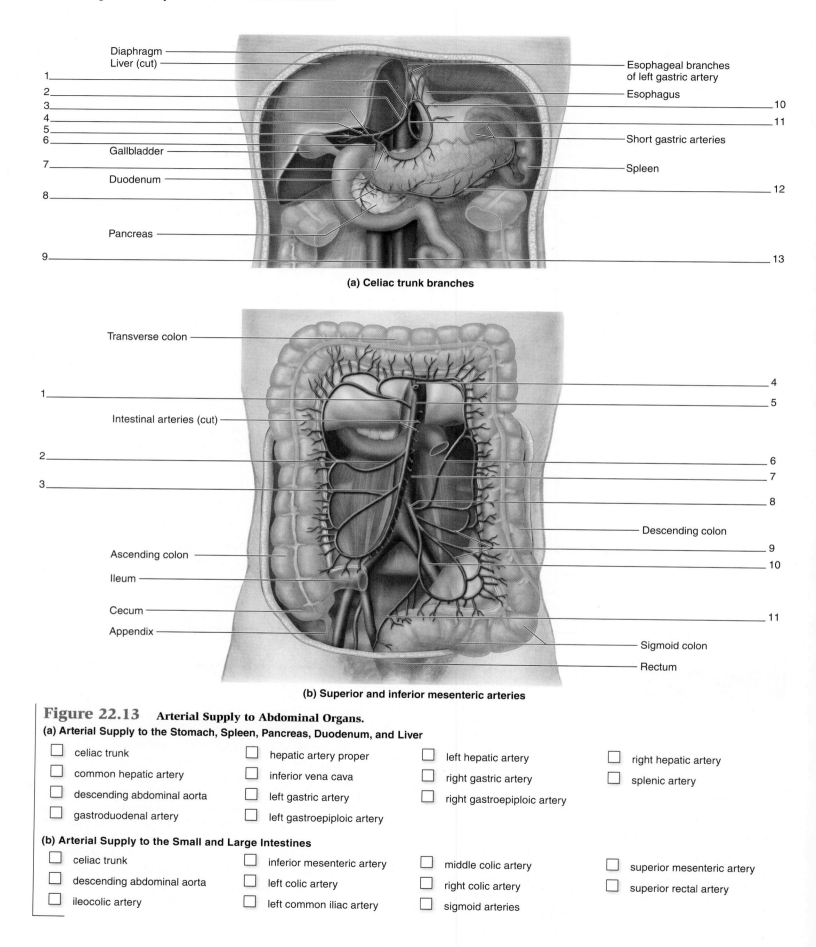

Diaphragm
Liver (cut)
1
2
3
4
5
6
Gallbladder
7
Duodenum
8
Pancreas
9

Esophageal branches
of left gastric artery
Esophagus
10
11
Short gastric arteries
Spleen
12
13

(a) Celiac trunk branches

Transverse colon
1
Intestinal arteries (cut)
2
3
Ascending colon
Ileum
Cecum
Appendix

4
5
6
7
8
Descending colon
9
10
11
Sigmoid colon
Rectum

(b) Superior and inferior mesenteric arteries

Figure 22.13   Arterial Supply to Abdominal Organs.
(a) Arterial Supply to the Stomach, Spleen, Pancreas, Duodenum, and Liver

☐ celiac trunk   ☐ hepatic artery proper   ☐ left hepatic artery   ☐ right hepatic artery

☐ common hepatic artery   ☐ inferior vena cava   ☐ right gastric artery   ☐ splenic artery

☐ descending abdominal aorta ☐ left gastric artery   ☐ right gastroepiploic artery

☐ gastroduodenal artery   ☐ left gastroepiploic artery

(b) Arterial Supply to the Small and Large Intestines

☐ celiac trunk   ☐ inferior mesenteric artery ☐ middle colic artery   ☐ superior mesenteric artery

☐ descending abdominal aorta ☐ left colic artery   ☐ right colic artery   ☐ superior rectal artery

☐ ileocolic artery   ☐ left common iliac artery   ☐ sigmoid arteries

4. Using your textbook as a guide, identify the *veins* listed in **figure 22.14** that compose the hepatic portal system. Then label them in figure 22.14.

5. The pathway a drop of blood takes to get from the abdominal aorta to the *spleen* and back to the right atrium of the heart is shaded in **figure 22.15**. Trace this flow of blood by writing in the names of the vessels in the figure. Label the vessels in order, starting at number 1, so you are figuratively tracing the pathway and naming the vessels encountered along the way.

6. The pathway a drop of blood takes to get from the abdominal aorta to the *duodenum* and back to the right atrium of the heart is shaded in **figure 22.16**. Trace this flow of blood by writing in the names of the vessels in the figure. Label the vessels in order, starting at number 1, so you are figuratively tracing the pathway and naming the vessels encountered along the way.

7. The pathway a drop of blood takes to get from the abdominal aorta to the *sigmoid colon* and back to the right atrium of the heart is shaded in **figure 22.17**. Trace this flow of blood by writing in the names of the vessels in the figure. Label the vessels in order, starting at number 1, so you are figuratively tracing the pathway and naming the vessels encountered along the way.

WHAT DO YOU THINK?

7 What type of capillaries are located within the liver? Given what you have just learned about the contents of the blood entering the liver from the hepatic portal system, why do you think it is advantageous for the liver capillaries to be of this type?

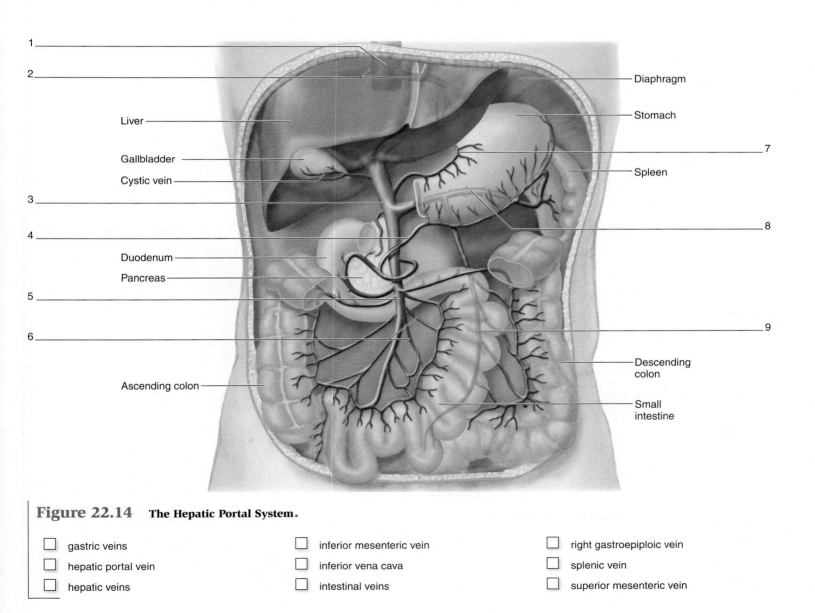

Figure 22.14   **The Hepatic Portal System.**

☐ gastric veins

☐ hepatic portal vein

☐ hepatic veins

☐ inferior mesenteric vein

☐ inferior vena cava

☐ intestinal veins

☐ right gastroepiploic vein

☐ splenic vein

☐ superior mesenteric vein

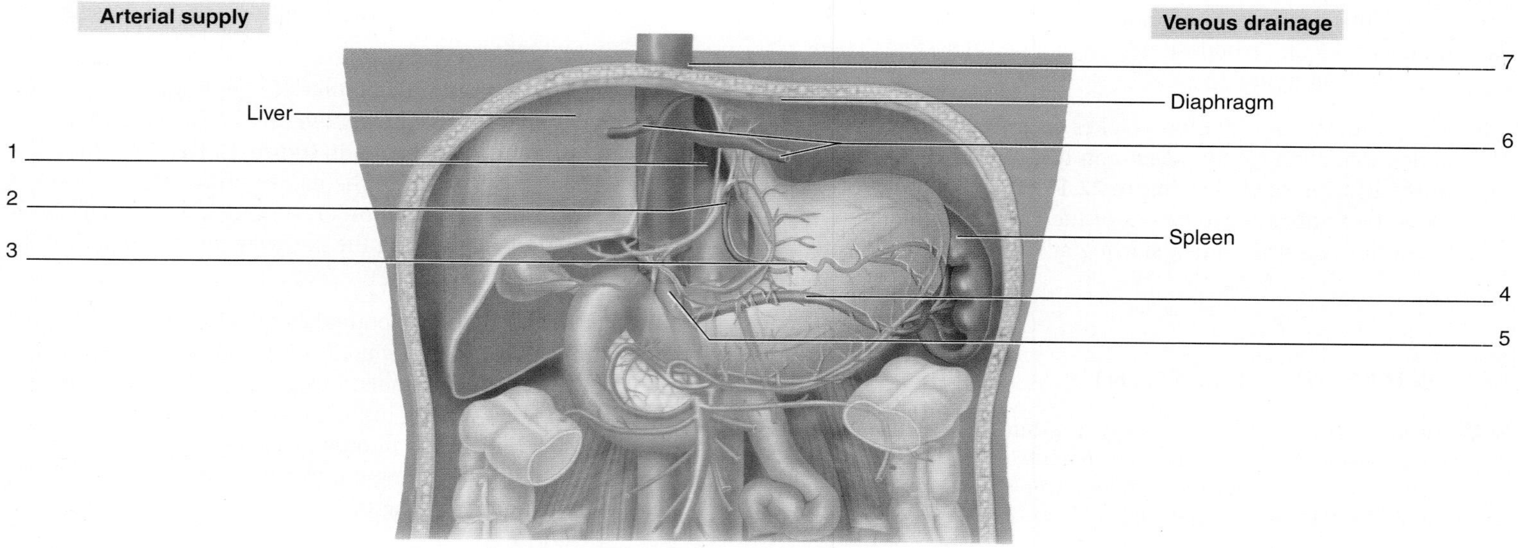

Arterial supply

Venous drainage

Liver

7

Diaphragm

1

6

2

3

Spleen

4

5

Figure 22.15 **Circulation from the Abdominal Aorta to the Spleen and Back to the Right Atrium of the Heart.**

☐ abdominal aorta

☐ celiac trunk

☐ hepatic portal vein

☐ hepatic veins

☐ inferior vena cava

☐ splenic artery

☐ splenic vein

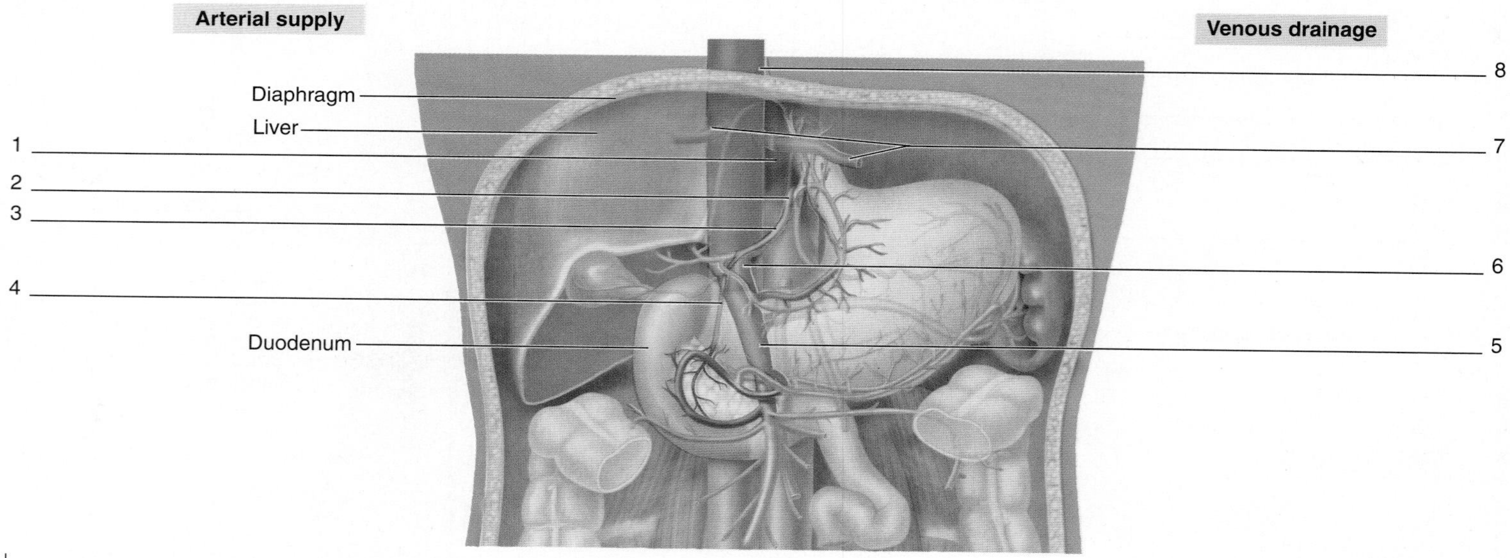

Arterial supply

Venous drainage

Diaphragm

Liver

8

1

7

2

3

6

4

Duodenum

5

Figure 22.16 **Circulation from the Abdominal Aorta to the Duodenum and Back to the Right Atrium of the Heart.**

☐ abdominal aorta

☐ celiac trunk

☐ common hepatic artery

☐ gastroduodenal artery

☐ hepatic portal vein

☐ hepatic veins

☐ inferior vena cava

☐ superior mesenteric vein

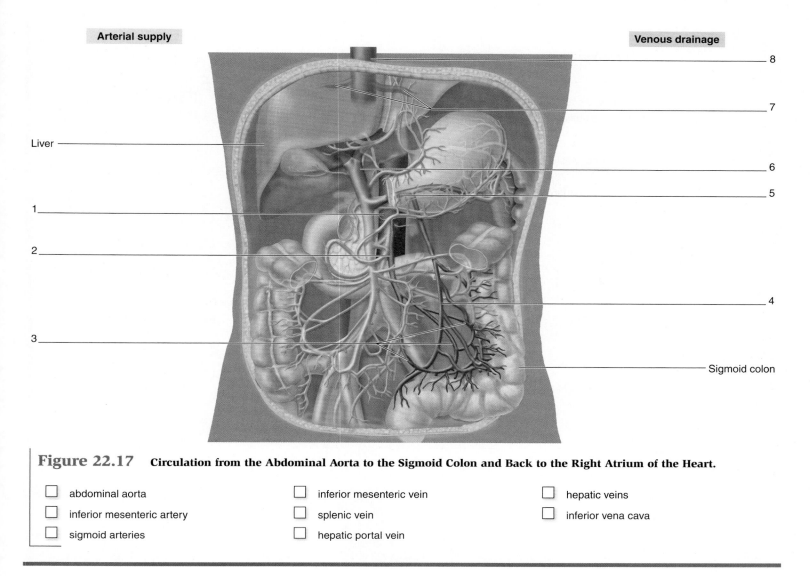

Arterial supply

Venous drainage

Liver

Sigmoid colon

Figure 22.17   **Circulation from the Abdominal Aorta to the Sigmoid Colon and Back to the Right Atrium of the Heart.**

☐ abdominal aorta

☐ inferior mesenteric artery

☐ sigmoid arteries

☐ inferior mesenteric vein

☐ splenic vein

☐ hepatic portal vein

☐ hepatic veins

☐ inferior vena cava

Focus | Portal-Systemic Anastomoses

Veins of the portal system are unique in that they have *no valves*. Thus, when blood backs up in this system, it can back up into systemic veins and will attempt to take an alternate route to the inferior vena cava through anastomoses (s. anastomosis). An **anastomosis** (*anastomo*, to furnish with a mouth) is a connection between two blood vessels. The major anastomoses of the portal circulation are found in veins of the esophagus and the rectum and surrounding the umbilicus. These vessels are common sites of **varicosities** (*varix*, a dilated vein). Varicosities of the esophageal veins are extremely serious because rupture of varicose veins of the esophagus can cause a patient to bleed to death. Varicosities of the rectal veins cause **hemorrhoids** (*haima*, blood, + *rhoia*, a flow), and varicosities of the umbilical veins form a *caput medusa*—because the varicose vessels radiating out from the umbilicus resemble the snakes on Medusa's head (from Greek mythology). Long time alcoholics often incur liver damage leading to **cirrhosis** (*Kirrhos*, yellow, + *–osis*, condition). With cirrhosis, normal liver tissue is replaced over time with scar tissue, decreasing the size of the liver and the number of capillaries within. This creates resistance to blood flow through the liver, which causes blood to back up into the veins of the hepatic portal system.

EXERCISE 22.12   Circulation to the Upper Limb

1. Observe the upper limb of a prosected human cadaver or observe classroom models or charts demonstrating blood vessels of the upper limb.

2. Using your textbook as a guide, identify the *arteries* listed in **figure 22.18a** that supply blood to the upper limb. Then label them in figure 22.18a.

3. Using your textbook as a guide, identify the *veins* listed in figure 22.18b that drain blood from the upper limb. Then label them in figure 22.18b.

4. *Superficial Trace*: The pathway a drop of blood takes to get from the aortic arch to the *anterior surface of the index finger* and back along a superficial route to the superior

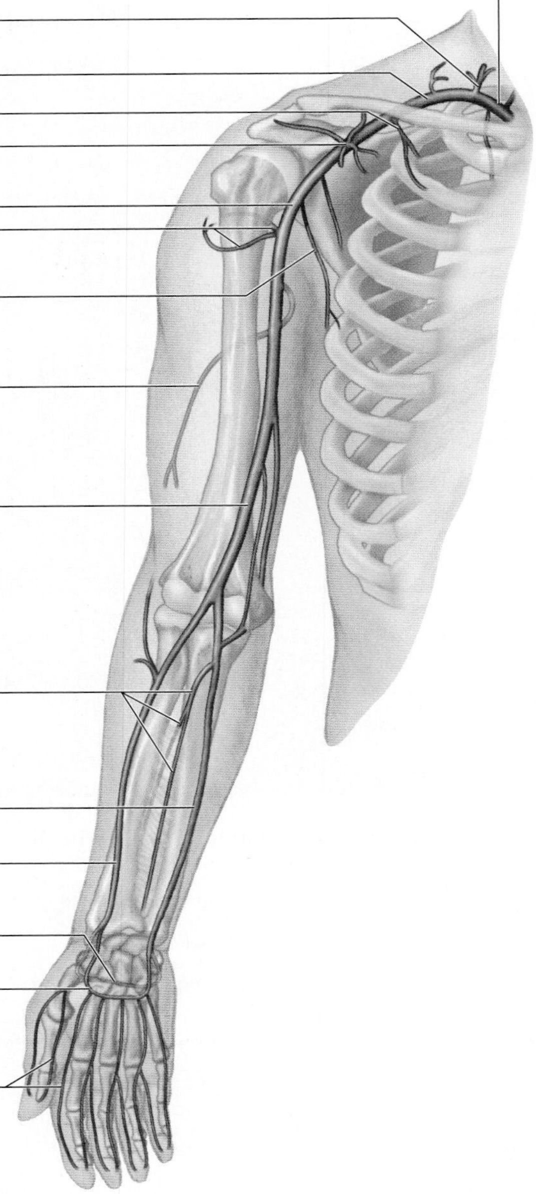

1 _____

Thyrocervical trunk _____

2 _____

Supreme thoracic artery _____

Thoracoacromial artery _____

3 _____

Anterior and posterior humeral _____
circumflex artery

Subscapular artery _____

4 _____

5 _____

Interosseous arteries _____

6 _____

7 _____

8 _____

9 _____

10 _____

(a) Arteries of right upper limb

Figure 22.18   Circulation to the Upper Limb.

(a) Arterial Supply

☐ axillary artery   ☐ deep brachial artery   ☐ radial artery   ☐ superficial palmar arch

☐ brachial artery   ☐ deep palmar arch   ☐ subclavian artery   ☐ ulnar artery

☐ brachiocephalic artery   ☐ digital arteries

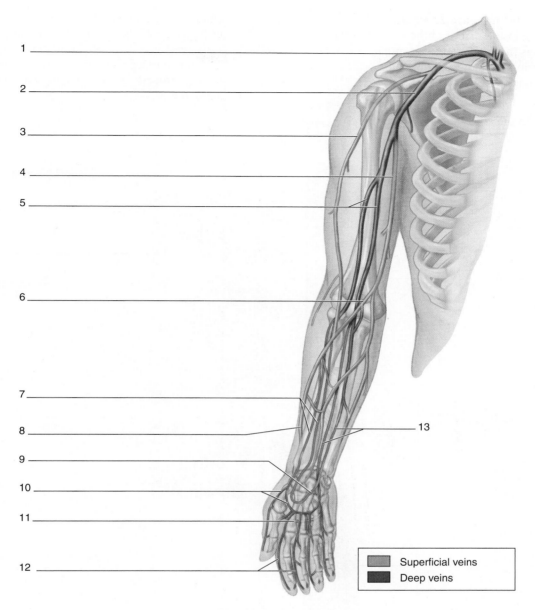

1 _____

2 _____

3 _____

4 _____

5 _____

6 _____

7 _____

8 _____

9 _____

10 _____

11 _____

12 _____

13 _____

Superficial veins
Deep veins

(b) Veins of right upper limb

(b) Venous Drainage

☐ axillary vein   ☐ cephalic vein   ☐ dorsal venous network   ☐ subclavian vein

☐ basilic vein   ☐ deep palmar venous arch ☐ median cubital vein   ☐ superficial palmar venous arch

☐ brachial veins   ☐ digital veins   ☐ radial veins   ☐ ulnar veins

vena cava is shaded in **figure 22.19**. Trace this flow of blood by writing in the names of the vessels in the figure. Label the vessels in order, starting at number 1, so you are figuratively tracing the pathway and naming the vessels encountered along the way.

5. *Deep Trace*: The pathway a drop of blood takes to get from the aortic arch to the *capitate bone in the wrist* and back

along a deep route to the superior vena cava is shaded in **figure 22.20**. Trace this flow of blood by writing in the names of the vessels in the figure. Label the vessels in order, starting at number 1, so you are figuratively tracing the pathway and naming the vessels encountered along the way.

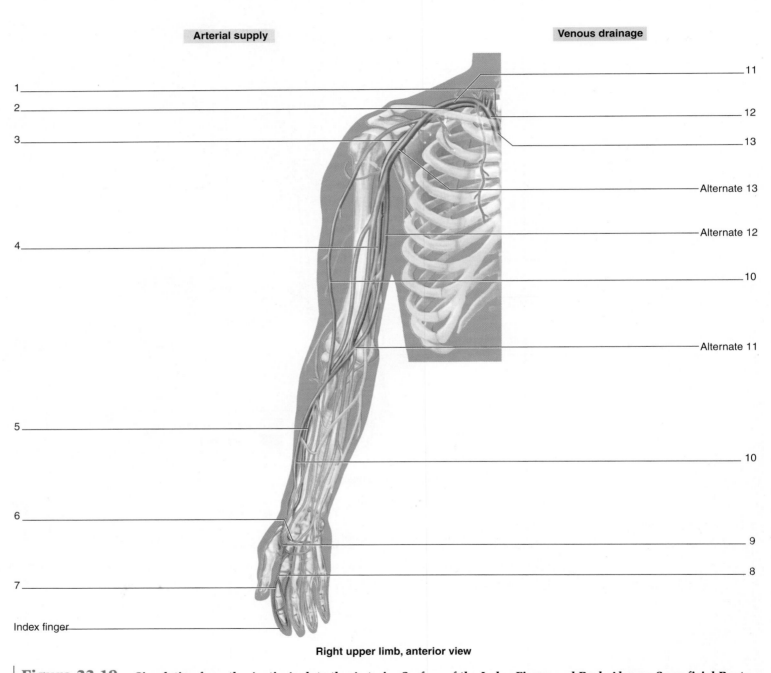

Arterial supply   **Venous drainage**

1
2
3
4
5
6
7
Index finger

11
12
13
Alternate 13
Alternate 12
10
Alternate 11
10
9
8

Right upper limb, anterior view

Figure 22.19   Circulation from the Aortic Arch to the Anterior Surface of the Index Finger and Back Along a Superficial Route to the Superior Vena Cava.

- ☐ axillary artery
- ☐ axillary vein
- ☐ basilic vein
- ☐ brachial artery
- ☐ brachiocephalic trunk (artery)
- ☐ brachiocephalic vein
- ☐ cephalic vein
- ☐ digital artery
- ☐ digital vein
- ☐ median cubital vein
- ☐ radial artery
- ☐ subclavian artery
- ☐ subclavian vein
- ☐ superficial palmar arch
- ☐ superficial palmar venous arch
- ☐ superior vena cava

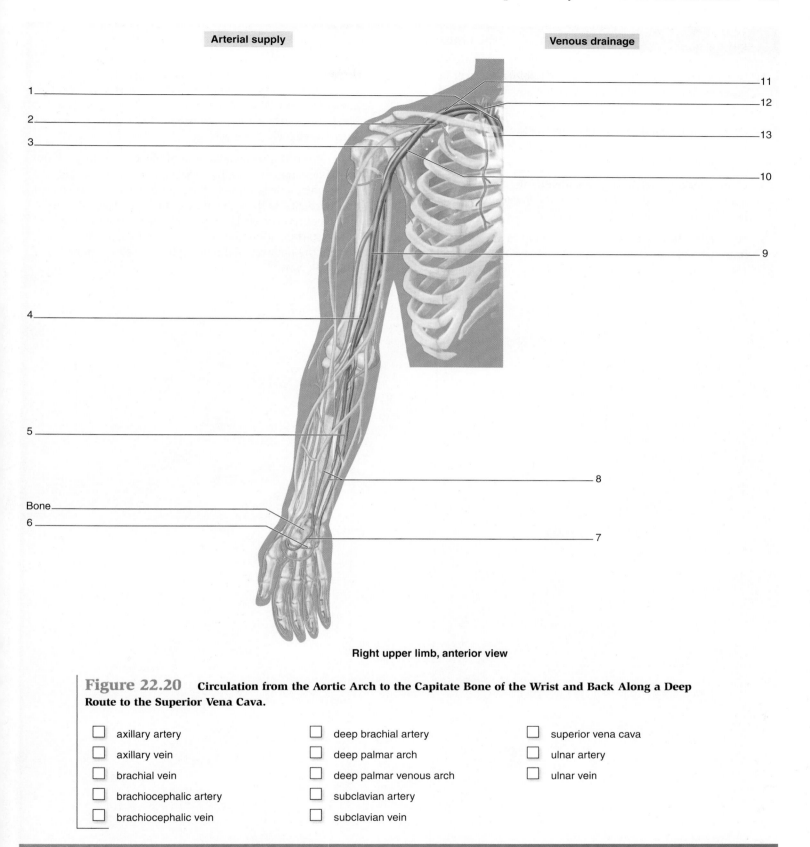

Arterial supply

Venous drainage

1

2

3

4

5

Bone

6

11

12

13

10

9

8

7

Right upper limb, anterior view

Figure 22.20   **Circulation from the Aortic Arch to the Capitate Bone of the Wrist and Back Along a Deep Route to the Superior Vena Cava.**

☐ axillary artery

☐ axillary vein

☐ brachial vein

☐ brachiocephalic artery

☐ brachiocephalic vein

☐ deep brachial artery

☐ deep palmar arch

☐ deep palmar venous arch

☐ subclavian artery

☐ subclavian vein

☐ superior vena cava

☐ ulnar artery

☐ ulnar vein

EXERCISE 22.13 Circulation to the Lower Limb

1. Observe the lower limb of a prosected human cadaver or observe classroom models or charts demonstrating blood vessels of the lower limb.

2. Using your textbook as a guide, identify the *arteries* listed in **figure 22.21a** that supply blood to the lower limb. Then label them in figure 22.21a.

3. Using your textbook as a guide, identify the *veins* listed in figure 22.21b that drain blood from the lower limb. Then label them in figure 22.21b.

4. *Superficial Trace*: The pathway a drop of blood takes to get from the abdominal aorta to the *dorsal surface of the big toe (hallux)* and back along a superficial route to the inferior vena cava is shaded in **figure 22.22**. Trace this flow of blood by writing in the names of the vessels in the figure. Label the vessels in order, starting at number 1, so you are figuratively tracing the pathway and naming the vessels encountered along the way.

5. *Deep Trace*: The pathway a drop of blood takes to get from the abdominal aorta to the *cuboid bone in the foot* and back along a deep route to the inferior vena cava is shaded in **figure 22.23**. Trace this flow of blood by writing in the names of the vessels in the figure. Label the vessels in order, starting at number 1, so you are figuratively tracing the pathway and naming the vessels encountered along the way.

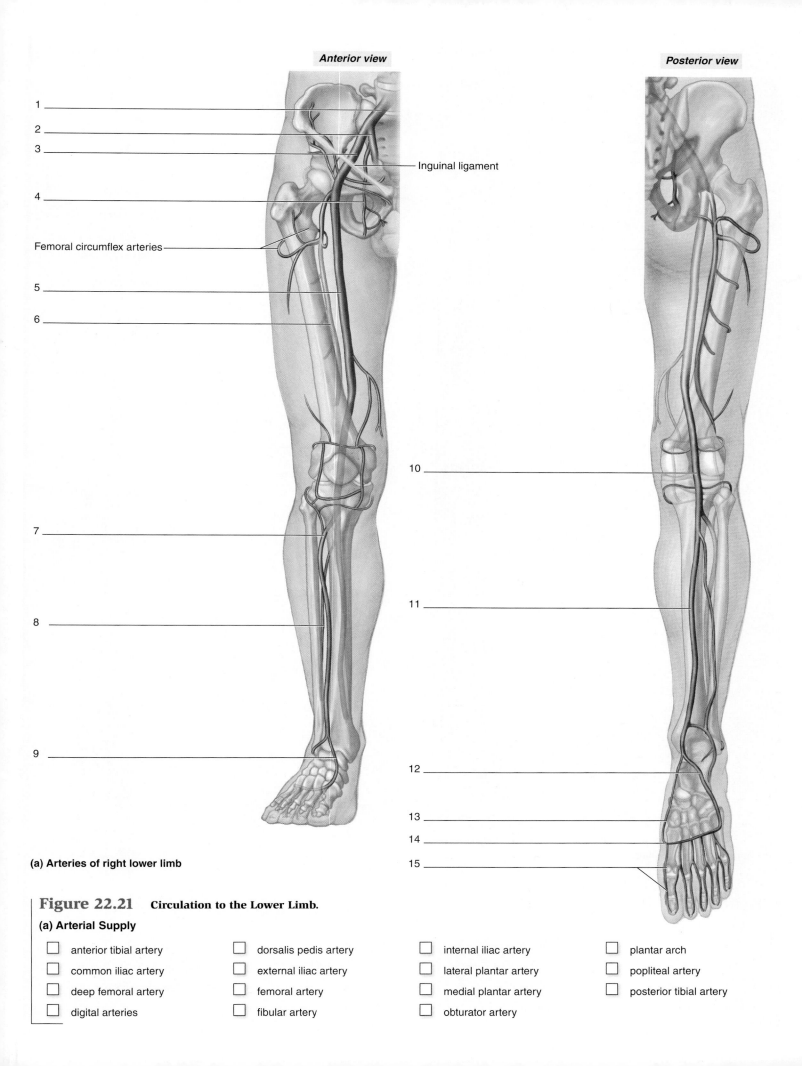

Anterior view

1 _____

2 _____

3 _____

— Inguinal ligament

4 _____

Femoral circumflex arteries —

5 _____

6 _____

7 _____

8 _____

9 _____

(a) Arteries of right lower limb

Posterior view

10 _____

11 _____

12 _____

13 _____

14 _____

15 _____

Figure 22.21   **Circulation to the Lower Limb.**

(a) Arterial Supply

- ☐ anterior tibial artery
- ☐ common iliac artery
- ☐ deep femoral artery
- ☐ digital arteries
- ☐ dorsalis pedis artery
- ☐ external iliac artery
- ☐ femoral artery
- ☐ fibular artery
- ☐ internal iliac artery
- ☐ lateral plantar artery
- ☐ medial plantar artery
- ☐ obturator artery
- ☐ plantar arch
- ☐ popliteal artery
- ☐ posterior tibial artery

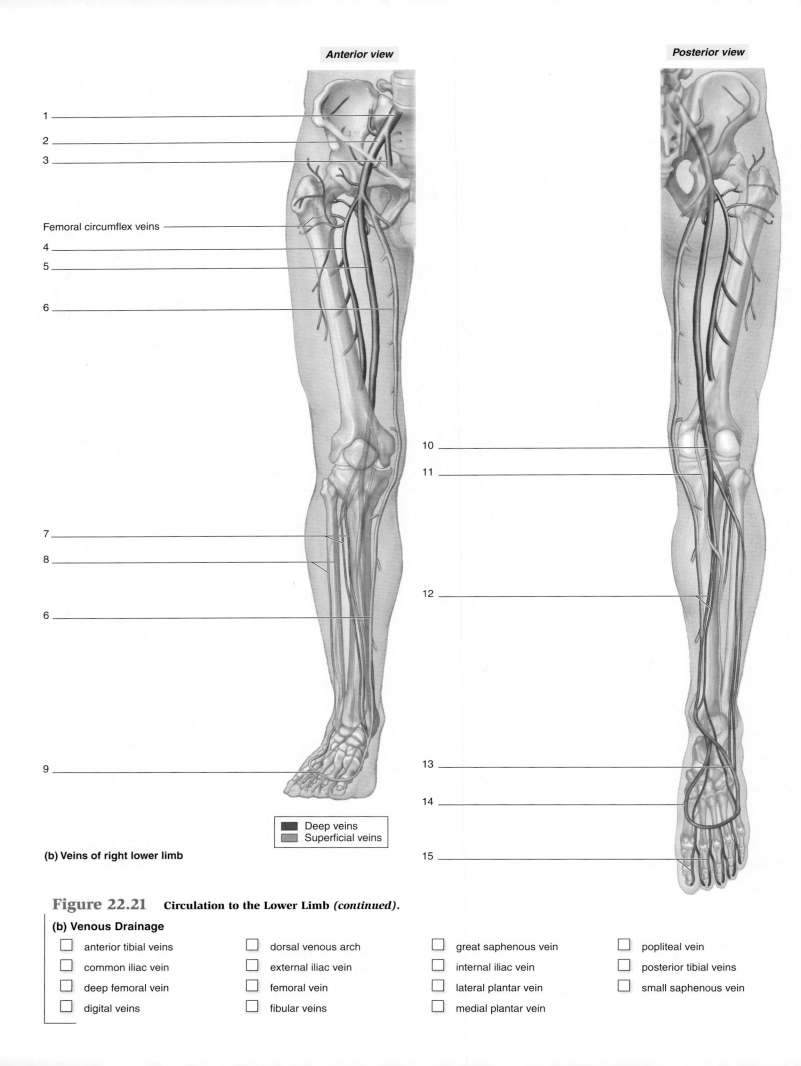

Anterior view

Posterior view

1 _____

2 _____

3 _____

Femoral circumflex veins _____

4 _____

5 _____

6 _____

10 _____

11 _____

7 _____

8 _____

6 _____

12 _____

9 _____

13 _____

14 _____

Deep veins
Superficial veins

(b) Veins of right lower limb

15 _____

Figure 22.21   **Circulation to the Lower Limb** *(continued)*.

(b) Venous Drainage

☐ anterior tibial veins   ☐ dorsal venous arch   ☐ great saphenous vein   ☐ popliteal vein

☐ common iliac vein   ☐ external iliac vein   ☐ internal iliac vein   ☐ posterior tibial veins

☐ deep femoral vein   ☐ femoral vein   ☐ lateral plantar vein   ☐ small saphenous vein

☐ digital veins   ☐ fibular veins   ☐ medial plantar vein

Arterial supply

Venous drainage

1

2

3

Inguinal ligament

4

Femoral vein

5

Popliteal vein

6

7

Hallux (great toe)

8

15

14

13

12

11

11

10

9

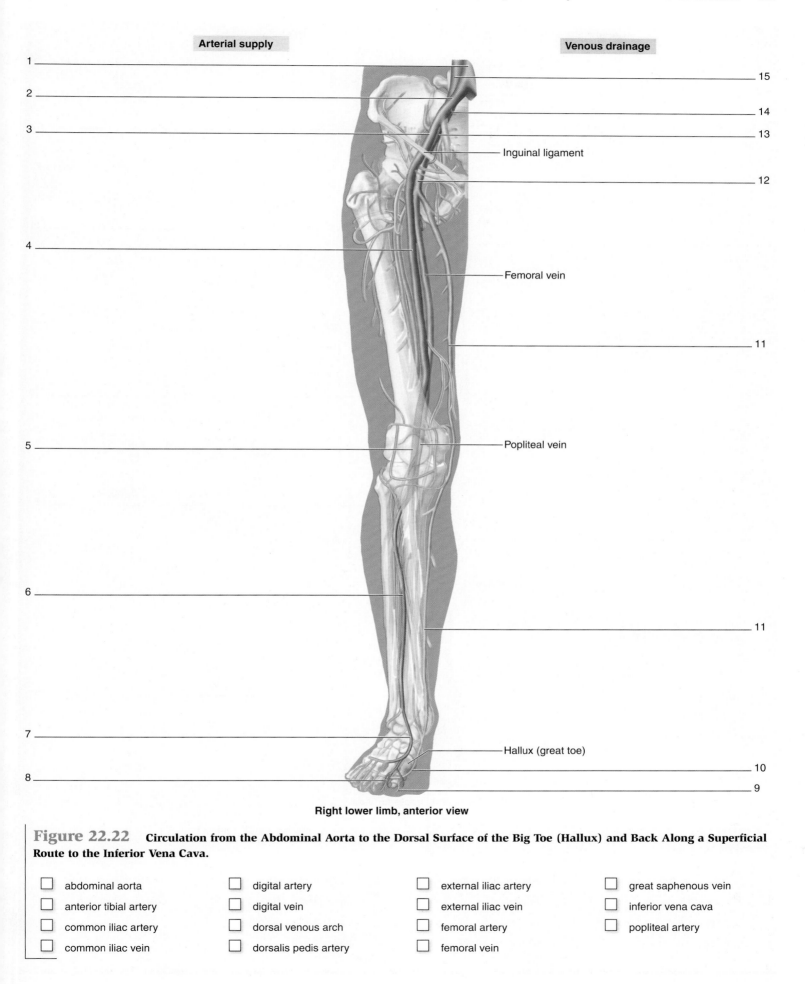

Right lower limb, anterior view

Figure 22.22   **Circulation from the Abdominal Aorta to the Dorsal Surface of the Big Toe (Hallux) and Back Along a Superficial Route to the Inferior Vena Cava.**

☐ abdominal aorta	☐ digital artery	☐ external iliac artery	☐ great saphenous vein
☐ anterior tibial artery	☐ digital vein	☐ external iliac vein	☐ inferior vena cava
☐ common iliac artery	☐ dorsal venous arch	☐ femoral artery	☐ popliteal artery
☐ common iliac vein	☐ dorsalis pedis artery	☐ femoral vein	

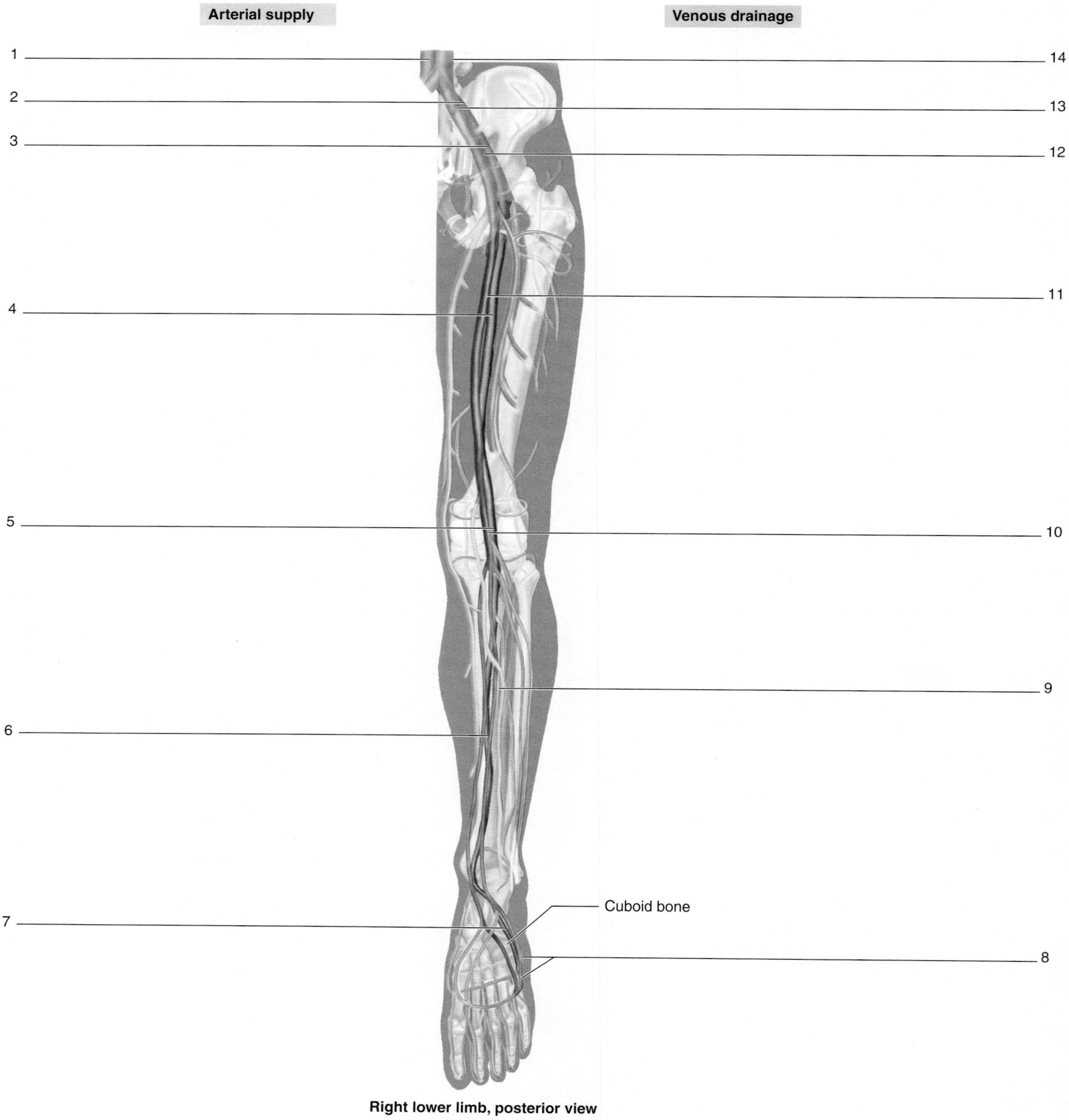

Arterial supply

Venous drainage

1

2

3

4

5

6

7

14

13

12

11

10

9

8

Cuboid bone

Right lower limb, posterior view

Figure 22.23   **Circulation from the Abdominal Aorta to the Cuboid Bone in the Foot and Back Along a Deep Route to the Inferior Vena Cava.**

☐ abdominal aorta

☐ common iliac artery

☐ common iliac vein

☐ external iliac artery

☐ external iliac vein

☐ femoral artery

☐ femoral vein

☐ inferior vena cava

☐ lateral plantar artery

☐ lateral plantar vein

☐ popliteal artery

☐ popliteal vein

☐ posterior tibial artery

☐ posterior tibial vein

Fetal Circulation

In the fetus the lungs are nonfunctional and need only a small amount of blood to support the developing lung tissue. This blood must be oxygenated blood coming from the fetal respiratory organ: the placenta. Once the fetus is born, the circulation must change and the lungs take over from the placenta as the respiratory organs.

Thus, there are a number of **shunts** present in the fetal circulation that direct blood away from the lungs or to and from the placenta. These shunts must close at birth to establish the normal postnatal circulatory pathways. In the following exercise you will identify the unique cardiovascular structures of the fetal circulation, trace the flow of blood through the fetal circulation, and identify the postnatal structures that are remnants of the fetal circulation.

EXERCISE 22.14 Fetal Circulation

1. Using your textbook as a guide, label the fetal circulatory system structures listed in **figure 22.24**.

2. In the space below, trace (in words) the flow of blood from the left ventricle of the fetal heart to the placenta and back to the right atrium of the heart.

_____

_____

_____

_____

_____

_____

_____

_____

_____

_____

_____

3. In **table 22.4**, write in the names of the postnatal structures that are remnants of the fetal circulation, and describe each structure's function in the fetus.

Table 22.4	Fetal Cardiovascular Structures and Associated Postnatal Structures	
Fetal Cardiovascular Structure	**Postnatal Structure**	**Function of Fetal Structure**
Ductus Arteriosus		
Ductus Venosus		
Foramen Ovale		
Umbilical Arteries		
Umbilical Vein		

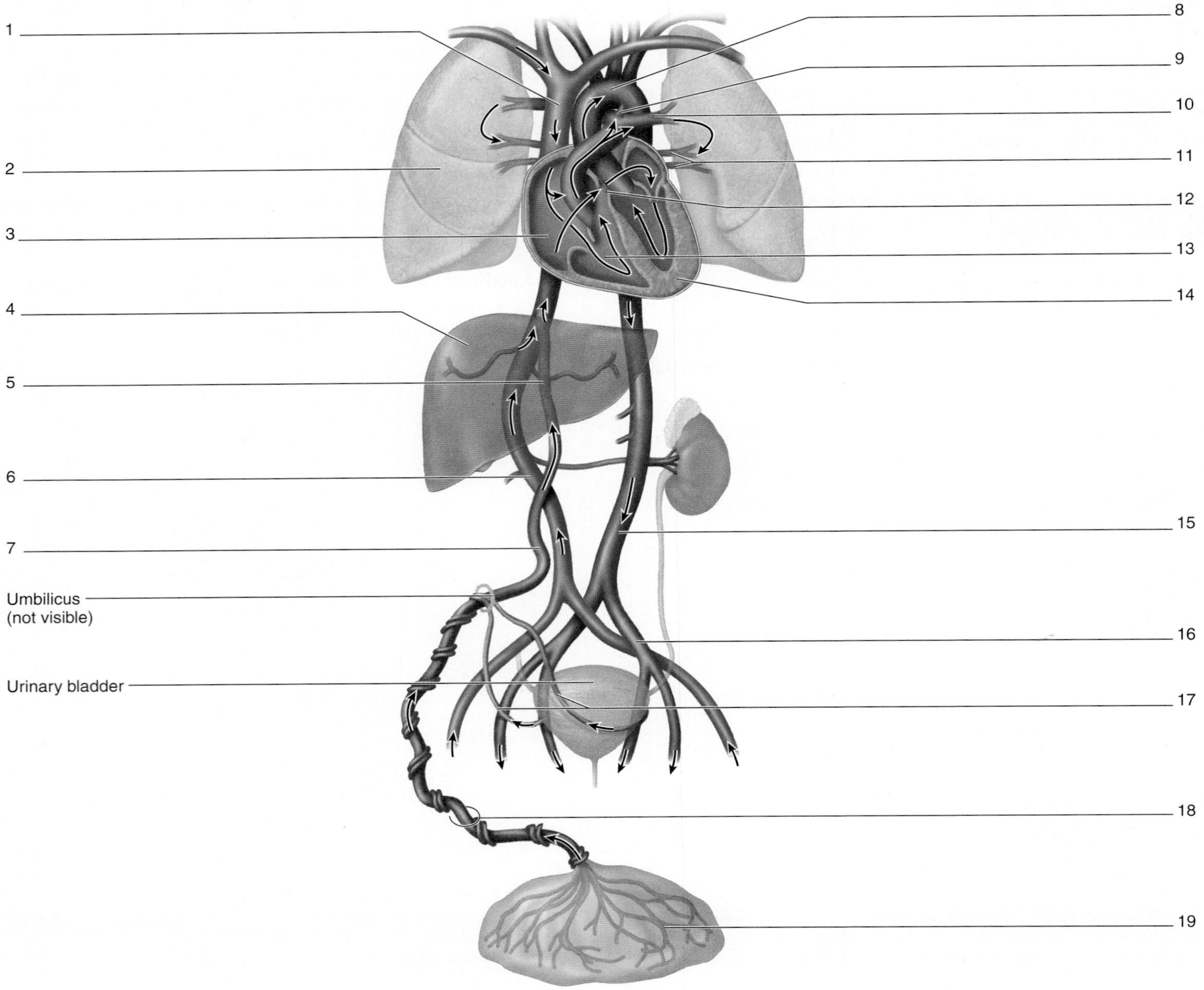

1 _____

2 _____

3 _____

4 _____

5 _____

6 _____

7 _____

Umbilicus
(not visible) _____

Urinary bladder _____

8 _____

9 _____

10 _____

11 _____

12 _____

13 _____

14 _____

15 _____

16 _____

17 _____

18 _____

19 _____

Figure 22.24 **Fetal Circulation.**

☐ aortic arch
☐ common iliac artery
☐ descending abdominal aorta
☐ ductus arteriosus
☐ ductus venosus

☐ foramen ovale
☐ heart
☐ inferior vena cava
☐ liver
☐ lung

☐ placenta
☐ pulmonary artery
☐ pulmonary veins
☐ right atrium
☐ right ventricle

☐ superior vena cava
☐ umbilical arteries
☐ umbilical cord
☐ umbilical vein

Chapter 22: Vessels and Circulation

POST-LABORATORY WORKSHEET

1. Label the diagram below with the appropriate artery names.

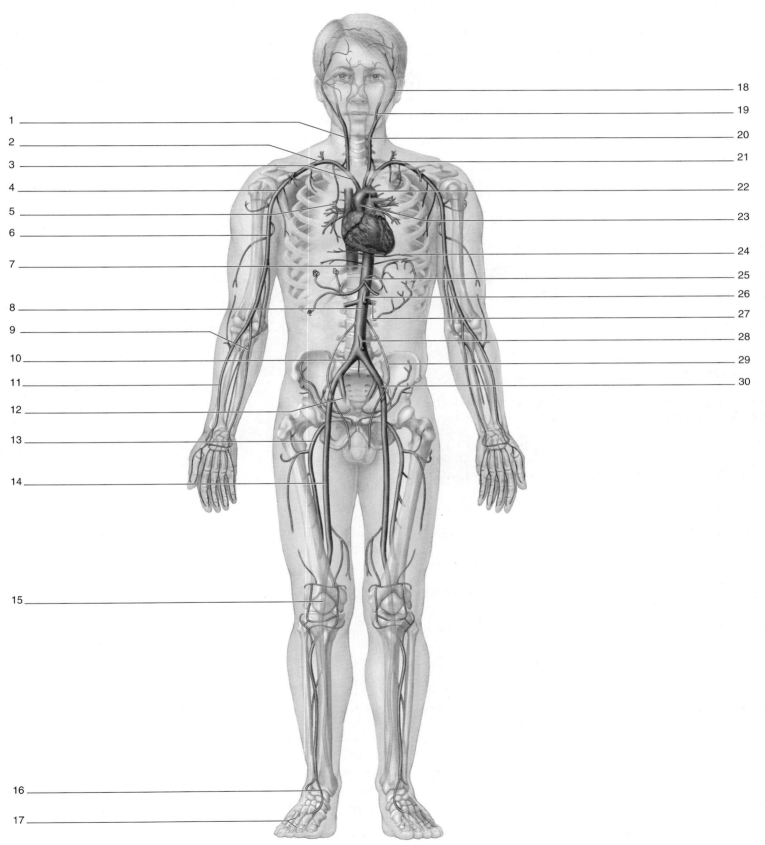

(a) Arteries, anterior view

1 _____
2 _____
3 _____
4 _____
5 _____
6 _____
7 _____
8 _____
9 _____
10 _____
11 _____
12 _____
13 _____
14 _____
15 _____
16 _____
17 _____

18 _____
19 _____
20 _____
21 _____
22 _____
23 _____
24 _____
25 _____
26 _____
27 _____
28 _____
29 _____
30 _____

2. Label the diagram below with the appropriate vein names.

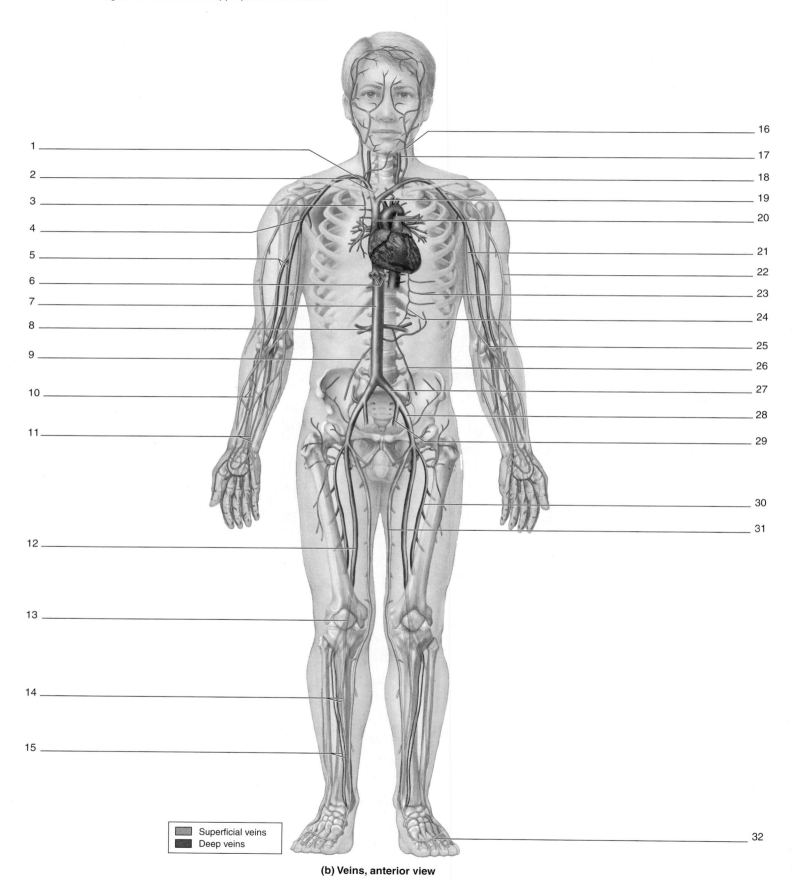

1 _____

2 _____

3 _____

4 _____

5 _____

6 _____

7 _____

8 _____

9 _____

10 _____

11 _____

12 _____

13 _____

14 _____

15 _____

16 _____

17 _____

18 _____

19 _____

20 _____

21 _____

22 _____

23 _____

24 _____

25 _____

26 _____

27 _____

28 _____

29 _____

30 _____

31 _____

32 _____

Superficial veins
Deep veins

(b) Veins, anterior view

3. When observing a histology slide demonstrating a cross section of an artery, how will you decide if the artery in question is an elastic artery or a muscular artery?

_____

_____

_____

4. Describe the role arterioles play in regulation of blood flow.

_____

_____

_____

5. A venous valve is an infolding of the tunica _____ of the vessel; the valve contains _____ cusps, and its function is to _____

_____.

6. Fill in the chart with examples of locations in the body where each type of capillary is located.

Type of Capillary	Locations	
Continuous		
Fenestrated		
Sinusoidal		

7. Which of the following organs does *not* receive oxygenated blood from an artery that is a direct, paired branch off of the abdominal aorta?

   a. Kidney

   b. Spleen

   c. Adrenal (suprarenal) gland

   d. Gonad (testis or ovary)

8. The _____ gonadal vein empties into the renal vein on the same side of the body.

9. Trace the flow of a drop of blood from the left ventricle of the heart to the left ovary and back to the left ventricle of the heart by filling in the blanks below:

left ventricle → _____ → _____ → _____ →

_____ → _____ → _____ → ovary →

_____ → _____ → _____ → right atrium →

_____ → _____ → pulmonary semilunar valve → _____ →

_____ → capillary bed in lung → _____ → _____ →

_____ → left ventricle

10. Trace the flow of a drop of blood from the left ventricle of the heart to the right styloid process of the radius (lateral wrist) and back to the left ventricle of the heart (use superficial vessels for venous return) by filling in the blanks below:

left ventricle → _____ → _____ → _____ →

_____ → _____ → _____ → _____ →

capillary bed in wrist → _____ → _____ → _____ →

_____ → _____ → _____ → right atrium →

_____ → _____ → _____ →

_____ → _____ → capillary bed in lung → _____ →

_____ → _____ → left ventricle

11. Explain the functional significance of the hepatic portal system.

_____

_____

_____

_____

_____

_____

_____

12. A physician wishes to place a balloon catheter into the left coronary artery of his patient. To do this, he will place the catheter into the femoral artery just inferior to the inguinal ligament and then string the catheter backward through the arterial system until it reaches the ascending aorta, and from there it will go into the left coronary artery. List the arteries, in order, the catheter will pass through as it travels from the femoral artery to the left coronary artery.

1. _____   5. _____

2. _____   6. _____

3. _____   7. _____

4. _____   8. Left coronary artery

13. A physician wishes to place a central line (a catheter used to repeatedly administer drugs such as chemotherapy drugs into a patient's circulatory system) into the right atrium of her patient's heart. To do this, she will place the catheter into the basilic vein just superior to where it branches from the median cubital vein, and then string the catheter along the venous system until it reaches the right atrium. List, in order, the veins through which the central line passes as it travels from the basilic vein to the right atrium of the heart.

    1. _____

    2 _____

    3. _____

    4. _____

    5. Right atrium

14. After the central line was placed, it pinched off and failed to work when the patient held a heavy weight in her hand and the inferior movement of her clavicle compressed the central line within the subclavian vein. The physician decided to remove the central line from the basilic vein and place another central line into the internal jugular vein, which provided a more direct route to the heart and avoided the problem of catheter pinch-off. List the veins, in order, through which the central line passes as it travels from the internal jugular vein to the right atrium of the heart.

    1. _____

    2. _____

    3. _____

    4. Right atrium

15. Circle the correct answer for each blank to complete the statement: In the fetus, the umbilical veins carry oxygenated/deoxygenated blood, while the umbilical arteries carry oxygenated/deoxygenated blood.

The Lymphatic System

23

OUTLINE and OBJECTIVES

Anatomy & Physiology REVEALED®
aprevealed.com

Module 10: LYMPHATIC SYSTEM

INTRODUCTION

The **lymphatic system** functions in close association with two other body systems: the **circulatory system** and the **immune system**. The lymphatic system's circulatory function comes from the role lymphatic vessels play in the circulation of body fluids. The lymphatic system's immune function comes from the roles that lymphocytes play in the body defenses. The lymphatic system has a number of functions, which include: returning interstitial fluid to the cardiovascular system, filtration of lymph by lymph nodes, destruction of bloodborne antigens, production of lymphocytes, and absorption of dietary fats from the digestive system (through lymphatic capillaries in the small intestine called **lacteals**).

The terms *lymph* and *lymphatic* come from the Latin word *lympha*, which means "pure spring water." Although lymph isn't exactly comparable to spring water in its composition, the fluid is relatively clear and free of suspended material because it contains no red blood cells or plasma proteins. Lymphatic vessels return approximately 1–3 L of fluid to the cardiovascular system each day. When lymphatic vessels become blocked or damaged so lymph cannot flow freely through them, severe **edema** (*oidema*, a swelling) results in the body part distal to the location of the obstruction.

Chapter 23: The Lymphatic System

PRE-LABORATORY WORKSHEET

1. What is lymph?

   _____

   _____

   _____

2. List three functions of the lymphatic system:

   a. _____

   b. _____

   c. _____

3. List the three major organs of the lymphatic system:

   a. _____

   b. _____

   c. _____

4. The basic unit of lymphatic tissue is called a _____.

5. *MALT* stands for _____

   _____.

6. Lymph draining from the left side of the head and neck, and the entire body below the thorax, drains into a lymphatic vessel called the _____

   _____.

7. A lymphatic organ that filters blood is the _____.

8. A lymphatic organ located within the thoracic cavity just deep to the sternum is the _____.

Because most lymphatic structures are anatomically small and difficult to see in a human cadaver or on isolated organ preparations, the majority of the exercises in this chapter involve histological observations of lymphatic tissues and organs. Your histological observations will be complemented by observations of classroom models demon-

strating lymphatic organs such as lymph nodes, the spleen, and the thymus. If human cadavers are used in your laboratory, most lymphatic structures will be studied on the cadaver in conjunction with other organ systems, such as the respiratory and digestive systems, rather than within this laboratory exercise.

Histology

Lymphatic Vessels

Lymphatic vessels are most similar in structure to veins because they carry fluid at low pressure and velocity from the tissues back toward the heart.

EXERCISE 23.1 Lymphatic Vessels

1. Obtain a *dissecting* microscope and a slide demonstrating lymphatic vessels.

2. Place the slide on the microscope stage. Bring the tissue sample into focus on low power and locate a lymphatic vessel (**figure 23.1**). Then move the microscope stage to

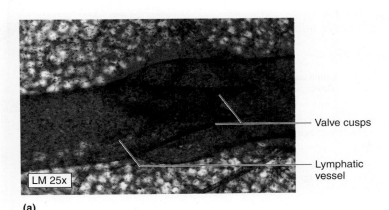

(a)

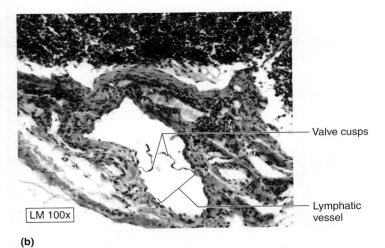

(b)

Figure 23.1   Lymphatic Vessels.   (a) View of lymphatic vessels and valves as seen through a dissecting microscope. (b) Cross section through a lymphatic vessel demonstrating a valve as seen through a compound microscope.

scan the length of the vessel until you locate a valve. If necessary, change to high power and focus in on the valve to see its structure more clearly.

3. Using figure 23.1*a* as a guide, identify the following structures:

☐ lymphatic vessel   ☐ valve cusps

4. Obtain a compound microscope and a slide demonstrating lymphatic vessels.

5. Place the slide on the microscope stage. Bring the tissue sample into focus on low power and locate a cross section through a lymphatic vessel (figure 23.1*b*). The walls of lymphatic vessels contain the same three tunics as the walls of blood vessels, but they are not as well defined. Like in the veins, the valves are extensions of the tunica interna of the vessel. Do you see any red blood cells within the lumen of the vessel? _____ Should you see any red blood cells within the lumen of the vessel?_____ Do you see any white blood cells within the lumen of the vessel? Should you see any white blood cells within the lumen of the vessel? _____

6. Using figure 23.1*b* as a guide, identify the following structures:

☐ lymphatic vessel   ☐ valve cusps

7. In the spaces below, draw brief sketches of the lymphatic vessel as viewed through both the dissecting and the compound microscopes.

_____ ×   _____ ×

Mucosa-Associated Lymphatic Tissue (MALT)

The basic functional unit of lymphatic tissue is a **lymphatic nodule (follicle)**. A lymphatic nodule is a ball of cells that are predominantly B-lymphocytes, with some macrophages and T-lymphocytes on its outer borders. **Figure 23.2** demonstrates a lymphatic nodule as seen under a compound microscope on high power. Notice that the central region stains lighter than the surrounding area. The central region, called the **germinal center** (*germen*, sprout), is where B-lymphocytes are most actively proliferating. Lymphatic nodules are found in hundreds of locations throughout the body. They are commonly found just deep to epithelial tissues and are even more common in locations where the epithelium changes from one type to another (such as where epithelium of the gut tube transitions from stratified squamous lining the esophagus to simple columnar lining the stomach). Lymphatic nodules are particularly abundant in the nasal and oral cavities, and in the walls of the digestive tract. Such tissue is collectively referred to as **MALT**, or **m**ucosa- **a**ssociated **l**ymphatic **t**issue. In several regions of the respiratory and digestive tracts, the aggregations of MALT are so consistent in structure and so regular in location that they are given

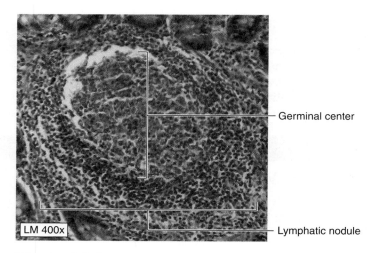

Figure 23.2 **Lymphatic Nodule.** The germinal center of the nodule consists of actively proliferating B-lymphocytes. The marginal zone consists of less actively proliferating B-lymphocytes as well as T-lymphocytes and macrophages.

names. Such named aggregations of MALT include **tonsils**, **Peyer patches**, and the **vermiform appendix**. **Table 23.1** summarizes the characteristics of MALT.

Table 23.1	Named Aggregations of MALT (Mucosa-Associated Lymphatic Tissue)		
Named Aggregation of MALT	**Description and Location**	**Function**	**Word Origins**
Peyer Patches	Lymphatic nodules in the submucosa (deep to the epithelium) of the ileum of the small intestine.	Protect against pathogens that enter the body through the intestinal mucosa.	*Peyer*, Johan K. Peyer, a Swiss anatomist (1653–1712)
Tonsils	Lymphatic nodules deep to the epithelium lining the pharynx. See table 23.2 for descriptions and locations of specific tonsils.	Protect against pathogens that enter the body through the mucosa of the pharynx.	*tonsilla*, a stake
Vermiform Appendix	A diverticulum (blind-ended pouch) extending from the cecum just past the ileocecal junction that contains lymphatic nodules in its walls.	Protect against pathogens that enter the body through the mucosa of the cecum.	*vermis*, wormlike, + *forma*, shape, + *appendix*, appendage

---

EXERCISE 23.2 Tonsils

1. Obtain a slide demonstrating tonsils and place it on the microscope stage.

2. Tonsils are lymphatic nodules located in the walls of the oral cavity and pharynx. They consist of lymphatic nodules interspersed between deep **crypts** (*crypt*, a pitlike depression) that open to the surface of the tonsil. The tonsils are covered superficially by the same epithelium that lines the part of the body in which they are located. There are three sets of tonsils: **pharyngeal tonsils** (called *adenoids* when they are swollen), **palatine tonsils**, and **lingual tonsils**. **Table 23.2** lists characteristics of the three kinds of tonsils.

3. Using **figure 23.3,** table 23.2, and your textbook as guides, identify the following on the microscope slide of a tonsil:

   ☐ crypt   ☐ lymphatic nodule
   ☐ epithelium

4. Based on the type of epithelium covering the tonsil and the number and length of the crypts, can you determine what type of tonsil (pharyngeal, palatine, or lingual) is shown on the slide you are observing? _____

Table 23.2	Tonsils			
Tonsil	**Description and Location**	**Crypts**	**Epithelium**	**Word Origins**
Lingual	Small, paired tonsils at the base of the tongue.	1–2 short crypts	Stratified squamous nonkeratinized	*lingua*, tongue
Palatine	Paired tonsils located in the fauces (the space between the mouth and pharynx) just posterior to the soft palate.	10–20 deep crypts	Stratified squamous nonkeratinized	*palatum*, palate
Pharyngeal (Adenoid)	Single tonsil projecting from the roof of the nasopharynx.	None	Respiratory (ciliated pseudostratified columnar)	*pharynx*, the throat, *adenos*, a gland, + *eidos*, appearance

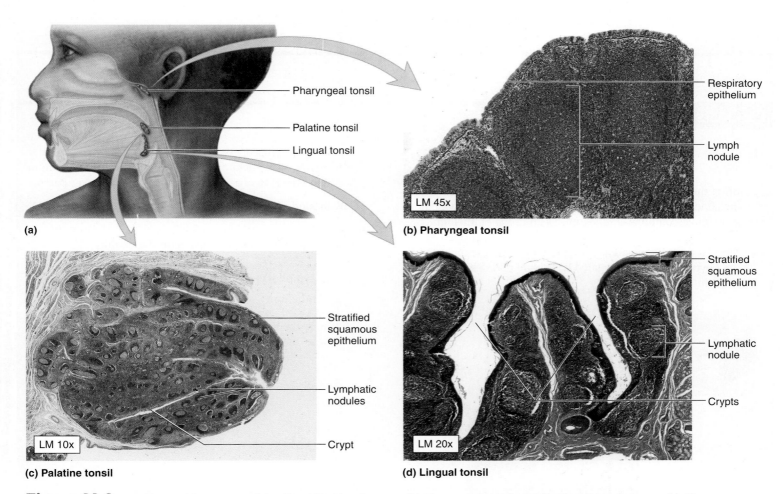

Figure 23.3 **Tonsils.** (a) Location of tonsils within the pharynx. (b) Pharyngeal tonsils are lined with respiratory epithelium (pseudostratified columnar with cilia and goblet cells). (c) Palatine tonsils contain numerous deep crypts and are lined with stratified squamous epithelium. (d) Lingual tonsils contain a few shallow crypts and are lined with stratified squamous epithelium.

5. In the space below, draw a simple sketch of the tonsil(s) that you observed under the microscope.

_____ ×

---

1. Obtain a slide of the ileum and place it on the microscope stage.

2. Bring the tissue sample into focus on low power and scan the slide to locate the epithelial lining of the ileum, which is a simple columnar epithelium. Deep to the epithelium, look for aggregations of purple-staining cells (**figure 23.4**). These are the aggregations of lymphocytes that compose the **Peyer patches** (aggregated lymphatic nodules of the small intestine).

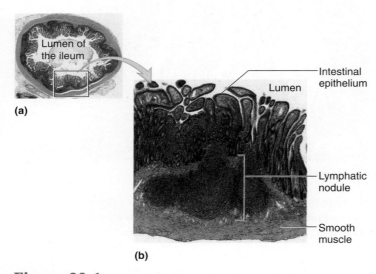

(a)

(b)

Figure 23.4   Peyer Patches. A cross-section of the ileum of the small intestine is shown in (a). Deep to the intestinal epithelium and superficial to the smooth muscle that surrounds the ileum are the lymphatic nodules called Peyer patches. Peyer patches are shown in detail in (b).

3. Move the microscope stage until you have a Peyer patch at the center of the field of view, and then change to medium or high power to observe the cells within the Peyer patch.

What kind of cells are these?_____

4. Using table 23.2 and figure 23.4, as guides, identify the following structures on the slide of the ileum:

☐ B-lymphocytes   ☐ lymphatic nodule

☐ epithelium   ☐ Peyer patch

5. In the space below, make a simple line drawing of a Peyer patch as seen through the microscope.

_____ ×

WHAT DO YOU THINK?

① Why do you think lymphatic nodules are so abundant in the wall of the small intestine?

EXERCISE 23.4 The Vermiform Appendix

The **vermiform appendix** (*vermiform*, shaped like a worm) projects from the inferior region of the cecum. It is usually about 2–4 centimeters in length and about half a centimeter in diameter. It is an evagination of the cecum and, as such, has a lumen that opens into the lumen of the cecum. Its epithelium is also a continuation of the epithelium of the cecum, and it contains numerous lymphatic nodules in its walls.

1. Obtain a slide demonstrating a cross section of the appendix and place it on the microscope stage.

2. Bring the tissue sample into focus on low power and scan the slide until you locate the lumen of the appendix (**figure 23.5**). Then change to medium power and bring the tissue sample into focus once again. Do you see

anything within the lumen of the appendix? _____ One identifying feature of the appendix is that its lumen usually appears to have a lot of "junk" inside it. This "junk" is typically cellular debris, bacteria, and other breakdown products of food that are left over after digestion in the small intestine. What type of epithelium lines the inside of the appendix? _____

_____

3. Observe the tissues located deep to the epithelial tissue to locate the lymphatic nodules, which are aggregations of purple-staining lymphocytes.

4. Using table 23.2 and figure 23.5 as guides, identify the following structures on the slide of the appendix:

☐ epithelium   ☐ lumen of the appendix

☐ lymphatic nodule

5. In the space below, make a simple line drawing of a cross section of the appendix as seen through the microscope.

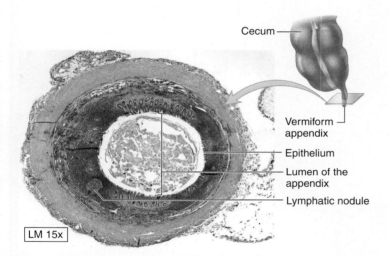

Cecum

Vermiform appendix

Epithelium

Lumen of the appendix

Lymphatic nodule

LM 15x

Figure 23.5   Vermiform Appendix.   A cross section through the appendix, demonstrating epithelium, lymphatic nodules, and a lumen that contains breakdown products of food.

_____ ×

Lymphatic Organs

Lymphatic organs differ from mucosa-associated lymphatic tissue (MALT) in that they are covered by a dense connective tissue capsule and are more highly organized. Lymphatic organs include lymph nodes, the thymus, and the spleen.

EXERCISE 23.5 Lymph Nodes

Lymph nodes are tiny lymphatic organs located along lymphatic vessels. They filter lymph that flows through the lymphatic vessels to identify and attack lymph borne foreign antigens.

1. Obtain a slide demonstrating a lymph node (**figure 23.6**) and place it on the microscope stage.

2. Bring the tissue sample into focus on low power. Identify the dense irregular connective tissue **capsule**, and notice the kidney-bean shape of the lymph node (if an entire node is on the slide). A lymph node is composed of an outer **cortex** and an inner **medulla**. The indented region of the

lymph node is the **hilum**. This is where blood vessels enter and leave the node, and where the **efferent lymphatic vessels** drain lymph from the lymph node. **Afferent lymphatic vessels** bring lymph into the lymph node on the regions that lie opposite the hilum of the lymph node.

3. Note the dense irregular connective tissue **trabeculae** that partition the cortex into smaller regions. Each region of the cortex contains one or two **lymphatic nodules**. What types of cells are found within lymphatic nodules?

_____

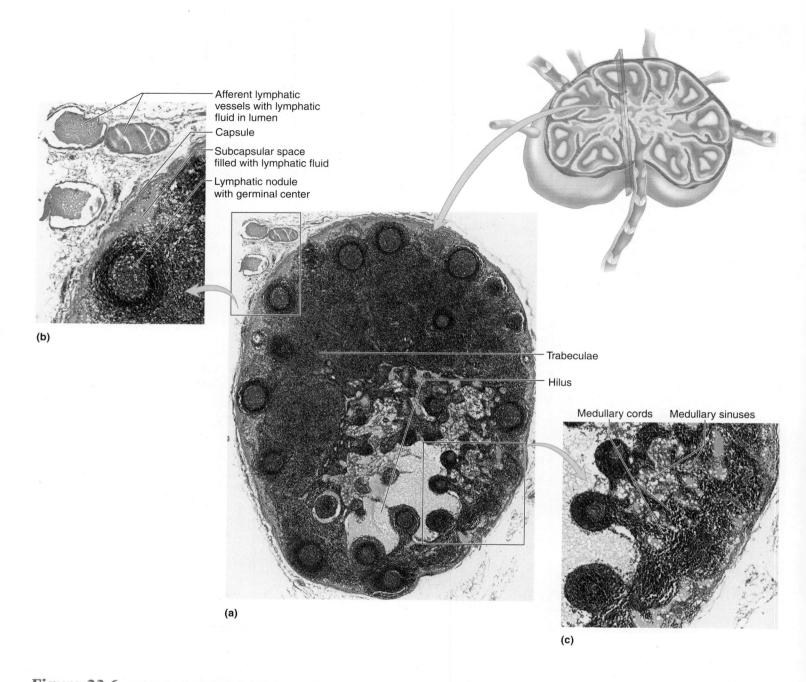

Afferent lymphatic vessels with lymphatic fluid in lumen

Capsule

Subcapsular space filled with lymphatic fluid

Lymphatic nodule with germinal center

(b)

Trabeculae

Hilus

Medullary cords   Medullary sinuses

(a)

(c)

Figure 23.6   Lymph Node   (a) Photomicrograph of a cross section of a lymph node that has been sectioned through the hilus. On the surface opposite the hilus three cross-sections of afferent lymphatic vessels are visible, as well as the capsule and lymphatic fluid within the subcapsular space (b). Medullary cords and spaces are visible in (c).

4. Scan the slide until the medulla is at the center of the field of view, and locate the dark-staining **medullary cords**. These contain mainly T-lymphocytes and macrophages. Can B- and T-lymphocytes be distinguished from each other using light microscopy? _____

5. Identify the following structures on the histology slide of the lymph node:

☐ capsule
☐ cortex
☐ hilum
☐ lymphatic nodules

☐ medulla
☐ medullary cords
☐ trabeculae

6. Observe the numerous spaces, called **medullary sinuses,** where lymph travels as it flows through the lymph node. Figure 23.7 demonstrates the pathway taken by lymph as it flows through a lymph node.

7. Using figure 23.6 and **table 23.3** as guides, identify the following spaces or vessels on the slide of the lymph node:

☐ afferent lymphatic vessels
☐ efferent lymphatic vessels

☐ medullary sinuses
☐ peritrabecular spaces
☐ subcapsular spaces

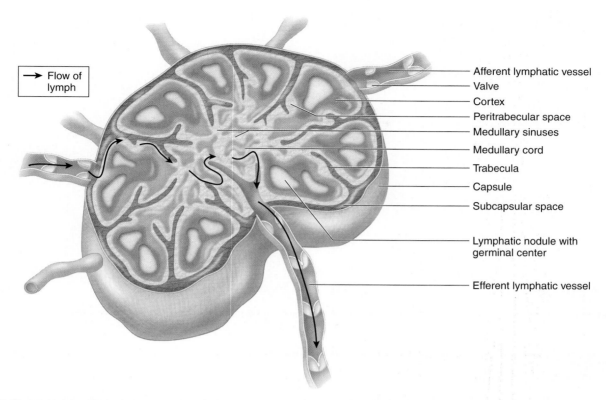

Flow of lymph

Afferent lymphatic vessel
Valve
Cortex
Peritrabecular space
Medullary sinuses
Medullary cord
Trabecula
Capsule
Subcapsular space
Lymphatic nodule with germinal center
Efferent lymphatic vessel

Figure 23.7   **Lymph Node.**   Pathway that lymph takes as it flows through a lymph node.

Table 23.3	Parts of a Lymph Node	
Structure	**Description**	**Word Origins**
Afferent Lymphatic Vessel	Vessel that carries lymph into the lymph node.	*af-fero*, to bring to
Capsule	Dense irregular connective tissue that surrounds the entire lymph node to support and protect it.	*capsula*, a box
Cortex	Outer portion of the lymph node that lies deep to the capsule and contains lymphatic nodules.	*cortex*, bark
Efferent Lymphatic Vessel	Vessel that carries lymph away from the lymph node; located at the hilum.	*effero*, to bring out
Hilum	Indented area where blood vessels enter and leave and where efferent lymphatic vessels also exit.	*hilum*, a small bit
Lymphatic nodule	Ball of lymphocytes containing a germinal center, where B-lymphocytes are actively proliferating.	*lympha*, clear spring water, + *nodulus*, a small knot
Medulla	Inner portion of the lymph node that contains medullary cords of B-lymphocytes and macrophages.	*medius*, middle
Medullary Cord	Strandlike cluster of B-lymphocytes and macrophages located in the medulla of a lymph node.	*medius*, middle, + *chorda*, a string
Medullary Sinuses	Spaces between the medullary cords in a lymph node through which lymphatic fluid flows.	*medius*, middle
Peritrabecular Space	Space located between a lymphatic nodule and a trabecula in the cortex of the lymph node.	*peri-*, around, + *trabecula*, trabeculae
Subcapsular Space	Space located between a lymphatic nodule and the capsule of the lymph node.	*sub-*, under, + *capsular*, capsule
Trabeculae	Invaginations of the dense irregular connective tissue capsule of a lymph node that partition the cortex into smaller compartments.	*trabs*, a beam

8. In the space to the right, make a simple line drawing of a lymph node as seen through the microscope. Label all of the structures listed in steps 5 and 7. Then, using arrows, indicate the pathway taken by lymph as it flows through the lymph node.

_____ ×

EXERCISE 23.6 The Thymus

The thymus is a small gland located in the superior mediastinum (**figure 23.8**). It is the site of T-lymphocyte maturation, and it is also an endocrine organ, secreting the hormone **thymosin**. The thymus is composed of two **lobes**, each separated into smaller **lobules** by connective tissue **septae** (singular: septa) (**figure 23.9**). Each lobule has a darker-staining outer cortex and a lighter-staining inner medulla. The majority of the cells within the thymus are T-lymphocytes, which are surrounded by cells of epithelial origin. Lymphocytes migrate to the thymus from the bone marrow early on in life. Once in the thymus, T-lymphocytes travel from the outer cortex to the inner medulla as they complete the process of **selection**. Selection is a process by which only those T-lymphocytes that are able to recognize self-antigen, and react to foreign antigen, are permitted to leave the thymus. Selection occurs in the outer cortex, and once selection of T-lymphocytes is complete the cells leave the thymus through venules located in the medulla. The majority of T-lymphocytes that migrate to the thymus from the bone marrow either fail to recognize self-antigen or fail to react to foreign antigen. These cells are destroyed through a process called **apoptosis** (programmed cell death), which occurs within the thymic cortex. After puberty the thymus begins a gradual decline in size with age. Epithelial-derived cells in the medulla come together and form small, onionlike masses of tissue called **thymic corpuscles**. Thymic corpuscles produce chemical signals that induce development of T-regulatory cells, which patrol the body looking for "bad" T-cells and destroying them so they cannot produce autoimmune disease. A large number of thymic corpuscles in the thymic medulla indicates an aging thymus.

1. Obtain a slide of the thymus and place it on the microscope stage.

2. Bring the tissue sample into focus at low power, then scan the slide to locate the connective tissue trabeculae that separate the gland into lobules (figure 23.8).

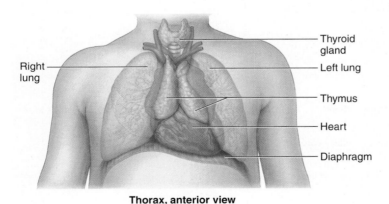

Thorax, anterior view

Figure 23.8   **Location of the Thymus.**   The thymus is located in the superior mediastinum, deep to the sternum.

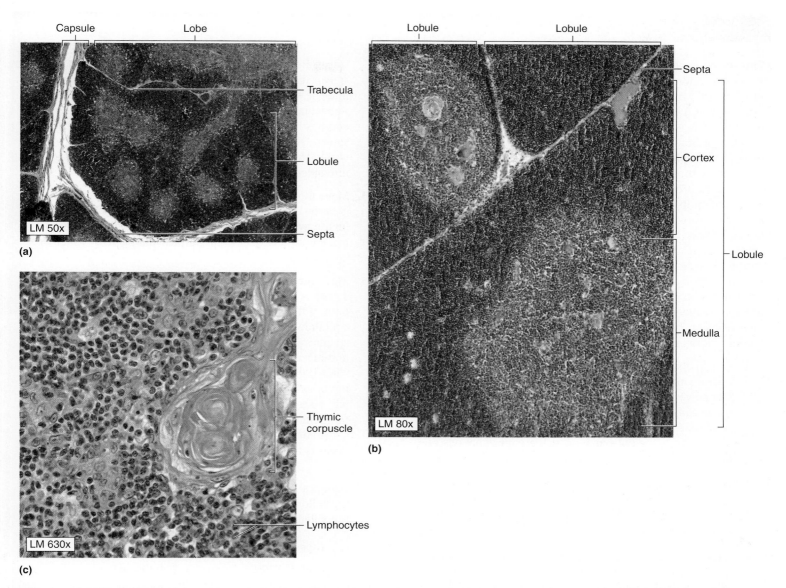

Capsule   Lobe

Trabecula

Lobule

Septa

LM 50x

(a)

Thymic corpuscle

Lymphocytes

LM 630x

(c)

Lobule   Lobule

Septa

Cortex

Lobule

Medulla

LM 80x

(b)

Figure 23.9   **Histology of the Thymus.**   (a) A thymic lobe, which is divided into lobules. (b) Each thymic lobule consists of an outer cortex and an inner medulla. (c) Thymic corpuscles are located within the medullary region of a thymic lobule, and are indicative of an aging thymus.

3. Using figure 23.9 as a guide, identify the following structures:

☐ cortex   ☐ thymic lobule

☐ medulla   ☐ trabeculae

☐ thymic corpuscle

4. In the space to the right, make a simple line drawing of the thymus as seen through the microscope. Be sure to label all the structures listed above.

✕

EXERCISE 23.7 The Spleen

The spleen is similar to a lymph node in many ways, with one major exception: the spleen filters *blood*, whereas a lymph node filters *lymph*. However, both organs filter their fluid for the purpose of identifying and fighting antigens that may be present in the fluid. The spleen receives blood through the **splenic artery**, which is a branch of the celiac trunk. Blood leaves the spleen through the **splenic vein**, which drains into the hepatic portal vein (see figures 22.13 and 22.14, pp. 456–57).

When a fresh spleen is cut, two types of tissue are clearly seen: red pulp and white pulp. **Red pulp**, which consists of splenic sinusoids, is red due to the large amount of blood contained within the sinusoids. It composes the majority of the splenic tissue. **White pulp**, which consists of aggregations of lymphocytes, is white due to the white blood cells (leukocytes) within each mass of tissue. Because you will be observing a stained histology slide of the spleen, red and white pulp will not appear red and white in color. The red pulp will be reddish-pink in color, but the white pulp will be darker purple, with a lighter-colored central region.

1. Obtain a slide of the spleen and place it on the microscope stage. Bring the tissue sample into focus on low power and locate the dense irregular connective tissue capsule of the spleen, if present in the section on your slide (figure 23.10). You may also notice several connective tissue trabeculae, which are invaginations of the capsule that separate the spleen into distinct regions.

2. Notice the circular purple aggregations of cells. These aggregations constitute the white pulp of the spleen, which consists mainly of B-lymphocytes, T-lymphocytes, and some macrophages.

3. Look for a **central artery**, which is found within each mass of white pulp. Central arteries receive blood from **trabecular arteries**, whose cross sections can

sometimes be seen in the surrounding red pulp of the spleen that surrounds the white pulp. Blood coming into the spleen travels from the splenic artery to the trabecular arteries to the central arteries, and finally empties into the sinusoids or surrounding tissues of the spleen (the exact nature of capillary-type blood flow within the spleen remains unknown) before entering into splenic veins.

4. Move the microscope stage until a mass of white pulp is in the center of the field of view and then change to high power. At this magnification you should be able to make out individual small purple-staining cells, which are lymphocytes.

5. Move the microscope stage so the red pulp of the spleen is in the center of the field of view. The red pulp of the spleen consists mainly of *splenic sinusoids* containing numerous erythrocytes. The blood-filled splenic sinusoids also contain numerous lymphocytes and macrophages. As blood travels slowly through the spleen, the lymphocytes and macrophages monitor the blood for **bloodborne antigens**, which, if present, will be removed or attacked by these immune cells. In addition, the structure and function of the splenic sinusoids is designed to destroy abnormal erythrocytes. What is the life span of a typical erythrocyte? _____ days. (Hint: the answer is in chapter 20.)

6. Using **table 23.4** and figure 23.10 as guides, identify the following structures on the histology slide of the spleen:

☐ capsule   ☐ sinusoids
☐ central artery   ☐ trabecula
☐ red pulp   ☐ white pulp

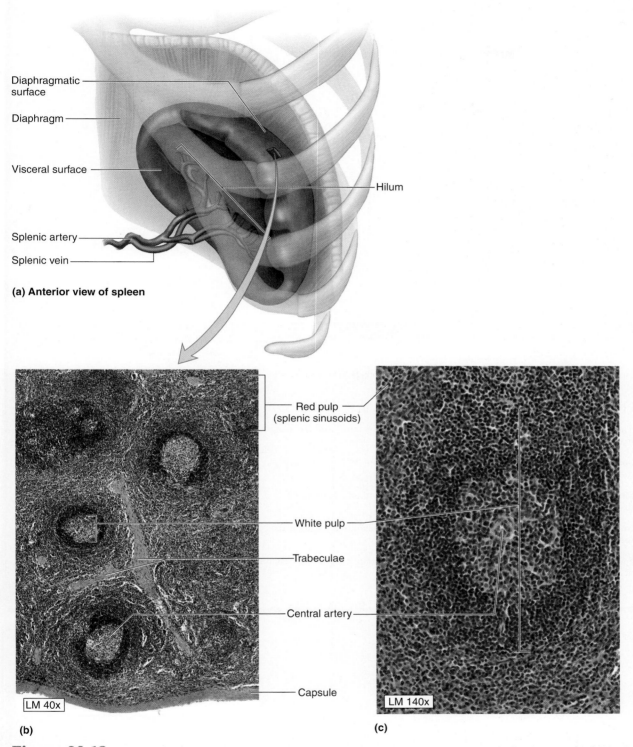

Diaphragmatic surface

Diaphragm

Visceral surface

Splenic artery

Splenic vein

Hilum

(a) Anterior view of spleen

Red pulp (splenic sinusoids)

White pulp

Trabeculae

Central artery

Capsule

LM 40x

(b)

LM 140x

(c)

Figure 23.10   The Spleen.   The spleen contains two predominant tissues: red pulp, which consists of splenic sinusoids, and white pulp, which contains lymphocytes. (a) Location of spleen. (b) Low-magnification and (c) high-magnification.

7. In the space below, make a simple line drawing of the spleen as seen through the microscope. Be sure to label all the structures listed in step 4.

_____ ×

Table 23.4	Parts of the Spleen	
Structure	**Description**	**Word Origins**
Capsule	Dense irregular connective tissue that supports and protects the outside of the spleen; surrounds the entire spleen.	capsula, a box
Central Arteries	Arteries located within the white pulp of the spleen.	centrum, center, + arteria, the windpipe
Hilum	The indented area of the spleen where the splenic artery enters and the splenic vein leaves.	hilum, a small bit
Red Pulp	Splenic tissue consisting of splenic sinusoids lined by T-lymphocytes and macrophages.	pulpa, flesh
Splenic Sinusoids	Sinusoidal capillaries, which consist of fenestrated endothelial cells and a discontinuous basement membrane.	sinus, cavity, + eidos, resemblance
Trabeculae	Invaginations of the connective tissue capsule of the spleen that partition the cortex into smaller compartments.	trabs, a beam
White Pulp	Splenic tissue consisting of lymphatic nodules that contain lymphocytes; each mass of white pulp contains a central artery.	pulpa, flesh

Gross Anatomy

EXERCISE 23.8 Gross Anatomy of Lymphatic Structures

1. *Major Lymph Vessels of the Body:* Obtain a model of the thorax and abdomen. Remove the heart and lungs from the thorax and the stomach and intestines from the abdomen. Using your textbook as a guide, identify the structures listed in **figure 23.11** on the model. Then label them in figure 23.11.

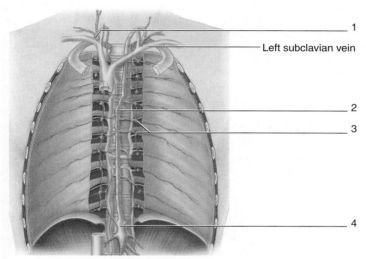

Thorax, anterior view

Figure 23.11   **Major Lymph Vessels of the Body.**

☐ cisterna chyli   ☐ right lymphatic duct
☐ lymph nodes   ☐ thoracic duct

> ## Study Tip!
>
> Identification of the thoracic duct within the thoracic cavity of a cadaver can be challenging because of its relatively small diameter and lack of color (because lymph does not contain erythrocytes). It travels between the esophagus and the azygos vein. Anatomists thus refer to it as the "duck between two gooses." The "duck" is the thoracic duct, and the "gooses" are the "azygoose" (azygos vein) and the "esophagoose" (esophagus).

2. *Mucosa-Associated Lymphatic Tissue (MALT):* Obtain a model demonstrating a midsagittal section of the head and neck, and a model of the abdomen. Using your textbook as a guide, identify the structures listed in **figure 23.12** on the model. Then label them in figure 23.12.

3. *Lymph Node:* Obtain a classroom model of a lymph node. Using table 23.3 and your textbook as a guide, identify the structures listed in **figure 23.13** on the model. Then label them in figure 23.13.

4. *The Spleen:* Obtain a classroom model of the abdominal cavity, or observe the abdominal cavity of a human cadaver. Locate the spleen and identify the structures listed in **figure 23.14** on the model or the cadaver. Then label them in figure 23.14.

5. *Optional Activity:* **AP|R** **Lymphatic System**—Watch the "Lymphatic System Overview" animation for a summary of the primary lymphatic structures.

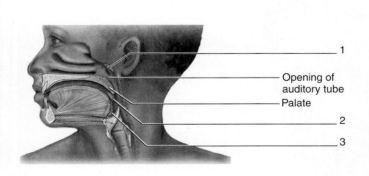

(a)

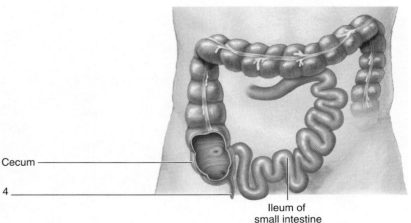

(b)

Figure 23.12   **Mucosa-Associated Lymphatic Tissue (MALT).**

☐ lingual tonsils   ☐ palatine tonsils   ☐ pharyngeal tonsils   ☐ vermiform appendix

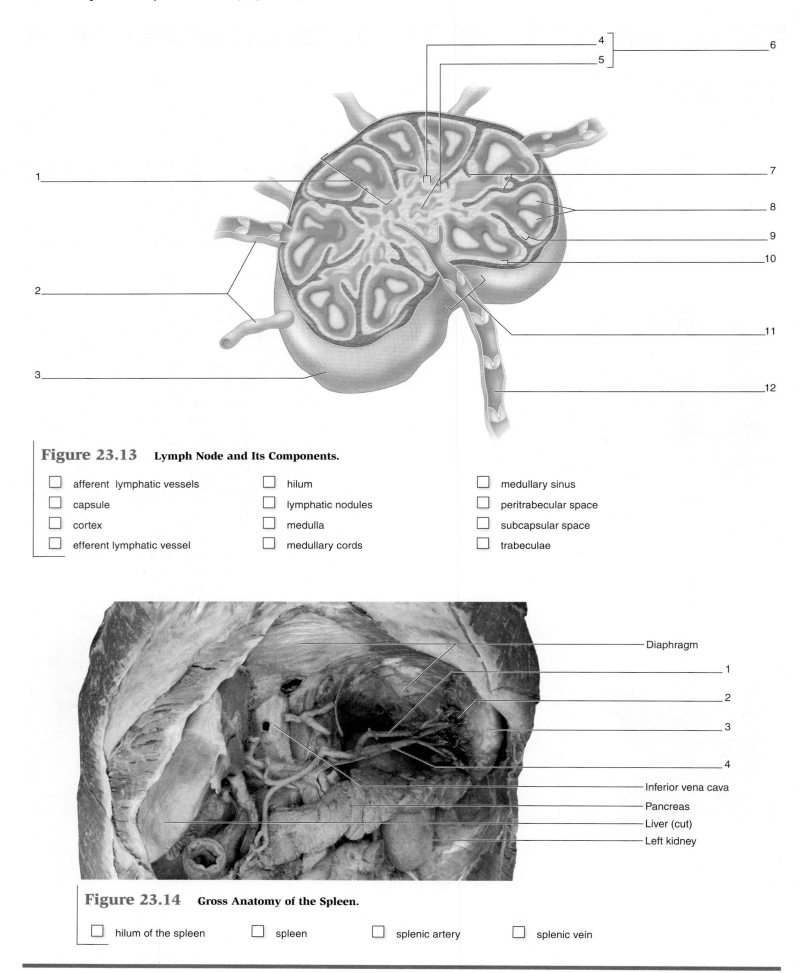

Figure 23.13   **Lymph Node and Its Components.**

☐ afferent lymphatic vessels   ☐ hilum   ☐ medullary sinus

☐ capsule   ☐ lymphatic nodules   ☐ peritrabecular space

☐ cortex   ☐ medulla   ☐ subcapsular space

☐ efferent lymphatic vessel   ☐ medullary cords   ☐ trabeculae

Figure 23.14   **Gross Anatomy of the Spleen.**

☐ hilum of the spleen   ☐ spleen   ☐ splenic artery   ☐ splenic vein

Chapter 23: The Lymphatic System

Name:_____

Date:_____Section:_____

POST-LABORATORY WORKSHEET

1. Imagine you are a bacterium that has entered a lymph capillary in the right arm of a human. Describe:

   a. the pathway you would take to travel from the lymph capillary you initially entered to a lymph node in the axilla:

   _____

   _____

   _____

   b. the pathway you would take as you traveled within the lymph node (include the types and locations of leukocytes you might encounter as you travel through the lymph node):

   _____

   _____

   _____

   _____

   _____

   _____

   c. the pathway you would take to get from the lymph node to the right atrium of the heart (assuming you are lucky enough to escape detection and make it through the lymph node!):

   _____

   _____

   _____

   _____

   _____

   _____

2. Lacteals are specialized lymphatic capillaries found in the small intestine that have the special function of absorbing dietary fats. Other nutrients that are absorbed by the digestive tract are absorbed into blood capillaries, not lymphatic capillaries. Because absorbed fats are relatively large structures, they must be absorbed into the lymphatic system instead of being absorbed directly into the blood vascular system. Using your knowledge of the flow of both lymph and blood in the body, describe the route a dietary lipid would take to get from its location of absorption in the small intestine to its entry into the right atrium of the heart.

   _____

   _____

   _____

   _____

   _____

3. At what age does the thymus reach its maximum size? _____

4. What is a thymic corpuscle, and what does the presence of many thymic corpuscles tell you about the thymus? _____
_____
_____

5. The white pulp of the spleen is composed of _____ and functions to _____
_____.

6. The red pulp of the spleen is composed of _____ and functions to _____
_____
_____.

7. Describe the anatomic location of each of the three types of tonsils:

   Lingual _____

   Palatine _____

   Pharyngeal _____

8. For each of the three types of tonsils, list the type of epithelium that is located superficial to the tonsil.

   Lingual _____

   Palatine _____

   Pharyngeal _____

9. When pharyngeal tonsils become inflamed, they are referred to as adenoids. Based on the location of the adenoids (pharyngeal tonsils), what kinds of symptoms might a patient experience due to the presence of swollen adenoids?
_____
_____
_____

10. An individual suffering from an inflamed appendix will most commonly have pain in which abdominopelvic quadrant?
_____

11. When an individual suffers from a ruptured spleen, the spleen is surgically removed so the patient does not die from internal bleeding. Using your knowledge of the circulation of the spleen, answer the following questions.

   a. Why does a ruptured spleen bleed so profusely? (Hint: What type of capillaries are found in the spleen?)
_____
_____

   b. Why can a patient survive without his/her spleen? (Hint: What other organs may take up the functions of the spleen?)
_____
_____

   c. A patient who has had his/her spleen removed will be more susceptible to certain types of infections. These would be infections that travel through the _____
_____.

The Respiratory System

24

OUTLINE and OBJECTIVES

Module 11: RESPIRATORY SYSTEM

INTRODUCTION

Breathe in deeply. What muscles did you contract to do this? As you may recall from chapter 11, the **external intercostal muscles** and the **diaphragm** are the key muscles used for inspiration. Contraction of these muscles increases the size of the thoracic cavity, which pulls on the **pleura**, causing the lungs to expand. As the lungs expand and the volume of the lungs increases, the pressure within the lungs becomes lower than atmospheric pressure and air flows into the lungs. This air brings with it a fresh supply of oxygen that is subsequently picked up by **erythrocytes** in the blood flowing through the lung capillaries. This freshly oxygenated blood will be transported back to the heart and pumped out to the working tissues of the body. Take another deep breath. This time focus on the process of **expiration** (breathing out). Normal expiration is a passive process, which does not require muscular effort. As soon as you stop contracting your external intercostal muscles and your diaphragm, the volume of the thoracic cavity (and the lungs) decreases, which increases the pressure. As the pressure in the lungs becomes greater than atmospheric pressure, air flows out of the lungs. This air is rich in carbon dioxide, a waste product from cellular respiration that must be removed from the body. Your study of the anatomy of the respiratory system will largely be focused on the anatomical structures used in this process of conveying air into and out of the lungs. It will also focus on the histology of the lung tissues themselves, whose structure is ideally suited to the task of exchanging gases between the lungs and the blood.

Chapter 24: The Respiratory System

Name:_____

Date:_____ Section:_____

PRE-LABORATORY WORKSHEET

1. Define the makeup (cell shape, number of layers, etc.) of *respiratory epithelium* (the epithelium found lining the nasal cavity and upper respiratory tract).

   _____

   _____

   _____

2. How many lobes compose the left lung? _____

3. Which of the following bones does *not* form a border of the nasal cavity?

   a. vomer

   b. ethmoid

   c. mandible

   d. palatine

   e. maxilla

4. The pulmonary arteries carry _____ blood from the _____ (chamber) of the heart to the lungs, and the pulmonary veins carry _____ blood from the lungs to the _____ (chamber) of the heart. This circuit is referred to as the _____ circuit.

5. True or False: Contraction of the diaphragm decreases the volume of the thoracic cavity. _____

6. The respiratory membrane is composed of three structures. List these structures in the spaces below.

   a. _____

   b. _____

   c. _____

7. Which lung cells produce surfactant? _____

8. Structures that compose the respiratory portion of the respiratory tree all contain _____ .

9. A _____ bronchus leads into each lobe of a lung.

   a. primary

   b. secondary

   c. tertiary

   d. quaternary

10. The lungs are contained within the _____ cavities.

In this laboratory exercise, you will observe the gross and histological structure of the upper respiratory tract and the lungs. You will begin by investigating the general layering pattern of the walls of the respiratory tract so that you will be able to identify specific histological structures located within each layer. This will be followed by analysis of the histology of major respiratory tract structures, including the lungs. Finally, you will observe the gross anatomy of the upper airways and the lungs, using either cadaveric specimens, fresh sheep specimens, models, or some combination of these.

Histology

Upper Respiratory Tract

The upper respiratory tract consists of the nose, nasal cavity, paranasal sinuses, and pharynx. These structures, along with the trachea and bronchi (which are part of the lower respiratory tract), compose the **conducting portion** of the respiratory system. The function of the conducting portion is to convey air into and out of the lungs. Structures within the lungs that are involved with gas exchange compose the **respiratory portion** of the respiratory system. Details of respiratory portion structures will be covered in exercise 24.4, "The Lungs."

In the conducting portion of the respiratory system, the walls of the airways are composed of three layers: mucosa, submucosa, and adventitia/serosa (**figure 24.1**). The **mucosa** (*mucosus*, mucus) consists of the epithelium and an underlying layer of loose connective tissue and blood vessels called the **lamina propria** (*lamina*, layer, + *proprius*, one's own). The **submucosa** (*sub-*, under, + mucosa) is the middle layer and consists of glands, blood vessels, nerves, smooth muscle, and connective tissues. The adventitia/serosa is the outermost layer. If the outer layer of the organ is composed of connective tissue, it is called an **adventitia** (*adventicius*, coming from abroad). If the outer layer of the organ is composed of a mesothelial membrane, it is called a **serosa** (*serosus*, serous). As you view slides of respiratory system structures, it is helpful to first identify these three layers, particularly the mucosa and submucosa, which will make it easier to identify specific cell types within each layer.

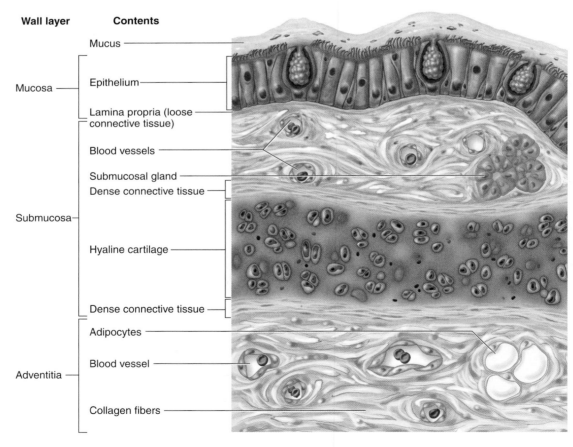

Figure 24.1 **Wall Layers of the Trachea.** The walls of all structures of the respiratory tract are composed of three layers: mucosa, submucosa, and adventitia/serosa. This drawing depicts the wall of the trachea, which has an outer covering of connective tissue referred to as an adventitia. If the outer covering was a mesothelial membrane, it would be referred to as a serosa.

EXERCISE 24.1 Olfactory Mucosa

The epithelium that lines the nasal cavity is highly specialized—not only does it assist with the cleansing and warming of incoming air, but it also acts as a special sensory organ to detect odors. **Figure 24.2** demonstrates the histological appearance of the olfactory epithelium, and **table 24.1** describes the appearance and function of the various cell types and structures within the olfactory epithelium.

1. Obtain a compound microscope and a slide of olfactory epithelium. Place the slide on the microscope stage and bring the tissue sample into focus on low power.

2. Identify the **mucosa** and **epithelium**. Move the microscope stage so the epithelium is at the center of the field of view, then change to high power.

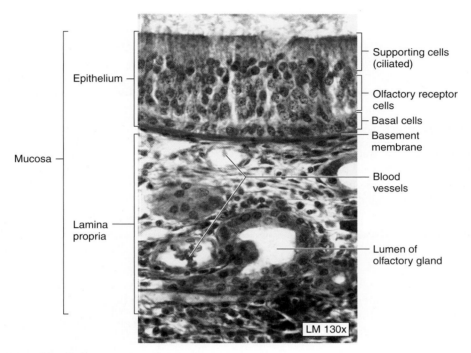

Figure 24.2 **Olfactory Epithelium.** Olfactory epithelium contains three cell types: basal cells, olfactory receptor cells, and supporting (ciliated) cells.

Table 24.1	Olfactory Epithelium	
Structure	**Description and Function**	**Word Origins**
Basal Cells	Small, triangular cells that lie on the basement membrane and do not reach the apical surface of the epithelium. Their *nuclei are located on the basal surface of the epithelium.* These cells are the precursor cells to the other cell types and are responsible for regeneration of the epithelium.	*basis*, bottom
Olfactory Glands	Serous-secreting glands located in the lamina propria of the olfactory epithelium that open to the surface. They secrete a serous fluid that "washes" the olfactory hairs, clearing the surface so that new odors can be sensed by the olfactory cells.	*ol-facio*, to smell
Olfactory Receptor Cells	Bipolar neurons that contain 6–8 cilia-like extensions on their apical surface called olfactory hairs. These hairs function to detect odorous substances. The cell *nuclei are located in the middle of the epithelium.* Axons of these cells travel through the cribriform foramina of the ethmoid bone to synapse with CNS neurons in the olfactory bulbs.	*ol-facio*, to smell
Supporting Cells (Ciliated)	The most prominent and abundant cells. Their *nuclei are located near the apical surface of the epithelium.* They are columnar epithelial cells with cilia on their apical surfaces. They function to propel mucus and particulate matter along the apical surface of the epithelium toward the pharynx.	*cilium*, eyelash

3. Using table 24.1 and figure 24.2 as guides, identify the following structures on the slide:

☐ basal cells ☐ olfactory glands

☐ olfactory receptor cells ☐ supporting cells (ciliated)

4. In the space to the right, draw a brief sketch of the olfactory epithelium as viewed under the microscope. Be sure to label all the structures listed in step 3 in your drawing.

_____ ×

Lower Respiratory Tract

The lower respiratory tract includes the **trachea** and the **bronchi** (_bronchos_, windpipe). As the airways extend progressively deeper into the lungs from the trachea, into the bronchi and bronchioles, the amount of cartilage in the submucosa decreases while the amount of smooth muscle increases. In addition, several changes occur in the epithelium as it transitions from the respiratory epithelium of the trachea to the simple squamous epithelium of the alveoli within the lungs. Table 24.2 describes the wall structure of the trachea, bronchi, and bronchioles.

EXERCISE 24.2 The Trachea

1. Obtain a slide of the trachea (_tracheia_, rough artery) and place it on the microscope stage. This slide will also contain the esophagus (_oisophagos_, gullet) because the two structures lie next to each other _in vivo_.

2. Bring the tissue sample into focus on low power and distinguish the trachea from the esophagus. Move the microscope stage so the lumen and epithelium of the trachea are in the center of the field of view. Then switch to high power.

3. As you observe the trachea, first identify the layers of the wall of the trachea (mucosa, submucosa, and adventitia), and then identify the structures located within each layer. **Table 24.2** describes the wall structure of the trachea and the structures located within each layer, and **figure 24.3** demonstrates the histological appearance of the wall of the trachea.

Table 24.2	Histology of the Trachea
Structure	**Description and Functions**
MUCOSA	Innermost layer that lines the tracheal wall; consists of respiratory epithelium and underlying lamina propria. The lamina propria is separated from the underlying submucosa by a layer of elastic connective tissue (the _elastic lamina_).
Epithelium	Pseudostratified columnar epithelium with cilia and goblet cells.
Basal Cells	Small triangular cells that lie on the basal lamina and do not reach the apical surface of the epithelium. These cells are the precursor cells to the other cell types and are responsible for regeneration of the epithelium.
Ciliated Cells	The most prominent and abundant cells. These are pseudostratified ciliated columnar cells of the epithelium. They function to propel mucus and particulate matter via the "mucus escalator" along the epithelial sheet toward the pharynx.
Goblet Cells	Large round cells that appear white or clear. They secrete mucus onto the surface of the epithelium. The mucus traps particulate matter that enters the trachea while the cilia move the mucus superiorly (a "mucus escalator") toward the pharynx so that it may be swallowed.
Lamina Propria	A layer of loose connective tissue that underlies the respiratory epithelium. In the trachea this layer can be seen in the area between the epithelial folds and the trachealis muscle or C-shaped cartilages.
SUBMUCOSA	The middle layer of the tracheal wall containing seromucus glands, blood vessels, the trachealis muscle, C-shaped cartilages, and nerves.
C-Shaped Tracheal Cartilages	C-shaped plates of hyaline cartilage located on the anterior surface of the trachea. There are 16–20 of them. They are C-shaped so there is room on the posterior aspect of the trachea for expansion of the esophagus during swallowing and to allow the trachialis muscle to alter the diameter of the trachea.

Table 24.2	Histology of the Trachea *(continued)*
Structure	**Description and Functions**
Trachealis Muscle	A layer of smooth muscle found on the posterior aspect of the trachea where the trachea lies against the esophagus. Laterally, the trachealis muscle is anchored to the ends of the cartilage C rings, and its contraction decreases the diameter of the trachea, which is important for coughing and sneezing. The decreased diameter of the trachea causes the air to exit more forcefully, which helps dislodge substances within the airways.
Submucous Glands	These produce a substance that is part watery (serous) and part viscous (mucus).
ADVENTITIA	Dense connective tissue on the outermost surface of the tracheal wall.

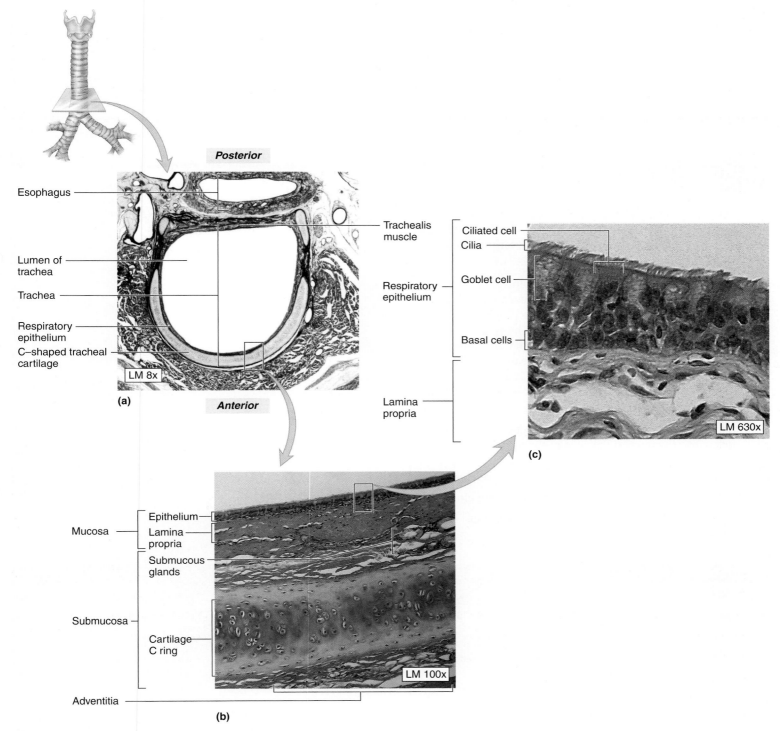

Figure 24.3 **Histology of the Trachea.** (a) Cross section through the trachea and esophagus. (b) Wall layers of the trachea. (c) Close-up view of the tracheal mucosa.

4. Using table 24.2 and figure 24.3 as guides, identify the following structures on the slide of the trachea:

- ☐ adventitia
- ☐ basal cells
- ☐ C-shaped tracheal cartilages
- ☐ ciliated cells
- ☐ goblet cells
- ☐ lamina propria
- ☐ mucosa
- ☐ respiratory epithelium
- ☐ submucosa
- ☐ submucous glands
- ☐ trachealis muscle

5. In the space below, draw a brief sketch of the cross section of the trachea as viewed under the microscope. Be sure to label all of the structures listed in step 4 in your drawing.

_____ ×

EXERCISE 24.3 The Bronchi and Bronchioles

1. Obtain a slide of the lungs and place it on the microscope stage. Bring the tissue sample into focus on low power and scan the slide to locate cross sections of **large bronchi**, **small bronchi**, and **bronchioles** (figure 24.4).

2. The different characteristics of the conducting zone structures are listed in **table 24.3**. In general, as the airways travel deeper into the lung and become progressively smaller, three changes take place: (1) The epithelium transitions from pseudostratified columnar to simple cuboidal, and the number of cilia decrease, (2) large plates of hyaline cartilage in the walls give way to smaller and smaller pieces of cartilage, and (3) the amount of smooth muscle in the airways increases.

3. Using figure 24.4 and table 24.3 as guides, identify the following structures of the slide of the lungs:

- ☐ branch of pulmonary artery
- ☐ bronchiole
- ☐ hyaline cartilage
- ☐ large bronchus
- ☐ small bronchus
- ☐ smooth muscle

4. In the space below, draw a cross-section of a bronchiole as viewed through the microscope.

Table 24.3	Histology of the Bronchial Tree		
Structure	**Epithelium**	**Hyaline Cartilage**	**Smooth Muscle**
Large Bronchi	Pseudostratified columnar with cilia	Large plates, which keep airway open	Encircles the lumen
Small Bronchi	Simple columnar with cilia	Small plates	Encircles the lumen
Bronchioles	Simple columnar to simple cuboidal	None	Encircles the lumen and is an important regulator of airway diameter

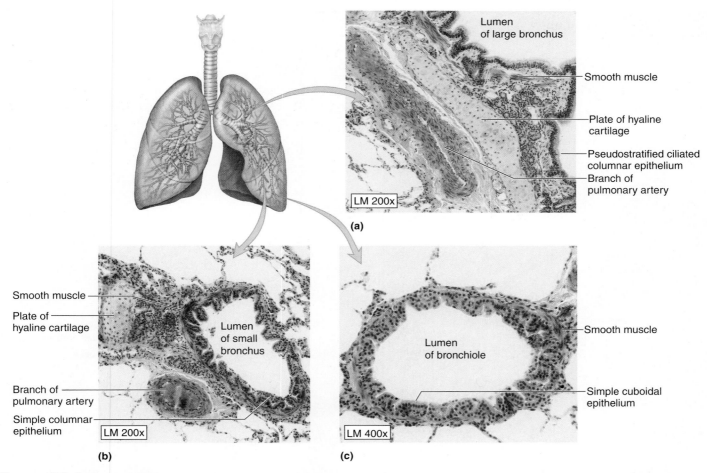

Figure 24.4 The Bronchial Tree.   When making observations of the lung, you will see cross sections of portions of the bronchial tree. (a) The larger bronchi are lined with a ciliated pseudostratified columnar epithelium, and there are large plates of hyaline cartilage in the walls, along with smooth muscle. (b) The smaller bronchi are lined with a simple columnar epithelium, and have no cilia, smaller plates of cartilage in their walls, and more smooth muscle. (c) Bronchioles have a simple cuboidal epithelium, no cartilage, and smooth muscle.

Lungs

The lungs consist of functional units called **alveoli** (*alveus*, hollow sac), which are the sites of gas exchange between the air within the lungs and the blood. The alveoli are lined with simple squamous epithelium, which provides the thinnest possible barrier to diffusion between the air within the alveoli and the blood within the capillaries that surround them. In a slide of the lung, you will see numerous alveoli, but you will also find cross sections of some of the smaller airways (such as respiratory bronchioles). The transition from conducting portion structures to respiratory portion structures occurs deep within the lungs, and the key feature in distinguishing them from each other is the presence or absence of alveoli. Any structure that has at least one alveolus coming off of it participates in gas exchange and, thus, is part of the respiratory portion.

EXERCISE 24.4 **The Lungs**

1. Obtain a slide of the lungs and place it on the microscope stage. Bring the tissue sample into focus on low power and scan the slide to locate cross sections of **bronchioles** (**figure 24.5**).

2. Increase the magnification to observe the smaller airways and the air sacs (alveoli) in greater detail. **Table 24.4** describes the structure and function of the airways that compose the respiratory portion. **Table 24.5** describes the

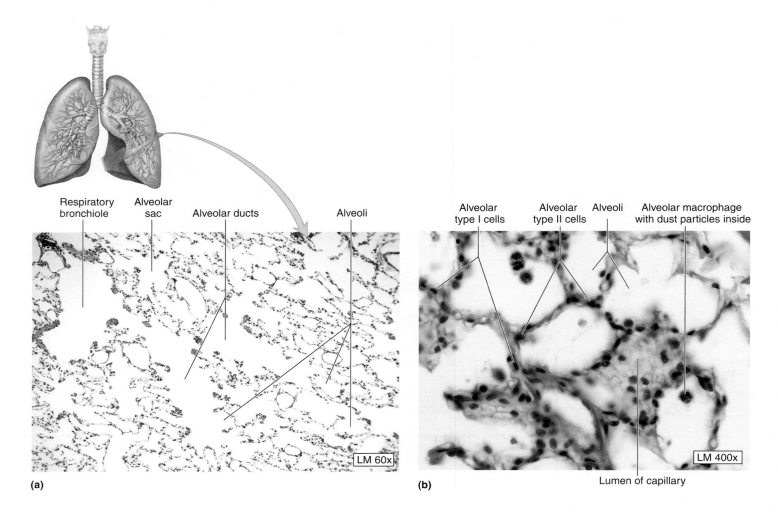

(a)

(b)

Figure 24.5 **Histology of the Lungs.** (a) Low-magnification view demonstrating respiratory portion structures. (b) High-magnification view demonstrating alveolar macrophages with particles of dust inside the cells.

Table 24.4	Structures Composing the Respiratory Zone	
Structure	**Description**	**Epithelium**
Respiratory Bronchioles	Ducts with alveoli scattered along the passageway.	Ciliated simple cuboidal
Alveolar Ducts	Ducts with alveoli lining the entire passageway.	Simple squamous
Alveolar Sacs	Shaped like a bunch of grapes with several alveoli (the "grapes") bunched together.	Simple squamous
Alveoli	A balloon-like structure that is the site of gas exchange.	Simple squamous

Table 24.5	Microscopic Structures Within the Lungs
Structure	**Description**
Alveolar Macrophages	Derived from monocytes. Engulf particulate matter or pathogenic organisms.
Alveolar Type I Cell	Simple squamous epithelial cells covering 97% of the alveolar surface area. They are connected to each other with tight junctions.
Alveolar Type II Cell	Compose about 3% of alveolar surface area. Joined to type I cells by desmosomes and tight junctions. Contain a large nucleus, foamy cytoplasm, and vesicles containing **pulmonary surfactant** (a mixture of phospholipids, glycosaminoglycans, and proteins that functions to reduce surface tension of alveoli).

structure and function of the cell types within the alveoli themselves. For gas exchange to occur, gas molecules must pass from the alveoli to the capillaries that surround them. The structures lining the alveoli and the capillaries collectively form the **respiratory membrane**. The respiratory membrane is composed of alveolar type I cells, the fused basement membrane of alveolar type I cells, and the endothelial cells lining the capillaries.

3. Using tables 24.4 and 24.5 and figure 24.5 as guides, identify the following structures on the slide of the lungs:

- ☐ alveolar ducts
- ☐ alveolar macrophages
- ☐ alveolar sacs
- ☐ alveolar type I cell
- ☐ alveolar type II cell
- ☐ alveoli
- ☐ capillary
- ☐ respiratory bronchioles

4. In the space below, draw a brief sketch of the respiratory membrane. Then label the three components that make up the respiratory membrane.

5. *Optional Activity*: **AP|R** **Respiratory System**—Watch the "Diffusion Across Respiratory Membrane" animation to visualize the microscopic structure and function of lung tissue.

Gross Anatomy

Upper Respiratory Tract

In the following exercises you will observe the gross anatomy of the respiratory structures that convey air into the lungs, the cavities that house the lungs (the pleural cavities), and the lungs themselves. Recall, however, that the process of breathing requires muscular action. You learned the details of the structure, location, and actions of the respiratory muscles in chapter 12.

This is a good point in time to go back to exercise 12.9 on p. 229 and review the actions of the respiratory musculature, paying particular attention to the location and actions of the diaphragm.

Structures of the upper respiratory tract are best seen through gross observation of a sagittal section of the head and neck. This view allows you to see the gross structures within the nasal cavity that warm and clean the air as it is brought into the airways.

EXERCISE 24.5 Sagittal Section of the Head and Neck

1. Obtain a classroom model of a sagittal section of the head and neck (or a cadaveric specimen that has been sectioned along the sagittal plane). **Figure 24.6** shows respiratory system structures that are visible in a sagittal section of the head and neck.

2. Using your textbook as a guide, identify the structures listed in **figure 24.7** on the classroom model or cadaveric specimen. Then label figure 24.7.

3. *Optional Activity*: **AP|R** **Respiratory System**—Watch the "Respiratory System Overview" to review the structures of the upper and lower respiratory tracts.

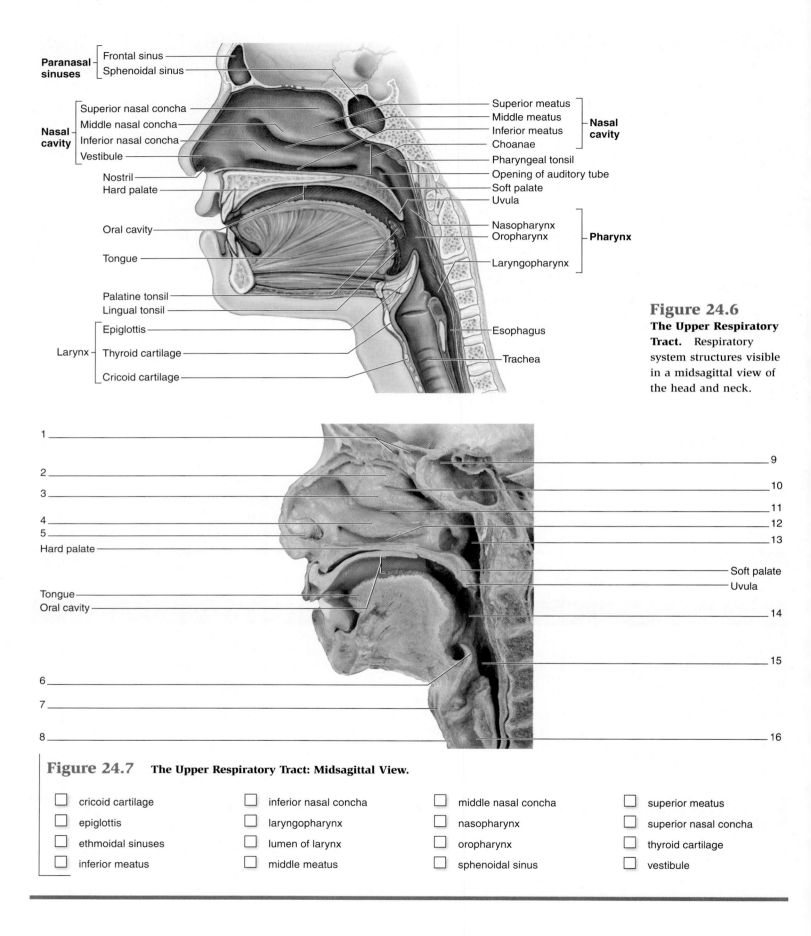

Paranasal sinuses
Frontal sinus
Sphenoidal sinus

Nasal cavity
Superior nasal concha
Middle nasal concha
Inferior nasal concha
Vestibule

Nostril
Hard palate

Oral cavity

Tongue

Palatine tonsil
Lingual tonsil

Larynx
Epiglottis
Thyroid cartilage
Cricoid cartilage

Superior meatus
Middle meatus
Inferior meatus
Choanae **Nasal cavity**
Pharyngeal tonsil
Opening of auditory tube
Soft palate
Uvula

Nasopharynx
Oropharynx **Pharynx**
Laryngopharynx

Esophagus

Trachea

Figure 24.6
The Upper Respiratory Tract. Respiratory system structures visible in a midsagittal view of the head and neck.

1
2
3
4
5
Hard palate

Tongue
Oral cavity

6
7
8

9
10
11
12
13

Soft palate
Uvula

14

15

16

Figure 24.7   **The Upper Respiratory Tract: Midsagittal View.**

☐ cricoid cartilage   ☐ inferior nasal concha   ☐ middle nasal concha   ☐ superior meatus

☐ epiglottis   ☐ laryngopharynx   ☐ nasopharynx   ☐ superior nasal concha

☐ ethmoidal sinuses   ☐ lumen of larynx   ☐ oropharynx   ☐ thyroid cartilage

☐ inferior meatus   ☐ middle meatus   ☐ sphenoidal sinus   ☐ vestibule

Lower Respiratory Tract

The lower respiratory tract is composed of structures of both the conducting portion (larynx, trachea, bronchi, and bronchioles) and the respiratory portion (respiratory bronchioles on down through alveoli). The **larynx** (*larynx*, organ of voice production) is much more than just an opening into the lower respiratory tract. It also houses the **vocal folds**, referred to as the "true vocal cords," which are responsible for phonation (sound production), and intrinsic muscles that control the length and tension of the vocal ligaments.

The structural framework of the larynx is composed of several paired and unpaired cartilages, which act as attachment sites for the intrinsic musculature. The muscles of the larynx are all innervated by branches of the vagus nerve (CN X). Thus, lesions of the vagus nerve result in problems with phonation, such as hoarseness. **Table 24.6** describes the cartilaginous structures of the larynx, and **table 24.7** describes the noncartilaginous structures of the larynx.

Table 24.6	Cartilaginous Structures of the Larynx		
Cartilage	**Description**		**Word Origins**
Arytenoid	Small paired pyramid-shaped cartilages found on the superior, posterior aspect of the cricoid cartilage. The vocal ligaments (true vocal cords) attach to them.		*arytania*, a ladle, + *eidos*, resemblance
Corniculate	Small paired cartilages found superior to the arytenoid cartilages. The vestibular folds ("false vocal cords") attach to them.		*cornicatus*, horned
Cricoid	The second largest laryngeal cartilage, this ring-shaped cartilage serves as an attachment point for muscles.		*krikos*, a ring, + *eidos*, resemblance
Cuneiform	Small paired cartilages found within the aryepiglottic fold.		*cuneus*, a wedge, + *forma*, form
Epiglottis	A plate of elastic cartilage at the superior aspect of the larynx that closes over the opening to the larynx during swallowing to prevent substances from entering the larynx. It is covered by stratified squamous epithelium on its superior aspect and by respiratory epithelium on its inferior aspect.		*epi-*, above, + *glottis*, the mouth of the windpipe
Thyroid	The largest of the laryngeal cartilages, located superior to the isthmus of the thyroid gland. The vocal ligaments (true vocal cords) attach to it.		*thyreos*, an oblong shield, + *eidos*, resemblance

Table 24.7	Noncartilaginous Structures of the Larynx		
Structure	**Description and Function**		**Word Origins**
Glottis	Consists of the rima glottis plus the vocal folds.		*glottis*, the mouth of the windpipe
Rima Glottidis	The space between the true vocal cords, also known as the true glottis.		*rima*, a slit, + *glottis*, the mouth of the windpipe
Vestibular Folds	The "false vocal cords," which are the vestibular ligaments plus the folds of mucous membrane that lie over them.		*vestibulum*, a small cavity at the entrance of a canal
Vestibular Ligaments	Ligaments that stretch between the angle of the thyroid cartilage to the corniculate cartilages.		*vestibulum*, a small cavity at the entrance of a canal, + *ligamentum*, a bandage
Vocal Folds ("True Vocal Cords")	The vocal ligaments plus the mucosa overlying them. Form the "true vocal cords." Involved directly in voice production. Alterations in tension of the cords affect the pitch of the sound.		*vocalis*, pertaining to the voice, + *chorda*, cord
Vocal Ligaments	Ligaments that stretch between the thyroid and arytenoid cartilages.		*vocalis*, pertaining to the voice, + *ligamentum*, a bandage

EXERCISE 24.6 The Larynx

1. Obtain a classroom model of the larynx (**figure 24.8**).

2. Using tables 24.6 and 24.7 and your textbook as guides, identify the structures listed in figure 24.8 on the model of the larynx. Then label figure 24.8.

3. *Optional Activity*: **AP|R** **Respiratory System**—Take the "Lower Respiratory Tract" test in the Quiz section to review the larynx and other lower respiratory structures.

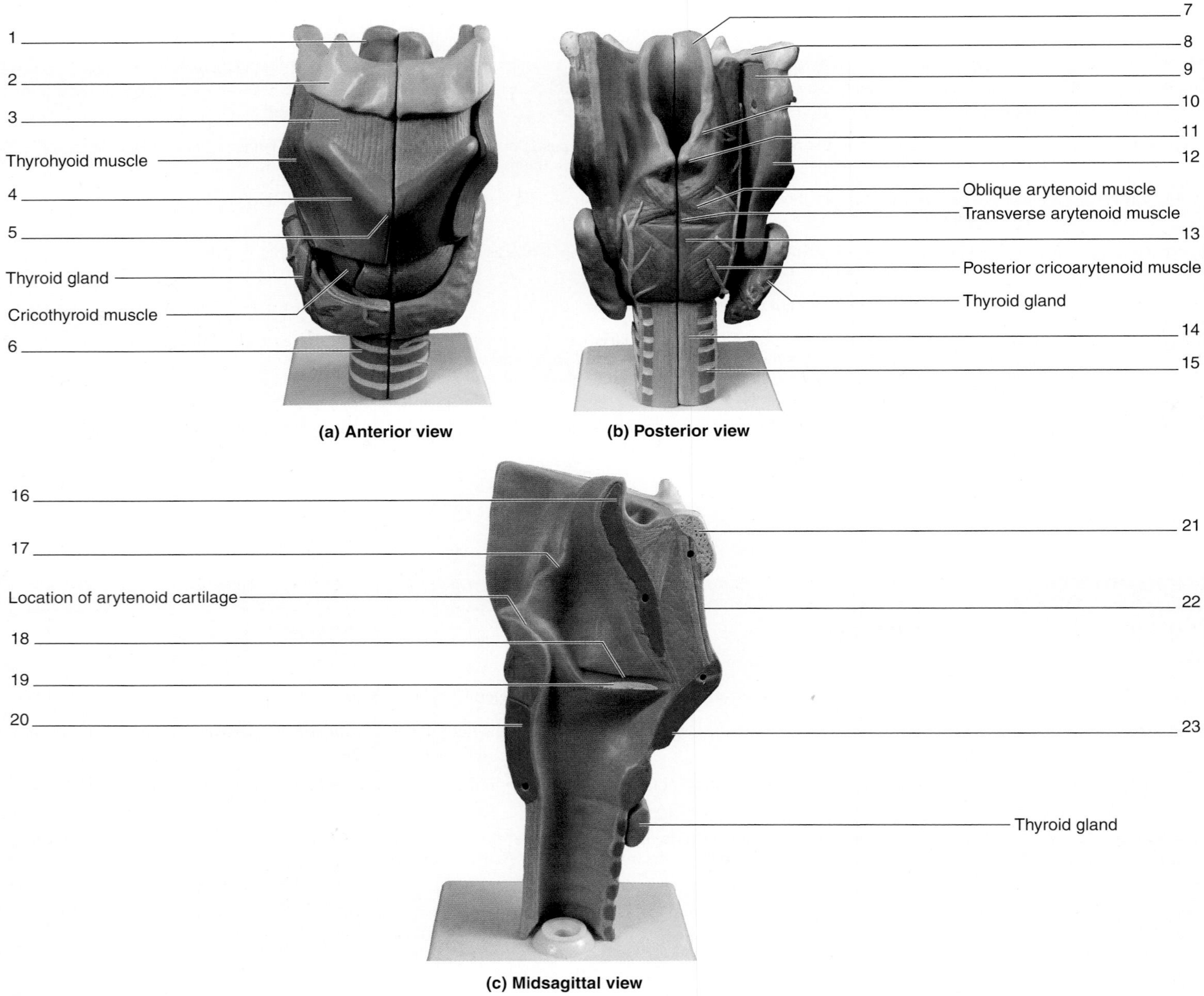

1 _____
2 _____
3 _____
Thyrohyoid muscle _____
4 _____
5 _____
Thyroid gland _____
Cricothyroid muscle _____
6 _____

(a) Anterior view

7
8
9
10
11
12
Oblique arytenoid muscle
Transverse arytenoid muscle
13
Posterior cricoarytenoid muscle
Thyroid gland
14
15

(b) Posterior view

16 _____
17 _____
Location of arytenoid cartilage _____
18 _____
19 _____
20 _____

21
22
23
Thyroid gland

(c) Midsagittal view

Figure 24.8   Classroom Model of the Larynx.

☐ arytenoid cartilage   ☐ cuneiform cartilage   ☐ laryngeal prominence   ☐ trachea   ☐ vestibular ligament

☐ corniculate cartilage   ☐ epiglottis   ☐ thyrohyoid membrane   ☐ tracheal C ring ☐ vocal ligament

☐ cricoid cartilage   ☐ hyoid bone   ☐ thyroid cartilage   ☐ trachealis muscle

The Pleural Cavities and the Lungs

Within the thoracic cavity, the lungs are located within separate **pleural cavities** (*pleura*, a rib). The space within the thoracic cavity between the two pleural cavities is the **mediastinum** (*medius*, middle). Because the mediastinum is simply the space between the two pleural cavities, if one of the lungs and its surrounding pleural cavity collapse, the mediastinum will shift toward the side of the collapsed lung. The pleural cavities form in much the same way as the pericardial cavity. Each pleural cavity is composed of

mesothelium (simple squamous epithelium), which is referred to as the **pleura**. The pleura are composed of two layers: visceral and parietal (**figure 24.9**). The **visceral pleura** is the inner layer and is tightly adhered to, and inseparable from, the outer surface of the lung. The **parietal pleura** is the outer layer and forms the wall of the pleural cavity. On a human cadaver the parietal pleura can often be seen as a shiny tissue attached to the innermost part of the rib cage or the superior surface of the diaphragm.

EXERCISE 24.7 The Pleural Cavities

1. Observe the thoracic cavity of a human cadaver or a model of the thorax (figure 24.9).

2. Using your textbook as a guide, identify the structures listed in figure 24.9 on the cadaver or model of the thorax. Then label figure 24.9.

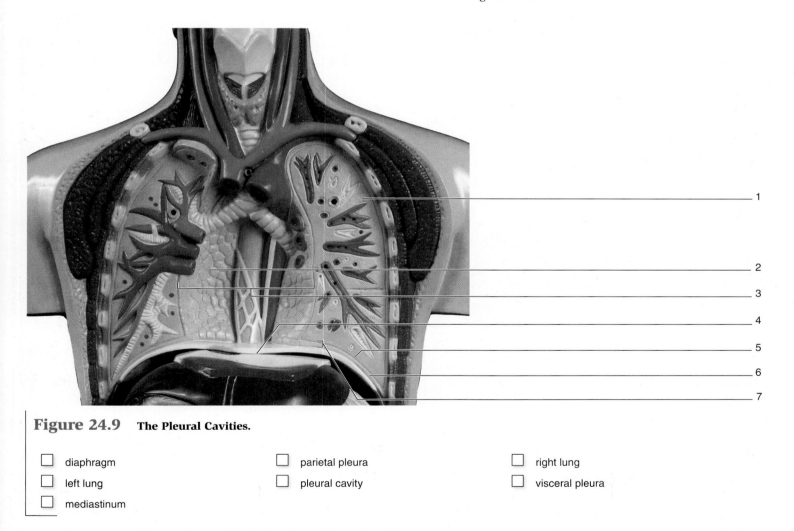

Figure 24.9 **The Pleural Cavities.**

☐ diaphragm   ☐ parietal pleura   ☐ right lung

☐ left lung   ☐ pleural cavity   ☐ visceral pleura

☐ mediastinum

EXERCISE 24.8 The Lungs

In this exercise you will compare and contrast the right and left lungs, and will observe the branching pattern of the respiratory tree. Although it is easy to distinguish the right and left lungs from each other based on the number of lobes (two for the left,

three for the right), locating the structures that enter the hilum of the lung (pulmonary arteries, pulmonary veins, and bronchi) is more difficult. There are patterns for recognizing these structures that you will learn in this laboratory exercise. In addition, you will observe several **impressions** made in the lungs by the structures that surround them. These impressions are visible in preserved human cadaver lungs and on classroom models of the lungs, but may not be visible in fresh lungs. This is because the process of fixing the lungs with preservative also fixes the impressions of adjacent organs. The impressions are not found in fresh lungs because they have not been fixed with preservative.

EXERCISE 24.8A: The Right Lung

1. Observe the lungs of a human cadaver, a fresh or preserved sheep pluck (a *pluck* contains the heart, lungs, and trachea), or a classroom model of the lungs.

2. Begin by observing the right lung (**figure 24.10**). How many lobes does the right lung have? _____

3. Turn the lung so the hilum is visible (medial view; figure 24.10*b*). The hilum of the right lung contains branches of the pulmonary arteries and veins, bronchi, and small bronchial arteries and veins, which represent

the systemic circulation to the lungs. Which of the vessels (pulmonary arteries or pulmonary veins) do you think will have thicker walls? Explain your answer:

_____

_____

_____

4. In general, the pulmonary arteries are located on the superior aspect of the hilum of the right lung, the pulmonary veins are located on the inferior and anterior aspect of the hilum of the right lung, and the bronchi are located on the superior and posterior aspect of the hilum of the right lung. If you are viewing cadaveric lungs, these structures will be more difficult to differentiate from each other because they are not color-coded, so you will need to rely on the texture of the vessels and their locations for identification.

5. Using your textbook as a guide, identify the structures listed in figure 24.10 on the lung. Then label figure 24.10.

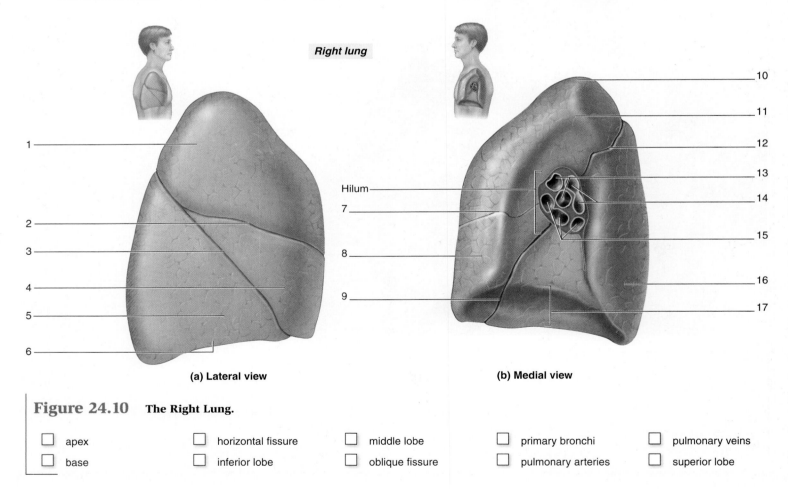

(a) Lateral view   (b) Medial view

Figure 24.10   **The Right Lung.**

| ☐ apex | ☐ horizontal fissure | ☐ middle lobe | ☐ primary bronchi | ☐ pulmonary veins |
| ☐ base | ☐ inferior lobe | ☐ oblique fissure | ☐ pulmonary arteries | ☐ superior lobe |

EXERCISE 24.8B: The Left Lung

1. Observe the lungs of a human cadaver, a fresh or preserved sheep pluck, or a classroom model of the lungs.

2. Observe the left lung (**figure 24.11**). How many lobes does the left lung have? _____ Is the left lung larger or smaller than the right lung? _____ Why do you think this is the case? _____
_____

3. Turn the lung so the hilum is visible (medial view; figure 24.11*b*). The hilum of the left lung contains branches of the pulmonary arteries and veins, bronchi, and small bronchial arteries and veins, which represent the systemic circulation to the lungs.

4. In general, the pulmonary arteries are located on the superior aspect of the hilum of the left lung, the pulmonary veins are located on the inferior and anterior aspect of the

hilum of the left lung, and the bronchi are located on the superior and posterior aspect of the hilum of the left lung. If you are viewing cadaveric lungs, these structures will be more difficult to differentiate from each other because they are not color-coded, so you will need to rely on the texture of the vessels and their locations for identification. The medial view of the left lung allows you to see several impressions in the lung that are made by adjacent structures in the living human. The most prominent of these impressions is the **cardiac impression** made by the heart.

5. How will you distinguish the left lung from the right lung?
_____
_____
_____

6. Using your textbook as a guide, identify the structures listed in figure 24.11 on the lung. Then label figure 24.11.

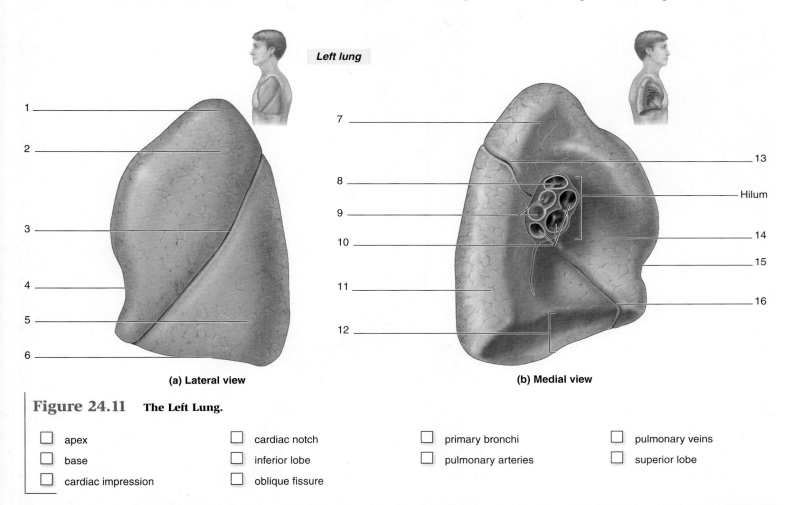

(a) Lateral view   (b) Medial view

Figure 24.11   **The Left Lung.**

☐ apex   ☐ cardiac notch   ☐ primary bronchi   ☐ pulmonary veins
☐ base   ☐ inferior lobe   ☐ pulmonary arteries   ☐ superior lobe
☐ cardiac impression   ☐ oblique fissure

1. Observe the lungs of a human cadaver, a fresh or preserved sheep pluck, or a classroom model of the lungs.

2. The lungs are segmental by nature and are subdivided, from larger to smaller units, into **lobes, bronchopulmonary segments,** and **lobules.** The branching pattern of the bronchial tree follows the segmentation of the lungs (**figure 24.12**). For example, secondary bronchi lead into lobes of the lung so there are three secondary bronchi leading into the right lung and its three lobes, and two secondary bronchi leading into the left lung and its two lobes. **Table 24.8** describes the levels of the bronchial tree that serve each segment of the lungs. This segmental nature of the lungs and bronchial tree makes it relatively easy to remove a segment of the lung that contains a tumor (for example) without interfering with the other parts of the lung. The **carina** is the *internal* ridge between the most inferior tracheal cartilage and the start of the primary bronchi. In the laboratory you will identify its location externally, where the primary bronchi originate from the trachea. However, it is important to realize that the actual structure is internal. It is an important landmark for physicians performing bronchoscopy.

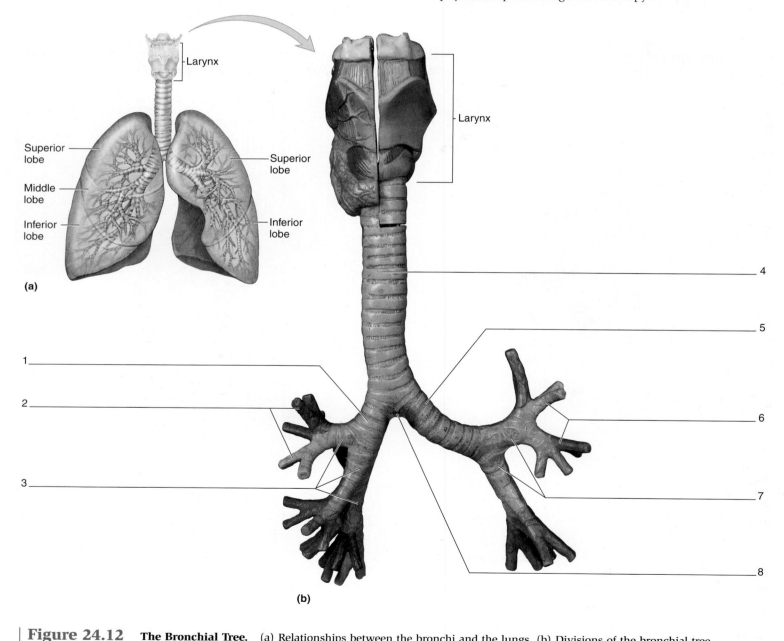

(a)

(b)

Figure 24.12   **The Bronchial Tree.**   (a) Relationships between the bronchi and the lungs. (b) Divisions of the bronchial tree.

| ☐ carina | ☐ left secondary bronchi | ☐ right primary bronchus | ☐ right tertiary bronchi |
| ☐ left primary bronchus | ☐ left tertiary bronchi | ☐ right secondary bronchi | ☐ trachea |

Table 24.8	The Bronchial Tree
Portion of Bronchial Tree	**Description**
Primary Bronchi	One to each lung. The right is more vertical than the left, so foreign objects are more likely to lodge in it.
Secondary (Lobar) Bronchi	One to each lobe (2 on the left, 3 on the right).
Tertiary (Lobular) Bronchi	One to each bronchopulmonary segment.
Quaternary Bronchi	One to each lobule.

3. Using table 24.8 and your textbook as guides, identify the structures listed in figure 24.12 on the cadaver lungs, sheep pluck, or classroom model of the lungs. Then label figure 24.12.

4. In the space below, sketch the bronchial tree. Be sure to label all of the structures listed in figure 24.12 in your drawing.

Chapter 24: The Respiratory System

POST-LABORATORY WORKSHEET

1. Fully classify the epithelium that lines the nasal cavity.

   _____

2. A fracture of the ethmoid bone, as might occur in an auto accident involving severe facial injuries, can cause an individual to lose his or her sense

   of _____. This is due to damage to the _____ nerves, which travel through the _____

   _____ of the ethmoid bone en route to the brain.

3. Which structures are located more superior in the larynx?

   a. the vestibular folds

   b. the vocal folds

4. Define *rima glottidis*: _____

   _____

5. Label the structures in the figure below. Then list, in order, the three structures through which a molecule of carbon dioxide must travel in order to

   diffuse from the blood into the alveoli.

   a. _____

   b. _____

   c. _____

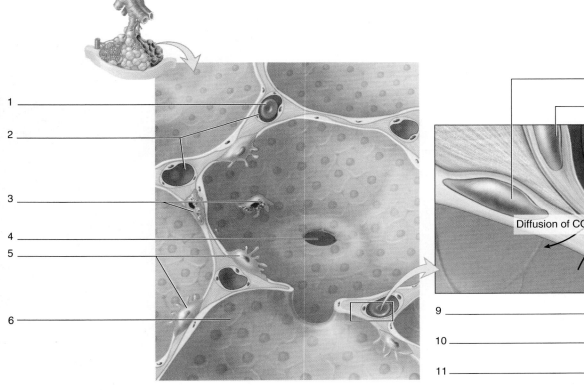

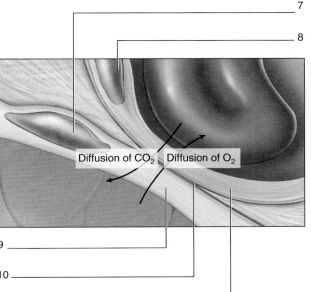

Diffusion of CO_2 Diffusion of O_2

6. How many secondary bronchi lead into the right lung? _____

7. A toddler coughs when she is attempting to swallow a bite of a hot dog, and the hot dog is directed into the respiratory tree instead of the esophagus. The piece of hot dog is most likely to become lodged in the airways leading to the _____ lung because the primary bronchus to this lung is more vertically oriented than the primary bronchus to the other lung. (Note: in 2010 the FDA listed hot dogs as a major choking hazard for children. They even suggested that hot dogs should have warning labels on them due to the large number of children who choke on them).

8. Hyaline membrane disease of the newborn, a disease in which the infant experiences great difficulty breathing, is characterized by a lack of pulmonary surfactant production. Which cells within the lung are responsible for producing surfactant? _____

9. In the space below trace the pathway a molecule of oxygen must take to travel from the nasal cavity to an alveolus in the inferior lobe of the right lung. Be sure to name all the conducting and respiratory portion structures the molecule will pass through along the way.

_____

_____

_____

_____

_____

_____

_____

10. The following questions refer to the micrograph shown below.

   a. What type of airway is indicated in this micrograph? (The star is in the lumen of the airway in question.)

   _____

   b. In the spaces below, explain your reasoning for your answer to part (a) of this question. That is, explain the histological features of the airway (epithelium, cartilage, etc.) that you used to determine your answer.

The Digestive System

25

OUTLINE and OBJECTIVES

Anatomy & Physiology REVEALED®
aprevealed.com

Module 12: DIGESTIVE SYSTEM

INTRODUCTION

The digestive (*digero*, to force apart, dissolve) system is concerned with the breakdown of food into usable molecules needed by the body for energy, maintenance, and ultimately survival. The digestive tract consists of a long tube (the gut tube), which is open to the environment at both ends (the oral cavity and the anus), which means that the lining of the digestive tract is open to the external environment. Thus, the interface between the lumen of the gut tube and the internal environment of the body presents a special problem. The digestive tract must transport needed substances from the lumen of the gut tube into the body, while at the same time it must prevent pathogens from entering. For this reason, the subepithelial tissues of the entire digestive tract are densely populated with lymphatic tissues.

In the previous chapter you were introduced to the layering pattern of the walls of the conducting tubes of the respiratory system. The walls of the gut tube have a similar layering pattern. In this laboratory exercise you will explore how the wall layers of the various parts of the digestive tract are modified to suit the particular needs of the organ. In addition, you will explore the structure and function of organs such as the liver and pancreas—highly specialized glands that develop from the lining of the gut tube. As you observe the structures of the digestive system, you will also review the various circulatory and lymphatic structures associated with these structures.

Chapter 25: The Digestive System

Name: _____

Date: _____ Section: _____

PRE-LABORATORY WORKSHEET

1. How many sets of salivary glands does a human have? _____

2. What are the three parts of the small intestine?

    a. _____

    b. _____

    c. _____

3. How many lobes is the liver composed of? _____

4. What is the function of the gallbladder? _____

5. Which portion of the pharynx does the oral cavity open into? _____

6. List the three unpaired branches of the abdominal aorta that deliver oxygenated blood to the organs of the digestive system.

    a. _____

    b. _____

    c. _____

7. What part of the small intestine do the pancreatic ducts empty into? _____

8. How many layers of smooth muscle are in the wall of the stomach? _____

9. A large pouch that represents the first part of the large intestine is the _____.

10. The epithelium that lines the small and large intestines is classified as _____

    _____.

IN THE LABORATORY

In this laboratory session you will explore both the histological structure and the gross structure of the digestive system. The materials in the 'Histology' and the 'Gross Anatomy' sections are organized so the order of structures studied begins at the mouth and ends at the anus. In other words, if you study the structures in the order in which they are described in this chapter, you will be encountering digestive tract structures in the same order that a bolus of food encounters them as it moves through the digestive tract. It is not imperative that you observe the structures in this order while you are in the laboratory. However, it is a good idea to view them in this order when you are reviewing the material upon completion of the laboratory exercises. This will allow you to reflect on how the different parts of the digestive tract work together to accomplish the goals of digestion.

Histology

Salivary Glands

The **salivary glands (figure 25.1)** are accessory digestive glands composed of modified epithelial tissue that produce **saliva**, a watery secretion that helps dissolve foodstuffs and contains an enzyme, **salivary amylase**, that begins the initial digestion of carbohydrates. There are three pairs of salivary glands: parotid, submandibular, and sublingual, and all of them empty their secretions into the oral cavity through ducts. Salivary glands contain two cell types: serous cells and mucous cells. **Serous cells** produce watery secretions containing proteins, electrolytes, and the enzymes salivary amylase and lysozyme, and **mucous cells** produce mucus, which lubricates the food as it passes into the esophagus. Definitions of structures related to salivary glands are described in **table 25.1**, and details of the structure and function of each of the salivary glands are listed in **table 25.2**.

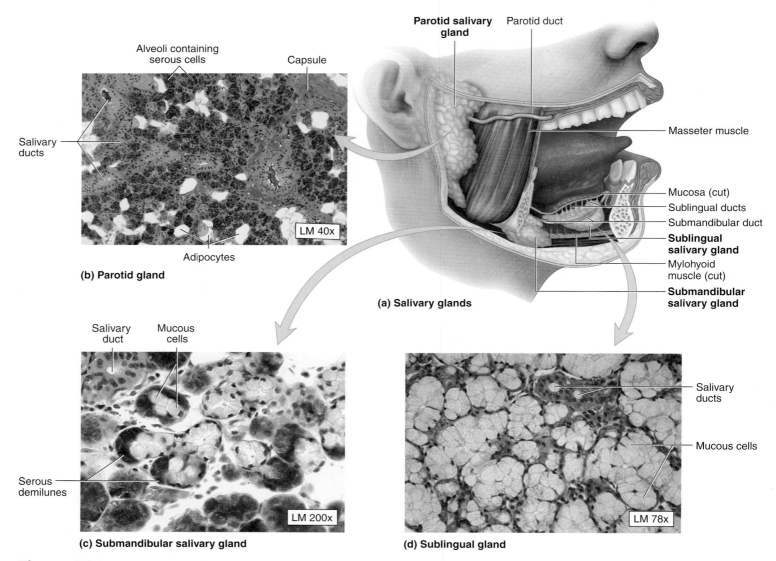

Figure 25.1   Salivary Glands. (a) Location of the salivary glands. Histology of the (b) parotid, (c) submandibular, and (d) sublingual glands.

Table 25.1	Salivary Gland Structures		
Structure	**Description and Location**	**Functions**	**Word Origins**
Alveolus	The grape-shaped secretory portion of a gland.	NA	*acinus*, grape
Serous Demilunes	Crescent- or moon-shaped groups of serous cells located at the periphery of a mucous alveolus.	Secrete proteins, electrolytes, and salivary amylase.	*serous*, having a watery consistency, + *demilune* half-moon
Serous Cells	Cells have round nuclei and contain numerous secretory granules. Located in the alveoli of the parotid gland and in the demilunes of the submandibular and sublingual glands.	Secrete proteins, electrolytes, and salivary amylase.	*serous*, having a watery consistency
Mucous Cells	Cells have flattened nuclei that are located on the basal surface. Mainly located along the tubules of salivary glands.	Secrete mucus.	*mucosus*, mucous
Myoepithelial Cells	Cells are not visible in light microscopy. Located around the alveoli and the long axes of the ducts.	Contraction of these cells expels the secretions from salivary glands.	*mys*, muscle, + *epithelial*, relating to epithelial tissues

Table 25.2	Histological Characteristics of Salivary Glands				
Gland	**Secretory Cells**	**Type of Secretion**	**% of Saliva**	**Opening**	**Word Origins**
Parotid	All **serous**—serous cells occupy about ~90% of the gland's volume (the rest is fat)	Mostly water; 25% of dissolved solutes are glycoproteins; high amylase content.	25%	Empties via the parotid duct opposite the second upper molar.	*para*, beside, + *ous*, ear
Submandibular	Mixed: 80% serous, 20% mucus, contains a few serous demilunes	40–60% of dissolved solutes are glycoproteins; low amylase content; contains lysozyme, an enzyme that inhibits the growth of bacteria.	70%	Empties via the submandibular ducts between the lingual frenulum and the mandible.	*sub-*, under, + *mandible*, the mandible
Sublingual	Mixed: 60–70% serous, 30–40% mucus; serous cells are located in serous demilunes	90% of solutes are glycoproteins (the most viscous secretion); low amylase content.	5%	Empties via multiple ducts into either the submandibular duct or directly into the oral cavity.	*sub-*, under, + *lingual*, the tongue

EXERCISE 25.1 Histology of the Salivary Glands

1. Obtain a compound microscope and histology slides of the parotid, submandibular, and sublingual salivary glands.

2. Place the slide of the **parotid gland** on the microscope stage. Bring the tissue sample into focus on low power and then switch to high power.

3. Using figure 25.1*a* and tables 25.1 and 25.2 as guides, identify the following structures on the slide of the parotid gland:

   ☐ adipocytes   ☐ serous cells

4. Place the slide of the **submandibular gland** on the microscope stage. Bring the tissue sample into focus on low power and then switch to high power.

5. In the submandibular and sublingual salivary glands, the serous cells are located surrounding the mucous cells and are shaped like half moons. Thus, they are referred to as

serous demilunes (*demi*, half, + *luna*, moon). Scan the slide to locate serous demilunes (figure 25.1*b* and 25.1*c*).

6. Using figure 25.1*c* and tables 25.1 and 25.2 as guides, identify the following structures on the slide of the submandibular gland:

   ☐ adipocytes   ☐ serous demilunes
   ☐ mucous cells

7. Place the slide of the **sublingual gland** on the microscope stage. Bring the tissue sample into focus on low power and then switch to high power.

8. The sublingual gland is similar to the submandibular gland in that it contains serous demilunes. There are very few adipocytes in the gland, so that is one characteristic that can be used to distinguish it from the submandibular gland. It also contains more mucous cells (table 25.2 and figure 25.1*d*).

9. Using figure 25.1c and tables 25.1 and 25.2 as guides, identify the following structures on the slide of the sublingual gland:

☐ mucous cells   ☐ serous demilunes

In the following spaces, draw brief sketches of the histological appearance of the parotid, submandibular, and sublingual salivary glands. Label mucous cells, serous cells, and serous demilunes in your drawings.

Parotid gland

Submandibular gland

_____ ×

Sublingual gland

_____ ×

_____ ×

The Stomach

The **stomach** is an organ that digests food by mixing it up mechanically, through the action of smooth muscle in its walls, and dissolves food molecules chemically, through the action of stomach enzymes and a highly acidic environment within. In this section you will first observe the wall layering pattern of the stomach and then you will observe regional characteristics of the epithelium lining the stomach wall.

EXERCISE 25.2 Wall Layers of the Stomach

The walls of the digestive tract, much like the walls of the respiratory tract, are composed of four layers, the **mucosa, submucosa, muscularis**, and **adventitia/serosa**. **Figure 25.2** demonstrates the general wall layering pattern of organs of the digestive tract, and **figure 25.3** demonstrates how these wall layers are modified in the stomach. **Table 25.3** describes the types of tissues that are located in each of the layers of the stomach wall.

1. Obtain a histology slide of the stomach (figure 25.3) and place it on the microscope stage. Bring the tissue sample into focus on low power.

2. Using figure 25.3 and table 25.3 as guides, identify the following layers of the stomach wall on the slide:

☐ blood vessels

☐ epithelium

☐ inner oblique muscle layer

☐ lamina propria

☐ middle circular muscle layer

☐ mucosa

☐ muscularis

☐ muscularis mucosa

☐ nerves

☐ outer longitudinal muscle layer

☐ serosa

☐ submucosa

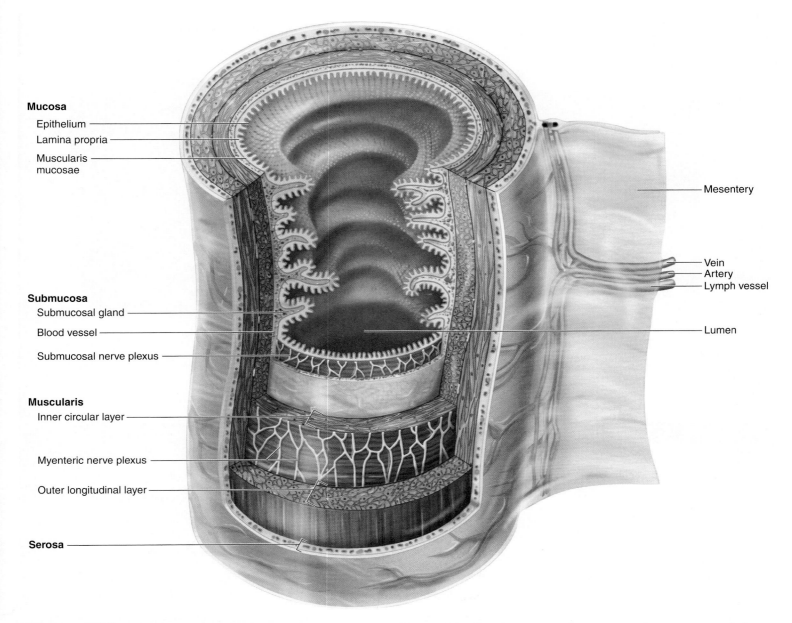

Mucosa
Epithelium
Lamina propria
Muscularis
mucosae

Submucosa
Submucosal gland
Blood vessel
Submucosal nerve plexus

Muscularis
Inner circular layer

Myenteric nerve plexus

Outer longitudinal layer

Serosa

Mesentery

Vein
Artery
Lymph vessel

Lumen

Figure 25.2   **Wall Layers of the Digestive Tract.**   The four wall layers of the digestive tract are the mucosa, submucosa, muscularis, and adventitia/serosa.

3. In the space to the right, draw a brief sketch of the histology of the stomach wall as seen through the microscope. Be sure to label the structures listed in step 2 in your drawing.

×

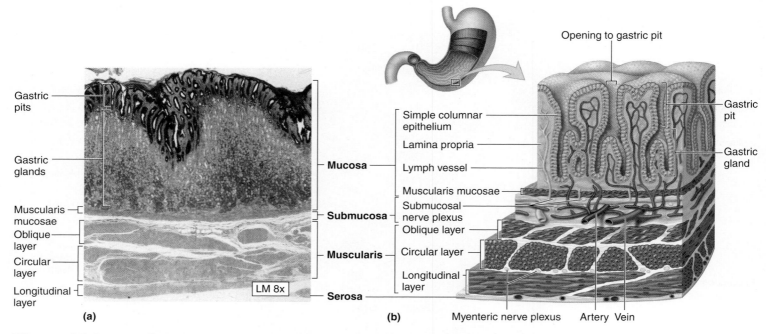

Figure 25.3   The Stomach Wall.   (a) Histology of the stomach wall. (b) Layers of the stomach wall.

Table 25.3	Wall Layers of the Stomach	
Layer	**Sublayer**	**Characteristics**
Mucosa	Epithelium	Simple columnar epithelium containing five distinct cell types (see table 25.5).
	Lamina propria	Highly cellular, contains reticular connective tissue below and in between glands. Free lymphocytes and lymphatic nodules are common.
	Muscularis mucosae	Composed of thin layer of smooth muscle.
Submucosa	Connective tissue	Composed of coarse bundles of collagen fibers, many elastic fibers, and adipose tissue.
	Vessels	Contains both blood and lymphatic vessels.
	Nerves	Submucosal nerve plexus.
Muscularis	Inner oblique	Contains muscle fibers responsible for creating the "twisting" action of the stomach.
	Middle circular	Forms the thickest layer of the three layers of muscle. Thickenings of this layer form the inferior esophageal (cardiac) and pyloric sphincters.
	Outer longitudinal	Found in the greater and lesser curvatures only.
	Nerves	Myenteric nerve plexus is located between the layers of smooth muscle.
Serosa	NA	Thin connective tissue and mesothelium.

EXERCISE 25.3 Histology of the Stomach

The epithelium of the stomach is modified to form **gastric pits**, invaginations of the epithelium that contain the openings of **gastric glands (figure 25.4)**. **Epithelial stem cells** are located at the junction between the gastric pits and the gastric glands. These cells continuously renew the epithelial lining of the stomach, the epithelium lining the gastric pits, and the epithelial cells that form the gastric glands. The epithelial cells are replaced approximately every three days. Stem cells that will line the gastric pits and the lumen of the stomach migrate *up* from the pit/gland junction, whereas stem cells that form the gastric glands migrate *down* from the pit/gland junction. The cell types in the gastric pits and glands vary in different regions of the stomach. **Table 25.4** describes the structure of the pits and glands in each of the major regions of the stomach (cardia, fundus/body, and pylorus), **table 25.5** describes the five cell types that compose the gastric pits and glands, and the type and function of the secretions produced by each cell.

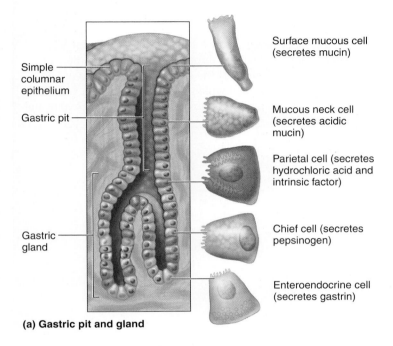

Simple columnar epithelium

Gastric pit

Gastric gland

Surface mucous cell (secretes mucin)

Mucous neck cell (secretes acidic mucin)

Parietal cell (secretes hydrochloric acid and intrinsic factor)

Chief cell (secretes pepsinogen)

Enteroendocrine cell (secretes gastrin)

(a) Gastric pit and gland

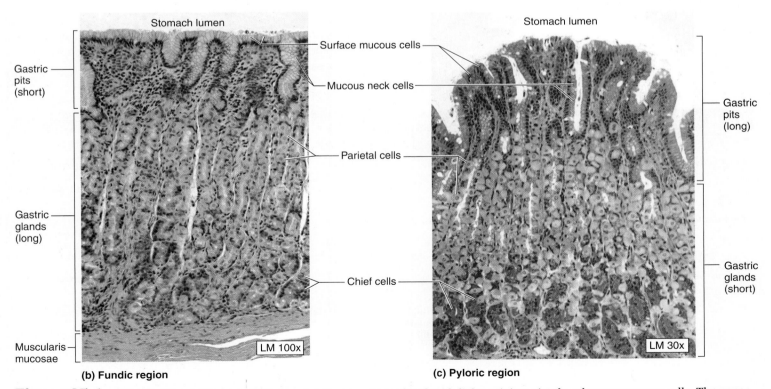

Stomach lumen

Stomach lumen

Gastric pits (short)

Gastric glands (long)

Muscularis mucosae

Surface mucous cells

Mucous neck cells

Parietal cells

Chief cells

Gastric pits (long)

Gastric glands (short)

LM 100x

LM 30x

(b) Fundic region

(c) Pyloric region

Figure 25.4   Gastric Pits and Gastric Glands. (a) Gastric pits are lined with light-staining, simple columnar mucous cells. The upper regions of the gastric pits contain large, light-staining parietal cells that resemble fried eggs, while the lower regions of the gastric pits contain smaller, dark-staining chief cells. (b) The fundic region of the stomach is characterized by short pits and long glands. (c) The pyloric region of the stomach is characterized by long pits and short glands.

Table 25.4	Regional Characteristics of the Gastric Pits and Glands of the Stomach		
Region	**Pit Structure**	**Main Cell Type(s)**	**Gland Structure**
Cardia	Short pits with long glands	All mucous	Simple branched tubular glands
Fundus/Body	Short pits with long glands	Parietal and chief cells with mucous neck cells	Branched, tubular glands
Pylorus	Long pits with short, coiled glands	Mostly mucous	Branched, tubular glands

1. Obtain a histology slide of the stomach and place it on the microscope stage. Bring the tissue sample into focus on low power and then scan the slide to locate the epithelial lining of the stomach (figure 25.4). Once you have the epithelium of the stomach at the center of the field of view, change to high power and bring the tissue sample into focus once again. How would you classify the epithelium that lines the stomach? _____

   _____

2. Once you have identified the epithelium lining the stomach wall, locate a gastric pit and gland and move the microscope stage so it is in the center of the field of view.

3. Identify the following structures on the slide:

   ☐ chief cells   ☐ mucous neck cells
   ☐ gastric glands   ☐ parietal cells
   ☐ gastric pits   ☐ surface mucous cells

4. In the space below, draw a brief sketch of the gastric glands and pits as seen through the microscope. Be sure to label the locations of the specialized cell types in your drawing. Can you tell which part of the stomach wall you are looking at based on the structure of the gastric pits and glands? _____

_____ ✕

5. *Optional Activity*: **AP|R** **Digestive System**—Watch the "Stomach" animation for a review of the stomach wall layers and their histology.

Table 25.5	Cell Types in the Gastric Pits and Glands of the Stomach		
Cells	**Location**	**Secretions**	**Action of Secretion**
Chief Cells (Zymogenic)	Lower 1/2 to 1/3 of the gastric glands. Contain numerous eosinophilic (red) granules.	Pepsinogen	A zymogen* that is converted to pepsin when it encounters the acidic environment of the stomach. Pepsin is a protease (it breaks down proteins).
Enteroendocrine Cells	Scattered throughout the gastric glands.	Gastrin	Hormone that stimulates chief cells and parietal cells to secrete their products, and stimulates the smooth muscle in the stomach walls to contract.
Mucous Neck Cells	Lining the interior of the gastric pits.	Mucin	Glycoprotein that protects the mucosa from HCl.
Parietal Cells	Upper 2/3 of the gastric glands in the fundus and body of the stomach. A few are found in the pylorus. None are found in the cardia.	HCl (hydrochloric acid) Intrinsic factor	Decreases the pH of the stomach to about 2 (very acidic). Necessary for vitamin B_{12} absorption in the small intestine.
Surface Mucous Cells	Covering the ridges between the gastric pits.	Mucin	Glycoprotein that protects the mucosa from HCl.

*Zymogen is a general term for an inactive protein. Generally, the names of zymogens end in -ogen, as in *pepsinogen*.

The Small Intestine

After initial digestion in the mouth and stomach, the mixture of digested food and gastric juices is collectively called **chyme.** When this chyme leaves the stomach, it enters the small intestine, where digestion of foodstuffs is completed and nutrients in the food are transported into the circulatory system. The walls of the small intestine contain several modifications that hugely increase the surface area of the small intestine for absorption: circular folds, villi, and microvilli. The mucosal and submucosal layers of the small intestine are thrown into folds called **circular folds** (plicae circularis). On top of the circular folds, there are folds of mucosa called **villi** (s., *villus*). Each villus is lined with a simple columnar epithelium containing **microvilli** and goblet cells. Goblet cells in the small intestine produce mucus, which helps protect the epithelial lining and lubricates the passage of chyme through the small intestine. **Figure 25.5** demonstrates histology slides of the three parts of the small intestine (duodenum, jejunum, and ileum), and **table 25.6** summarizes the distinguishing histological features of each part.

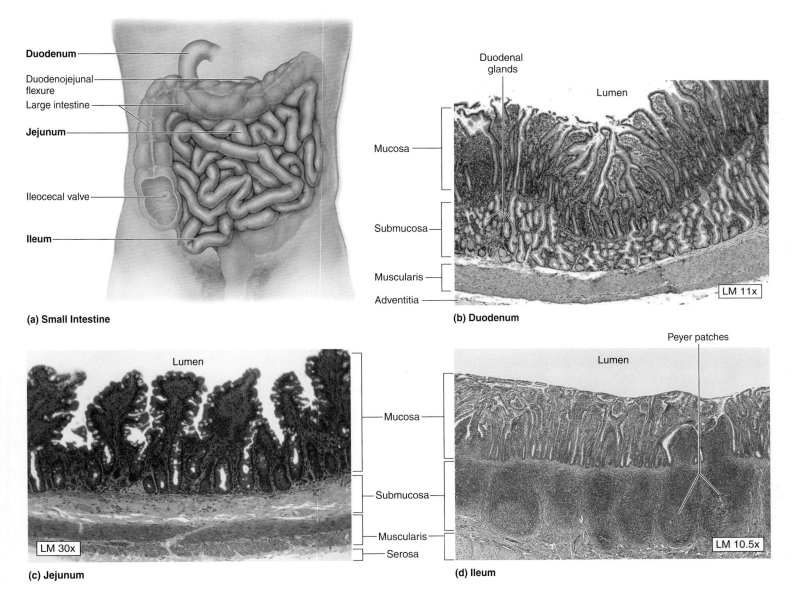

Figure 25.5   The Small Intestine.   (a) Parts of the small intestine. (b) Histology of the duodenum, demonstrating duodenal glands in the submucosa. (c) Histology of the jejunum, which lacks duodenal glands and Peyer patches. (d) Histology of the ileum, demonstrating Peyer patches in the submucosa.

Table 25.6	Regional Differences in the Small Intestine	
Region	**Distinguishing Histological Characteristics**	**Word Origins**
Duodenum	Contains duodenal glands, which produce an alkaline secretion that protects the duodenum from stomach acids.	*duodeno-*, breadth of twelve fingers
Jejunum	Identified by prominent circular folds and a lack of duodenal glands and Peyer patches.	*jejunus*, empty
Ileum	Peyer patches are scattered throughout the submucosa. Goblet cells increase in number closer to the the iliocecal valve.	*eileo*, to roll up, twist

EXERCISE 25.4 Histology of the Small Intestine

1. Obtain a histology slide of the small intestine and place it on the microscope stage. If your laboratory is equipped with slides of each section of the small intestine (duodenum, jejunum, and ileum), be sure to view all three. If not, use figure 25.5 and table 25.6 to decide which part of the small intestine is on the slide.

2. Bring the tissue sample into focus at low power and move the microscope stage until the epithelium is at the center of the field of view. Then change to high power. What type of

   epithelium lines the small intestine? _____

   _____

   What surface modifications are present?_____

   _____

   What is the purpose of these surface modifications? _____

   _____

3. Using figure 25.5 and table 25.6 as guides, identify the following structures on the slide(s):

   ☐ blood vessels   ☐ muscularis
   ☐ duodenal glands   ☐ muscularis mucosa
   ☐ epithelium   ☐ Peyer patches
   ☐ goblet cells   ☐ serosa
   ☐ lamina propria   ☐ submucosa
   ☐ microvilli   ☐ villi
   ☐ mucosa

4. In the spaces to the right, draw brief sketches of the histology of the duodenum, jejunum, and ileum of the small intestine (if you only looked at one slide, just draw that one). Identify the histological features that will allow you to differentiate between these three portions of the small intestine when viewing histology slides of the small intestine (refer to table 25.6 for reference).

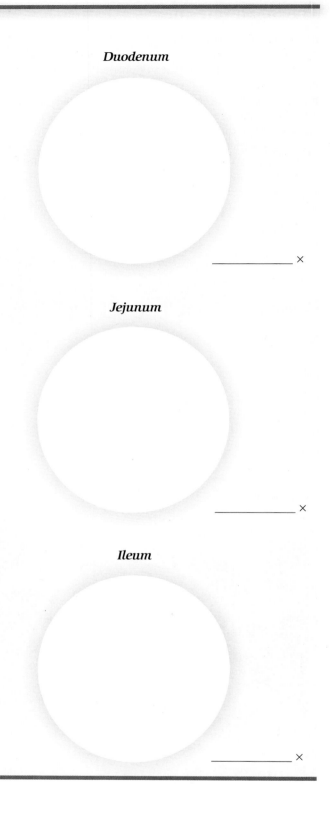

Duodenum

_____ ×

Jejunum

_____ ×

Ileum

_____ ×

The Large Intestine

Chyme leaving the small intestine enters the **large intestine** at the iliocecal junction. A valve, the **iliocecal valve**, controls the passage of chyme from the small intestine to the large intestine. Because the vast majority of nutrients have been absorbed in the small intestine, the function of the large intestine is mainly to absorb water from the chyme that remains and compact the waste products as **feces** for elimination from the body. Because the feces get more solid and compacted in the large intestine, the epithelium of the large intestine contains many goblet cells, which help lubricate the epithelium to ease the passage of feces through the large intestine.

EXERCISE 25.5 Histology of the Large Intestine

1. Obtain a histology slide of the large intestine (colon) and place it on the microscope stage.

2. Bring the tissue sample into focus on low power and identify the epithelial layer (figure 25.6). Move the microscope stage so the epithelium is at the center of the field of view and then change to high power.

3. Using figure 25.6 and your textbook as guides, identify the following structures in the slide of the large intestine:

☐ blood vessels   ☐ muscularis
☐ epithelium   ☐ muscularis mucosae
☐ goblet cells   ☐ nerves
☐ inner circular   ☐ outer longitudinal
   muscle layer   muscle layer
☐ lamina propria   ☐ serosa
☐ mucosa   ☐ submucosa

4. In the space below, draw a brief sketch of the histology of the large intestine as seen through the microscope. Label the four major layers of the wall of the large intestine in your drawing.

_____ ✕

WHAT DO YOU THINK?

1 What special cell type is prominent in the epithelium of the large intestine? Why do you think this cell type is particularly abundant in the large intestine?

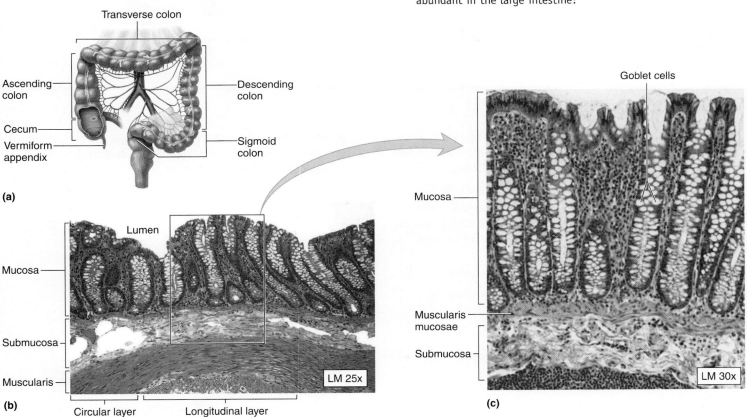

(a)

Transverse colon

Ascending colon

Descending colon

Cecum

Sigmoid colon

Vermiform appendix

(b)

Lumen

Mucosa

Submucosa

Muscularis

Circular layer   Longitudinal layer

Goblet cells

Mucosa

Muscularis mucosae

Submucosa

LM 25x

LM 30x

(c)

Figure 25.6   The Large Intestine. (a) Parts of the large intestine. (b) Histology of the wall of the large intestine. (c) Close-up of the epithelium lining the large intestine; numerous goblet cells are present.

The Liver

The **liver** is the largest accessory organ in the digestive system. Indeed, it is one of the largest organs in the human body and performs numerous vital functions that include producing bile, detoxifying the blood, storing nutrients, and producing plasma proteins. The structural and functional unit of the liver is a **hepatic lobule**, which is a hexagonally shaped structure consisting of strands of **hepatocytes** (the strands of hepatocytes are **hepatic cords**) with a **central vein** in the middle (**figure 25.7**). In the areas where the outer edges of the hepatic lobules come together there are **portal triads**, which consist of a branch of the hepatic artery, a branch of the hepatic portal vein, and a branch of a bile duct. Within the hepatic lobules, in between the hepatic cords, are **hepatic sinusoids**: capillaries that carry blood from the branches of the hepatic artery and hepatic portal vein to the central veins. Along the sinusoids are several macrophage-like cells, **reticuloendothelial** (Kupffer) cells. These cells engulf bacteria that enter the liver from the portal circulation.

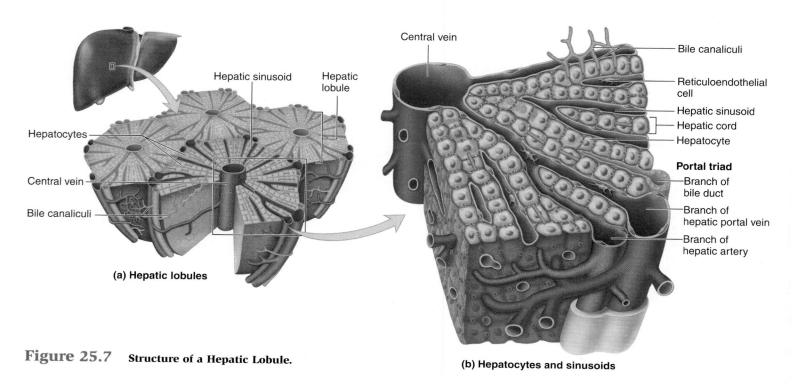

(a) Hepatic lobules

(b) Hepatocytes and sinusoids

Figure 25.7   **Structure of a Hepatic Lobule.**

EXERCISE 25.6 | Histology of the Liver

1. Obtain a histology slide of the liver and place it on the microscope stage. Bring the tissue sample into focus on low power and locate a hepatic lobule (**figure 25.8a**).

2. Move the microscope stage until the hepatic lobule is at the center of the field of view, and then change to high power (figure 25.8b). Using figures 25.7 and 25.8 as guides, identify the following structures on the slide of the liver:

   ☐ branch of bile duct
   ☐ branch of hepatic artery
   ☐ branch of hepatic portal vein
   ☐ central vein
   ☐ hepatic cords
   ☐ hepatic lobule
   ☐ hepatic sinusoids
   ☐ hepatocyte
   ☐ portal triad
   ☐ reticuloendothelial cells

3. In the space below, draw a brief sketch of the liver as seen with the microscope. Be sure to label all of the structures listed in step 2 in your drawing.

_____ ×

4. *Optional Activity:* 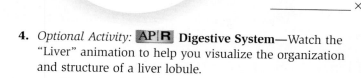 **Digestive System**—Watch the "Liver" animation to help you visualize the organization and structure of a liver lobule.

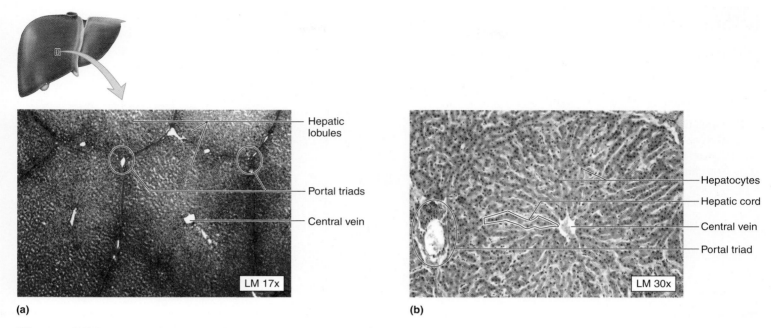

Figure 25.8 **The Liver.** (a) Low-magnification histology slide demonstrating multiple hepatic lobules with portal triads in the spaces between lobules. (b) High-magnification histology slide demonstrating a central vein, hepatocytes arranged into hepatic cords, and a portal triad, which contains a branch of the hepatic portal vein, a branch of the hepatic artery, and a bile duct.

The Pancreas

The **pancreas** is the second-largest accessory organ in the digestive system after the liver. It is both an endocrine and an exocrine gland. The histology of the endocrine part of the pancreas (the pancreatic islets) was covered in chapter 19 (The Endocrine System). The **exocrine** portion of the pancreas consists of grapelike bunches of cells called **acini** (s., *acinus*), which are similar in many ways to the cells that compose the salivary glands. These **acinar cells** produce many substances important for digestion, including pancreatic amylase and bicarbonate. The products of acinar cells are collectively referred to as **pancreatic juice**. Pancreatic juice is transported from the acinar cells into small ducts that become larger ducts and eventually dump the secretions into the duodenum via the **main pancreatic duct** (see **figure 25.9**).

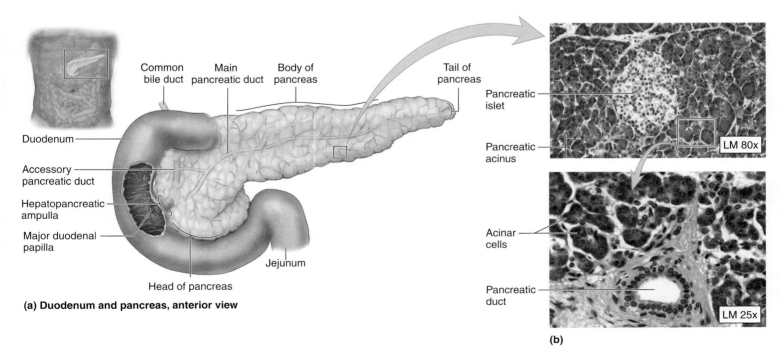

Figure 25.9 **The Pancreas.** (a) Location of the pancreas. (b) Histology of the pancreas demonstrating acinar cells, which secrete digestive enzymes and pancreatic islets, which contain the hormone-secreting cells of the pancreas.

Histology of the Pancreas

1. Obtain a histology slide of the **pancreas** and place it on the microscope stage. Bring the tissue sample into focus on low power. Scan the slide until you locate a pancreatic islet surrounded by acinar cells (figure 25.9a).

2. Move the microscope stage until the acinar cells are in the center of the field of view, and then change to high power. Using figure 25.9 as a guide, identify the following structures on the slide:

   ☐ acinar cells   ☐ pancreatic duct
   ☐ pancreatic acinus   ☐ pancreatic islet

3. In the space below, draw a brief sketch of the pancreas as seen through the microscope. Be sure to label all of the structures listed in step 2 in your drawing.

_____ ✕

Gross Anatomy

The Oral Cavity, Pharynx, and Esophagus

The oral cavity contains a number of digestive system structures, including the teeth, salivary glands, lips, and tongue. These structures are important for wetting and manipulating food as it enters the digestive tract. Food leaves the oral cavity, travels through the **oropharynx** (*oris*, mouth, + *pharynx*, the throat), and enters the **esophagus** (*oisophagos*, gullet), which transports the food to the stomach. The esophagus enters the stomach just after it pierces through the **esophageal hiatus** in the diaphragm.

Gross Anatomy of the Oral Cavity, Pharynx, and Esophagus

1. Observe a cadaver specimen or a classroom model demonstrating the head and neck.

2. Using your textbook as a guide, identify the structures listed in **figure 25.10** on the cadaver or classroom model. Then label them in figure 25.10.

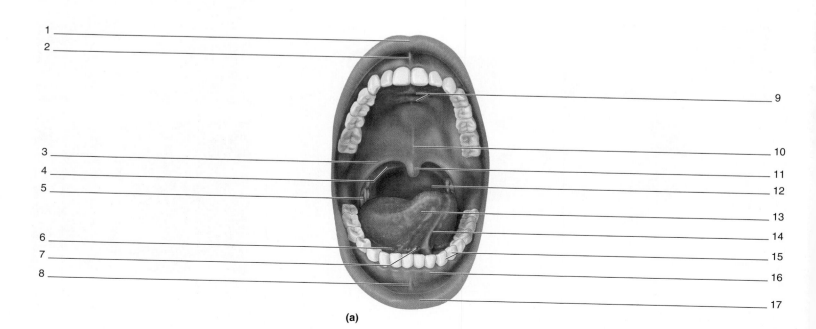

(a)

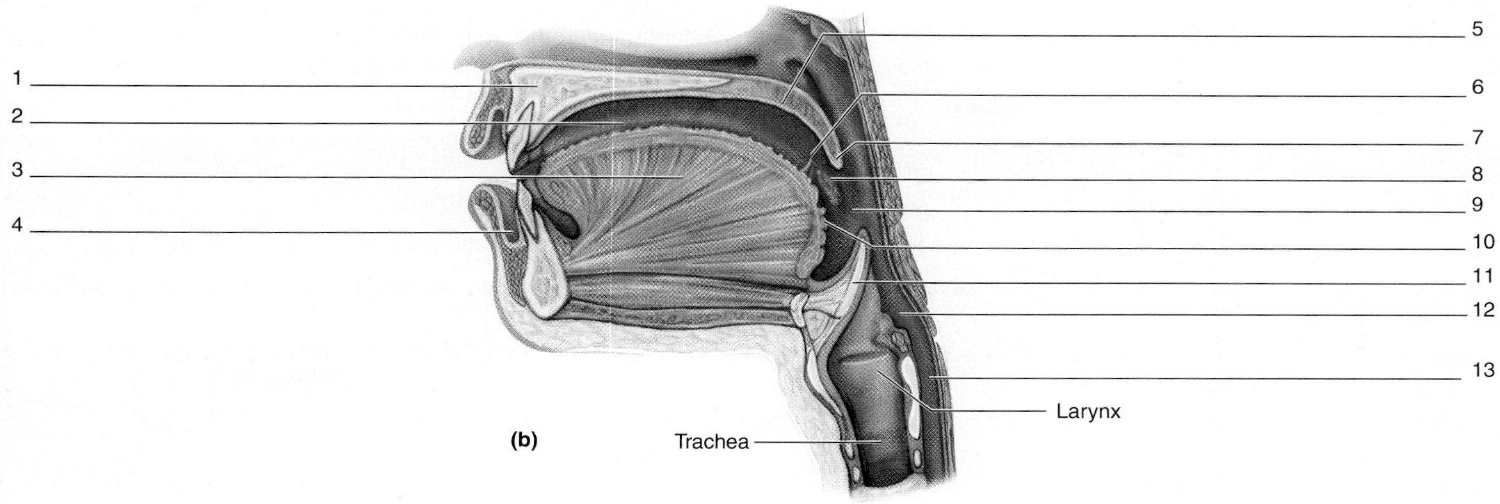
1
2
3
4
5
6
7
8
9
10
11
12
13
Larynx
Trachea
(b)

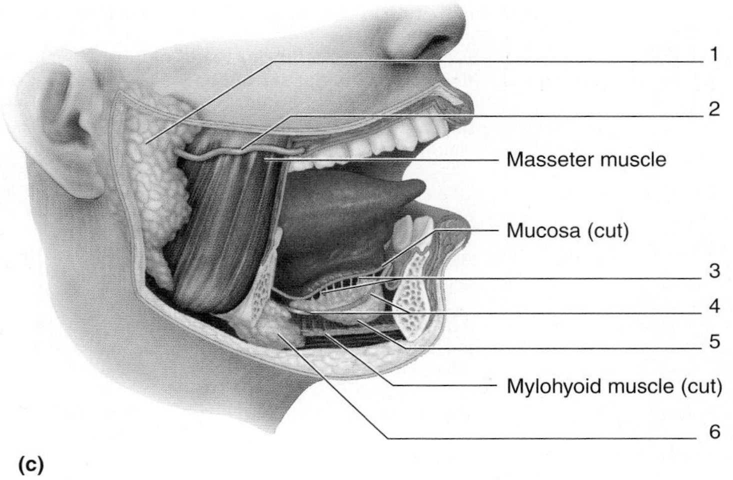
1
2
Masseter muscle
Mucosa (cut)
3
4
5
Mylohyoid muscle (cut)
6
(c)

Figure 25.10 Upper Digestive Tract.

(a) Anterior View of the Oral Cavity

- ☐ fauces
- ☐ gingivae
- ☐ glossopalatine arch
- ☐ hard palate/transverse palatine folds
- ☐ inferior labial frenulum
- ☐ inferior lip
- ☐ lingual frenulum
- ☐ palatine tonsil
- ☐ pharyngopalatine arch
- ☐ soft palate
- ☐ sublingual duct orifice
- ☐ submandibular duct orifice
- ☐ superior labial frenulum
- ☐ superior lip
- ☐ teeth
- ☐ tongue
- ☐ uvula

(b) Midsagittal View of the Oral Cavity and Pharynx

- ☐ epiglottis
- ☐ esophagus
- ☐ hard palate
- ☐ laryngopharynx
- ☐ lingual tonsil
- ☐ oral cavity
- ☐ oropharynx
- ☐ palatine tonsil
- ☐ palatoglossal arch
- ☐ soft palate
- ☐ tongue
- ☐ uvula
- ☐ vestibule

(c) Salivary Glands

- ☐ parotid duct
- ☐ parotid salivary gland
- ☐ sublingual ducts
- ☐ sublingual salivary gland
- ☐ submandibular duct
- ☐ submandibular salivary gland

The Stomach

The stomach is a large, sac-like organ that both stores and breaks down ingested foodstuffs. The stomach is located in the epigastric abdominopelvic region, superficial to the pancreas, and deep to the anterior abdominal wall. **Table 25.7** lists the major features of the stomach and describes their functions.

Table 25.7	Gross Anatomical Features of the Stomach	
Structure	**Description**	**Word Origins**
Body	Main part of the stomach located between the fundus and the pylorus.	body, the principal mass of a structure
Cardia	The region of the stomach that the esophagus opens into.	kardia, heart; relating to the part of the stomach nearest the heart
Cardiac Notch	A deep notch located between the fundus of the stomach and the esophagus.	kardia, heart; relating to the part of the stomach nearest the heart, + notch, indentation
Fundus	The dome-shaped part of the stomach that lies superior to the cardiac notch.	fundus, bottom
Greater Curvature	The large, inferior curvature of the stomach. It is one attachment point for the greater omentum.	greater, larger
Greater Omentum	A fold of 4 layers of peritoneum that stretches from the greater curvature of the stomach to the transverse colon.	omentum, the membrane that encloses the bowels
Inferior Esophageal (Cardiac) Sphincter	A physiological sphincter (band of muscle), composed of the part of the diaphragm that surrounds the esophagus. When the diaphragm contracts, it closes off this opening, preventing reflux of stomach contents back into the esophagus. Some circular smooth muscle in the wall of the esophagus also contributes to this sphincter, but its contribution is weak.	kardia, heart; relating to the sphincter of the stomach nearest the heart
Lesser Curvature	The small, superior curvature of the stomach. It is one attachment point of the lesser omentum.	lesser, smaller
Lesser Omentum	A fold of 4 layers of peritoneum that stretches from the liver to the lesser curvature of the stomach.	omentum, the membrane that encloses the bowels
Pyloric Sphincter	An anatomic sphincter (band of muscle), composed of smooth muscle in the wall of the pylorus. It controls passage of chyme from the stomach to the duodenum (and vice versa).	pyloros, a gatekeeper, + sphinkter, a band
Pylorus	The region of the stomach that opens into the duodenum.	pyloros, a gatekeeper
Rugae	Folds of the mucosal lining of the stomach.	ruga, a wrinkle

EXERCISE 25.9 Gross Anatomy of the Stomach

1. Observe a cadaver specimen or a classroom model demonstrating the stomach.

2. Using table 25.7 and your textbook as a guide, identify the structures listed in **figure 25.11** on the cadaver or classroom model. Then label them in figure 25.11.

3. In the space to the right, draw a brief sketch of the stomach. Be sure to label all of the structures listed in figure 25.11 in your drawing.

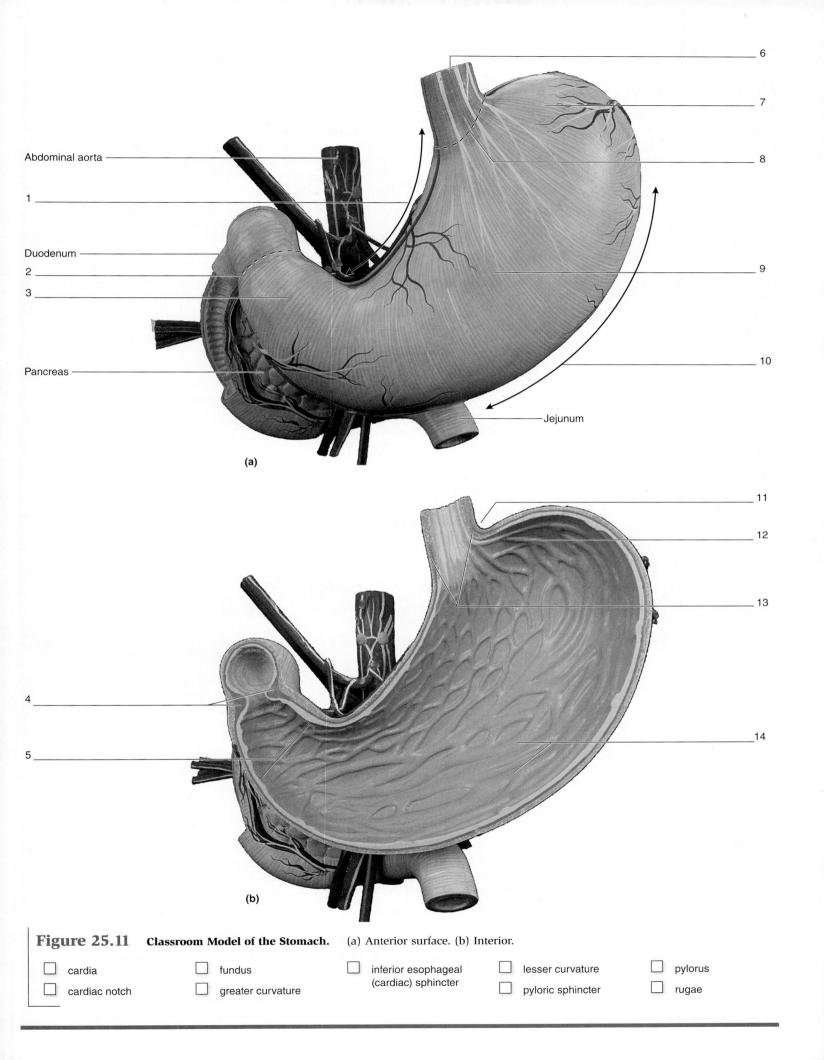

Abdominal aorta

1

Duodenum

2

3

Pancreas

6

7

8

9

10

Jejunum

(a)

11

12

13

4

5

14

(b)

Figure 25.11   **Classroom Model of the Stomach.**   (a) Anterior surface. (b) Interior.

☐ cardia

☐ cardiac notch

☐ fundus

☐ greater curvature

☐ inferior esophageal (cardiac) sphincter

☐ lesser curvature

☐ pyloric sphincter

☐ pylorus

☐ rugae

The Duodenum, Gallbladder, Liver, and Pancreas

The **duodenum** (duodeno-, breadth of twelve fingers) is the first part of the small intestine. It is C-shaped, and mostly retroperitoneal, which allows it to be anchored to the posterior abdominal wall. This is advantageous because a number of ducts coming from the gallbladder, pancreas, and liver empty their contents into the duodenum. The relationships between the duodenum and the gallbladder, liver, and pancreas are critically important for the process of digestion. The liver produces **bile**, a substance that emulsifies fats, which is temporarily stored within the **gallbladder**. The pancreas produces **bicarbonate**, which neutralizes stomach acids, and **pancreatic juice**, which contains bicarbonate and enzymes that break down proteins, fats, and carbohydrates. When these organs dump their secretions into the duodenum, the acidity of the chyme (chymos, juice) that has entered the duodenum from the stomach is neutralized and the digestion of proteins and carbohydrates continues. Here the digestion of fats and nucleic acids also begins. In addition to their functions in the breakdown of food, the duodenum, liver, and pancreas all produce hormones that are important in signaling processes of digestion. **Table 25.8** lists the major features of the duodenum, gallbladder, liver, and pancreas, and describes each of their functions.

Table 25.8	Gross Anatomical Structures of the Duodenum, Liver, Gallbladder, and Pancreas	
Structure	**Description**	**Word Origins**
Accessory Pancreatic Duct	Excretory duct located in the head of the pancreas. Empties into duodenum at the minor duodenal papilla.	pankreas, the sweetbread
Body of the Pancreas	The main portion of the pancreas extending between the head and the tail.	pankreas, the sweetbread
Caudate Lobe of the Liver	A small lobe of the liver located between the right and left lobes and on the posterior, inferior part of the liver.	caudate, possessing a tail, + lobos, lobe
Common Bile Duct	The bile duct formed from the union of the common hepatic duct and the cystic duct. Empties into the duodenum at the major duodenal papilla.	bilis, a yellow/green fluid produced by the liver
Common Hepatic Duct	The bile duct formed from the union of the right and left hepatic ducts. Drains bile into the common bile duct.	hepatikos, liver
Cystic Duct	The bile duct that transports bile from the gallbladder to the junction of the common hepatic duct and common bile duct.	cystic, relating to the gallbladder
Falciform Ligament	A fold of peritoneum that extends from the diaphragm and anterior abdominal wall to the liver. Its free inferior border contains the round ligament of the liver.	falx, sickle, + forma, form
Gallbladder	A sac-like appendage of the liver that stores and concentrates bile.	gealla, bile, + blaedre, a distensible organ
Head of the Pancreas	The portion of the pancreas that sits in the depression formed by the curvature of the duodenum.	pankreas, the sweetbread
Hepatopancreatic Ampulla	The space within the major duodenal papilla that contains the common bile duct and the main pancreatic duct.	hepatikos, relating to the liver, + pancreatic, relating to the pancreas, + ampulla, a two-handled bottle
Left Lobe of the Liver	The second largest lobe of the liver. Extends from the falciform ligament toward the midline of the body.	lobos, lobe
Main Pancreatic Duct	The main excretory duct of the pancreas. Runs longitudinally in the center of the gland and empties into the duodenum at the major duodenal papilla.	pankreas, the sweetbread
Major Duodenal Papilla	A raised "nipple-like" bump located on the posterior wall of the descending part of the duodenum. The main pancreatic duct and the common bile duct empty their contents here.	major, great, + papilla, a nipple
Minor Duodenal Papilla	A small raised "nipple-like" bump located superior to the major duodenal papilla. Contains the opening of the accessory pancreatic duct.	minor, smaller, + papilla, a nipple
Porta Hepatis	A depression on the inferomedial part of the liver that contains the hepatic artery, hepatic portal vein, and common bile duct.	porta, gate, + hepatikos, liver
Quadrate Lobe of the Liver	A small lobe of the liver located between the right and left lobes and on the anterior, inferior part of the liver between the gallbladder and the round ligament.	quadratus, square, + lobos, lobe
Right Lobe of the Liver	The largest lobe of the liver, it is on the right side of the abdomen and composes over half of the mass of the liver.	lobos, lobe
Round Ligament of the Liver (Ligamentum Teres)	A remnant of the fetal umbilical vein, which connects to the umbilicus. Located within the free edge of the falciform ligament on the anterior abdominal wall.	ligamentum, a bandage, + teres, round
Tail of the Pancreas	The tapered, right end of the pancreas located near the hilum of the spleen.	pankreas, the sweetbread

EXERCISE 25.10 Gross Anatomy of the Duodenum, Liver, Gallbladder, and Pancreas

1. Obtain a classroom model demonstrating the relationships between the duodenum, liver, gallbladder, and pancreas (**figure 25.12**), or view these structures in the superior abdominal cavity of a prosected human cadaver.

2. Using table 25.8 and your textbook as guides, identify structures listed in figure 25.12 on the classroom model or human cadaver. Then label them in figure 25.12.

3. Using Table 25.8 and your textbook as guides, draw a brief sketch of the ducts coming from the gallbladder, liver, and pancreas in the space to the right. Show how the ducts merge to eventually empty their contents into the pancreas. Label each duct and organ in your drawing.

4. Obtain a classroom model demonstrating the liver (**figure 25.13**), or view the liver from a human cadaver.

5. Using table 25.8 and your textbook as guides, identify structures listed in figure 25.13 on the classroom model or human cadaver. Then label them in figure 25.13.

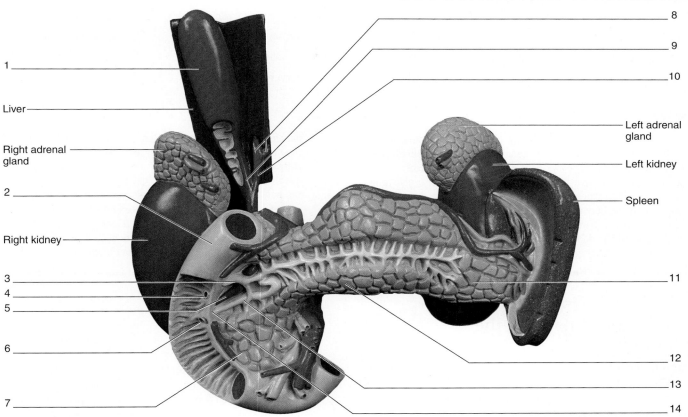

Figure 25.12   **Classroom Model of the Duodenum, Gallbladder, Liver, and Pancreas.**   Anterior view.

- ☐ accessory pancreatic duct
- ☐ body of pancreas
- ☐ common bile duct
- ☐ common hepatic duct
- ☐ cystic duct
- ☐ duodenum
- ☐ gallbladder
- ☐ head of pancreas
- ☐ hepatopancreatic ampulla
- ☐ left and right hepatic ducts
- ☐ main pancreatic duct
- ☐ major duodenal papilla
- ☐ minor duodenal papilla
- ☐ tail of pancreas

Figure 25.13 **Classroom Model of the Liver.** Anteroinferior view. In this view, the anterior surface of the liver has been rotated away from your point of view so you can see the structures on the inferior surface of the liver.

☐ caudate lobe of liver   ☐ falciform ligament   ☐ inferior vena cava   ☐ quadrate lobe of liver

☐ common bile duct   ☐ gallbladder   ☐ left hepatic duct   ☐ right hepatic duct

☐ common hepatic duct   ☐ hepatic artery proper   ☐ left lobe of liver   ☐ right lobe of liver

☐ cystic duct   ☐ hepatic portal vein   ☐ porta hepatis

WHAT DO YOU THINK?

2 Gallstones are hard stones that form from cholesterol and bile deposits in the gallbladder. The condition of having gallstones is called *cholelithiasis* (*chole*, bile, + *lithos*, stone, + *-iasis*, condition). One of the most serious complications of a gallstone occurs when the stone passes into the cystic duct and makes its way toward the duodenum, but gets lodged somewhere en route to the duodenum. This cuts off the flow of bile from the liver and gallbladder to the duodenum. However, if it lodges in a particular location, it can also cut off the flow of pancreatic juice from the pancreas. Normally pancreatic enzymes are not activated until they enter the duodenum. However, when these enzymes build up within the pancreas because of a blockage of flow of pancreatic juice out of the pancreas, they get activated and begin to digest the pancreas itself. This causes *pancreatitis* (inflammation of the pancreas), which can be life-threatening. Using your knowledge of the relationships between ducts draining bile from the liver and gallbladder and the pancreatic ducts, where do you think a gallstone would have to lodge to lead to this condition?

The Jejunum and Ileum of the Small Intestine

The **jejunum** and **ileum** compose the longest part of the small intestine. The two parts can be distinguished from each other both histologically and grossly. Distinguishing the two via gross anatomic features involves comparing four anatomic features: circular folds, encroaching fat, arterial arcades, and vasa recta (**figure 25.14**). **Circular folds** are the mucosal folds found within the lumen of the small intestine. **Encroaching fat** is mesenteric fat that "rides up" upon the wall of the intestine.

Arterial arcades are arching branches of the mesenteric arteries, and **vasa recta** (*vasa*, vessel, + *rectus*, straight) are straight vessels that come off of the arterial arcades and enter the small intestine proper. In general the jejunum is located in the upper left part of the abdominal cavity and the ileum is located in the lower right part of the abdominal cavity. **Table 25.9** summarizes the gross anatomical features that distinguish the jejunum from the ileum.

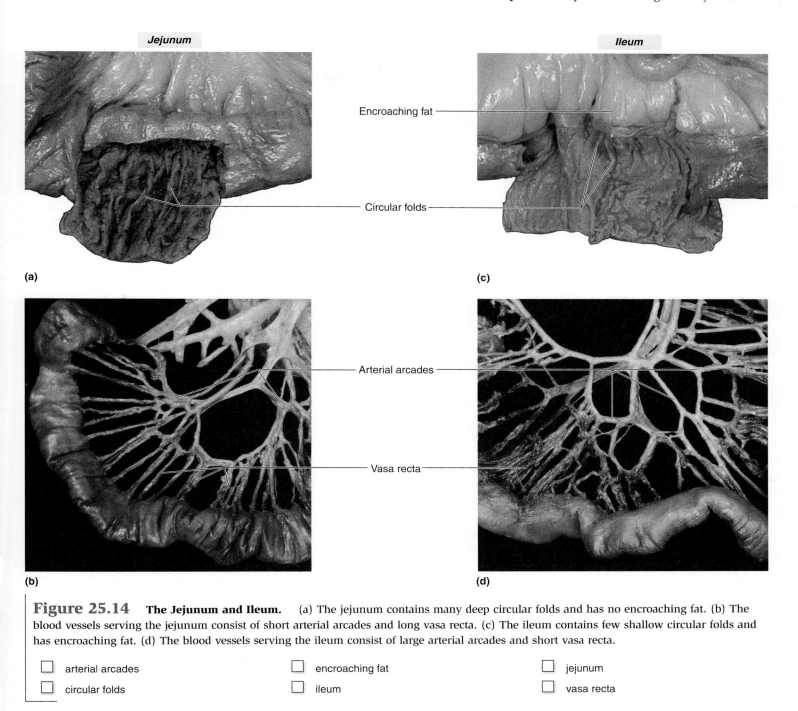

Figure 25.14 **The Jejunum and Ileum.** (a) The jejunum contains many deep circular folds and has no encroaching fat. (b) The blood vessels serving the jejunum consist of short arterial arcades and long vasa recta. (c) The ileum contains few shallow circular folds and has encroaching fat. (d) The blood vessels serving the ileum consist of large arterial arcades and short vasa recta.

☐ arterial arcades   ☐ encroaching fat   ☐ jejunum

☐ circular folds   ☐ ileum   ☐ vasa recta

Table 25.9	Gross Anatomical Differences Between the Jejunum and the Ileum			
Part of the Small Intestine	**Circular Folds**	**Encroaching Fat**	**Arterial Arcades**	**Vasa Recta**
Jejunum	Deep, many	No	Fewer, larger	Longer
Ileum	Shallow, few	Yes	More, smaller, stacked upon each other	Shorter

EXERCISE 25.11 Gross Anatomy of the Jejunum and Ileum of the Small Intestine

1. Observe a classroom model of the abdominal cavity or the abdominal cavity of a prosected human cadaver in which the small intestine is intact.

2. Using table 25.9 and your textbook as guides, identify the gross structures listed in figure 25.14 on the classroom model of the abdomen or in the abdominal cavity of the human cadaver.

The Large Intestine

The large intestine begins as a large sac called the **cecum**, which is located in the right lower abdominopelvic quadrant. Exiting the cecum, the **colon** runs along the borders of the abdominal cavity (ascending, transverse, and descending portions) before entering the pelvic cavity via the sigmoid colon. The sigmoid colon then becomes the **rectum**, which is located within the pelvic cavity proper. The rectum ends at the **anus**. **Table 25.10** summarizes these structures.

Table 25.10	The Cecum, Large Intestine, Rectum, and Anus	
Structure	**Description**	**Word Origins**
Anus	Inferior opening of the digestive tract.	anus, the lower opening of the digestive tract
Ascending Colon	Part of the colon that extends from the cecum to the liver.	kolon, the part of the large intestine from cecum to rectum
Cecum	Blind-ended sac located at the junction between the ileum of the small intestine and the ascending colon.	caecus, blind
Haustra	Pouches of the colon formed when the taenia coli (longitudinal smooth muscle) contract.	haustus, to draw up
Ileocecal Valve	Smooth muscle sphincter located where the ileum opens into the cecum.	ileo-, ileum, + cecal, cecum
Left Colic (Splenic) Flexure	A curve of the colon medial to the spleen, where the transverse colon becomes the descending colon.	splenic, relating to the spleen, + flexura, a bend
Omental (Epiploic) Appendices	Small, fatty appendages that hang off of the colon.	omentum, the membrane that encloses the bowels; epiploic, related to the omentum
Rectum	Final portion of the digestive tract, located within the pelvic cavity and extending from the sigmoid colon to the anus.	rectus, straight
Right Colic (Hepatic) Flexure	Curve of the colon medial to the liver, where the ascending colon becomes the transverse colon.	hepatikos, relating to the liver, + flexura, a bend
Sigmoid Colon	The S-shaped part of the colon that extends from the descending colon to the rectum.	sigma, the letter S, + eidos, resemblance
Taenia Coli	Three small bands of longitudinal smooth muscle of the muscularis externa of the colon; contraction of this muscle creates pouches (haustra) in the colon.	tainia, a band, + coli, colon
Transverse Colon	The part of the colon that extends between the liver and the spleen, and connects to the greater omentum.	transversus, crosswise

EXERCISE 25.12 Gross Anatomy of the Large Intestine

1. Observe a classroom model of the abdominal cavity demonstrating the large intestine or the large intestine of a prosected human cadaver (**figure 25.15**).

2. Using table 25.10 and your textbook as guides, identify the gross structures listed in figure 25.15 on the classroom model of the abdomen or in the abdominal cavity of the human cadaver.

3. *Optional Activity*: **AP|R** **Digestive System**—Review the locations and functions of the major organs of the digestive system by watching the "Digestive System Overview" animation.

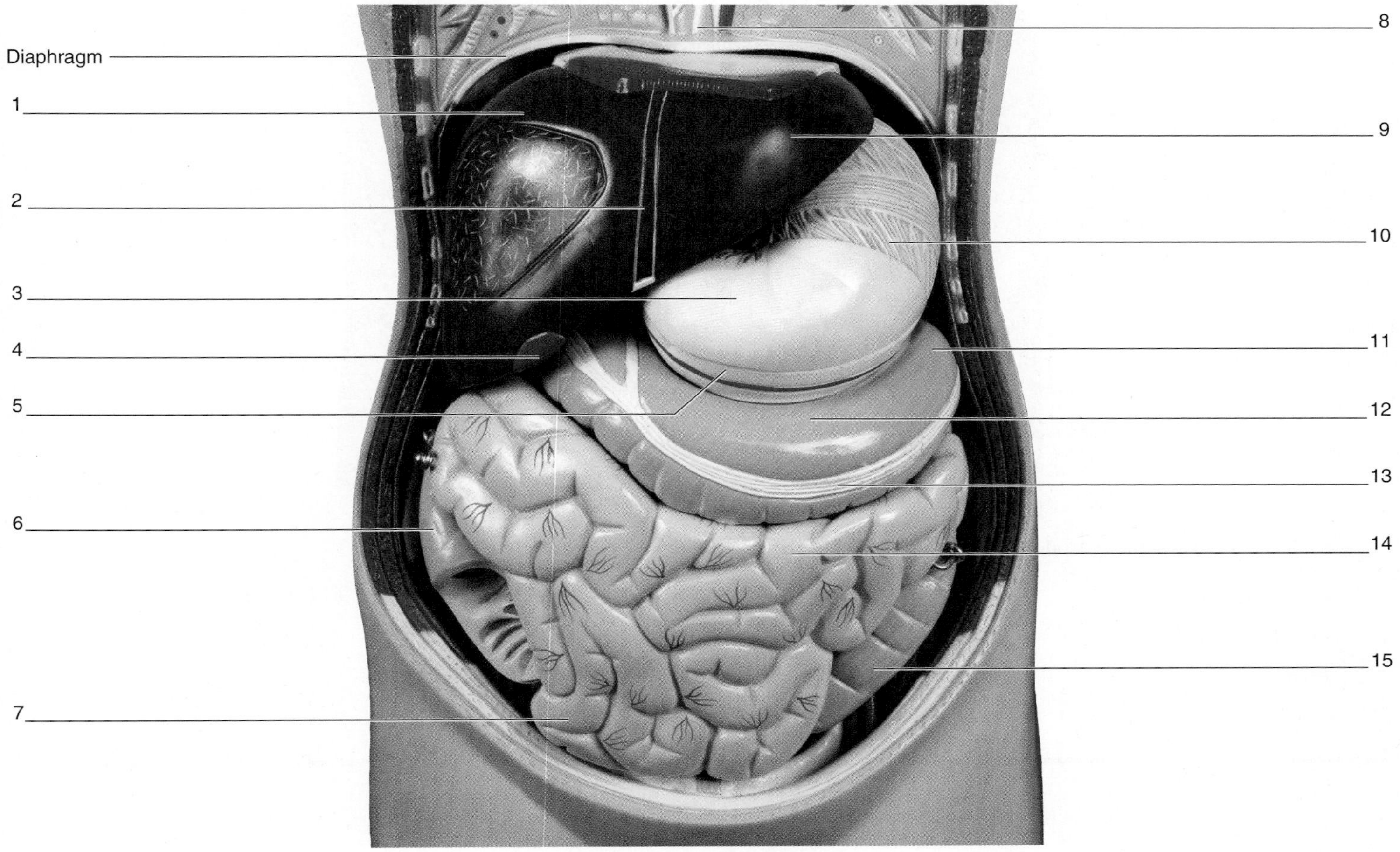

(a)

Figure 25.15 **Classroom Model of the Abdominal Cavity and Large Intestine.** (a) Superficial view. (b) Deep view with the liver, stomach, and small intestine removed.

☐ ascending colon

☐ body of stomach

☐ descending colon

☐ esophagus

☐ falciform ligament

☐ gallbladder

☐ greater curvature of stomach

☐ ileum of small intestine

☐ jejunum of small intestine

☐ left colic (splenic) flexure of colon

☐ left lobe of liver

☐ pylorus of stomach

☐ right lobe of liver

☐ taenia coli

☐ transverse colon

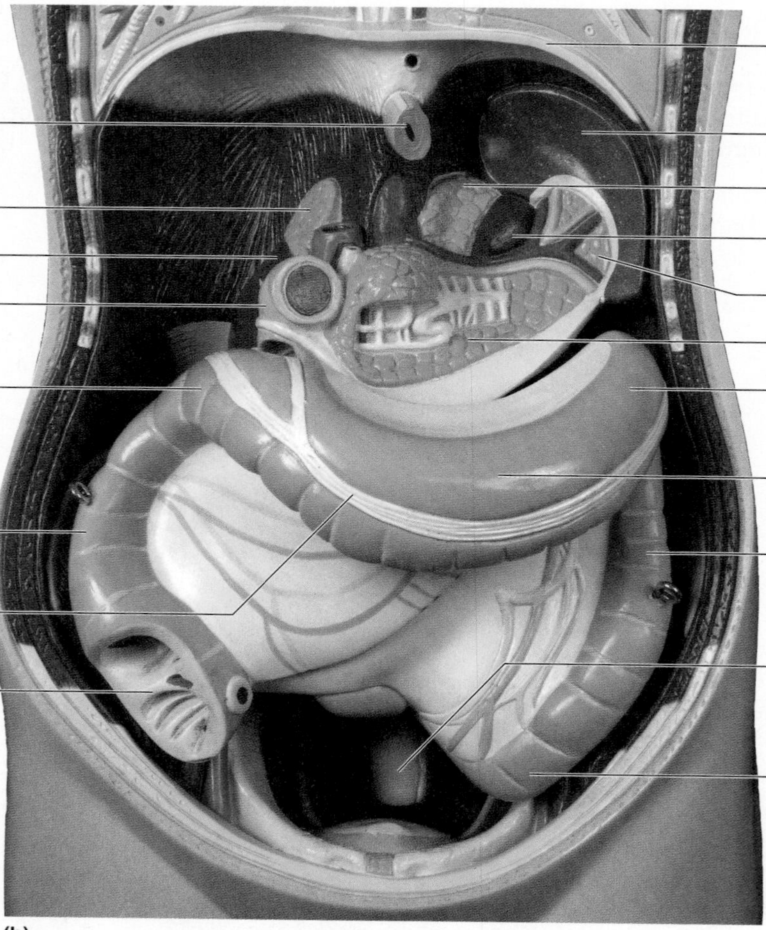

16

Right adrenal gland

Right kidney

17

18

19

20

21

Diaphragm

Spleen

Left adrenal gland

Left kidney

22

23

24

25

26

27

28

(b)

Figure 25.15   **Classroom Model of the Abdominal Cavity and Large Intestine,** *continued.*

- ☐ ascending colon
- ☐ body of pancreas
- ☐ cecum
- ☐ descending colon

- ☐ duodenum
- ☐ esophagus
- ☐ left colic (splenic) flexure of colon

- ☐ rectum
- ☐ right colic (hepatic) flexure of colon
- ☐ sigmoid colon

- ☐ taenia coli
- ☐ tail of pancreas
- ☐ transverse colon

Chapter 25: The Digestive System

POST-LABORATORY WORKSHEET

Name:_____

Date:_____Section:_____

1. Mary likes sour foods, and she decided to eat a slice of lemon. As she bit down on the lemon, she felt an uncomfortable squeezing-type sensation in her cheek as one of her salivary glands emptied its secretions into her mouth. Which salivary gland did she feel?

   _____

2. A patient is suffering from gastroesophageal reflux disease (GERD). When she lies down, she feels a burning sensation in her esophagus caused by the reflux of stomach acids into the esophagus. This occurs because one of the sphincters in her stomach is not working properly. Which sphincter is not

   functioning properly? _____. Is this sphincter considered an anatomic or physiologic sphincter? _____

   _____. What type of muscle composes this sphincter? _____.

3. A student is observing a histology slide of the stomach. He notices in the slide he is observing that there are short gastric pits and long gastric glands, and he can easily identify chief cells, parietal cells, and mucous neck cells. What part of the stomach did the slide most likely come from?

   _____.

4. Pepsinogen is secreted by _____ cells, and HCl is secreted by _____ cells.

5. What histological feature is unique to the duodenum of the small intestine?

   _____

6. What histological feature is unique to the ileum of the small intestine?

   _____

7. What epithelial modification is particularly abundant in the large intestine?

   _____

8. Label the parts of a hepatic lobule in the figure below:

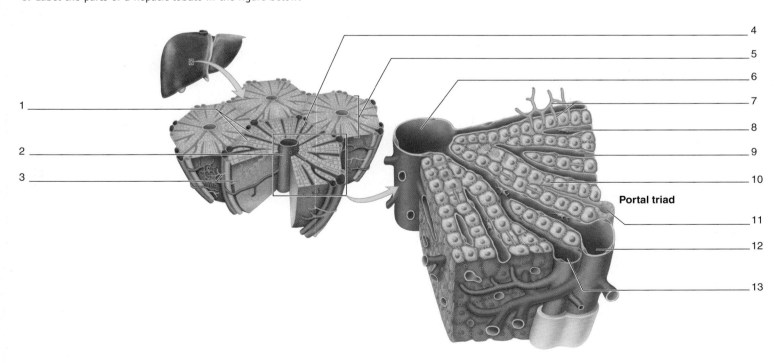

9. Trace a drop of bile from its site of production in the liver to its entry into the duodenum. Assume that it does not enter the gallbladder.

_____

_____

_____

_____

_____

10. Describe how to distinguish the jejunum from the ileum of the small intestine, using only gross anatomic features.

_____

_____

_____

_____

11. A patient presented to his physician with pain in the left lower quadrant of his abdomen. The source of the pain was an adhesion* between the visceral peritoneum covering part of the patient's colon and the parietal peritoneum lining his anterolateral abdominopelvic wall in that region. What part of the colon was most likely adhered to the abdominopelvic wall?

_____.

12. The image to the right is a cross section through part of the small intestine.

   a. What part of the small intestine is it (duodenum, jejunum, or ileum)?

   _____

   b. In the space below, explain how you came to your answer for part (a). That is, what characteristic(s) did you use to determine which part of the small intestine this sample was taken from?

_____

_____

_____

_____

_____

_____

*An adhesion (adhaereo, to stick to) in the abdominopelvic cavity is an area where two layers of peritoneum are stuck to each other with connective tissue. It usually is the result of some sort of injury or inflammation.

The Urinary System

26

OUTLINE and OBJECTIVES

Anatomy & Physiology REVEALED®
aprevealed.com

Module 13: URINARY SYSTEM

INTRODUCTION

The urinary system is responsible for maintaining blood volume and composition. It accomplishes this task through the process of **ultrafiltration,** which involves forcing fluid out of the blood across a membrane called the **filtration membrane.** The fluid thus formed is called **filtrate.** As the filtrate flows through the structural and functional units of the kidney, the **nephrons,** it is further processed. Many substances, including over 90% of the water in the filtrate, are **reabsorbed** back into the bloodstream so they are not lost from the body. Other substances, such as **urea** (a breakdown product of protein metabolism), are **secreted** into the filtrate so they can be removed from the body. Ultimately, after the blood and filtrate have made their way through the kidney, a small (relative to the volume of blood that is filtered) amount of filtrate leaves the kidney as **urine,** which will be transported via the **ureters** to the **urinary bladder** for storage. At a time that is convenient for the individual, the urine is emptied from the bladder and exits the body through the **urethra.**

The structural and functional unit of the kidney is the **nephron.** Each kidney contains more than 1.25 million nephrons. Remarkably, the kidneys can maintain their function even when 85–90% of their nephrons have been destroyed through disease. However, further losses will result in **kidney failure**. If an individual's kidneys fail, he or she must be placed on **dialysis** (*dialyo*, to separate). This involves filtering the blood using a dialysis machine, or artificial kidney. A patient on dialysis must undergo three to four sessions a week, each session lasting approximately 4 hours. Without dialysis, the individual cannot survive because the balance of fluid, electrolytes, and waste products in the blood cannot be maintained at appropriate levels. The consequences of kidney failure underscore the enormous role the kidneys play in maintaining health.

Chapter 26: The Urinary System

Name: _____

Date: _____ Section: _____

PRE-LABORATORY WORKSHEET

1. List four functions of the urinary system:

   a. _____

   b. _____

   c. _____

   d. _____

2. List the four components of the urinary system.

   a. _____

   b. _____

   c. _____

   d. _____

3. The structural and functional unit of the kidney is the _____.

4. Describe the structure of transitional epithelium. _____

   _____

5. The three components of the urinary trigone are:

   a. _____

   b. _____

   c. _____

6. Define *retroperitoneal*: _____

   _____

7. Which of the following is lined with simple cuboidal epithelium with microvilli?

   a. collecting duct

   b. distal convoluted tubule

   c. renal corpuscle

   d. proximal convoluted tubule

8. The glomerulus of the kidney consists of _____ capillaries.

9. How many layers compose the capsule of the kidney? _____.

10. The renal artery is a branch off of the _____ and the renal vein drains into the _____.

IN THE LABORATORY

In this laboratory exercise you will explore the structures of the urinary system with a special emphasis on the kidney. Nearly all functions of the urinary system are functions of the kidney. The remaining structures of this system (ureter, urinary bladder, and urethra) transport or store the urine formed by the kidney. Because the structural and functional units of the kidney—the nephrons—are actively involved in altering the composition of the blood, it is critical to understand the pattern of blood flow through the kidney and how this pattern of blood flow parallels the parts of the nephron. You will accomplish this task by observing the gross anatomy of the kidney and its blood supply, and correlating gross structures with histological observations of the parts of the nephron (renal corpuscles, proximal convoluted tubules, distal convoluted tubules, and nephron loops).

Histology

The Kidney

Each kidney is composed of two major regions: an outer **renal cortex** and an inner **renal medulla** (**figure 26.1**). The arrangement of nephrons along the **corticomedullary junction** means that some nephron components fall predominantly in the renal cortex and others in the renal medulla. Thus, the two regions exhibit distinct histological features. The renal cortex contains the renal corpuscles, proximal convoluted tubules (PCTs), distal convoluted tubules (DCTs), and peritubular capillaries. The renal medulla contains the nephron loops, collecting ducts (CDs), and vasa recta. Most structures within the kidney can be identified histologically by recognition of both the region (cortex or medulla) and the type of epithelium lining the structure. **Table 26.1** summarizes the type of epithelium that lines each of the structures and lists the major functions of each structure.

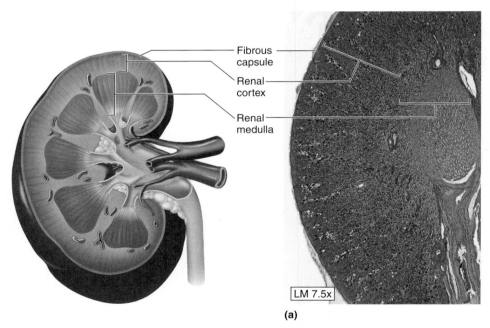

Fibrous capsule

Renal cortex

Renal medulla

LM 7.5x

(a)

Figure 26.1   Histology of the Kidney.   (a) Histological appearance of the kidney at low power demonstrates the fibrous capsule, outer cortex, and inner medulla. (b) Placement of nephron structures within the renal cortex and renal medulla.

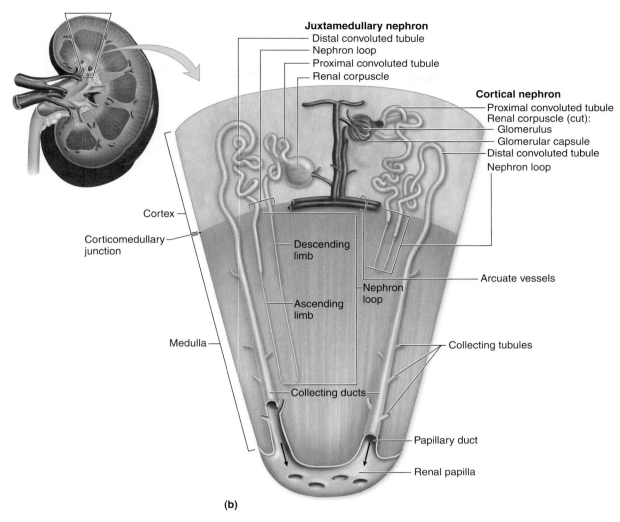

Juxtamedullary nephron
Distal convoluted tubule
Nephron loop
Proximal convoluted tubule
Renal corpuscle

Cortical nephron
Proximal convoluted tubule
Renal corpuscle (cut):
Glomerulus
Glomerular capsule
Distal convoluted tubule
Nephron loop

Cortex

Corticomedullary junction

Descending limb

Nephron loop

Ascending limb

Arcuate vessels

Medulla

Collecting tubules

Collecting ducts

Papillary duct

Renal papilla

(b)

Figure 26.1 **Histology of the Kidney,** *continued.*

Table 26.1	Histological Features of the Kidney		
Structure	**Epithelium**	**Function**	**Region Where Structure Is Predominantly Located**
Glomerulus	Fenestrated endothelium (simple squamous).	Filtration.	Cortex
Visceral Layer of Glomerular Capsule	Simple squamous modified to form podocytes.	Secondary processes of podocytes have pedicels, which contain actin filaments. The spaces between the pedicels, called filtration slits, participate in the filtration process.	Cortex
Parietal Layer of Glomerular Capsule	Simple squamous.	Forms an outer wall to the capsule.	Cortex
Proximal Convoluted Tubule (PCT)	Simple cuboidal with long, dense microvilli. Nuclei are located near the basal surface of the cells.	Reabsorbs glucose, amino acids, Ca^{2+}, PO_4, HCO_3^-, and 80% of the water and NaCl present in the filtrate. Secretes substances like penicillin and toxins after they have undergone modification by the liver. Also secretes organic acids and bases.	Cortex
Descending Limb of the Nephron Loop	Begins as simple cuboidal, then transitions to simple squamous.	Epithelial cells are impermeable to sodium, but water is drawn out into the interstitial spaces. Thus, the filtrate becomes more concentrated as it moves down the descending nephron loop.	Medulla
Ascending Limb of the Nephron Loop—Thin Segment	Simple squamous.	Epithelial cells impermeable to water, but sodium passively diffuses out.	Medulla

(continued on next page)

Table 26.1	Histological Features of the Kidney *(continued)*		
Structure	**Epithelium**	**Function**	**Region Where Structure Is Predominantly Located**
Ascending Limb of the Nephron Loop—Thick Segment	Simple cuboidal. Cells are darker than in the collecting duct.	The epithelial cells are impermeable to water, and they actively transport sodium out of the tubule. Thus, the filtrate becomes less concentrated as it moves up the ascending nephron loop.	Medulla
Distal Convoluted Tubule (DCT)	Simple cuboidal with few, short microvilli. Nuclei are located near the apical surface of the cells.	Secretes H^+ and K^+. Reabsorbs Na^+ and water. Contains the macula densa of the juxtaglomerular apparatus, which is involved with the sensation and regulation of blood pressure.	Cortex
Collecting Duct (CD)	Simple cuboidal epithelium. Cells have very precise boundaries. Overall tube diameter is the same as the PCT, but the CD has a larger lumen and no microvilli. Cells are paler than those of the thick segment of the nephron loop.	Concentrates urine under the influence of antidiuretic hormone (ADH). ADH causes CD epithelial cells to transport aquaporins (membrane proteins that transport water) to their apical surface, allowing water to be reabsorbed into the vasa recta.	Medulla

EXERCISE 26.1 Histology of the Renal Cortex

The renal cortex contains **renal corpuscles**, which are the site of filtration. Each renal corpuscle is composed of a ball of capillaries, the **glomerulus**, surrounded by the first part of the tubular system of the nephron, the **glomerular capsule** (also known as Bowman's capsule). The glomerular capsule itself is composed of an inner **visceral layer,** which consists of modified simple squamous epithelial cells called **podocytes**, and an outer **parietal layer**, which consists of unmodified simple squamous epithelium. The renal cortex also contains **PCTs**, which are the site of most reabsorption, and **DCTs**, which are involved in both reabsorption and secretion and compose part of the **juxtaglomerular apparatus**.

EXERCISE 26.1A: The Renal Corpuscle

1. Obtain a compound microscope and a histology slide of the kidney and place the slide on the microscope stage.

2. Bring the tissue sample into focus at low power and scan the slide to distinguish between the outer cortex and inner medulla (figure 26.1). Next, move the microscope stage so the renal cortex is in the center of the field of view.

3. Scan the slide in the region of the cortex to locate a circular **renal corpuscle** (**figure 26.2**). Bring the renal corpuscle into the center of the field of view and then change to high power. Although you won't be able to distinguish the visceral layer of the glomerular capsule from the glomerular capillaries, identify the tissue that contains both the visceral layer of the glomerular capsule and the glomerular capillaries (figure 26.2). Next, identify the parietal layer of the glomerular capsule. What type of epithelium composes the parietal layer of the glomerular

   capsule?_____

   What is the name of the space located between the visceral and parietal layers of glomerular capsule?

   _____

   The space you just identified becomes continuous with the lumen of which of the following? (Circle the appropriate response.)

   PCT   nephron loop   DCT   collecting duct

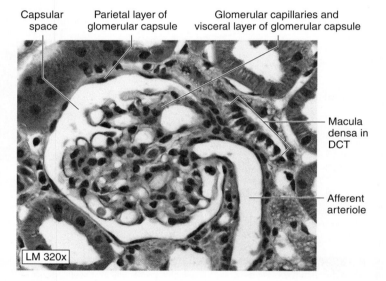

Capsular space | Parietal layer of glomerular capsule | Glomerular capillaries and visceral layer of glomerular capsule | Macula densa in DCT | Afferent arteriole

LM 320x

Figure 26.2 **The Renal Corpuscle.** The renal corpuscle consists of the glomerulus and the glomerular capsule.

4. In the space below, draw a brief sketch of the renal corpuscle as seen under the microscope. Be sure to label the visceral and parietal layers of the glomerular capsule on your drawing.

5. *Optional Activity:* **AP|R** **Urinary System**—Review the "Kidney—microscopic anatomy" animation to visualize the parts of the nephron and their placement in the renal cortex and medulla.

EXERCISE 26.1B: Proximal and Distal Convoluted Tubules

1. After completing exercise 26.1A, you should have a slide of the kidney on the microscope stage that has the outer cortex in the center of the field of view. If you did not perform exercise 26.1A, go through steps 1–2 of exercise 26.1A and then continue with this exercise.

2. Scan the slide in the region of the cortex to locate **proximal** and **distal convoluted tubules** (PCTs and DCTs) (**figure 26.3**). When viewing a histology slide of the cortex of the kidney, you will mostly see cross sections of PCTs and DCTs, with a few renal corpuscles scattered throughout. Although both PCTs and DCTs are lined with simple cuboidal epithelium, the *proximal* convoluted tubules have long, dense microvilli, which make the lumens of the PCTs appear "fuzzy." The distal convoluted tubules have very few, short microvilli, which make the lumens appear to be clear. What does the presence of microvilli indicate about the function of an epithelium? _____

_____

3. In the space to the right, sketch cross sections of proximal and distal convoluted tubules.

4. Using figures 26.1 to 26.3 and table 26.1 as guides, identify the following structures on the slide of the cortex of the kidney:

☐ capsular space   ☐ renal corpuscle

☐ distal convoluted tubule   ☐ visceral layer of glomerular capsule and glomerular capillaries

☐ microvilli

☐ parietal layer of glomerular capsule

☐ proximal convoluted tubule

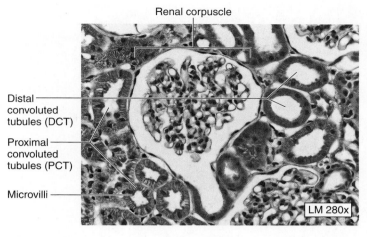

Figure 26.3   **The Renal Cortex.**   The renal cortex contains renal corpuscles, proximal convoluted tubules (PCT), and distal convoluted tubules (DCT). PCTs have "fuzzy" lumens because of the presence of microvilli.

_____ ×

EXERCISE 26.2 Histology of the Renal Medulla

The renal medulla contains **nephron loops** (loops of Henle) and **collecting ducts**, with surrounding capillaries called the **vasa recta**. These structures are all elongated tubules that lie next to each other and function together to concentrate the urine formed by the nephrons.

1. Obtain a compound microscope and a histology slide of the kidney, and place the slide on the microscope stage.

2. Bring the tissue sample into focus at low power and scan the slide to distinguish between the outer cortex and inner medulla. Next, move the microscope stage so the renal medulla is in the center of the field of view (figure 26.1).

3. With the medulla in the center of the field of view, switch to high power and bring the tissue sample into focus once again. In this region of the kidney you will identify thick and thin limbs of the nephron loops, collecting ducts, and possibly vasa recta (table 26.1, **figure 26.4**). Notice how, in longitudinal section, these structures appear as row after row of cells lined up next to each other. Your main objective in viewing this part of the kidney is to gain an appreciation for the way these structures line up next to each other, which is critical to their function. You should also be able to distinguish thick limbs of the nephron loops from the collecting ducts based on the diameter of the lumens. Both

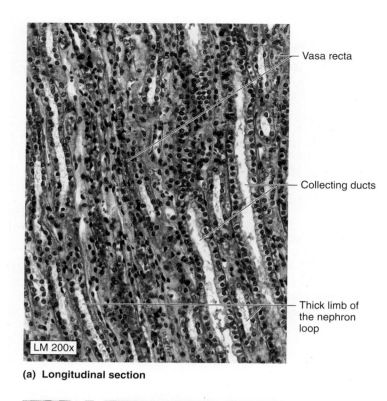

Vasa recta

Collecting ducts

Thick limb of the nephron loop

LM 200x

(a) Longitudinal section

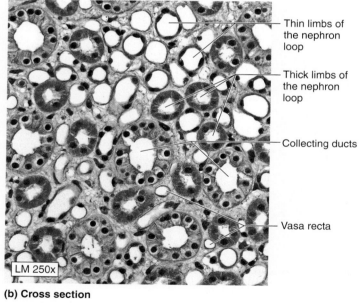

Thin limbs of the nephron loop

Thick limbs of the nephron loop

Collecting ducts

Vasa recta

LM 250x

(b) Cross section

Figure 26.4   **The Renal Medulla.**

structures are lined with simple cuboidal epithelium, but the collecting ducts have lumens with very large diameters compared to those of the thick nephron loops. It will be difficult to identify the vasa recta because of the extreme thinness of their walls and because they cannot easily be distinguished from thin limbs of the nephron loops (unless erythrocytes can be seen inside the lumen of the vessels, in which case you'll know the structures must be vasa recta).

4. Using figure 26.4 and table 26.1 as guides, identify the following on the histology slide of the medulla of the kidney.

☐ collecting duct   ☐ thin limbs of the
☐ thick limbs of the   nephron loop
    nephron loop   ☐ vasa recta

5. In the space below, draw a brief sketch of the histology of the renal medulla as observed through the microscope. Be sure to label the structures listed in step 4 in your drawing.

――――――― ×

The Urinary Tract

The **urinary tract** consists of structures that function to transport or store urine: the ureters, urinary bladder, and urethra. In this section you will observe the histological structure of the walls of the ureters and urinary bladder. **Table 26.2** lists the type of epithelium that lines each of the structures of the urinary tract and the calyces of the kidney.

Table 26.2	Urine-Draining Structures			
Structure	**Epithelium**		**No. of Cell Layers**	**Word Origins**
Calyces	Transitional		2–3	*calyx*, cup of a flower
Ureter	Transitional		4–5	*oureter*, urinary canal
Urinary Bladder	Transitional		>6	*urinary*, relating to urine
Urethra	In males, begins as transitional epithelium and ends with stratified squamous of the external body surface. In females, is primarily stratified squamous.		NA	*ourethra*, canal leading from the bladder

EXERCISE 26.3 Histology of the Ureters

The **ureters** are long, muscular tubes that transport urine from the hilum of the kidney to the urinary bladder. They are lined with **transitional epithelium**, which has the ability to stretch when urine is being transported through the ureters. Like other tubular structures in the body, the wall of the ureter is composed of multiple layers. The walls of the ureters, however, do not have a submucosa. There are three layers to the walls of the ureters: the mucosa, muscularis, and adventitia (**figure 26.5**). Contraction of smooth muscle in the muscularis layer of the ureters transports urine from the renal pelvis to the urinary bladder. The arrangement of smooth muscle layers in the muscularis

of the ureters is just the opposite of that of the digestive tract organs. The inner layer is composed of **longitudinal** smooth muscle, whereas the outer layer is composed of **circular** smooth muscle. In addition, because the ureters are **retroperitoneal** (they lie behind the peritoneal cavity and are not covered by peritoneum), their outer layer is an **adventitia**.

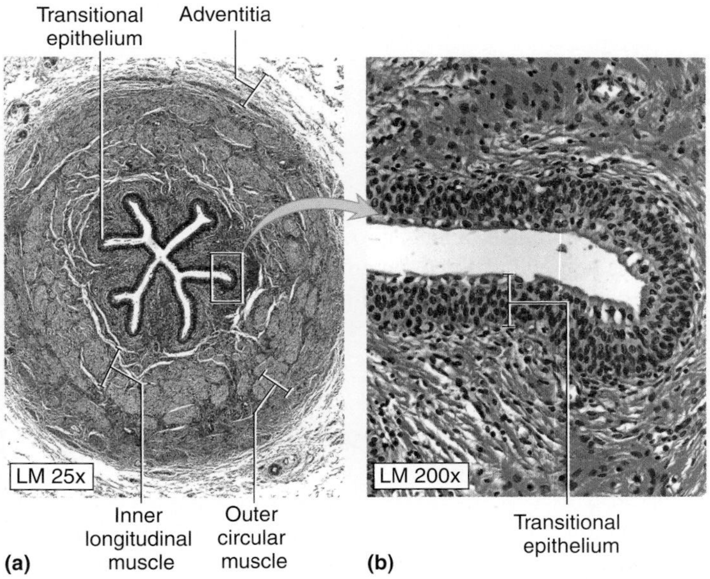

1. Obtain a histology slide demonstrating a cross-sectional view of a ureter. Place it on the microscope stage and bring the tissue sample into focus on low power.

2. Using figure 26.5 as a guide, identify the following structures on the slide of the ureter:

   ☐ adventitia   ☐ outer circular muscle
   ☐ inner longitudinal   ☐ transitional epithelium
      muscle

3. In the space below, draw a brief sketch of a cross section of a ureter as seen through the microscope. Be sure to label all the structures listed in step 2 in your drawing.

Figure 26.5   **The Ureter.**   (a) Cross section through entire ureter. (b) Close-up of the mucosa of the ureter.

EXERCISE 26.4 Histology of the Urinary Bladder

The wall of the urinary bladder has a more typical arrangement of layers, similar to that of other organs in the body: the mucosa, submucosa, muscularis, and adventitia (with serosa only on the superior surface of the urinary bladder) (**figure 26.6**). Similar to the ureters, the urinary bladder is lined with **transitional epithelium**,

which allows it to stretch as it fills with urine, and its outermost layer is an adventitia because the bladder lies outside the peritoneal cavity. The muscularis layer of the bladder wall is composed of several individual layers of smooth muscle that are collectively referred to as the **detrusor muscle** (*detrudo*, to drive away).

1. Obtain a histology slide of the wall of the urinary bladder. Place it on the microscope stage and observe at low power. Using figure 26.6 as a guide, identify the following structures:

   ☐ adventitia   ☐ serosa

   ☐ detrusor muscle   ☐ submucosa

   ☐ mucosa   ☐ transitional epithelium

   ☐ muscularis

2. Move the microscope stage so the epithelium is at the center of the field of view, then change to high power. Look for rounded epithelial cells on the apical surface of the epithelium, which are often binucleate. These cells are sometimes referred to as **dome cells** because of their shape.

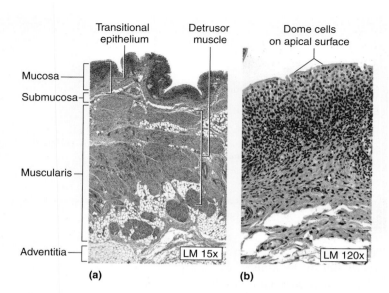

Figure 26.6   **Histology of the Urinary Bladder.** (a) Section of the wall of the bladder. (b) Close-up of the transitional epithelium.

Gross Anatomy

In this set of exercises you will observe the gross anatomy of the kidneys, ureters, urinary bladder, and urethra, paying particular attention to the location of these structures within the abdominopelvic cavity and their relationships with other organs.

The Kidney

The kidneys are located along the posterior body wall from vertebral levels T12 – L3, with the right kidney slightly lower than the left kidney because of the location of the liver. The kidneys, like the ureters and urinary bladder, are retroperitoneal structures (*retro*, behind, + *peritoneal*, referring to the peritoneum). In the following exercises you will observe the gross anatomy of the kidney, the blood supply to the kidney, and the urine draining structures within the kidney.

EXERCISE 26.5 Gross Anatomy of the Kidney

1. Each kidney is encased by a **capsule**, which is composed of three layers: the renal fascia, perinephric fat, and fibrous capsule. The first two of these layers, the **renal fascia** and **perinephric fat**, are not often visualized in the gross anatomy laboratory unless an intact cadaver is available because the kidney must be extracted from these two layers of its capsule to be removed from the body for observation of its internal structure.

2. Obtain a kidney that has been sectioned along a coronal plane or a classroom model of a coronal section of the kidney (**figure 26.7**). Whether you are viewing an actual kidney or a model of the kidney, the outermost part of what you are looking at is actually the innermost layer of the capsule: the **fibrous capsule**.

3. Identify the structures listed in figure 26.7 on the kidney or classroom model of the kidney. Then label them in figure 26.7.

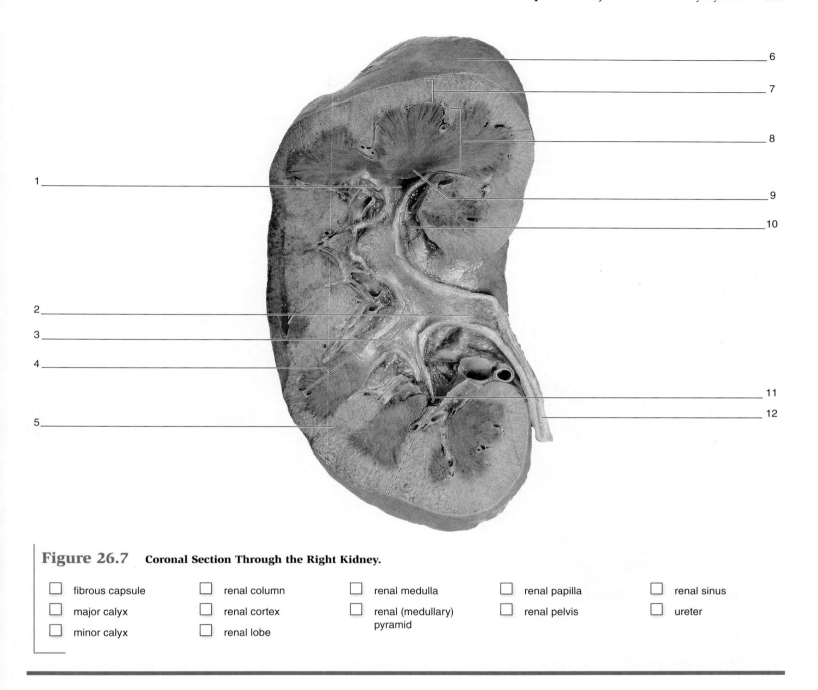

Figure 26.7 **Coronal Section Through the Right Kidney.**

☐ fibrous capsule   ☐ renal column   ☐ renal medulla   ☐ renal papilla   ☐ renal sinus

☐ major calyx   ☐ renal cortex   ☐ renal (medullary)   ☐ renal pelvis   ☐ ureter
                                                                 pyramid

☐ minor calyx   ☐ renal lobe

EXERCISE 26.6   Blood Supply to the Kidney

The kidneys receive approximately 25% of the blood leaving the heart each minute. This is not because they have a huge demand for oxygen or nutrients, but instead because their major function is to filter the blood to alter its volume and composition. The kidney accomplishes this through the processes of filtration, absorption, and secretion, which all involve transport of fluids and other substances between the functional units of the kidney (nephrons) and the blood. Thus, an understanding of blood flow through the kidney is critical to understanding the function of the kidney.

1. Observe a classroom model of the kidney that demonstrates blood vessels of the kidney.

2. Blood flow through the kidney is unique, and is similar to a portal system such as the hepatic portal system. **Figure 26.8** diagrams the flow of blood through the kidney. There are three capillary beds in the kidney: the glomerulus (within the renal corpuscle), the peritubular capillaries, and the vasa recta. The **glomerulus,** the first capillary bed, is the site of *filtration*: movement of substances from the blood across the filtration membrane to the capsular space of the renal corpuscle (figure 26.8a,b). Blood enters the glomerulus through an afferent arteriole and leaves through an efferent arteriole. The efferent arteriole then feeds into the second capillary

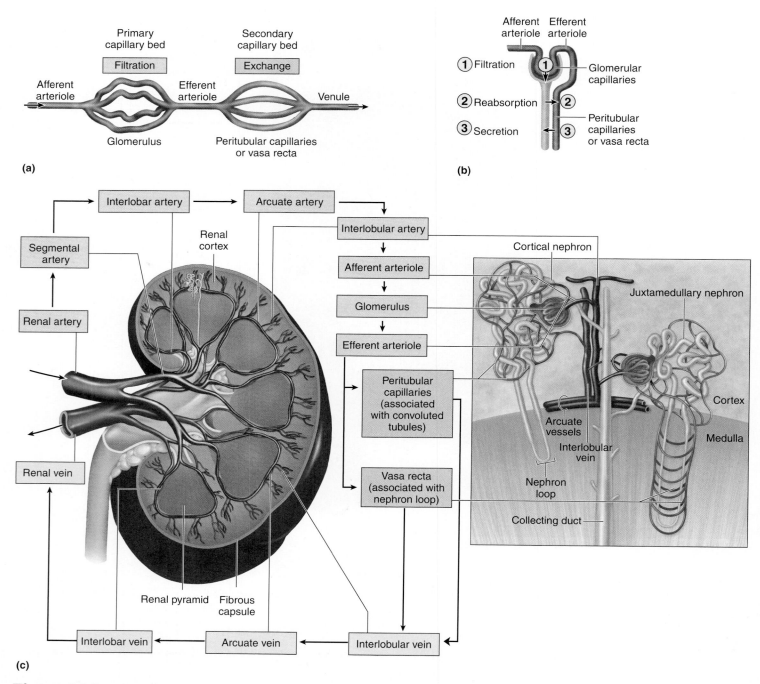

Figure 26.8 **Blood Supply to the Kidney.** (a) Blood flows through two capillary beds in the kidney. The first capillary bed, the glomerulus, is the site of filtration. The second capillary bed, peritubular capillaries in the cortex or vasa recta in the medulla, is the site of exchange (reabsorption or secretion). (b) The basic renal processes are filtration, reabsorption, and secretion. (c) Blood flow through the kidney.

bed, which can be either the **peritubular capillaries** (within the cortex) or the **vasa recta** (within the medulla). The second capillary bed is the site of *exchange* of fluids, electrolytes, respiratory gases, and nutrients between the tubular portions of the nephron and the blood. Exchange in these capillaries involves either *reabsorption* (movement of substances from the tubular spaces into the blood), or *secretion* (movement of substances from the blood into the tubular spaces). Finally, blood leaving the peritubular capillaries and vasa recta drains into veins that carry the blood back toward the inferior vena cava.

3. Using your textbook as a guide, identify the blood vessels listed in **figure 26.9** on the classroom model of the kidney. Then label them in figure 26.9. What parts of the nephron do the peritubular capillaries surround?

_____

What parts of the nephron do the vasa recta surround?

_____

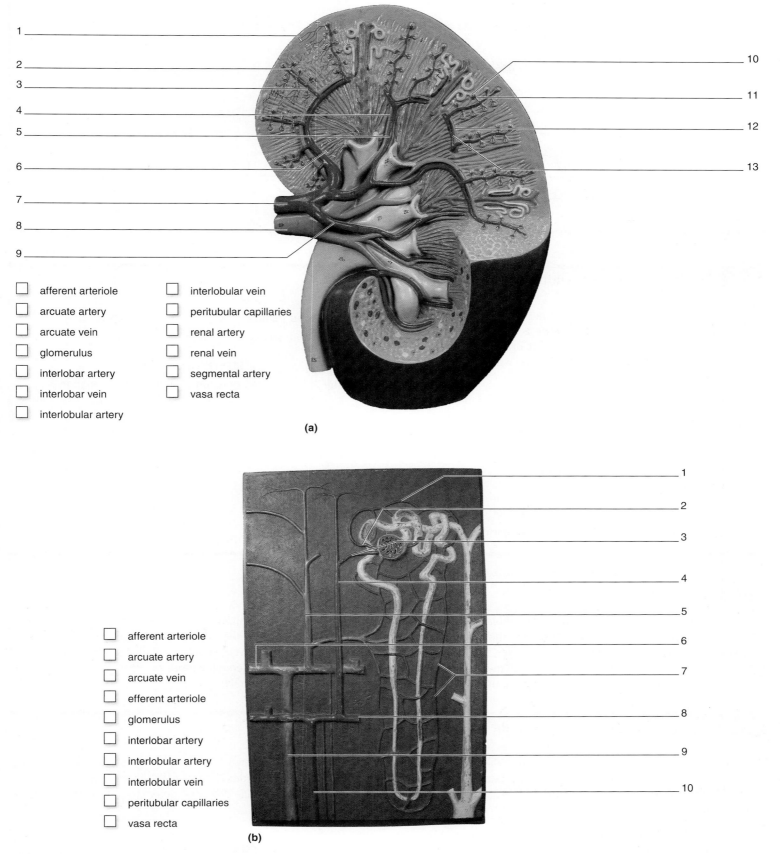

1 _____

2 _____

3 _____

4 _____

5 _____

6 _____

7 _____

8 _____

9 _____

10 _____

11 _____

12 _____

13 _____

☐ afferent arteriole

☐ arcuate artery

☐ arcuate vein

☐ glomerulus

☐ interlobar artery

☐ interlobar vein

☐ interlobular artery

☐ interlobular vein

☐ peritubular capillaries

☐ renal artery

☐ renal vein

☐ segmental artery

☐ vasa recta

(a)

1 _____

2 _____

3 _____

4 _____

5 _____

6 _____

7 _____

8 _____

9 _____

10 _____

☐ afferent arteriole

☐ arcuate artery

☐ arcuate vein

☐ efferent arteriole

☐ glomerulus

☐ interlobar artery

☐ interlobular artery

☐ interlobular vein

☐ peritubular capillaries

☐ vasa recta

(b)

Figure 26.9 **Models of the Kidney Demonstrating the Blood Supply to the Kidney.** (a) Coronal section. (b) Close-up of the renal cortex and renal medulla.

4. In the space below, trace the flow of blood in words through the kidney starting at the abdominal aorta and ending at the inferior vena cava.

_____

_____

_____

_____

_____

_____

_____

EXERCISE 26.7 Urine-Draining Structures Within the Kidney

Once filtrate has been formed by the nephrons, the DCTs empty it into a collecting duct, and the filtrate is subsequently referred to as _urine_. The urine thus formed will drain through a series of structures located in the kidney that will conduct it into the **ureter**.

1. Observe a classroom model of a coronal section of a kidney, or a gross sample of a kidney that has been sectioned along a coronal plane.

2. Using figures 26.7 and 26.9, and table 26.2 as guides, identify the following urine-draining structures of the kidney:

☐ collecting duct   ☐ minor calyx

☐ papillary duct   ☐ major calyx

☐ renal papilla   ☐ renal pyramid

☐ renal pelvis   ☐ ureter

3. _Optional Activity_: **AP|R** **Urinary System**—Watch the "Kidney—gross anatomy" and "Urine formation" animations for an overview of kidney anatomy, vasculature, and function.

The Urinary Tract

The urinary tract consists of the ureters, urinary bladder, and urethra. Within the kidney, urine draining from the calyces begins to collect in the renal pelvis. Every 2 to 3 minutes, peristaltic contractions of the smooth muscle lining the ureters conveys the urine into the ureters, which transport it into the urinary bladder.

Urine is then stored within the urinary bladder until a time when it is convenient to allow the urine to exit the body through the urethra. The process of voiding urine is called **micturition**. In this section you will observe the gross anatomy of the structures composing the urinary tract.

EXERCISE 26.8 Gross Anatomy of the Ureters

The ureters are long, thin, muscular tubes that run from the hilum of the kidney to the posterior, inferior surface of the urinary bladder (**figure 26.10**). Like the kidney, they are located retroperitoneally, behind the peritoneal cavity.

1. Observe a human cadaver or a classroom model of the abdomen.

2. Identify the structures listed in figure 26.10 on the cadaver or classroom model. Then label figure 26.10.

3. Follow the pathway of the ureter from its site of origin near the hilum of the kidney to its entry into the urinary bladder. What major muscle does the ureter cross over to get to the urinary bladder? _____

☙ WHAT DO YOU THINK?

❶ Why do you think muscular contractions are used to transport urine through the ureters?

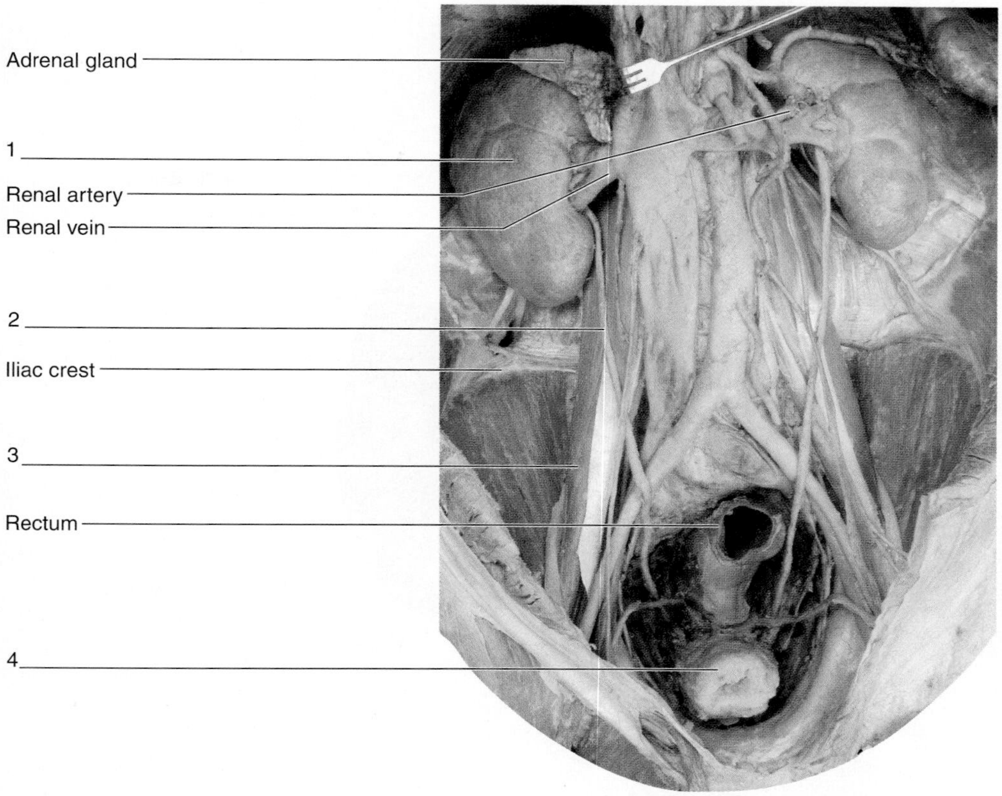

Adrenal gland

1 _____

Renal artery

Renal vein

2 _____

Iliac crest

3 _____

Rectum

4 _____

Figure 26.10 **Location of the Ureters Within the Abdominopelvic Cavity.**

☐ psoas major muscle
☐ right kidney
☐ ureter
☐ urinary bladder

EXERCISE 26.9 Gross Anatomy of the Urinary Bladder and Urethra

Like the kidneys and ureters, the **urinary bladder** is a retroperitoneal structure. The superior surface of the bladder is covered with peritoneum, but the rest is not. The urinary bladder lies in the pelvic cavity, just posterior to the pubic symphysis. The **urethra** is a simple passageway running from the inferior surface of the urinary bladder to the external urethral orifice. It is much longer in males than in females because of its dual purpose in the male (it also serves as the passageway for sperm during ejaculation) and location within the penis. In both males and females there are both internal and external urethral sphincters. The **internal urethral sphincter** is composed of smooth muscle in the wall of the urinary bladder that encloses the entrance into the urethra, and is under involuntary control. The **external urethral sphincter** is composed of skeletal muscle of the **urogenital diaphragm**, and is under voluntary control.

1. Observe the abdominopelvic cavity of a male human cadaver or a classroom model of the abdominopelvic cavity of a male (**figure 26.11**). The area in the floor of the urinary bladder that is bounded by the openings of the two ureters and the urethra is called the **urinary trigone**. This area is clinically significant because urinary tract infections often establish themselves in this location. If the urinary bladder is cut open on the cadaver or if you can observe a classroom model that demonstrates the interior of the urinary bladder, identify the urinary trigone in the floor of the urinary bladder.

2. Using your textbook as a guide, identify the structures listed in figure 26.11 on the cadaver or classroom model. Then label them in figure 26.11.

3. In the space below, draw a brief sketch of the structures composing the urinary trigone as seen from the interior of the urinary bladder.

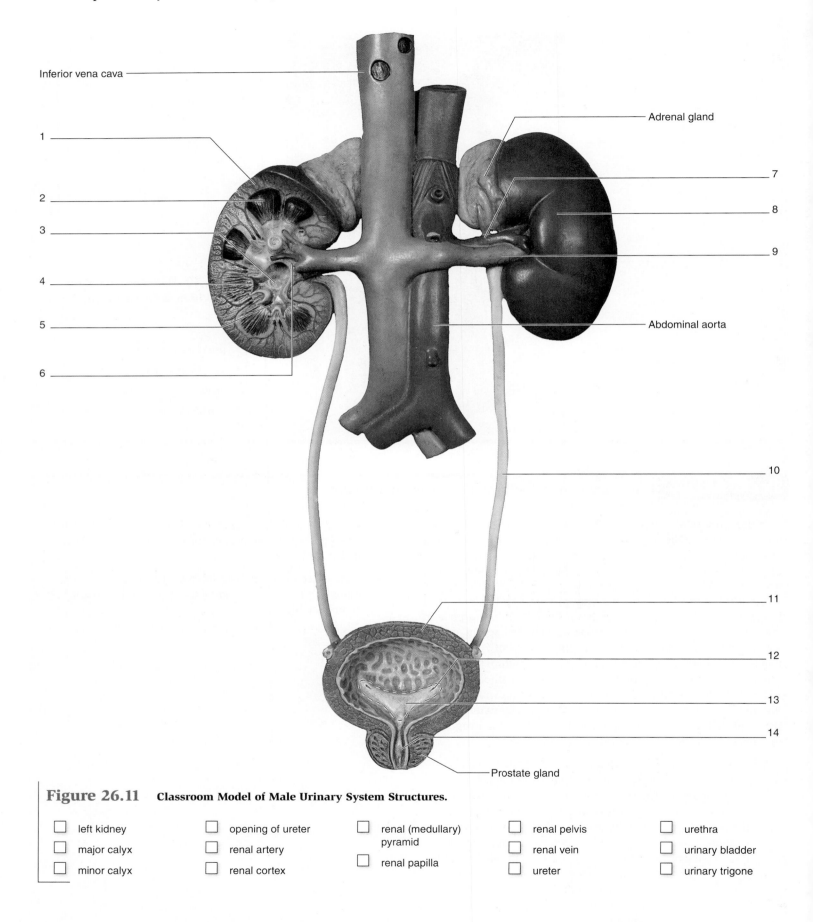

Inferior vena cava

Adrenal gland

Abdominal aorta

Prostate gland

1

2

3

4

5

6

7

8

9

10

11

12

13

14

Figure 26.11 **Classroom Model of Male Urinary System Structures.**

- [ ] left kidney
- [ ] major calyx
- [ ] minor calyx
- [ ] opening of ureter
- [ ] renal artery
- [ ] renal cortex
- [ ] renal (medullary) pyramid
- [ ] renal papilla
- [ ] renal pelvis
- [ ] renal vein
- [ ] ureter
- [ ] urethra
- [ ] urinary bladder
- [ ] urinary trigone

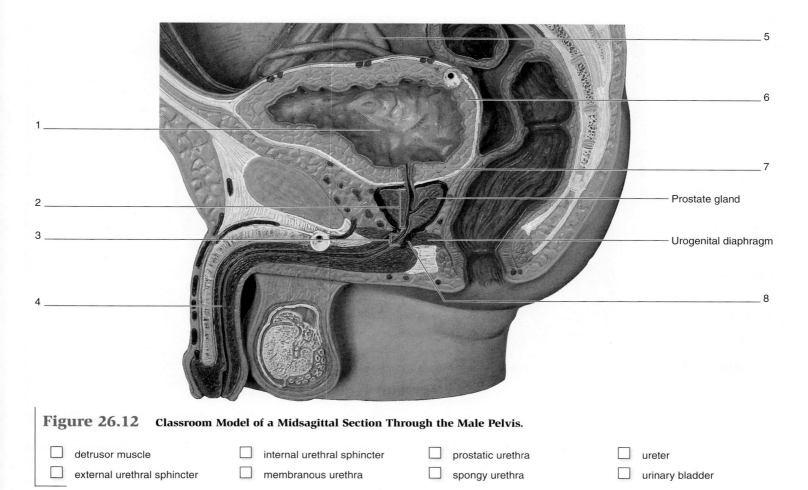

Figure 26.12   **Classroom Model of a Midsagittal Section Through the Male Pelvis.**

☐ detrusor muscle   ☐ internal urethral sphincter   ☐ prostatic urethra   ☐ ureter

☐ external urethral sphincter ☐ membranous urethra   ☐ spongy urethra   ☐ urinary bladder

4. Observe the male urethra on a classroom model demonstrating a midsagittal section through the male pelvis (**figure 26.12**). The male urethra has three parts: the **prostatic urethra**, the **membranous urethra**, and the **spongy urethra**. The internal urethral sphincter of the male is located between the urinary bladder and the prostate gland, and the external urethral sphincter is the skeletal muscle that composes the urogenital diaphragm (which surrounds the membranous urethra).

5. Using your textbook as a guide, identify the structures listed in figure 26.12 on the classroom model. Then label them in figure 26.12.

6. Observe the female urethra on a classroom model demonstrating a midsagittal section through the female pelvis (**figure 26.13**). The female urethra is short, and the external urethral orifice is located between the clitoris and the vagina. The entire urethra is surrounded by the internal urethral sphincter. The external urethral sphincter is skeletal muscle of the urogenital diaphragm, just as it is in the male.

7. Using your textbook as a guide, identify the structures listed in figure 26.13 on the classroom model. Then label them in figure 26.13.

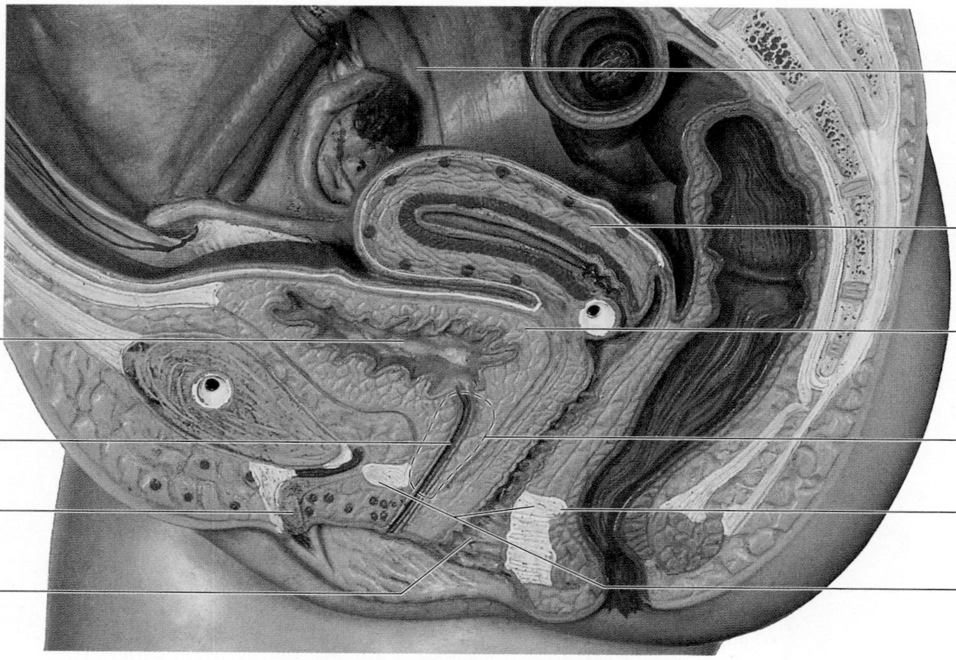

Figure 26.13   **Classroom Model of a Midsagittal Section Through the Female Pelvis.**

☐ detrusor muscle

☐ external urethral sphincter

☐ internal urethral sphincter

☐ ureter

☐ urethra

☐ urinary bladder

Chapter 26: The Urinary System

Name:_____

Date:_____ Section:_____

POST-LABORATORY WORKSHEET

1. Which of the two kidneys is located more superiorly than the other (and why)?

_____

_____

_____

2. In the space below, trace the path of a drop of filtrate from its origin in the capsular space to its exit (as urine) from the kidney through the renal pelvis.

_____

_____

_____

_____

_____

3. In the space below, trace the path of a drop of urine from its exit from the kidney (through the renal pelvis) to its removal from the body through the urethra.

_____

_____

_____

_____

_____

4. Label the parts of a nephron and the blood supply to the nephron in the figure below:

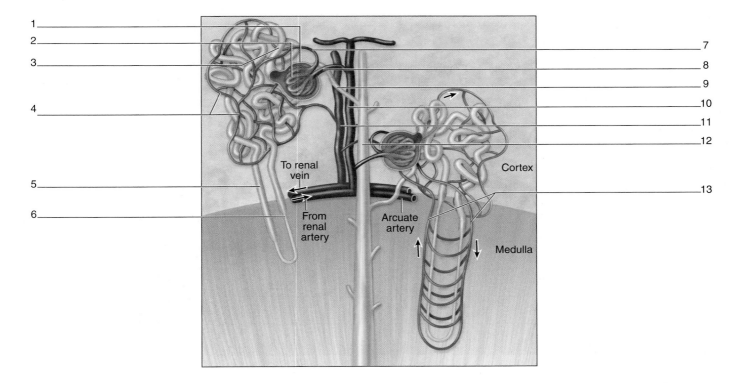

1_____

2_____

3_____

4_____

5_____

6_____

To renal vein

From renal artery

Arcuate artery

Cortex

Medulla

7

8

9

10

11

12

13

5. Fill in the table below with the type of epithelium lining each structure, and a brief description of the function of each structure.

Structure	Epithelium	Function
Glomerulus		
Visceral Layer of Glomerular Capsule		
Parietal Layer of Glomerular Capsule		
PCT		
Descending Limb of Nephron Loop		
Ascending Limb of Nephron Loop—Thin		
Ascending Limb of Nephron Loop—Thick		
DCT		
Collecting Duct		
Minor and Major Calyces		
Ureter		
Urinary Bladder		
Urethra		

6. A surgeon has the job of removing a patient's kidney. His approach to the kidney will be to first cut through skin and adipose tissue of the back, remove a section of the lower ribs, and then enter the abdominal cavity. Once the surgeon has entered the abdominal cavity, what are the layers of the capsule of the kidney the surgeon will encounter, from superficial to deep?

_____

_____

_____

7. Define *retroperitoneal:*

_____

_____

8. Describe the wall-layering pattern of the ureter and explain how urine is transported by the ureter.

_____

_____

_____

9. Describe the wall-layering pattern of the urinary bladder.

_____

_____

_____

10. What is the detrusor muscle?

_____

11. What three structures compose the urinary trigone?

    a. _____

    b. _____

    c. _____

12. Explain the structural and functional differences between the internal and external urethral sphincters (that is, what type of muscle is each composed of and what type of nervous control each sphincter is under).

_____

_____

_____

_____

13. Label the figure below.

14. For each photomicrograph below, identify the organ. Then explain the characteristics you used to determine what organ it is.

(a)

Organ: _____

Identifying characteristics: _____

_____

_____

_____

_____

_____

(b)

Organ: _____

Identifying characteristics: _____

_____

_____

_____

_____

_____

(c)

Organ: _____

Identifying characteristics: _____

_____

_____

_____

_____

The Reproductive System

27

OUTLINE and OBJECTIVES

Anatomy & Physiology REVEALED
aprevealed.com

Module 14: REPRODUCTIVE SYSTEM

INTRODUCTION

Of all the systems of the human body, the **reproductive system** (*re-*, again, + *productus*, to lead forth) is the only system that is not required for the survival of the human body. Instead, its function is to ensure the survival of the species by allowing for the creation of another human being. Thus far in this course you have discovered much about the amazing complexity and beauty of the human body. We will now turn our attention to the process by which this amazingly complex human comes into being. It requires two separate **sexes**: male and female. Each sex produces **gametes** (sperm in males; ova in females), which are haploid cells that combine to form a **zygote**, the first step in the development of a new human. Male and female gametes are both formed by **meiosis**, which occurs in the **gonads** (*gone*, seed). The human body invests a great deal of energy in the formation of gametes and in the maintenance of characteristics that are meant to attract members of the opposite sex. This much energy expenditure may seem counterintuitive when one thinks about homeostasis as the primary driving force for most organ system functions. However, when you consider the consequences of humans failing to come together for reproducing the species (that is, extinction of the species), the amount of energy invested makes a great deal more sense.

Because the reproductive structures in males and females are different, the structures can seem to be more challenging to learn (because there are more of them). However, all human embryos develop from the same overall plan. Certain embryonic structures degrade in males but not in females. On the other hand, many embryonic structures persist in both males and females—forming apparently very different adult structures. Such structures are called **homologues** (*homos*, the same, + *logos*, relation). Ovaries and testes are examples of homologous structures. As you explore the male and female reproductive systems, pay particular attention to homologous structures. This will not only simplify the task of learning reproductive system structures, it will also help you to appreciate the similarities between the male and female reproductive systems.

Chapter 27: The Reproductive System

PRE-LABORATORY WORKSHEET

1. List three functions of the reproductive system:

   a. _____

   b. _____

   c. _____

2. The female gonad is the _____, while the male gonad is the _____.

3. The female gamete is the _____, while the male gamete is the _____.

4. In the space below, make a drawing depicting the stages of meiosis.

Stages of Meiosis

In this laboratory session you will explore the histological and gross structure of the male and female reproductive systems, with a focus on the histology of the primary reproductive organs of both females and males, the *ovaries* and *testes*. Observing the histology of the gonads allows us to best see and understand the stages of gametogenesis. The focus of your observations of the gross anatomy of the reproductive systems will be to understand the regional relationships between gross reproductive structures and structures of other body systems, as well as to understand homologous relationships between male and female reproductive structures.

Histology

Female Reproductive System

The female reproductive system has two important functions: produce the female gametes (ova) and support and protect the developing embryo/fetus that is created when an ovum becomes fertilized by a sperm. Components of the female reproductive system include the ovaries, uterine tubes, uterus, vagina, and mammary glands. Exercises 27.1 to 27.4 explore the internal detail and wall structures of the ovary, uterine tube, uterus, and vagina. Mammary glands will be covered in the gross anatomy portion of this chapter.

EXERCISE 27.1 Histology of the Ovary

The **ovary** is the primary reproductive organ in the female and functions in the production of the ova (eggs) and the female sex steroid hormones **estrogen** and **progesterone**. The ovaries contain several **ovarian follicles**, all at various stages of development. Each follicle contains a developing **oocyte** (**table 27.1**). The structure and function of the ovary is best viewed at the histological level so you can appreciate the characteristics of ovarian follicles at each stage of development.

1. Obtain a histology slide of an ovary and place it on the microscope stage.

2. Bring the tissue sample into focus on low power and identify the outer **cortex** and inner **medulla** of the ovary (**figure 27.1**). Next, move the microscope stage so the ovarian cortex is at the center of the field of view and then change to high power.

3. Observe the outermost region of the ovarian cortex. Locate the two layers of tissue that compose the outer coverings of the ovary: the germinal epithelium and the tunica albuginea (**figure 27.2**). The **germinal epithelium** (*germen*, a sprout) is a simple cuboidal epithelium that forms the outermost covering of the ovary. The name *germinal* refers to the fact that scientists once believed the germ cells (oocytes) were formed from this epithelial layer. Most ovarian cancers arise in the germinal epithelium. Deep to the germinal epithelium is the **tunica albuginea** (*tunica*, a coat, + *albugineus*, white spot), which is composed of dense irregular connective tissue.

4. Focus on the cortex and identify follicles in each stage of follicular development listed in **table 27.2**.

Table 27.1	Developmental Stages of an Oocyte			
Cell Name/Oocyte Stage	**Description and Function**	**Stage of Mitosis/ Meiosis**	**Ploidy**	**Word Origins**
Oogonia	Primitive germs cells that undergo mitosis in the second to fifth months of embryonic life to form approximately 7 million oogonia that will subsequently develop into primary oocytes. Approximately 70% of the primary oocytes will degenerate in a proces called atresia.	Formed by mitosis; undergo mitosis to form primary oocytes.	Diploid (2n)	*oon*, egg, + *gonia*, generation
Primary Oocyte	Oogonia that begin the process of meiosis and arrest in prophase of meiosis I. Primary oocytes will not undergo subsequent meiotic divisions unless the follicle is stimulated to mature by-follicle-stimulating hormone (FSH) and luteinizing hormone (LH).	Arrested in Prophase I of meiosis.	Diploid (2n)	*primary*, first, + *oon*, egg, + *kytos*, cell
Secondary Oocyte	Under the influence of FSH and LH, primary oocytes complete meiosis I to become secondary oocytes, which will not undergo subsequent meiotic divisions unless fertilization occurs.	Arrested in Metaphase II of meiosis.	Haploid (1n)	*secondary*, second, + *oon*, egg, + *kytos*, a cell
Definitive Oocyte	Upon fertilization, the secondary oocyte undergoes the second meiotic division to become an ovum.	Formed after secondary oocyte is fertilized and completes meiosis.	Haploid (1n)	*oon*, egg, + *kytos*, cell

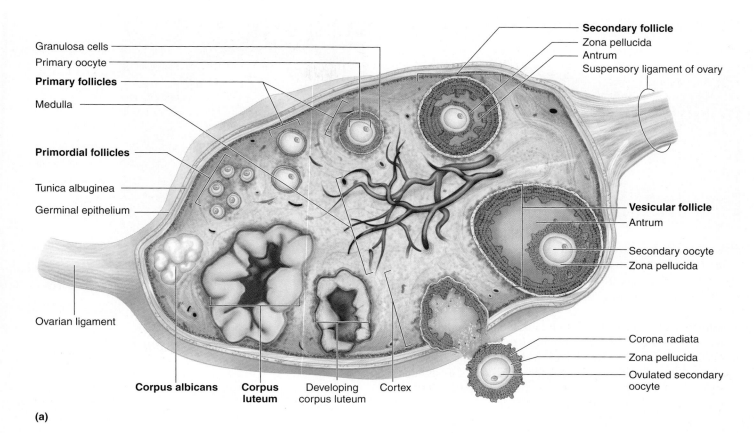

Granulosa cells
Primary oocyte
Primary follicles
Medulla
Primordial follicles
Tunica albuginea
Germinal epithelium
Ovarian ligament

Secondary follicle
Zona pellucida
Antrum
Suspensory ligament of ovary

Vesicular follicle
Antrum
Secondary oocyte
Zona pellucida

Corona radiata
Zona pellucida
Ovulated secondary oocyte

Corpus albicans   **Corpus luteum**   Developing corpus luteum   Cortex

(a)

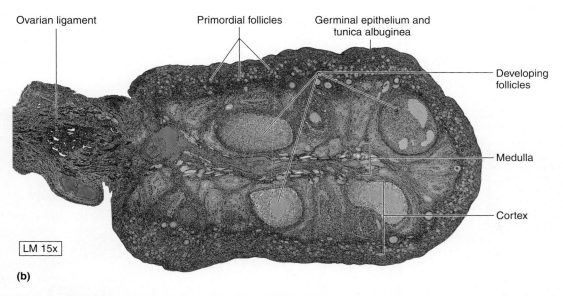

Ovarian ligament   Primordial follicles   Germinal epithelium and tunica albuginea

Developing follicles

Medulla

Cortex

LM 15x

(b)

Figure 27.1   **The Ovary.**   (a) The ovary is the primary reproductive organ of the female and produces the female gametes, the ova. (b) Histological structure of the ovary demonstrating the outer cortex containing follicles in various stages of development, and the inner medulla, which consists mainly of blood vessels and nerves.

5. Using tables 27.1 and 27.2 and figures 27.1 and 27.2 as guides, locate the following structures on the slide of the ovary:

☐ corpus albicans   ☐ germinal epithelium
☐ corpus luteum   ☐ primary follicle

☐ primary oocyte   ☐ secondary oocyte
☐ primordial follicle   ☐ tunica albuginea
☐ secondary follicle   ☐ vesicular (Graafian) follicle

6. Next move the microscope stage so a vesicular follicle is in the center of the field of view. A **vesicular (Graafian)**

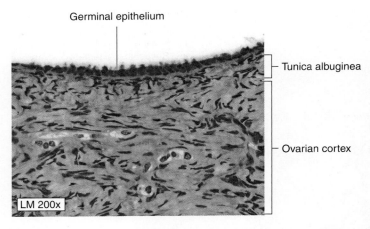

Figure 27.2 contents:
- Germinal epithelium
- Tunica albuginea
- Ovarian cortex
- LM 200x

Figure 27.2   Germinal Epithelium and Tunica Albuginea of the Ovary. The germinal epithelium is the most common origin site for ovarian cancers. The tunica albuginea is a "white coat" of dense irregular connective tissue that surrounds the entire ovary.

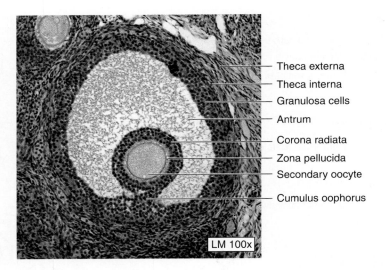

Figure 27.3 contents:
- Theca externa
- Theca interna
- Granulosa cells
- Antrum
- Corona radiata
- Zona pellucida
- Secondary oocyte
- Cumulus oophorus
- LM 100x

Figure 27.3   Vesicular (Graafian) Follicle. A vesicular follicle contains a secondary oocyte and a single antrum.

follicle (**figure 27.3**) is a mature follicle that is ready to be released during ovulation, after stimulation by a surge in luteinizing hormone (LH) secreted by the anterior pituitary gland. **Table 27.3** lists the histological features of a vesicular follicle. Using table 27.3 and figure 27.3 as guides, identify the following parts of the vesicular follicle:

☐ antrum   ☐ theca externa

☐ corona radiata   ☐ theca interna

☐ cumulus oophorus   ☐ zona pellucida

☐ granulosa cells

7. In the spaces to the right, draw brief sketches of primordial, primary, secondary, and vesicular follicles as seen under the microscope.

Primordial follicle

_____ ×

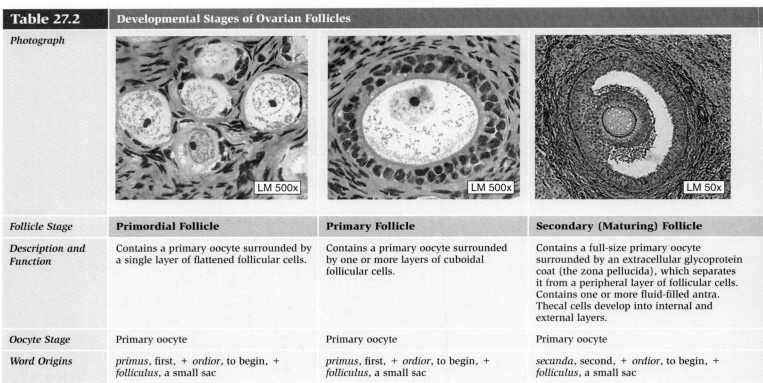

Table 27.2	Developmental Stages of Ovarian Follicles		
Photograph	LM 500x	LM 500x	LM 50x
Follicle Stage	**Primordial Follicle**	**Primary Follicle**	**Secondary (Maturing) Follicle**
Description and Function	Contains a primary oocyte surrounded by a single layer of flattened follicular cells.	Contains a primary oocyte surrounded by one or more layers of cuboidal follicular cells.	Contains a full-size primary oocyte surrounded by an extracellular glycoprotein coat (the zona pellucida), which separates it from a peripheral layer of follicular cells. Contains one or more fluid-filled antra. Thecal cells develop into internal and external layers.
Oocyte Stage	Primary oocyte	Primary oocyte	Primary oocyte
Word Origins	*primus*, first, + *ordior*, to begin, + *folliculus*, a small sac	*primus*, first, + *ordior*, to begin, + *folliculus*, a small sac	*secunda*, second, + *ordior*, to begin, + *folliculus*, a small sac

Table 27.3	Components of a Vesicular Follicle		Word Origins
Antrum	The fluid-filled space in the center of the follicle.		antron, a cave
Corona Radiata	A single layer of columnar cells derived from the cumulus oophorus that attach to the zona pellucida of the oocyte.		*corona*, a crown, + *radiatus*, to shine
Cumulus Oophorus	A "mound" of granulosa cells that supports and surrounds the ovum within the follicle.		*cumulus*, a heap, + *oophoron*, ovary
Granulosa Cells	Epithelial cells lining the follicle that will become the luteal cells of the corpus luteum after ovulation. They secrete the liquor folliculi, which is the fluid that fills the antrum.		*granulum*, a small grain
Theca Externa	The external fibrous layer of a well-developed vesicular follicle. The cells and fibers are arranged in concentric layers.		*theca*, a box, + *externus*, external
Theca Interna	The inner cellular layer of the vesicular follicle; these cells secrete androgen that is converted to estrogen by granulosa cells.		*theca*, a box, + *internus*, internal
Zona Pellucida	A thick coat of glycoproteins that surrounds the oocyte.		*zona*, zone, + *pellucidus*, clear

Primary follicle

_____ ×

Secondary follicle

_____ ×

Vesicular follicle

_____ ×

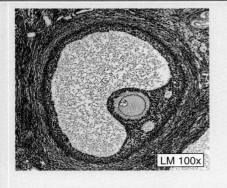

LM 100x

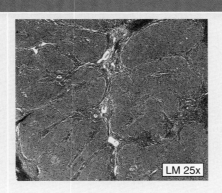

LM 25x

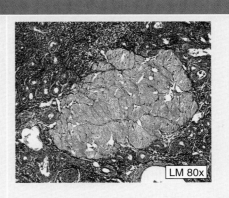

LM 80x

Vesicular (Graafian) Follicle	**Corpus Luteum**	**Corpus Albicans**
Has nearly the same structure and function as a secondary follicle, but contains only a single large antrum.	A "yellow body" that gets its color from the steroid hormones it secretes, which are lipids. After ovulation, the theca interna cells enlarge and continue to secrete the steroid hormones estrogen and progesterone. In the center of the corpus luteum is a large blood clot.	If fertilization does not occur, the corpus luteum stops secreting hormones after two weeks and becomes a smaller "white body" consisting mainly of scar tissue.
Secondary oocyte	NA	NA
vesicular, a blister, + *folliculus*, a small sac	*corpus*, body, + *luteus*, yellow	*corpus*, body, + *albus*, white

8. Using table 27.2 as a guide, locate a corpus luteum and corpus albicans on the slide.

9. In the spaces below, draw a brief sketch of a corpus luteum and a corpus albicans as seen under the microscope.

WHAT DO YOU THINK?

1 How can you distinguish between a secondary follicle and a vesicular follicle?

2 During which phase of the ovarian cycle is the corpus luteum present?

3 What is the function of the corpus luteum?

Corpus luteum

_____ ×

Corpus albicans

_____ ×

EXERCISE 27.2 Histology of the Uterine Tubes

The **uterine tubes** (fallopian tubes, or oviducts) are long muscular tubes that transport the ovum from the ovary to the uterus. The wall of the uterine tube is composed of three layers: mucosa, muscularis, and serosa.

1. Obtain a histology slide demonstrating a cross-sectional view of a uterine tube.

2. Place the slide on the microscope stage and bring the tissue sample into focus on low power. Note the highly folded **mucosa** of the tube (**figure 27.4**).

3. Move the microscope stage until you have an area of folded mucosa at the center of the field of view and then switch to high power. The epithelium of the uterine tube contains two cell types, **ciliated epithelial cells** and **secretory cells**. **Table 27.4** describes the histological components of the uterine tubes. Focus on the epithelial cells lining the uterine tube, and distinguish ciliated cells from secretory cells (figure 27.4b).

4. Using table 27.4 and figure 27.4 as guides, identify the following structures on the slide of the uterine tube:

☐ ciliated cells   ☐ muscularis   ☐ serosa
☐ mucosa   ☐ secretory cells

5. In the space below, draw a brief sketch of a cross section of the uterine tube, as seen under the microscope. Be sure to label the structures listed above in your drawing.

_____ ×

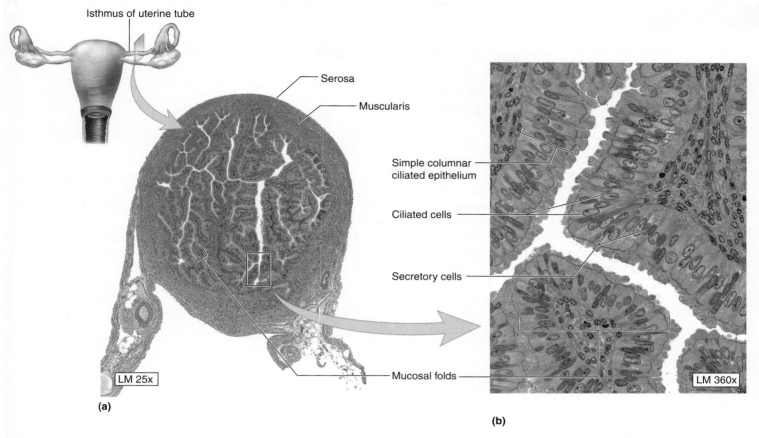

Figure 27.4   Histology of the Uterine Tubes.   (a) Cross section through the isthmus of the uterine tubes demonstrating the many folds of the mucosa. (b) The epithelial cells lining the uterine tubes contain two cell types: ciliated cells and secretory cells. Ciliated cells have light nuclei, whereas secretory cells have dark nuclei.

Table 27.4	Components of the Uterine Tube	
Uterine Tube Structure	**Description and Function**	**Word Origins**
Ampulla	The wide part of the uterine tube located medial to the infundibulum. Its mucosa is highly folded and lined with simple columnar epithelium containing ciliated cells and secretory cells. Fertilization most commonly occurs in the ampulla.	*ampulla*, a two-handled bottle
Ciliated Cells	Columnar epithelial cells that contain cilia that beat toward the uterus to transport the ovum to the uterus. Nuclei stain lighter than those of the secretory cells.	*cilium*, eyelash
Infundibulum	The funnel-like expansion of the ovarian end of the uterine tube. Fingerlike extensions of the infundibulum are fimbriae (*fimbria*, fringe).	*infundibulum*, a funnel
Isthmus	The narrow part of the uterine tube located right next to the uterus.	*isthmos*, a constriction
Mucosa	The inner layer of the wall of the uterine tube. It is highly folded in the infundibulum and ampulla of the uterine tube. Contains ciliated columnar epithelium and a layer of areolar connective tissue.	*mucosus*, mucous
Muscularis	The middle layer of the wall of the uterine tube. It consists of smooth muscle whose contraction assists in transporting the ovum toward the uterus.	*musculus*, mouse (referring to muscle)
Secretory Cells	Nonciliated columnar epithelial cells that promote the activation of spermatozoa (capacitation) and provide nourishment for the ovum. Nuclei stain darker than those of the ciliated cells.	*secretus*, to separate
Serosa	The outer layer of the uterine tube. It consists of simple squamous epithelium (a fold of peritoneum).	*serosus*, serous
Uterine Part (Interstitial Segment)	The part of the uterine tube that penetrates the wall of the uterus.	*intra −*, within, + *muralis*, wall (*inter*, between, + *sisto*, to stand)

The **uterus** is a muscular structure whose primary function is to house the developing fetus.

1. Obtain a histology slide demonstrating a portion of the uterine wall. Place it on the microscope stage and bring the tissue sample into focus on low power.

2. The uterine wall is composed of three layers: endometrium, myometrium, and perimetrium. The inner lining of the uterus, the **endometrium**, goes through its own cycle of growth throughout the ovarian cycle. Each time a woman's ovary undergoes a single ovarian cycle, the endometrium becomes prepared for the possibility that a fertilized egg will become implanted. If no implantation occurs, the innermost layer of the endometrium, the **functional layer**, sloughs off in the process of **menstruation** (*menstruus*,

monthly). **Table 27.5** describes the layers of the uterine wall and the components that make up each layer. **Table 27.6** summarizes three phases of the menstrual cycle and demonstrates the appearance of the endometrium during each phase.

3. Using tables 27.5 and 27.6 as guides, identify the following structures on the slide of the uterus:

   ☐ endometrium (functional layer)   ☐ myometrium
   ☐ endometrium (basal layer)   ☐ perimetrium

4. Move the microscope stage so the functional layer of the endometrium is in the center of the field of view. Switch to high power and identify **uterine glands**.

Table 27.6	**Phases of the Menstrual Cycle**	
Photograph	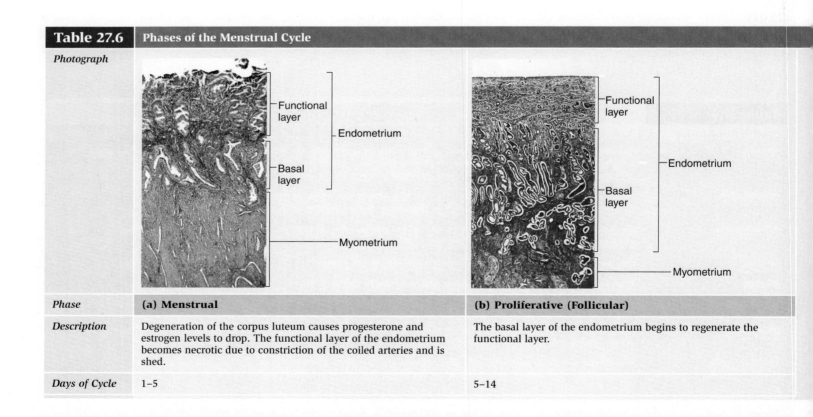	
Phase	**(a) Menstrual**	**(b) Proliferative (Follicular)**
Description	Degeneration of the corpus luteum causes progesterone and estrogen levels to drop. The functional layer of the endometrium becomes necrotic due to constriction of the coiled arteries and is shed.	The basal layer of the endometrium begins to regenerate the functional layer.
Days of Cycle	1–5	5–14

Table 27.5	Wall Layers of the Uterus	
Wall Layer	**Description and Function**	**Word Origins**
Endometrium	The mucous membrane composing the inner layer of the uterine wall. Consists of simple columnar epithelium and a lamina propria with simple tubular uterine glands. The structure, thickness, and state of the endometrium undergo marked change during the menstrual cycle.	*endon*, within, + *metra*, uterus
Functional Layer (Stratum Functionalis)	The apical layer of the endometrium. Most of this layer is shed during menstruation.	*stratum*, a layer, + *functus*, to perform
Basal Layer (Stratum Basalis)	The basal layer of the endometrium. It undergoes minimal changes during the menstrual cycle and serves as the basis for regrowth of the more apical stratum functionalis.	*stratum*, a layer, + *basalis*, basal
Myometrium	The muscular wall of the uterus composed of three layers of smooth muscle.	*mys*, muscle, + *metra*, uterus
Perimetrium	The outermost covering of the uterus; a serous membrane formed from peritoneum.	*peri –*, around, + *metra*, uterus

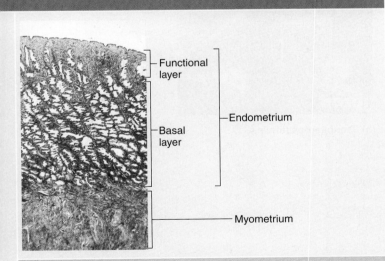

(c) Secretory (Luteal)

Begins at ovulation. Progesterone secreted by the corpus luteum stimulates the uterine glands to begin secretion.

14–28

The **vagina** (*vagina*, sheath) is a muscular tube that receives the penis during intercourse and serves as the birth canal during parturition (*parturio*, to be in labor).

1. Obtain a histology slide demonstrating a portion of the vaginal wall. Place it on the microscope stage and bring the tissue sample into focus on low power. The walls of the vagina are composed of a mucosa and muscularis, and the mucosa is lined with a nonkeratinized stratified squamous epithelium (**figure 27.5**).

2. Using figure 27.5 as a guide, identify the following structures on the slide of the vagina:

   ☐ lamina propria   ☐ nonkeratinized
   ☐ mucosa   stratified squamous
   ☐ muscularis   epithelium

3. *Optional Activity*: **AP|R Reproductive System**—Watch the "Female reproductive system overview" animation to review the female reproductive organs and their histological features.

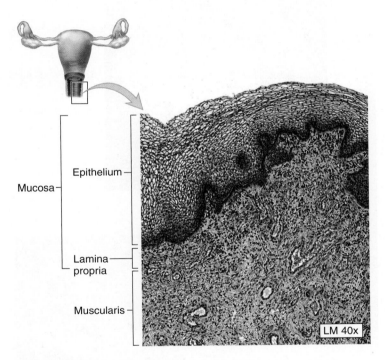

Figure 27.5 **Histology of the Vaginal Epithelium.** The epithelium is nonkeratinized stratified squamous.

Male Reproductive System

The male reproductive system consists of the testes, epididymis, ductus deferens, seminal vesicles, prostate gland, bulbourethral glands, and penis. Unlike the female reproductive system, which must support and nourish a developing fetus, the only function of the male reproductive system is to produce the male gamete (sperm) and provide nourishment for the sperm and a mechanism for the sperm to be delivered to the female reproductive tract. Like female reproductive organs, male reproductive structures have unique histological features that will help you identify them under the microscope.

EXERCISE 27.5   Histology of Seminiferous Tubules

1. Obtain a slide demonstrating seminiferous tubules and place it on the microscope stage. The **testes** are the primary reproductive organ in the male, and function in the production of sperm and the male sex steroid hormone **testosterone**. Within each testis are several hundred long, coiled **seminiferous tubules (figure 27.6)**, which are the site of sperm production, or **spermatogenesis** (*sperma*, seed, + *genesis*, origin). Within each tubule, **spermatogonia** undergo successive meiotic divisions as they move from the basal to the apical surface of the tubule epithelium. **Table 27.7** describes the developmental stages of the sperm, and **table 27.8** describes the structure

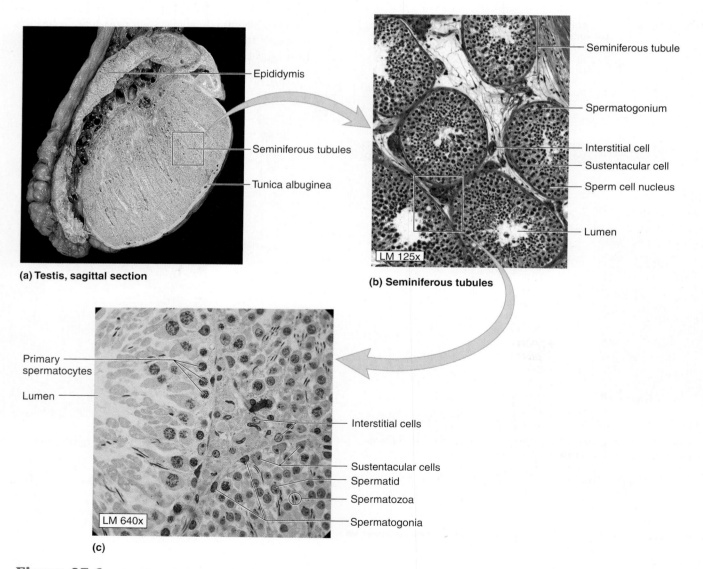

(a) **Testis, sagittal section**

— Epididymis
— Seminiferous tubules
— Tunica albuginea

(b) **Seminiferous tubules**

— Seminiferous tubule
— Spermatogonium
— Interstitial cell
— Sustentacular cell
— Sperm cell nucleus
— Lumen

LM 125x

(c)

Primary spermatocytes
Lumen
Interstitial cells
Sustentacular cells
Spermatid
Spermatozoa
Spermatogonia

LM 640x

Figure 27.6   Testes and Seminiferous Tubules.   (a) Testes and epididymis showing seminiferous tubules within the testes and tunica albuginea surrounding the testes. (b) Medium-power histological appearance of seminiferous tubules. (c) High-power histological view of seminiferous tubules demonstrating stages of spermatogenesis and sustentacular cells.

Table 27.7	Developmental Stages of Sperm			
Cell Name	**Description and Function**	**Stage of Mitosis/Meiosis**	**Ploidy**	**Word Origins**
Spermatogonia	Primitive male gametes that are formed by mitosis from male germline stem cells.	Formed by mitosis; undergo mitosis to form primary spermatocytes.	Diploid (2n)	*sperma*, seed, + *gonia*, generation
Primary Spermatocyte	Male gamete that has replicated its DNA in preparation for meiosis. It will subsequently undergo meiosis I to form two secondary spermatocytes.	DNA is replicated in prophase of meiosis I.	Diploid (2n)	*primary*, first, + *sperma*, seed, + *kytos*, a cell
Secondary Spermatocyte	Male gamete that has completed meiosis I and will subsequently undergo meiosis II to form two spermatids.	Formed after the first meiotic division.	Haploid (1n)	*secondary*, second, + *sperma*, seed, + *kytos*, a cell
Spermatid	Male gamete that has completed meiosis, but has not undergone spermiogenesis, the process of becoming a mature spermatozoon.	Formed after the second meiotic division.	Haploid (1n)	*sperma*, seed, + *-id*, a young specimen
Spermatozoon (Sperm)	A mature male gamete that has undergone spermiogenesis.	Formed by maturation of spermatids.	Haploid (1n)	*sperma*, seed, + *zoon*, animal

Table 27.8	Accessory Cells of the Testis	
Cell Name	**Description and Function**	**Word Origins**
Sustentacular (Sertoli) Cells	Surround multiple developing spermatocytes to provide them with support and nourishment. Phagocytize excess cytoplasm from developing spermatocytes. Form the blood-testis barrier, which protects the developing spermatocytes from antigens that circulate in the blood. Produce androgen-binding protein (ABP), which concentrates testosterone around the developing spermatocytes.	*sustento*, to hold upright
Interstitial (Leydig) Cells	Produce the steroid hormone testosterone, which is required for proper sperm development and for the development of male secondary sex characteristics.	*inter−*, between, + *sisto*, to stand

and function of the accessory cell types located within the testes.

2. Observe the slide on low power and identify several cross sections of the seminiferous tubules. Move the microscope stage so that one tubule is at the center of the field of view and change to high power.

3. Using tables 27.7 and 27.8 and figure 27.6 as a guide, identify the following structures:

☐ interstitial cell   ☐ spermatids

☐ primary spermatocyte   ☐ spermatogonia

☐ secondary spermatocyte   ☐ sustentacular cell

☐ seminiferous tubule

4. In the space to the right, draw a brief sketch demonstrating the histology of a seminiferous tubule as seen through the

microscope. Be sure to label the structures listed in step 3 in your drawing.

_____ ×

5. *Optional Activity*: **AP|R** **Reproductive System**—Watch the "Spermatogenesis" animation to visualize the formation of sperm in the seminiferous tubules.

After spermatozoa are formed in the seminiferous tubules, they travel through the efferent ductules, straight tubules, and the rete testis to enter the long, coiled tube that lies upon the testis, the **epididymis** (*epi* −, upon, + *didymos*, a twin [related to *didymoi*, testes]). The epididymis is the first part of the duct system in the male reproductive tract. Here the sperm undergo the process of maturation and are stored until ejaculation takes place. One of the most characteristic features of the epididymis is the presence of thousands of sperm cells in the lumen of the tube (**figure 27.7**). **Table 27.9** describes the structure and function of the male accessory reproductive structures, including the epididymis.

1. Obtain a slide of the epididymis and place it on the microscope stage. Bring the tissue sample into focus on low power and identify several cross sections of the epididymis. Move the microscope stage so that one part of the lumen of the epididymis is at the center of the field of view and then change to high power. Notice the sperm inside the lumen of the tubule, and how they do not come right up against the apical surface of the columnar epithelial cells. This is because of the stereocilia on the apical surface of the epithelial cells. **Stereocilia** (*stereo*, solid, + *cilium*, eyelid) are single, long microvilli (ironically, they are not cilia at all!) that increase the surface area of the epithelial cells for the purpose of secreting substances that nourish the sperm and absorbing substances from the sperm as they undergo maturation.

2. Using table 27.9 and figure 27.7 as guides, identify the following structures on the slide of the epididymis:

   ☐ columnar epithelial cells ☐ stereocilia
   ☐ sperm cells

3. In the space below, draw a brief sketch of the histology of the epididymis as seen through the microscope. Be sure to label the structures listed above in your drawing.

_____ ×

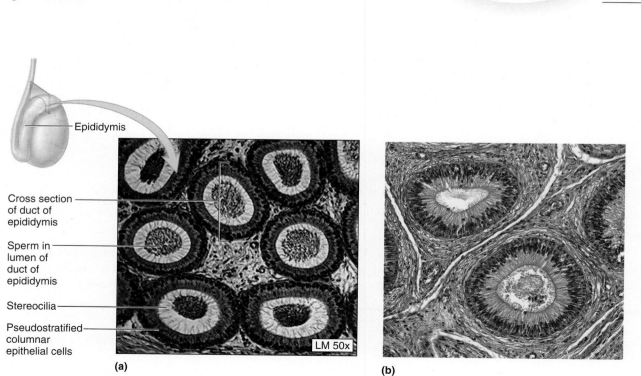

(a)   (b)

Figure 27.7 Cross Section Through the Epididymis. The epididymis is a long, coiled tube. In these sections you can see several cross sections through the duct. (a) Note the numerous spermatozoa filling the lumen. (b) At higher magnification you can better view the pseudostratified columnar epithelium with stereocilia.

Table 27.9	Male Accessory Reproductive Structures		
Structure	**Epithelium**	**Function**	**Word Origins**
Epididymis	Pseudostratified columnar epithelium with cilia	Site of storage and maturation of sperm. Walls contain some smooth muscle that will propel sperm into the ductus deferens during ejaculation.	*epi* −, upon, + *didymos*, a twin (related to *didymoi*, testes)
Ductus Deferens (vas deferens)	Pseudostratified columnar epithelium with stereocilia	A thick, muscular tube whose walls undergo peristaltic contractions during ejaculation to propel sperm into the urethra.	*ductus*, to lead, + *defero*, to carry away
Prostate Gland	Glandular epithelium resembling pseudostratified or simple columnar	Accessory reproductive gland that contributes approximately 25% of the volume of semen. The fluid produced is rich in Vitamin C (citric acid) and enzymes (such as PSA, prostate-specific antigen) that are important for proper sperm function.	*prostates*, one standing before
Seminal Vesicles	Pseudostratified columnar	Accessory reproductive gland that contributes approximately 60% of the volume of semen. The fluid produced is rich in fructose, which is a source of energy for the sperm. They also produce prostaglandins, which are important in promoting sperm motility and may stimulate uterine contractions.	*seminal*, relating to sperm, + *vesicula*, a blister
Bulbourethral Glands	Glandular epithelium consisting of simple and pseudostratified columnar	Accessory reproductive gland that produces a mucus-like lubricating substance during sexual arousal that lubricates the urethra and neutralizes the acidity of the urine, thus preparing the way for the spermatozoa to pass.	*bulbus*, bulb, + *urethral*, relating to the urethra

EXERCISE 27.7   Histology of the Ductus Deferens

The **ductus deferens** (vas deferens) is a continuation of the epididymis and is the route by which sperm travel from the epididymis to the urethra during ejaculation (*ejaculo*, to shoot out). It is a highly muscular tube lined with pseudostratified columnar epithelium with cilia (**figure 27.8**).

1. Obtain a slide demonstrating the ductus deferens and place it on the microscope stage. Bring the tissue sample into focus on low power and identify the cross section of the ductus deferens (figure 27.8). Move the microscope stage so the lumen of the ductus deferens is at the center of the field of view and then change to high power.

2. Using table 27.9 and figure 27.8 as guides, identify the following structures on the slide of the ductus deferens:

☐ cilia

☐ lumen

☐ mucosa

☐ pseudostratified columnar epithelial cells

☐ smooth muscle

3. In the space below, draw a brief sketch of the histology of a cross-section of the ductus deferens as seen through the microscope. Be sure to label the structures listed in step 2 in your drawing.

×

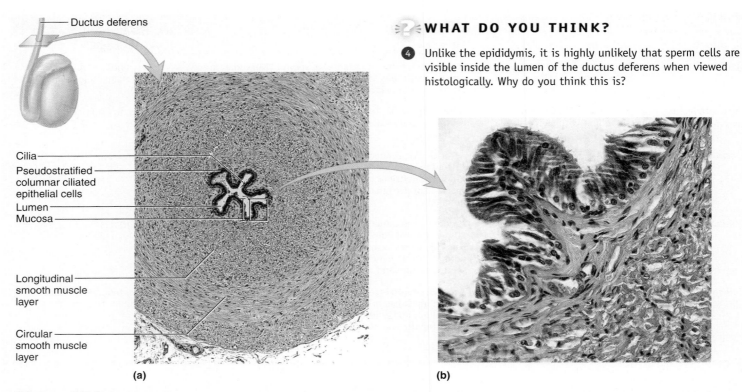

Figure 27.8   Ductus Deferens.   (a) Cross section through the entire ductus deferens, demonstrating thick layers of smooth muscle surrounding the lumen. (b) Close-up view of the pseudostratified columnar ciliated epithelium.

EXERCISE 27.8 Histology of Seminal Vesicles

As each ductus deferens approaches the posterior wall of the bladder, each comes together with a **seminal vesicle** (**figure 27.9**). These accessory reproductive glands produce substances (like fructose, which is a source of energy for the sperm) that are an important component of **semen**, the fluid expelled from the penis during ejaculation (table 27.9).

1. Obtain a slide of the seminal vesicles and place it on the microscope stage. Bring the tissue sample into focus on low power. Identify cross sections of the seminal vesicles. Move the microscope stage so the lumen of a portion of the seminal vesicle is at the center of the field of view, and then change to high power.

2. Using table 27.9 and figure 27.9 as guides, identify the following structures on the slide of the seminal vesicles:

☐ lumen   ☐ muscular wall
☐ mucosal folds

3. In the space below, draw a brief sketch of the histology of the seminal vesicles as seen through the microscope. Be sure to label the structures listed in step 2 in your drawing.

_____ ✕

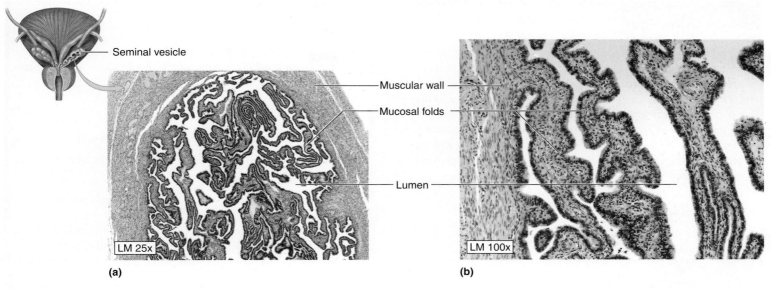

Seminal vesicle

Muscular wall

Mucosal folds

Lumen

LM 25x

(a)

LM 100x

(b)

Figure 27.9   **The Seminal Vesicles.**   (a) Low magnification. (b) Medium magnification.

Histology of the Prostate Gland

The **prostate** gland is an accessory reproductive gland located immediately inferior to the urinary bladder. Like the seminal vesicles, the prostate produces substances that will become part of semen and that are important for proper sperm function.

1. Obtain a slide demonstrating the prostate gland and place it on the microscope stage. Bring the tissue sample into focus on low power and identify the prostatic urethra and the prostate gland itself (**figure 27.10**). Move the microscope stage so a part of the gland relatively far away from the urethra is at the center of the field of view, and then change to high power.

2. Using table 27.9 and figure 27.10 as guides, identify the following structures on the slide of the prostate:

   ☐ tubuloalveolar glands

3. As men age, calcifications often form in the prostate. Such calcifications are called **prostatic calculi** (corpora aranacea) (figure 27.10). Scan the slide and see if any of these are present in your specimen.

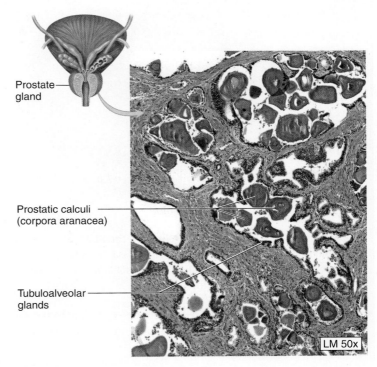

Prostate gland

Prostatic calculi (corpora aranacea)

Tubuloalveolar glands

LM 50x

Figure 27.10   **The Prostate Gland.**   Note the many prostatic calculi (corpora aranacea) seen at high magnification.

EXERCISE 27.10 | Histology of the Penis

The **penis** is the male copulatory organ and is homologous to the female clitoris. It is composed of three erectile bodies, the paired **corpora cavernosa** and the single **corpus spongiosum**, which contains the male urethra (**figure 27.11**). In the center of each corpus cavernosa is a central artery that supplies blood to the penis. The erectile tissues of the corpus cavernosa are the equivalent of the deep venous drainage accompanying the central arteries. During an erection, the veins draining the penis become constricted and venous blood fills up the erectile tissues.

1. Obtain a slide demonstrating a cross section of the penis and place it on the microscope stage. Observe at low power.

2. Using figure 27.11 as a guide, identify the following structures on the slide of the penis.

   ☐ central artery   ☐ dorsal vein
   ☐ corpus cavernosum   ☐ superficial fascia (CT)
   ☐ corpus spongiosum   ☐ urethra
   ☐ dorsal surface   ☐ ventral surface

3. In the space below, draw a brief sketch of the histology of a cross section of the penis as seen through the microscope. Be sure to label the structures listed above in your drawing.

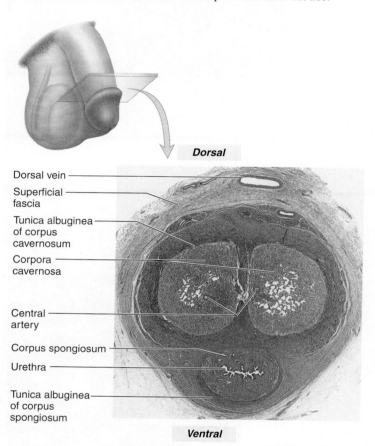

Dorsal

Dorsal vein

Superficial fascia

Tunica albuginea of corpus cavernosum

Corpora cavernosa

Central artery

Corpus spongiosum

Urethra

Tunica albuginea of corpus spongiosum

Ventral

_____ ×

4. *Optional Activity:* **AP|R** **Reproductive System**—Watch the "Male reproductive system overview" animation to review the male reproductive structures and their relationships.

WHAT DO YOU THINK?

⑤ The homologue to the male penis is the female clitoris. The clitoris consists of two paired erectile tissues, the corpora cavernosa, but it does not contain a corpus spongiosum. What structure is the clitoris "missing" as compared to the penis?

Figure 27.11   Cross Section Through the Shaft of the Penis. The penis contains two paired corpora cavernosa dorsally, and a single corpus spongiosum ventrally. Note the male urethra in the center of the corpus spongiosum.

Gross Anatomy

Female Reproductive System

The internal detail, wall structures, and functions of most of the female reproductive organs were covered in the histology section of this chapter. In this section, you will observe the gross structure of these organs to gain an appreciation of their location within the female pelvic cavity, and their relationships with each other. You will also observe the numerous suspensory ligaments that anchor the ovaries, uterine tubes, and uterus in place within the pelvic cavity, and the female breast.

EXERCISE 27.11 Gross Anatomy of the Ovary, Uterine Tubes, Uterus, and Suspensory Ligaments

The ovary, uterine tubes, and uterus are all contained within the pelvic cavity of the female and are held in place by a number of suspensory ligaments, many of which form from folds of the peritoneum as it drapes over these structures. **Table 27.10** summarizes the structure, function, and location of the suspensory ligaments.

1. Observe a female human cadaver or classroom models of the female pelvis and female reproductive organs. Also observe a classroom model demonstrating a sagittal section through a female pelvis.

2. Using table 27.10 and your textbook as guides, identify the structures listed in **figures 27.12** and **27.13** on the cadaver or on classroom models. Then label them in figures 27.12 and 27.13.

3. *Optional Activity*: **AP|R** Reproductive System—In the Quiz section, select the "Structures: Female" option to test yourself on gross anatomy of female structures.

Table 27.10	Suspensory Ligaments of the Ovary, Uterine Tubes, and Uterus	
Suspensory Ligament	**Description and Function**	**Word Origins**
Broad Ligament	Fold of peritoneum draped over the superior surface of the uterus. Portions of the broad ligament form the mesovarium and the mesosalpinx.	*broad*, wide, + *igamentum*, a bandage
Mesosalpinx	The mesentery of the uterine tube, which is formed as a fold of the most superior part of the broad ligament.	*meso* −, a mesentery-like structure, + *salpinx*, trumpet (tube)
Mesovarium	The mesentery of the ovary, which is formed as a posterior extension of the broad ligament.	*meso* −, a mesentery-like structure, + *ovarium*, ovary
Ovarian Ligament	Ligament contained within folds of the broad ligament. It extends from the medial part of the ovary to the superolateral surface of the body of the uterus.	*ovarium*, ovary, + *ligamentum*, bandage
Round Ligament of the Uterus	Ligament attached to the superolateral surface of the body of the uterus. It extends laterally to the deep inguinal ring, passes through the inguinal canal, and attaches to the skin of the labia majora.	*ligamentum*, a bandage, + *uterus*, uterus
Suspensory Ligament of the Ovary	A fold of peritoneum draping over the ovarian artery and vein superolateral to the ovary. Anchors the ovary to the lateral body wall.	*suspensio* −, to hang up, + *ligamentum*, a bandage
Transverse Cervical (Cardinal) Ligament	Ligament extending laterally from the cervix and vagina, connecting them to the pelvic wall.	*transverse* −, across, + *cervix*, neck, + *ligamentum*, a bandage
Uterosacral Ligaments	Ligament connecting the inferior part of the uterus to the sacrum posteriorly.	*utero* −, the uterus, + *sacral*, the sacrum

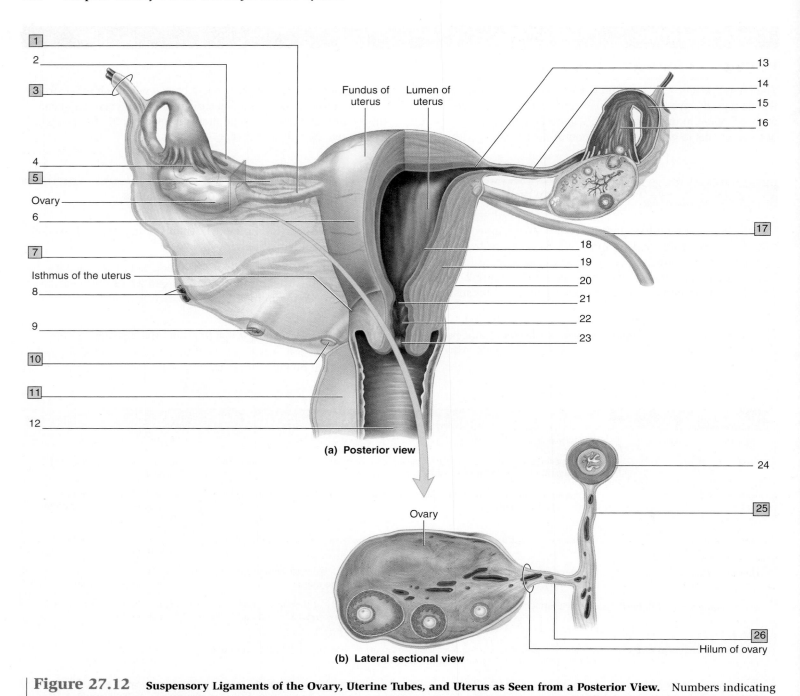

1

2

3

Fundus of uterus   Lumen of uterus

13

14

15

16

4

5

Ovary

6

7

Isthmus of the uterus

8

9

10

11

12

17

18

19

20

21

22

23

(a) Posterior view

Ovary

24

25

26

Hilum of ovary

(b) Lateral sectional view

Figure 27.12 **Suspensory Ligaments of the Ovary, Uterine Tubes, and Uterus as Seen from a Posterior View.** Numbers indicating suspensory ligaments are highlighted in green. Numbers of lines to reproductive and other associated structures have no highlight. (a) Posterior view. (b) Lateral view.

☐ ampulla of uterine tube

☐ body of uterus

☐ broad ligament

☐ cervical canal

☐ endometrium

☐ external os

☐ fimbria

☐ infundibulum of uterine tube

☐ internal os

☐ isthmus of uterine tube

☐ mesosalpinx

☐ mesovarium

☐ myometrium

☐ ovarian ligament

☐ perimetrium

☐ round ligament of the uterus

☐ suspensory ligament of the ovary

☐ transverse cervical ligament

☐ ureter

☐ uterine blood vessels

☐ uterine part of uterine tube

☐ uterine tube

☐ uterosacral ligament

☐ vagina

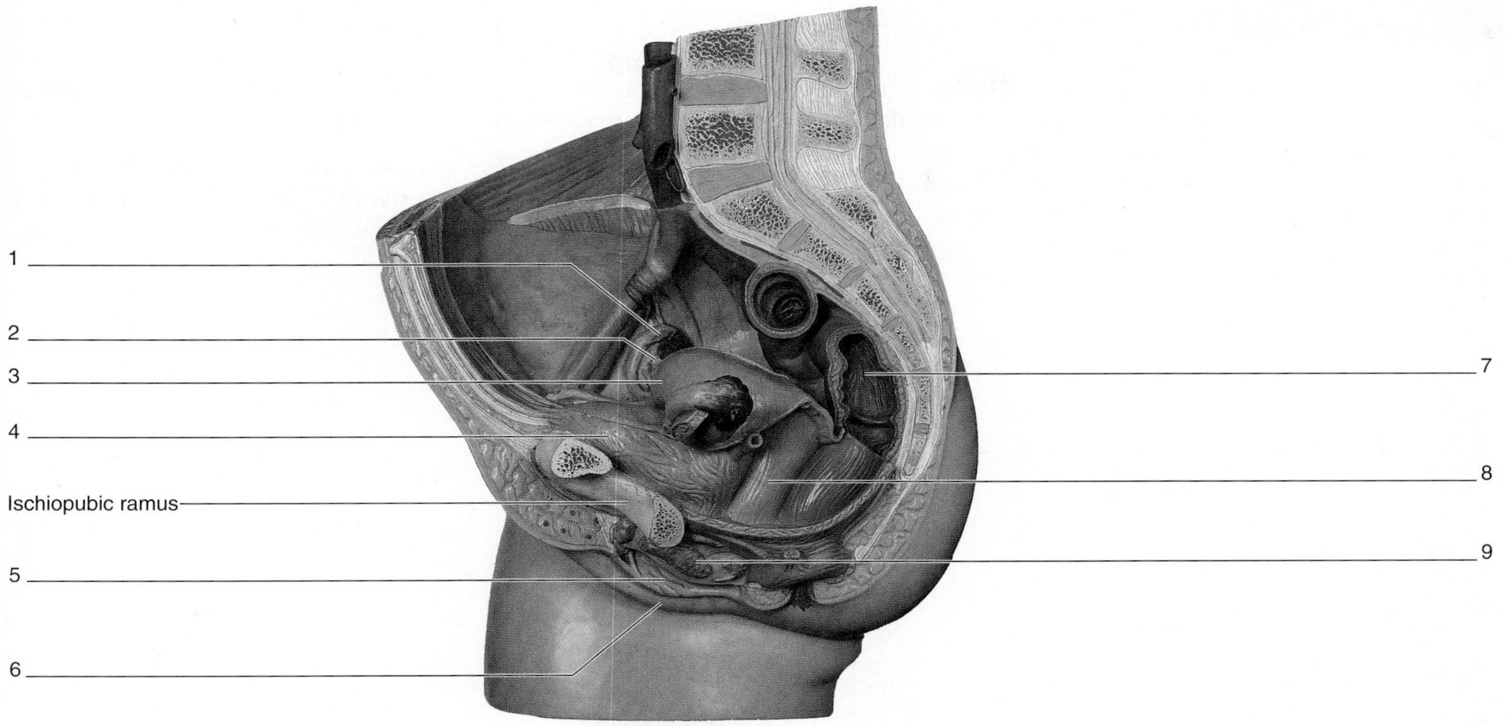

1
2
3
4
Ischiopubic ramus
5
6

7
8
9

(a) Midsagittal view

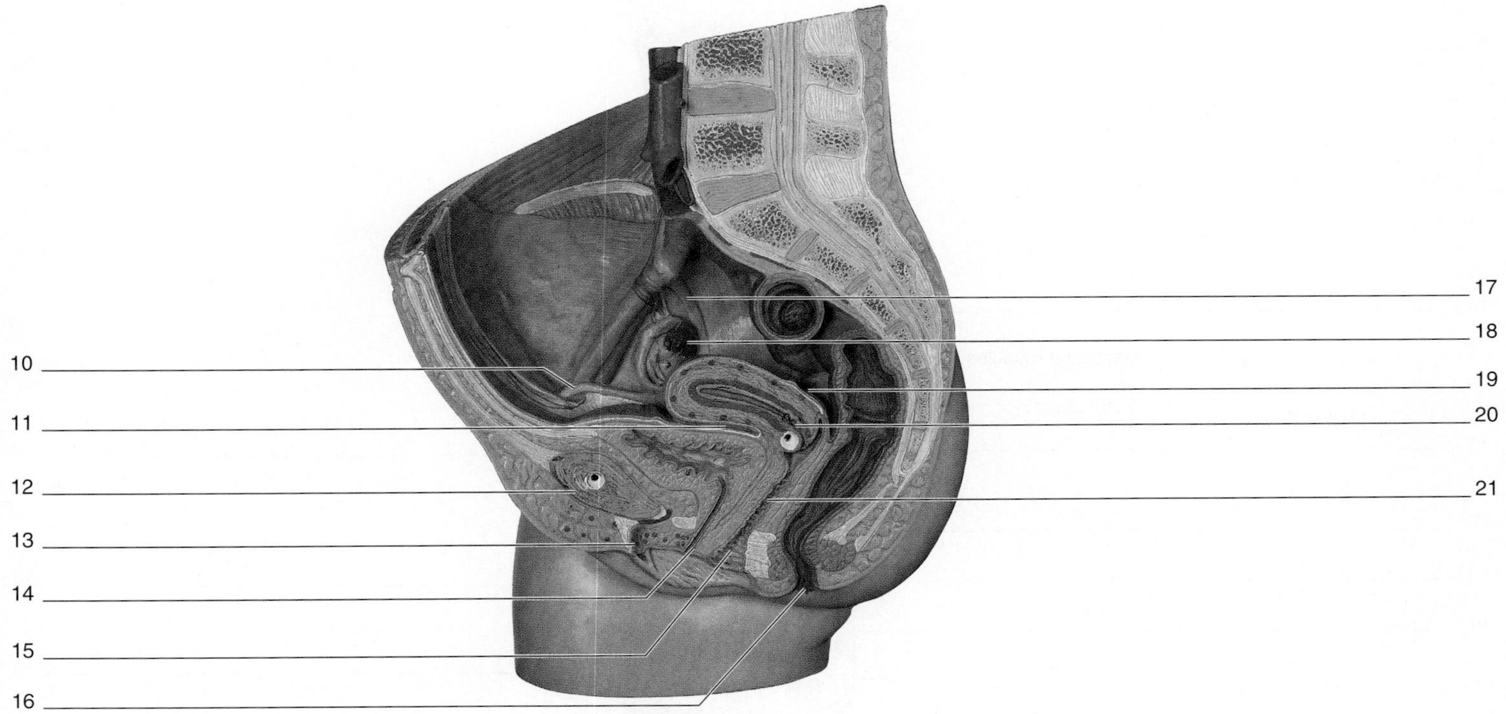

10
11
12
13
14
15
16

17
18
19
20
21

(b) Midsagittal view

Figure 27.13   **Classroom Model of the Female Pelvic Cavity.**   (a) Midsagittal view with pelvic structures intact. (b) Midsagittal view.

- [ ] anus
- [ ] bulb of the vestibule
- [ ] cervix of uterus
- [ ] clitoris
- [ ] external urethral orifice

- [ ] fimbria of uterine tube
- [ ] labia majora
- [ ] labia minora
- [ ] ovary
- [ ] pubic symphysis

- [ ] rectouterine pouch
- [ ] rectum
- [ ] round ligament of the uterus
- [ ] ureter
- [ ] urinary bladder

- [ ] uterine tube
- [ ] uterus
- [ ] vagina
- [ ] vaginal orifice
- [ ] vesicouterine pouch

EXERCISE 27.12   Gross Anatomy of the Female Breast

The female breast consists largely of fatty tissue and suspensory ligaments. Imbedded within are numerous modified sweat glands, the **mammary glands** (*mamma*, breast), which are compound tubuloalveolar exocrine glands. These glands enlarge greatly during pregnancy, enabling them to produce milk to nourish the new baby. Table 27.11 describes the structures that compose the female breast.

1. Observe the breast of a prosected female human cadaver or a classroom model of the female breast.

2. Using Table 27.11 and your textbook as guides, identify the structures listed in **figure 27.14** on the cadaver or the classroom model of the female breast. Then label them in figure 27.14.

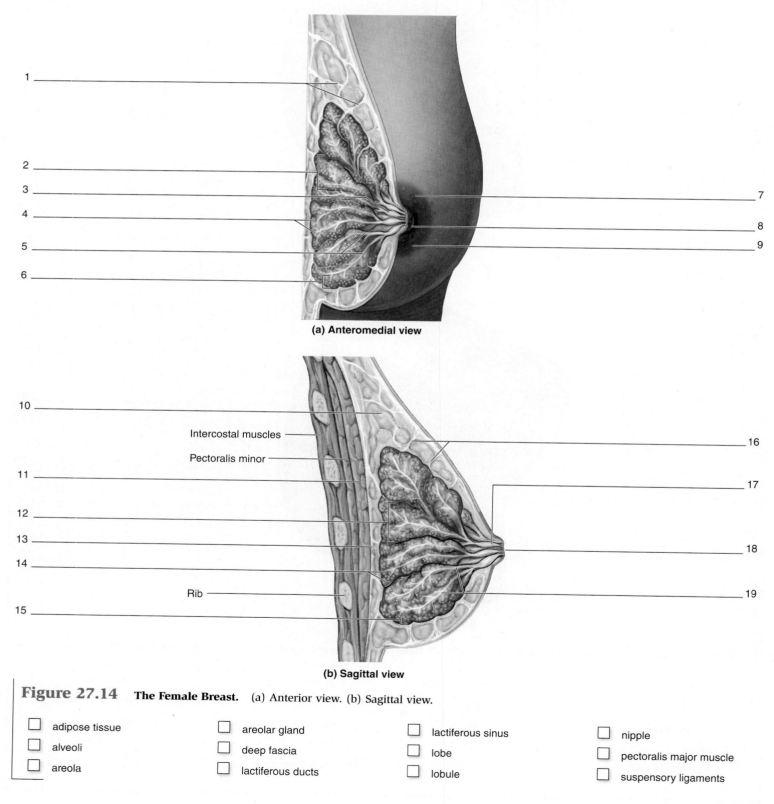

(a) Anteromedial view

(b) Sagittal view

Figure 27.14   **The Female Breast.**   (a) Anterior view. (b) Sagittal view.

☐ adipose tissue   ☐ areolar gland   ☐ lactiferous sinus   ☐ nipple

☐ alveoli   ☐ deep fascia   ☐ lobe   ☐ pectoralis major muscle

☐ areola   ☐ lactiferous ducts   ☐ lobule   ☐ suspensory ligaments

Focus | Breast Cancer and Breast Self Examination (BSE)

Incidence

In females, breast cancer is the second most commonly diagnosed cancer after skin cancer, and the second most common cause of death from cancer after lung cancer. Due to increased screening for breast cancers, currently only about 20% of patients diagnosed with breast cancer are likely to die from breast cancer. Luckily, breast cancer is a highly curable disease if caught early and there are several screening mechanisms that can be used to identify early lesion: clinical breast exam (CBE), mammography, and breast self exam (BSE). Of these, breast self exam is one of the most common ways that breast cancers are first noticed.

What is Breast Cancer?

The most common forms of breast cancers are carcinomas, which are tumors that originate in epithelial tissues. In the breast, the mammary glands and ducts consist of modified epithelial tissue, and most breast cancers originate in these tissues. The most common types of breast cancer are ductal carcinoma in situ (DCIS) and lobular carcinoma in situ (LCIS), which are tumors that originate in cells lining the ducts (DCIS) or in the glands themselves (LCIS). These tumors remain limited to the epithelial tissue (*in situ* means a tumor has not crossed the basement membrane of the epithelium). More serious forms of cancer are invasive carcinomas, which are tumors that have spread beyond the basement membrane of the epithelial tissues.

Screening

Clinical Breast Exam (CBE)

All women of reproductive age need to make annual visits to their gynecologist for a clinical breast exam and a pap smear to test for early cervical changes that are risk factors for cervical cancer. The physician performing the CBE is palpating for abnormal lumps. However, they can also help teach a woman how to perform breast self exam (BSE) so she can monitor the condition of her breasts monthly instead of just during yearly visits. The vast majority of breast cancers are discovered by the patient herself.

Mammography

A mammogram is an x-ray of the breast that is used to look for areas of extra density or calcifications, which can be early signs of breast cancer. Most women are advised to start having mammograms at age 40, but those with a family history of the disease are often advised to have their first mammogram at age 35. Mammography is good at locating very small tumors (especially DCIS), which may not be palpable on self examinations or clinical examinations.

Breast Self Exam (BSE)

Women should perform a breast self exam at least once a month to look for any unusual bumps. Upon palpation, a typical breast feels a bit lumpy, with harder areas and softer areas. The harder areas can be thickenings of the suspensory ligaments, or fibrocystic changes to the breast. The softer areas are typically just adipose tissue. There are a number of benign (non-cancerous) changes to the breast that can make breasts feel "lumpy." Specifically, fibrocystic changes in the breast consist of fluid-filled cysts and are often surrounded by dense fibrous tissue. This tissue, though often large and dense, is typically mobile (moves around easily), and often mirrors itself on the opposite breast. That is, if you find it in the lower right quadrant of the right breast, you might also find it on the lower right quadrant of the left breast. On the contrary, breast cancers tend to feel much harder, like a kernel of unpopped popcorn, and they are immobile (they do not move when you touch them). They are generally painless and sometimes cause dimpling of the skin overlying the tumor.

The following are three procedures for performing a breast self exam:

1. In the shower: Use soapy hands to palpate the breast while raising the arm on the same side of the breast you are palpating. Starting at the periphery of the breast, use two or three fingers to make small circles that progressively move toward the nipple. Squeeze the nipple to look for discharge. Be sure to palpate all the way up into the axilla, as the majority of tumors arise in the outer/upper quadrant toward the axillary region of the breast.

2. Lying down: Use the same procedure as in the shower. It's good to do an exam lying down because it makes the breasts flatter and thus it may be easier to feel certain lumps.

3. Standing in front of mirror: Place your hands on your hips, lean your elbows forward, and look for any indentations in the skin of the breast or "orange-peel" looking skin, which can be indicative of underlying tumor.

If you discover anything that concerns you, make an appointment with your gynecologist as soon as possible to discuss your findings so he or she can help determine if what you felt was something benign or something that needs further exploration. Remember, always, that breast cancer is *highly curable* when caught early.

Table 27. 11	The Female Breast	
Structure	**Description and Function**	**Word Origins**
Alveoli	Secretory units of the mammary glands, which produce milk.	alveolus, a concave vessel, a bowl
Areola	The pigmented area of skin surrounding the nipple.	areola, area
Areolar glands	Sebaceous glands deep to the skin of the areola; produce sebum, which keeps the skin of the areola moist, particularly during lactation.	areolar, relating to the areola of the breast
Lactiferous duct	Ducts that form from the confluence of small ducts draining milk from the alveoli and lobules.	lacto-, milk + ductus, to lead
Lactiferous sinus	10-20 large channels that form from the confluence of several lactiferous ducts; the spaces where milk is stored prior to release from the nipple.	lacto-, milk + sinus, cavity
Lobes	Large subdivisions of the mammary glands.	lobos, lobe
Lobules	Smaller subdivisions of the mammary glands, which contain the alveoli.	lobulus, a small lobe
Nipple	A cylindrical projection in the center of the breast that contains the openings of the lactiferous ducts.	neb, beak or nose
Suspensory ligaments	Bands of connective tissue that anchor the breast skin and tissue to the deep fascia overlying the pectoralis major muscle.	suspensio, to hang up, + ligamentum, a band

Male Reproductive System

The **testes**, the primary male reproductive organs, function in the production of sperm and testosterone. Sperm produced in the testes are stored in the **epididymis** and transported via the **ductus deferens** to the male urethra during ejaculation. The **seminal vesicles** and **prostate gland** are accessory reproductive glands that produce substances that nourish and protect the sperm and also compose the vast majority of semen. The **bulbourethral glands** are small glands located within the urogenital diaphragm that produce a mucous-like substance during sexual arousal that neutralizes the acidity of the male urethra and provides lubrication to ease the passage of semen during ejaculation.

In exercises 27.5 to 27.10 you examined the histological appearance of many of these structures. Now we will observe the male reproductive structures at the gross level.

EXERCISE 27.13 Gross Anatomy of the Scrotum, Testis, Spermatic Cord, and Penis

1. Observe a prosected male human cadaver or classroom models of the male reproductive organs.

2. Using your textbook as guides, identify the structures listed in **figures 27.15** and **27.16** on the cadaver or on classroom models. Then label them in figures 27.15 and 27.16.

3. *Optional Activity*: AP|R **Reproductive System**—In the Quiz section, select the "Structures: Male" option to test your knowledge of male reproductive gross anatomy.

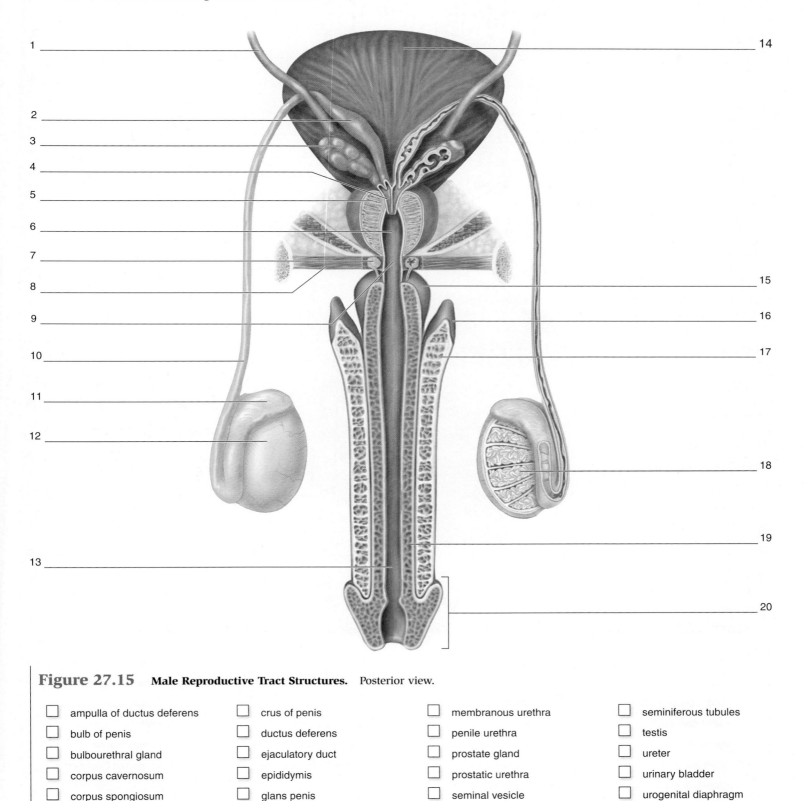

Figure 27.15 **Male Reproductive Tract Structures.** Posterior view.

☐ ampulla of ductus deferens

☐ bulb of penis

☐ bulbourethral gland

☐ corpus cavernosum

☐ corpus spongiosum

☐ crus of penis

☐ ductus deferens

☐ ejaculatory duct

☐ epididymis

☐ glans penis

☐ membranous urethra

☐ penile urethra

☐ prostate gland

☐ prostatic urethra

☐ seminal vesicle

☐ seminiferous tubules

☐ testis

☐ ureter

☐ urinary bladder

☐ urogenital diaphragm

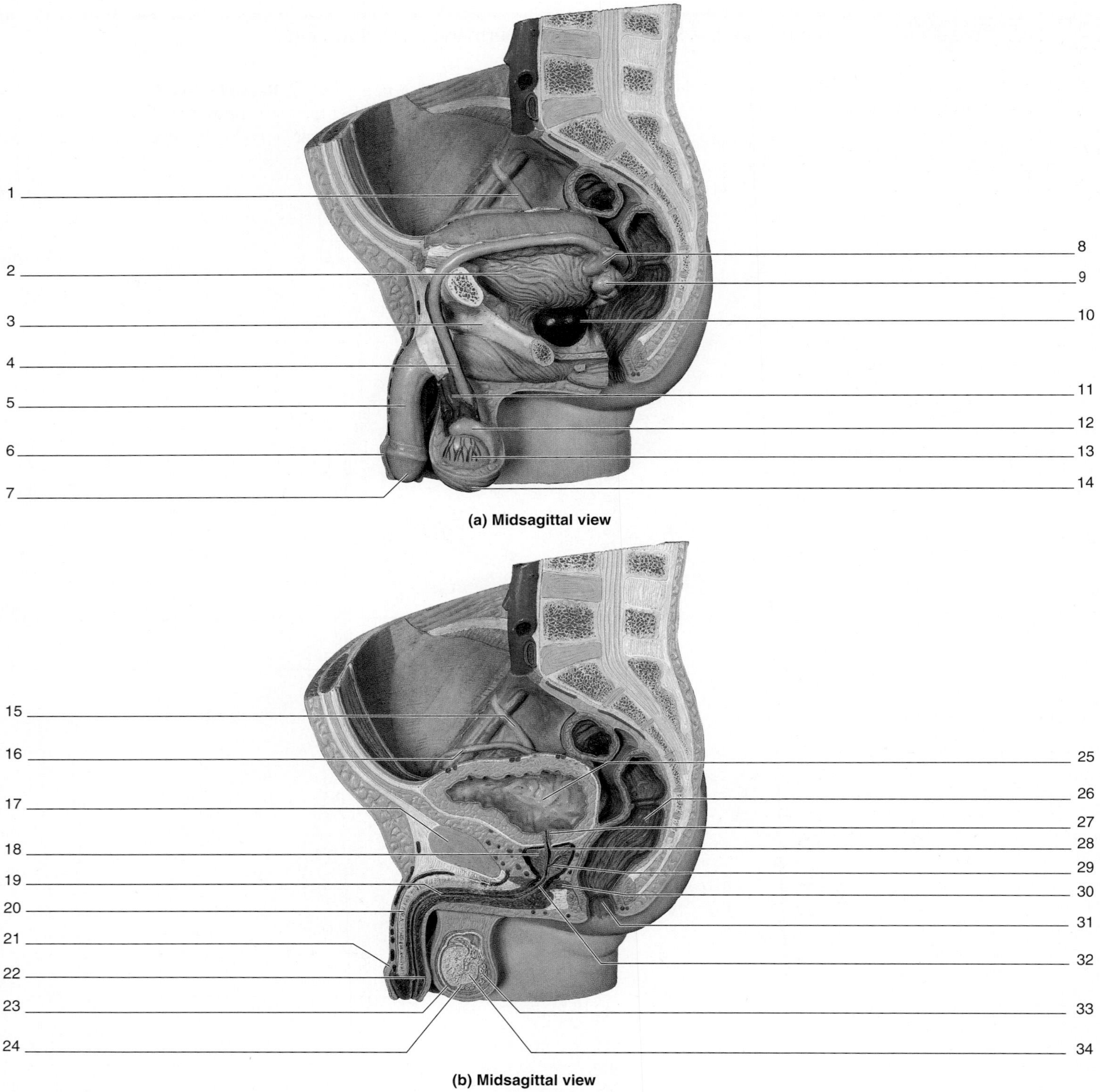

1

2

3

4

5

6

7

8

9

10

11

12

13

14

(a) Midsagittal view

15

16

17

18

19

20

21

22

23

24

25

26

27

28

29

30

31

32

33

34

(b) Midsagittal view

Figure 27.16 **Classroom Model of the Male Pelvic Cavity.** (a) Midsagittal view with pelvic structures intact. (b) Midsagittal view.

☐ anus
☐ corpus cavernosum
☐ ductus deferens
☐ ejaculatory duct
☐ epididymis
☐ glans penis
☐ internal urethral sphincter

☐ ischiopubic ramus
☐ membranous urethra
☐ penile urethra
☐ penis
☐ prepuce
☐ prostate gland
☐ prostatic urethra

☐ pubic symphysis
☐ rectum
☐ scrotum
☐ seminal vesicle
☐ seminiferous tubules
☐ testicular artery and vein

☐ testis
☐ tunica albuginea of testis
☐ tunica vaginalis of testis
☐ ureter
☐ urinary bladder
☐ urogenital diaphragm

1. Compare and contrast the following:

   a. primary follicle   vs.   primary oocyte

   _____

   _____

   _____

   _____

   b. secondary follicle   vs.   secondary oocyte

   _____

   _____

   _____

   _____

2. How can you distinguish histologically between a secondary follicle and a vesicular (Graafian) follicle?

   _____

   _____

   _____

3. Put the following terms in the correct order to describe the pathway that an ovum takes as it travels from the ovary to the uterus in the female.

   ovarian cortex   infundibulum of uterine tube

   uterus   uterine part of uterine tube

   isthmus of uterine tube   ampulla of uterine tube

   _____

   _____

   _____

4. What type of epithelium lines the uterine tube?

   _____

5. The figure below demonstrates a _____ follicle, which contains a _____ oocyte. The oocyte is in the _____ stage of meiosis.

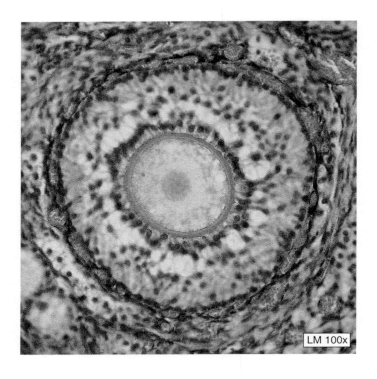

LM 100x

6. The peritoneal cavity in the female is considered an open cavity. Based on your knowledge of the female reproductive organs, explain why.

_____

_____

_____

_____

_____

7. Does the inguinal canal exist in females? _____ If so, what structure(s) run through it?

_____

_____

8. You are a nurse and your job is to insert a catheter into the urethra of a female patient.
   a. Where will you look for the opening to the female urethra (be specific)?

   _____

   b. Approximately how long must the catheter be in order to reach into the bladder of the female?

   _____

9. The two layers of the uterine endometrium are the _____ layer and the _____ layer. The _____ layer is shed during menstruation.

10. What type of glands are mammary glands?

   _____

11. List the following terms in the correct sequence to describe the correct maturation sequence for a developing sperm cell.

spermatid   primary spermatocyte   spermatozoon

secondary spermatocyte   spermatogonia

_____

_____

_____

_____

12. List three functions of sustentacular cells.

a. _____

b. _____

c. _____

13. Label the following diagram of the testis.

1 _____

2 _____

3 _____

4 _____

5 _____

6

7

8

9

10

11

12

13

14

14. What type of epithelium lines the epididymis?

_____

15. What are cilia? (Describe their structure, location, and function.)

_____

_____

16. What is the function of the epididymis?

_____

_____

17. What type of epithelium lines the ductus deferens?

_____

18. Trace the pathway a sperm travels, starting at the epididymis, as it travels through the male reproductive tract during ejaculation. Name all accessory reproductive glands encountered along the way.

_____

_____

_____

_____

_____

_____

19. Based on your observations of male reproductive structures (and using table 27.9 as a guide), list the main constituents of semen and the relative percentages of each constituent.

_____

_____

_____

20. Complete the table below, which lists homologous female and male reproductive structures.

Female Structure	Male Homologue
Ovary	
	Penis
Labia majora	
	Bulbourethral glands

21. Identify the structure shown in each of the photomicrographs below. Then describe the characteristic features of each structure that allowed you to come to your answer.

a. Structure:_____

Description: _____

_____

_____

_____

b. Structure:_____

Description: _____

_____

_____

_____

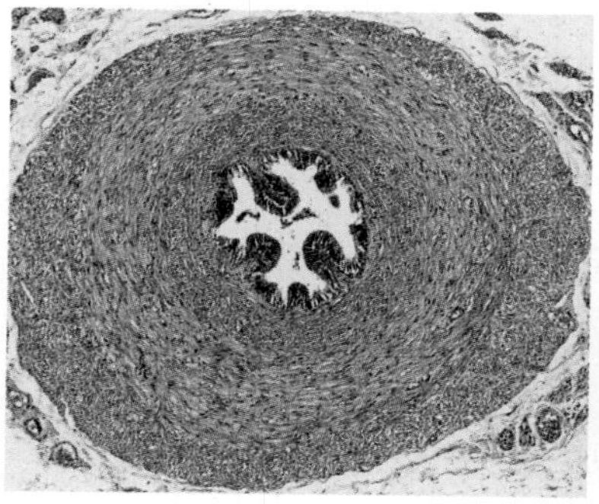

(a)

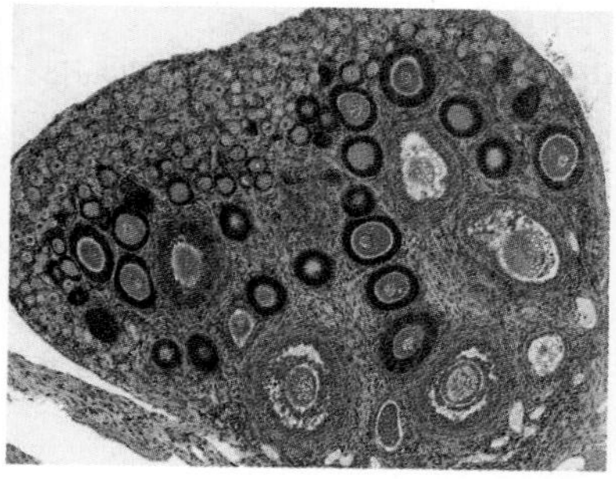

(b)

Appendix

This appendix includes answers to the What Do You Think? questions, labeling exercises, table completion activities, and calculation questions found in each chapter. Answers to in-exercise questions prompting students to record their personal observations and drawings are not provided due to their variability.

Chapter 1

Exercise 1.1
1 Forceps
2 Scalpel blades
3 Scalpel (disposable)
4 Scalpel blade handle (#3)
5 Scalpel blade handle (#4)
6 Scissors (curved)
7 Scissors (pointed)
8 Dissecting pins
9 Blunt probe
10 Dissecting needle
11 Blunt probe

Exercise 1.2
1 A
2 B
3 A
4 A
5 B
6 B
7 A

Chapter 2

What Do You Think?
1. The **spleen** was likely the source of the bleeding/injured organ. The spleen is located in the **left upper quadrant** and the **left hypochondriac region.** Because the spleen is located deep and inferior to the ribs, it can be injured by the sharp edges of a broken rib.

Chapter Activities
Figure 2.2   Sections Through a Human Brain
1 coronal (frontal)
2 midsagittal (median)
3 transverse (horizontal)
4 sagittal (or parasagittal)

Figure 2.3   Posterior View of an Individual with Three Reference Locations Marked
1 thoracic
2 antebrachial
3 femoral

Figure 2.4   Regional Terms
1 oral (mouth)
2 cervical (neck)
3 axillary (armpit)
4 brachial (arm)
5 antebrachial (forearm)
6 carpal (wrist)
7 digital (finger)
8 femoral (thigh)
9 crural (leg)
10 frontal (forehead)
11 orbital (eye)
12 buccal (cheek)
13 mental (chin)
14 mammary (breast)
15 pelvic
16 inguinal (groin)
17 tarsal (ankle)
18 otic (ear)
19 vertebral (spinal column)
20 sacral
21 sural (posterior leg)
22 calcaneal (heel)
23 occipital (back of head)
24 lumbar (lower back)
25 perineal
26 popliteal (back of knee)

Figure 2.5   Body Cavities
1 posterior aspect
2 ventral cavity
3 cranial cavity
4 vertebral canal
5 thoracic cavity
6 abdominopelvic cavity
7 diaphragm
8 abdominal cavity
9 pelvic cavity
10 thoracic cavity
11 abdominopelvic cavity
12 mediastinum
13 pleural cavity
14 pericardial cavity
15 abdominal cavity
16 pelvic cavity

Exercise 2.5, #2

Organ	Quadrant(s)	Region(s)
Left kidney	left upper	left hypochondriac, left lumbar
Liver	right upper, left upper	right hypochondriac, right lumbar, epigastric
Pancreas	right and left upper	epigastric, left hypochondriac
Small intestine	right and left lower	umbilical, hypogastric, right and left lumbar and right and left iliac
Spleen	left upper	left hypochondriac
Stomach	right and left upper	epigastric, left hypochondriac
Urinary bladder	right and left lower	hypogastric

Chapter 3

What Do You Think?
1. As total magnification increases, working distance decreases. In a practical sense, this means you have less room to move the stage without having the objective lens run into the slide. With this in mind, remember you should *never* use the coarse adjustment knob with the high-power objective in place because you can easily crack or break the slide or the objective lens.
2. You can answer this question using common sense and get pretty close to the correct answer, or you can be more specific and calculate the diameters of the field of view at both magnifications, estimate cell diameter at 200× total magnification, and then calculate the number of cells that could fit within the field of view at 500× total magnification (see calculations below):

Diameter of the field of view at 200×: 0.6 mm
Total magnification at starting power: 200×
Diameter of field of view at 500×: *unknown*
Total magnification at higher power: 500×

Calculation:

$$0.6 \text{ mm} * 200× = \underline{unknown} * 500×$$
$$120 = \underline{unknown} * 500×$$
$$120/500 = 0.24 \text{ mm diameter of field of view at 500× total magnification}$$
$$\text{Four cells wide} = 0.6 \text{ mm}/4 \text{ cells} = .15 \text{ mm/cell at 200× (diameter of one cell)}$$

Final answer: Field of view at 500× is **0.24 mm,** so approximately **1.6 cells** could be seen ([0.24 mm diameter /(0.15 mm/cell)] = 1.6 cells).

Chapter Activities
Figure 3.5
a. First, note that approximately 5 "objects" would fit side-by-side at the widest part of the circle. Thus, to calculate the length of the object, divide the diameter of the field of view by 5.
$$0.6 \text{ mm} / 5 = .12 \text{ mm}$$
$$\text{Length of object} = 0.12 \text{ mm}$$
b. For this question, use the same procedure described for What Do You Think? question 2 (see above).
$$\text{magnification 1} * \text{diameter 1} = \text{magnification 2} * \text{unknown diameter of field}$$
$$200 * 0.6 = 40 * \text{unknown diameter of field}$$
$$120/40 = \text{unknown diameter of field}$$
$$\text{Diameter of field} = 3 \text{ mm}$$
Then decide how many "objects" would fit side-by-side at the widest part of the circle, which looks like about 25. Finally, divide the diameter of the field (3 mm) by the number of objects (25) to get the diameter of one object: 3/25 = **0.12 mm**
c. For this question, use the same procedure described for What Do You Think? question 2 (see above).
$$\text{Magnification 1} * \text{diameter 1} = \text{magnification 2} * \text{unknown diameter of field}$$
$$200 * 0.6 = 500 * \text{unknown diameter of field}$$
$$120/500 = \text{unknown diameter of field}$$
$$\text{Diameter of field} = 0.24 \text{ mm}$$

Then decide how many "objects" would fit side-by-side at the widest part of the circle, which looks like about 2. Finally, divide the diameter of the field (0.24 mm) by the number of objects (2) to get the diameter of one object: .24/2 = **0.12 mm.**

Notice that your results for diameter of the object should be the same, even though the magnifications are different.

Chapter 4

What Do You Think?

1. Rough endoplasmic reticulum (ER) functions to **synthesize proteins destined for the plasma membrane** of the cell. Nerve cells have a lot of protein channels in their cell membranes, which are used to transport ions such as Na^+ and K^+ in and out of the cell as they generate electrical signals. Thus, nerve cells have an abundance of rough ER, which is used in the synthesis of these protein channels.

2. Microtubules are necessary for forming the mitotic spindle and moving chromosomes during mitosis. A drug that interferes with the lengthening or shortening of microtubules will interfere with prophase, metaphase, anaphase, and telophase of mitosis. Thus, **mitosis will fail to occur.** This is advantageous for a chemotherapy drug because the objective is to stop the tumor cells from dividing and multiplying to slow or halt the growth of the tumor.

Chapter 5

What Do You Think?

1. For each of the examples given below, is the structure likely to fit *in its entirety* on a microscope slide? (Circle yes or no for each.)
   1. yes; 2. no; 3. yes; 4. no 5. yes

2. The function of the uterine tube is to transport an ovum (egg) from the ovary to the uterus in the female reproductive tract (see chapter 26 for more information on this). Thus, the function of the cilia is to **propel the ovum toward the uterus** within the uterine tube.

3. Cuboidal and columnar cells contain many more **mitochondria** and **rough endoplasmic reticulum** than squamous cells. Mitochondria, the site of cellular respiration, produce most of the ATP in the cell. Rough endoplasmic reticulum functions to synthesize the carrier proteins—or "pumps"—that are used for membrane transport. To transport substances across an epithelial lining, the pumps use a lot of energy in the form of ATP. Because squamous cells are so thin, they generally allow substances to be transported via diffusion or pinocytosis rather than active transport. Simple diffusion does not require energy input, so the cell does not have the same energy needs or requirements for pumps. Thus, it does not need as many mitochondria or as much rough endoplasmic reticulum.

4. Avascular means the tissue does not contain blood vessels. Without a blood supply, cartilage will be **slow to grow and repair.** In addition, because the chondrocytes must rely on diffusion from the blood vessels that surround the cartilage to obtain nutrients, there will be a **limitation on growth in thickness** of the cartilage.

5. Most brain tumors arise from **glial cells** and are referred to as **gliomas.** Because glial cells retain the capacity for cell division and are constantly being replaced, they are more likely to experience damage to their DNA, which is the first step involved in the transformation of normal cells into cancerous cells (which subsequently form tumors).

Chapter 6

What Do You Think?

1. The layer of the epidermis that represents the transition from living to dead epithelial cells is the **stratum granulosum.** Keratinocytes begin to die within this layer because the epidermis lacks a blood supply and the keratinocytes **are too far away from their blood supply** (which is in the dermis) to receive adequate nutrients for survival.

2. The keratinocytes closest to the stratum basale are constantly undergoing cell division, or **mitosis.** These keratinocytes concentrate melanin granules just apical to their nuclei so the melanin can absorb UV radiation and prevent it from hitting the nuclei of the cells. UV radiation can cause errors to occur in DNA replication. When errors of DNA replication accumulate over time, it causes the cells to become abnormal and/or cancerous. For this reason, cells that are actively undergoing mitosis are most in need of the protection provided by melanin.

3. **Thick skin has more dermal papillae and epidermal ridges** than thin skin, and the epidermal/dermal layer is thrown into deep folds. In thin skin, the epidermal/dermal junction is flatter, and there are fewer dermal papillae and epidermal ridges. This is because **thick skin,** which is found on the palms of the hands and soles of the feet, **experiences far greater frictional forces** than thin skin. The greater number of epidermal ridges **helps keep the epidermis and dermis from separating.** This is why it is more difficult to develop a *blister* (an accumulation of fluid between the epidermis and dermis) in thick skin than in thin skin.

4. Placing a high concentration of tactile corpuscles on the surface of the skin allows us to sense very fine fluctuations in vibration, pressure and texture. There are a great deal more sensory receptors on the skin of your hand than on your back, which is why you can discriminate objects much better with your hands than your back.

Chapter 7

What Do You Think?

1. When a bone fractures and subsequently undergoes the process of repair, **osteoblast** cells are involved in laying down new bone matrix and **osteoclast** cells are involved in remodeling the new bone that is formed as the repair process continues.

2. For an epiphyseal plate to close, the **formation of new cartilage** on the epiphyseal side of the plate must stop. Then, as the process of bone replacing cartilage continues over time, the bone eventually replaces all the cartilage and closure of the plate becomes complete.

3. The medullary cavity of an adult long bone is filled with adipose connective tissue because, as ossification of the bone occurs, red bone marrow is converted into yellow bone marrow (adipose connective tissue). The **adipose tissue fills up what is a potentially large, open space within the bone** that, if left open, would make it easy for any infection within the bone to spread easily.

4. Because the **collagen fibers of the knee ligaments are continuous with the collagen fibers within the fibrocartilage of the meniscus,** most individuals who tear knee ligaments also have a torn meniscus. This is particularly true of tears to the anterior cruciate ligament, which are often associated with tears to the medial or lateral meniscus (or both).

Chapter Activities

Figure 7.8   The Human Skeleton

1 cranium	9 ilium	17 femur
2 mandible	10 ischium	18 patella
3 clavicle	11 radius	19 fibula
4 scapula	12 ulna	20 tibia
5 sternum	13 pubis	21 tarsals
6 humerus	14 carpals	22 metatarsals
7 ribs	15 metacarpals	23 phalanges
8 vertebrae	16 phalanges	

Chapter 8

What Do You Think?

1. Fontanelles are present after the birth of the infant **so the cranial cavity can continue to expand** to allow for the massive growth that occurs in the brain.

2. The bifid spinous processes of a superior cervical vertebra allow it to fit *over* the spinous process of the cervical vertebra below it. This **increases the range of motion in extension** of the neck.

3. The superior and inferior articular processes of the atlas are **oriented in the horizontal plane** instead of vertically. This allows for the gliding, fore-aft "yes" movement allowed at the atlanto-occipital joint, and the rotational "no" movement allowed at the atlantoaxial joint.

4. Little to no lateral rotation is allowed in the lumbar region because the function of the lumbar region of the vertebral column is to **provide solid support for the rest of the vertebral column,** which lies above it.

5. Whenever needles are inserted into the thoracic cavity they are always placed along the superior border of a rib **so as not to injure the intercostal nerve, artery, and vein,** which run in the costal groove on the inferior surface of a rib.

Chapter Activities

Figure 8.1   Anterior View of the Skull

1 frontal bone	13 maxilla
2 parietal bone	14 superciliary arch
3 glabella	15 supraorbital margin
4 frontonasal suture	16 supraorbital foramen (notch)
5 sphenoid bone	17 optic foramen (canal)
6 superior orbital fissure	18 inferior orbital fissure
7 lacrimal bone	19 perpendicular plate of ethmoid
8 nasal bone	20 vomer
9 temporal process of zygomatic bone	21 inferior nasal concha
10 infraorbital foramen	22 mandible
11 zygomatic bone	23 mental foramen
12 mastoid process	24 mental protuberance

Figure 8.2   The Orbit

1 nasal bone
2 supraorbital foramen
3 ethmoid bone
4 frontal bone
5 zygomatic process of frontal bone
6 greater wing of sphenoid
7 optic foramen
8 superior orbital fissure
9 lacrimal bone
10 inferior orbital fissure
11 zygomatic bone
12 infraorbital foramen
13 maxilla
14 frontal process of maxilla

Figure 8.3   The Nasal Cavity

1 frontal sinus
2 nasal bone
3 superior nasal concha
4 middle nasal concha
5 lacrimal bone
6 inferior nasal concha
7 maxilla
8 palatine process of maxilla
9 crista galli
10 cribriform plate of ethmoid
11 sella turcica
12 sphenoid sinus
13 palatine bone
14 sphenoid bone
15 horizontal plate
    of palatine bone
16 perpendicular plate
    of ethmoid bone
17 vomer

Figure 8.4   The Mandible

1 mandibular condyle
2 coronoid process
3 mandibular foramen
4 condylar process
    (mandibular condyle)
5 mandibular notch
6 mylohyoid line
7 ramus
8 alveolar process
9 mental foramen
10 angle
11 body
12 mental protuberance

Figure 8.5   Lateral View of the Skull

1 parietal eminence
2 parietal bone
3 inferior temporal line
4 squamosal suture
5 lambdoid suture
6 squamous part of temporal bone
7 occipital bone
8 external acoustic meatus
9 mastoid process of temporal bone
10 styloid process of temporal bone
11 head of mandible
12 zygomatic process of temporal bone
13 temporal process
    of zygomatic bone
14 body of mandible
15 coronal suture
16 frontal bone
17 superior temporal line
18 pterion
19 greater wing of sphenoid bone
20 nasal bone
21 lacrimal bone
22 ethmoid bone
23 lacrimal groove
24 zygomatic bone
25 maxilla
26 mental foramen
27 mental protuberance

Figure 8.6   Posterior View of the Skull

1 sagittal suture
2 parietal foramina
3 parietal bone
4 parietal eminence
5 lambdoid suture
6 sutural (wormian) bone
7 occipital bone
8 temporal bone
9 external occipital protuberance
10 mastoid process
11 mandible

Figure 8.7   Superior View of the Skull

1 frontal bone
2 coronal suture
3 sagittal suture
4 parietal bone
5 parietal foramina
6 lambdoid suture
7 sutural (wormian) bone
8 occipital bone

Figure 8.8   Inferior View of the Skull

1 maxilla
2 palatine bone
3 palatine foramina
4 vomer
5 sphenoid bone
6 foramen ovale
7 foramen spinosum
8 foramen lacerum
9 jugular foramen
10 carotid canal
11 mastoid foramen
12 inferior nuchal line
13 superior nuchal line
14 incisive foramen
15 temporal process of zygomatic
    bone
16 lateral pterygoid plate
17 zygomatic process of temporal
    bone
18 mandibular fossa
19 styloid process
20 temporal bone
21 mastoid process
22 occipital condyle
23 hypogossal canal
24 basilar region of occipital
    bone
25 foramen magnum
26 external occipital crest
27 lambdoid suture
28 external occipital
    protuberance

Figure 8.9   Cranial Fossae

1 anterior cranial fossa
2 middle cranial fossa
3 posterior cranial fossa

Figure 8.11   Superior View of the Cranial Floor

1 frontal sinus
2 frontal bone
3 optic canal (foramen)
4 lesser wing of sphenoid
5 foramen rotundum
6 greater wing of sphenoid
7 temporal bone
8 petrous part of temporal bone
9 parietal bone
10 internal auditory meatus
11 foramen magnum
12 occipital bone
13 internal occipital protuberance
14 frontal crest
15 crista galli
16 cribriform plate of the ethmoid
17 sella turcica
18 foramen lacerum
19 foramen ovale
20 foramen spinosum
21 jugular foramen
22 hypoglossal canal
23 groove for sigmoid sinus
24 clivus
25 groove for transverse sinus
26 internal occipital crest
27 anterior clinoid process
28 tuberculum sellae
29 hypophyseal fossa
30 posterior clinoid process
31 dorsum sellae

Figure 8.12   The Hyoid Bone

1 greater cornu
2 lesser cornu
3 body

Figure 8.13   The Fetal Skull

1 frontal bone
2 parietal bone
3 sphenoidal fontanelle
4 sphenoid bone
5 mandible
6 temporal bone
7 mastoid fontanelle
8 occipital bone
9 frontal bone
10 anterior fontanelle
11 parietal bone

Figure 8.14   Lateral View of the Vertebral Column

1 cervical vertebrae
    Number of vertebrae: 7
2 thoracic vertebrae
    Number of vertebrae: 12
3 lumbar vertebrae
    Number of vertebrae: 5
4 sacral vertebrae/sacrum
    Number of vertebrae: 5
5 coccygeal vertebrae
    Number of vertebrae: 4
6 cervical curvature
7 thoracic curvature
8 lumbar curvature
9 sacral curvature

Figure 8.15   A Typical Vertebra

1 lamina
2 transverse process (costal facet)
3 superior articular process
4 body
5 spinous process
6 vertebral arch
7 transverse process
8 pedicle
9 vertebral (spinal) foramen
10 transverse process (costal facet)
11 superior articular process
12 body
13 vertebral notch
14 inferior articular process

Figure 8.16   Cervical Vertebra

1 body
2 transverse process
3 superior articular process
    (and facet)
4 transverse foramen
5 pedicle
6 vertebral (spinal) foramen
7 lamina
8 spinous process (bifid)
9 spinous process
10 superior articular process
11 transverse foramen
12 body
13 inferior articular process
    (and facet)

Figure 8.17   The Atlas (C1)

1 anterior tubercle
2 anterior arch
3 superior articular facet
4 transverse process
5 transverse foramen
6 posterior tubercle
7 posterior arch
8 articular facet for dens
9 vertebral foramen
10 groove for vertebral artery

Figure 8.18   The Axis (C2)

1 dens (odontoid process)
2 superior articular process
3 transverse foramen
4 transverse process
5 pedicle
6 vertebral (spinal) foramen
7 lamina
8 spinous process

Figure 8.19   Thoracic Vertebra

1 body
2 vertebral (spinal) foramen
3 superior articular process
4 spinous process
5 superior costal facet
6 pedicle
7 costal facet
8 transverse process
9 lamina

10 costal facet
11 spinous process
12 inferior articular process
13 superior articular process
14 superior costal facet
15 body
16 inferior costal facet
17 inferior vertebral notch

Figure 8.20   Lumbar Vertebra

1 body
2 vertebral (spinal) foramen
3 superior articular process
4 spinous process
5 pedicle
6 transverse process

7 lamina
8 superior articular facet
9 transverse process
10 spinous process
11 inferior articular process
12 body

Figure 8.21   Sacrum and Coccyx

1 superior articular process
2 ala
3 sacral promontory
4 anterior sacral foramina
5 transverse lines (ridges)
6 coccygeal cornu
7 coccyx
8 sacral canal

9 superior articular facet
10 median sacral crest
11 auricular surface
12 posterior sacral foramina
13 sacral hiatus
14 coccygeal cornu
15 coccyx

Figure 8.22   The Sternum

1 suprasternal notch
2 manubrium
3 sternal angle

4 body
5 xiphoid process
6 second rib

Figure 8.23   A Typical Rib

1 superior articular facet
2 inferior articular facet
3 shaft
4 head
5 neck

6 tubercle
7 articular facet
   for transverse process
8 angle
9 costal groove

Figure 8.24   The First Rib

1 shaft
2 groove for subclavian vein
3 groove for subclavian artery

4 head
5 neck
6 tubercle

Chapter 9

What Do You Think?

1. The **surgical neck** of the humerus is more likely to fracture in an accident, particularly in younger individuals. The surgical neck is so named *because* of the propensity of this portion of the humerus to fracture over that of the "regular", or anatomical, neck. The joint between the scapula and humerus is the **glenohumeral joint.**
2. The **scaphoid** bone is most likely to fracture when someone falls on an outstretched hand.
3. It is functionally important that the bones of the os coxae are fused together, rather than remaining as independent bones so that they can **provide solid support between the lower limb and the axial skeleton,** because the weight of the entire body must be supported by the lower limb when standing.
4. A fracture to the **tibia,** a bone that must support the weight of the entire body, would result in the greatest loss of function of the lower limb. Because the fibula is not a weight-bearing bone, fractures of the fibula create fewer functional problems than fractures of the tibia.

Chapter Activities

Figure 9.1   The Clavicle

1 acromial end
2 sternal end
3 acromial end

4 conoid tubercle
5 sternal end
6 costal tuberosity

Figure 9.2   The Right Scapula

1 acromion
2 coracoid process
3 glenoid cavity
4 lateral border

5 inferior angle
6 suprascapular notch
7 spine
8 superior border

9 superior angle
10 subscapular fossa
11 medial border
12 acromion
13 spine
14 glenoid cavity
15 infraglenoid tubercle
16. lateral border
17 superior angle
18 supraglenoid tubercle
19 coracoid process
20 subscapular fossa

21 inferior angle
22 coracoid process
23 suprascapular notch
24 superior border
25 supraspinous fossa
26 spine
27 infraspinous fossa
28 medial border
29 acromion
30 glenoid cavity
31 lateral border
32 inferior angle

Figure 9.3   The Right Humerus

1 anatomical neck
2 greater tubercle
3 lesser tubercle
4 intertubercular sulcus
5 surgical neck
6 deltoid tuberosity
7 shaft
8 coronoid fossa
9 radial fossa
10 lateral epicondyle
11 capitulum
12 trochlea
13 head
14 medial epicondyle
15 head
16 greater tubercle
17 anatomical neck

18 surgical neck
19 deltoid tuberosity
20 medial epicondyle
21 olecranon fossa
22 lateral epicondyle
23 trochlea
24 lateral epicondyle
25 capitulum
26 radius
27 medial epicondyle
28 trochlea
29 ulna
30 medial epicondyle
31 ulna
32 humerus
33 lateral epicondyle
34 radius

Figure 9.4   The Radius

1 head
2 neck
3 radial tuberosity

4 shaft
5 styloid process of radius
6 ulnar notch

Figure 9.5   The Ulna

1 trochlear notch
2 coronoid process
3 radial notch
4 tuberosity of ulna

5 shaft of ulna
6 styloid process of ulna
7 olecranon
8 coronoid process

Exercise 9.2D, #3

Carpal Bone	Word Origin	Bone Shape/Appearance
Scaphoid	*skaphe,* boat	shaped like a boat
Lunate	*luna,* moon	moon-shaped
Triquetrum	*triquetrus,* 3-cornered	triangular
Pisiform	*pisum,* pea	pea-shaped
Trapezium	*trapezion,* a table	table-shaped
Trapezoid	*trapezion,* a table	table-shaped
Capitate	*caput,* head	head-shaped
Hamate	*hamus,* a hook	hook-shaped

Figure 9.7   The Metacarpals and Phalanges

1 phalanges
2 metacarpals
3 carpals
4 distal phalanx
5 middle phalanx
6 proximal phalanx
7 distal phalanx of pollex
8 proximal phalanx of pollex

9 metacarpal V
10 metacarpal IV
11 metacarpal III
12 pollex (metacarpal I)
13 metacarpal II
14 head
15 body
16 base

Figure 9.8   Surface Anatomy of Upper Limb

1 clavicle
2 acromion
3 deltoid tuberosity
4 medial epicondyle
5 suprasternal notch

6 styloid process of ulna
7 olecranon
8 styloid process of radius
9 spine of scapula
10 medial boarder of scapula

Figure 9.9   The Right Os Coxae

1 ala
2 anterior gluteal line
3 posterior gluteal line
4 posterior superior iliac spine

5 posterior inferior iliac spine
6 greater sciatic notch
7 ischial body
8 ischial spine

9 lesser sciatic notch
10 ischial tuberosity
11 iliac crest
12 anterior superior iliac spine
13 inferior gluteal line
14 anterior inferior iliac spine
15 lunate surface
16 acetabulum
17 superior pubic ramus
18 pubic crest
19 pubic tubercle
20 inferior pubic ramus
21 obturator foramen
22 ischial ramus
23 iliac fossa
24 anterior superior iliac spine
25 anterior inferior iliac spine

26 arcuate line
27 pectineal line
28 superior pubic ramus
29 pubic tubercle
30 symphysial surface of pubic bone
31 obturator foramen
32 inferior pubic ramus
33 iliac crest
34 posterior superior iliac spine
35 auricular surface
36 posterior inferior iliac spine
37 greater sciatic notch
38 ischial spine
39 lesser sciatic notch
40 ischial tuberosity
41 ischial ramus

Figure 9.11   The Right Femur

1 head
2 greater trochanter
3 fovea
4 neck
5 intertrochanteric crest
6 lesser trochanter
7 shaft
8 head
9 shaft
10 patellar articular surface
11 intercondylar notch
12 lateral epicondyle
13 medial epicondyle
14 head
15 greater trochanter
16 neck
17 intertrochanteric crest
18 lesser trochanter
19 shaft
20 adductor tubercle
21 lateral epicondyle
22 medial epicondyle

23 patellar articular surface
24 lateral condyle
25 medial condyle
26 head
27 fovea
28 greater trochanter
29 neck
30 intertrochanteric crest
31 lesser trochanter
32 pectineal line
33 gluteal tuberosity
34 linea aspera
35 medial supracondylar line
36 lateral supracondylar line
37 lateral epicondyle
38 popliteal surface
39 adductor tubercle
40 medial epicondyle
41 medial condyle
42 intercondylar notch
43 lateral condyle

Figure 9.12   The Tibia

1 lateral condyle
2 intercondylar eminence
3 medial condyle
4 tibial tuberosity
5 anterior border
6 medial malleolus

7 medial condyle
8 intercondylar eminence
9 lateral condyle
10 lateral tibial condyle
11 soleal line
12 medial malleolus

Figure 9.13   The Right Fibula

1 head
2 neck

3 shaft
4 lateral malleolus

Figure 9.14   The Tarsals

1 phalanges
2 metatarsals
3 medial cuneiform
4 navicular
5 intermediate cuneiform

6 lateral cuneiform
7 cuboid
8 talus
9 calcaneus

Exercise 9.5D, #5

Tarsal Bone	Word Origin	Bone Shape/Appearance
Talus	talus, ankle	convex, triangular
Calcaneus	calcaneous, heel	elongated
Navicular	navis, ship	shaped like a boat
Medial Cuneiform	cuneus, wedge	wedge-shaped
Intermediate Cuneiform	cuneus, wedge	wedge-shaped
Lateral Cuneiform	cuneus, wedge	wedge-shaped
Cuboid	kybos, cube	cube-shaped

Figure 9.15   The Metatarsals and Phalanges

1 lateral cuneiform
2 cuboid
3 distal phalanx
4 middle phalanx

5 proximal phalanx
6 hallux (metatarsal I)/head
7 shaft
8 base

9 medial cuneiform
10 intermediate cuneiform
11 navicular

12 talus
13 calcaneus

Figure 9.16   Surface Anatomy of Lower Limb

1 iliac crest
2 anterior superior iliac spine
3 iliac crest
4 sacrum
5 greater trochanter of femur
6 ischial tuberosity
7 lateral condyle of femur
8 head of fibula
9 lateral malleolus

10 medial condyle of femur
11 patella
12 tibial tuberosity
13 shaft of tibia
14 medial malleolus
15 calcaneus
16 metatarsals
17 phalanges

Chapter 10

What Do You Think?

1. The ACL is commonly ruptured during contact sports because **it is relatively weak compared to other knee ligaments.** In addition, contact sports like football often cause a player to be hit from behind. If the player who is hit doesn't contract his muscles to protect the knee joint, the resulting movement of the femur with respect to the tibia can cause the ACL to tear.

2. Rupture of the tibial collateral ligament often coincides with a tear of the medial meniscus because **connective tissue fibers that compose the tibial collateral ligament are attached to the bundles of collagen fibers that compose the medial meniscus.**

Chapter Activities

Figure 10.1   Fibrous Joints

1 syndesmosis
2 gomphosis
3 suture

Figure 10.2   Cartilaginous Joints

1 synchondrosis
2 symphysis
3 synchondrosis
4 symphysis

Figure 10.3   General Structure of Synovial Joints

1 periosteum
2 yellow bone marrow
3 fibrous layer of articular capsule
4 synovial membrane

5 synovial (joint) cavity
6 articular cartilage
7 ligament

Figure 10.4   Structural Classifications of Synovial Joints

1 saddle
2 hinge
3 plane

4 ball-and-socket
5 condylar
6 pivot

Figure 10.5   A Representative Synovial Joint: Right Knee Joint

1 posterior cruciate ligament
2 lateral condyle of femur
3 lateral meniscus
4 fibular collateral ligament
5 anterior cruciate ligament
6 fibula
7 medial condyle of femur
8 medial meniscus
9 tibial collateral ligament
10 tibia
11 anterior cruciate ligament
12 medial condyle of femur
13 medial meniscus
14 posterior cruciate ligament
15 tibial collateral ligament
16 tibia

17 femur
18 lateral condyle of femur
19 fibular collateral ligament
20 lateral meniscus
21 fibula
22 femur
23 quadriceps femoris tendon
24 suprapatellar bursa
25 patella
26 prepatellar bursa
27 articular cartilage
28 patellar ligament
29 infrapatellar bursa
30 meniscus
31 tibia

Chapter 11

What Do You Think?

1. **quadriceps longus** (*quadriceps* = four heads; *longus* = long)
2. The muscles in the posterior compartment will have the opposite action as those in the anterior compartment (they are *antagonists*). Therefore, the common action will be **extension.**

Chapter Activities
Figure 11.6 Muscles of the Human Body

Muscle Number	Muscle Name	Muscle Architecture
1	orbicularis oculi	circular
2	deltoid	multipennate
3	pectoralis major	convergent
4	sartorius	parallel
5	rectus femoris	bipennate
6	trapezius	convergent
7	triceps brachii	bipennate
8	extensors of wrist/hand	parallel
9	gastrocnemius	bipennate (individual heads multipennate (entire muscle)

Chapter 12

What Do You Think?

1. The vertebral artery travels through the **transverse foramina of the cervical vertebrae** en route to the posterior region of the neck. When it reaches the atlas, it bends medially and runs in the **groove for the vertebral artery** on the superior surface of the atlas.
2. In a male, contents of the spermatic cord pass through the inguinal canal. In a female, only a small ligament, the round ligament of the uterus, passes through. Because the **round ligament of the uterus is much smaller than the combined contents of the spermatic cord, the defect in the abdominal wall is smaller in a female than in a male.** Thus, indirect inguinal hernias are more common in males than in females.

Chapter Activities
Figure 12.1   Muscles of Facial Expression

1 epicranius (occipitofrontalis)
2 epicranial aponeurosis
3 frontal belly of occipitofrontalis
4 procerus
5 orbicularis oculi
6 levator labii superioris
7 zygomaticus minor
8 zygomaticus major
9 depressor anguli oris
10 depressor labii inferioris
11 platysma
12 corrugator supercilii
13 nasalis (dilator naris)
14 levator anguli oris
15 masseter
16 orbicularis oris
17 mentalis
18 epicranial aponeurosis
19 frontal belly of occipitofrontalis
20 epicranius (occipitofrontalis)
21 buccinator
22 orbicularis oculi
23 levator labii superioris
24 zygomaticus minor
25 zygomaticus major
26 orbicularis oris
27 levator anguli oris
28 depressor labii inferioris
29 depressor anguli oris
30 platysma

Figure 12.2   Muscles of Mastication

1 temporalis
2 masseter
3 temporalis
4 lateral pterygoid
5 medial pterygoid

Figure 12.3   Extrinsic Eye Muscles

1 superior rectus
2 lateral rectus
3 inferior oblique
4 inferior rectus
5 superior oblique
6 medial rectus
7 lateral rectus
8 inferior oblique

Figure 12.4   Muscles That Move the Tongue

1 palatoglossus
2 styloglossus
3 hyoglossus
4 genioglossus

Figure 12.5   Muscles of the Pharynx

1 tensor veli palatini
2 levator veli palatini
3 superior constrictor
4 stylopharyngeus
5 middle constrictor
6 inferior constrictor

Focus: Understanding Actions of Agonists, Synergists, and Antagonists (p. 220)

The right sternocleidomastoid rotates the neck to the left.
The right splenius capitis rotates the neck to the right.
In summary: to rotate your neck to the right, you use the sternal head of the sternocleidomastoid on the left side of the neck, and the splenius capitis muscle on the right side of the neck.

Figure 12.6   Muscles of the Head and Neck

1 mylohyoid
2 stylohyoid
3 digastric (anterior belly)
4 digastric (posterior belly)
5 omohyoid
6 sternohyoid
7 sternocleidomastoid
8 thyrohyoid
9 sternothyroid
10 scalenes
11 digastric (posterior belly)
12 mylohyoid
13 digastric (anterior belly)
14 sternothyroid
15 omohyoid (superior belly)
16 sternohyoid
17 splenius capitis
18 sternocleidomastoid
19 scalenes
20 omohyoid (inferior belly)
21 semispinalis capitis
22 sternocleidomastoid
23 splenius capitis
24 rectus capitis posterior minor
25 rectus capitis posterior major
26 superior oblique (obliquus capitis superior)
27 inferior oblique (obliquus capitis inferior)

Figure 12.7   Suboccipital Muscles

1 Rectus capitis posterior minor
2 Rectus capitis posterior major
3 Obliquus capitis superior (superior oblique)
4 Obliquus capitis inferior (inferior oblique)

Figure 12.8   Muscles of the Vertebral Column

1 splenius capitis
2 splenius cervicis
3 Iliocostalis
4 longissimus
5 Spinalis
6 semispinalis cervicis
7 semispinalis thoracis
8 multifidus
9 quadratus lumborum
10 intertransversarii
11 rotatores
12 interspinales

Figure 12.9   Muscles of Respiration

1 external intercostals
2 internal intercostals
3 transverse thoracis
4 diaphragm
5 internal intercostals
6 external intercostals
7 diaphragm

Figure 12.11   Muscles of Abdominal Wall

1 tendinous intersections
2 rectus abdominis
3 transverse abdominis
4 internal oblique
5 external oblique

Figure 12.15   Muscles of Pelvic Floor

1 bulbospongiosus
2 ischiocavernosus
3 superficial transverse perineal muscle
4 bulbospongiosus
5 ischiocavernosus
6 levator ani
7 external urethral sphincter
8 deep transverse perineal muscle
9 external anal sphincter
10 external urethral sphincter
11 deep transverse perineal muscle
12 external anal sphincter
13 coccygeus
14 iliococcygeus
15 pubococcygeus

Chapter 13

Chapter Activities
Figure 13.1   Muscles that Move the Pectoral Girdle and Glenohumeral Joint

1 trapezius
2 deltoid
3 pectoralis major
4 biceps brachii (long head)
5 biceps brachii (short head)
6 subscapularis
7 tendon of biceps brachii (long head)
8 coracobrachialis
9 pectoralis minor
10 serratus anterior
11 trapezius
12 rhomboid minor
13 rhomboid major
14 deltoid
15 rhomboid major
16 latissimus dorsi
17 levator scapulae
18 supraspinatus
19 infraspinatus
20 teres minor
21 teres major
22 serratus anterior

Figure 13.2   Anterior (Flexor) Compartment of the Arm
1 coracobrachialis
2 biceps brachii (short head)
3 biceps brachii (long head)
4 tendon of the long head of biceps brachii
5 coracobrachialis
6 brachialis

Figure 13.3   Posterior (Extensor) Compartment of the Arm
1 lateral head of triceps brachii
2 long head of triceps brachii
3 olecranon process of ulna

Figure 13.4   Anterior (Flexor) Muscles of the Forearm
1 pronator teres
2 brachioradialis
3 flexor retinaculum
4 medial epicondyle of humerus
5 common flexor tendon
6 flexor carpi radialis
7 palmaris longus
8 flexor carpi ulnaris
9 flexor digitorum superficialis
10 palmar aponeurosis
11 supinator
12 flexor pollicis longus
13 pronator quadratus
14 flexor digitorum profundus

Figure 13.5   Posterior (Extensor) Compartment of the Forearm
1 anconeus
2 extensor carpi ulnaris
3 extensor digiti minimi
4 extensor retinaculum
5 extensor digitorum tendons
6 brachioradialis
7 extensor carpi radialis longus
8 extensor carpi radialis brevis
9 extensor digitorum
10 abductor pollicis longus
11 extensor pollicis brevis
12 olecranon process of ulna
13 extensor pollicis longus
14 extensor indicis
15 dorsal interossei
16 supinator
17 abductor pollicis longus
18 extensor pollicis brevis

Figure 13.8   Intrinsic Muscles of the Hand
1 lumbricals
2 flexor digiti minimi brevis
3 abductor digiti minimi
4 first dorsal interosseous
5 lateral lumbrical
6 adductor pollicis
7 flexor pollicis brevis
8 abductor pollicis brevis
9 thenar group
10 hypothenar group

Figure 13.9   Muscles that Move the Pelvic Girdle
1 gluteus maximus
2 piriformis
3 superior gemellus
4 obturator internus
5 inferior gemellus
6 gluteus mediu
7 gluteus minimus
8 gluteus mediu
9 quadratus femoris
10 iliacus
11 psoas major
12 tensor fasciae latae
13 iliotibial tract
14 psoas minor
15 iliopsoas

Figure 13.11   Anterior Compartment of the Thigh
1 inguinal ligament
2 tensor fasciae latae
3 iliotibial tract
4 rectus femoris
5 vastus lateralis
6 pectineus
7 adductor longus
8 gracilis
9 sartorius
10 vastus medialis
11 quadriceps tendon
12 patella

Figure 13.12   Medial Compartment of the Thigh
1 pectineus
2 adductor brevis
3 adductor longus
4 gracilis

Figure 13.13   Posterior Compartment of the Thigh
1 semimembranosus
2 semitendinosus
3 biceps femoris, long head
4 biceps femoris, short head
5 gluteus maximus
6 adductor magnus
7 iliotibial tract

Figure 13.14   Anterior Compartment of the Leg
1 extensor digitorum longus
2 extensor hallucis longus
3 fibularis tertius tendon
4 tibialis anterior

Figure 13.15   Lateral View of the Leg
1 gastrocnemius
2 soleus
3 fibularis longus
4 fibularis brevis
5 fibularis tertius
6 extensor digitorum brevis
7 fibularis tertius tendon
8 tibialis anterior
9 extensor digitorum longus
10 extensor hallucis longus
11 extensor hallucis brevis
12 extensor hallucis longus tendon
13 extensor digitorum longus tendons

Figure 13.16   Right Leg, Deep Posterior View
1 plantaris
2 popliteus
3 tibialis posterior
4 flexor digitorum longus
5 flexor hallucis longus
6 calcaneal tendon

Figure 13.17   Intrinsic Muscles of the Foot
1 flexor digitorum brevis
2 abductor hallucis
3 abductor digiti minimi
4 lumbricals
5 tendon of flexor hallucis longus
6 tendons of flexor digitorum longus
7 quadratus plantae
8 adductor hallucis
9 flexor hallucis brevis
10 flexor digiti minimi brevis
11 quadratus plantae
12 abductor digiti minimi
13 abductor hallucis
14 plantar interossei
15 dorsal interossei

Chapter 14

What Do You Think?

1. **Gray matter** is more involved in integration of information because it contains the cell bodies and dendrites of neurons. Synapses are abundant in such areas, and each synapse involves the transfer of information from one neuron to the next. White matter, on the other hand, consists of myelinated axons, which are involved with the transmission of information long distances from one area of the nervous system to another.
2. The compression of a peripheral nerve that compresses blood vessels within the nerve will **cut off the nerve's blood supply** and cause the axons to fail to send signals properly. This will result in nerve damage if blood flow is not restored soon. The lack of blood flow causes the area served by the nerve to become weak and go numb. Restoration of blood flow is associated with a prickly sensation along the sensory distribution of the nerve.

Chapter 15

What Do You Think?

1. If the passage of fluid was blocked at the confluence of sinuses, fluid would back up into the **inferior sagittal sinus, straight sinus,** and the **superior sagittal sinus,** because these three sinuses all drain into the confluence of sinuses.
2. The pineal gland functions in the control of circadian or annual rhythms. Its size in sheep indicates that **it exerts a stronger influence on brain function in sheep** than humans. Thus, a sheep's reproductive cycle, mating cycle, etc., is under much tighter control by its nervous system than is a human's.

Chapter Activities

Figure 15.3   Meningeal Structures

15.1a
1 superior sagittal sinus
2 falx cerebri
3 pia mater
4 subarachnoid space
5 arachnoid mater
6 falx cerebri
7 periosteal dura
8 dura mater
9 superior sagittal sinus
10 arachnoid villi
11 meningeal dura

15.1b
1 cranium
2 superior sagittal sinus
3 inferior sagittal sinus
4 tentorium cerebelli
5 diaphragma sellae
6 dura mater
7 falx cerebri
8 straight sinus
9 confluence of sinuses
10 falx cerebelli

Figure 15.5   Cast of the Ventricles of the Brain
1 lateral ventricles
2 interventricular foramen
3 third ventricle
4 cerebral aqueduct
5 fourth ventricle
6 lateral ventricles
7 interventricular foramen
8 third ventricle
9 cerebral aqueduct
10 fourth ventricle

Figure 15.6   Superior View of the Brain
1 frontal lobes
2 precentral gyrus
3 central sulcus
4 postcentral gyrus
5 longitudinal fissure
6 parietal lobe
7 occipital lobes

Figure 15.7   Lateral View of the Brain
1 frontal lobe
2 parietal lobe
3 central sulcus
4 precentral gyrus
5 postcentral gyrus
6 lateral fissure
7 temporal lobe
8 pons
9 medulla oblongata
10 occipital lobe
11 transverse fissure
12 cerebellum
13 spinal cord

Figure 15.8   Inferior View of the Brain

1 frontal lobe
2 infundibulum
3 mammillary bodies
4 temporal lobe
5 pons
6 occipital lobe
7 olfactory bulb
8 olfactory tract
9 optic chiasm
10 optic nerve
11 optic tract
12 mesencephalon
13 cerebral peduncles
14 cerebellum
15 medulla oblongata
16 spinal cord

Figure 15.9   Midsagittal View of the Brain

1 frontal lobe
2 cingulate gyrus
3 corpus callosum
4 septum pellucidum
5 interthalamic adhesion
6 thalamus
7 hypothalamus
8 tegmentum
9 mesencephalon
10 cerebral peduncle
11 temporal lobe
12 pons
13 medulla oblongata
14 spinal cord
15 central sulcus
16 parietal lobe
17 thalamus
18 parieto-occipital sulcus
19 occipital lobe
20 pineal body (gland)
21 tectal plate (corpora quadrigemina)
22 mammillary body
23 cerebral (mesencephalic) aqueduct
24 fourth ventricle
25 cerebellum

Chapter 16

What Do You Think?

1. Because the **facial nerve (CN VII)** travels alongside the vestibulocochlear nerve (CN VIII) en route to the face, it is also commonly affected by the pressure of an enlarging acoustic neuroma.
2. The **optic nerve (CN II)** carries sensory information from the light toward the brain. On the other hand, the oculomotor nerve (CN III) sends motor information to the pupil to alter its diameter. If the consensual light reflex is absent it can either indicate a disorder of the optic nerve of the eye in which the light was shone, a problem with the control centers within the brainstem that are involved in the consensual light reflex, or a disorder of the oculomotor nerve of the eye in which the light was *not* shone.

Chapter Activities

Exercise 16.1, #4

Point of Exit Cranial Nerve	
Midbrain	III, IV
Pons	V, VI, VII, part of VIII
Medulla oblongata	part of VIII, IX, X, XI, XII

Exercise 16.2, #4

Foramen	Nerves Traveling Through
olfactory foramina	CN I
optic canal	CN II
superior orbital fissure	CN III, IV, V$_1$, VI
foramen rotundum	CN V2
foramen ovale	CN V3
internal acoustic meatus	CN VII, VIII
jugular foramen	CN IX, X, XI (accessory division)
hypoglossal canal	CN XII

Chapter 17

What Do You Think?

1. If a posterior root was severed there would be a **loss of sensation** from the area of the body served by the spinal nerve that posterior root contributed to. There would be no loss of motor function because all motor nerves run in the anterior root.
2. Because the phrenic nerve arises from the cervical plexus instead of arising from of the thoracic region of the spinal cord, **the innervation to the diaphragm will remain intact in the event of a spinal cord injury below C5.** This is hugely advantageous because while injuries to the lower cervical or upper thoracic region of the spinal cord will cause paralysis of upper limb muscles (not life-threatening), they will not cause an individual to stop breathing, which would be life-threatening.

Chapter Activities

Figure 17.2   Regional Gross Anatomy of the Spinal Cord

1 brain (cerebellum)
2 posterior rootlets
3 posterior median sulcus
4 posterior roots*
5 anterior roots*
6 posterior median sulcus
7 pia mater
8 denticulate ligament
9 dura mater
10 conus medullaris
11 dura mater
12 cauda equina
13 filum terminale
14 posterior root ganglion

Figure 17.3   Model of a Cervical Vertebra and Spinal Cord in cross section

1 posterior median sulcus
2 subarachnoid space
3 epidural space
4 anterior rootlets
5 posterior rootlet
6 posterior root ganglion
7 anterior root
8 spinal nerve
9 anterior median fissure
10 dura mater
11 arachnoid mater
12 white matter
13 pia mater
14 posterior horn
15 lateral horn
16 anterior horn
17 central canal

Figure 17.5   The Phrenic Nerves in the Thoracic Cavity

1 right phrenic nerve
2 diaphragm
3 left phrenic nerve

Figure 17.7   Major Nerves of the Brachial Plexus

1 lateral cord
2 posterior cord
3 musculocutaneous nerve
4 axillary nerve
5 medial cord
6 radial nerve
7 median nerve
8 ulnar nerve
9 long thoracic nerve
10 superior trunk
11 middle trunk
12 inferior trunk
13 axillary artery

Figure 17.8   Nerves of the Lumbosacral Plexus Within the Gluteal Region

1 gluteus maximus muscle
2 inferior gluteal nerve
3 sciatic nerve
4 pudendal nerve
5 gluteus medius muscle
6 gluteus minimus muscle
7 superior gluteal nerve
8 piriformis muscle
9 posterior femoral cutaneous nerve
10 gluteus maximus muscle

Figure 17.9   Nerves of the Lumbosacral Plexus

1 subcostal nerve
2 iliohypogastric nerve
3 ilioinguinal nerve
4 lateral femoral cutaneous nerve
5 genitofemoral nerve
6 obturator nerve
7 femoral nerve
8 medial sural cutaneous nerve
9 tibial nerve
10 common fibular nerve
11 lateral sural cutaneous nerve

Chapter 18

What Do You Think?

1. Tactile corpuscles are sensitive to **light touch.** They are located relatively close to the surface of the skin so they can be sensitive to small disturbances occurring on the surface of the skin. If tactile corpuscles were located deeper within the dermis, it would take a stronger stimulus, such as deep pressure, to stimulate the receptors adequately.
2. Lamellated corpuscles are sensitive to **deep pressure and high frequency vibration.** They are located deep within the dermis of the skin. This location requires relatively strong pressure to stimulate the sensory receptor.
3. The optic disc is where the axons of ganglion cells exit the retina. There are **no photoreceptor cells (rods and cones) in this location,** which is why it is called the "blind spot" of the eye. Without photoreceptors, no light information can be sensed by the retina when light waves hit the optic disc.
4. A detached retina occurs because **there is no connective tissue to hold the retina in place.** The only thing holding the retina in place is the **fluid pressure of the vitreous humor** in the posterior cavity (vitreous chamber) of the eye.

Chapter Activities

Figure 18.12   External Eye Structures

1 eyelashes
2 lacrimal caruncle
3 medial palpebral commissure
4 palpebral fissure
5 eyebrow
6 superior eyelid

7 pupil
8 sclera
9 iris

10 lateral palpebral commissure
11 inferior eyelid

Chapter 19

What Do You Think?

1. Individuals with pituitary gland tumors often experience visual disorders because of the **close proximity of the pituitary gland to the optic chiasm.** The optic chiasm, where fibers from the two optic nerves cross over to the opposite side of the brain, is located directly anterior to the infundibulum. The pituitary gland is located within the sella turcica of the sphenoid bone, directly inferior to the infundibulum. As a pituitary tumor grows, the bone tissue of the sella turcica somewhat limits the inferior movement of the tumor. Instead, the tumor tends to project superiorly where it begins to put pressure on the optic chiasm and the optic tracts, which lie next to the infundibulum. The pressure on these nerves is responsible for the visual disturbances.

Chapter Activities

Figure 19.9   Classroom Models Demonstrating Major Endocrine Glands of the Body

1 thyroid gland
2 adrenal gland
3 pancreas
4 hypothalamus

5 anterior pituitary gland
6 pineal gland
7 posterior pituitary gland

Chapter 21

What Do You Think?

1. The **volume of blood pumped by the two ventricles must be the same** because ours is a closed circulatory system and all blood pumped to the pulmonary circuit returns to the systemic circuit and vice versa. If the volumes pumped by the two ventricles were different, then blood would back up in the circuit behind the ventricle that was pumping a lower volume. For instance, if the left ventricle pumped a lower volume of blood than the right ventricle, blood would back up in the pulmonary circuit, causing congestion within the lungs. This condition often occurs when an individual suffers from left-sided heart failure. If the right ventricle pumped a lower volume of blood than the left ventricle, blood would back up in the systemic circuit, causing congestion in the peripheral tissues, which is often seen as edema (swelling) of the extremities. This condition occurs when an individual suffers from right-sided heart failure.

Chapter Activities

Figure 21.1   Cardiac Muscle Tissue

1 striations
2 intercalated disc

3 nucleus
4 cardiac muscle cell

Figure 21.3   Location of Heart within the Thoracic Cavity

1 right lung
2 pericardial cavity
3 diaphragm
4 trachea

5 mediastinum
6 left lung
7 heart

Figure 21.4   The Pericardial Sac

1 fibrous pericardium
2 parietal layer of serous pericardium
3 pericardial cavity
4 visceral layer of serous pericardium (epicardium)
5 myocardium
6 pericardial cavity

7 myocardium
8 endocardium
9 visceral layer of serous pericardium (epicardium)
10 fibrous pericardium
11 diaphragm

Figure 21.12   Circulation to and from the Heart Wall

1 right coronary artery
2 marginal artery
3 left coronary artery
4 circumflex artery
5 great cardiac vein

6 anterior interventricular artery
7 coronary sinus
8 right coronary artery
9 posterior interventricular artery
10 middle cardiac vein

Chapter 22

What Do You Think?

1. Large blood vessels have thick walls composed of many cells (particularly smooth muscle cells), which need oxygen and nutrients just like any other cells of the body. The blood vessel walls are thick enough that they must have their own blood vessels, the vasa vasorum, to supply the tissues with oxygen and nutrient-rich blood because the **blood flowing within the lumen of the vessel is not adequate to supply these tissues.** In addition, blood flowing within the lumen of the vessels is generally flowing at high velocity, which is disadvantageous when it comes to exchange of substances between the blood and the tissues.

2. An increase in carbon dioxide in the tissues of smooth muscle results in the local vasodilation in order to supply the region with adequate blood supply and glucose, as well as an avenue for the removal of metabolic wastes. In the case of a highly metabolically active organ, like the brain, vasodilation is necessary for the survival of the organ under conditions of hypoxia (low oxygen and high carbon dioxide). Elevated oxygen levels are associated with vasoconstriction, owing to the toxic nature of oxygen on tissues of the body. Vasoconstriction allows for the high oxygen concentration in the blood to be dispersed throughout the body, avoiding any injury to local tissue.

3. The blood-brain barrier is composed of astrocytes and **continuous capillaries.** Continuous capillaries form the tightest barrier between the blood and the tissues because substances to be exchanged between the blood and the tissues are forced to pass through an endothelial cell to do so, giving the endothelial cells control over which substances pass through and which do not.

4. The blood-cerebrospinal fluid barrier of the choroid plexus is composed of ependymal cells and **fenestrated capillaries.** Fenestrated capillaries have small holes (fenestrations) in the endothelial cells, which makes them more "leaky" than non-fenestrated endothelial cells. This "leakiness" in the capillaries of the choroid plexus allows fluid to escape from the blood and enter the ventricles of the brain as cerebrospinal fluid (CSF).

5. Continuous capillaries form the tightest barrier between blood and tissues. Because of this, **the blood-brain barrier is a more complete barrier between the blood and the tissues** than is the blood-cerebrospinal fluid barrier.

6. The right parietal lobe of the brain functions in logical reasoning and stereognosis (the ability to recognize by touch). Thus, a stroke of the middle cerebral artery that results in damage to the right parietal lobe of the brain would result in neurological deficits such as **difficulties with problem-solving (particularly math problems) and an inability to recognize objects based on touch (as when trying to identify an object placed inside a paper bag).**

7. The liver contains **sinusoidal capillaries.** Sinusoidal capillaries allow maximum exchange between the blood and the tissues. Because the liver's main functions involve processing of the blood, it is advantageous to **allow maximum exchange between the hepatocytes of the liver and the substances within the blood,** which is what sinusoidal capillaries are specialized for.

Chapter Activities

Exercise 22.7, #4

right ventricle ⟶ pulmonary semilunar valve ⟶ pulmonary trunk ⟶ pulmonary arteries ⟶ lungs ⟶ pulmonary veins ⟶ left atrium

Figure 22.6   Pulmonary Circulation

1 right atrium
2 right AV valve
3 right ventricle
4 pulmonary semilunar valve
5 pulmonary trunk
6 right pulmonary artery
7 branch of pulmonary artery
8 pulmonary capillaries

9 branch of pulmonary vein
10 right pulmonary vein
11 left atrium
12 left AV valve
13 left ventricle
14 aortic semilunar valve
15 aorta

Figure 22.7   Circulation to the Head and Neck
(a) Arterial supply

1 internal carotid artery
2 external carotid artery
3 carotid sinus
4 common carotid artery
5 vertebral artery
6 thyrocervical trunk
7 subclavian artery
8 superficial temporal artery

9 posterior auricular artery
10 occipital artery
11 maxillary artery
12 facial artery
13 superior laryngeal artery
14 superior thyroid artery
15 brachiocephalic trunk

(b) Venous drainage

1 vertebral vein
2 external jugular vein
3 internal jugular vein
4 subclavian vein
5 right brachiocephalic vein
6 superficial temporal vein
7 posterior auricular vein
8 maxillary vein
9 pharyngeal vein
10 facial vein
11 lingual vein
12 superior thyroid vein

Figure 22.8   Circulation from the Aortic Arch to the Anterior Part of the Right Parietal Bone and Back to the Superior Vena Cava

1 right brachiocephalic trunk
2 right common carotid artery
3 right external carotid artery
4 right superficial temporal artery
5 right superficial temporal vein
6 right internal jugular vein
7 right brachiocephalic vein

Figure 22.9   Circulation to the Brain
(a) Arterial Supply

1 middle cerebral artery
2 internal carotid artery
3 posterior cerebral artery
4 anterior communicating artery
5 anterior cerebral artery
6 internal carotid artery
7 posterior communicating artery
8 posterior cerebral artery
9 basilar artery
10 vertebral artery

(b) Venous Drainage

1 straight sinus
2 transverse sinus
3 sigmoid sinus
4 internal jugular vein
5 superior sagittal sinus
6 inferior sagittal sinus
7 cavernous sinus
8 superior petrosal sinus
9 inferior petrosal sinus

Figure 22.10   Circulation from the Aortic Arch to the Right Parietal Lobe of the Brain and Back to the Right Brachiocephalic Vein

1 right brachiocephalic artery
2 right common carotid artery
   a. Alternate 2: right subclavian artery
3 right internal carotid artery
   b. Alternate 3: right vertebral artery
4 middle cerebral artery
   c. Alternate 4: basilar artery
d. Alternate 5: posterior cerebral artery
e. Alternate 6: posterior communicating artery
5 superior sagittal sinus
6 transverse sinus
7 sigmoid sinus
8 internal jugular vein
9 right brachiocephalic vein

Figure 22.11   Circulation to the Thoracic and Abdominal Walls
(a) Arterial Supply

1 right common carotid artery
2 vertebral artery
3 right subclavian artery
4 brachiocephalic trunk
5 internal thoracic artery
6 anterior intercostal arteries
7 posterior intercostal arteries
8 superior epigastric artery
9 right inferior suprarenal artery
10 right renal artery
11 descending abdominal aorta
12 right gonadal artery
13 inferior mesenteric artery
14 inferior epigastric artery
15 left common carotid artery
16 left subclavian artery
17 aortic arch
18 descending thoracic aorta
19 left gastric artery
20 celiac trunk
21 left superior suprarenal artery
22 left renal artery
23 superior mesenteric artery
24 left gonadal artery
25 left common iliac artery
26 left internal iliac artery
27 left external iliac artery
28 left femoral artery

(b) Venous Drainage

1 right subclavian vein
2 right brachiocephalic vein
3 superior vena cava
4 anterior intercostal veins
5 azygos vein
6 internal thoracic vein
7 right posterior intercostal vein
8 inferior vena cava
9 hepatic veins
10 right superior epigastric vein
11 right suprarenal vein
12 right renal vein
13 right gonadal vein
14 right lumbar veins
15 right inferior epigastric vein
16 left subclavian vein
17 left brachiocephalic vein
18 accessory hemiazygos vein
19 left posterior intercostal vein
20 hemiazygos vein
21 left suprarenal vein
22 left renal vein
23 left gonadal vein
24 left common iliac vein
25 left external iliac vein
26 left internal iliac vein
27 left femoral vein
28 inferior vena cava

Figure 22.12   Circulation from the Left Ventricle of the Heart to the Right Kidney and Back to the Right Atrium of the Heart

1 ascending aorta
2 aortic arch
3 descending thoracic aorta
4 abdominal aorta
5 right renal artery
6 right renal vein
7 inferior vena cava

Figure 22.13   Arterial Supply to Abdominal Organs
(a) Arterial Supply to the Stomach, Spleen, Pancreas, Duodenum, and Liver

1 celiac trunk
2 common hepatic artery
3 hepatic artery proper
4 left hepatic artery
5 right hepatic artery
6 gastroduodenal artery
7 right gastric artery
8 right gastroepiploic artery
9 inferior vena cava (not part of arterial supply)
10 left gastric artery
11 splenic artery
12 left gastroepiploic artery
13 descending abdominal aorta

(b) Arterial Supply to the Small and Large Intestines

1 middle colic artery
2 right colic artery
3 ileocolic artery
4 celiac trunk
5 superior mesenteric artery
6 left colic artery
7 descending abdominal aorta
8 inferior mesenteric artery
9 sigmoid arteries
10 left common iliac artery
11 superior rectal artery

Figure 22.14   The Hepatic Portal System

1 inferior vena cava
2 hepatic veins
3 hepatic portal vein
4 right gastroepiploic vein
5 superior mesenteric vein
6 intestinal veins
7 gastric veins
8 splenic vein
9 inferior mesenteric vein

Figure 22.15   Circulation from the Abdominal Aorta to the Spleen and Back to the Right Atrium of the Heart

1 abdominal aorta
2 celiac trunk
3 splenic artery
4 splenic vein
5 hepatic portal vein
6 hepatic veins
7 inferior vena cava

Figure 22.16   Circulation from the Abdominal Aorta to the Duodenum and Back to the Right Atrium of the Heart

1 abdominal aorta
2 celiac trunk
3 common hepatic artery
4 gastroduodenal artery
5 superior mesenteric vein
6 hepatic portal vein
7 hepatic veins
8 inferior vena cava

Figure 22.17   Circulation from the Abdominal Aorta to the Sigmoid Colon and Back to the Right Atrium of the Heart

1 abdominal aorta
2 inferior mesenteric artery
3 sigmoid arteries
4 inferior mesenteric vein
5 splenic vein
6 hepatic portal vein
7 hepatic veins
8 inferior vena cava

Figure 22.18   Circulation to the Upper Limb
(a) Arterial Supply

1 brachiocephalic artery
2 subclavian artery
3 axillary artery
4 deep brachial artery
5 brachial artery
6 ulnar artery
7 radial artery
8 deep palmar arch
9 superficial palmar arch
10 digital arteries

(b) Venous Drainage

1 subclavian vein
2 axillary vein
3 cephalic vein
4 basilic vein
5 brachial veins
6 median cubital vein
7 radial veins
8 cephalic vein
9 deep palmar venous arch
10 superficial palmar venous arch
11 dorsal venous network
12 digital veins
13 ulnar veins

Figure 22.19   Circulation from the Aortic Arch to the Anterior Surface of the Index Finger and Back along a Superficial Route to the Superior Vena Cava

1 brachiocephalic trunk (artery)	9 superficial palmar venous arch
2 subclavian artery	10 cephalic vein
3 axillary artery	a. Alternate 11: median cubital vein
4 brachial artery	b. Alternate 12: basilic vein
5 radial artery	c. Alternate 13: axillary vein
6 superficial palmar arch	11 subclavian vein
7 digital artery	12 brachiocephalic vein
8 digital vein	13 superior vena cava

Figure 22.20   Circulation from the Aortic Arch to the Capitate Bone of the Wrist and Back along a Deep Route to the Superior Vena Cava

1 brachiocephalic artery	8 ulnar veins
2 subclavian artery	9 brachial vein
3 axillary artery	10 axillary vein
4 brachial artery	11 subclavian vein
5 ulnar artery	12 brachiocephalic vein
6 deep palmar arch	13 superior vena cava
7 deep palmar venous arch	

Figure 22.21   Circulation to the Lower Limb
(a) Arterial Supply

1 common iliac artery	9 dorsalis pedis artery
2 internal iliac artery	10 popliteal artery
3 external iliac artery	11 posterior tibial artery
4 obturator artery	12 lateral plantar artery
5 femoral artery	13 medial plantar artery
6 deep femoral artery	14 plantar arch
7 anterior tibial artery	15 digital arteries
8 fibular artery	

(b) Venous Drainage

1 common iliac vein	9 dorsal venous arch
2 external iliac vein	10 popliteal vein
3 internal iliac vein	11 small saphenous vein
4 deep femoral vein	12 posterior tibial veins
5 femoral vein	13 lateral plantar vein
6 great saphenous vein	14 medial plantar vein
7 anterior tibial veins	15 digital veins
8 fibular veins	

Figure 22.22   Circulation from the Abdominal Aorta to the Dorsal Surface of the Big Toe and Back along a Superficial Route to the Inferior Vena Cava

1 abdominal aorta	9 digital vein
2 common iliac artery	10 dorsal venous arch
3 external iliac artery	11 great saphenous vein
4 femoral artery	12 femoral vein
5 popliteal artery	13 external iliac vein
6 anterior tibial artery	14 common iliac vein
7 dorsalis pedis artery	15 inferior vena cava
8 digital artery	

Figure 22.23   Circulation from the Abdominal Aorta to the Cuboid Bone of the Foot and Back along a Deep Route to the Inferior Vena Cava

1 abdominal aorta	8 lateral plantar vein
2 common iliac artery	9 posterior tibial vein
3 external iliac artery	10 popliteal vein
4 femoral artery	11 femoral vein
5 popliteal artery	12 external iliac vein
6 posterior tibial artery	13 common iliac vein
7 lateral plantar artery	14 inferior vena cava

Exercise 22.14, #2

Left ventricle ⟶ aorta ⟶ common iliac arteries ⟶ internal iliac arteries ⟶ umbilical arteries ⟶ placenta ⟶ umbilical vein ⟶ ductus venosus ⟶ inferior vena cava ⟶ right atrium

Exercise 22.4, #3

Table 21.4	Fetal Cardiovascular Structures and Associated Postnatal Structures	
Fetal Cardiovascular Structure	**Postnatal Structure**	**Function of Fetal Structure**
Ductus arteriosus	Ligamentum arteriosum	Shunt blood from the pulmonary trunk to the aorta, thereby bypassing the pulmonary circuit and nonfunctional lungs
Ductus venosus	Ligamentum venosum	Carry oxygenated blood from umbilical vein to the inferior vena cava
Foramen ovale	Fossa ovalis	Shunt blood from the right atrium to the left atrium, thereby bypassing the nonfunctional lungs.
Umbilical arteries	Medial umbilical ligaments	Carry deoxygenated blood from the fetus to the placenta so it can obtain oxygen from the mother's blood.
Umbilical vein	Round ligament of the liver (ligamentum teres)	Carry oxygenated blood from the placenta to the fetal circulation

Figure 22.24   The Fetal Circulation

1 superior vena cava	11 pulmonary veins
2 right lung	12 foramen ovale
3 right atrium	13 right ventricle
4 liver	14 heart
5 ductus venosus	15 descending abdominal aorta
6 inferior vena cava	16 common iliac artery
7 umbilical vein	17 umbilical arteries
8 aortic arch	18 umbilical cord
9 ductus arteriosus	19 placenta
10 pulmonary artery	

Chapter 23

What Do You Think?

1. The small intestine functions to absorb nutrients from the food we eat. Unfortunately, our mouths are not only a large entryway into the body for food and drink, but also for microorganisms and pathogens. In addition, the epithelium of the small intestine is thin relative to the epithelium of our skin. Thus, **an abundance of lymphatic nodules within the walls of the small intestine provides a robust "second-line defense" against pathogens that might enter our bodies through the wall of the small intestine.**

Chapter Activities

Figure 23.10   Major Lymphatic Vessels of the Body

1 right lymphatic duct	3 lymph nodes
2 thoracic duct	4 cisterna chyli

Figure 23.11   Mucosa-Associated Lymphatic Tissue (MALT)

1 pharyngeal tonsils	3 lingual tonsils
2 palatine tonsils	4 vermiform appendix

Figure 23.12   Lymph Node and Its Components

1 lymphatic nodule	8 lymphatic nodules
2 afferent lymphatic vessels	9 peritrabecular space (cortical sinus)
3 capsule	10 subcapsular space
4 medullary sinuses	11 hilum
5 medullary cords	12 efferent lymphatic vessels (cortical sinus)
6 medulla	
7 trabeculae	

Figure 23.13   Gross Anatomy of the Spleen

1 splenic artery	3 spleen
2 hilum of the spleen	4 splenic vein

Chapter 24

Chapter Activities

Figure 24.7   The Upper Respiratory Tract: Midsagittal View

1 ethmoidal sinuses
2 superior nasal concha
3 middle nasal concha
4 inferior nasal concha
5 vestibule
6 epiglottis
7 thyroid cartilage
8 cricoid cartilage
9 sphenoidal sinus
10 superior meatus
11 middle meatus
12 inferior meatus
13 nasopharynx
14 oropharynx
15 laryngopharynx
16 lumen of larynx

Figure 24.8   Classroom Model of the Larynx

1 epiglottis
2 hyoid bone
3 thyrohyoid membrane
4 thyroid cartilage
5 laryngeal prominence
6 tracheal "C" ring
7 epiglottis
8 hyoid bone
9 thyrohyoid membrane
10 cuneiform cartilage
11 corniculate cartilage
12 thyroid cartilage
13 cricoid cartilage
14 trachealis muscle
15 tracheal C-Ring/trachea
16 epiglottis
17 corniculate cartilage
18 vestibular ligament
19 vocal ligament
20 cricoid cartilage
21 hyoid bone
22 thyrohyoid membrane
23 thyroid cartilage

Figure 24.9   The Pleural Cavities

1 left lung
2 right lung
3 mediastinum
4 diaphragm
5 visceral pleura
6 parietal pleura
7 pleural cavity

Figure 24.10   The Right Lung

1 superior lobe
2 horizontal fissure
3 oblique fissure
4 middle lobe
5 inferior lobe
6 base
7 horizontal fissure
8 middle lobe
9 oblique fissure
10 apex
11 superior lobe
12 oblique fissure
13 pulmonary arteries
14 primary bronchi
15 pulmonary veins
16 inferior lobe
17 base

Figure 24.11   The Left Lung

1 apex
2 superior lobe
3 oblique fissure
4 cardiac notch
5 inferior lobe
6 base
7 superior lobe
8 pulmonary arteries
9 primary bronchi
10 pulmonary veins
11 inferior lobe
12 base
13 oblique fissure
14 cardiac impression
15 cardiac notch
16 oblique fissure

Figure 24.12   The Bronchial Tree

1 right primary bronchus
2 right tertiary bronchi
3 right secondary bronchi
4 trachea
5 left primary bronchus
6 left tertiary bronchi
7 left secondary bronchi

Chapter 25

What Do You Think?

1. The epithelium of the large intestine has many **goblet cells,** which produce mucus. **Mucus is important for lubricating the large intestine to ease the passage of the feces as they solidify within the large intestine.**
2. For a gallstone to result in pancreatitis, at least one of the ducts must be blocked, preventing pancreatic juices from entering the duodenum. The hepatopancreatic ampulla is the location where the common bile duct and the main pancreatic duct come together to enter the duodenum. Thus, a gallstone lodged in the **common bile duct within the hepatopancreatic ampulla** could result in obstruction of the main pancreatic duct, which could then lead to pancreatitis.

Chapter Activities

Figure 25.10   Upper Digestive Tract
(a) Anterior View of the Oral Cavity

1 upper lip (superior lip)
2 superior labial frenulum
3 glossopalatine arch
4 pharyngopalatine arch
5 palatine tonsil
6 sublingual duct orifice
7 submandibular duct orifice
8 inferior labial frenulum
9 transverse palatine folds/ hard palate
10 soft palate
11 uvula
12 fauces
13 tongue
14 lingual frenulum
15 teeth
16 gingivae
17 lower lip (inferior lip)

(b) Midsagittal View of the Oral Cavity and Pharynx

1 hard palate
2 oral cavity
3 tongue
4 vestibule
5 soft palate
6 palatoglossal arch
7 uvula
8 palatine tonsil
9 oropharynx
10 lingual tonsil
11 epiglottis
12 laryngopharynx
13 esophagus

(c) Salivary Glands

1 parotid salivary gland
2 parotid duct
3 sublingual ducts
4 submandibular duct
5 sublingual salivary gland
6 submandibular salivary gland

Figure 25.11   Classroom Model of the Stomach

1 lesser curvature
2 pyloric sphincter
3 pylorus
4 pyloric sphincter
5 pylorus
6 esophagus
7 cardia of stomach
8 inferior esophageal (cardiac)
sphincter
9 body of stomach
10 greater curvature
11 cardiac notch
12 cardia of stomach
13 inferior esophageal sphincter
15 rugae in body of stomach

Figure 25.12   Classroom Model of the Duodenum, Gallbladder, Liver, and Pancreas

1 gallbladder
2 duodenum
3 accessory pancreatic duct
4 minor duodenal papilla
5 common bile duct
6 major duodenal papilla
7 head of pancreas
8 left and right hepatic ducts
9 cystic duct
10 common hepatic duct
11 tail of pancreas
12 body of pancreas
13 main pancreatic duct
14 hepatopancreatic ampulla

Figure 25.13   Classroom Model of the Liver

1 gallbladder
2 right lobe of liver
3 common hepatic duct
4 cystic duct
5 common bile duct
6 hepatic portal vein
7 hepatic artery proper
8 inferior vena cava
9 quadrate lobe of liver
10 falciform ligament
11 right hepatic duct
12 left hepatic duct
13 left lobe of liver
14 caudate lobe of liver
15 porta hepatis

Figure 25.15   Classroom Model of the Abdominal Cavity and Large Intestine

1 right lobe of liver
2 falciform ligament
3 pylorus of stomach
4 gallbladder
5 greater curvature of stomach
6 ascending colon
7 ileum of small intestine
8 esophagus
9 left lobe of liver
10 body of stomach
11 left colic (splenic) flexure
12 transverse colon
13 taenia coli
14 jejunum of small intestine
15 sigmoid colon
16 esophagus
17 duodenum
18 right colic (hepatic) flexure of colon
19 ascending colon
20 taenia coli
21 cecum
22 tail of pancreas
23 body of pancreas
24 left colic (splenic) flexure of colon
25 transverse colon
26 descending colon
27 rectum
28 sigmoid colon

Chapter 26

What Do You Think?

1. To answer this question, first think of what would happen if gravity alone were the force that moved urine from the kidneys to the urinary bladder. What would happen to the flow of urine when you were lying down or doing a handstand? You guessed it; urine would no longer flow from the kidney to the urinary bladder. Thus, muscular contractions are necessary to actively propel urine from the kidney to the urinary bladder.

Chapter Activities

Figure 26.7   Coronal Section Through the Right Kidney

1 Minor calyx	7 renal cortex
2 renal pelvis	8 renal medulla
3 major calyx	9 renal papilla
4 renal pyramid	10 renal sinus
5 renal column	11 Liver lobe
6 fibrous capsule	12 ureter

Figure 26.9   Models of the Kidney Demonstrating the Blood Supply to the Kidney

(a) Coronal Section
1 peritubular capillaries
2 interlobular vein
3 interlobular artery
4 interlobar vein
5 interlobar artery
6 vasa recta
7 renal artery
8 renal vein
9 segmental artery
10 glomerulus
11 afferent arteriole
12 arcuate vein
13 arcuate artery
14 efferent arteriole

(b) Close-up of the Renal Cortex and Renal Medulla
1 afferent arteriole
2 peritubular capillaries
3 efferent arteriole
4 interlobular artery
5 interlobular vein
6 arcuate vein
7 vasa recta
8 arcuate artery
9 interlobar vein
10 interlobar artery

Exercise 26.6, #4

Abdominal aorta ⟶ renal artery ⟶ segmental artery ⟶ interlobar artery ⟶ arcuate artery ⟶ interlobular artery ⟶ afferent arteriole ⟶ glomerulus ⟶ efferent arteriole ⟶ peritubular capillaries or vasa recta ⟶ interlobular vein ⟶ arcuate vein ⟶ interlobar vein ⟶ renal vein ⟶ inferior vena cava

Figure 26.10   Location of the Ureters Within the Abdominopelvic Cavity

1 right kidney	3 psoas major muscle
2 ureter	4 urinary bladder

Figure 26.11   Classroom Model of Male Urinary System Structures

1 renal cortex	8 left kidney
2 renal pyramid	9 renal vein
3 major calyx	10 ureter
4 renal papilla	11 urinary bladder
5 minor calyx	12 opening of ureter
6 renal pelvis	13 urinary trigone
7 renal artery	14 urethra

Figure 26.12   Classroom Model of a Midsagittal Section Through the Male Pelvis

1 urinary bladder	5 ureter
2 prostatic urethra	6 detrusor muscle
3 membranous urethra	7 internal urethral sphincter
4 spongy urethra	8 external urethral sphincter

Figure 26.13   Classroom Model of a Midsagittal Section Through the Female Pelvis

1 urinary bladder	4 detrusor muscle
2 urethra	5 internal urethral sphincter
3 ureter	6 external urethral sphincter

Chapter 27

What Do You Think?

1. A vesicular follicle is distinguished from a secondary follicle by the **presence of a single antrum.** Thus, if the antrum is divided into more than one region, the follicle is a secondary follicle, not a vesicular follicle.
2. The corpus luteum is present during the **luteal phase** of the ovarian cycle. Hence the name of the phase (corpus luteum; luteal phase).
3. The corpus luteum **secretes the hormones estrogen and progesterone,** which prepare the uterine lining for implantation of a fertilized egg, should fertilization occur.
4. The ductus deferens is a muscular tube that functions to propel sperm toward the urethra during ejaculation. **Because sperm are only propelled through the ductus deferens during ejaculation, at other times the lumen of the ductus deferens will not have many, if any, sperm within its lumen.** On contrast, sperm are commonly visible inside the lumen of the epididymis when viewed histologically because the epididymis is the site for storage and maturation of sperm before ejaculation.
5. The clitoris has no corpus spongiosum, which is the erectile tissue that surrounds the urethra in the male. Thus, the **clitoris also has no urethra.** The corpus spongiosum in the female surrounds the labia minora, and the female urethral opening is located posterior to the clitoris and within the folds of the labia minora. Thus, while the female urethra *is* surrounded by the corpus spongiosum (just as in the male), neither the corpus spongiosum nor the urethra are part of the clitoris.

Chapter Activities

Figure 27.12   Suspensory Ligaments of the Ovary, Uterine Tubes, and Uterus as Seen from a Posterior View

1 ovarian ligament	14 isthmus of uterine tube
2 uterine tube	15 ampulla of uterine tube
3 suspensory ligament of the ovary	16 infundibulum of uterine tube
4 fimbria	17 round ligament of the uterus
5 mesosalpinx	18 endometrium
6 body of uterus	19 myometrium
7 broad ligament	20 perimetrium
8 uterine blood vessels	21 internal os
9 ureter	22 cervical canal
10 uterosacral ligament	23 external os
11 transverse cervical ligament	24 uterine tube
12 vagina	25 mesosalpinx
13 uterine part of uterine tube	26 mesovarium

Figure 27.13   Classroom Model of the Female Pelvic Cavity

1 uterine tube	12 pubic symphysis
2 ovary	13 clitoris
3 uterus	14 external urethral orifice
4 urinary bladder	15 vaginal orifice
5 labia minora	16 anus
6 labia majora	17 ureter
7 rectum	18 fimbria of uterine tube
8 vagina	19 rectouterine pouch
9 bulb of vestibule	20 cervix of uterus
10 round ligament of the uterus	21 vagina
11 vesicouterine pouch	

Figure 27.14   The Female Breast

1 suspensory ligaments	11 pectoralis major muscle
2 lobe	12 lobe
3 lactiferous sinus	13 deep fascia
4 alveoli	14 alveoli
5 lactiferous ducts	15 lobule
6 lobule	16 suspensory ligaments
7 areolar gland	17 lactiferous sinus
8 nipple	18 nipple
9 areola	19 lactiferous ducts
10 adipose tissue	

Figure 27.15   Male Reproductive Tract Structures

1 ureter
2 ampulla of ductus deferens
3 seminal vesicle
4 ejaculatory duct
5 prostate gland
6 prostatic urethra
7 bulbourethral gland
8 urogenital diaphragm
9 membranous urethra
10 ductus deferens
11 epididymis
12 testis
13 spongy (penile) urethra
14 urinary bladder
15 bulb of penis
16 crus of penis
17 corpus cavernosum
18 seminiferous tubules
19 corpus spongiosum
20 glans penis

Figure 27.16   Classroom Model of the Male Pelvic Cavity

1 ureter
2 urinary bladder
3 ischiopubic ramus
4 ductus deferens
5 penis
6 prepuce
7 glans penis
8 ductus deferens (ampulla)
9 seminal vesicle
10 prostate gland
11 testicular artery and vein
12 epididymis
13 testis
14 scrotum
15 ureter
16 ductus deferens
17 pubic symphysis
18 prostate gland
19 spongy (penile) urethra
20 corpus cavernosum
21 prepuce
22 tunica vaginalis of testis
23 tunica albuginea of testis
24 tunica vaginalis of testis
25 urinary bladder
26 rectum
27 internal urethral sphincter
28 prostatic urethra
29 ejaculatory duct
30 urogenital diaphragm
31 anus
32 membranous urethra
33 epididymis
34 seminiferous tubules

Credits

Chapter 22

Figure 22.1a–22.3: © Christine Eckel; **22.4:** © The McGraw-Hill Companies, Inc./Photo by Dr. Alvin Telser; **22.5:** © Christine Eckel; **Table 22.3a–b:** © Dr. Don W. Fawcett/Visuals Unlimited; **Table 22.3c:** Image from Dr. Jastrow's electron microscopic atlas (http://www.uni-mainz.de/FB/Medizin/Anatomie/workshop/EM/EMAtlas.html).

Chapter 23

Figure 23.1a: © Ed Reschke; **23.1b:** © The McGraw-Hill Companies, Inc./Photo by Dr. Alvin Telser; **23.2:** © Christine Eckel; **23.3b:** © The McGraw-Hill Companies, Inc./Photo by Dr. Alvin Telser; **23.3c:** © Biophoto Associates/Photo Researchers, Inc.; **23.3d:** © The McGraw-Hill Companies, Inc./Photo by Dr. Alvin Telser; **23.4a–b:** © Christine Eckel; **23.5:** © The McGraw-Hill Companies, Inc./Photo by Dr. Alvin Telser; **23.6(all):** © Christine Eckel; **23.9a–b:** © The McGraw-Hill Companies, Inc./Photo by Dr. Alvin Telser; **23.9c:** © Christine Eckel; **23.10b–c:** © The McGraw-Hill Companies, Inc./Photo by Dr. Alvin Telser; **23.14:** © The McGraw-Hill Companies, Inc./Photo and Dissection by Christine Eckel.

Chapter 24

Figure 24.2: © Victor Eroschenko; **24.3a:** © Science VU/Visuals Unlimited; **24.3b–24.4c:** © Christine Eckel; **24.5(all):** © The McGraw-Hill Companies, Inc./Photo by Dr. Alvin Telser; **24.7:** © The McGraw-Hill Companies, Inc./Photo and Dissection by Christine Eckel; **24.8a–p. 516:** © Christine Eckel.

Chapter 25

Figure 25.1b: © Carolina Biological Supply Company/Phototake, NYC; **25.1c:** © Dr. F. C. Skvara/Visuals Unlimited; **25.1d:** © The McGraw-Hill Companies, Inc./Photo by Dr. Alvin Telser; **25.3a:** © Victor Eroschenko; **25.4b:** © Dr. John D. Cunningham/Visuals Unlimited; **25.4c–25.5b:** © Victor Eroschenko; **25.5c–d:** © Carolina Biological Supply Company/Phototake, NYC; **25.6b–25.9b(top):** © Victor Eroschenko; **25.9b(bottom):** © Dr. Alvin Telser/Visuals Unlimited; **25.11a–25.14a:** © Christine Eckel; **25.14b:** Courtesy of David A. Morton and Chris Steadman, University of Utah School of Medicine; **25.14c:** © Christine Eckel; **25.14d:** Courtesy of David A. Morton and Chris Steadman, University of Utah School of Medicine; **25.15a–p. 544:** © Christine Eckel.

Chapter 26

Figure 26.1a–26.6b: © The McGraw-Hill Companies, Inc./Photo by Dr. Alvin Telser; **26.7:** © Ralph T. Hutchings/Visuals Unlimited; **26.9a–p. 566(all):** © Christine Eckel.

Chapter 27

Figure 27.1b–27.2: © The McGraw-Hill Companies, Inc./Photo by Dr. Alvin Telser; **27.3:** © Christine Eckel; **Table 27.2(1–2):** © The McGraw-Hill Companies, Inc./Photo by Dr. Alvin Telser; **Table 27.2(3–4):** © Ed Reschke/Peter Arnold, Inc.; **Table 27.2(5)–27.4b:** © The McGraw-Hill Companies, Inc./Photo by Dr. Alvin Telser; **Table 27.6a:** © Educational Images/Custom Medical Stock Photo; **Table 27.6b–c:** © Biophoto Associates/Photo Researchers, Inc.; **27.5:** © The McGraw-Hill Companies, Inc./Photo by Dr. Alvin Telser; **27.6a:** From Anatomy & Physiology Revealed, © The McGraw-Hill Companies, Inc./The University of Toledo, photography and dissection; **27.6b:** © The McGraw-Hill Companies, Inc./Photo by Dr. Alvin Telser; **27.6c:** © Christine Eckel; **27.7a:** © Ed Reschke/Peter Arnold, Inc.; **27.7b:** © Christine Eckel; **27.8a:** © The McGraw-Hill Companies, Inc./Photo by Dr. Alvin Telser; **27.8b:** © Christine Eckel; **27.9a–27.10:** © The McGraw-Hill Companies, Inc./Photo by Dr. Alvin Telser; **27.11–p. 596(all):** © Christine Eckel.

Index

Page numbers followed by an *f* or *t* indicate figures and tables.